Neuroanatomical Tract-Tracing 3

Molecules, Neurons, and Systems

Neuroanatomical Tract-Tracing 3

Molecules, Neurons, and Systems

Edited by

Laszlo Zaborszky
Rutgers University
Newark, NJ, USA

Floris G. Wouterlood
Vrije University
Amsterdam, The Netherlands

José Luis Lanciego
University of Navarra
Pamplona, Spain

 Springer

Laszlo Zaborszky
Center for Molecular and
    Behavioral Neuroscience
Rutgers University
Newark, NJ, USA

Floris G. Wouterlood
Department of Anatomy
Vrije University Medical Center
Amsterdam, The Netherlands

José Luis Lanciego
Neurosciences Division
    Center for Applied
    Medical Research
    (CIMA)
University of Navarra
Pamplona, Spain

Cover illustration: Reconstruction of functional connectivity between neurons based on the temporal coherence of spiking activity recorded extracellulary from the somatosensory cortex of a rat. See Fig. 20.8 on page 666.

Printed on acid-free paper.

e-ISBN: 0-387-28942-9
ISBN: 978-1-4419-3963-0   e-ISBN: 978-0-387-28942-7

9 8 7 6 5 4 3 2 1

springer.com

To Sarah Z for her 18th birthday

Preface

Between the first edition of *Neuroanatomical Tract-Tracing Methods* in 1981 and the current, third edition, neuroscience has witnessed a total transition into the information age. Scientists, whether they wanted it or not, have turned digital. Today, everyone is linked up with worldwide computer communication networks, with local and worldwide digital environments offering vastly increased speed and accuracy of data acquisition and processing. Communication and exchange of information between scientists worldwide is a matter of seconds. The electronic dissemination of research data has become routine. Publication of scientific results has changed from the typewritten manuscript to electronic online submission. Search engines, PubMed-like services, and electronic notification and delivery services are making life more convenient for scientists. *What is not on the Web does not exist.* Do we still need books?

We think positively about books. In the first place, it is common sense to have at hand a printed technical protocol in the setting of a laboratory engaged in experimental neuroscience. Although the workbench protocol does not necessarily have to be a book, it is nonetheless helpful to have a book at hand that not only provides the technical protocol, but also explains why the protocol is designed as it is, what the alternatives are, and what their consequences are. The Web is a wonderful yet particularly fluid medium in which things change very quickly. Data that were here today are gone tomorrow. A book has a longer time constant, which sometimes is beneficial.

The first two editions of the book (Heimer and Robards, 1981; Heimer and Zaborszky, 1989—both published by Plenum Press) had a tremendous impact on neuroscience. They are still among the frequently consulted books in the laboratory. We feel that the moment has arrived to pursue a third, thoroughly updated version of this landmark book, in order to continue the line originated by *Neuroanatomical Tract-Tracing Methods*. The target audience remains the graduate students and young investigators working in the laboratory, seeking fast, complete, up-to-date, and immediately applicable information about techniques, written by acknowledged experts in the field.

Since the last edition, several methods that were in their infancy 15 years ago have become routine, older methods have experienced a renaissance, and newly emerging techniques need validation. Molecular techniques, such as genomics and proteomics, have become established methods, which allow for the study of gene expression of recorded and traced neurons (chapters by Ginsberg, Griffith, Stornetta, and their colleagues). The simultaneous development of new fluorescence probes, single- and multiphoton confocal laser scanning microscopes, and vastly increased computer processing power have contributed to a renaissance of fluorescence methods (chapters by Lanciego, Molnar, Reiner, and Wouterlood). In vivo tractography as well as structural and functional imaging techniques allow for the study of the living human brain with a degree of detail never dreamed possible (chapter by Amunts and Zilles). Immunocytochemistry using pre- and postembedding electron microscopy for identifying neuroactive substances still remains an art which has to be tailored to one's needs (chapters by Sesack, Ottersen, and their coworkers). Viral tracers for the analysis of neural circuits have become established in some laboratories (chapter by Geerling and his colleagues). Several chapters briefly discuss how databases help in acquisition, analysis, modeling, and integration of complex cross-scale data sets (chapters by Zilles, Bjaalie, and Nadasdy). Readers interested in these topics are referred to a recently published textbook edited by Koslow and Subramaniam (2005) that summarizes efforts in this field led by the Human Brain Project of the National Institutes of Health.

As neuroscience research progresses, we witness drastic changes in how methods are used. The previous editions of this series reflected the reductionist approach to study neuroscience which was characteristic for the previous century. Even in the second edition, only 2 out of 13 chapters combined techniques that crossed the traditional borders of anatomy and physiology. In the present edition, most of the chapters describe methods, which allow for the integration of molecular, cellular, and system level data, reflecting a holistic-integrative approach to neuroscience in the twenty-first century. Specifically, using sophisticated combinations of tracing methods, the molecular and genetic identity of a neuron (chapters by Ginsberg and his colleagues), as well as the synaptology of any circuitry can be accurately determined (chapter by Sesack and her coworkers). Using extracellular, juxtacellular, and intracellular recordings (chapters by Duque and Zaborszky, and Sik), anatomical features can be correlated with electroencephalographic (EEG), multiunit activity (MUA), local field potentials, and intrinsic membrane characteristics. Recent advances in voltage-sensitive dye imaging (chapter by Petersen) and two-photon calcium imaging (chapter by Goldberg et al.) are promising techniques for studying the spatiotemporal dynamics of hundreds of neurons in the living brain. Sophisticated statistical designs (Avendaño) and expanding computational approaches have the potential to capture full three-dimensional (3D) relations of neuronal and architectonic features of entire brain systems (chapters by Ascoli, Bjaalie, and their colleagues). The last chapter (by Nadasdy *et al.*) predicts that within the next 10 years the complete 3D vectorial database of the rat brain will be available to address specific questions about hidden organization principles of the nervous system. However, these authors also address the gap that exists in our understanding between "structural" and "functional" connectivity (e.g., Friston *et al.*, 1993). We hope that students of this book will bridge this gap eventually, leading to a better understanding of how the human brain functions in health, aging, and disease. This is our ultimate goal.

ACKNOWLEDGMENT

It has been an honor as well as a pleasure to select contributors for the current edition of *Neuroanatomical Tract-Tracing Methods* and to produce with them this book. However, this book would never have been written without several generations of scientists designing and optimizing the methods discussed in the chapters. Among all those whose shared legacy is the current technological standard, we would like to specially mention the recently deceased **Sanford Palay** (Sandy) (1992), who saw the first synapse using electron microscopy and **Theodor Blackstad** (see, e.g., Blackstad and Bjaalie, 1988) whose contribution was instrumental to computational neuroanatomy as we understand it today. We had the great privilege to work or interact with them. However, we learned the most from our mentors, **Lennart Heimer** and **Enrico Mugnaini**, pioneers in tract-tracing and cellular neuroscience. They were not only mentors and teachers, but friends as well. We dedicate this book to them.

We thank the authors for their contributions and patience during the somewhat lengthy editorial process. They exerted great effort in writing their chapters, and also provided essential feedback by cross-reviewing each manuscript. We are also indebted to Drs. James Tepper (Rutgers University), Harry Uylings (The Netherlands Brain Research Institute), and Rolf Kötter (C. and O. Vogt Brain Research Institute, Düsseldorf) who as external reviewers read earlier versions of some of the chapters. We have the fortune to have on our side Kathleen Lyon, Senior Biosciences Editor of Springer, who spearheaded this edition with great enthusiasm. Last, but not least, we would like to acknowledge Professors Ian Creese and Paula Tallal, codirectors of the Center for Molecular and Behavioral Neuroscience, Rutgers University, for encouragement. The National Institutes of Health (L.Z.) gave generous financial support over many years. Elizabeth Hur, helping the editorial process at Rutgers University, was the first Graduate Student who read and benefited from this book. We hope many will follow.

Laszlo Zaborszky
Floris G. Wouterlood
Jose L. Lanciego

REFERENCES

Blackstad, T. W., and Bjaalie, J. G., 1988, Computer programs for neuroanatomy: three-dimensional reconstruction and analysis of populations of cortical neurons and other bodies with a laminar distribution. *Comput. Biol. Med.* **18**:321–340.

Friston, K. J., Frith, C. D., Liddle, P. F., and Frackowiak, R. S., 1993, Functional connectivity: the principal-component analysis of large (PET) data sets. *J. Cereb. Blood Flow Metab.* **13**:5–14.

Heimer, L., and Robards, M., 1981, *Neuroanatomical Tract-Tracing Methods*, New York: Plenum Press, p. 567.

Heimer, L., and Zaborszky, L., 1989, *Neuroanatomical Tract-Tracing Methods 2. Recent Progress*, New York: Plenum Press, p. 408.

Koslow, S. H., and Subramaniam, S., 2005, *Databasing the Brain: From Data to Knowledge*, Hoboken: Wiley-Liss, p. 466.

Palay, S. L., 1992, A concatenation of accidents, In: Samson, F. S., Adelman, G. (eds.), *The Neurosciences: Paths of Discovery*, II, Boston: Birkhauser, pp. 191–212.

Contents

Short Retrospection

LENNART HEIMER

When Laszlo Zaborszky and I dedicated the 1989 tract-tracing book to Alf Brodal, Walle J. H. Nauta, and János Szentágothai, we did it in part to emphasize the maxim that tract-tracing is a crucial part of the overall effort to understand the functions of the central nervous system. We also wanted to highlight the fact—repeatedly illustrated in the history of neuroscience—that breakthroughs in neurobiology are often the result of histotechnical advances. The scientists saluted above began their careers in neuroanatomy with efforts to improve techniques for tracing neuronal connections. Although they were standing on the shoulders of giants who had grabbled with histotechnical problems of tract-tracing long before they arrived on the scene, I would like to begin this short retrospection in the middle of the twentieth century with reference to the book *New Research Techniques of Neuroanatomy*, edited by William Windle and published by Charles C. Thomas in 1957. Here, Nauta presented his silver staining method by summarizing his continuing efforts to improve the silver technique for the tracing of degenerating axons. What was often referred to as the "suppressive" Nauta–Gygax method (Nauta and Gygax, 1954) was by far the most popular tract-tracing method at the time. The word "suppressive" underscored its ability to suppress the staining of normal fibers while still revealing the main parts of degenerating fibers, an attractive feature which greatly facilitated the tracing of axons undergoing degeneration as a result of an experimental lesion.

When the next tract-tracing book: *Contemporary Research Methods in Neuroanatomy*, edited by Nauta and Ebbesson and published by Springer Verlag, appeared more than a decade later in 1970, the silver methods for the staining of degenerating axons were still the methods of choice for most scientists interested in tracing neuronal connections. In fact, one third of the chapters in the Nauta–Ebbesson book dealt with the Nauta–Gygax method and its modifications, whereas another third described various modifications of the Golgi technique, which at that time experienced a revival as a tract-tracing method. Many of the major pathways of the brain and spinal cord were convincingly described for the first time by the aid of experimental silver methods during this period. The secret of Nauta's success in tract-tracing is

LENNART HEIMER • Departments of Neurosurgery and Neuroscience, University of Virginia Health Science Center, P.O. Box 800212, Charlottesville, VA 22908 most easily explained by reference to his unfailing persistence and ability to spend a large part of his career to improving and promoting the art of silver staining. To those unfamiliar with the tinker-and-toil process of developing new techniques, Nauta's never-ending quest for histotechnical perfection might understandably have evoked the image of an obsessive mind. Nauta vividly described the ups and downs of developing tract-tracing methods in his "farewell" article in the *Journal of Neuroscience* (Nauta, 1993).

An important reason for the success of the silver method was the fact that its most sensitive modifications, which became available at the end of the 1960s, made it possible to trace pathways throughout the central nervous system and to visualize the entire axonal projections including their terminations at the light microscopic level. The areas of terminations could be conveniently confirmed and investigated with the aid of the electron microscope, provided the sampling of the tissue was guided by a silver method sensitive enough to stain the terminal fields. With some justification, therefore, the silver methods can be called the first modern methods for the tracing of pathways in the central nervous system. This is in large part Walle Nauta's legacy.

When I was introduced to the silver methods around 1960, it seemed reasonable to suggest that we needed a new anatomical technique every 10 or 15 years in order to keep good progress. Little did I know what was about to happen. If the 1960s are remembered for the introduction of the Falck–Hillarp fluorescence method for the tracing and mapping of monoamine neurons in the central nervous system (Falck *et al.*, 1962)—in itself a milestone on the road toward establishing the field of chemical neuroanatomy—the 1970s were a time of unprecedented renewal in the field of tract-tracing with anterograde and retrograde axonal transport of tracers replacing the silver techniques as the methods of choice in tract-tracing (Cowan and Cuenod, 1975). The novel axonal transport methods could be effectively exploited on the backdrop of an already established basic anatomical framework, thanks in large part to earlier tract-tracing studies aided by silver methods. Although the silver methods lost their dominant role as tract-tracing methods during the last quarter of the twentieth century, they experience a dramatic revival as a tool for neurotoxic assessment (Beltramino *et al.*, 1993).

Then followed in the 1980s in rapid succession a host of new methods for the anatomical and chemical mapping of pathways and neuronal microcircuits, including immunohistochemical methods, in situ hybridization histochemistry, transneuronal tracing using live viruses, and the identification of successive links in chains of neurons using a combination of different techniques at both the light and electron microscopic levels. Computational neuroanatomy is gaining momentum, and as reflected in several chapters in this volume, sophisticated combinations of anatomical and physiological techniques have become routine. The application of voltage-sensitive dye imaging has opened up the possibility to study the function and architecture of neuronal networks in living animals, and the genetic expression of recorded and traced neurons can be studied. We may now attempt to elucidate Vogt's classic theory of Pathoklise (Vogt, 1925), or "selective vulnerability," with molecular biological techniques. Many of the recent pioneering discoveries in neuroscience are in large part the result of this unprecedented development of neuroscience methods, and for the neuroscientists who came of age in the last quarter of the twentieth century, no problem of neuronal connectivity seems too difficult. For those willing to venture beyond the mere routine application of standard methods, the possibilities are almost unlimited, as they are determined primarily by the scientist's own imagination and technical skills.

The successes in experimental tract-tracing during the second half of the last century did not translate into similar gains in the study of connections in the human brain, although the phenomenal achievements in the field of chemical neuroanatomy have to some extent compensated for this deficiency in tract-tracing per se. Methods based on postmortem diffusion of different chemical substances have been tried (Haber, 1988; Sparks *et al.*, 2000), but the slow rate of diffusion has limited the usefulness of these methods in the large human brain. The use of postmortem silver techniques or myelin stains in patients with focal lesions (Grafe and Leonard, 1981; Mesulam, 1979; Miklossy and Van der Loos, 1991) is another approach, but the inability to control the location and size of human brain lesions has so far restricted the value of these methods. Nonetheless, it appears that postmortem silver staining of human brains with focal lesions or degenerative disorders is underutilized, maybe in part because the art of silver staining is gradually disappearing as the younger generation of experimental neuroanatomists have embraced the axonal transport methods. But the situation is not without its silver lining. The expertise of silver staining has been uniquely preserved in some laboratories, especially by Robert Switzer (Switzer, 2000), whose activities and writings suggest that it would be worthwhile to more actively try to introduce modern silver methods on a routine basis in neuropathology laboratories.

In vivo *tractography* through the method of diffusion tensor imaging (DTI; Basser *et al.*, 2000), based on the tendency of water molecules to diffuse in the direction of myelinated fiber bundles, has become the new buzzword in clinical neuroscience. Results obtained by this method in the living human brain so far are spectacular (Behrens *et al.*, 2003; Catani *et al.*, 2002). In vivo tractography will in all likelihood have a great impact in clinical neuroscience. In spite of the great expectations surrounding in vivo fiber tractography, however, it is unrealistic to expect that new pathways, defined by origin and termination, will be discovered by DTI alone. Only white matter bundles can be visualized, and there is currently no possibility to reconstruct intracortical connections. The axons of many myelinated pathways, furthermore, often intermingle with axons belonging to other pathways, which makes it difficult to disentangle individual myelinated pathways, even if efforts to do just that have had some success (Catani *et al.*, 2002).

Whatever limitations DTI may have in discovering new pathways, the technique has served as a powerful incentive to study fiber tracts in the human brain; neurosurgeons and neuroradiologists, in particular, are promoting blunt dissection of the white substance as part of their education and daily activities (Kier *et al.*, 2004; Türe *et al.*, 2000; Yaşargil, 2004). However, dissection of fiber tracts, even if aided by the operation microscope, has some obvious drawbacks as soon as the fiber bundles in question are intermingling with other fiber tracts, and this, needless to say, is the rule rather than the exception in the CNS.

DTI has caught the imagination of basic and clinical neuroscientists alike. It is already yielding dividends in clinical neuroscience, and it has reinforced the value of brain dissection as an integral part of studying the major pathways in the human brain. It may also have raised the awareness among clinical neuroscientists that a naturally occurring lesion may sometimes provide a golden opportunity to trace connections in the human brain (see earlier discussion of silver methods). Maybe one day an even more powerful method, or a combination of methods, will fulfill Mesulam's dream of a new frontier in the field of tract-tracing in the human brain (Mesulam, 2005).

Let us return to experimental tract-tracing methods in the broadest sense of the term. This includes studying all aspects of neuronal circuitry using anatomical, physiological, histochemical, molecular biological, and neuroimaging methods. The histotechnical revolution, which gained momentum in the last quarter of the twentieth century, is reflected by dramatic progress in the basic and clinical neurosciences. This in turn, has raised the hope that some day in the not too distant future, the most crippling brain disorders might be understood and successfully treated. For those chosen to carry the torch of neuroscience in the early part of the twenty-first century, this book is a godsend considering the often difficult problems of choosing and applying the right technique or combination of techniques. Great ambitions were needed to produce a coherent volume on tract-tracing methods in the broadest sense of the term. The editors, the authors, and the publisher deserve our congratulations.

REFERENCES

Basser, P. J., Pajevic, S., Pierpaoli, C., Duda, J., and Aldroubi, A., 2000, In vivo fiber tractography using DT-MRI data, *Magn. Reson. Med.* **44**:625–632.

Behrens, T. E. J., Johansen-Berg, H., Woolrich, M. W., Smith, S. M., Wheeler-Kinshott, C. A. M., Boulby, P. A., Barker, G. J., Sillery, S. L., Sheehan, K., Ciccarelli, O., Thompson, A. J., Brady, J. M., and Matthews, P. M., 2003, Non-invasive mapping of connections between human thalamus and cortex using diffusion imaging, *Nat. Neurosci.* **6**:750–757.

Beltramino, C. A., DeOlmos, J., Gallyas, F., Heimer, L., and Zaborszky, L., 1993, Silver impregnation as a tool for neurotoxic assessment, In: Erinoff, L. (ed.), *Assessing Neurotoxicity of Drugs of Abuse.* Rockville, Maryland: NIDA Monograph series, Vol. 136, pp. 101–132.

Catani, M., Howard, R. J., Pajevic, S., and Jones, D. K., 2002, Virtual in vivo interactive dissection of white matter fasciculi in the human brain, *Neuroimage* **17**:77–94.

Cowan, W. M., and Cuenod, M., 1975, *The Use of Axonal Transport for Studies of Neuronal Connectivity*, Amsterdam: Elsevier.

Falck, B., Hillarp, N. -Å., Thieme, G., and Torp, A., 1962, Fluorescence of catecholamines and related compounds condensed with formaldehyde, *J. Histochem. Cytochem.* **10**:348–354.

Grafe, M. F., and Leonard, C. M., 1980, Successful silver impregnation of degenerating axons after long survivals in the human brain, J. Neuropathol. Exp. Neurol. **39**:555–574.

Haber, S., 1988, Tracing intrinsic fiber connections in postmortem human brain with WGA-HRP, *J. Neurosci. Methods* **23**:15–22.

Kier, E. L., Staib, L. H., Davis, L. M., and Bronen, R. A., 2004, MR imaging of the temporal stem: anatomic dissection tractography of the uncinate fasciculus, inferior occipitofrontal fasciculcus, and Meyer's loop of the optic radiation, *Am. J. Neuroradiol.* **25**:677–691.

Mesulam, M. M., 1979, Tracing neural connections of human brain with selective silver impregnation. Observations on geniculocalcarine, spinothalamic, and entorhinal pathways, *Arch. Neurol.* **36**:814–818.

Mesulam, M. M., 2005, Imaging connectivity in the human cerebral cortex: the next frontier? *Ann. Neurol.* **57**:5–7.

Miklossy, J., and Van Der Loos, H., 1991, The long-distance effects of brain lesions: visualizing of myelinated pathways in the human brain using polarizing and fluorescence microscopy, *J. Neuropathol. Exp. Neurol.* **50**:1–15.

Nauta, W. H. J., 1993, Some early travails of tracing axonal pathways in the brain, *J. Neurosci.* **13**:1337–1345.

Nauta, W. H. J., and Gygax, P. A., 1954, Silver impregnation of degenerating axons in the central nervous system: a modified technic, *Stain Technol.* **29**:91–93.

Sparks, D. L., Lue, L. -F., Martin, T. A., and Rogers, J., 2000, Neural tract tracing using Di-I: a review and a new method to make fast Di-I faster in human brain, *J. Neurosci. Methods* **103**:3–10.

Switzer, R. C., III, 2000, Application of silver degeneration stains for neurotoxicity testing, *Toxicol. Pathol.* **28**:70–83.

Türe, U., Yaşargil, M. G., Friedman, A. H., Al-Mefty, O., 2000, Fiber Dissection technique: lateral aspect of the brain, *Neurosurgery* **47**:417–427.

Vogt, O., 1925, Der Begriff der Pathoklise, *J. Psychol. Neurol. (Lpz)* **31**:245–255.

Yaşargil, M. G., 2004, Impact of temporal lobe surgery, *J. Neurosurg.* **101**:725–738.

2

Preembedding Immunoelectron Microscopy: Applications for Studies of the Nervous System

SUSAN R. SESACK, LEEANN H. MINER, and NATALIA OMELCHENKO

SUSAN R. SESACK, LEEANN H. MINER, AND NATALIA OMELCHENKO • Department of Neuroscience, University of Pittsburgh, 446 Crawford Hall, Pittsburgh, PA 15260

Abstract: This chapter addresses the basic applications of tract-tracing and preembedding immunoperoxidase and immunogold–silver labeling for transmission electron microscopy, focusing primarily on identifying the cellular and subcellular localization of proteins of relevance to neurotransmission and on defining synaptic connectivity within neuronal circuits. Information is provided regarding the use of preembedding immunoperoxidase and immunogold techniques to identify the cellular and subcellular localization of neuronal receptors and transporters. The chapter also describes in detail a triple-labeling approach designed by our laboratory for identifying synaptic inputs to neuronal cell populations defined both by their projection targets and by their transmitter phenotype. Protocols presented in the Appendix are designed to enable researchers trained in small animal surgery, immunocytochemistry, electron microscopy, and appropriate laboratory safety procedures to perform ultrastructural investigations similar to those described here.

Keywords: electron microscopy, immunocytochemistry, immunogold, immunoperoxidase, preembedding, tract-tracing, ultrastructure

I. INTRODUCTION

Having been available for 70 years, it seems reasonable to ask what transmission electron microscopy (TEM) can add to the investigation of the nervous system in the new millennium. Today, the basic synaptic organization of most brain regions is known, as is the morphological detail of most neurons and support cells. Chapters in previous volumes of this series have addressed the applications of TEM for studies of brain/neuronal structure, patterns of degeneration, and identification of synaptic connectivity using tract-tracing, immunocytochemistry, and electrophysiological cell filling (Heimer and Robards, 1981; Heimer and Zaborszky, 1989). In the chapter by Wouterlood of the current volume, a sophisticated technique for using confocal microscopy to define synaptic contacts is described, raising further questions about how long TEM will endure as a staple approach for neuroscience investigation. Is it the case that TEM will increasingly become a legacy method used only by aesthete scholars eager to have quality photomicrographic evidence at the highest magnifications possible?

Obviously, we believe that this is not the case and that TEM, in combination with procedures to identify discrete pathways and proteins at cellular and subcellular levels, is as powerful and up to date as ever. Of course, TEM is not without its disadvantages; it is expensive, time-consuming, requires considerable technical skill, and is capable of generating false-positive and false-negative results, the latter even in the hands of experienced researchers. In the current chapter, we will present a general assessment of the progress made in TEM studies of the nervous system since the last volume of this series and provide technical and interpretational guidelines that are critical for planning modern TEM experiments. We will briefly characterize relatively common procedures and refer readers to other chapters where these are described in greater detail. In addition, we will supply sometimes hard-to-find information that has not yet been described in this series or needs to be updated. This chapter focuses on preembedding immunocytochemical methods for TEM. The reader is referred to the chapter by Mathisen *et al.* (this volume) for a description of postembedding techniques that are also used for immunoelectron microscopy.

II. APPLICATIONS

A. General Applications and Appraisal of the Methods

The higher magnification and resolution afforded by the TEM method of analysis presents distinct advantages compared to light microscopic (LM) techniques. TEM allows the observation of cell ultrastructure and how this morphology changes with natural or extrinsically imposed events such as aging, drug treatment, or stress. For such questions, the application of LM and Golgi impregnation is also quite useful, particularly for measuring dendritic branching and for counting spine density on identified cell types (Irwin *et al.*, 2002; Kolb *et al.*, 2003). However, TEM in combination with unbiased stereological measurements can more accurately and more quickly provide counts of overall spine number in a given brain region, as well as the number of axon terminals, dendrites, and synapses in that area. As such, TEM application of the physical disector method has become an important tool for analyzing how structure and synaptology change with experience. This methodology can also be combined with immunocytochemistry (Beaulieu *et al.*, 1994). Such applications will not be addressed here, and the reader is directed to the existing literature on physical disector methods for counting neuronal elements (Howard, 1990; Mayhew, 1992; von Bartheld, 2002) and to the chapter by Avendaño in this volume.

With relatively few exceptions, most antigens can be immunolabeled by both LM and TEM, although the latter allows for a more precise assessment of the subcellular position of antigens, and the relationship of labeled structures to each other can be determined more precisely with TEM techniques. This makes the latter method essential for the most definitive determination of the cellular and subcellular distribution of neurotransmitters, receptors, transporters, enzymes, metabolites, and so on (see section "Ultrastructural Immunolocalization of Neurobiological Proteins"). TEM is also suitable for more recent studies of protein trafficking (Bloch *et al.*, 2003; Ravary *et al.*, 2001), although LM approaches have the advantage of being able to follow these intracellular processes in real or near-real time.

TEM allows the visualization of the fine characteristics of synapses occurring between neuronal structures identified by tract-tracing or immunocytochemistry. Hence, it continues to find utility as a tool for studying communication between functional neural systems. For this particular application, TEM approaches have gained in analytical power since the last release of neuroanatomical tract-tracing methods, due to the ability to recognize at least three markers on the same sections. Although LM techniques such as that described in the chapter by Wouterlood (this volume) may eventually provide comparable information, TEM will continue to represent the principal method for detailed synaptic connectivity studies of neuronal circuitry. Below, we address in more detail the important technical issues that deal with the studies of this type.

B. Ultrastructural Immunolocalization of Neurobiological Proteins

The functional impact of any given neurotransmitter is determined by the activation of receptor subtypes with particular affinities and activities and often by the removal of the transmitter from the extracellular space by plasma membrane transporters. Although these general principles of neurotransmission apply to most neuronal systems, differences in the extent to which receptors and transporters are inserted in the plasma membrane can cause distinct cellular responses to neurochemicals that are highly regulated. Hence, a powerful application of ultrastructural immunocytochemistry is the localization of neurobiologically relevant proteins (Cornea-Hebert *et al.*, 1999; Ferguson *et al.*, 2003; Garzón *et al.*, 1999; Glass *et al.*, 2004; Hanson and Smith, 1999; Huang and Pickel, 2002; Pickel and Chan, 1999; Pickel *et al.*, 2004; Miner *et al.*, 2000, 2003a, 2003c; Riad *et al.*, 2000). The information obtained from ultrastructural descriptions of their distribution is crucial to our understanding of the operation of these systems.

1. General Issues of Pre- and Postembedding
Immunocytochemical Methods

The main techniques used for immunologically based electron-dense staining of these proteins are preembedding immunoperoxidase, preembedding immunogold–silver, and postembedding immunogold. These are "indirect" methods of labeling that involve a "primary" unconjugated antibody directed against the antigen of interest. The primary antibody is an unmodified protein that cannot be visualized by TEM. Therefore, the presence of the primary antibody is detected by using a secondary antibody that is directed against species-specific antigenic sites on the primary antibody. This secondary antibody is conjugated to a molecule that will allow the creation of an electron-dense product that can be readily visualized. Some of the labeling molecules are enzymes whose reaction product can be detected by the addition of a substrate, whereas others are electron-dense heavy metals that can be visualized directly. Because the end reaction products from the various immunocytochemical techniques possess distinguishing characteristics, a combination of the labeling methods can be used to identify multiple proteins and determine their cellular or subcellular relationships to each other. For instance, preembedding immunoperoxidase labeling of one antigen can be followed by preembedding immunogold–silver (Fig. 2.1A) or postembedding immunogold labeling of another antigen. Also, both

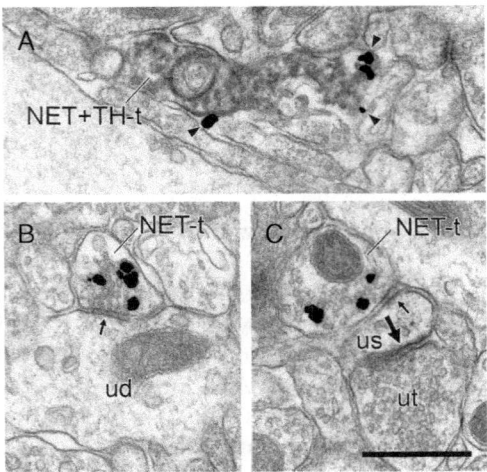

Figure 2.1. Electron micrographs showing axon terminals in the rat PFC singly labeled by immunogold–silver for NET (NET-t) or dually labeled for NET and by immunoperoxidase for TH (NET + TH-t). Axons dually labeled for NET and TH contain the majority of gold–silver particles along the plasma membrane (arrowheads in A). Axons singly labeled for NET contain immunogold–silver particles mainly within the cytoplasm and occasionally form symmetric synapses (small arrows) onto unlabeled dendrites (ud in B) or spines (us in C). In (C), the spine also receives an asymmetric synapse (large arrow) from an unlabeled terminal (ut). Scale bar represents 0.5 μm.

preembedding (Yi *et al.*, 2001) and postembedding (see chapter by Mathisen *et al.*, this volume) immunogold procedures can make use of different-sized gold or gold–silver particles to perform dual-labeling studies.

For the preembedding techniques, the immunocytochemical labeling is performed on thick sections (50 μm), prior to embedding in the plastic resin necessary for cutting the ultrathin (60 nm) sections used for TEM. These preembedding techniques allow satisfactory visualization of most proteins and are becoming somewhat standardized, as described in detail in the Appendix. However, proteins situated in areas that may be inaccessible to antibodies are not consistently recognized with these methods. Such locations include dense membrane barriers like the synaptic complex (e.g., many receptors and transporters) or within vesicles (e.g., the synthetic enzyme dopamine β-hydroxylase).

For the postembedding procedure, immunocytochemistry is performed after the thick sections are embedded in plastic and sliced into ultrathin sections. Due to the thinness of the sections, the experimenter can be confident that antibodies contact most portions of any given cell. Therefore, the major advantage of the postembedding technique is that it allows the visualization of proteins within areas that are typically not accessible with the preembedding procedures. For instance, receptors located within pre- and postsynaptic densities can be visualized (Hanson and Smith, 1999; Nusser *et al.*, 1995). However, many antigens become denatured during the embedding process, rendering this technique unfeasible for the localization of many proteins. Another potential disadvantage is that postembedding gold reagents may interact nonspecifically with the surface of ultrathin sections, resulting in background staining and interfering with unequivocal identification of specifically labeled structures (Van Haeften and Wouterlood, 2000). For a complete review of the applications and technical considerations regarding postembedding methods, the reader is referred to the chapter by Mathisen *et al.* (in this volume).

2. Preembedding Immunoperoxidase Labeling

The preembedding immunoperoxidase technique involves localizing horseradish peroxidase enzyme activity. The horseradish peroxidase molecule is easily conjugated to antibodies, and because it is relatively small, it can readily penetrate membranes of cells fixed with aldehydes. When the peroxidase molecule is exposed to a hydrogen peroxide substrate and the capturing agent 3,3'-diaminobenzidine (DAB), the oxidation products of the DAB will form a brown reaction product. This insoluble product becomes electron dense following chelation by osmium tetroxide (Fig. 2.1A).

a. Strengths and Weaknesses

The major benefit of the immunoperoxidase technique is its superior level of sensitivity. This is particularly true given the various techniques for signal amplification with this method (Hsu *et al.*, 1981; Ordronneau *et al.*, 1981). Compared to the preembedding immunogold approach (described below), immunoperoxidase maximizes the labeling of antigen (Chan *et al.*, 1990), which presents a considerable advantage if the amount of the protein of interest in a given structure is low. Hence, if the desired end point is to demonstrate that a particular neuronal element contains a particular protein, this is the method of choice.

The main disadvantage of immunoperoxidase labeling is that the catalytic activity of the enzyme can result in the accumulation of reaction product that is capable of diffusion away from the original site of activity and, thus, the location of the antigen (Novikoff *et al.*, 1972). In this case, immunoperoxidase is not the best choice if the goal of the study is to determine the exact subcellular location of a protein. Another potential disadvantage of immunoperoxidase is that the accumulation of excessive amounts of reaction product can obscure the visualization of cytoplasmic organelles and synaptic specializations, which limits the determination of cellular structure or the presence of synaptic contacts, and challenges the identification of synapse type. In some cases, the crystalline reaction product can become sufficiently excessive to break through membranes and leak into surrounding structures. Consequently, one must take great care when considering structures lying immediately adjacent to heavily labeled profiles to ensure that all membranes that separate the profiles are intact.

Another potential drawback of immunoperoxidase staining is that some cellular elements contain proteins with endogenous peroxidase-like activity. Such activity is present in red blood cells. As a result, suboptimal perfusions that leave behind many blood cells within brain capillaries can be problematic, in that they have the capacity to use up reagents intended for immunoperoxidase labeling with subsequent reduction in the density of specific labeling. In addition, some glial cells and even some neurons appear to have endogenous peroxidase-like activity (Conradi, 1981; Srebro, 1972; Svensson *et al.*, 1984). Therefore, one may wish to incubate tissue sections in hydrogen peroxide (see Appendix, section "Sodium Borohydride and Hydrogen Peroxide Treatments") prior to immunocytochemical procedures in order to quench this endogenous peroxidase activity.

b. Applications and Analysis

The immunoperoxidase technique is well suited for qualitative analysis of the distribution of proteins. In other words, the *presence* or *absence* of a protein can be readily determined using this labeling procedure (Fig. 2.1A). Moreover, if the aim of an experiment is to establish the general localization of a protein (e.g., expression in soma, dendrites, or axons), this procedure is optimal, given its sensitivity. Quantitative analysis of immunoperoxidase-labeled tissue is typically limited to counting the number of labeled structures within a given area or volume. Because the peroxidase reaction product is capable of spreading over some distance, determination of the subcellular localization of proteins is imprecise with this technique. Therefore, if knowledge of exact protein location is desired, this procedure is not the most viable choice. Likewise, the density of antigens within labeled structures cannot be discerned with precision, although estimates of light or heavy localization are possible. It should be noted that this method is valuable as a first step in localizing proteins. Because immunoperoxidase labeling is the most sensitive immunocytochemical technique and the least costly in terms of time and money, it is practical to use this technique to gain knowledge about the overall amount of an antigen and its general distribution in order to guide further labeling with less sensitive techniques. Some proteins, particularly those in low abundance, often do show "hot spots" of immunoreactivity that strongly suggest a certain subcellular pattern of localization (Aoki *et al.*, 2001; Doly *et al.*, 2004; Mi *et al.*, 2000). Hence, this may be the best first approximation that is available in the event that the less sensitive immunogold–silver methods do not produce consistently detectable labeling.

3. Preembedding Immunogold–Silver Labeling

Immunogold labeling exploits the conjugation of colloidal gold particles to immunoglobulins (DeMey and Moeremans, 1986). Therefore, the tissue is exposed to the primary antibody and then to a secondary antibody that has been conjugated to a gold particle. Colloidal gold probes may be prepared in a variety of sizes (from ~1 to 150 nm). For electron microscopy, gold probes below 30 nm are typically used. For example, it has been shown that gold particles in the range of 10 nm can be used without silver enhancement for preembedding subcellular immunolocalization (Zaborszky *et al.*, 2004). However, the use of ultrasmall gold particles (~1 nm) conjugated to secondary antibodies has the advantage of greater tissue penetration for preembedding studies (Chan *et al.*, 1990; Pickel *et al.*, 1993). Because particle of this size are not readily visualized by electron microscopy, the gold is then enlarged by allowing silver to complex with it. It should be noted that, because antibody penetration is not an issue for postembedding immunolabeling, the gold particles conjugated to the secondary antibodies used for that method are much larger (10–20 nm), and can be easily distinguished by electron microscopy without further enhancement procedures.

a. Strengths and Weaknesses

In contrast to the precipitates formed in immunoperoxidase staining, silver-enhanced gold particles are more electron dense, more discrete, and spherical (Figs. 2.1–2.3). Hence, they are quite easily identified. Their discrete nature and the fact that the gold particle remains attached to the secondary antibody means that diffusion away from the source of the antigen is not a problem as it can be for immunoperoxidase. Thus, the predominant strength of this technique is that it permits a more precise view of antigen localization (Figs. 2.1–2.3). For instance, this method will distinguish whether a protein is inserted within the plasma membrane or within the cytoplasm (see especially Fig. 2.1) (Ferguson *et al.*, 2003; Garzón *et al.*, 1999; Miner *et al.*, 2000, 2003c; Pickel and Chan, 1999). It is possible that the silver-enhanced gold particles will settle at areas that are slightly removed from the antigen. However, the distance in question is typically in the range of 20–25 nm (Paspalas and Goldman-Rakic, 2004) (e.g., along the inner or the outer plasma membrane). If a more refined level of protein localization is required, postembedding immunogold labeling may be necessary (see the chapter by Mathisen *et al.*). Having discrete particles also allows for a certain degree of quantification with the immunogold–silver method. Particles can be counted, measured as number per unit area or per unit volume, and assigned to different compartments such as cytoplasmic or plasmalemmal.

Although the preembedding immunogold–silver technique is a powerful method for localizing proteins, it has limitations. One weakness is that the level of labeling achieved for a given amount of antigen is generally lower than with immunoperoxidase. It has been estimated that the immunogold–silver approach is perhaps one order of magnitude less sensitive than immunoperoxidase (Chan *et al.*, 1990). This is likely due to decreased penetration into the tissue section of the secondary antibody with the gold particle conjugated to it and the absence of methods for signal amplification.

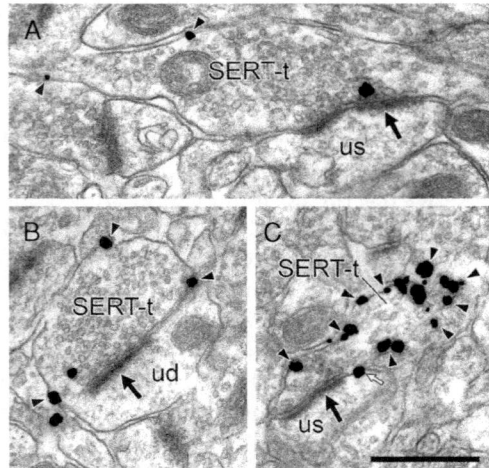

Figure 2.2. Electron micrographs of the rat PFC showing immunogold–silver labeling for SERT in axon terminals (SERT-t) forming asymmetric synapses (large arrows) onto unlabeled dendrites (ud) or spines (us). The majority of gold–silver particles for SERT are located along the plasma membrane (arrowheads), although some are also distributed within the cytoplasm. Occasionally, gold–silver particles for SERT are found in close proximity to sites of synaptic contact (white arrow). Scale bar represents 0.5 μm.

Another possible weakness is that preembedding immunogold may be better suited to detecting plasmalemmal antigens when they are located at extrasynaptic than at synaptic sites (Figs. 2.2 and 2.3) (Bernard *et al.*, 1997). Some studies indicate that preembedding immunogold is unable to detect

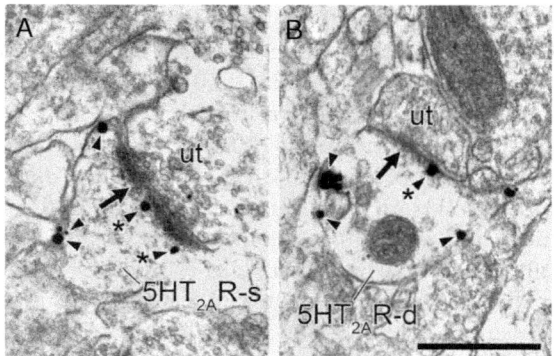

Figure 2.3. Electron micrographs from the rat PFC illustrating immunogold–silver labeling for 5HT$_{2A}$ receptors (5HT$_{2A}$ R) in dendritic shafts (5HT$_{2A}$ R-d) and spines (5HT$_{2A}$ R-s) receiving synaptic input (arrows) from unlabeled terminals (ut). Arrowheads indicate immunogold–silver particles associated with the plasma membrane. Those with asterisks are within or near the postsynaptic density. Scale bar represents 0.5 μm.

proteins that are embedded with in the synaptic complex, including the synaptic cleft (Baude *et al.*, 1995; Bernard *et al.*, 1997). However, other published studies show acceptable labeling of synaptic proteins (Chen *et al.*, 2004; Kulik *et al.*, 2003; Wong *et al.*, 2002) (see also Fig. 2.3). Conversely, the postembedding immunogold method is considered the preferred method for visualizing proteins within synaptic specializations (Bernard *et al.*, 1997). Most likely, this is due to the fact that the synaptic complex has been sectioned at 60 nm, exposing antigenic sites within the complex. Another potential weakness of the preembedding immunogold–silver technique is that the silver enhancement of the gold particles is sensitive to several experimental factors, and therefore care must be taken to obtain reliable gold–silver labeling (see section "Choice of Markers").

b. Applications and Analysis

The ultrastructural immunolocalization of neurobiological proteins is a powerful method for identifying the subcellular distribution of receptors and transporters that may or may not be situated at predicted locations (Ferguson *et al.*, 2003; Hanson and Smith, 1999; Masson *et al.*, 1999). Regardless of whether such studies match or deviate from expectations, they elucidate the anatomical substrates for the actions of neurotransmitters.

For example, we have utilized the preembedding immunogold–silver method to investigate the subcellular localization of monoamine transporters in the rat cortex (Miner *et al.*, 2000, 2003c). In the case of the serotonin transporter (SERT), immunogold–silver particles are typically found on the plasma membrane in extrasynaptic locations (Fig. 2.2). Occasionally, particles are localized close to the synaptic specialization (Fig. 2.2C). Surprisingly, the norepinephrine transporter (NET) is localized predominantly in the cytoplasm (~75%), with only a minority of gold–silver particles found along the plasma membrane (Fig. 2.1). We have also performed a preliminary investigation (unpublished data) of the subcellular localization of the serotonin (5-hydroxytryptamine, 5HT) 2A receptor subtype ($5HT_{2A}R$) in cortical cells utilizing an antibody that was used previously to localize this receptor by immunoperoxidase (Miner *et al.*, 2003a). The preembedding gold–silver method shows that this receptor is localized to both extrasynaptic and synaptic plasmalemmal sites (Fig. 2.3).

In addition to characterizing the typical distribution of neurobiologically relevant proteins, these techniques allow the provocative documentation of changes following physiological or pathological treatments. For instance, alterations in the subcellular distribution of various proteins have been observed in response to genetic, pharmacological, or environmental manipulations (Dumartin *et al.*, 1998, 2000; Glass *et al.*, 2004; Miner *et al.*, 2003b, 2004; Nusser *et al.*, 1999; Riad *et al.*, 2004). The findings from such studies provide valuable insights into the function and trafficking of these important proteins.

For analysis of tissue in which the subcellular localization of gold–silver particles, and hence antigen, is the dependent measure, it is customary to count the total number of gold particles in the structure of interest, and then to assign each gold particle to a particular compartment (e.g., cytoplasm, smooth endoplasmic reticulum, vesicle, plasma membrane). One can then assess as a percentage the tendency of gold particles to accumulate within a particular compartment. It is also possible to determine whether gold particles are localized to the inside or outside edge of a membrane, although as mentioned earlier, the exact position of the relatively large preembedding immunogold–silver particles may not reflect the antigen distribution with perfect fidelity. More refined localization requires postembedding immunogold methods.

4. Dual-Labeling Procedures

A major strength of the immunocytochemical procedures discussed here is that several procedures may be combined in order to visualize two antigens in the same tissue section. The type of dual-labeling method used most frequently in our laboratory is preembedding immunoperoxidase and immunogold–silver. The immunoperoxidase reaction product is dense, dark, and flocculent, whereas the silver-enhanced gold particles are fully black and particulate. Therefore, the two types of markers can be readily distinguished, even in the same structure. In our experience, we prefer to use a "parallel" labeling method in which the tissue is incubated simultaneously in the primary antibodies for both antigens. Then, the immunoperoxidase labeling is performed, followed by the immunogold–silver procedures. The order of manipulations can also be switched so that immunogold–silver is performed first (Katona *et al.*, 2001). However, in this case, an additional step of gold toning may be needed to stabilize the immunogold–silver prior to peroxidase treatments (Arai *et al.*, 1992). Either dual-labeling combination necessitates that the primary antibodies be raised in different species, but it allows the experimenter to determine the presence of one antigen and the precise subcellular localization of the other. For example, in our studies of the localization of NET in the cortex, we found that axons containing immunoperoxidase labeling for the catecholamine synthetic enzyme, tyrosine hydroxylase (TH), contain preembedding immunogold–silver labeling for NET primarily along the plasma membrane (Fig. 2.1A) (Miner *et al.*, 2003c). This is in distinct contrast to axons that contain NET and not TH (Fig. 2.1B,C), where NET is predominantly cytoplasmic.

Other TEM methods for labeling two antigens in the same tissue sections include combinations of immunoperoxidase chromogens with different physical characteristics (Charara *et al.*, 1996; Norgren and Lehman, 1989; Smith *et al.*, 1994; Zaborszky and Heimer, 1989; Zhou and Grofova, 1995) (see section "Choice of Markers") and combinations of preembedding immunogold–silver techniques (Yi *et al.*, 2001). For the latter procedure, silver intensification of one antigen is performed, followed by silver intensification of a second antigen. The labeling for each can be distinguished based on the size of the gold–silver particles, with the particles for the first antigen being larger, because they are exposed to two enhancement sessions. This procedure allows the subcellular localization of two antigens within the same tissue.

Finally, two antigens can be visualized simultaneously by combining preembedding immunoperoxidase or immunogold–silver with postembedding immunogold procedures. An advantage of this technique is that two primary antibodies that were generated within the same species can be used, because the osmication and dehydration performed prior to plastic embedding destroys the antigenicity of the first primary antibody. However, it should be noted that this is feasible only if those same steps do not denature the antigen to be labeled by the postembedding method (see above).

C. Analyses of Synaptic Connections

In order to determine whether neurons of interest are synaptically connected, it is necessary to distinguish particular populations of cells based on their specific protein content and/or on their topographic projections. Although neurochemical phenotypes can typically be defined in naive animals, the identification of neuronal connections requires surgical intervention to introduce tract-tracing agents, at least at present (see section "Prospects for the Future"). Here, we discuss the principles of study design that combine tract-tracing and neurochemical definitions with TEM analysis.

It is beyond the scope of this chapter to compare the effectiveness and methodology for all the available anterograde and retrograde tracing approaches, many of which have been described in previous volumes of this series (Gerfen et al., 1989; Pickel and Milner, 1989; Skirboll et al., 1989; Steward, 1981; Warr et al., 1981; Zaborszky and Heimer, 1989). Relevant information regarding different methods of intraparenchymal injections, including iontophoretic and pressure injections have also been covered earlier (Alheid et al., 1981). Here, we will focus on the tracers that are used most commonly in ultrastructural studies, and particularly those that combine well with each other and with techniques to label phenotypic markers. The main requirements for such tracers are (1) sensitivity in labeling as many of the neurons and processes that contribute to a pathway as possible, (2) specificity in the direction of transport, (3) recognizable filling of the parts of neurons that will be examined by TEM, (4) compatibility with ultrastructural preservation, and (5) compatibility with other markers. If multiple tracers need to be combined, the choices may become more limited. In order to avoid the performance of multiple survival surgeries, we use tracers with similar optimal survival times. The clear understanding of individual tracer competence is a critical issue for planning TEM experiments, and the inability of any particular tracer or combination of tracers to fulfill all of the above requirements is the main reason why the results may need to be verified by alternative tracer approaches.

1. Anterograde Tract-Tracing

a. Anterograde Degeneration

The induction of anterograde degeneration following brain lesion is the oldest tracing method available, but it still has excellent utility for some investigations. The three main lesion approaches are physical disruption (e.g., suction), cytotoxic injection, and electrolytic lesion (Moore, 1981), with cytotoxic injections having the advantage of being directed only to cell bodies and dendrites in the region while sparing fibers of passage. Physical disruption and electrolytic lesions provide a more certain disruption of all cellular elements at the same time, and the electrolytic approach can be calibrated to control the degree of current spread. After a survival time that must be determined empirically for each pathway, the axon terminals of the lesioned neurons will show signs of degeneration that can be readily distinguished in the electron microscope (Mugnaini and Friedrich, 1981). In the best case, the terminals will demonstrate the characteristics of electron-dense degeneration (Fig. 2.4A): shrunken and distorted boundaries, darkened cytoplasm, disrupted organelles like mitochondria and synaptic vesicles, and frequent engulfment on nonsynaptic sides by glial processes (Mugnaini and Friedrich, 1981; Sesack and Pickel, 1992). Despite this extensive disruption of structure, axon terminals will retain their synaptic connections with target cells until they are completely removed by glia, and this provides the degeneration method its utility in studies of synaptic connections.

Anterograde degeneration has two major advantages and several disadvantages of note. First, in our experience, it is the most effective method of anterograde tracing for TEM in the rodent, assuming that disruption of fibers of passage is not an issue for the pathway under study (Sesack and Pickel, 1992). Depending on the size of the lesion, many of the axons of interest will be affected. Moreover, the method does not require any chemical or immunolabeling, and affected terminals will be observed throughout the thickness of ultrathin sections. This is in contrast to tracers that require immunocytochemical detection, whose recognition will be limited to the tissue plastic interface (see section "General Sampling Issues"). Second, anterograde degeneration can be combined with immunocytochemical methods, including sensitive immunoperoxidase approaches (e.g., avidin–biotin peroxidase; ABC (Hsu *et al.*, 1981) to label target neurons or convergent structures so that the two methods together provide excellent likelihood of finding hypothesized synaptic connections (i.e., low false-negative rate). In this regard, it should be noted that the two methods can be further combined with preembedding immunogold–silver for triple-labeling studies.

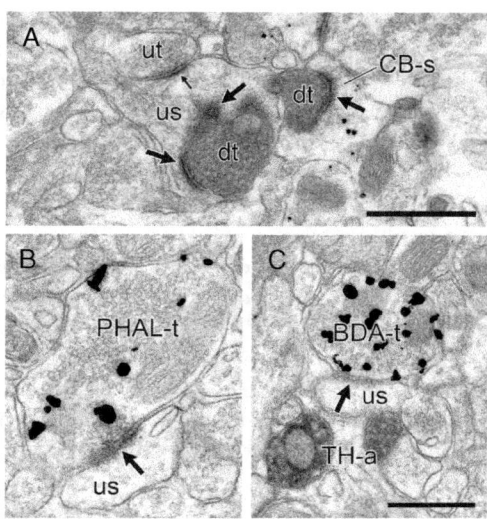

Figure 2.4. Electron micrographic examples of anterograde tract-tracing in the rat forebrain. (A) Two degenerating axon terminals (dt) in the PFC following an electrolytic lesion in the mediodorsal thalamus form asymmetric synapses (large arrows) onto spines either labeled by immunogold–silver for calbindin (CB-s) or unlabeled (us). The latter receives an additional symmetric synapse (small arrow) from an unlabeled terminal (ut). (B, C) Axon terminals in the amygdala (B) or NAc (C) containing immunogold–silver labeling for PHA-L (PHA-L-t in B) or BDA (BDA-t in C) transported anterogradely from the PFC or paraventricular thalamus, respectively, synapse onto unlabeled spines. In (C) immunoperoxidase labeling for TH is evident within axons (TH-a) in the adjacent neuropil. Scale bar in (A) represents 0.5 μm for (A) and (B); scale bar in (C) represents 0.5 μm for (C).

There are several disadvantages to the anterograde degeneration method:

1. The correct survival time after lesion is critically important. At short times, it will be hard to recognize terminals undergoing degeneration, whereas at long times, synaptic contacts will be disrupted and the affected terminals will be eliminated. Deviations of a half day can make a difference, so that the optimal timing for a particular pathway must be established empirically.

2. The method is not applicable for all connections or in all species. Monoamine pathways are particularly difficult to label with this approach, and early studies that attempted to use the method produced erroneous results. A detailed analysis of the difficulties associated with anterograde degeneration for detection of monoamine axons is available (Zaborszky *et al.*, 1979). In addition, not all pathways appear to undergo a process of anterograde degeneration that can be easily detected by TEM. For example, in a prior study of efferents from the prefrontal cortex (PFC), we were able to label the projection to the nucleus accumbens (NAc) using anterograde degeneration but not

the fibers known to innervate the ventral tegmental area (VTA), despite examination of lengthy survival times (Sesack and Pickel, 1992). With regard to other species, the rat is particularly amenable to the anterograde degeneration method because of its small size and the ease of lesioning the majority of a pathway nearly simultaneously. In primates, the anterograde degeneration process can proceed at different rates in different neurons, making the selection of an optimal survival time nearly impossible.

3. False-positive results can be generated if the lesion disrupts fibers not originating from but passing through the lesioned area en route to the target of interest.

4. Anterograde degeneration cannot be used for phenotypic characterization of afferents. Such features as size, normal morphology, and immunoreactivity for particular markers will be compromised.

b. *Phaseolus vulgaris* Leucoagglutinin (PHA-L)

PHA-L is a plant lectin that has many superior properties for anterograde tract-tracing, including specificity for the anterograde direction and the ability to fill the processes of neurons that take it up in a Golgi-like fashion (Gerfen and Sawchenko, 1984; Wouterlood and Groenewegen, 1985; Wouterlood and Jorritsma-Byham, 1993). A detailed description of the application of PHA-L tract-tracing can be found in Gerfen *et al.* (1989), and combination of this method with transmitter identification in postsynaptic targets has been described in a previous chapter in this series (Zaborszky and Heimer, 1989). Here, we will focus primarily on the advantages and disadvantages of this tracer for TEM studies.

The requirement of PHA-L for iontophoretic injection to stimulate uptake is both an advantage and a disadvantage of this method. Iontophoresis is advantageous because it contributes to the selective uptake into soma and dendrites and, hence, the anterograde specificity. It also helps to create the smaller injection sites that are characteristic of this tracer, which may be desirable or undesirable, depending on the target of interest. Moreover, the equipment for iontophoretic application is somewhat expensive, and so this can limit the availability of the PHA-L method.

Regarding other advantages, PHA-L rarely undergoes retrograde transport and only in certain systems (Shu and Peterson, 1988), making it an otherwise exclusive anterograde tracer for most pathways. Another important advantage is that it is not taken up by fibers of passage. To date, there are no reports of PHA-L failing to label any particular pathway. Furthermore, PHA-L remains within the neurons that take it up for considerable periods of time (up to 10 weeks) (Wouterlood *et al.*, 1990), facilitating studies in which the introduction of the tracer and sacrifice of the animal must be separated by some period of time. In the early stages after injection of PHA-L, the extent of anterograde transport increases with time, so that for TEM studies, it is generally advisable to wait 10–14 days following injection.

PHA-L is transported into all branches of an axon, making it possible to use this tracer to study clearly identified local collaterals within the region of injection. Although biotinylated dextran amine (BDA) has the same ability (Melchitzky et al., 2001), its greater potential to undergo retrograde transport and trafficking into the collateral branches of retrogradely labeled cells (Wouterlood and Jorritsma-Byham, 1993) (see below) may compromise its usefulness for studying local collaterals. Finally, the PHA-L method can be combined with: BDA as a second anterograde tracer (Dolleman-Van der Weel et al., 1994; French and Totterdell, 2002, 2003), retrograde tract-tracing agents, or immunocytochemistry to identify target phenotype. In this regard, PHA-L can be localized by either immunoperoxidase or immunogold–silver methods (Fig. 2.4B), with the other marker being used to label additional tracers or transmitters.

The PHA-L method has two principal disadvantages:

1. It is reported to have less sensitivity compared to BDA (Wouterlood and Jorritsma-Byham, 1993), although we have not found this difference to be particularly remarkable in our own studies of brain connectivity. Lower sensitivity is, in part, due to the relatively small injection sites achieved with the iontophoretic technique. However, the major reason for lower sensitivity is the requirement for immunolabeling to visualize PHA-L. This lowers the sensitivity for TEM studies in which ultrastructural preservation and penetration of the tissue surface must be carefully balanced (see section "General Sampling Issues").

2. Small injection sites limit the usefulness of the method for studies in primates. Of course, PHA-L is the preferred method when injections must be confined to a small target.

c. Biotinylated Dextran Amine (BDA)

BDA (10,000 MW) is another example of a highly successful method for anterograde tract-tracing (Brandt and Apkarian, 1992; Reiner et al., 2000; Veenman et al., 1992), as described in more detail in the chapter by Reiner and Honig. The tracer can be injected as a 10% solution in 0.01 M phosphate buffer, pH 7.4, either by iontophoresis or by pressure, enabling the investigator to make large or small injections as desired. The ability to make large injections is particularly important for tract-tracing in primates. BDA utilizes a simple detection method, in which brain sections are simply incubated in ABC reagents (Hsu et al., 1981) without the need for immunolabeling procedures. This provides BDA with greater sensitivity than PHA-L (Wouterlood and Jorritsma-Byham, 1993). However, when needed, BDA can be labeled by immunogold–silver methods (Fig. 2.4C) using an antibody directed against biotin (Pinto et al., 2003).

At the same time, BDA presents with some rather significant disadvantages. Specifically, it can be transported retrogradely and via fibers of passage (Reiner et al., 2000; Van Haeften and Wouterlood, 2000). Moreover, the

BDA that undergoes retrograde transport can subsequently travel into the collaterals of these neurons and give rise to false-positive results when used for anterograde tracing (Chen and Aston-Jones, 1998; Reiner *et al.*, 2000). This requires that investigators check carefully for retrograde transport of BDA into any brain area that is also afferent to the region of interest.

2. Retrograde Tract-Tracing

For LM studies, it is usually sufficient that retrograde tracers transport into the soma and proximal dendrites of labeled cells. However, for TEM studies of synaptic inputs to identified cell populations, there may be the added requirement that a retrograde tract-tracing agent also traffic into the distal dendrites of neurons, if the afferent under study synapses distally. This poses one of the most difficult challenges in studies of synaptic connectivity. To our knowledge, only a few tracers possess this ability, and then only in certain cells or under special circumstances. Here, we review three tracers that have been reported to infiltrate distal dendrites following retrograde transport. A fourth retrograde tracer, cholera toxin subunit B, has also been reported in the literature to show usefulness in ultrastructural analyses, and interested readers are directed to several reviews on this subject (Bruce and Grofova, 1992; Ericson and Blomqvist, 1988; Llewellyn-Smith *et al.*, 1990). In our hands, we have observed no obvious advantages with cholera toxin over the tracers that follow.

a. BDA

A smaller molecular form of BDA (3000 MW) is available for use as a retrograde tracer that can be delivered either by pressure or by iontophoresis using a different buffer: 10% solution in 0.1 M sodium citrate–HCl (Reiner *et al.*, 2000). Like the use of BDA for anterograde tract-tracing, large injections of BDA can be made, making this technique useful for primate studies. Retrogradely transported BDA can be detected with a simple ABC method, and there is no evidence that it undergoes transneuronal transport (Brandt and Apkarian, 1992; Rajakumar *et al.*, 1993). However, the ability of BDA (3000 MW) to also undergo anterograde transport limits the usefulness of this tracer in multitracing experiments, and the tracer can also be taken up by fibers of passage (Reiner *et al.*, 2000; Veenman *et al.*, 1992).

With regard to the need to visualize the distal dendrites of labeled cells, a method has been introduced for improving axonal uptake of BDA by first applying NMDA *N*-methyl-D-aspartate to lesion the soma and dendrites at the injection site that might compete for uptake of BDA (Jiang *et al.*, 1993). However, in our experience, this procedure does not work for all pathways. For example, when we attempted to use BDA as a retrograde tracer in the PFC to VTA pathway, the prior injection of NMDA into the VTA resulted in BDA diffusion throughout a considerable portion of the midbrain with minimal retrograde transport (unpublished data).

b. FluoroGold (FG)

FG has been utilized for some time as an effective retrograde tracer for LM studies (Schmued and Fallon, 1986; Skirboll *et al.*, 1989). More recently, FG has also been adapted as a highly effective tracer for TEM (Chang *et al.*, 1990; Naumann *et al.*, 2000; Van Bockstaele *et al.*, 1994). FG can be delivered as a 2% solution in 0.1 M cacodylate buffer, pH 7.5, by iontophoretic or pressure injections. FG is a highly effective and sensitive retrograde tracer that, to our knowledge, has never failed to label any particular cell population. Neither does this tracer diffuse out of retrogradely labeled cells into adjacent neurons (Novikova *et al.*, 1997; Schmued and Fallon, 1986; Van Bockstaele *et al.*, 1994). Moreover, FG appears to be an exclusively retrograde tracer, at least in the systems for which we have used it to date. Following forebrain or midbrain injections of FG, we have systematically examined sections in the midbrain or pons by TEM and found no evidence for anterograde transport of FG into axon terminals, despite the known existence of available pathways for such transport. The inherent fluorescence of FG also provides a nice benefit to this method, in that the appropriateness of the injection sites and transport can be checked immediately after the brain is sectioned. Misplaced injections or cases of poor transport can then be discarded without loss of more costly reagents.

At the TEM level, FG is seen as being diffusely distributed within the cytoplasm of retrogradely labeled cells or concentrated within lysosomes (Fig. 2.5). In certain neurons, the passage of FG into distal dendrites is excellent, particularly for cells with modest dendritic branching such as monoamine neurons (Carr and Sesack, 2000b; Van Bockstaele and Pickel, 1995). Within these populations, FG appears to undergo equivalent penetration into the distal dendrites of cells that have different axonal targets but share similar dendritic morphology. For example, we have analyzed random ultrastructural samples of dopamine (DA) neurons in the VTA that were retrogradely labeled with FG from the NAc or from the PFC and found no difference in the mean or distribution of the cross-sectional diameters of the dendrites containing FG (unpublished observations). This suggests that FG was transported to an equivalent extent into large and small caliber dendrites in the two populations.

Unfortunately, FG fails to be transported into the distal most dendrites of cell classes with complex dendritic trees, in particular, those with extensive dendritic spines (e.g., cortical pyramidal cells). This difficulty is compounded by the need for immunocytochemistry to detect FG for TEM studies (Chang *et al.*, 1990), with the reduced antibody penetration that accompanies the requirement for ultrastructural preservation. To some extent, this limitation is mitigated by the fact that FG induces the formation

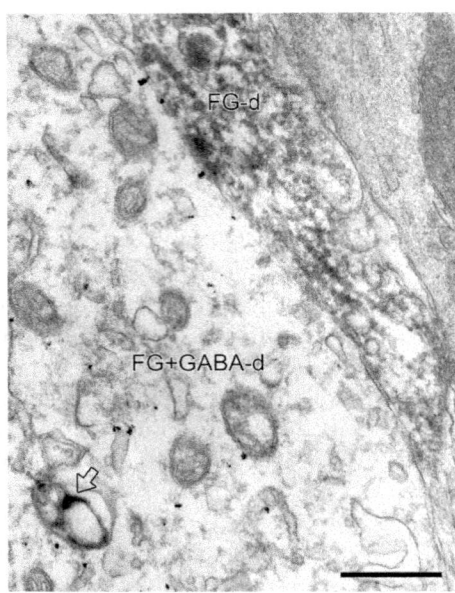

Figure 2.5. Electron micrograph of the rat VTA showing dendrites containing immunoperoxidase for FG (FG-d) retrogradely transported from the NAc. The labeling is diffusely distributed to the cytoplasm in one case, but is concentrated in a lysosome (open arrow) in the other. The latter dendrite is dually labeled by immunogold—silver for GABA (FG + GABA-d). Scale bar represents 0.5 μm.

of lysosomes in labeled cells, which can aid in the detection of this tracer (Schmued *et al.*, 1989). However, labeled lysosomes do not always appear in a particular plane of section. Moreover, individual dendrites can exhibit markedly different densities of FG content (Figs. 2.5 and 2.14), with the lowest densities falling near the limit of detection. Another disadvantage of FG is that it can be taken up by fibers of passage (Dado *et al.*, 1990). The iontophoretic application method reduces this drawback (Pieribone and Aston-Jones, 1988), and in our experience, this problem is less with FG than with other retrograde tracers such as cholera toxin (see also Llewellyn-Smith *et al.*, 1990). Nevertheless, it is an important limitation that must be taken into account when interpreting the results of studies using FG. Finally, FG has only limited usefulness for long-term studies. After 1 week, the intensity of labeling begins to diminish progressively (Novikova *et al.*, 1997), and FG can be cytotoxic in some cases (Garrett *et al.*, 1991; Naumann *et al.*, 2000).

c. Pseudorabies Virus (PRV)

The use of live viruses that replicate within neurons (Fig. 2.6) is an essential tool for retrograde tracing of multisynaptic connections in the nervous system, as detailed in the chapter by Geerling *et al.* and as previously

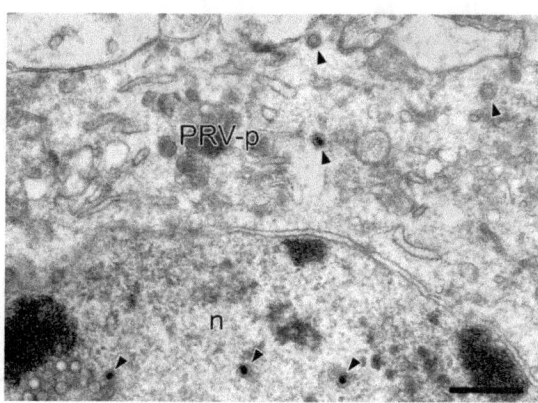

Figure 2.6. Electron micrograph of the rat PFC showing a neuronal perikaryon that is infected with PRV (PRV-p) following transport of this tracer from the NAc. Numerous viral particles (arrowheads) are evident in the cytoplasm and nucleus (n). Scale bar represents 0.5 μm.

described in reviews and the primary literature (Aston-Jones and Card, 2000; Card and Enquist, 1994; Card *et al.*, 1993; McLean *et al.*, 1989; Strick and Card, 1992). In collaboration with J. Patrick Card, we have adapted this technique to label the distal dendrites of neurons infected with virus through a first-order process (Fig. 2.7) in order to study identified synaptic inputs to these dendrites in dual-labeling TEM studies (Figs. 2.8–2.10) (Carr *et al.*, 1999; Carr and Sesack, 2000a). The technique utilizes an attenuated Bartha strain of PRV and has a number of advantages for retrograde tract-tracing.

PRV is an exclusively retrograde tracer, although viral strains that undergo anterograde transport are also available (Aston-Jones and Card, 2000). PRV has a high affinity for brain proteins and is rapidly sequestered by neurons at intraparenchymal injection sites, which are typically rather small. The latter may be disadvantageous if large injections are needed to fill a region of interest. Uptake of PRV by fibers passing through the injection site is lower than typically observed with most retrograde tracers (Aston-Jones and Card, 2000; Chen *et al.*, 1999), although such transport can occur and must be taken into consideration for each study. Viral transport is the only tract-tracing method in which the sensitivity is not limited by the amount of tracer incorporated into cells at the time of injection. The ability of the virus to replicate progeny (Fig. 2.6), to traffic viral particles into the dendritic tree (Fig. 2.7), and to cause infected neurons to synthesize viral-specific proteins that concentrate in the soma and dendrites (Figs. 2.8–2.10) provides a degree of signal amplification that is Golgi-like in nature and unique among tracers (Card *et al.*, 1990). In essence, these processes are driven by the necessity for the virus to repeat its life cycle in newly infected cells. Hence, replicated virus is trafficked into the dendrites to facilitate transneuronal passage at sites of synaptic afferent contact (Aston-Jones and Card, 2000; Card *et al.*, 1993). Reactive astrocytes that are drawn to infected neurons take up any

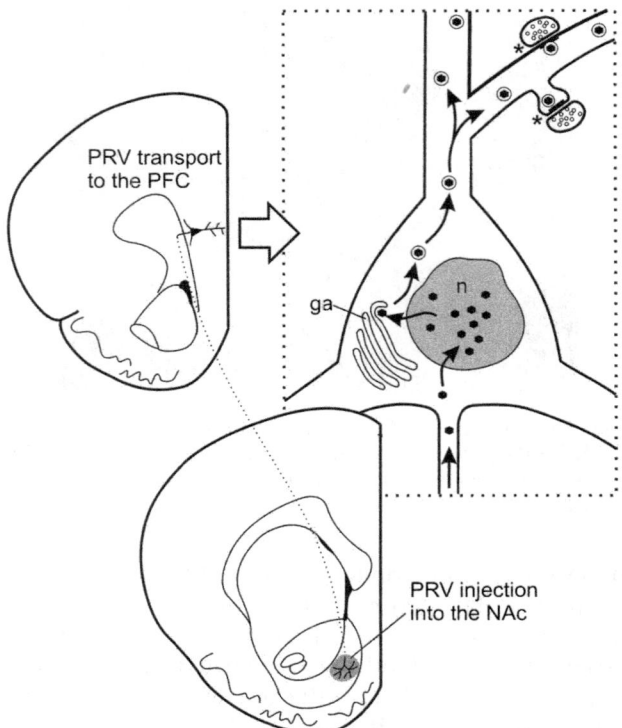

Figure 2.7. Schematic diagram illustrating the application of PRV for first-order retrograde transport that labels the distal dendrites of neurons with complex dendritic trees. PRV is taken up by axon terminals at the site of intraparenchymal injection (the NAc in this example) and transported retrogradely (to the PFC in this case). The virus then enters the nucleus (n), in which the virus replicates itself and manufactures viral-specific proteins. Viral particles then acquire two membrane coats from the Golgi apparatus (ga; only one coat is shown here) and then travel into the dendritic tree seeking synaptic sites (*) at which to pass out of the cell to infect other neurons.

infectious particles that are not transported into axon terminals, and thereby limit nonsynaptic spread. The astrocytes themselves are unable to replicate infectious virus (Aston-Jones and Card, 2000; Card *et al.*, 1993).

For retrograde tracing in cells with simple dendritic branching patterns, like VTA neurons, we have shown (Carr and Sesack, 2000a) that the ability of PRV to label distal dendrites (Fig. 2.8) is comparable to FG (Fig. 2.5). However, for the identification of the thin distal dendrites and dendritic spines of cortical pyramidal neurons, infection with PRV (Figs. 2.9 and 2.10) is superior to any other tracing method in our experience (Carr *et al.*, 1999; McLean *et al.*, 1989). Moreover, the PRV method is readily amenable to dual-labeling studies of the synaptic inputs to these dendrites (Fig. 2.10) (Carr *et al.*, 1999). Although detection of PRV is based on immunocytochemistry,

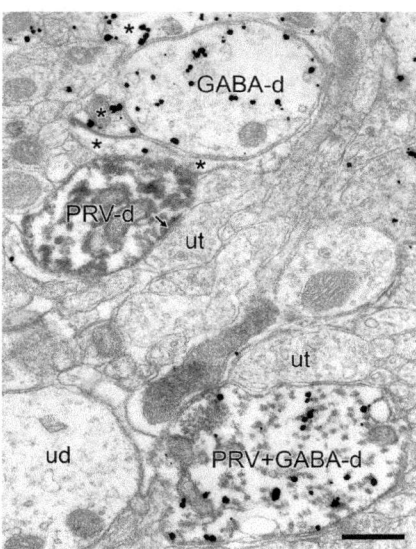

Figure 2.8. Electron micrograph of the rat VTA showing distal dendrites that are: singly labeled by immunoperoxidase for the retrograde tracer PRV (PRV-d) transported from the NAc, singly labeled by immunogold–silver for GABA (GABA-d), or dually labeled for both markers (PRV + GABA-d). An unlabeled dendrite (ud) is shown for comparison. The PRV-d receives a symmetric synapse (small arrow) from an unlabeled terminal (ut). A second unlabeled terminal is apposed to the PRV + GABA-d without synapsing in this section. Immunogold–silver labeling for GABA is also evident in glial processes (*). Scale bar represents 0.5 μm.

the primary antibodies are directed against multiple epitopes of viral proteins and so are quite sensitive, even for TEM studies.

Despite the decided advantages of retrograde tract-tracing with PRV for TEM studies of synaptic connectivity, there are substantial limitations that are important to note. It is sometimes difficult to precisely localize PRV injection sites, due to rapid sequestration and transport of the virus (Aston-Jones and Card, 2000). Viruses cause progressive infections that eventually kill the subjects. However, the survival times that are used for most tracing experiments are typically earlier than this period (Kelly and Strick, 2000), and the application of viral tracing for TEM requires only short times needed for first-order infection. Viruses are living elements and so can have considerable differences in their neuroinvasiveness that reflect growth methods, storage conditions, titer, and other factors (Aston-Jones and Card, 2000; Card *et al.*, 1991, 1999). Another significant limitation to viral tracing is the need for biosafety containment facilities in which to inject and house animals. PRV is a swine pathogen that is not infectious to humans, but infected rodents need to be kept isolated from other animals (Strick and Card, 1992). Herpes simplex virus type 1 and rabies virus are infectious to humans and therefore require even more extensive biosafety precautions and facilities

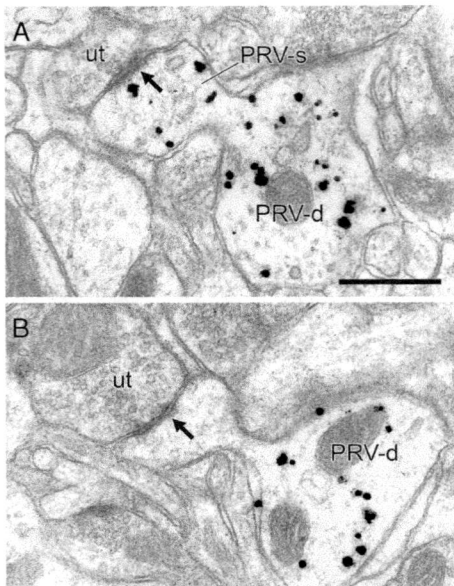

Figure 2.9. Electron micrograph of the rat PFC showing immunogold–silver for PRV in spines (PRV-s) and distal dendrites (PRV-d) of pyramidal neurons retrogradely labeled from the contralateral PFC. Immunoreactivity for PRV extends from the parent dendrite into the spine in (A) but not in (B). Both spines receive asymmetric synapses from unlabeled terminals (ut). Scale bar represents 0.5 μm.

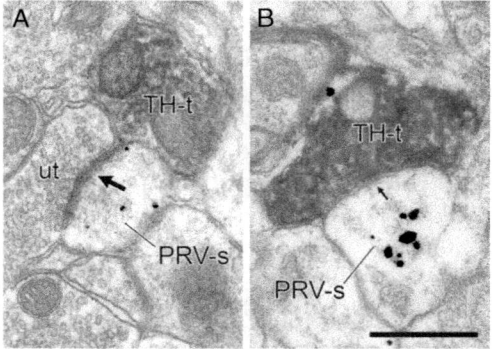

Figure 2.10. Electron micrographs of the rat PFC showing immunogold–silver for PRV in the spines of pyramidal neurons (PRV-s) retrogradely labeled from the contralateral PFC and immunoperoxidase labeling for tyrosine hydroxylase (TH) in axon terminals (TH-t). (A). The PRV-s receives an asymmetric synapse (large arrow) from an unlabeled terminal (ut) and is apposed by a TH-t that does not form a synapse in this section. (B). The TH-t forms a symmetric synapse (small arrow) onto the PRV-s. Scale bar represents 0.5 μm.

that may be prohibitive for some laboratories (Kelly and Strick, 2000; Strick and Card, 1992).

One of the most attractive aspects of viral tracing, namely its transneuronal transport, may lead to false-positive results when this method is used for TEM studies. Avoiding such outcomes requires careful attention to survival times (Aston-Jones and Card, 2000) and control experiments to ensure that second-order infection has not occurred within the region of interest at the time when TEM studies are performed (Carr *et al.*, 1999). For example, in our prior analysis of cortical pyramidal neurons infected by first-order transport, we monitored second-order infection of GABA local circuit neurons whose synapses onto the pyramidal cells are proximal and therefore likely to be the earliest to pass virus. Survival times at which substantial infection of GABA cells was detected were not used in the TEM analysis of synaptic inputs to pyramidal neurons.

Another notable drawback to PRV tract-tracing is the fact that some neurons may not label with this virus. For example, our repeated attempts to label the known projection from the PFC to the VTA with this method were unsuccessful. The most probable explanation is competition for uptake of the virus at the injection site (Card *et al.*, 1999), with afferents having the largest terminal density being most likely to transport quantities of virus sufficient to mount an infection. Indeed, our injections of PRV into the VTA lead to extensive retrograde transport into brainstem sites, seemingly at the expense of forebrain areas.

As a final consideration of the viral tracing method for use in studying synaptic connectivity, it is worth considering why we have emphasized TEM investigation of first-order transport rather than the better known and more easily applied method of LM analysis of transsynaptic transport. For certain experiments, we agree that carefully timed LM studies of viral trafficking are sufficient to provide evidence of synaptic connections between given cell populations. However, for other circuits, there may be multiple pathways that can transport virus, making it difficult without further manipulations to be certain which course viruses took in moving between synaptically connected cells (Fig. 2.11). For example, if investigating whether the PFC synapses onto DA neurons in the VTA that project to the NAc, one might inject PRV into the NAc and wait for second-order transport into the PFC by way of the VTA. However, following NAc injections, the labeling of PFC neurons could arise from connections within the basolateral amygdala or the paraventricular thalamus rather than the VTA (Problem 1). Indeed, our inability to label the PFC by first-order viral uptake from the VTA (see above) suggests that these alternate pathways would be the more probable routes for virus to reach the PFC. Another likely circuit would involve transneuronal transport of virus into PFC pyramidal neurons via intrinsic synapses onto neighboring pyramidal cells that underwent first-order infection from the NAc (Problem 2). Finally, and perhaps most importantly, even if selective lesions could confine the virus into passing transsynaptically via the VTA, the method would not reveal the phenotype of VTA cells involved in this

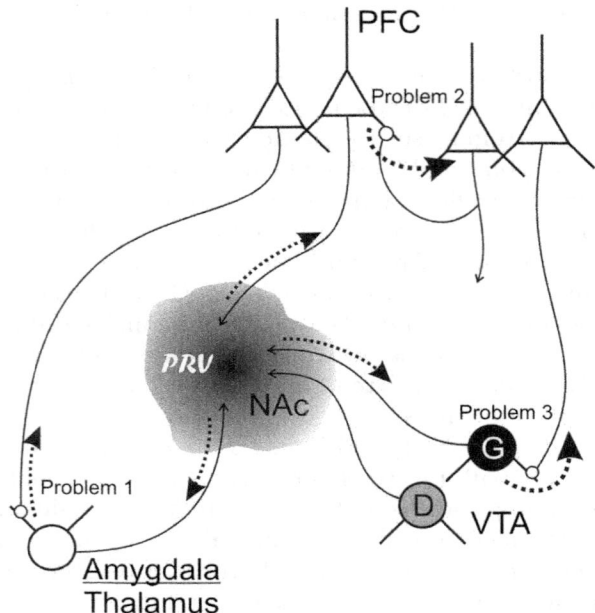

Figure 2.11. Schematic diagram illustrating the use of PRV for multisynaptic tract-tracing to test whether PFC neurons synapse onto DA cells (D) in the VTA that project to the NAc. Interpretational problems arise (see section "Retrograde Tract-Tracing") when the virus passes transneuronally through alternate pathways or via GABA cells (G).

transneuronal passage. For example, second-order infection of PFC neurons would most likely occur due to uptake by mesoaccumbens GABA neurons that receive PFC synaptic input (Problem 3) (Carr and Sesack, 2000a). Recent technological developments are beginning to solve the latter problem, for example in creating a form of PRV whose replication is dependent on Cre-mediated recombinant events that can be isolated to neurons with specific phenotype (DeFalco *et al.*, 2001). However, if such an approach is not available for the complex circuit being investigated, the use of TEM to analyze first-order retrograde transport of virus remains a viable option for examining synaptic inputs to identified cell populations.

3. Identification of Neurochemical Phenotype

For tract-tracing experiments conducted for TEM analysis, it is often important to identify the neurochemical phenotype of the neurons/dendrites targeted by the traced pathway. For example, we wished to determine the extent to which projections from the PFC innervate DA or GABA neurons in the VTA (Carr and Sesack, 2000a; Sesack and Pickel, 1992). For such studies, antibodies directed against transmitters, enzymes, receptors, transporters, calcium-binding proteins, or peptides that are unique identifiers of the target cells can be used. Combinations of immunocytochemical markers are then needed to label both the tracer and the phenotypic marker. As described above in section "Dual-Labeling Procedures," the most typical combination we use is preembedding immunoperoxidase and immunogold–silver, although other combinations are also available. It is important to consider that the peroxidase method is more sensitive than gold–silver by approximately one order of magnitude (Chan *et al.*, 1990). Hence, we typically use peroxidase or immunoperoxidase to label the least abundant antigen (e.g., BDA or PHA-L) and immunogold–silver to label the more plentiful antigen (e.g., TH or GABA).

In other cases, it may be desirable to identify the neurochemical phenotype of neurons containing retrograde tracer from a known injection site. Often such studies are conducted at the LM level, taking advantage of the many available fluorescent markers for double and triple labeling (Skirboll *et al.*, 1989) (see also the chapter by Wouterlood). However, some phenotypes are difficult to detect by LM because a particular marker is not present in sufficient quantity to clearly define the soma. In our experience, this has been a particular problem with GABA in projection neurons. Although the levels of GABA are high in local circuit neurons, they are markedly less robust in projection neurons, such as those in midbrain and basal ganglia structures. Often it is difficult to define these cells using LM methods. Use of GAD antibodies can help with an LM definition of this population, although even this approach requires colchicine treatment to boost GAD levels, and colchicine can interfere with transport of retrograde tracer (Ford *et al.*, 1995). Hence, TEM becomes a useful alternative for determining the extent to which GABA cells contribute to particular projections, as GABA levels are sufficient to detect by TEM. For this purpose, we have followed the protocol originated by Van Bockstaele (Van Bockstaele *et al.*, 1994; Van Bockstaele and Pickel, 1995) to demonstrate the strength of the GABA projection from the VTA to the PFC (Carr and Sesack, 2000b).

Finally, some experiments may call for the identification of neurochemical phenotype in axons that innervate particular brain regions as determined by anterograde tract-tracing. In our experience, this is a particularly difficult issue to resolve using dual-labeling preembedding methods. Although evidence of dually labeled axons is often obtained, it appears that the incidence of such antigen coexpression is underestimated by the combination of immunoperoxidase and immunogold–silver that our laboratory typically uses. This issue is further discussed below in section "Limitations: Sources of False-Negative Errors".

4. Triple-Labeling Studies

TEM studies can be designed that analyze several combinations of tract-tracing agents and neurochemical markers. The general methodological

principles for combining tracing with neurotransmitter phenotype identification for LM and TEM have been described in a previous volume of
this series (Zaborszky and Heimer, 1989). Here, we deal specifically with
experiments in which investigators seek to define the synaptic target of a
particular anterogradely labeled pathway both in terms of its neurochemical phenotype and in terms of its own projection targets. This requires the
combination of two tracers, one anterograde and one retrograde, with a
phenotypic marker.

As an example of such an application, we will refer to our findings with
regard to the afferent regulation of midbrain DA neurons. The major inputs
and outputs of the substantia nigra (SN) and VTA have been known for some
time (Oades and Halliday, 1987; Phillipson, 1979; Swanson, 1982), although
more minor projections are still the subject of investigation. It is known that
the major forebrain projections of the VTA arise from separate populations of neurons, both DA and non-DA, presumably GABA cells (Carr and
Sesack, 2000b; Swanson, 1982; Van Bockstaele and Pickel, 1995). DA cells
innervating the NAc regulate locomotion and motivated behaviors, whereas
DA neurons projecting to the PFC modulate cognitive and affective functions. To understand the morphological basis of behavioral control of DA
cell activity, it is important to develop a detailed picture of the specific afferents that synapse onto different populations of VTA DA neurons. Afferents
can be identified by anterograde tract-tracing or, in some cases, phenotypic
markers, if these derive from single sources (e.g., acetylcholine). Cell populations can be defined on the basis of retrograde tract-tracing from known
target areas (e.g., PFC and NAc), and neurotransmitter identity of these populations can be delineated based on immunocytochemistry for transmitter
markers. In the first study of its kind, our laboratory used a combination
of anterograde and retrograde tract-tracing with immunocytochemistry to
show that excitatory projections from the PFC synapse selectively onto DA
and not GABA neurons that project back to the PFC and onto GABA but
not DA cells that innervate the NAc (Carr and Sesack, 2000a). Such synaptic
specificity may explain some of the unique functional properties of mesopre-
frontal and mesoaccumbens neurons and help to advance understanding
of the circumstances and mechanisms for their behavioral activation.

More recently, we have performed a similar study of the inputs to the
VTA from the brainstem laterodorsal tegmentum (LDT), which, unlike the
PFC, synapses onto DA neurons that innervate the NAc (Omelchenko and
Sesack, 2005a). The LDT projection contains a mixed neurochemical phe-
notype, and with this in mind, we have also completed an analysis of cholin-
ergic inputs to different VTA cell populations (Omelchenko and Sesack,
2005b), as the cholinergic innervation derives predominantly from the LDT
(Hallanger and Wainer, 1988; Oakman et al., 1995).

The success of these experiments depends on the combination of immunoperoxidase and immunogold–silver labeling methods, and on the use
of one of these markers, typically immunoperoxidase, to label two of the
three desired components that are known to be segregated in different

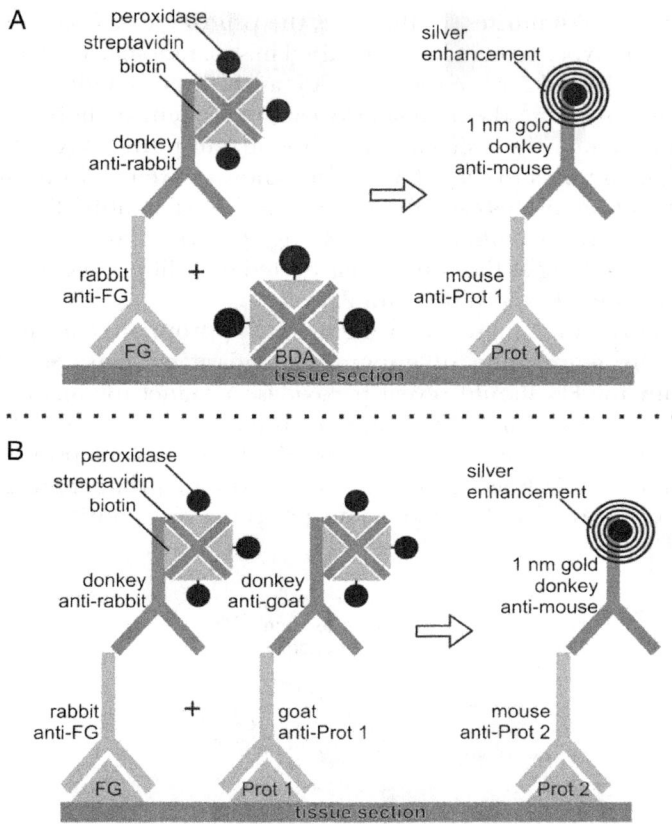

Figure 2.12. Schematic diagrams showing the labeling procedures used for triple labeling in the experiments described here. (A) Two primary antibodies raised in different species against FG and phenotypic protein 1 (Prot 1) are co-applied, and then a biotinylated secondary antibody against the first species is added, followed by the avidin–biotin peroxidase complex. The latter will also bind to the biotin in the anterograde tracer BDA. Following peroxidase histochemistry, a gold-conjugated secondary antibody against the second species is added, and the bound gold particles are silver enhanced. (B) Three primary antibodies raised in different species against FG, a protein labeling the afferent pathway (Prot 1; e.g., PHA-L or a unique phenotypic marker such as VAChT) and a phenotypic protein labeling the target (Prot 2) are co-applied. A mixture of biotinylated secondary antibodies against the first two species is then applied, followed by the avidin–biotin peroxidase complex. Following peroxidase histochemistry, a gold-conjugated secondary antibody against the third species is added, and the bound gold particles are silver enhanced.

neuronal compartments. In the most typical case, BDA is used as the anterograde tracer, and FG is the retrograde tracer of choice. Immunoperoxidase labeling for FG in soma and dendrites is accomplished using the ABC method, which will by design also label the axons that contain BDA (Fig. 2.12A). Immunogold–silver is then used to label the phenotypic marker in the soma and dendrites. In this case, the primary antibodies against FG and the phenotypic protein must be raised in separate species. The method can also be adapted for the use of PHA-L as the anterograde tracer or the use of a phenotypic marker that labels a class of afferent axons from a known source, for example the vesicular acetylcholine transporter (VAChT) to label cholinergic afferents to the VTA that are known to derive from the brainstem tegmentum. In this case, there are three primary antibodies, and each must either be from a different species (Fig. 2.12B) or two can be from the same species as long as they will be segregated into different compartments (e.g., rabbit anti-FG and rabbit anti-PHA-L).

It is important for these experiments that appropriate controls are run to verify the segregation of immunoperoxidase markers. Sections processed only for FG should reveal peroxidase product for this tracer only in dendrites, either diffused in the cytoplasm or concentrated in lysosomes (Figs. 2.5 and 2.13), and not in axons. Similarly, sections processed only for the anterograde tracer should reveal peroxidase product for this tracer diffusely distributed within axons and not soma or dendrites (Fig. 2.13).

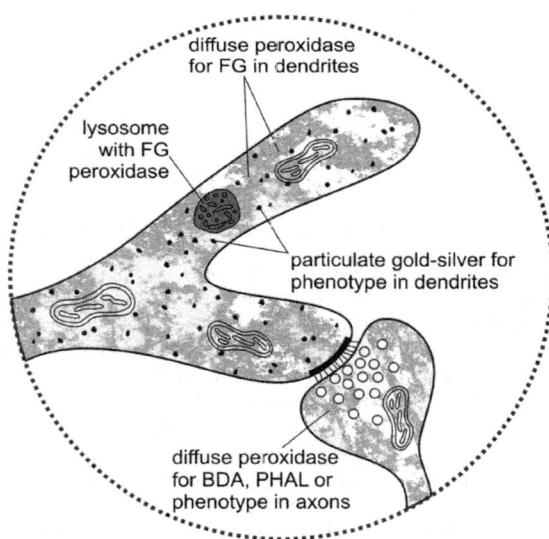

Figure 2.13. Schematic diagram illustrating the differential distribution of tracers and phenotypic markers within different compartments in an experiment in which anterograde and retrograde tract-tracing are combined. Peroxidase or immunoperoxidase is used to label both the retrograde tracer FG within soma and dendrites and the anterograde tracer BDA or PHA-L (or a unique phenotypic marker) within axon terminals. Control experiments are needed for each system under study to ensure that the tracers label only their respective compartments. Immunoperoxidase is typically diffusely distributed within these compartments but is occasionally concentrated within lysosomes in the case of FG. Finally, preembedding immunogold–silver is used to label antigens unique for different neuronal phenotypes in the retrogradely labeled cell population(s).

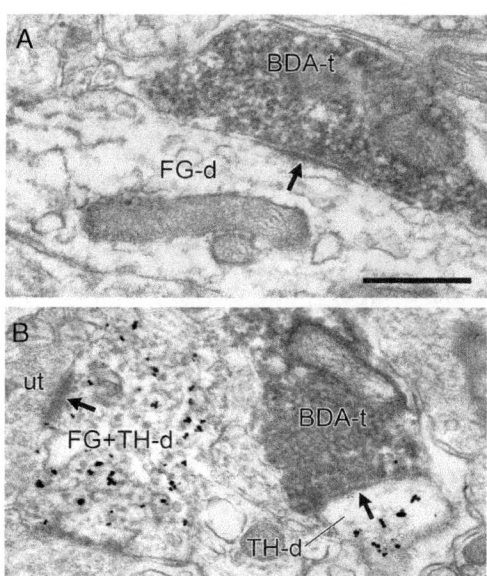

Figure 2.14. Electron micrographs of the rat VTA showing the juxtaposition of axon terminals labeled by peroxidase for BDA (BDA-t) anterogradely transported from the LDT, dendrites singly labeled by immunoperoxidase for FG (FG-d) retrogradely transported from the NAc, singly labeled by immunogold–silver for TH (TH-d), or dually labeled for both markers (FG + TH-d). (A) The BDA-t forms an asymmetric synapse (large arrow) onto the FG-d. In (B) the BDA-t synapses onto the TH-d, while the FG + TH-d receives synaptic input from an unlabeled terminal (ut). Scale bar represents 0.5 μm.

This is especially important for BDA, which is capable of some degree of retrograde transport in certain neuronal pathways. The segregation of the immunogold–silver is less of a concern, as it may appear in soma and dendrites (Fig. 2.13) as well as axons, for example in the case of GABA.

Using this approach, it is possible to see synaptic contacts on several different dendrite populations. For example, the anterograde tracer may occur within axon terminals that synapse onto dendrites containing only the retrograde tracer and not the phenotypic marker (Fig. 2.14A) or onto dendrites that are phenotypically labeled but do not contain the retrograde tracer (Fig. 2.14B). However, in fortuitous cases, the axons containing anterograde tracer are observed to synapse onto dendrites that contain both the retrograde tracer and the phenotypic marker (Fig. 2.15A). Alternatively, axons labeled by a phenotypic marker (e.g., VAChT) will synapse onto dendrites that have these characteristics (Fig. 2.15B).

For this method to provide a useful estimate of the extent to which synapses occur between identified inputs and outputs of a region, it is necessary to perform a certain degree of analysis in serial sections. In many cases, labeled axons may be closely apposed to labeled dendrites without synapsing

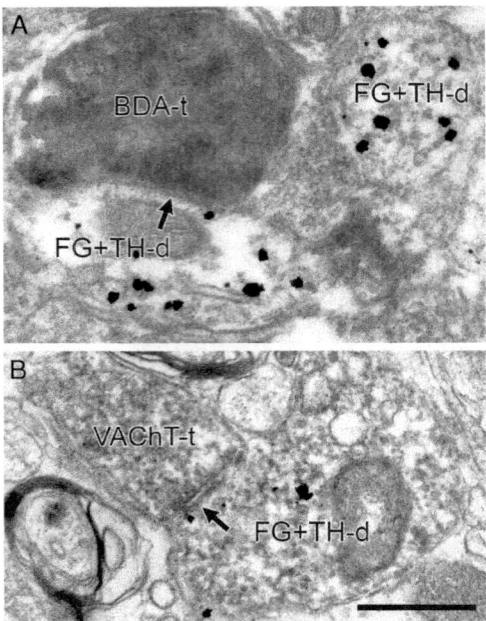

Figure 2.15. Electron micrographs of the rat VTA showing axon terminals labeled by peroxidase for BDA (BDA-t in A) anterogradely transported from the LDT or for VAChT (VAChT-t in B) forming asymmetric synapses (large arrows) onto dendrites dually labeled by immunoperoxidase for FG retrogradely transported from the PFC or the NAc, respectively, and immunogold–silver for TH (FG + TH-d). In (A) a second FG + TH-d does not receive synaptic input in this section. Scale bar represents 0.5 μm.

onto them in one plane of section. In this case, synapses may be revealed in immediately adjacent sections. Serial sections also assist the determination of whether low levels of gold–silver particles are repeated over the same structure in adjacent sections and therefore likely to be specific. However, in this case, it would be preferable to examine sections closer to rather than further from the surface (see Fig. 2.16). In order to thoroughly investigate the possible presence of synapses onto each identified cell population, it is also necessary to sample extensive amounts of tissue. This is particularly true if the investigator uses mesh grids as recommended in section "General Sampling Issues," as the metal obscures part of the tissue being examined. As further discussed in that section below, extensive sampling is necessary to address the possibility that failure to find a particular synapse type is not due simply to underrepresentation in the sample.

Other investigators have developed a triple-labeling approach that utilizes two different anterograde tract-tracing agents and examines whether both inputs converge onto a common target neuron (French and Totterdell, 2002, 2003). In this case, the target neuron is defined not by neurochemical phenotype but by morphological type, as assessed by its uptake of locally administered BDA. The success of this method relies on careful comparative analyses between LM and TEM in order to identify different qualities of immunoperoxidase for the two anterograde tracers and identify regions of probable synaptic input to the target neuron. These regions are then analyzed by TEM to verify the presence of synapses.

III. PRINCIPLES OF THE METHODS

Although tract-tracing and immunocytochemical methods can be combined in a variety of ways, the major steps of the TEM preembedding procedures are usually the same. First, the animal is deeply anesthetized until all pain reflexes are gone. The animal then undergoes transcardial perfusion with aldehydes to fix proteins. The brain blocks are sliced with a Vibratome; some method is used to enhance antibody penetration; and free-floating brain sections are incubated in antibody solutions. Tissue lipids are then fixed with osmium tetroxide, and sections are dehydrated and embedded in plastic resin. After ultrathin sectioning and counterstaining, the tissue is then examined by a TEM. Many of these procedures are common to LM immunolabeling and so involve similar technical concerns: antibody specificity, prevention of nonspecific labeling, penetration enhancement, etc. However, it is important to appreciate that TEM studies are not simply LM experiments with some extra steps. The experiments must be designed with TEM in mind, which typically involves altering procedures in order to balance maximal preservation of ultrastructure with optimal detection of the desired antigens.

A. Animals

The methods presented here are designed primarily for small rodents, although they can be adapted for larger animals. Moreover, many of the immunocytochemical and sampling procedures are fully applicable to other species once fixation is completed and brain sections are cut. An important question to consider for TEM studies is how many animals are typically needed. The answer varies depending on the details of the experiment. If tract-tracing is part of the design, additional animals are often required to allow for misplaced injections. If a new antibody is being tested, pilot studies of optimal fixatives and dilutions are needed. Once all the parameters are optimized, the number of animals required may depend on the results that are obtained. Abundant proteins or highly robust synaptic connections will typically be observed in multiple samples from within and across animals. In that case, three animals each showing similar protein localization or a similar frequency of synapse detection may be sufficient. The more the data seem to vary within or between animals, the greater the sampling required to ensure that the conclusion being developed accurately represents reality.

A question that relates to the issue of variability is the extent to which animals with different densities of labeling should be represented in the overall sample. Obviously, one would not be expected to produce a large sample from an animal in which the immunodetection procedures were clearly suboptimal, just to say that animals were sampled equivalently. Such an approach is sure to produce false-negative results. Conversely, if the tissue from one particular animal seems to have been blessed with the most optimal immunolabeling and ultrastructural preservation, it is tempting to overrepresent this animal in the sample, arguing that it is actually more representative of reality. Of course, a balance needs to be struck between the desire for absolute rigor and the search for "truth" in neuroanatomy. Animals with poor immunolabeling should be excluded from quantitative analyses, and conclusions should not be based on data from single animals. Moreover, clearly stating in published work how the sample was collected and how extensively each animal was represented in the sample should allow colleagues to make their own conclusions regarding the validity of the findings.

B. Tract-Tracing

The principles of experimental design for combining anterograde and retrograde tract-tracing with immunochemical markers have been extensively described in the sections above. Here, we wish to present some methodological issues associated with animal welfare. The surgery necessary to introduce multiple tract-tracing agents can be rather long. However, we chose to combine tracers that have similar survival times (e.g., BDA and FG) specifically so that they can be injected during a single survival surgery. Most institutions require extensive justification for performing multiple survival surgeries, and for the best interests of the animals, we endeavor to avoid such designs. Investigators must choose an anesthetic regimen that can be maintained over several hours. We prefer to use a mixture of ketamine (34 mg/kg), xylazine (7 mg/kg), and acepromazine (1 mg/kg) that is injected i.m. For long surgeries, it is also imperative that the animals' temperature be monitored with a rectal probe and maintained at 36.5°C using a thermostatically regulated heating pad. It is also important to monitor animals closely during recovery from surgery and to administer analgesic if they show any signs of pain or discomfort. We recommend butorphanol, 2 mg/kg, s.c. every 8 h as needed.

C. Phenotypic Labeling

In the sections above, we have also extensively described the principles of immunocytochemistry as related to detection of neurochemical phenotype. One issue that was not addressed is the use of intracerebral (typically intraventricular) injections of colchicine to enhance the content of certain proteins or peptides so that they may be more readily detected by LM or TEM (Dube and Pelletier, 1979; Ford et al., 1995; Graybiel and Chesselet, 1984;

Ribak *et al.*, 1978). By disrupting axonal transport (Paulson and McClure, 1975), colchicine leads to accumulation of peptide in soma and sometimes in dendrites, thereby improving the sensitivity of antigen detection. Most commonly, colchicine is injected into the lateral ventricles and allowed to perfuse the brain through the ventricular system. However, local injections of colchicine within a brain region can reveal the most likely sources of peptide-containing afferents that can then be verified with retrograde tract-tracing (Arluison *et al.*, 1994). In either case, colchicine interferes with axonal transport of tracing agents (Monti-Graziadei and Berkley, 1991), and so if the goal of an experiment is to examine phenotypic markers in retrogradely labeled cells, it is necessary to inject the tracer first and perform intracranial injection of colchicine in a second survival surgery. Because colchicine treatment is distressful for animals, it is recommended that the second survival time be no more than 24–48 h and that butorphanol is given at 2 mg/kg, s.c. every 8 h to relieve suffering. It should also be borne in mind that the full consequences of colchicine treatment are not yet understood. Data indicate that the drug may evoke abnormal gene expression, protein synthesis, and morphological changes within neurons (Pirnik *et al.*, 2003; Rho and Swanson, 1989; Yan and Ribak, 1999).

D. Intracardial Perfusion

1. Pretreatments

Certain neuronal elements contain endogenous metals (e.g., zinc in glutamate nerve terminals) that can complex with silver during the silver enhancement steps for preembedding immunogold–silver labeling. Hence, animals are first treated with a zinc chelator to minimize this source of spurious labeling (Veznedaroglu and Milner, 1992).

2. Choice of Fixative

Immunocytochemical staining methods necessitate labeling antigens in fixed cellular material. The goal of tissue fixation for immunocytochemical processing is the preservation of tissue in a state as close to natural as possible while maintaining the ability of the antigen to react with the antibody. Rapid, thorough preservation of the brain is required. A detailed description of the fixation procedures used in our laboratory is provided in the Appendix. However, it should be noted that the characteristics of some antibodies allow optimal labeling only following certain types of fixation. For instance, the antibody against the neurotransmitter DA is generated against a peptide sequence that is conjugated to glutaraldehyde (Chagnaud *et al.*, 1987). Because of this trait, labeling with this antibody is feasible only following fixation using high amounts of glutaraldehyde. On the other hand, certain fixatives may denature antigens, resulting in little to no staining.

Therefore, the optimal fixative for each antigen–antibody combination must be determined empirically.

The standard fixative for TEM is typically some combination of 4% formaldehyde and glutaraldehyde in concentrations from 0.05 to 1% depending on compatibility with primary antibodies. However, in our experience, certain antibodies actually produce better immunolabeling in tissue fixed with 2% formaldehyde and 3.75% acrolein, a related aldehyde that gives excellent tissue preservation for TEM. This can be true even for antibodies that label only poorly in glutaraldehyde-fixed sections, such as those generated against the monoamine plasma membrane transporters described above. Hence, we consider it worth the effort to attempt acrolein fixation in testing new antibodies, although many laboratories avoid this chemical because the hazardous risks associated with it are considered greater than for glutaraldehyde. The basic procedure for acrolein perfusion was presented in Leranth and Pickel (1989). Here in the Appendix, we provide information on how to safely mix acrolein solutions and perform intracardial perfusions in rodents without undue risk to personnel. Of course, each laboratory should become familiar with the material safety data sheet for acrolein as for any hazardous chemical.

Regardless of the fixative employed, the speed at which fixation of brain tissue is accomplished is essential for the best ultrastructural preservation. In our experience, the time between the opening of the animal's diaphragm and the introduction of fixative should be as short as possible (20–30 s). Extensive saline rinsing to remove blood cells only delays the time at which fixative is introduced. Hence, we recommend the inclusion of heparin with the initial saline rinse to quickly remove blood cells and prevent clotting within the vasculature. It is also important to avoid introducing air bubbles into the system, as these might also block vessel perfusion.

3. Postfixation and Sectioning

A well-fixed brain should contain no visible blood and be firm to the touch. It should be ready for sectioning on a Vibratome after a short postfixation period (30–60 min). Postfixation is usually performed in the final fixative that was perfused through the animal. Inappropriately fixed tissue will often be difficult to section, and although the brain may be further hardened by postfixation overnight, such prolonged exposure to fixative can render antigens of interest inaccessible to primary antibodies. In our opinion, a brain that does not section well should be abandoned unless it is quite valuable. Cutting on a Vibratome is the recommended method of sectioning brain material, as it avoids any artifacts that would be associated with frozen sectioning. As presented previously in this series (Leranth and Pickel, 1989), it is recommended that sections be treated with a brief incubation in 1% sodium borohydride in order to stop fixation (by reducing any aldehydes still exposed to the tissue) and reduce background labeling.

Finally, in the event that immunocytochemical labeling cannot proceed immediately, or the investigator wishes to save sets of tissue for future analysis (not unusual in the case of tract-tracing studies), it is possible to store sections in a cryoprotectant solution (see Appendix, section "Cryoprotection and Tissue Storage"). However, pilot studies should be run to determine whether this storage alters antigenicity for any given protein.

E. Immunocytochemistry

1. Penetration Enhancement

For many immunocytochemical experiments, materials or procedures are introduced to the tissue sections to disrupt cellular membranes in order to potentiate antibody penetration, and thus increase the level of labeling. However, because the membranes must remain visibly intact for electron microscopic examination, only the mildest techniques for enhancing antibody penetration can be employed. The typical compounds used for TEM are detergents such as Triton X-100 (0.04%) or surfactants like PhotoFlo (0.1%). Some investigators use a slightly higher concentration of these reagents but only briefly expose tissue to them during the normal serum blocking procedure. Alternatively, the lower concentrations can be used throughout the primary antibody incubation. In our experience, the use of detergents like Triton tends to reduce background nonspecific labeling, which is advantageous. On the other hand, we have found a few proteins whose antigenicity is actually reduced by detergent treatment (Luedtke *et al.*, 1999; Sesack and Snyder, 1995), most likely due to disruption of transmembrane domains upon solubilization. Hence, penetration enhancement can also be accomplished by treating tissue with a cryoprotectant and subjecting it to rapid freezing and thawing using either liquid nitrogen or a −80°C freezer. The recipe for the latter procedure is given in the Appendix.

2. Choice of Markers

TEM involves passing a beam of electrons through the tissue specimen. Therefore, only markers that can trap electrons and thus appear "electron dense" can be visualized with this method. In contrast to the use of varied colors for LM techniques, there are only a few types of markers that can be distinguished from each other at the TEM level. These include autoradiography (i.e., silver grains developed in a photographic emulsion), gold with or without silver enhancement, and peroxidase chromogens. Autoradiographic immunolabeling has been discussed in detail in a previous chapter in this series (Pickel and Milner, 1989). This method has high sensitivity, and the product is easy distinguishable from immunogold and immunoperoxidase markers. However, it can take 3–12 months to develop autoradiographic material for TEM. Moreover, it is not entirely suited for subcellular localization studies, as radioactivity can spread to produce silver grains a short distance away from antigen sites. In recent years, immunoautoradiography has generally been replaced by methods that are quicker to develop and provide more discrete localization.

In the sections above, we have provided general information regarding the advantages and disadvantages of using immunoperoxidase versus immunogold–silver for preembedding studies of protein localization or synaptic connectivity. Here, we discuss more specific details of the procedures within these two general classes.

a. Preembedding Immunoperoxidase

One of the earliest types of immunoperoxidase labeling involved the application of a soluble complex of peroxidase enzyme bound to antiperoxidase antibodies, termed PAP (for peroxidase antiperoxidase) (Sternberger, 1974). This procedure, which has been previously discussed in this series (Pickel, 1981), involves primary antibody binding to tissue antigens, followed by application of a secondary antibody raised in another species and directed against the species in which the primary antibody was raised, and finally addition of the PAP complex for which the antiperoxidase is raised in the same species as the primary antibody. Following exposure of the peroxidase enzyme to a chromogen substrate (e.g., DAB) and H_2O_2, an electron-dense reaction product is formed. The density of the peroxidase product can then be further enhanced by osmication of the tissue (Johansson and Backman, 1983). Moreover, an additional round of exposure to the secondary antibody and the PAP complex produces amplification of the peroxidase signal (Ordronneau et al., 1981). Subsequent efforts to enhance even further the incorporation of peroxidase, and thus the sensitivity of the method, lead to the introduction of the ABC technique (Hsu et al., 1981). For this procedure, the secondary antibody is conjugated to several molecules of biotin. The tissue is then exposed to a solution containing avidin, which binds to biotin with high affinity, and biotinylated horseradish peroxidase. Similar to the other types of immunoperoxidase procedures, exposure of the enzyme to a chromogen results in a flocculent, electron-dense reaction product. Although most laboratories now use the ABC method as their sole immunoperoxidase technique, we continue to find applications for the PAP method, for example when using immunogold–silver to localize compounds like BDA that would otherwise produce false dual labeling if exposed to ABC (Pinto and Sesack, 1998). In addition, the presence of endogenous biotin in some glial cells (Yagi et al., 2002) may present a circumstance in which the PAP method would be preferred to ABC.

As previously reviewed in this series, there are several chromogens that can be used for immunoperoxidase, the most common being DAB, benzidine dihydrochloride, tetramethylbenzidine, VIP peroxidase, and SG peroxidase (both from Vector Laboratories Burlingame, CA) (Warr et al., 1981;

Zaborszky and Heimer, 1989; Zhou and Grofova, 1995). DAB is the most commonly used chromogen for TEM studies, as it produces the smallest crystalline size and therefore the most diffuse, flocculent reaction product that can fill even small neuronal processes. It is also more stable for the TEM processing steps than tetramethylbenzidine. More recently, VIP peroxidase has shown excellent sensitivity and stability for TEM studies, and its distinctive rosette-like appearance may provide easier recognition within dendrites where the diffuse nature of the DAB reaction product may sometimes be difficult to discern (Zhou and Grofova, 1995). Benzidine dihydrochloride appears to be less sensitive than DAB or VIP peroxidase and its precipitates are not always uniform in appearance (Zhou and Grofova, 1995). Nevertheless, it is still a useful marker for abundant antigens in large processes such as soma and proximal dendrites (Charara et al., 1996). Tetramethylbenzidine is reported to have greater sensitivity than other chromogens, but its instability in aqueous solutions and alcohol requires that it be stabilized for use in TEM (Llewellyn-Smith et al., 1993; Marfurt et al., 1988; Rye et al., 1984). Such stabilization procedures can result in some loss of sensitivity. Because these different immunoperoxidase chromogens produce precipitates of different size, texture, and appearance, they can be used in combination for dual-labeling studies (Norgren and Lehman, 1989; Smith et al., 1994; Zaborszky and Heimer, 1989; Zhou and Grofova, 1995). However, it may not always be possible to distinguish the presence of two different peroxidase products within the same neuronal structure, especially axon terminals (Zhou and Grofova, 1995) (see also section "Limitations: Sources of False-Negative Errors").

Another important consideration for TEM studies is the strength of the immunoperoxidase reaction. We have experimented with changing the concentration of the DAB chromogen, the concentration of the H_2O_2, and the duration of the incubation. In our experience, the main determinant of the amount of peroxidase product generated is the incubation time. The time in solution should be chosen empirically: inadequate exposure will not label antigens deep in the tissue, whereas overexposure will obscure ultrastructural detail. The peroxidase product in overstained profiles may also pierce the plasma membrane and spread into adjacent profiles. This is usually readily detectable in the TEM as peroxidase product in the vicinity of broken membranes. In the event that the sensitivity of immunoperoxidase is needed for a study but a clear view of cellular detail is also critical, the DAB precipitate can be intensified by metallic silver grains through an argyrophil III reaction (Gallyas, 1982). This procedure has been applied with high sensitivity in TEM studies (Liposits et al., 1984) and has been further modified to allow enhancement of DAB reaction product that is deliberately produced at a low level so as not to obscure subcellular detail (Smiley and Goldman-Rakic, 1993). A further variant of this approach is to perform silver–gold enhancement of a nickel-DAB precipitate, and this method produces a reaction product that can be distinguished from nonenhanced DAB for dual-labeling TEM studies (Hajszan and Zaborszky, 2002). In this regard, it should be noted that it is not uncommon to find light silver labeling of immunoperoxidase in tissue that has been dually labeled by the standard ABC and preembedding immunogold–silver methods. However, the size of these particles is generally quite small and therefore readily distinguished from specific immunogold–silver particles.

b. Preembedding Immunogold–Silver

The protocols for immunogold–silver labeling are rather different from the immunoperoxidase steps, and the distinction begins at the time of sacrifice. As noted above, it is recommended that the animal should be treated with a zinc chelator prior to perfusion in order to prevent silver intensification of endogenous zinc (Veznedaroglu and Milner, 1992). Another approach using L-cysteine exposure to reduce tissue argyrophilia in fixed sections has also been developed (Smiley and Goldman-Rakic, 1993).

Following incubation of sections in primary antibody, they are exposed to secondary antibodies conjugated to "ultrasmall" gold particles in the range of 1 μm. There are several commercial sources of these antibodies as well as the silver enhancement solutions that are matched to them. The most common sources of silver reagents used for preembedding immunogold–silver are Amersham Biosciences Corps (Piscataway, NJ), Nanoprobes Inc. (Yaphank, NY), and Aurion (Electron Microscopy Sciences, Fort Washington, PA). Regarding dual- or triple-labeling studies such as those described here (sections "Dual-Labeling Procedures" and "Triple-Labeling Studies"), it is important to note that Electron Microscopy Sciences offers a line of gold-conjugated secondary antibodies raised in donkey. We have found that having all secondary antibodies from donkey helps to avoid species cross-reaction.

A critical determinant for the success of preembedding immunogold–silver is the size of the gold–silver particles, and hence the total incubation time in silver reagents. This issue has been extensively covered in the original reviews of the method (Chan *et al.*, 1990; Pickel *et al.*, 1993). Briefly, at short incubation times, labeled structures appear light gold at the LM level. As silver intensification proceeds, the color moves increasingly toward brown and finally black. In addition, the levels of nonspecific gold–silver deposit increase with time. The LM appearance of specific gold–silver labeling that is considered optimal for TEM is generally in the range of brown. The total incubation time to achieve this is empirical and based largely on the primary antibody and the density of the antigen. It is recommended that each experiment involve a timed series (4–12 min) on a few test sections and then bulk processing of the remaining tissue sections at two different time points, one that is deemed optimal by LM (clear detection of the structures of interest with minimal evident background) and one that is 1–2 min shorter.

The silver processing steps also require extensively clean glassware, which usually involves acid washing. In our experience, it is more convenient, though admittedly more costly, to use disposable cultured cell well plates for this purpose. In addition, there must be no metal ions in the tissue at the time when the enhancement procedure is performed. Any metal ions present in the tissue will be silver intensified in addition to the gold particles, resulting in nonspecific labeling. Therefore, the tissue must be manipulated using nonmetal instruments. For this purpose, our laboratory employs wooden applicator sticks.

Another important issue is the temperature at which the silver reaction is conducted. The amount of time necessary for the silver enhancement is highly dependent on the temperature of the silver solution (i.e., the warmer the solution, the shorter the silver enhancement time required). In our laboratory, we store the silver solutions in the refrigerator, but allow them to reach room temperature prior to use. Furthermore, once the silver enhancement has been performed, the solutions to which the tissue is subsequently exposed must be at room temperature. In other words, the osmium tetroxide solution must be at an ambient temperature prior to osmication of silver-intensified immunogold-labeled tissue. Otherwise, it is possible that tissue expansion and contraction upon temperature changes will dislodge silver-enhanced gold particles from the tissue.

3. Parallel Versus Serial Antibody Incubations

In our experience, the most efficient way to perform dual or triple immunocytochemical labeling is to incubate sections in primary antibodies raised in different species in a parallel manner, followed by serial applications of secondary antibodies for immunoperoxidase and immunogold–silver respectively. Of course, it is also possible to apply primary antibodies in a sequential manner, performing all of the immunoperoxidase procedures before incubation in the second primary antibody. However, we have observed at least one case in which low levels of peroxidase reaction product were washed out during subsequent lengthy antibody incubations. Hence, we continue to prefer the parallel versus the serial approach.

4. Antibody Dilutions

The optimal dilution of primary antibody will vary considerably and must be empirically determined for each antigen–antibody combination and for each brain region. For example, in the VTA where TH immunoreactivity in soma and dendrites is abundant, anti-TH antibodies can be used at higher dilution than in the forebrain where the levels of TH in axons are lower. Moreover, the concentration of primary antibody often has to be several times higher for immunogold–silver than for immunoperoxidase due to the lower sensitivity of the former method. For initial pilot studies, a range of antibody dilutions should be tested that include concentrations published in the literature. Additional recommendations for dilution testing have recently been published (Saper, 2003).

5. Antibody Controls

It is essential for any laboratory using immunocytochemical methods to establish the specificity of the antibodies they are employing. The appropriate control experiments for immunocytochemical detection of antigens for TEM are similar to those required for LM or for confocal microscopy and have been discussed elsewhere (Saper, 2003). Briefly, lack of immunolabeling must be observed following: (1) omission of the primary antiserum, (2) exclusion of the secondary antiserum, (3) preadsorption of the primary antibody with the antigen prior to exposure to the tissue, and, where possible, (4) incubation of tissue sections from transgenic mice in which the antigen has been "knocked out."

When performing double- and triple-labeling procedures, special attention should be paid to ensure absence of cross-reaction between secondary antibodies. Obviously, the best results can be achieved if all secondary antibodies used in the study are obtained from the same species. However, the immunological similarity between certain species (e.g., sheep and goat) can still lead to problems, and we have sometimes noted cross-reaction of certain antibodies against immunologically different species (e.g., anti-rabbit IgG labeling rat primary antibodies). For biotinylated IgG, we recommend the use of antibodies with minimal species cross-reaction, for example those obtained from Jackson ImmunoResearch Laboratories, West Grove, PA. However, as gold-conjugated antibodies are unlikely to be developed for minimal species cross-reaction, it is still necessary to perform the following control experiment. Incubate tissue sections in a mixture of primary antibodies that are known to have distinct compartmentalization or are known not to colocalize (e.g., rabbit anti-TH and mouse anti-GABA in striatal sections). Then divide the tissue into two sets and perform dual immunoperoxidase and immunogold–silver with one of the two secondary antibodies left out of each set. Evidence of markers in inappropriate compartments or unexpected dual labeling of structures will indicate that a particular secondary antibody labels more than its respective species.

F. Tissue Preparation for TEM

1. Osmication

Treatment of tissue sections with osmium tetroxide provides fixation of lipids by rendering them insoluble prior to dehydration and plastic embedding. Osmium also imparts a heavy metal stain to the lipids, enhancing the contrast of membranes. However, osmium is a strong oxidizing agent that can alter the antigenicity of many proteins, making it a challenge to immuno-cytochemically label such proteins following plastic embedding (Hemming *et al.*, 1983). Consequently, investigators have developed methods to substitute other chemicals for osmium to achieve lipid fixation without loss of antigenicity (Phend *et al.*, 1995).

For preembedding procedures, the greatest challenge with osmium is that it can oxidize silver metal to silver salt with resultant loss of the silver used to enhance immunogold labeling. In our experience, loss of silver during the osmication step is primarily associated with the presence of chloride ions in the buffering solution. The use of PBS (phosphate-buffered saline) or PB buffers that have been pH adjusted with HCl introduces chloride anions that can form a salt with silver cations, and this seems to greatly speed the oxidation of silver metal to silver salt. Indeed, when inadvertently using solutions that contained chloride, we have witnessed the oxidation of silver in brain sections to nondetectable levels within 1–2 min. Conversely, scrupulous avoidance of chloride ions (or other halides that can readily form silver salts) can reduce this problem and slow the rate of oxidation so that it is minimal within the typical times used for osmication. Hence, the buffer mixed with osmium tetroxide should be PB that has been pH adjusted with phosphoric acid and contains no chloride ions. Using this reagent, it should not be necessary to shorten the osmication time.

2. Dehydration

Preparation of tissue for electron microscopy requires fixation of lipids followed by extensive dehydration. During these procedures, the volume of tissue decreases (Hillman and Deutsch, 1978), suggesting that a portion of the extracellular space has been lost. Therefore, caution must be exercised in determining some measurements, such as distance between labeled structures, and it should always be acknowledged that such measurements are only semi-quantitative in nature.

3. Counterstaining

Tissue to be examined by electron microscopy is typically stained with heavy metals (first osmium and then uranyl and lead) in order to increase the electron density of membranes and enhance contrast. Many procedures call for application of uranyl acetate en bloc, meaning on sections during the alcohol dehydration steps (usually with the 70%). However, such treatment can sometimes make it difficult to identify sparse immunoperoxidase labeling, such as that associated with low abundance proteins. For this reason, we routinely apply heavy metal stains only after ultrathin sectioning and only on a portion of the grids. In this case, the investigator has the option to omit uranyl acetate and stain ultrathin sections only with lead citrate in order to visualize low amounts of immunoperoxidase.

G. Tissue Sampling

1. General Sampling Issues

In preembedding tissue labeled by immunoperoxidase or immunogold–silver, it is crucial to examine only the surface of sections, which is represented by the interface between the tissue and the plastic embedding medium. Because large amounts of penetration enhancers are not feasible for immunolabeling of tissue for TEM, antibodies do not have access to the full extent of the section thickness. Thus, maximal antibody penetration is limited to a few microns of the tissue surface. Moreover, even if one restricts sampling to the tissue surface, the actual protein levels will be underestimated with these techniques, due to the chemical fixation and low concentrations of penetration enhancers. Hence, immunocytochemistry for TEM involves a compromise that balances immunodetection with morphological integrity. Immunoreactivity that is excessive on the surface may not penetrate more than a few microns (Fig. 2.16). Conversely, morphological preservation may be excellent within the depths of the tissue section but compromised at the surface where antibody penetration is greatest. Moreover, the extent to which immunoreagents penetrate the tissue is not equal between methods, with preembedding immunogold–silver generally penetrating less well than immunoperoxidase (Fig. 2.16) (Chan *et al.*, 1990).

Knowing that immunoreactivity is confined to the outer surface of flat-embedded sections, the investigator must decide what approach to take regarding ultramicrotomy. Some investigators actually turn their thick sections on edge and cut ultrathin sections perpendicular to (i.e., at 90° from) the original plane of sectioning. This approach has the advantage of allowing the experimenter to measure precisely the distance from the tissue surface where immunolabeling becomes weak or nondetectable. However, in addition to being a considerable technical challenge, this method has the disadvantage of causing lost perspective regarding how the area being sectioned relates to typical LM views of the region of interest. It is also possible to cut the tissue surface at an oblique angle so that the zone containing optimal antibody penetration and tissue morphology is "stretched" over many sections. However, in our experience, this reduces the size of the useful zone within each ultrathin section and creates the need to collect many more sections in order to obtain a sufficient sample size. Hence, our approach has always been to embed the thick sections in plastic as flat as possible (by placing them under glass slides and heavy lead bricks) and to perform ultramicrotomy en face.

In order to accomplish the goals of analysis at the tissue surface, we further recommend collecting ultrathin sections on grids that contain at least three sections, each with part tissue and part embedding resin. Focusing the analysis on the middle section allows serial examination in sections both deeper and more superficial to the central one (Fig. 2.16). Using mesh grids allows the creation of a grid map for estimating the area analyzed and relocating

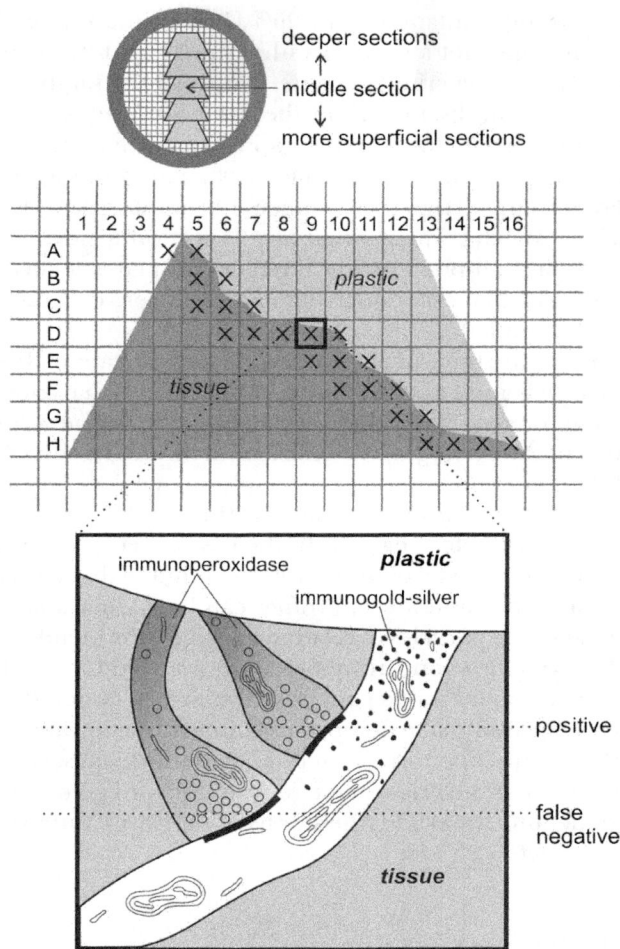

Figure 2.16. Schematic diagrams showing the sampling strategy used for preembedding immunoelectron microscopic studies (see text, section "General Sampling Issues").

regions of interest during serial section analysis or in the event that better photomicrographs are required. Mesh grids are also generally tougher and stand up well to repeat handling. Of course, loss of desired regions under the metal mesh is inevitable with these grids. If this becomes a problem for a study, then slot grids are recommended so that all of the tissue is supported on a nonobscuring film like formvar. However, slot grids are generally more fragile and it can be difficult to relocate specific areas of interest if it becomes necessary after a viewing session.

At low magnification, it is advisable to take a photomicrograph or make a drawing of the ultrathin section and assign letter and number coordinates to each grid square. At 3000× magnification, our laboratory chooses grid squares that (a) contain at least 25%, but no more than 75% tissue or (b) share an edge (not a corner) with a square that meets the first criterion (Fig. 2.16). These criteria ensure that the TEM sample is collected at a relatively uniform distance from the tissue surface. Recording which grid squares are to be analyzed and the approximate amount of tissue that each contains (25, 50, 75, or 100%) allows estimation of the area of tissue analyzed. For example, thin bar copper 400 mesh grids (Electron Microscopy Sciences) have squares that are 55 μm on a side, or 3025 μm^2 area. This is then multiplied by the number of squares analyzed and the proportion of tissue that each square contained to derive an estimate of the total tissue area sampled.

Selected squares are then analyzed at higher magnification (10–30,000×). For each square, it is good to record the number of fields that contain specific peroxidase and/or specific gold–silver labeling. We define a "field" as the area delimited by the photographic brackets on the microscope. In this case, either fields that are photographed or simply analyzed for content without photo documentation can be recorded for the number of events per unit area (area of the bracketed region). In the case of receptor or transporter studies, this gives an estimate of the density of immunolabeling (i.e., number of labeled profiles per unit area). Profiles can be further analyzed for the position of gold–silver particles in relation to the plasma membrane or other structures. For synaptic connectivity, fields are analyzed for whether labeled processes (usually axons) contact unlabeled or labeled targets (usually soma, dendrites, or spines) or make no obvious contacts. A coordinate relocation system on the microscope can be used to examine serial sections to verify whether labeling of profiles is specific (e.g., number of gold–silver particles per profile or per unit area) and whether synaptic specializations are present at points of contact.

2. Criteria for Immunoperoxidase

In general, specific labeling using immunoperoxidase staining is easy to discern as a flocculent, dense precipitate within labeled structures. Occasionally, nonspecific or "background" labeling may also be present. Unfortunately, there are no established principles for estimating the amount of background staining in immunoperoxidase-labeled tissue. In fact, in some instances, nonspecific labeling may be somewhat difficult to determine, because it is typically related to the primary antibody. In other words, if the primary antibody is omitted, the background staining disappears. It is for this reason that simply omitting the primary antibody is not an acceptable test for specificity; rather, it is merely a control for the specificity of the secondary antibody. Specific immunoperoxidase staining should be confined to structures that have the potential to synthesize the antigen. For instance, labeling in cells that do *not* express the mRNA for the protein has a high likelihood of indicating nonspecific or background reactivity and must be examined with caution. Background labeling also tends to be more diffuse and less dense than specific peroxidase product. Nevertheless, we have often noted that diffuse cellular immunoperoxidase labeling that is visualized by LM is not always evident by TEM. Conversely, TEM may detect immunoreactivity that is too weak to produce a visible signal in LM. In any event, the presence of suspected background labeling necessitates rigorous tests for specificity of the primary antibody.

3. Criteria for Immunogold–Silver

Immunogold–silver labeling is sometimes associated with a higher probability of background labeling than is immunoperoxidase, necessitating the establishment of consistent criteria for determining specific immunoreactivity. Such criteria will depend on the localization and density of the antigen as well as the level of background immunogold–silver labeling. Basically, experimenters should be confident that their criterion for specific immunogold–silver labeling would not be met by randomly distributed gold–silver particles throughout the tissue, and this in turn is affected by the size of the structure of interest. For example, the probability that three gold–silver particles would distribute randomly within a proximal dendrite is high, but the presence of three random particles within an axon terminal is a low probability event. Similarly, the probability that two gold–silver particles within an axon terminal would both be randomly distributed to the plasma membrane is a low probability event compared to the same particles localized to the cytoplasm. Hence, for our published studies of plasma membrane transporters, we have set our criteria for specific immunogold–silver labeling within axons as at least two particles on the plasmalemma or three in total (including those in the cytoplasm) (Miner *et al.*, 2000, 2003c). These criteria are somewhat more conservative than other laboratories (Garzón *et al.*, 1999; Pickel and Chan, 1999), and so we must acknowledge a higher likelihood of false-negative outcomes in our studies. On the other hand, we can more confidently assert the absence of false-positive results.

IV. SUMMARY OF ADVANTAGES AND LIMITATIONS

A. Advantages

Preembedding immunoperoxidase methods involving signal amplification have the advantage of superior sensitivity for the localization of sparse antigens. This is particularly true for TEM, as the greater resolving power of electron microscopy allows detection of low levels of peroxidase reaction product that may appear too diffuse for LM identification. The more discrete and nondiffusible marker associated with the preembedding immunogold–silver technique provides the advantage of indicating the precise subcellular localization of antigens, including neurotransmitter receptors and transporters. Discrete gold particles, with or without silver enhancement, are also advantageous because they are more readily quantified than the precipitate formed by peroxidase reaction. When combined, the immunoperoxidase and immunogold methods provide a powerful means for localizing two or more antigens in relation to each other. With TEM detection, these approaches allow the identification of specific synaptic relationships formed between cellular elements that are labeled by tract-tracing and/or neurochemical phenotype. If two antigens are known to be localized to separate neuronal compartments (e.g., anterograde and retrograde tracing agents in dendrites and axons, respectively), then the combination of preembedding immunoperoxidase and immunogold methods can be used for triplelabeling studies, specifically the determination of whether afferents into a region of interest synapse onto populations of cells identified both by their neurotransmitter phenotype and major axonal target.

B. Limitations: Sources of False-Positive Errors

For any TEM study involving tract-tracing and immunocytochemistry, consideration must be given to potential sources of false-positive and falsenegative results. Common sources of false-positive errors are cross-reaction of the primary antibody with unknown proteins, cross-reaction of secondary antibodies with inappropriate species, nonspecific immunolabeling, or erroneous transport of tract-tracing agents. The evidence supporting specificity of immunoreagents for TEM is the same as for LM studies and should be demonstrated with appropriate controls prior to the experiment. Much has been written on this subject and on sources of nonspecific/background labeling. The reader is referred elsewhere for a consideration of these issues (Saper, 2003) (see also section "Antibody Controls").

For experiments involving tract-tracing, the investigator should be familiar with the LM literature on afferents and efferents of each region of interest and hence the potential sources of false-positive results if tracer injections spread beyond the target. Such cases should be eliminated if tracer spread would involve an adjacent pathway that is not the one under study. Another common source of false-positive labeling in tract-tracing studies of the type described here is uptake of retrograde tracer by fibers of passage. The extent of this problem varies with different retrograde tracers, but the tracer with the least amount of uptake into passing fibers is PRV. If performing retrograde tract-tracing from a region in which uptake by fibers of passage is likely to produce false-positive results, it is recommended that PRV be used, and that potential contributors to false positives are systematically checked in control experiments. Finally, and as described above, use of BDA for anterograde tracing has the potential to produce false-positive results if it undergoes retrograde transport and subsequent anterograde trafficking into collateral axons. When using BDA, care should be taken to examine the most likely brain regions in which retrograde transport might occur. Evidence of retrograde transport indicates that collateral transport is a possibility; such cases should be discarded, and PHA-L should then be used as the preferred anterograde tracer.

C. Limitations: Sources of False-Negative Errors

False-negative results are a common concern in studies using TEM immunocytochemistry due to the limited penetration of immunochemicals in tissue prepared with minimal or no detergent. This drawback especially affects the detection of immunogold reagents, which typically penetrate less deeply into the tissue than immunoperoxidase compounds (Chan *et al.*, 1990). Although these limitations cannot be completely avoided, their impact can be minimized by utilizing the less sensitive immunogold–silver method to label antigens in high abundance, confining the analysis to the surface of the sections where both peroxidase and gold–silver markers are present, and analyzing serial sections for all profiles with sparse labeling. Anterograde and retrograde tract-tracing can also contribute to false-negative results, because all neurons in a population of interest are unlikely to be labeled in any given study. The use of multiple animals with slightly different injection sites can help to overcome this limitation, as can the use of large injections where possible. Finally, for retrograde tract-tracing with FG, an additional source of false-negative results discussed earlier (section "Retrograde Tract-Tracing") is that the tracer may be confined to lysosomes and not spread diffusely within dendrites. Extensive sampling, examination of dendrites in serial section, and following dendrites cut longitudinally to see whether lysosomes are present can help to minimize this problem.

An additional methodological consideration for the methods presented here is the ability to detect the presence of multiple markers within one profile. Unfortunately, the limitations of preembedding immunocytochemistry are such that it may not be possible to detect the colocalization of markers in every structure in which these antigens are present, particularly when the neuronal structures are small. This limitation may result from imperfect antibody penetration, spatial interference when relatively large reagents compete for access to antigens within confined spaces, differential location of antigens (e.g., cytoplasmic versus vesicular), disproportional concentration of antigens, and unequal sensitivity of immunoperoxidase and immunogold–silver methods. In our experience, such impediments typically have a greater impact on the dual labeling of axon terminals as opposed to dendrites. Sometimes, the problem can be addressed by switching the order in which immunoperoxidase and immunogold labeling are performed (Katona *et al.*, 2001) or using a sequential method of incubation in primary antibodies rather than the parallel method recommended here. Adjusting the concentrations of the primary antibody may also help, for example, reducing the concentration of the antibody directed against the antigen in greatest abundance. However, this issue remains a limitation of preembedding methods and one for which postembedding methodologies (Charara *et al.*, 1996; Smith *et al.*, 1996) may be superior for avoiding false-negative outcomes.

V. PROSPECTS FOR THE FUTURE

It is hoped that microarray studies (see chapter by Ginsberg *et al.*) will eventually identify unique phenotypic markers for neuronal pathways and so render obsolete the need for tract-tracing. Having such protein markers would avoid the need for survival surgery in animals. Moreover, identifying pathways based on unique protein signature is likely to have greater sensitivity compared to tract-tracing, in that antibodies have the potential to detect all the neuronal structures that contain that protein, whereas tracer injections virtually never include all the cells and/or processes that contribute to a given projection. Of course, limited antibody penetration will still present difficulties for interpretation of TEM studies of protein localization or synapse identification. A second potential solution to the limitations of tract-tracing would be the introduction of genetically altered animals in which marker proteins (e.g., green fluorescent protein) are expressed in cells of interest (Zhao *et al.*, 2004) that are otherwise difficult to study by standard methods, for example GABA neurons that project from the VTA to the PFC. The ability to be certain that the marker protein is expressed in all such cells and their processes will allow a more complete analysis of their synaptic organization at both cell body and nerve terminal levels.

APPENDIX

Here, we will present a full protocol for the immunoperoxidase and immunogold–silver procedures for TEM. This can be followed for single-, dual-, or triple-labeling experiments.

A. Recipes for Standard Buffers

1. 0.2 M Sodium Phosphate Buffer

1000 ml distilled water
21.8 g sodium phosphate dibasic
6.4 g sodium phosphate monobasic
pH to 7.3 with phosphoric acid, not HCl (see section "Osmication")
Dilute 1:1 with water to make 0.1 M sodium phosphate buffer (PB)

2. 0.01 M Phosphate-Buffered Saline (PBS)

1000 ml distilled water
50 ml 0.2 M PB
9 g sodium chloride

3. 0.1 M Tris-Buffered Saline (TBS)

1000 ml distilled water
12.1 g trizma base
9 g sodium chloride
pH to 7.6 with HCl

B. Fixation

1. Rat Preparation for Immunogold Procedure

Animals are first anesthetized with 60 mg/kg pentobarbital i.p. and then given 1 g/kg i.p. of diethyldithiocarbamic acid (Sigma, St. Louis, MO) for 15 min prior to aldehyde perfusion. During this treatment, animals should be carefully monitored for seizures, which can be induced by chelation of zinc. In the event that seizure activity is detected, animals should be given supplemental doses of anesthetic.

2. Perfusion and Fixatives

Correctly prepared fixatives for perfusion should be filtered and then checked to ensure that they are clear and colorless.

a. 3.8% Acrolein, 2% Formaldehyde in 0.1 M PB

Rats are perfused with 10 ml heparin saline (1000 U/ml; Elkins-Sinn, NJ), followed by 50 ml of 3.8% acrolein and 2% formaldehyde, followed by 200–400 ml of 2% formaldehyde in 0.1 M PB. Coronal blocks of brain are postfixed in 2% formaldehyde for 30–60 min.

In a well-ventilated hood, prepare 2% formaldehyde in 0.1 M phosphate buffer as follows. Heat 500 ml of ultrapure water in a 1-l glass beaker to 60–65°C. Do not exceed 65°C. Turn off the heat and add 20 g of EM grade granular paraformaldehyde (Electron Microscopy Sciences; Fort Washington, PA), stirring constantly. Stir for several minutes and then add small volumes of 1 N NaOH, stirring for several minutes after each addition until the solution is mostly clear. A few granules of paraformaldehyde might still be present. Filter through a Buchner funnel and an aspiration flask using #3 filter paper. Also filter 500 ml of 0.2 M PB. Transfer the solution to a beaker.

For one rat, measure 48.1 ml of the 2% solution of freshly depolymerized paraformaldehyde (i.e., formaldehyde) into a 100 ml graduated cylinder. We recommend the use of acrolein from Electron Microscopy Sciences because it is supplied in 2 ml single-use glass ampoules and so does not require storage of opened containers of acrolein. Wearing gloves and eye goggles use a 5-cm^3 syringe and an 18-G needle (the large gauge is needed because of the high vapor pressure) to remove 1.9 ml of acrolein from the glass

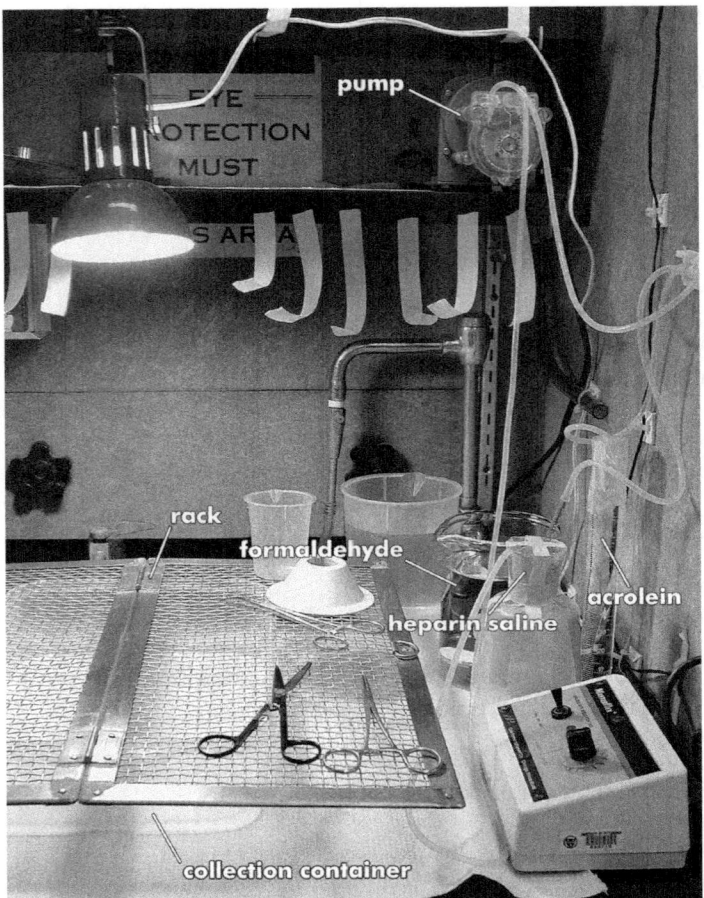

Figure 2.17. Equipment setup for intracardial perfusion with acrolein by using a peristaltic pump. The rack above a collection container allows recovery of perfusate for safe disposal, and the entire assembly is placed within a laminar flow hood to contain acrolein vapors.

ampoule. Add acrolein to the graduated cylinder; seal tightly with parafilm and invert several times to mix. For two rats, use two ampoules of acrolein and add 3.8 ml of acrolein to 96.2 ml of 2% formaldehyde.

The perfusion system that is needed for this fixative consists of a peristaltic pump and tubes that are attached to a three-way stopcock with two inlets and one outlet, all assembled in a well-ventilated laminar flow hood (Figs. 2.17 and 2.18). This system allows the delivery of two to three different solutions without the introduction of air bubbles that might block brain capillaries. Although not shown in the figure, we recommend using a metal clamp stand to ensure that the graduated cylinder containing acrolein does not inadvertently tip over.

The rat is placed on a perfusion rack that is set over a plastic container in order to collect the blood with acrolein perfusate. Full descriptions of

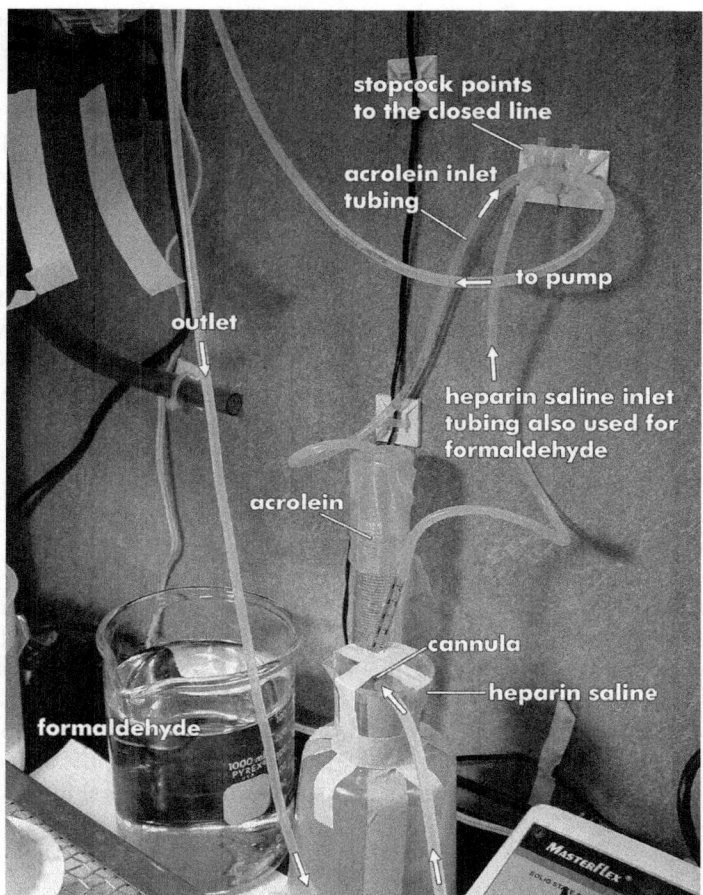

Figure 2.18. Tubing arrangement for acrolein perfusion. The use of a three-way stop-cock allows for two inlet lines and one outlet. The first inlet tube is primed for acrolein, and the stopcock is then switched to shut off the acrolein and allow flow through the second inlet tube. This is first rinsed with water to remove residual acrolein and then primed with heparin saline. The outlet tube is run into the heparin saline, which circulates while the animal's abdomen and thorax are exposed. The cannula from the outlet tube is then inserted through the base of the left ventricle into the aorta and clamped in place. The stopcock is then switched to allow acrolein to perfuse the animal. This shuts off the flow of heparin saline, making it safe to move the inlet tube from the heparin saline into the formaldehyde. Once the correct volume of acrolein is perfused, the stopcock is switched one final time to complete the perfusion with formaldehyde.

the surgical procedure can be found elsewhere (Friedrich and Mugnaini, 1981; Leranth and Pickel, 1989). Briefly, once the anesthetized rat's pain reflexes have ceased completely, tape down the arms and tail. Turn the peristaltic pump to run at a flow rate of approximately 50 ml/min. Cut a wide opening in the lower abdomen; find the xiphoid process and clamp with a regular hemostat. Holding the hemostat cut up either side of the rib cage, cut open the diaphragm and remove any membrane surrounding the heart. Slit open the right atrium and then the base of the left ventricle. Push the outflow tubing with the cannula through the left ventricle and up into the aorta; clamp the cannula into the aorta with a vascular hemostat. The cannula can also be clamped onto the left ventricle if preferred, but in either case the hemostat should be propped to keep it from twisting. Immediately switch the stopcock to introduce the acrolein with 2% formaldehyde and turn up the perfusion speed to 90 ml/min; pump through 50 ml of acrolein. In the meantime, transfer the heparin inflow tubing into the plain 2% formaldehyde. After acrolein, switch the stopcock to pump through 200 ml of formaldehyde and turn down the pump speed to 80 ml/min.

When handling acrolein, always wear gloves and eye protection. Vials should only be opened and the perfusion should only be performed in a well-ventilated hood with the shield lowered. Acrolein remaining in the ampoules and any fluids containing acrolein after the perfusion is completed should be collected into a labeled glass waste container that is stored in a flammable liquids cabinet until it can be properly disposed. Use a funnel to empty the blood/perfusate into the glass storage bottle and rinse several times with water until the collection container is safe to remove from the hood for final cleaning. Any items that contact acrolein (vials, ampoules, syringes, etc.) should be kept in the hood at least overnight until the solution has evaporated (even a small amount of acrolein put in the trash will soon become evident to anyone in the room). If a spill occurs outside the hood, evacuate the room immediately and call chemical safety. If possible, turn on the hood, as this may help to clear the acrolein vapors and contain them to the affected room. Use a safety shower to wash any skin or clothing that contacts acrolein.

b. 0.05–1% Glutaraldehyde, 4% Formaldehyde in 0.1 M PB

Rats are perfused first with heparin saline as above, followed by 500 ml of the para/glut fixative. Coronal blocks are postfixed in 4% formaldehyde for 30–60 min.

In a well-ventilated hood, prepare 4% formaldehyde in 0.1 M phosphate buffer as follows and as described in detail above. Heat 500 ml of ultrapure water. Add 40 g of EM grade granular paraformaldehyde, followed by small volumes of 1 N NaOH until clear. Filter solution, followed by 500 ml of 0.2 M PB. Transfer the solution to a beaker. For the standard 0.2% glutaraldehyde, add 8 ml of 25% EM grade glutaraldehyde to a liter of the 4% freshly depolymerized paraformaldehyde.

3. Vibratome Sectioning

After perfusion, remove the brain from the skull and cut it into thick blocks that contain the brain regions of interest. Postfix for the times described above, and then transfer to 0.1 M PB and section on a Vibratome. For the best sectioning, use well-fixed brains, fill the Vibratome with 0.1 M PB that has been cooled to 4°C, use fresh, sharp blades, and set the slicer to low speed and wide amplitude. Section at 40–60 μm and collect sections in serial order into cell wells containing 0.1 M PB.

4. Sodium Borohydride and Hydrogen Peroxide Treatments

Rinse sections in 0.1 M PB and divide into multiple conditions as desired. Incubate sections for 30 min in 1% sodium borohydride in PB. As sections will float to the top, maintain hydration by mixing them back down occasionally. Rinse sections extensively in PB until all bubbles are gone. If endogenous peroxidase activity has the potential to confound interpretation in the study, treat sections for 15 min with 3% H_2O_2 in 0.1 M TBS and then rinse extensively in this buffer.

5. Cryoprotection and Tissue Storage

For long-term storage in a solution that is compatible with later EM analysis, place sections into the following cryoprotectant (generously provided by Darlene Melchitzky) and freeze at -20°C. Other potential cryoprotectant solutions may also serve this purpose (Rosene *et al.*, 1986).

Storage cryoprotectant

> 300 ml ethylene glycol
> 300 ml glycerol
> 100 ml 0.2 M PB
> 300 ml ultrapure water

C. Immunolabeling

Immunolabeling procedures are performed on free-floating sections at room temperature with constant shaking unless otherwise specified.

1. Primary Antibody Steps and Penetration Enhancement

a. Optional, Freeze–Thaw Procedure (protocol generously provided
  by Dr. Yoland Smith)

Place sections in cryoprotectant for 20 min.
Freeze thaw cryoprotectant

200 ml 0.2 M PB
520 ml distilled water
80 ml glycerol
200 g sucrose
qs final volume to 1000 ml
Solution can be stored in −20°C freezer.

Place sections into −80°C freezer for 20 min.

Thaw sections at room temperature for 10 min each in the following
solutions: cryoprotectant at 100, 70, 50, and 30% diluted in PBS.

Rinse sections in PBS (3 × 5 min).

Rinse sections in TBS (3 × 5 min).

b. Blocking Solution

87 ml TBS
10 ml 0.4% Triton X-100 (Sigma) (optional)
Final concentration is 0.04% Triton.
3 ml normal serum
1 g BSA

Place sections in blocking solution for 30 min.

Incubate sections in primary antibody made up in blocking solution
overnight at room temperature or over two nights at 4°C.

Rinse sections in 0.1 M TBS (1 min then 3 × 10 min).

If single labeling for immunogold–silver is being performed, proceed to
section "Immunogold Labeling." Otherwise, follow the next steps.

2. Immunoperoxidase Labeling

Mix biotinylated secondary antibody in blocking solution. Typical final
concentrations are 1:100 to 1:400.

Incubate sections in secondary antibody for 30 min.

Mix ABC complex solution using the Vectastain Elite kit (Vector Laboratories) by adding two drops each of A and B solutions to 10 ml of
0.1 M TBS. Allow to stand at least 30 min before use. Take care not
to contaminate the dropper bottles and do not overmix or vortex the
solution.

Rinse sections in 0.1 M TBS (1 min then 3 × 5 min).

Incubate sections in ABC solution for 30–120 min.

Rinse sections in 0.1 M TBS (1 min then 3 × 5 min).

Prepare the DAB (Sigma) immediately before use. To 100 ml of 0.1 M
TBS, add 22 mg of DAB and 10 μl of 30% H_2O_2. Filter the solution.

Incubate sections in DAB for 3–6 min, depending on the desired strength
of the reaction, keeping in mind that copious labeling in the light
microscope may appear overblown by TEM.

Stop the reaction by rinsing sections in 0.1 M TBS (1 min then 3 × 5 min).

For tissue to be plastic embedded after immunoperoxidase labeling, transfer sections to 0.1 M PB (2 × 5 min) and proceed to section "Tissue Preparation for Electron Microscopy." Otherwise, follow the next steps.

3. Immunogold Labeling

Rinse sections in 0.01 M PBS (1 min then 3 × 5 min).
Place sections in washing buffer for 30 min.
Washing buffer

> 93.5 ml 0.01 M PBS
> 0.8 g BSA
> 0.5 ml fish gelatin (comes with the gold secondary antibody)
> 6 ml normal serum

Incubate sections in gold-conjugated secondary antibody diluted 1:50 in washing buffer for 2–4 h or overnight.
Rinse sections in washing buffer (1 min then 3 × 5 min).
Rinse sections in 0.01 M PBS (1 min then 3 × 5 min).
In a ventilated hood, fix sections in 2% glutaraldehyde in PBS for 10 min.
Remove silver enhancement solutions from the refrigerator (see below).
Rinse sections in 0.01 M PBS (1 min, then 3 × 5 min) until odor of glutaraldehyde is gone.

0.2 M sodium citrate buffer

> 100 ml ultrapure water
> 5.88 g sodium citrate dihydrate
> pH to 7.4 using citric acid

0.2 M citric acid

> 100 ml ultrapure water
> 4.2 g citric acid monohydrate
> Keep refrigerated.

4. Silver Enhancement

Fill a 12-well cultured cell plate and a 24-well plate as shown in Figure 2.19. For the 0.1 M PB in the last two rows of wells, make sure to use solution that is not contaminated by chloride ions, as these are the final steps before osmication. Divide up the tissue in batches in the first row of 0.01 M PBS depending on the number of silver enhancement times to be tested. Do not proceed until all the wells (except silver) are filled with solution and the silver enhancement solutions are at room temperature. Do not allow sections to sit for long periods in the citrate buffer, as this has only weak buffering capacity that is not optimal for morphological preservation.

Once all is ready, fill the second row of the 24-well culture plate with equal drops of IntenSE M kit A and B (Amersham, Arlington Heights, IL)

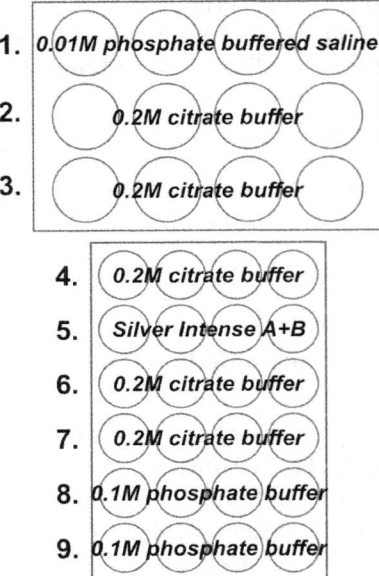

1. 0.01M phosphate buffered saline
2. 0.2M citrate buffer
3. 0.2M citrate buffer
4. 0.2M citrate buffer
5. Silver Intense A+B
6. 0.2M citrate buffer
7. 0.2M citrate buffer
8. 0.1M phosphate buffer
9. 0.1M phosphate buffer

Figure 2.19. Schematic illustration of cell culture wells showing the order of reagents used for silver intensification at different time points.

solutions. The exact volume will depend on the number and size of the sections, but usually 10 drops each of A and B are sufficient for 4–8 sections.

Using a wooden applicator stick with a pointed end (snap in half from the ends to get a clean break), transfer sections to 0.2 M sodium citrate buffer in the second row of the 12-well cell plate. Rinse sections for 1 min. Using a new wooden applicator stick (to minimize contamination by phosphate), transfer sections to the third row of the 12-well cell plate and rinse for 1 min. Now transfer sections to the citrate buffer in the first row of the 24-well cell plate. Rinse for 1 min. Then transfer sections to the silver solution (second row of the 24-well cell plate) and start a timer.

Gently swirl the plates and time the silver enhancement reaction carefully, as differences of 30 s are significant. If needed, use a dissecting microscope to watch the progress of silver intensification. Stop the silver enhancement reaction by rinsing in buffer. Do not return sections to the silver once they are removed. Transfer sections into the third through sixth rows of the 24-well cell plate in succession (1 min each for citrate buffer and longer for PB). Sections can remain in the final row of 0.1 M PB until dishes for osmication are ready.

In practice, we have found it useful to perform the steps above for a pilot determination of optimal silver enhancement times on a small number of sections (see also section "Choice of Markers"). We usually run test sections at wide intervals (e.g., 4, 8, and 12 min), mount these on slides, and examine them by LM. The most optimal silver enhancement time is that which produces clearly labeled neuronal elements, typically with a golden-brown color, with minimal background (e.g., few silver-enhanced gold particles in the white matter). Pale gold labeling is probably underdeveloped; black labeling is considered overdeveloped and likely to be associated with greater background particles (see also Chan *et al.*, 1990). For the actual silver enhancement of the experimental sections, we recommend dividing the sections into half and running them at two different times, 1–2 min apart. Subsequent TEM analysis may reveal that one of these tissue sets has better labeling characteristics with regard to specific and nonspecific labeling.

D. Tissue Preparation for Electron Microscopy

Rinse sections in 0.1 M PB (2×5 min) in Coors dishes. Place dishes in a well-ventilated hood. Prepare 2% OsO_4 in 0.1 M PB by mixing equal volumes of 4% OsO_4 and 0.2 M PB. Make sure sections are lying flat in the Coors wells then slowly draw off the phosphate buffer without disturbing the sections. Gently add OsO_4 taking care not to twist or fold the sections. Incubate in OsO4 for 1 h in the hood.

The remaining steps for tissue preparation are standard in electron microscopy (Friedrich and Mugnaini, 1981) and will not be presented in detail here. Briefly, after rinsing several times with PB, sections are dehydrated at 10-min intervals through a standard series of increasing strength ethanol solutions (30%, 50%, 70%, 95%, 100%, 100%), then twice in 100% propylene oxide or acetone, followed by embedding resin equally mixed with propylene oxide or acetone. The sections are left for several hours or overnight and then infiltrated with pure resin for 2–4 h. Sections are then embedded between sheets of commercial plastic. Once polymerized at 60°C for at least 18–24 h, the resin-embedded sections are trimmed and cut on an ultramicrotome, and ultrathin sections are collected onto mesh or coated slot grids.

REFERENCES

Alheid, G. F., Edwards, S. B., Kitai, S. T., Park, M. R., and Switzer, R. C. I., 1981, Methods for delivering tracers, In: Heimer, L., and RoBards, M. J. (eds.), *Neuroanatomical Tract-Tracing Methods*, New York: Plenum Press, pp. 91–116.

Aoki, C., Miko, I., Oviedo, H., Mikeladze-Dvali, T., Alexandre, L., Sweeney, N., and Bredt, D. S., 2001, Electron microscopic immunocytochemical detection of PSD-95, PSD-93, SAP-102, and SAP-97 at postsynaptic, presynaptic, and nonsynaptic sites of adult and neonatal rat visual cortex, *Synapse* **40**:239–257.

Arai, R., Kojima, Y., Geffard, M., Kitahama, K., and Maeda T., 1992, Combined use of silver staining of the retrograde tracer WGAapoHRP-Au and pre-embedding immunocytochemistry for electron microscopy: demonstration of dopaminergic terminals in synaptic contact with striatal neurons projecting to the substantia nigra in the rat, *J. Histochem. Cytochem.* **40**:889–892.

Arluison, M., Brochier, G., Vankova, M., Leviel, V., Villalobos, J., and Tramu G., 1994, Demonstration of peptidergic afferents to the bed nucleus of the stria terminalis using local injections of colchicine. A combined immunohistochemical and retrograde tracing study, *Brain Res. Bull.* **34:**319–337.

Aston-Jones, G., and Card, J. P., 2000, Use of pseudorabies virus to delineate multisynaptic circuits in brain: opportunities and limitations, *J. Neurosci. Methods* **103:**51–61.

Baude, A., Nusser, Z., Molnár, E., McIlhinney, R. A. J., and Somogyi, P., 1995, High-resolution immunogold localization of AMPA type glutamate receptor subunits at synaptic and non-synaptic sites in rat hippocampus, *Neuroscience* **69:**1031–1055.

Beaulieu, C., Campistron, G., and Crevier, C., 1994, Quantitative aspects of the GABA circuitry in the primary visual cortex of the adult rat, *J. Comp. Neurol.* **339:**559–572.

Bernard, V., Somogyi, P., and Bolam, J. P., 1997, Cellular, subcellular, and subsynaptic distribution of AMPA-type glutamate receptor subunits in the neostriatum of the rat, *J. Neurosci.* **17:**819–833.

Bloch, B., Bernard, V., and Dumartin, B., 2003, "In vivo" intraneuronal trafficking of G protein coupled receptors in the striatum: regulation by dopaminergic and cholinergic environment, *Biol. Cell* **95:**477–488.

Brandt, H. M., and Apkarian, A. V., 1992, Biotin-dextran: a sensitive anterograde tracer for neuroanatomic studies in rat and monkey, *J. Neurosci. Methods* **45:**35–40.

Bruce, K., and Grofova, I., 1992, Notes on a light and electron microscopic double-labeling method combining anterograde tracing with *Phaseolus vulgaris* leucoagglutinin and retrograde tracing with cholera toxin subunit B, *J. Neurosci. Methods* **45:**23–33.

Card, J. P., and Enquist, L. W., 1994, Use of pseudorabies virus for definition of synaptically linked populations of neurons, In: Adolph, K. W. (ed.), *Methods in Molecular Genetics*, New York: Academic Press, pp. 363–382.

Card, J. P., Enquist, L. W., and Moore, R. Y., 1999, Neuroinvasiveness of pseudorabies virus injected intracerebrally is dependent on viral concentration and terminal field density, *J. Comp. Neurol.* **407:**438–452.

Card, J. P., Rinaman, L., Lynn, R. B., Lee, B.-H., Meade, R. P., Miselis, R. R., and Enquist, L. W., 1993, Pseudorabies virus infection of the rat central nervous system: ultrastructural characterization of viral replication, transport, and pathogenesis, *J. Neurosci.* **13:**2515–2539.

Card, J. P., Rinaman, L., Schwaber, J. S., Miselis, R. R., Whealy, M. E., Robbins, A. K., and Enquist, L. W., 1990, Neurotropic properties of pseudorabies virus: uptake and transneuronal passage in the rat central nervous system, *J. Neurosci.* **10:**1974–1994.

Card, J. P., Whealy, M. E., Robbins, A. K., Moore, R. Y., and Enquist, L. W., 1991, Two α-herpesvirus strains are transported differently in the rodent visual system, *Neuron* **6:**957–969.

Carr, D. B., O'Donnell, P., Card, J. P., and Sesack, S. R., 1999, Dopamine terminals in the rat prefrontal cortex synapse on pyramidal cells that project to the nucleus accumbens, *J. Neurosci.* **19:**11049–11060.

Carr, D. B., and Sesack, S. R., 2000a, Projections from the rat prefrontal cortex to the ventral tegmental area: target specificity in the synaptic associations with mesoaccumbens and mesocortical neurons, *J. Neurosci.* **20:**3864–3873.

Carr, D. B., and Sesack, S. R., 2000b, GABA-containing neurons in the rat ventral tegmental area project to the prefrontal cortex, *Synapse* **38:**114–123.

Chagnaud, J. L., Mons, N., Tuffet, S., Grandier-Vazeilles, X., and Geffard, M., 1987, Monoclonal antibodies against glutaraldehyde-conjugated dopamine, *J. Neurochem.* **49:**487–494.

Chan, J., Aoki, C., and Pickel, V. M., 1990, Optimization of differential immunogold–silver and peroxidase labeling with maintenance of ultrastructure in brain sections before plastic embedding, *J. Neurosci. Methods* **33:**113–127.

Chang, H. T., Kuo, H., Whittaker, J. A., and Cooper, N. G. F., 1990, Light and electron microscopic analysis of projection neurons retrogradely labeled with Fluoro-Gold: notes on the application of antibodies to Fluoro-Gold, *J. Neurosci. Methods* **35:**31–37.

Charara, A., Smith, Y., and Parent, A., 1996, Glutamatergic inputs from the pedunculopontine nucleus to midbrain dopaminergic neurons in primates: *Phaseolus vulgaris* leucoagglutinin anterograde labeling combined with postembedding glutamate and GABA immunohisto-chemistry, *J. Comp. Neurol.* **364:**254–266.

Chen, S., and Aston-Jones, G., 1998, Axonal collateral–collateral transport of tract tracers in brain neurons: false anterograde labelling and useful tool, *Neuroscience* **82:**1151–1163.

Chen, L., Boyes, J., Yung, W. H., and Bolam, J. P., 2004, Subcellular localization of GABAB receptor subunits in rat globus pallidus, *J. Comp. Neurol.* **474:**340–352.

Chen, S., Yang, M., Miselis, R. R., and Aston-Jones, G., 1999, Characterization of transsynaptic tracing with central application of pseudorabies virus, *Brain Res.* **838:**171–183.

Conradi, N. G., 1981, Endogenous peroxidatic activity in the cerebral and cerebellar cortex of normal adult rats, *Acta Neuropathol.* S7:3–6.

Cornea-Hebert, V., Riad, M., Wu, C., Singh, S. K., and Descarries, L., 1999, Cellular and subcellular distribution of the serotonin 5-HT2A receptor in the central nervous system of the adult rat, *J. Comp. Neurol.* **409:**187–209.

Dado, R. J., Burstein, R., Cliffer, K. D., and Giesler, G. J., 1990, Evidence that Fluoro-Gold can be transported avidly through fibers of passage, *Brain Res.* **553:**329–333.

DeFalco, J., Tomishima, M., Liu, H., Zhao, C., Cai, X., Marth, J. D., Enquist, L., and Friedman, J. M., 2001, Virus-assisted mapping of neural inputs to a feeding center in the hypothalamus, *Science* **291:**2608–2613.

DeMey, J., and Moeremans, M., 1986, The preparation of colloidal gold probes and their use as a marker in electron microscopy, In: Koehler, J. K. (ed.), *Advanced Techniques in Biological Electron Microscopy III*, New York: Springer-Verlag, pp. 229–271.

Dolleman-Van der Weel, M. J., Wouterlood, F. G., and Witter, M. P., 1994, Multiple anterograde tracing, combining *Phaseolus vulgaris* leucoagglutinin with rhodamine- and biotin-conjugated dextran amine, *J. Neurosci. Methods* **51:**9–21.

Doly, S., Madeira, A., Fischer, J., Brisorgueil, M. J., Daval, G., Bernard, R., Verge, D., and Conrath, M., 2004, The 5-HT2A receptor is widely distributed in the rat spinal cord and mainly localized at the plasma membrane of postsynaptic neurons, *J. Comp. Neurol.* **472:**496–511.

Dube, D., and Pelletier, G., 1979, Effect of colchicine on the immunohistochemical localization of somatostatin in the rat brain: light and electron microscopic studies, *J. Histochem. Cytochem.* **27:**1577–1581.

Dumartin, B., Caillé, I., Gonon, F., and Bloch, B., 1998, Internalization of D1 dopamine receptor in striatal neurons *in vivo* as evidence of activation by dopamine agonists, *J. Neurosci.* **18:**1650–1661.

Dumartin, B., Jaber, M., Gonon, F., Caron, M. G., Giros, B., and Bloch, B., 2000, Dopamine tone regulates D1 receptor trafficking and delivery in striatal neurons in dopamine transporter-deficit mice, *Proc. Natl. Acad. Sci.* **97:**1879–1884.

Ericson, H., and Blomqvist, A., 1988, Tracing of neuronal connections with cholera toxin subunit B: light and electron microscopic immunohistochemistry using monoclonal antibodies, *J. Neurosci. Methods* **24:**225–235.

Ferguson, S. M., Savchenko, V., Apparsundaram, S., Zwick, M., Wright, J., Heilman, C. J., Yi, H., Levey, A. I., and Blakely, R. D., 2003, Vesicular localization and activity-dependent trafficking of presynaptic choline transporters, *J. Neurosci.* **23:**9697–9709.

Ford, B., Holmes, C. J., Mainville, L., and Jones, B. E., 1995, GABAergic neurons in the rat pontomesencephalic tegmentum: codistribution with cholinergic and other tegmental neurons projecting to the posterior lateral hypothalamus, *J. Comp. Neurol.* **363:**177–196.

French, S. J., and Totterdell, S., 2002, Hippocampal and prefrontal cortical inputs monosynaptically converge with individual projection neurons of the nucleus accumbens, *J. Comp. Neurol.* **446:**151–165.

French, S. J., and Totterdell, S., 2003, Individual nucleus accumbens-projection neurons receive both basolateral amygdala and ventral subicular afferents in rats, *Neuroscience* **119:**19–31.

Friedrich, V. L., and Mugnaini, E., 1981, Preparation of neural tissues for electron microscopy, In: Heimer, L., and Robards, M. J. (eds.), *Neuroanatomical Tact-Tacing Methods*, New York: Plenum Press, pp. 345–375.

Gallyas, F., 1982, Suppression of the argyrophil III reaction by mercapto compounds (a prerequisite for the intensification of certain histochemical reactions by physical developers), *Acta Histochem.* **70:**99–105.

Garrett, W. T., McBride, R. L., Williams, J. K. J., and Feringa, E. R., 1991, Fluoro-Gold's toxicity makes it inferior to True Blue for long-term studies of dorsal root ganglion neurons and motoneurons, *Neurosci. Lett.* **128**:137–139.

Garzón, M., Vaughan, R. A., Uhl, G. R., Kuhar, M. J., and Pickel, V. M., 1999, Cholinergic axon terminals in the ventral tegmental area target a subpopulation of neurons expressing low levels of the dopamine transporter, *J. Comp. Neurol.* **410**:197–210.

Gerfen, C. R., and Sawchenko, P. E., 1984, An anterograde neuroanatomical tracing method that shows the detailed morphology of neurons, their axons and terminals: immunohistochemical localization of an axonally transported plant lectin, *Phaseolus vulgaris* leucoagglutinin (PHA-L), *Brain Res.* **290**:219–238.

Gerfen, C. R., Sawchenko, P. E., and Carlsen, J., 1989, The PHA-L anterograde axonal tracing method, In: Heimer, L., and Zaborszky, L. (eds.), *Neuroanatomical Tract-Tracing Methods 2: Recent Progress*, New York: Plenum Press, pp. 19–47.

Glass, M. J., Kruzich, P. J., Kreek, M. J., and Pickel, V. M., 2004, Decreased plasma membrane targeting of NMDA-NR1 receptor subunit in dendrites of medial nucleus tractus solitarius neurons in rat self-administering morphine, *Synapse* **53**:191–201.

Graybiel, A. M., and Chesselet, M. F., 1984, Compartmental distribution of striatal cell bodies expressing [Met]enkephalin-like immunoreactivity, *Proc. Natl. Acad. Sci.* **81**:7980–7984.

Hajszan, T., and Zaborszky, L., 2002, Direct catecholaminergic-cholinergic interactions in the basal forebrain. III. Adrenergic innervation of choline acetyltransferase-containing neurons in the rat, *J. Comp. Neurol.* **449**:141–157.

Hallanger, A. E., and Wainer, B. H., 1988, Ascending projections from the pedunculopontine tegmental nucleus and the adjacent mesopontine tegmentum in the rat, *J. Comp. Neurol.* **274**:483–515.

Hanson, J. E., and Smith, Y., 1999, Group I metabotropic glutamate receptors at GABAergic synapses in monkeys, *J. Neurosci.* **19**:6488–6496.

Heimer, L., and Robarts, M., 1981, *Neuroanatomical Tract-Tracing Methods*, New York: Plenum Press, p. 567.

Heimer, L., and Zaborszky, L., 1989, *Neuroanatomical Tract-Tracing Methods 2. Recent Progress*, New York: Plenum Press, p. 408.

Hemming, F. J., Mesguich, P., Morel, G., and Dubois, P. M., 1983, Cryoultramicrotomy versus plastic embedding: comparative immunocytochemistry of rat anterior pituitary cells, *J. Microsc.* **131**:25–34.

Hillman, H., and Deutsch, K., 1978, Area changes in slices of rat brain during preparation for histology or electron microscopy, *J. Microsc.* **114**:77–84.

Howard, V., 1990, Stereological techniques in biological electron microscopy, In: Hawkes, P. W., and Valdrè, U. (eds.), *Biophysical Electron Microscopy. Basic Concepts and Modern Techniques.* Bologna, Italy: Academic Press, pp. 479–508.

Hsu, S.-M., Raine, L., and Fanger, H., 1981, Use of avidin–biotin-peroxidase complex (ABC) in immunoperoxidase techniques: a comparison between ABC and unlabeled antibody (PAP) procedures, *J. Histochem. Cytochem.* **29**:577–580.

Huang, J., and Pickel, V. M., 2002, Serotonin transporters (SERTs) within the rat nucleus of the solitary tract: subcellular distribution and relation to 5HT2A receptors, *J. Neurocytol.* **31**:667–679.

Irwin, S. A., Idupulapati, M., Gilbert, M. E., Harris, J. B., Chakravarti, A. B., Rogers, E. J., Crisostomo, R. A., Larsen, B. P., Mehta, A., Alcantara, C. J., Patel, B., Swain, R. A., Weiler, I. J., Oostra, B. A., and Greenough, W. T., 2002, Dendritic spine and dendritic field characteristics of layer V pyramidal neurons in the visual cortex of fragile-X knockout mice, *Am. J. Med. Genet.* **111**:140–146.

Jiang, X., Johnson, R. R., and Burkhalter, A., 1993, Visualization of dendritic morphology of cortical projection neurons by retrograde axonal labeling, *J. Neurosci. Methods* **50**:45–60.

Johansson, O., and Backman, J., 1983, Enhancement of immunoperoxidase staining using osmium tetroxide, *J. Neurosci. Methods* **7**:185–193.

Katona, I., Rancz, E. A., Acsady, L., Ledent, C., Mackie, K., Hajos, N., and Freund, T. F., 2001, Distribution of CB1 cannabinoid receptors in the amygdala and their role in the control of GABAergic transmission, *J. Neurosci.* **21**:9506–9518.

Kelly, R. M., and Strick, P. L., 2000, Rabies as a transneuronal tracer of circuits in the central nervous system, *J. Neurosci. Methods* **103**:63–71.

Kolb, B., Gorny, G., Li, Y., Samaha, A. N., and Robinson, T. E., 2003, Amphetamine or cocaine limits the ability of later experience to promote structural plasticity in the neocortex and nucleus accumbens, *Proc. Natl. Acad. Sci.* **100**:10523–10528.

Kulik, Á., Vida, I., Luján, R., Haas, C. A., López-Bendito, G., Shigemoto, R., and Frotscher, M., 2003, Subcellular localization of metabotropic GABA$_B$ receptor subunits GABA$_{B1a/b}$ and GABA$_{B2}$ in the rat hippocampus, *J. Neurosci.* **23**:11026–11035.

Leranth, C., and Pickel, V. M., 1989, Electron microscopic pre-embedding double immunostaining methods, In: Heimer, L., and Zaborsky, L. (eds.), *Neuroanatomical Tract Tracing 2*, New York: Plenum Press, pp. 129–172.

Liposits, Z., Setalo, G., and Flerko, B., 1984, Application of the silver-gold intensified 3,3′-diaminobenzidine chromogen to the light and electron microscopic detection of the luteinizing hormone-releasing hormone system of the rat brain, *Neuroscience* **13**:513–525.

Llewellyn-Smith, I. J., Minson, J. B., Wright, A. P., and Hodgson, A. J., 1990, Cholera toxin B-gold, a retrograde tracer that can be used in light and electron microscopic immunocytochemical studies, *J. Comp. Neurol.* **294**:179–191.

Llewellyn-Smith, I. J., Pilowsky, P. M., and Minson, J. B., 1993, The tungstate-stabilized tetramethylbenzidine reaction for light and electron microscopic immunocytohemistry and for revealing biocytin filled neurons, *J. Neurosci. Methods* **46**:27–40.

Luedtke, R. R., Griffin, S. A., Conroy, S. S., Jin, X., Pinto, A., and Sesack, S. R., 1999, Immunoblot and immunohistochemical comparison of murine monoclonal antibodies specific for the rat D1a and D1b dopamine receptor subtypes, *J. Neuroimmunol.* **101**:170–187.

Marfurt, C. F., Turner, D. F., and Adams, C. E., 1988, Stabilization of tetramethylbenzidine (TMB) reaction product at the electron microscopic level by ammonium molybdate, *J. Neurosci. Methods* **25**:215–223.

Masson, J., Riad, M., Chaudhry, F., Darmon, M., Aidouni, Z., Conrath, M., Giros, B., Hamon, M., Storm-Mathisen, J., Descarries, L., and El Mestikaw, S., 1999, Unexpected localization of the Na+-dependent-like orphan transporter, Rxt1, on synaptic vesicles in the rat central nervous system, *Eur. J. Neurosci.* **11**:1349–1361.

Mayhew, T. M., 1992, A review of recent advances in stereology for quantifying neural structure, *J. Neurocytol.* **21**:313–328.

McLean, J. H., Shipley, M. T., and Bernstein, D. I., 1989, Golgi-like, transneuronal retrograde labelling with CNS injections of herpes simplex virus type 1, *Brain Res. Bull.* **22**:867–881.

Melchitzky, D. S., González-Burgos, B., Barrionuevo, G., and Lewis, D. A., 2001, Synaptic targets of the intrinsic axon collaterals of supragranular pyramidal neurons in monkey prefrontal cortex, *J. Comp. Neurol.* **430**:209–221.

Mi, Z. P., Jiang, P., Weng, W. L. F. P., Narayanan, V., and Lagenaur, C. F., 2000, Expression of a synapse-associated membrane protein, P84/SHPS-1, and its ligand, IAP/CD47, in mouse retina, *J. Comp. Neurol.* **416**:335–344.

Miner, L. A. H., Backstrom, J. R., Sanders-Bush, E., and Sesack, S. R., 2003a. Ultrastructural localization of serotonin2A receptors in the middle layers of the rat prelimbic prefrontal cortex, *Neuroscience* **116**:107–117.

Miner, L. A. H., Benmansour, S., Moore, F. W., Blakely, R. D., Morilak, D. A., Frazer, A., and Sesack, S. R., 2004, Chronic treatment with a selective serotonin reuptake inhibitor reduces the total and plasmalemmal serotonin transporter immunoreactivity in axon terminals within the rat prefrontal cortex, *Soc. Neurosci. Abstr.* **29**:54.55.

Miner, L. A. H., Moore, F. W., Jedema, H. P., Grace, A. A., and Sesack, S. R., 2003b. Ultrastructural localization of the norepinephrine transporter in the prefrontal cortex of chronically stressed rats, *Soc. Neurosci. Abstr.* **28**:506.505.

Miner, L. A. H., Schroeter, S., Blakely, R. D., and Sesack, S. R., 2000, Ultrastructural localization of the serotonin transporter in superficial and deep layers of the rat prefrontal cortex and its spatial relationship to dopamine terminals, *J. Comp. Neurol.* **427**:220–234.

Miner, L. A. H., Schroeter, S., Blakely, R. D., and Sesack, S. R., 2003c. Ultrastructural localization of the norepinephrine transporter in superficial and deep layers of the rat prelimbic prefrontal cortex and its spatial relationship to probable dopamine terminals, *J. Comp. Neurol.* **466**:478–494.

Monti-Graziadei, A. G., and Berkley, K. J., 1991, Effects of colchicine on retrogradely-transported WGA-HRP, *Brain Res.* **565**:162–166.

Moore, R. Y., 1981, Methods for selective, restricted lesion placement in the central nervous system, In: Heimer, L., and Robards, M. J. (eds.), *Neuroanatomical Tract-Tracing Methods,* New York: Plenum Press, pp. 55–90.

Mugnaini, E., and Friedrich, V. L., Jr., 1981, Electron microscopy: identification and study of normal and degenerating neural elements by electron microscopy, In: Heimer, L., and Robards, M. J. (eds.), *Neuroanatomical Tract-Tracing Methods,* New York: Plenum Press, pp. 377–406.

Naumann, T., Härtig, W., and Frotscher, M., 2000, Retrograde tracing with Fluoro-Gold: different methods of tracer detection at the ultrastructural level and neurodegenerative changes of back-filled neurons in long-term studies, *J. Neurosci. Methods* **103**:11–21.

Norgren, R. B. J., and Lehman, M. N., 1989, A double-label pre-embedding immunoperoxidase technique for electron microscopy using diaminobenzidine and tetramethylbenzidine as markers, *J. Histochem. Cytochem.* **37**:1283–1289.

Novikoff, A., Novikoff, P., Quintana, N., and Davis, C., 1972, Diffusion artifacts in 3,3'-diaminobenzidine cytochemistry, *J. Histochem. Cytochem.* **20**:745–749.

Novikova, L., Novikov, L., and Kellerth, J. O., 1997, Persistent neuronal labeling by retrograde fluorescent tracers: a comparison between Fast Blue, Fluoro-Gold and various dextran conjugates, *J. Neurosci. Methods* **74**:9–15.

Nusser, Z., Ahmad, Z., Tretter, V., Fuchs, K., Wisden, W., Sieghart, W., and Somogyi, P., 1999, Alterations in the expression of GABAA receptor subunits in the cerebellar granule cells after the disruption of the alpha6 subunit gene, *Eur. J. Neurosci.* **11**:1685–1697.

Nusser, Z., Roberts, J. D., Baude, A., Richards, J. G., Sieghart, W., and Somogyi, P., 1995, Immunocytochemical localization of the α1 and β2/3 subunits of the GABAA receptor in relation to specific GABAergic synapses in the dentate gyrus, *Eur. J. Neurosci.* **7**:630–646.

Oades, R. D., and Halliday, G. M., 1987, Ventral tegmental (A10) system: neurobiology. 1. Anatomy and connectivity, *Brain Res. Rev.* **12**:117–165.

Oakman, S. C., Faris, P. L., Kerr, P. E., Cozzari, C., and Hartman, B. K., 1995, Distribution of pontomesencephalic cholinergic neurons projecting to substantia nigra differs significantly from those projecting to ventral tegmental area, *J. Neurosci.* **15**:5859–5869.

Omelchenko, N., and Sesack, S. R., 2005, Laterodorsal tegmental projections to identified cell populations in the rat ventral tegmental area, *J. Comp. Neurol.* **483**:217–235.

Omelchenko, N., and Sesack, S. R., 2005b, cholinergic axons in the rat ventral tegmental are synapse preferentially onto mesoaccumbens dopamine neurons, J. Comp. Neurol.

Ordronneau, P., Lindström, P. B.-M., and Petrusz, P., 1981, Four unlabeled antibody bridge techniques: a comparison, *J. Histochem. Cytochem.* **29**:1397–1404.

Paspalas, C. D., and Goldman-Rakic, P. S., 2004, Microdomains for dopamine volume neurotransmission in primate prefrontal cortex, *J. Neurosci.* **24**:5292–5300.

Paulson, J. C., and McClure, W. O., 1975, Inhibition of axoplasmic transport by colchicine, podophyllotoxin, and vinblastine: an effect on microtubules, *Ann. N. Y. Acad. Sci.* **253**:517–527.

Phend, K. D., Rustioni, A., and Weinberg, R. J., 1995, An osmium-free method of epon embedment that preserves both ultrastructure and antigenicity for post-embedding immunocytochemistry, *J. Histochem. Cytochem.* **43**:283–292.

Phillipson, O. T., 1979, Afferent projections to the ventral tegmental area of Tsai and interfascicular nucleus: a horseradish peroxidase study in the rat, *J. Comp. Neurol.* **187**:117–144.

Pickel, V. M., 1981, Immunocytochemical methods, In: Heimer, L., and RoBards, M. (eds.), *Neuroanatomical Tract-Tracing Methods,* New York: Plenum Press, pp. 483–509.

Pickel, V. M., and Chan, J., 1999, Ultrastructural localization of the serotonin transporter in limbic and motor compartments of the nucleus accumbens, *J. Neurosci.* **19**:7356–7366.

Pickel, V. M., Chan, J., and Aoki, C., 1993, Electron microscopic immunocytochemical labelling of endogenous and/or transported antigens in rat brain using silver-intensified one-nanometre colloidal gold, In: Cuello, A. (ed.), *Immunohistochemistry II*, Chichester: John Wiley & Sons, pp. 265–280.

Pickel, V. M., Chan, J., Kash, T. L., Rodriguez, J. J., and Mackie, K., 2004, Compartment-specific localization of cannabinoid (CB1) and mu-opiod receptors in rat nucleus accumbens, *Neuroscience* **127:**101–112.

Pickel, V. M., and Milner, T. A., 1989, Interchangeable uses of autoradiographic and peroxidase markers for electron microscopic detection of neuronal pathways and transmitter-related antigens in single sections, In: Heimer, L., and Zaborszky, L. (eds.), *Neuroanatomical Tract-Tracing Methods 2: Recent Progress*, New York: Plenum Press, pp. 97–127.

Pieribone, V. A., and Aston-Jones, G., 1988, The iontophoretic application of Fluoro-Gold for the study of afferents to deep brain nuclei, *Brain Res.* **475:**259–271.

Pinto, A., Jankowski, M., and Sesack, S. R., 2003, Projections from the paraventricular nucleus of the thalamus to the rat prefrontal cortex and nucleus accumbens shell: ultrastructural characteristics and spatial relationships with dopamine afferent, *J. Comp. Neurol.* **459:**142–155.

Pinto, A., and Sesack, S. R., 1998, Paraventricular thalamic afferents to the rat prefrontal cortex and nucleus accumbens shell: synaptic targets and relation to dopamine afferents, *Soc. Neurosci. Abstr.* **24:**1595.

Pirnik, Z., Mikkelsen, J. D., and Kiss, A., 2003, Fos induction in the rat deep cerebellar and vestibular nuclei following central administration of colchicine: a qualitative and quantitative time-course study, *Brain Res. Bull.* **61:**63–72.

Rajakumar, N., Elisevich, K., and Flumerfelt, B. A., 1993, Biotinylated dextran: a versatile anterograde and retrograde neuronal tracer, *Brain Res.* **607:**47–53.

Ravary, A., Muzerelle, A., Darmon, M., Murphy, D. L., Moessner, R., Lesch, K. P., and Gaspar, P., 2001, Abnormal trafficking and subcellular localization of an N-terminally truncated serotonin transporter protein, *Eur. J. Neurosci.* **13:**1349–1362.

Reiner, A., Veenman, C. L., Medina, L., Jiao, Y., Del Mar, N., and Honig, M. G., 2000, Pathway tracing using biotinylated dextran amines, *J. Neurosci. Methods* **103:**23–37.

Rho, J. H., and Swanson, L. W., 1989, A morphometric analysis of functionally defined subpopulations of neurons in the paraventricular nucleus of the rat with observations on the effects of colchicine, *J. Neurosci.* 9:1375–1388.

Riad, M., Garcia, S., Watkins, K. C., Jodoin, N., Doucet, E., Langlois, X., El Mestikaw, S., Hamon, M., and Descarries, L., 2000, Somatodendritic localization of the 5-HT1A and preterminal axonal localization of 5-HT1B serotonin receptors in the adult rat brain, *J. Comp. Neurol.* **417:**181–194.

Riad, M., Zimmer, L., Rbah, L., Watkins, K. C., Hamon, M., and Descarries, L., 2004, Acute treatment with the antidepressant fluoxetine internalizes 5-HT1A autoreceptors and reduces the in vivo binding of the PET radioligand [18F]MPFF in the nucleus raphe dorsalis of rat, *J. Neurosci.* **24:**5420–5426.

Ribak, C. E., Vaughn, J. E., and Saito, K., 1978, Immunocytochemical localization of glutamic acid decarboxylase in neuronal somata following colchicine inhibition of axonal transport, *Brain Res.* **140:**315–332.

Rosene, D. L., Roy, N. J., and Davis, B. J., 1986, A cryoprotection method that facilitates cutting frozen sections of whole monkey brains for histological and histochemical processing without freezing artifact, *J. Histochem. Cytochem.* **34:**1301–1315.

Rye, D. B., Saper, C. B., and Wainer, B. H., 1984, Stabilization of the tetramethylbenzidine (TMB) reaction product: application for retrograde and anterograde tracing, and combination with immunohistochemistry, *J. Histochem. Cytochem.* **32:**1145–1153.

Saper, C. B., 2003, Magic peptides, magic antibodies: guidelines for appropriate controls for immunohistochemistry, *J. Comp. Neurol.* **465:**161–163.

Schmued, L. C., and Fallon, J. H., 1986, Fluoro-Gold: a new fluorescent retrograde axonal tracer with numerous unique properties, *Brain Res.* **377:**147–154.

Schmued, L. C., Kyriakidis, K., Fallon, J. H., and Ribak, C. E., 1989, Neurons containing retrogradely transported Fluoro-Gold exhibit a variety of lysosomal profiles: a combined brightfield, fluorescence, and electron microscopic study, *J. Neurocytol.* **18:**333–343.

Sesack, S. R., and Pickel, V. M., 1992, Prefrontal cortical efferents in the rat synapse on unlabeled neuronal targets of catecholamine terminals in the nucleus accumbens septi and on dopamine neurons in the ventral tegmental area, *J. Comp. Neurol.* **320:**145–160.

Sesack, S. R., and Snyder, C. L., 1995, Cellular and subcellular localization of syntaxin-like immunoreactivity in the rat striatum and cortex, *Neuroscience* **67:**993–1007.

Shu, S. Y., and Peterson, G. M., 1988, Anterograde and retrograde axonal transport of *Phaseolus vulgaris* leucoagglutinin (PHA-L) from the globus pallidus to the striatum of the rat, *J. Neurosci. Methods* **25:**175–180.

Skirboll, L. R., Thor, K., Helke, C., Hökfelt, T., Robertson, B., and Long, R., 1989, Use of retrograde fluorescent tracers in combination with immunohistochemical methods, In: Heimer, L., and Zaborszky, L. (eds.), *Neuroanatomical Tract-Tracing Methods 2: Recent Progress,* New York: Plenum Press, pp. 5–18.

Smiley, J. F., and Goldman-Rakic, P. S., 1993, Silver-enhanced diaminobenzidine-sulfide (SEDS): a technique for high-resolution immunoelectron microscopy demonstrated with monoamine immunoreactivity in monkey cerebral cortex and caudate, *J. Histochem. Cytochem.* **41:**1393–1404.

Smith, Y., Bennett, B. D., Bolam, J. P., Parent, A., and Sadikot, A. F., 1994, Synaptic relationship between dopaminergic afferents and cortical or thalamic input in the sensorimotor territory of the striatum in monkey, *J. Comp. Neurol.* **344:**1–19.

Smith, Y., Charara, A., and Parent, A., 1996, Synaptic innervation of midbrain dopaminergic neurons by glutamate-enriched terminals in the squirrel monkey, *J. Comp. Neurol.* **364:**231–253.

Srebro, Z., 1972, Ultrastructural localization of peroxidase activity in Gomori-positive glia, *Acta Anat.* **83:**388–397.

Sternberger, L. A., 1974, *Immunocytochemistry.* Englewood Cliffs, NJ: Prentice-Hall.

Steward, O., 1981, Horseradish peroxidase and fluorescent substances and their combination with other techniques, In: Heimer, L., and Robards, M. J. (eds.), *Neuroanatomical Tract-Tracing Methods,* New York: Plenum Press, pp. 279–310.

Strick, P. L., and Card, J. P., 1992, Transneuronal mapping of neural circuits with alpha herpesviruses, In: Bolam, J. (ed.), *Experimental Neuroanatomy: A Practical Approach,* New York: IRL Press, pp. 81–101.

Svensson, B. A., Rastad, J., and Westman, J., 1984, Endogenous peroxidase-like activity in the feline dorsal column nuclei and spinal cord, *Exp. Brain Res.* **55:**325–332.

Swanson, L. W., 1982, The projections of the ventral tegmental area and adjacent regions: a combined fluorescent retrograde tracer and immunofluorescence study in the rat, *Brain Res. Bull.* **9:**321–353.

Van Bockstaele, E. J., and Pickel, V. M., 1995, GABA-containing neurons in the ventral tegmental area project to the nucleus accumbens in rat brain, *Brain Res.* **682:**215–221.

Van Bockstaele, E. J., Wright, A. M., Cestari, D. M., and Pickel, V. M., 1994, Immunolabeling of retrogradely transported Fluoro-Gold: sensitivity and application to ultrastructural analysis of transmitter-specific mesolimbic circuitry, *J. Neurosci. Methods* **55:**65–78.

Van Haeften, T., and Wouterlood, F. G., 2000, Neuroanatomical tracing at high resolution, *J. Neurosci. Methods* **103:**107–116.

Veenman, C. L., Reiner, A., and Honig, M. G., 1992, Biotinylated dextran amine as an anterograde tracer for single- and double-labeling studies, *J. Neurosci. Methods* **41:**239–254.

Veznedaroglu, E., and Milner, T. E., 1992, Elimination of artifactual labeling of hippocampal mossy fibers seen following pre-embedding immunogold–silver technique by pretreatment with zinc chelator, *Microsc. Res. Tech.* **23:**100–101.

von Bartheld, C., 2002, Counting particles in tissue sections: choices of methods and importance of calibration to minimize biases, *Histol. Histopathol.* **17:**639–648.

Warr, W. B., de Olmos, J. S., and Heimer, L., 1981, Horseradish peroxidase: the basic procedure, In: Heimer, L., and Robards, M. J. (eds.), *Neuroanatomical Tract-Tracing Methods,* New York: Plenum Press, pp. 207–262.

Wong, H. K., Liu, X. B., Matos, M. F., Chan, S. F., Perez-Otano, I., Boysen, M., Cui, J., Nakanishi, N., Trimmer, J. S., Jones, E. G., Lipton, S. A., and Sucher, N. J., 2002, Temporal and regional expression of NMDA receptor subunit NR3A in the mammalian brain, *J. Comp. Neurol.* **450**:303–317.

Wouterlood, F. G., Goede, P. H., and Groenewegen, H. J., 1990, The in situ detectability of the neuroanatomical tracer *Phaseolus vulgaris*-leucoagglutinin (PHA-L), *J. Chem. Neuroanat.* **3**:11–18.

Wouterlood, F. G., and Groenewegen, H. J., 1985, Neuroanatomical tracing by use of *Phaseolus vulgaris* leucoagglutnin (PHA-L): electron microscopy of PHA-L filled neuronal somata, dendrites, axons and axon terminals, *Brain Res.* **326**:188–191.

Wouterlood, F. G., and Jorritsma-Byham, B., 1993, The anterograde neuroanatomical tracer biotinylated dextran-amine: comparison with the tracer *Phaseolus vulgaris* leucoagglutinin in preparations for electron microscopy, *J. Neurosci. Methods* **48**:75–87.

Yagi, T., Terada, N., Baba, T., and Ohno, S., 2002, Localization of endogenous biotin-containing proteins in mouse Bergmann glial cells, *Histochem. J.* **34**:567–572.

Yan, X. X., and Ribak, C. E., 1999, Alteration of GABA transporter expression in the rat cerebral cortex following needle puncture and colchicine injection, *Brain Res.* **816**:317–328.

Yi, H., Leunissen, J. L. M., Shi, G.-M., Gutekunst, C.-A., and Hersch, S. M., 2001, A novel procedure for pre-embedding double immunogold–silver labeling at the ultrastructural level, *J. Histochem. Cytochem.* **49**:279–284.

Zaborszky, L., and Heimer, L., 1989, Combinations of tracer techniques, especially HRP and PHA-L, with transmitter identification for correlated light and electron microscopic studies, In: Heimer, L., and Zaborszky, L. (eds.), *Neuroanatomical Tract-Tracing Methods 2: Recent Progress*, New York: Plenum Press, pp. 49–96.

Zaborszky, L., Léránth, C., and Palkovits, M., 1979, Light and electron microscopic identification of monoaminergic terminals in the central nervous system, *Brain Res. Bull.* **4**:99–117.

Zaborszky, L., Rosin, D. L., and Kiss, J., 2004, Alpha-adrenergic receptor (α2A) is colocalized in basal forebrain cholinergic neurons: a light and electron microscopic double immunolabeling study, *J. Neurocytol.* **33**:265–276.

Zhao, S., Maxwell, S., Jimenez-Beristain, A., Vives, J., Kuehner, E., Zhao, J., O'Brien, C., de Felipe, C., Semina, E., and Li, M., 2004, Generation of embryonic stem cells and transgenic mice expressing green fluorescence protein in midbrain dopaminergic neurons, *Eur. J. Neurosci.* **19**:1133–1140.

Zhou, M., and Grofova, I., 1995, The use of peroxidase substrate Vector VIP in electron microscopic single and double antigen localization, *J. Neurosci. Methods* **62**:149–158.

3

Postembedding Immunogold Cytochemistry of Membrane Molecules and Amino Acid Transmitters in the Central Nervous System

THOMAS MISJE MATHIISEN, ERLEND ARNULF NAGELHUS, BAHAREH JOULEH, REIDUN TORP, DIDRIK SØLIE FRYDENLUND, MARIA-NIKI MYLONAKOU, MAHMOOD AMIRY-MOGHADDAM, LUCIENE COVOLAN, JO KRISTIAN UTVIK, BJØRG RIBER, KAREN MARIE GUJORD, JORUNN KNUTSEN, ØIVIND SKARE, PETTER LAAKE, SVEND DAVANGER, FINN-MOGENS HAUG, ERIC RINVIK, and OLE PETTER OTTERSEN

THOMAS MISJE MATHIISEN, ERLEND ARNULF NAGELHUS, BAHAREH JOULEH, REIDUN TORP, DIDRIK SØLIE FRYDENLUND, MARIA-NIKI MYLONAKOU, MAHMOOD AMIRY-MOGHADDAM, LUCIENE COVOLAN, JO KRISTIAN UTVIK, BJØRG RIBER, KAREN MARIE GUJORD, JORUNN KNUTSEN, ØYVIND SKARE, PETTER LAAKE, SVEND DAVANGER, FINN-MOGENS HAUG, ERIC RINVIK, AND OLE PETTER OTTERSEN • Centre for Molecular Biology and Neuroscience, and Nordic Centre for Research on Water Imbalance Related Disorders (WIRED), Institute of Basic Medical Sciences, University of Oslo, PO Box 1105 Blindern, N-0317 Oslo, Norway

Abstract: This chapter deals with procedures for postembedding labeling of brain sections embedded in epoxy or methacrylate resins and focuses on protocols that are based on freeze substitution of chemically fixed tissue. When optimized for the target epitope, such protocols offer a high labeling efficiency and allow simultaneous visualization of several antigens by use of different-sized gold particles. Postembedding labeling can be combined with anterograde tracing, permitting the identification of transmitter and postsynaptic receptors of identified axons. By use of tailor-made model systems, antibody selectivity can be monitored in a quantitative manner and under conditions that are representative of the immunocytochemical procedure. Such model systems also allow the generation of calibration curves for assessment of the cellular and subcellular concentration of soluble antigens. When used in conjunction with computer programs for automated acquisition and analysis of gold particles, the postembedding immunogold procedure provides an accurate representation of the cellular and subcellular distribution of proteins and small compounds such as transmitter amino acids. The present chapter provides a quantitative analysis and critical discussion of how changes in incubation parameters influence the labeling intensity. Postembedding immunogold cytochemistry stands out as a powerful technique for analysis of the chemical architecture of the central nervous system and has proved useful for investigating disease processes at the molecular level.

Keywords: aquaporins, glutamate, glutamate receptors, quantitation, resolution, specificity testing

I. INTRODUCTION

The ultimate goal in immunocytochemistry is to be able to determine the exact number and position of a given target molecule in a biological tissue.

A priori, this requires access to monospecific antibodies that bind with a 1:1 stoichiometry to the target antigen, and a reporter system that allows accurate localization and quantitation of the primary antibodies. These are ideal conditions that cannot be met in practice. However, they can be approached by use of markers that are amenable to quantitative electron microscopic analysis.

Colloidal gold particles (Faulk and Taylor, 1971; Roth, 1996; van den Pol, 1989) have proven to be the most versatile markers for this purpose. They are electron dense, allowing easy identification and quantitation in the electron microscope, and can be prepared in many different sizes, permitting simultaneous detection of several different antigens. Most importantly, colloidal gold particles can be coupled directly to the primary or secondary antibody so as to afford a close spatial relation to the target antigen. These features set immunogold cytochemistry apart from the peroxidase–antiperoxidase method and other enzyme-based immunocytochemical techniques. The latter techniques typically rely on the analysis of an electron-dense reaction product that is difficult to quantify and that may diffuse away from the site of formation.

Colloidal gold particles may, in principle, be applied in two different ways: in preembedding or postembedding mode. In the preembedding mode, the antibodies and immunogold reagents are applied to permeabilized tissue that is subsequently embedded in a resin suitable for electron microscopic analysis. In the postembedding mode, the immunoreagents are applied directly onto ultrathin sections of resin-embedded tissue or cells. The latter approach allows immunodetection only of those antigen molecules that are exposed at the surface of the section. This implies that the proportion of antigen molecules that is available for antibody binding is severely restricted when compared with the preembedding mode.

So why use the postembedding mode? The major advantage offered by the postembedding mode is that each antigen molecule that occurs at the surface of the section should stand the same chance of being immunodetected, regardless of its cellular or subcellular localization. This contrasts with the situation in the preembedding mode, where diffusion barriers may constrain the labeling and distort the relationship between antigen concentration and gold particle density. Thus, for the purpose of quantitation, the postembedding procedure is generally considered as the superior of the two modes of immunogold cytochemistry. Preembedding procedures have their own set of advantages that will not be considered here (see Sesack *et al.*, this volume). We also need to emphasize that cryo-electron microscopy is outside the scope of the present chapter, which deals exclusively with postembedding immunogold labeling of resin-embedded sections. The chapter is focused on experience gained in our own laboratory and is not intended to provide a balanced overview of the historical development of the technique. A description of the pioneering work is found elsewhere (Griffiths, 1993; Maunsbach and Afzelius, 1999; Roth, 1996).

II. RESOLUTION

As stated in the Introduction, the resolution of the postembedding immunogold technique exceeds by far the resolution offered by enzyme-based techniques. Thus, with the former technique, the marker is a well-defined particle that is attached to the antigen through an antibody bridge, rather than a reaction product that may be deposited at a distance from its site of formation.

Indeed, it is the length of the antibody bridge and the dimension of the colloidal particle that restrict the resolution of the postembedding immunogold procedure. Theoretically, the distance between the epitope and the center of the gold particle should correspond to the radius of the particle plus the diameters of the interposed immunoglobulins (IgGs) (Fig. 3.1). Using 15-nm gold particles and a primary and secondary IgG (each with an efficient diameter of $\sim$8 nm), this theoretical distance should be $\sim$23 nm. Obviously, the distance will be shorter if colloidal gold is coupled to the primary antibody directly or by way of a secondary Fab complex rather than a secondary IgG.

The theoretical prediction as to lateral resolution is borne out by experiments based on the use of tailor-made model antigens (Fig. 3.2). The model antigens were embedded in the same resin as that used for the tissue, ensuring identical conditions. Further, the antigens were prepared to form discrete bodies with a distinct demarcation from the surrounding resin. Hence, the distance between the margin of the body and the centers

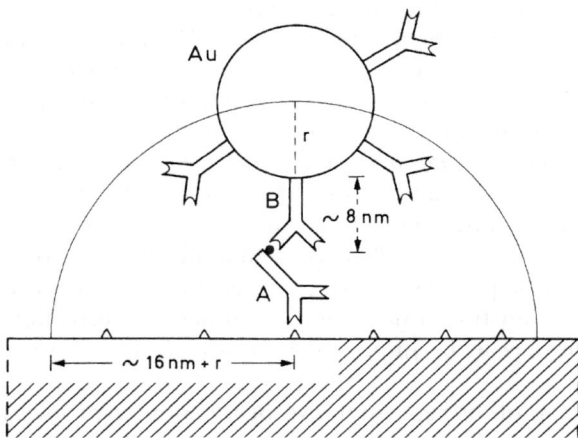

Figure 3.1. Simplified diagram of a two-step, postembedding immunogold procedure. Triangles represent the antigen against which the primary antibody (A) was raised. The secondary antibody (B) is coupled to a colloidal gold particle (Au). The radius (r) of the gold particle is 7.5 nm. This adds up to a maximum distance of about 23 nm between the epitope and the gold particle center (given an effective IgG diameter of 8 nm). (From Ottersen, 1989a.)

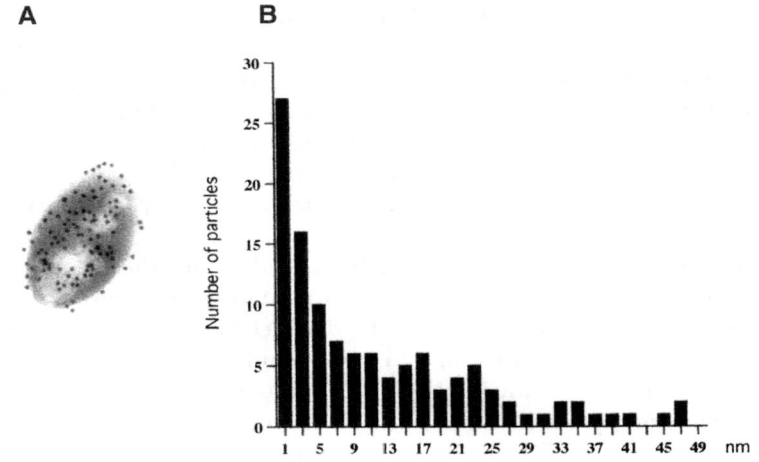

Figure 3.2. Lateral resolution of current immunogold procedure (15-nm gold particles). Values along the *x*-axis in B denote distance from centers of gold particles to the margin of antigen-containing bodies (A). Background level of labeling is reached ∼28 nm off the bodies. The data were based on the analysis of 30 bodies containing glutaraldehyde-fixed L-aspartate as a model antigen. Scale bar: 0.3 μm. (From Matsubara *et al.*, 1996.)

of the gold particles should be representative of the lateral resolution of the postembedding immunogold procedure. The gold particle density was found to reach background level at ∼28 nm off the margin of the test body (Fig. 3.2B) in good agreement with the theoretical prediction.

Obviously, the above-mentioned theoretically and experimentally determined values represent the *maximum* distance between an epitope and the respective gold particle. In practice, many particles are likely to end up closer to the epitope, due to a restricted rotational freedom. It must also be remembered that the *projected* distance between the gold particle and the epitope may be considerably shorter than the real distance.

This notwithstanding, the fact that the size of the antibody bridge significantly limits the resolution of the postembedding immunogold technique has practical consequences in several experimental settings. Two examples will be used to illustrate this. These examples will also show how the effective resolution can be improved by resorting to tailor-made statistical procedures.

The first example is representative of a common problem in neurobiology: the need to distinguish between two plasma membranes that are closely apposed to one another. A close membrane apposition is typical of central synapses where the pre- and postsynaptic membranes are separated by a synaptic cleft of ∼20 nm. This distance is less than the theoretical and experimental maximum values between the epitope and the gold particle center. In other words, a gold particle overlying the presynaptic membrane might reflect antibody binding to an epitope in the postsynaptic membrane,

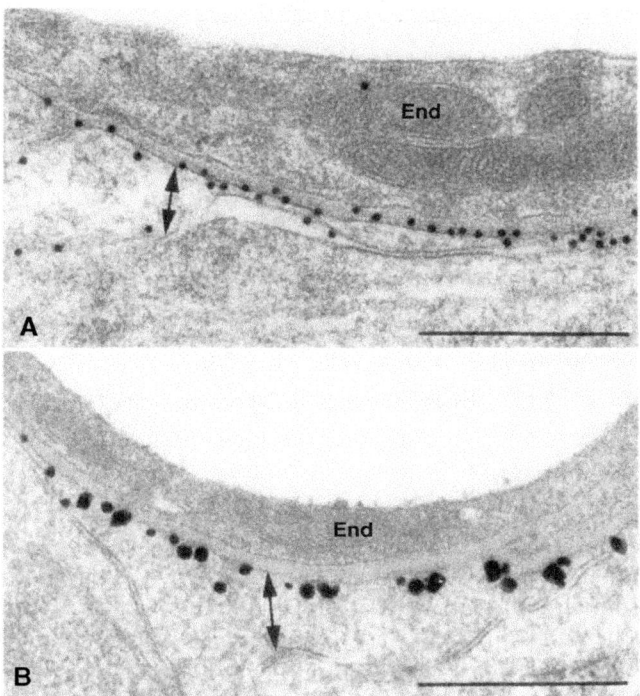

Figure 3.3. Comparison of postembedding (A) and preembedding (B) labeling of AQP4 in perivascular astrocyte end feet membranes (double arrows indicate width of end foot). End, endothelial cell. (A) In the postembedding mode, gold particles are deposited on either side of the membrane—the maximum distance from the membrane reflecting the size of the antibody bridge (cf. Figs. 3.1 and 3.2). (B) In preembedding mode (ultrasmall gold particles enhanced by silver), all particles end up at the cytoplasmic aspect, due to the constraints imposed by the membrane (the epitope is located at the intracellular tail of AQP4). Note variable size and confluence of particles, typical of the preembedding mode. (A, B: From Nielsen *et al.*, 1997.) (C) Analysis of AQP4 immunogold labeling by recording the distribution of gold particles along an axis perpendicular to the labeled plasma membrane (postembedding labeling as in A). The *ordinate* indicates number of gold particles per bin (bin width, 5 nm). The peak coincides with the plasma membrane (0 corresponds to midpoint of membrane) and the particle density approaches background level at ~50 nm from the membrane (inside negative). This section was labeled from both sides, explaining why some particles are located further off than the theoretically and experimentally determined maximum distance between epitope and gold particle (see text). (C: From Nagelhus *et al.*, 1998.)

and vice versa. So how is it possible to determine whether a given antigen is localized pre- and/or postsynaptically?

This question can be resolved by recording the distribution of gold particles along an axis perpendicular to the synaptic membranes (here defined as z-axis). Depending, i.a., on the rotational freedom of the epitope and the number of gold particles recorded, the average position of an antigen along this axis can be determined with a precision of ~1 nm,

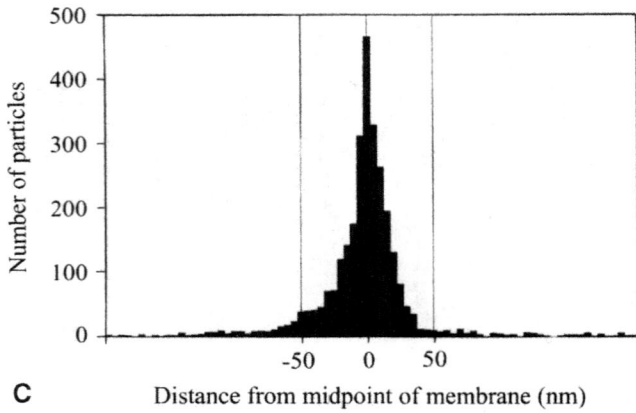

C

Figure 3.3. (*cont*).

defined by the standard error of the z-value of the peak particle density (Fig. 3.3; also see Nagelhus *et al.*, 1998, 2004). By this approach, it was possible to demonstrate that the BK potassium channel is expressed in presynaptic but not in postsynaptic membranes of hippocampal synapses (Hu *et al.*, 2001; Fig. 3.4). The same approach was exploited recently to distinguish between closely apposed membranes in the olfactory bulb (Panzanelli *et al.*, 2004) and has also been used, in a different context, to identify the relative positions of molecules engaged in glutamate receptor complexes (Valtschanoff and Weinberg, 2001). The common practice of labeling both sides of the section may decrease the precision of this approach, as the intersections of a membrane with the two surfaces of the 50- to 100-nm-thick tissue section are rarely superimposed in the image (Fig. 3.5).

The second example of a biological problem that requires due attention to the size of the antibody bridge relates to the analysis of synaptic vesicles (particularly the small, clear vesicles that have a diameter of ~50 nm). The question was whether glutamate is enriched in the synaptic vesicles of granule cell dendritic spines of the olfactory bulb (Didier *et al.*, 2001). These spines are presynaptic to the mitral cell dendrites and display a high density of gold particles signaling glutamate. But this signal could reflect metabolic glutamate, rather than a vesicular pool of transmitter glutamate. Due to the small dimensions of the clear synaptic vesicles, one cannot attach significance to individual gold particles: even a particle located at the center of a vesicle could theoretically depend on an epitope external to the vesicle in question.

To circumvent this problem, measurements were made of the intercenter distances between each gold particle and the nearest synaptic vesicle. It turned out that short distances were overrepresented compared with random distributions of gold particles, supporting the idea that glutamate is associated with synaptic vesicles (Didier *et al.*, 2001).

The two examples discussed above show that statistical analyses of large numbers of gold particles can partly compensate for the inaccuracy

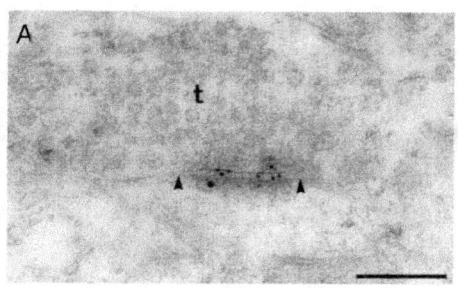

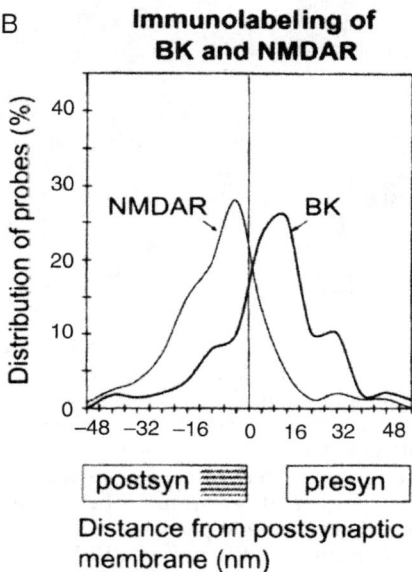

Figure 3.4. Analysis of gold particle distribution in closely apposed synaptic membranes. (A) Electron micrograph showing the distribution of BK channels (10-nm particles) and NMDA receptors (NMDARs; 15-nm particles) in double-labeled sections from the stratum radiatum of CA1. t, terminal; arrowheads indicate extent of postsynaptic density. (B) To determine whether the two epitopes were associated with the pre- and/or postsynaptic membranes, the vertical distribution of particles was analyzed by the approach described in Fig. 3.3. The peak density of particles coincided with the presynaptic membrane in the case of BK channels and with the postsynaptic membrane in the case of NMDARs. The dimensions of the synaptic cleft and postsynaptic density are indicated below the abscissa. (From Hu *et al.*, 2001.)

introduced by the antibody bridge. Other factors that affect the effective resolution of the postembedding technique are discussed in Appendix II. How could resolution be further improved? There is a marginal gain by coupling small gold particles directly to the primary antibodies (or Fab fragments of primary antibodies), rather than to the secondary ones. A substantial increase in resolution would be achieved by visualizing (by negative staining) the antibody bridge between the epitope and the gold particle. But an even larger step toward the "ultimate goal" of defining the precise

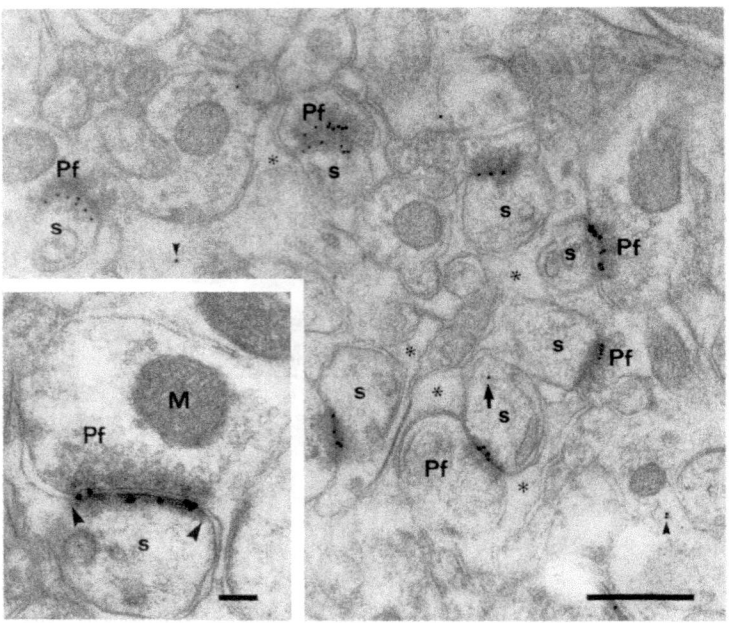

Figure 3.5. Distribution of gold particles signaling δ2-glutamate receptors at synapses between parallel fibers (Pf) and Purkinje cell spines (s) of rat cerebellum. The synapse at top center is obliquely cut, the two rows of particles representing receptors exposed at opposite surfaces of the section (the section was labeled from both sides). Asterisks indicate glial lamellae. The labeling is highly selective: only seven particles in this field are not associated with any postsynaptic density. Of these, one is found within a spine (arrow) and two within other intracellular compartments (arrowheads). The linear density of gold particles (particles per micrometer membrane) in parallel fiber–Purkinje cell synapses was 19.7, compared to 0 in all other types of cerebellar synapse. Inset: Higher magnification of an immunolabeled parallel fiber synapse (postsynaptic density delimited by arrowheads). *M*, mitochondrion. Gold particles, 15 nm. Scale bars: 0.5 μm; 0.1 μm (inset). (From Landsend *et al.*, 1997.)

position of the target antigen is provided by the combination of freeze fracture and immunogold labeling (Fujimoto, 1995; Rash *et al.*, 2004, Hagiwara *et al.*, 2005). With this combination of techniques, the immunogold procedure is employed to determine the molecular identity of intramembrane particles (IMPs) visualized in metal replicas. This approach has proved particularly useful in the case of proteins that are clustered in the plasma membranes, such as connexins and aquaporins. For molecules with a more scattered distribution, the analysis may be hampered by limited sensitivity. This notwithstanding, the freeze fracture–immunogold labeling technique has enormous potential for determining the exact position and number of receptors, transporters, or other molecules in the plasma membranes of neural cells. The extent to which this potential can be realized depends on the ability to classify IMPs according to their sizes and structural features. A further refinement of the freeze fracture approach is required to define a sufficiently broad range of membrane molecules.

III. QUANTITATION

"Quantitative immunocytochemistry" is considered by many as a contradiction in terms. It is true that we are still far away from the "ultimate goal" of being able to determine the accurate numbers of molecules by means of immunocytochemistry. However, immunogold procedures have brought us closer to this goal and should open avenues for further advances, particularly if combined with appropriate calibration systems or with freeze fracture techniques. A thorough discussion of quantitative aspects of immunocytochemistry is provided by Griffiths (1993).

As set out in the Introduction, the use of gold particles facilitates quantitation. Gold particles represent an "all or none signal," setting them apart from the less easily quantifiable reaction product of enzyme-based immunocytochemistry. The particles can be readily identified and counted in the electron microscope, and computer programs have been designed for automated acquisition and analysis of their distribution (Blackstad *et al.*, 1990; Haug *et al.*, 1994, 1996; Monteiro-Leal *et al.*, 2003; Ruud and Blackstad, 1999). In most cases, each gold particle deposit can be regarded as the result of an independent antigen–antibody reaction, permitting the use of simple statistics. Hence, one would predict a linear relationship between the gold particle density and the number of available antigen molecules (Ottersen, 1989b; also see Posthuma *et al.*, 1988).

The major obstacle to quantitation resides not in the counting and analysis of gold particles but in the nature of the underlying event: the antibody–antigen coupling. A *conditio sine qua non* for a meaningful quantitation of a sample of antigen molecules is that each molecule in the sample faces the same likelihood of encountering and binding to an antibody molecule. In practice, this requirement is difficult to fulfill. Preembedding procedures pose particular problems, as diffusion barriers imposed by membranes and other tissue constituents will bias the access of immunoreagents to the target molecules. This bias remains even with optimum permeabilization of the tissue in question. With the postembedding immunogold technique, the problem of diffusion constraints is eliminated, as the sample of target antigens is restricted to those molecules that are available at the cut surface of the section. Diffusion to the interior of the section is effectively hindered by the resin. Thus, an attractive feature of the postembedding approach is that all molecules in the sample are equally likely to be visualized by the immunocytochemical procedure.

The situation in practice is probably not as simple (Griffiths, 1993). Access to antibodies may be skewed by the surface relief of the section, and epitopes may be obscured by protein–protein interactions. Also, sterical hindrance between the rather bulky immunoreagents may reduce the probability of

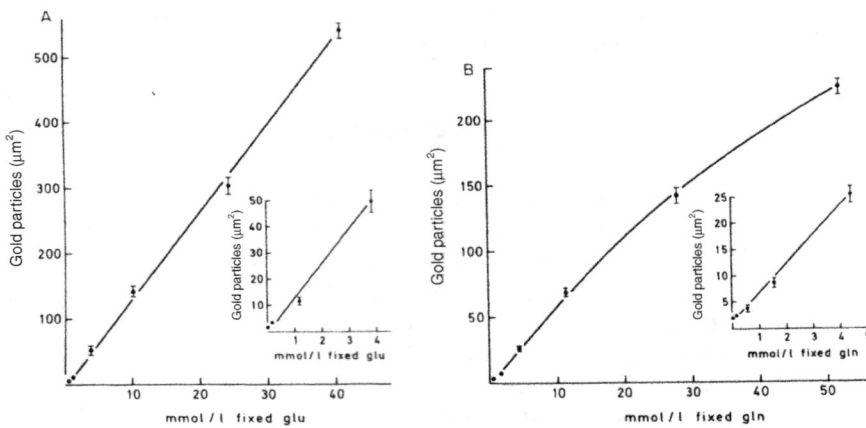

Figure 3.6. A calibration system tailor-made for postembedding immunogold cytochemistry (Ottersen, 1989b) demonstrates linear relationship between antigen concentration (mmol/l fixed glutamate in A; glutamine in B) and gold particle density (recorded over antigen-containing bodies such as those shown in Fig. 3.7). The linear relationship tends to break down at very high (biologically irrelevant) antigen concentrations (B). The lower ends of the plots are shown at larger scale in insets. The slopes and correlation coefficients for the two regression lines were 12.4 and 0.99 (glutamate, all data points included) and 5.4 and 1.00 (glutamine, estimated for data points shown in inset). (From Ottersen *et al.*, 1992.)

labeling at sites of very high antigen concentrations. This notwithstanding, by use of model antigens it has been shown that a linear relationship indeed can be obtained between antigen concentration and gold particle density.

We have used conjugated amino acids as model antigens (Ottersen, 1987, 1989b). Incorporation of radiolabeled amino acids allowed the concentration of antigen to be determined. Antigens in known concentrations were embedded in resin and sectioned for postembedding immunogold labeling and electron microscopy. For a series of different model antigens, it could be shown that the gold particle density was positively and linearly correlated with the calculated concentration of antigen molecules (Ottersen, 1989b; Ottersen *et al.*, 1992; Fig. 3.6). As predicted, in some cases the linearity was found to break down at very high antigen levels, probably as a result of steric hindrance.

The sections that were used to establish the correlation between the antigen concentration and the gold particle density could be incubated together with tissue sections and under exactly the same conditions as these. We thus had in our hands a calibration system that allowed us to determine the approximate concentration of antigen in different cells and organelles. This approach was used, inter alia, to assess the concentration of glutamate in nerve terminals, neuronal cell bodies, and astrocytes (Ottersen, 1989b; Ottersen *et al.*, 1992), and in organelles such as mitochondria and synaptic vesicles (Shupliakov *et al.*, 1992).

A prerequisite for an adequate use of calibration curves is knowledge of the fraction of antigens that are retained in the tissue after fixation. In the case of small molecules such as neurotransmitter amino acids, this is not a trivial problem, as substantial amounts may be lost from the tissue during the fixation procedure. With optimum fixation, using 1% glutaraldehyde or more, the retention of free amino acids exceeds 80%. This could be shown by equilibrating the tissue with tracer amounts of radiolabeled amino acids prior to fixation by immersion or perfusion (Storm-Mathisen and Ottersen, 1990).

When used in conjunction with appropriate model systems such as the one described above, postembedding immunocytochemistry permits an assessment of amino acid concentrations in cell compartments down to the level of synaptic vesicles and other organelles. In principle, this approach is applicable to all antigens that are available in pure form and that can be incorporated in calibration systems that are representative of the mode of antigen expression in vivo (see Griffiths, 1993, for a detailed discussion).

Plasma membrane proteins represent a more difficult case in regard to quantitation than small organic molecules or proteins that are distributed in the aqueous interior of the cell. For membrane proteins, a representative calibration system should be based on model membranes containing known concentrations of the protein in question. Ideally, the model membranes should have a composition similar to that of the membrane in which the protein is expressed in vivo. Obviously, these antigen-containing model membranes would have to be embedded and sectioned in parallel with the tissue and subjected to simultaneous immunoprocessing. It has proved difficult to develop model systems that meet all of these requirements. Unfortunately, therefore, postembedding immunogold cytochemistry of membrane proteins remains semiquantitative rather than quantitative.

To circumvent the difficulties entailed in developing a calibration system for membrane proteins, one may take advantage of quantitative immunoblotting and stereological data. Quantitative immunoblotting provides an estimate of the amount of a given protein per volume unit of tissue, whereas stereological analyses can be used to determine the total distribution surface for that protein (such as the total astrocytic surface in the case of a membrane protein that is restricted to astrocytes). In this way, one may calculate the average number of protein molecules per unit area of plasma membrane, as has been done for astroglial glutamate transporters (Lehre and Danbolt, 1998). This value can be correlated to the linear density of gold particles signaling the molecule in question. In principle, one ends up with a "correction factor" that can be used subsequently to translate gold particle densities into densities of target proteins.

Instead of calibrating the immunogold signal to biochemical data, one can relate the number of gold particles to functional parameters, such as the magnitude of synaptic currents in the case of postsynaptic receptors. Nusser *et al.* (1998) used this approach to assess the number of alpha-amino-3-hydroxy-5-methyl-4-isoxazole-propionic acid (AMPA) receptors at hippocampal synapses. One gold particle in postembedding-labeled methacrylate sections was found to correspond to 2.3 functional receptors, allowing estimates of synaptic receptor content.

As pointed out above, the major challenge in quantitation is to provide an accurate conversion factor between the number of gold particles and the number of tissue antigens. However, the analysis of particle distribution is not trivial, even if one refrains from assumptions regarding the underlying pattern of antigen distribution. Computer-based methods are indispensable if one embarks on large projects that involve sampling of several animals, blocks, sections, and images. In early studies in our laboratory, we digitized photographic prints by means of a digitizing table and counted particles with an electronic pen under control of the computer programs Morforel (Blackstad *et al.*, 1990) and Palirel (Ruud and Blackstad, 1999). Today, our main tool for quantifying immunogold labeling is IMGAP (IMmuno-Gold-Analysis-Program), created in our laboratory (Haug *et al.*, 1994, 1996). This program was developed in collaboration with SIS (Soft Imaging Systems Gmbh, Münster, Germany) as an extension to their product analySIS. IMGAP permits automated acquisition of gold particles, taking advantage of their high electron density and defined sizes. Appendix II describes the typical workflow when using IMGAP. The web site http://www.med.uio.no/imb/stat/immunogold/index.html is a service for methods and programs for statistical analysis of immunogold data.

Simplified methods are applicable when the biological problem at hand is limited to that of comparing immunogold labeling patterns in the same sets of compartments across different cells (Mayhew *et al.*, 2004). In this situation, no information may be required about compartment size or membrane length.

Rapid assessments of gold particle distributions can be obtained by use of simple stereological methods. One example from our own laboratory is described in Landsend *et al.* (1997), based on preparations such as that shown in Fig. 3.5. Using a stereological approach, Lucocq *et al.* (2004) claimed that counting 100–200 particles on each of two grids may be sufficient to produce a rough estimate of the gold particle distribution over as many as 10–16 different compartments. This should be kept in mind when designing an immunogold analysis so as to avoid excessive and pointless counting of gold particles.

Finally, it should be emphasized that the freeze fracture–immunogold technique represents a quite different approach to quantitation, holding great promise for the future (for review, see Rash *et al.*, 2004). Accurate quantitation should be feasible for any protein that can be immunodetected in freeze fracture replicas and that has structural features that set it apart from other IMPs in the same membrane domain. The limited ability to differentiate between different IMPs, based on their morphological appearance, remains a problem in this regard.

Closely related to the issue of quantitation is the term *labeling efficiency*. This term refers to the ratio between the number of gold particles and the number of antigen molecules that is available for immunolabeling. Labeling efficiency depends on many factors that have been addressed in studies of cryosections (e.g., Griffiths and Hoppeler, 1986). As pointed out by Griffiths (1993), it is probable that the labeling efficiency is always below 100% in postembedding-labeled sections. A rough estimate of the labeling efficiency can be obtained by analysis of membrane domains that contain known densities of the target proteins (derived, e.g., from correlative freeze fracture analyses). In a postembedding immunogold analysis of AMPA receptors in hippocampal synapses, it was shown that the number of gold particles increased by one particle per ~15-nm increment in the length of synaptic profile (Takumi *et al.*, 1999). Based on the known size of AMPA receptors and conservative estimates of their spacing, these data suggest that there is a close to 1:1 stoichiometry between gold particles and AMPA receptors at the section surface, given optimum experimental conditions.

Factors that affect labeling efficiency include quality of antibody, fixation procedure, embedding medium, and incubation parameters (Griffiths, 1993; Matsubara *et al.*, 1996). Several of these factors are addressed below, in the discussion of our "standard" postembedding immunogold procedure. Optimization of the postembedding immunogold procedure for a given antigen is very much a question of obtaining maximum labeling efficiency. High labeling efficiency is of critical importance in postembedding immunogold analyses because of the restricted sample of accessible epitopes, and becomes a decisive factor in analyses of membrane proteins that are expressed at low densities. Indeed, the choice of immunocytochemical procedure (pre- or postembedding) should always be based on available information on the prevalence of the antigen at hand.

IV. CONTROLS

The need for appropriate controls in immunocytochemistry can hardly be overemphasized. As a discussion of the general principles for assessing antibody selectivity is outside the scope of the present chapter, we will restrict ourselves to procedures that are specific for postembedding immunogold cytochemistry.

Testing of selectivity must be done in conditions that are representative of the conditions of the immunocytochemical procedure. This is because the conformation of the epitope and the nature of the antibody–epitope interaction may be influenced by a number of factors, such as the choice of fixative and resin and the selection of incubation parameters. In other words, showing that the antibody identifies a single band at the appropriate molecular weight in immunoblots cannot be taken to imply that the antibody provides selective labeling in the postembedding mode. Immunocytochemistry and antibody testing should be performed in parallel and under identical conditions.

This criterion can be met in the case of small molecules such as amino acids and larger antigens that are available in pure form (Davanger *et al.*, 1994; Ottersen, 1987). The target antigen and structurally related molecules can be embedded in resin, sectioned, and immunoincubated together with

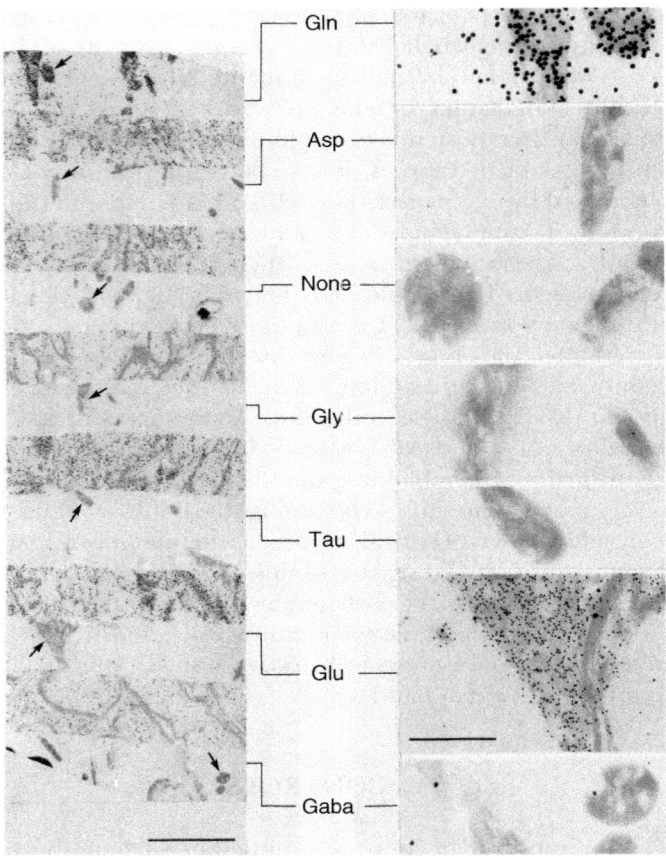

Figure 3.7. Test system designed to monitor antibody selectivity under the conditions of the immunocytochemical procedure. The target antigens and structurally homologous molecules were embedded in resin and incorporated in a test sandwich with alternating brain sections used as spacers. Ultrathin cross sections of this sandwich were incubated together with the tissue sections. The test antigens appear as dense bodies. Bodies identified by arrows are enlarged in the right part of the figure. In this case, the test antigens were prepared by coupling glutamate (Glu), glutamine (Gln), and structurally related amino acids (standard abbreviations) to brain macromolecules in the presence of glutaraldehyde. The test section shown here and the accompanying tissue section (Fig. 3.8) were double labeled for glutamate (15-nm particles) and glutamine (30-nm particles) using a modification of the procedure of Wang and Larsson (1985) (see Ottersen *et al.*, 1992, for details). Quantitative analysis of this test section (Ottersen *et al.*, 1992) confirmed that the antisera react selectively with the target antigen in the actual conditions used for fixation, embedding, and immunoincubation. The quantitative analysis also showed that the present double-labeling procedure (using two antibodies from the same species) distinguishes between the two antigens with negligible interference. (From Ottersen *et al.*, 1992)

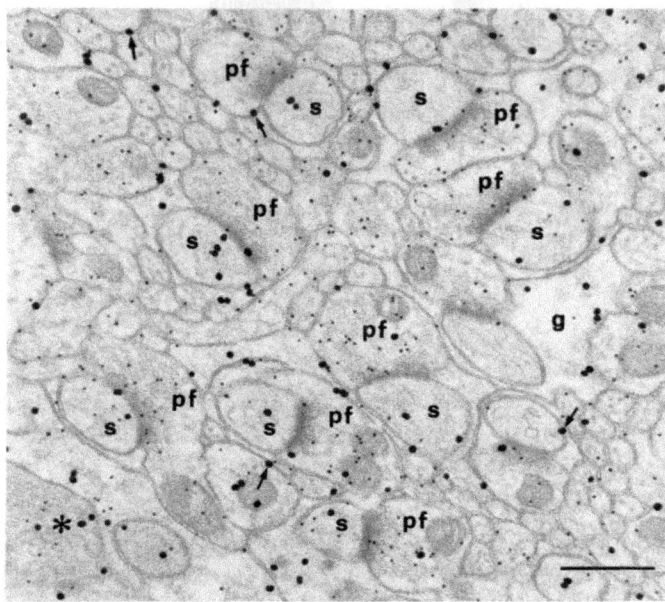

Figure 3.8. Section from the molecular layer of rat cerebellum incubated together with the test section shown in Fig. 3.7. The high selectivity obtained in the test section should be representative of the selectivity in the tissue section. Small gold particles signaling fixed glutamate are enriched in parallel fiber terminals (pf), whereas large gold particles signaling glutamine mainly decorate glial processes (g) and Purkinje cell dendritic spines (s). Asterisk, possible climbing fiber; arrows, intercellular clefts. Scale bar: 0.4 μm. (From Ottersen *et al.*, 1992.)

the tissue sections. This approach provides a direct and reliable validation of antibody selectivity. In the example shown here (Figs. 3.7 and 3.8), it could be documented that antibodies to glutamate and glutamine distinguish between the respective amino acids, despite their close structural similarity. The degree of cross-reactivity could be determined quantitatively (Ottersen *et al.*, 1992; Fig. 3.7). It is important that the test antigens are exposed to the same fixative as the target antigen in the tissue. Specifically, in the case of amino acids, these must be conjugated to brain proteins by glutaraldehyde before embedding and testing (Ottersen, 1987; Storm-Mathisen *et al.*, 1983). In this way, the test antigens will mimic the complexes that are formed in the tissue during perfusion fixation, when the fixative (glutaraldehyde) cross-links free amino acids to brain macromolecules.

Positive controls such as that discussed above document the ability of the antibody to differentiate between structurally similar epitopes. However, negative controls are required to ascertain that the immunogold signal represents antibody binding to the target antigen rather than unspecific labeling. Such controls are particularly important when the target antigen is believed to reside in nuclei, mitochondria, postsynaptic densities, or other sites that promote unspecific binding due to high protein concentrations. The most powerful negative control is provided by the availability of animals with a selective knockout of the gene encoding the target protein. Pending knockout animals, transfection experiments (comparing cells with and without the antigen in question) constitute a useful substitute. One must not put too much emphasis on standard absorption experiments (involving neutralization of the primary antibody by application of an excess of the immunizing peptide), as these do not differentiate between specific and unspecific binding of the antibody clone in question (for a comprehensive discussion of specificity controls, see Holmseth et al., 2005).

V. APPLICATIONS

It is outside the scope of the present chapter to provide a comprehensive discussion of the range of biological problems to which the postembedding immunogold technique can be successfully applied. In our own laboratory, we have found this technique to be particularly useful for the following purposes:

1. Demonstration of protein colocalization by use of double labeling with two different-sized gold particles (Fig. 3.9; also see Fig. 3.15). This application takes advantage of the discrete sizes of gold particles and the fact that double labeling can be successfully performed, with minimum cross-reactivity (Fig. 3.7), even when the two primary antisera are derived from the same species (Ottersen *et al.*, 1992; Wang and Larsson, 1985).
2. Demonstration of receptors and amino acid transmitter (e.g., glutamate) in the same synapses by use of double labeling (Matsubara *et al.*, 1996; Takumi *et al.*, 1999; Fig. 3.10). This procedure requires the use of glutaraldehyde in the fixative to retain the amino acid in question (see section "Postembedding Procedures").
3. Combination of postembedding labeling and anterograde tracing to identify transmitters and receptors in specific fiber projections (Ji *et al.*, 1991; Ragnarson *et al.*, 1998, 2003; Rinvik and Ottersen, 1993; Fig. 3.11).
4. Investigation of disease mechanisms at high resolution. Examples: analysis of glutamate redistribution in experimental stroke to explore the mechanisms of excitotoxic cell death (Torp *et al.*, 1993), and analysis of the mechanisms of β-amyloid generation in Alzheimer's disease and relevant animal models (Torp *et al.*, 2000, 2003; Fig. 3.12).
5. Phenotypic analyses of transgene animals to demonstrate changes in subcellular expression of neuronal or glial proteins (Amiry-Moghaddam *et al.*, 2003a,b; Amiry-Moghaddam and Ottersen, 2003; Kohr *et al.*, 2003; Neely *et al.*, 2001; Rossi *et al.*, 2002).

VI. POSTEMBEDDING PROCEDURES

The hallmark of postembedding immunocytochemistry is that the tissue is embedded in a resin prior to the immunocytochemical procedure. As most

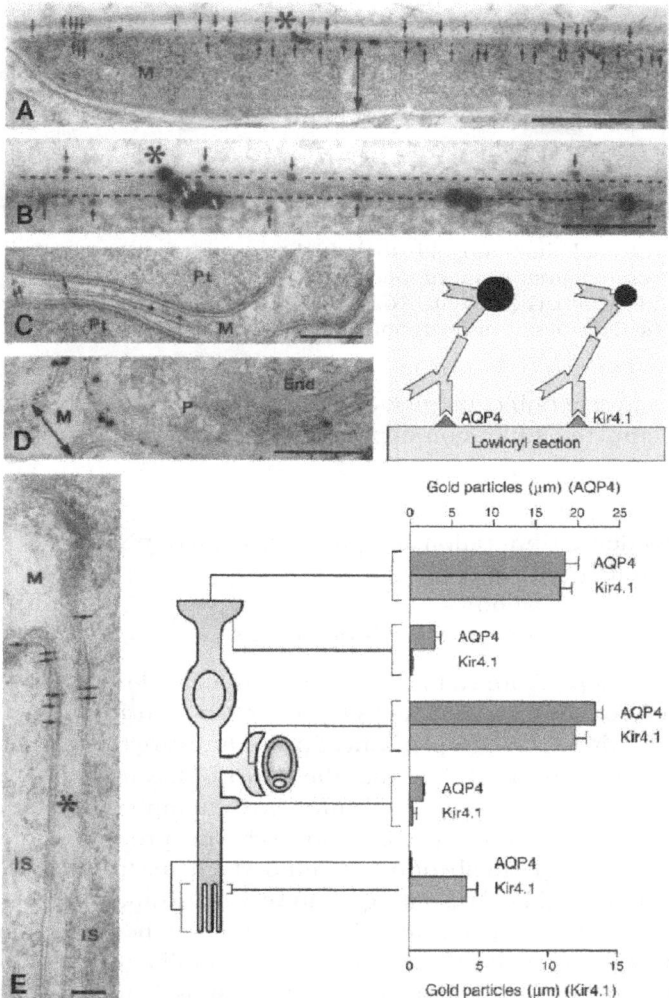

Figure 3.9. Semiquantitative analysis of subcellular expression patterns of two protein antigens. Double immunogold labeling of Kir4.1 (10-nm particles) and AQP4 (20-nm particles) in vitreal (A, B), perisynaptic (C), and perivascular (D) Müller cell (M) membranes. (A, B) The M end foot (M) is selectively labeled at its vitreal aspect (both aspects indicated by double arrow). The vitreal plasma membrane (between dashed lines in B) is obliquely cut, allowing labeling at the two sides of the section to be distinguished. Small arrows indicate Kir4.1 labeling. Asterisks indicate corresponding points at the vitreal surface (part of A is enlarged in B). (C) Weak Kir4.1 (arrows) and AQP4 labeling is found in the thin Müller cell processes (M) that surround photoreceptor terminals (Pt). (D) Perivascular end feet (M) shows a polarized distribution of gold particles (double arrow, compare with A). End, endothelial cell; P, pericyte. (E) Kir4.1 immunolabeling of an M microvillus (asterisk). The gold particles (arrows) are restricted to its basal part which is devoid of large particles signaling AQP4. IS, inner segment of photoreceptors. Scale bars: 0.5 μm (A); 0.1 μm (B, E); 0.25 μm (C, D). The graph shows linear densities of gold particles in different membrane domains of M. (Modified from Nagelhus *et al.*, 1999.)

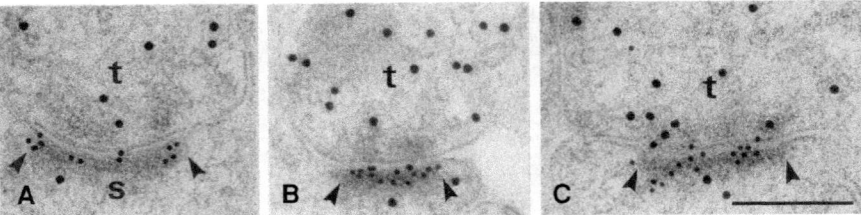

Figure 3.10. Double immunogold labeling of transmitter and receptor in synapses from stratum radiatum of rat hippocampus. Large gold particles signal fixed glutamate, and small particles signal NMDA receptors. Arrowheads indicate extent of postsynaptic densities. t, nerve terminal. Scale bar: 200 nm. (From Takumi *et al.*, 1999.)

resins are hydrophobic, the tissue water has to be replaced by an organic solvent before the infiltration step. Thus, the sequence of steps is as follows:

1. Fixation
2. Dehydration
3. Embedding (infiltration and polymerization of resin)
4. Sectioning
5. Immunoincubation
6. Electron microscopic analysis of sections

Each of the steps above can be performed in many different ways. In fact, there are about as many recipes as there are laboratories in the field. The reasons why there is such a plethora of procedures are that each laboratory has a unique set of needs and that the protocol has to be tailored to the antigen at hand. The latter point cannot be overemphasized: for each new target antigen, one must be prepared to modify the procedure for optimum results. This is why postembedding immunocytochemistry is oftentimes challenging and sometimes frustrating—and very rewarding when it works.

Any researcher who enters into the field of postembedding immunocytochemistry will soon realize that many antibodies will never work, regardless of fixation, embedding, and incubation conditions. Unfortunately, the performance of a given antibody in postembedding immunogold analyses cannot be predicted from its performance in immunoblots or immunofluorescence. An element of "trial and error" is inevitable.

Appendix I provides an outline of the protocol that is currently being used in our own laboratory for the initial screening of a novel protein antigen. Optimization of this protocol for a given protein may require substantial modifications, as emphasized below.

We will first provide a general discussion of the major steps of the procedure, starting with tissue fixation.

A. Fixation

As for immunocytochemistry in general, the choice of fixative is a trade-off between the need to preserve tissue ultrastructure and the need to

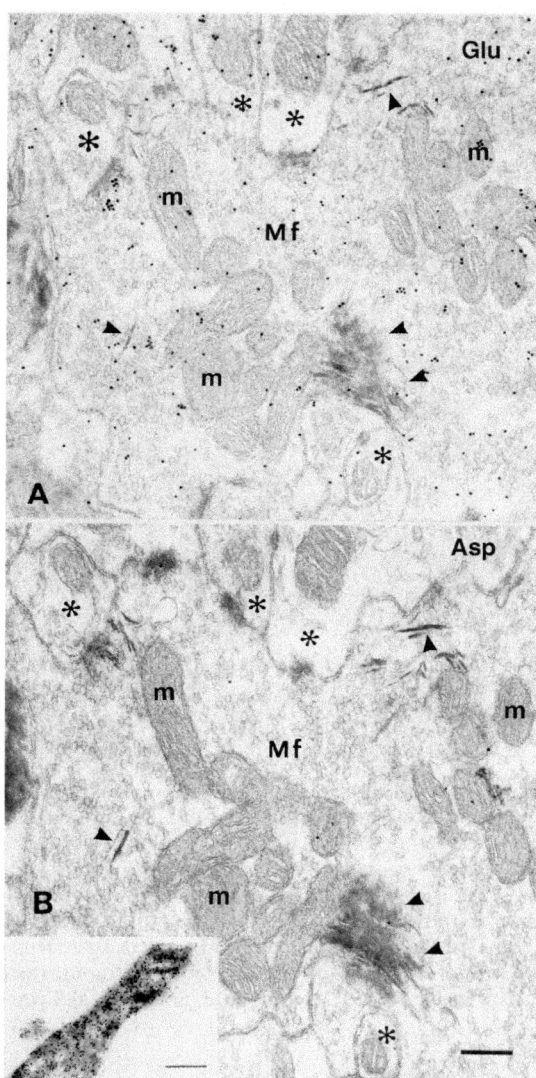

Figure 3.11. Combination of anterograde labeling and postembedding immunogold cytochemistry. Spinocerebellar terminals were identified by anterograde transport of horseradish peroxidase conjugated to wheat germ agglutinin (HRP-WGA). The HRP reaction product is indicated by arrowheads. The anterogradely labeled mossy fiber terminal (Mf) is immunopositive for glutamate (A) but immunonegative for aspartate (B; section adjacent to that in A). The absence of aspartate immunolabeling is not a false-negative result as aspartate-containing model conjugates, incubated together with the tissue section, showed strong immunogold labeling (inset in B). m, mitochondria; asterisks, granule cell dendritic digits. Scale bars: 0.2 μm (A, B); 0.6 μm (inset). (From Ji *et al.*, 1991.)

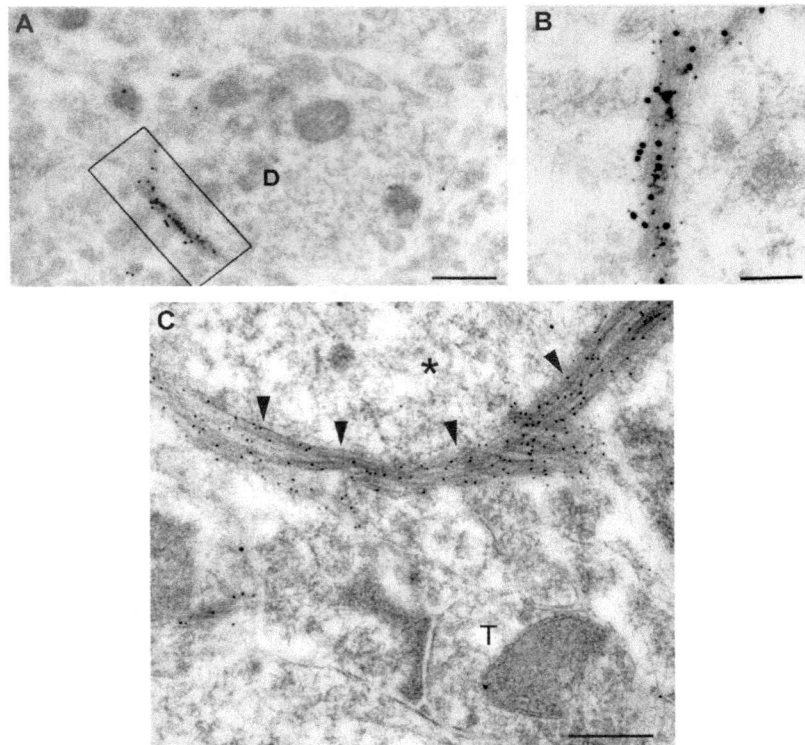

Figure 3.12. (A, B) Plasma membrane domains show coexpression of presenilin 1 (PS1) and amyloid (Aβ 42). Double immunogold labeling reveals colocalization of Aβ (large particles) and PS1 (small particles) in a discrete patch of neuronal plasma membrane in somatosensory cortex of an aged canine. Framed area in A is enlarged in B. D, dendrite. Scale bars: 0.3 μm (A); 0.2 μm (B). (From Torp *et al.*, 2000.) (C) Postembedding immunogold labeling of amyloid plaque in the hippocampus of a transgenic mouse model of Alzheimer's disease (3XTg-AD). Arrowheads point to extracellular deposit of amyloid. Intracellular compartments (asterisk) are devoid of gold particles. Antibody to Aβ 42. T, terminal. Scale bar: 0.4 μm (R. Torp, 2006).

preserve the antigen in a form that can be recognized by the specific antibodies. Strong fixatives, with a high concentration of glutaraldehyde, provide the best ultrastructure, whereas weak fixatives, with little or no glutaraldehyde, provide optimum preservation of antigenicity and hence the strongest immunocytochemical signal. The concentration of glutaraldehyde is more critical than the concentration of formaldehyde, as glutaraldehyde is bivalent (it has two reactive aldehyde groups). As such, glutaraldehyde is a very efficient cross-linker. Cross-linking of tissue macromolecules affords good ultrastructure but may severely distort or mask the target epitopes. In our hands, it is impossible to predict the amount of glutaraldehyde that a given antigen will tolerate. For some membrane proteins, glutaraldehyde concentrations as high as 1% have proved to be compatible with a strong immunogold signal. Our standard protocol (i.e., the protocol used for the first screening of a novel protein target) is based on the use of 0.1% glutaraldehyde in combination with 4% formaldehyde.

Visualization of neuroactive amino acids (such as glutamate, γ-amino butyric acid (GABA), and glycine) represents a special case (Ottersen, 1987; Somogyi et al., 1986; Somogyi and Hodgson, 1985), as these small molecules are lost from the tissue unless they are irreversibly cross-linked to tissue macromolecules (Dale et al., 1986; Ottersen, 1989a; Storm-Mathisen et al., 1983). Thus, immunocytochemical analysis of amino acids and other small molecules with free amino groups (including glutathione; Hjelle et al., 1994) requires the inclusion of significant amounts of glutaraldehyde. We have shown that 1% glutaraldehyde is sufficient to retain the major proportion of free amino acids. Concentrations as low as 0.1% may yield a good signal, permitting double labeling of a transmitter amino acid and its respective receptor (e.g., visualization of presynaptic glutamate and postsynaptic AMPA receptors in the same synapses; Takumi et al., 1999). However, the lower the glutaraldehyde concentration, the greater the risk for a skewed retention of amino acids across the different cell compartments.

B. Dehydration and Embedding

For protein antigens in the central nervous system, the combination of freeze substitution and embedding in methacrylate resins has become very popular (Griffiths, 1993; Humbel and Schwarz, 1989; Matsubara et al., 1996; Nusser et al., 1998; van Lookeren Campagne et al., 1991). This is also the standard procedure in our own laboratory. This combination of techniques is designed to preserve the original conformation of the protein antigen. For analyses of brain, we prefer to fix the tissue before freezing rather than at later stages in the process.

The first step is to cryoprotect the fixed tissue specimen by immersion in glycerol or sucrose. The specimen is then frozen, usually by plunging it into liquid propane. Freezing must be obtained as quickly as possible, in order to avoid loss of ultrastructure due to the formation of ice crystals (Griffiths, 1993). Liquid propane has a higher temperature than liquid helium but is considered superior to the latter since it allows a faster dissipation of heat from the tissue specimen. Obviously, the size of the specimen must be restricted in order to permit rapid freezing. In practice, we usually refrain from exceeding 1 mm in any dimension.

Once frozen, the tissue specimen is transferred from the cryofixation unit to a cryosubstitution unit. This unit supports three sequential processes:

1. Substitution of methanol for ice
2. Substitution of resin for methanol (i.e., infiltration)
3. Polymerization of resin by ultraviolet (UV) light

The overall effect of processing the tissue in the cryosubstitution unit is to replace ice with polymerized resin. The substitutions occur at very low temperatures and are therefore very slow, requiring several days to complete. In the cryosubstitution unit, contrast is conferred by exposure of the tissue to uranyl acetate, which has affinity for biological membranes. Uranyl acetate takes the place of osmium tetroxide, commonly used to provide contrast when the tissue is embedded in epoxy resins. Osmium tetroxide absorbs light and can be used only in low concentrations in combination with UV polymerization.

The freeze substitution procedure outlined above differs significantly from the classical procedure for dehydration and embedding. The classical procedure is based on postfixation of the tissue in osmium tetroxide, followed by dehydration, infiltration in an epoxy resin, and polymerization. These steps typically take place at room temperature (RT), except for the polymerization process which is run at 60°C, depending on the type of epoxy resin.

Although procedures based on the use of epoxy resins can be modified for visualization of proteins in the postembedding mode (Phend *et al.*, 1995; Salio *et al.*, 2005), the freeze substitution procedure is usually considered as superior. Our own experience with a number of protein antigens is that the latter procedure provides a much stronger immunosignal and higher signal-to-noise ratio than do procedures based on epoxy resins. Several factors have been proposed to explain why freeze substitution provides such an advantage:

1. All steps, including polymerization, are run at low temperatures, reducing the risk of protein denaturation.
2. Methacrylate resins can accommodate a certain amount of water, possibly allowing for a partial preservation of the proteins' hydration shells.
3. The polymerization of methacrylate resins does not engage the proteins to the same extent as does the polymerization of epoxy resins, reducing the risk of distorting or masking the target epitopes.
4. Due in part to their unique polymerization features, methacrylate resins provide for a more pronounced surface relief than do epoxy resins. In other words, proteins at the cut surface tend to stick out of the plane of section rather than being bisected.

In sum, these mechanisms may help retain the original conformation of the target protein and preserve its antigenicity (Griffiths, 1993). The freeze substitution procedure is also well suited for nonproteinaceous antigens such as neuroactive amino acids, allowing for double labeling analyses (Fig. 3.10; also see Matsubara *et al.*, 1996).

A range of methacrylate resins is available. The different resins differ primarily by their ability to accommodate water and by their fluidity. In principle, given the fact that proteins are normally expressed in an aqueous environment, high hydrophilicity should translate into a better preservation of protein conformation. For glutamate receptors and a number of plasma membrane transporters in the central nervous system, we have observed rather minor differences between different methacrylate resins when it comes to the strength of the immunosignal. Our standard procedure is based on the use of Lowicryl HM20, which is one of the more hydrophobic methacrylates. Very hydrophilic methacrylates may be difficult to work with due to inadvertent swelling of the sections once they are exposed to water.

C. Immunoincubation

Postembedding immunogold labeling of ultrathin sections usually comprises the following steps:

1. "Etching" of section to increase the availability of epitopes at the section surface
2. Blocking of unspecific binding of antibodies
3. Application of primary antibody
4. Application of secondary IgG, Fab, or protein A coupled to colloidal gold

The goal is to end up with a distribution of gold particles that truly reflects the distribution of target epitopes, given the constraints imposed by the size of the antibody bridge (Fig. 3.1). To minimize variability, the sections and grids to be analyzed should be included in the same batch for simultaneous incubation.

Appendix I shows the immunoincubation procedure that is in current use as our standard laboratory protocol. The standard protocol has been modified over the years (Chaudhry *et al.*, 1995; Hjelle *et al.*, 1994; Nagelhus *et al.*, 2005) and has been influenced strongly by protocols published from other laboratories (e.g., Nusser *et al.*, 1998; van Lookeren Campagne *et al.*, 1991). In fact, as is the case for immunocytochemical procedures in general, postembedding procedures evolve continuously as parameters are tested out and as resins and reagents improve.

The outcome of an immunoincubation depends on the combined effect of a number of different steps. In fact, the enormous number of permutations that can be obtained by combining variations of the different steps precludes a bona fide scientific approach to optimization of the procedure. This is unfortunate, as several steps in current use may have been carried over from other experimental settings, which may or may not be representative of the procedure in question.

For the purpose of the present chapter, we have performed a systematic variation of key parameters to assess their relative importance for the strength of the immunogold signal (measured as the number of gold particles per micrometer membrane following immunogold visualization of the water channel protein AQP4). As shown in Fig. 3.13, the choice of secondary IgG/colloidal gold conjugate profoundly affects the signal strength

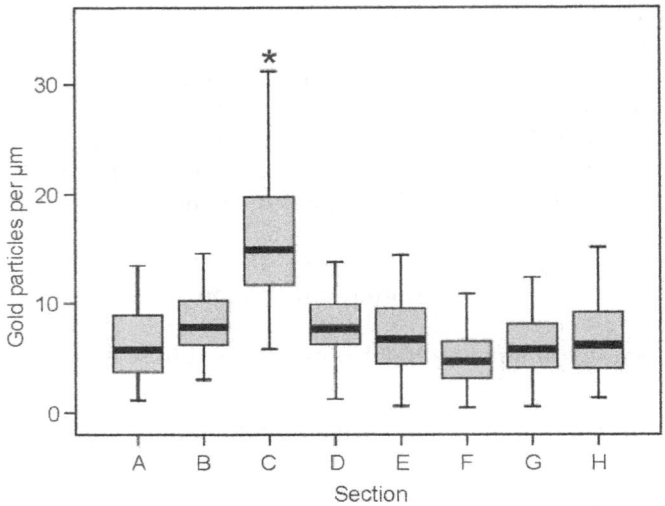

Figure 3.13. Quantitative assessment of how incubation parameters influence labeling efficiency. Preparations similar to that showed in Fig. 3.3A, postembedding labeled with an antibody to AQP4, were subjected to automated analysis of gold particle density by the procedure outlined in Appendix II. The labeling intensity (particles per micrometer perivascular membrane) depended on the combination of incubation parameters (A–H). Reference conditions (A) were identical to those of the "standard" protocol in Appendix I. These were sodium ethanolate (etching), 2% HSA (blocking), 0.04 M NaCl (in buffer), 0.1% Triton (T), 15-nm gold particles. B–H differed from A in the following ways. B: no sodium ethanolate; C: 10-nm gold particles; D: no HSA; E: 0.01% T; F: 0.2% milk powder in lieu of HSA; G: 0.12 M NaCl; H: identical to G, but analyzed by a different person (to check for consistency). Sixty membrane segments were analyzed for each combination of parameters.

(compare A and C). Secondary IgG coupled to 10-nm particles leads to a higher linear density of gold particles than does a secondary IgG coupled to 15-nm particles. This effect relates primarily to the size of gold particle (rather than differences between the two IgGs) as qualitative analyses of a wider range of particle sizes point to a negative correlation between gold particle size and signal strength. Steric hindrance may be one of the several factors that underlie this effect.

A logical extension of the latter observations would be to minimize the gold particle size to maximize the labeling efficiency. Gold particles that are less than 5 nm in diameter are not easily discerned in postembedding-labeled sections, calling for silver enhancement if smaller particles are used. Application of nanogold particles in combination with silver enhancement indeed leads to a significant increase in gold particle density, as shown in postembedding immunogold labeling of glutamate receptors (Matsubara *et al.*, 1996; also see Fig. 3.17). However, such preparations pose difficulties in quantitative analyses: The silver-enhanced particles are less uniform than gold particles in regard to size and shape, and their stoichiometric relationship to the target antigens is less well defined.

Another parameter that may affect signal strength is the concentration of NaCl in the buffer of the primary and secondary antibody steps. It is well known that antibody–antigen binding depends on a number of different forces, including ion bonds between oppositely charged residues in the antibody and antigen molecules. Thus, ions in the buffer (primarily Na^+ and Cl^- ions) would be expected to compete with antigen–antibody binding. This prediction was borne out in our postembedding analysis of AMPA glutamate receptors (Matsubara *et al.*, 1996) and led us to lower the NaCl concentration of the buffer (typically to one third of that of physiological saline) whenever an increased signal strength was wanted. In the case of AQP4, the effect of reducing the salt concentration is negligible (compare A with G in Fig. 3.13), underlining the fact that the significance of the different parameters depends on the nature of the target antigen.

Blocking of unspecific binding is an essential step in postembedding immunocytochemistry as in immunocytochemistry in general. Unspecific binding may be caused by a wide range of mechanisms. Obviously, when aldehydes are used for fixation, there is a risk that reactive aldehyde groups remain in the tissue and that these groups attach the immunoreagents to the section. For this reason, we always include glycine and TRIS in the first blocking buffer. Both of these molecules have free amino groups that would bind to and neutralize any free aldehyde groups at the section surface.

We also regularly use human serum albumin (HSA) to prevent unspecific binding of IgG. Fortunately, the inclusion of up to 2% of HSA does not cause any inadvertent reduction in the strength of the specific signal (compare A with D in Fig. 3.13). Milk powder is also in common use as a blocking agent, but our experience is that it is difficult to adjust its concentration so as to provide a reduction of background labeling without affecting the specific signal. In fact, in our quantitative analysis of AQP4 immunolabeling we recorded a reduction in gold particle density when exchanging 2% HSA with 0.2% milk powder (compare A with F in Fig. 3.13). The use of well-defined blocking agents (such as HSA) is encouraged for the purpose of standardization and reproducibility. Very low background labeling can be obtained with adequate blocking procedures (cf. Figs. 3.5 and 3.14).

The etching step is probably the most problematic step to justify in the postembedding immunogold protocol. As alluded to above, the purported aim of this step is to improve access to the target epitopes. However, it is not clear to what extent the commonly used etching procedure (a short immersion in sodium ethanolate; see Appendix I) removes resin and unmasks surface epitopes. In fact, our quantitative analysis of AQP4 suggests that there is no gain in immunogold signal by including this step (compare A with B in Fig. 3.13). Leaving out the etching steps also allows for excellent labeling of glutamate receptors and intracellular enzymes (Figs. 3.15 and 3.16). It should be emphazised that the comparison in Fig. 3.13 was done in the presence of Triton. Triton might help unmask epitopes at the section surface (by dissolving lipids) and could easily obscure any positive effect of the etching procedure. Pending a systematic analysis of the interaction

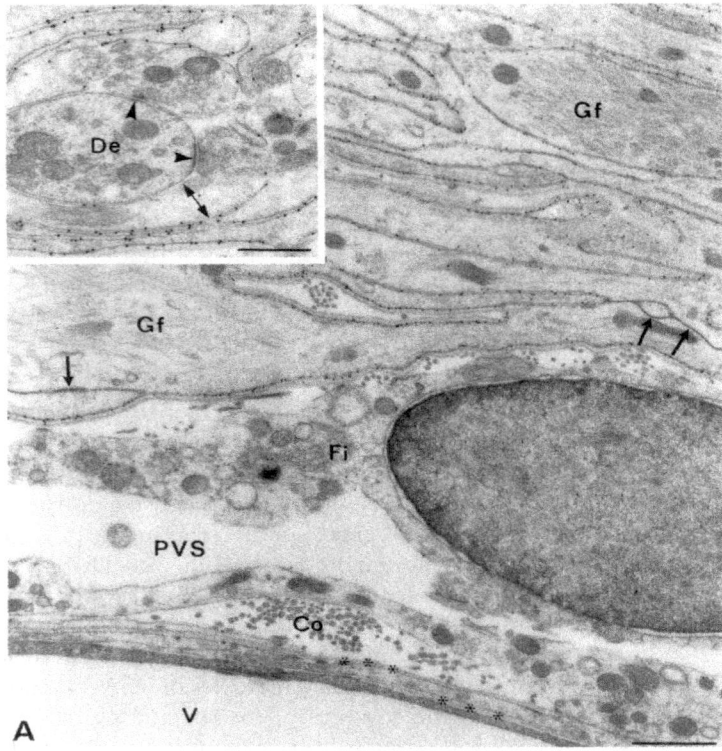

Figure 3.14. High signal-to-noise ratio afforded by optimization of postembedding immunogold procedure. Gold particles signal expression of AQP4 in the subfornical organ of rat. Immunogold particles identify AQP4 along the entire glial lamellae except at the membrane domains engaged in gap or adhaerens type junctions (arrows) or contacting neuronal elements (double-headed arrow in inset). The vessel (V) and associated basal laminae (asterisks) are devoid of AQP4 immunolabeling. Co, collagen; Fi, fibroblast; Gf, glial filaments; PVS, perivascular space. Inset: unlabeled synapses (arrowheads) sandwiched between glial lamellae. The adjacent glial processes are polarized with respect to AQP4 expression (double-headed arrow). De, dendrite. Scale bars: 1 μm. (Micrographs by E. Nagelhus taken From Nielsen *et al.*, 1997).

between etching and Triton application, the possibility exists that the etching step represents an unjustified adoption to methacrylate resins of a step that has been shown to work well with epoxy resins. In the absence of a proven effect, the etching step should be omitted as it significantly detracts from the quality of the ultrastructure. Again it should be recalled that the effect of etching (or lack thereof) may not be the same for all antigens.

The above discussion amply documents that immunocytochemistry is still an art and not a science. Quantitative analyses such as that shown in Fig. 3.13 can be used to monitor the effect of changing an individual parameter, but the magnitude of effect, if any, may depend on the other parameters and on the nature of the antigen. The take-home message is that the postembedding immunogold procedure must be tailored to one's needs and to

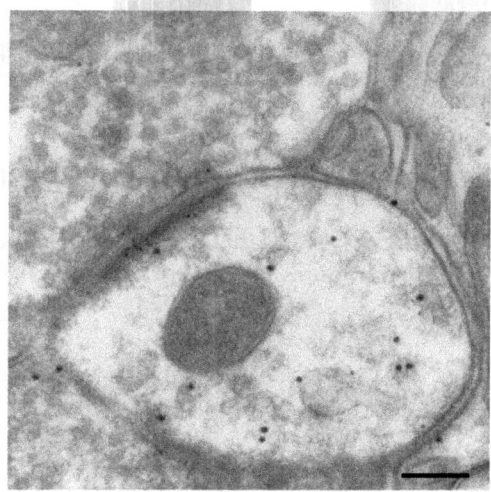

Figure 3.15. Omission of etching (conditions as in Fig. 3.13B but with 0.01% Triton) is compatible with high labeling efficiency. The sections were first immunolabeled with an antibody against tyrosine hydroxylase (20-nm gold particles), and after 1 h in formaldehyde vapor at 80°C (procedure of Wang and Larsson, 1985) the sections were immunolabeled with antibodies recognizing NMDA receptor subunits A and B (10-nm gold particles). From ventral tegmental area of rat. (Micrograph by E. Rinvik, 2006.)

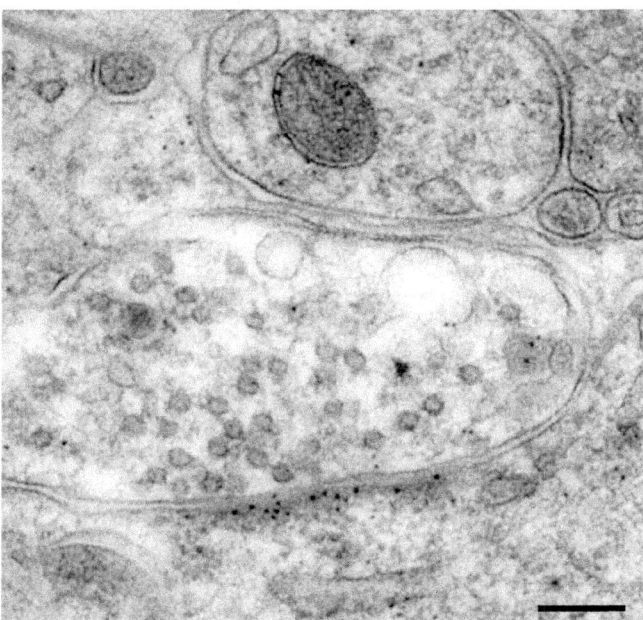

Figure 3.16. Omission of etching allows high labeling efficiency and good ultrastructural preservation (compare with Fig. 3.10). Same procedure as in Fig. 3.15, but single labeling with antibodies to AMPA receptor subunits 2 and 3. From ventral tegmental area of rat. (Micrograph by E. Rinvik, 2006.)

one's targets. This is often a major challenge but a challenge well worth taking because of the wealth of information that can be gained when an immunogold experiment succeeds.

APPENDIX I: A POSTEMBEDDING IMMUNOGOLD PROCEDURE FOR MEMBRANE PROTEINS

This procedure permits double labeling with antibodies to glutamate, GABA, or other neuroactive amino acids (Takumi *et al.*, 1999).

A. Tissue Preparation

Note. Steps 5–10 are carried out in a computer-controlled cryosubstitution unit.

1. Anesthetize the animal. Perfuse transcardially with 2% dextran (MW 70,000) in 0.1 M sodium phosphate buffer (PB; pH 7.4, 15 s at RT) followed by a mixture of glutaraldehyde (0.1%) and formaldehyde (4%; freshly depolymerized from paraformaldehyde) in the same buffer (for rats, 50 ml/min for 20 min at RT).
2. Leave the brain in situ overnight (4°C).
3. Isolate tissue specimens from brain, cryoprotect by immersion in increased concentrations of glycerol (10, 20, and 30% in PB), 0.5 h for each concentration (RT), and then overnight in 30% (4°C).
4. Place the tissue on the specimen pin and plunge into propane cooled to −170°C by LN2 in a cryofixation unit (Reichert KF80, Vienna, Austria).
5. Transfer the specimens to 1.5% uranyl acetate dissolved in anhydrous methanol (- 90°C) in a cryosubstitution unit (AFS; Reichert). After 30 h, raise the temperature stepwise (4°C increment per hour) from −90 to −45°C.
6. Wash the samples three times with anhydrous methanol.
7. Infiltrate with Lowicryl HM 20 resin (Polysciences, Inc., Warrington, PA 18976. Cat# 15924) at −45°C.

   (a) Lowicryl/methanol: 1:1, 2 h
   (b) Lowicryl/methanol: 2:1, 2 h
   (c) Pure Lowicryl: 2 h
   (d) Pure Lowicryl: overnight

8. Change to freshly prepared Lowicryl and move the Reichert capsules with the specimens to the Lowicryl-filled gelatin capsules in the G-chamber.
9. Transfer the capsules to a container filled with ethanol.
10. Polymerize with UV light. Start at −45°C (24 h), and then increase temperature to 0°C (increment 5°C/h). Complete the polymerization at 0°C (35 h).

11. Prepare ultrathin sections at 80- to 100-nm thickness and mount on nickel rids or gold-coated grids for immunogold cytochemistry.

B. Immunoincubation

1. Etch sections in saturated sodium ethanolate (can be omitted; cf. text) for 2–5 s.
2. Rinse well in distilled water for 3× short and 1 × 10 min. Let dry and check that the sections are in place.
3. Place grids into a grid support plate (Leica cat. no. 705698) and immerse in 50 mM glycine in TBST (Tris-buffered saline with 0.1% Triton X-100 or 0.01% Triton; cfr. text) for 10 min.
4. Preincubate in TBST containing 2% HSA for 10 min.
5. Incubate in primary antibody diluted in TBST containing 2% HSA for 2 h, overnight in RT.
6. Rinse in TBST for 3× short, 1 × 10 min and 3× short, 1 × 10 min.
7. Preincubate in TBST containing 2% HSA for 10 min.
8. Incubate in secondary antibody (IgGs coupled to colloidal gold particles). Dilute as recommended from the company and with 2% HSA and 0.05% polyethyleneglycol (PEG) for 1 h.
9. Rinse briefly 6× in distilled water, dry sections.
10. Incubate in 5% uranyl acetate in 40% ethanol for 90 s.
11. Rinse briefly 3× in distilled water, dry sections.
12. Incubate in lead citrate for 90 s.
13. Rinse briefly 3× in distilled water, dry sections.

C. Solutions

Sodium ethanolate: Add 100 g NaOH in 700 ml 100% ethanol.
TBST: 100 ml 0.05 M Tris buffer, pH 7.4 (pH adjusted with HCl); 900 ml ultrafiltered (UF) water containing 0.9% NaCl; 1 g Triton X-100.
Lead citrate: Dissolve 1.33 g lead nitrate and 1.76 g sodium citrate in 30 ml UF water. Stir for 30 min. Add 8 ml 1 M NaOH and fill up to 50 ml with UF water. Aliquot in 10 ml syringes and store protected from light in refrigerator.

D. Protocol for Postembedding Immunogold Labeling Using Ultrasmall Gold Particles Coupled to Fab Fragments (Secondary Antibodies) and Silver Intensification (Fig. 3.17)

1. Rinse grids in MilliQ filtered (18.2 MΩ) water for 1 min
2. TBST containing 0.1% sodium borohydride and 50 mM glycine
3. Blocking buffer 10 min
4. Primary antiserum 2 h or overnight

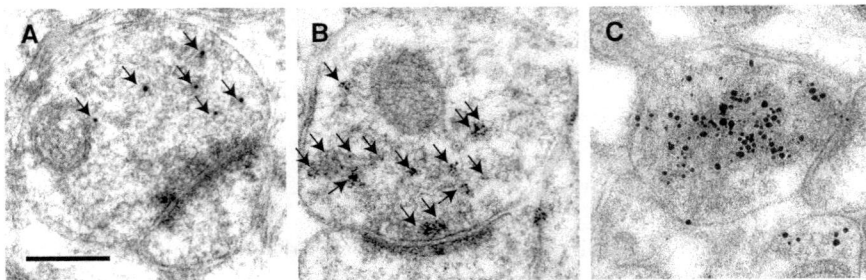

Figure 3.17. Silver enhancement provides increased labeling intensity of synaptic vesicle proteins. (A) Postembedding labeling of SV2 using 10-nm colloidal gold coupled to the secondary antibody. (B) Same as in A, but with 1.4-nm gold particles ("Nanogold") coupled to a secondary Fab fragment and followed by silver enhancement. (C) Same as in B, but with antibody to synaptophysin. Note clusters of small silver deposits in B, and larger silver deposits due to longer reaction time in C. Silver intensification gives silver deposits of varying sizes, prohibiting double-labeling experiments. (Micrographs by S. Davanger, unpublished data.)

 5. TBST 1 min
 6. Blocking buffer 3× 1 min
 7. Nanogold Fab 1:40 in blocking buffer containing PEG 2000
 8. Phosphate-buffered saline (PBS) 3× 1 min
 9. Glutaraldehyde 1% in PBS 3 min
10. Water 2 × 1 min
11. HQ silver Nanoprobes (N.B. safelight!)
12. Water 3 × 5 min
13. Dry sections
14. Uranyl acetate 2% 40 min
15. Water 4 × 1 min
16. Dry sections
17. Lead citrate 0.3% 3 min
18. Water 4 × 1 min
19. Dry sections

TBST (1000 ml): 100 ml 0.05 M Tris, pH 7.4; 900 ml MilliQ water with 0.9% NaCl; 1 g Triton X-100.
Blocking Buffer (1500 µl): 2% (30 µl) normal goat serum; 1% (125 µl of 12% stock solution) bovine serum albumin; 0.5% (7.5 µl) Tween 20; TBST 1337.5 µl.

APPENDIX II: COUNTING IMMUNOPARTICLES
BY DIGITAL IMAGE ANALYSIS

Computer support may substantially reduce the time to completion of large projects, as illustrated by the IMGAP software referred to in the text. The latter combines interactive and automated procedures for image storage and retrieval, particle counting over organelles and membranes, calculation of areal and linear densities, and converson of the final data set to a statistics program for further analyses. The following workflow does not distinguish between native analySIS functions and in-house extensions.

1. "Define" an "IMGAP project": Create a project folder with section folders to hold all images, ancillary information, and results. Define project-wide pick lists of Field types (see point 6 for a definition of Field in this context) and Object types ("objects" are transected spines, postsynaptic densities, etc.). Select measurements (native to analySIS or defined in IMGAP) to be automatically carried out on each object.

2. Choose the electron optic magnification that provides the best compromise between digital resolution, required field of view, and the size of the digital image file. Since lossy compression methods should be avoided, projects with many large images may run into several gigabytes, not a problem with today's storage facilities.

3. Capture TIFF images with any suitable camera and software and subsequently import them into IMGAP, or capture images with analySIS while keeping IMGAP active for online access to the project-wide pick list of image labels and possibly other services.

4. Freehand draw transected organelles as Regions of Interest (ROIs), and membranes as curves, attaching a type label to each such graphic object.

5. After background smoothing and global intensity thresholding, detect and classify particles in up to three size classes.

6. Inspect the result and revise if necessary, by changing parameters for automated procedures or by interactively deleting, adding, or splitting particles. In particular, although particle detection is partly automated, its quality depends 100% on the operator.

   Each ROI, curve, and particle is now uniquely identified and geometrically described by analySIS. Together with analySIS, IMGAP ensures that they are automatically stored with the corresponding image in "revisable format" and automatically displayed in the overlay whenever an image is reloaded. In IMGAP, an image with associated ROIs, curves, and particles, and other associated information, is termed a field (i.e., "field of view").

7. Automatically aggregate and export particle measurements (data on each particle in the project) to SPSS. Use to create transverse histograms for each "type" of membrane and for "quality control" of the detected particles (cf. point 6).

8. Automatically calculate areal density of particles per ROI, aggregate over the project, and export to SPSS.

9. Prior to calculating linear densities along membranes, set a distance filter to eliminate particles too far off the curve.

10. Export ROI and curve measurements for further graphical and statistical analyses.

11. In the SPSS data set, each object (ROI, curve, or particle) has one line and each measurement or other characteristic one column. Check the data set for errors as recommended in standard statistics textbooks and software manuals. Each measurement is traceable to the underlying image, and so errors may be diagnosed and corrected.

12. For images in need of corrections, repeat steps 4–6. Then go on to steps 7 and 8.

13. Further analyses are performed in SPSS or other software (the SPSS formats are compatible with a range of other statistics software, but more manual calculations may also be required). Also see web site http://www.med.uio.no/imb/stat/immunogold/index.html.

An important source of error in the above procedures should be noted. Due to simplistic particle-detection and quality-control algorithms, *small* falsely positive particles may easily go unnoticed unless the operator is alert to this possibility in step 6 and has been duly trained to optimize the parameters for steps 2, 3, and 5. Increased intensity resolution (14 rather than 12 bits) may improve on this, as emphasized by Monteiro-Leal *et al.* (2003), and in addition, the algorithms for particle detection and quality control should be improved.

Limited digital resolution and operator error during interactive drawing (step 4) may add error, presumably random. Obviously, both the automatically detected particles and the interactively drawn curves will be more accurate with increasing resolution. Compared to the "basic uncertainty zone" which surrounds immunogold attached via primary and secondary antibodies, the present inaccuracy may be thought insignificant. However, when planning histograms with 1-nm bins, select the electron optic magnification with a view to the resulting object pixel dimensions (the size of a pixel projected to the specimen). To illustrate this, our MegaView III camera (trademark of Soft Imaging Systems) has a nominal xy resolution just above 1280×1024 pixels. Mounted on our Tecnai 12 it renders a specimen area of 2.1 µm × 1.6 µm, with an object pixel size of 1.5 nm at a nominal electron optical magnification of 49,000×. (As an aside, with these parameters, 10-nm gold particles come out with diameters around 4–6 pixels and the particle detection algorithm still works satisfactorily.)

While a digital resolution of 1280×1024 pixels is small in relation to digital cameras in this year's (2005) consumer market, the price–performance ratio on cameras for electron microscopy remains two orders of magnitude higher. If you need higher digital resolution, automated image montage, in analySIS represented by the MIA module, may be an alternative, although in our hands its success rate is often less than 100%.

ACKNOWLEDGMENTS. Support is acknowledged from the Norwegian Research Council, the Nordic Council (the centre of excellence programme in molecular medicine), and EU (projects QLG3-CT-2001-02089 and LSHM-CT-2005-005320).

REFERENCES

Amiry-Moghaddam, M., Otsuka, T., Hurn, P. D., Traystman, R. J., Haug, F. M., Froehner, S. C., Adams, M. E., Neely, J. D., Agre, P., Ottersen, O. P., and Bhardwaj, A., 2003a, An alpha-syntrophin-dependent pool of AQP4 in astroglial end-feet confers bidirectional water flow between blood and brain, *Proc. Natl. Acad. Sci. USA* **100**(4):2106–2111.

Amiry-Moghaddam, M., and Ottersen, O. P., 2003, The molecular basis for water transport in brain, *Nat. Rev. Neurosci.* **4**(12):991–1001.

Amiry-Moghaddam, M., Williamson, A., Palomba, M., Eid, T., de Lanerolle, N. C., Nagelhus, E. A., Adams, M. E., Froehner, S. C., Agre, P., and Ottersen, O. P., 2003b, Delayed K^+ clearance associated with aquaporin-4 mislocalization: phenotypic defects in brains of alpha-syntrophin-null mice, *Proc. Natl. Acad. Sci. USA* **100**(23):13615–13620.

Blackstad, T. W., Karagulle, T., and Ottersen, O. P., 1990, MORFOREL, a computer program for two-dimensional analysis of micrographs of biological specimens, with emphasis on immunogold preparations, *Comput. Biol. Med.* **20**(1):15–34.

Chaudhry, F. A., Lehre, K. P., van Lookeren Campagne, M., Ottersen, O. P., Danbolt, N. C., and Storm-Mathisen, J., 1995, Glutamate transporters in glial plasma membranes: highly differentiated localizations revealed by quantitative ultrastructural immunocytochemistry, *Neuron* **15**(3):711–720.

Dale, N., Ottersen, O. P., Roberts, A., and Storm-Mathisen, J., 1986, Inhibitory neurones of a motor pattern generator in *Xenopus* revealed by antibodies to glycine, *Nature (Lond)* **324**:255–257.

Davanger, S., Hjelle, O. P., Babaie, E., Larsson, L. I., Hougaard, D., Storm-Mathisen, J., and Ottersen, O. P., 1994, Colocalization of gamma-aminobutyrate and gastrin in the rat antrum: an immunocytochemical and in situ hybridization study, *Gastroenterology* **107**(1):137–148.

Didier, A., Carleton, A., Bjaalie, J. G., Vincent, J. D., Ottersen, O. P., Storm-Mathisen, J., and Lledo, P. M., 2001, A dendrodendritic reciprocal synapse provides a recurrent excitatory connection in the olfactory bulb, *Proc. Natl. Acad. Sci. USA* **98**:6441–6446.

Faulk, W. P., and Taylor, G. M., 1971, An immunocolloid method for the electron microscope, *Immunochemistry* **11**:1081–1083.

Fujimoto, K., 1995, Freeze-fracture replica electron microscopy combined with SDS digestion for cytochemical labeling of integral membrane proteins. Application to the immunogold labeling of intercellular junctional complexes, *J. Cell Sci.* **108**:3443–3449.

Griffiths, G., 1993, *Fine Structure Immunocytochemistry*, Berlin: Springer-Verlag, p. 459.

Griffiths, G., and Hoppeler, H., 1986, Quantitation in immunocytochemistry: correlation of immunogold labeling to absolute number of membrane antigens, *J. Histochem. Cytochem.* **34**(11):1389–1398.

Hagiwara, A., Fukazawa, Y., Deguchi-Tawarada, M., Ohtsuka, T., and Shigemoto, R., 2005, Differential distribution of release-related proteins in the hippocampal CA3 area as revealed by freeze-fracture replica labeling, *J. Comp. Neurol.* **489**(2): 195–216.

Haug, F. M., Desai, V. D., Nergaard, P. O., Laake, J., and Ottersen, O. P., 1994, Particle-counting in immunogold labelled ultrathin sections by transmission electron microscopy and image analysis, *Anal. Cell. Pathol.* **6**:197.

Haug, F. M. S., Desai, V., Nergaard, P. O., and Ottersen, O. P., 1996, Quantifying immunogold labelled neurotransmitters and receptors by image analysis, *Soc. Neurosci. Abstr.* **22**(1):581.

Hjelle, O. P., Chaudhry, F. A., and Ottersen, O. P., 1994, Antisera to glutathione: characterization and immunocytochemical application to the rat cerebellum, *Eur. J. Neurosci.* **6**(5):793–804.

Holmseth, S., Dehnes, Y., Bjornsen, L. P., Boulland, J. L., Furness, D. N., Bergles, D., and Danbolt, N. C., 2005, Specificity of antibodies: unexpected cross-reactivity of antibodies direct against the excitatory amino acid transporter 3(EAAT3), *Neuroscience* **136**: 649–660.

Hu, H., Shao, L. R., Chavoshy, S., Gu, N., Trieb, M., Behrens, R., Laake, P., Pongs, O., Knaus, H. G., Ottersen, O. P., and Storm, J. F., 2001, Presynaptic Ca^{2+}-activated K^+ channels in glutamatergic hippocampal terminals and their role in spike repolarization and regulation of transmitter release, *J. Neurosci.* **21**(24):9585–9597.

Humbel, B. M., and Schwartz, H., 1989, Freeze-substitution for immunochemistry, In: Verkleij, A. J., and Leunissen, J. L. M. (eds.), *Immunogold Labelling in Cell Biology*, Boca Raton, FL: CRC Press, pp. 115–134.

Ji, Z. Q., Aas, J. E., Laake, J., Walberg, F., and Ottersen, O. P., 1991, An electron microscopic, immunogold analysis of glutamate and glutamine in terminals of rat spinocerebellar fibers, *J. Comp. Neurol.* **307**(2):296–310.

Kohr, G., Jensen, V., Koester, H. J., Mihaljevic, A. L., Utvik, J. K., Kvello, A., Ottersen, O. P., Seeburg, P. H., Sprengel, R., and Hvalby, O., 2003, Intracellular domains of NMDA receptor subtypes are determinants for long-term potentiation induction, *J. Neurosci.* **23**(34):10791–10799.

Landsend, A. S., Amiry-Moghaddam, M., Matsubara, A., Bergersen, L., Usami, S., Wenthold, R. J., and Ottersen, O. P., 1997, Differential localization of delta glutamate receptors in the rat cerebellum: coexpression with AMPA receptors in parallel fiber-spine synapses and absence from climbing fiber-spine synapses, *J. Neurosci.* **17**(2):834–842.

Lehre, K. P., and Danbolt, N. C., 1998, The number of glutamate transporter subtype molecules at glutamatergic synapses: chemical and stereological quantification in young adult rat brain, *J. Neurosci.* **18**(21):8751–8757.

Lucocq, J. M., Habermann, A., Watt, S., Backer, J. M., Mayhew, T. M., and Griffiths, G., 2004, A rapid method for assessing the distribution of gold labeling on thin sections, *J. Histochem. Cytochem.* **52**(8):991–1000.

Matsubara, A., Laake, J. H., Davanger, S., Usami, S., and Ottersen, O. P., 1996, Organization of AMPA receptor subunits at a glutamate synapse: a quantitative immunogold analysis of hair cell synapses in the rat organ of Corti, *J. Neurosci.* **16**(14):4457–4467.

Maunsbach, A. B., and Afzelius, B. A., 1999, *Biomedical Electron Microscopy: Illustrated Methods and Interpretations*, San Diego: Academic Press, 548 p.

Mayhew, T. M., Griffiths, G., and Lucocq, J. M., 2004, Applications of an efficient method for comparing immunogold labelling patterns in the same sets of compartments in different groups of cells, *Histochem. Cell Biol.* **122**(2):171–177.

Monteiro-Leal, L. H., Troster, H., Campanati, L., Spring, H. F., and Trendelenburg, M., 2003, Gold finder: a computer method for fast automatic double gold labeling detection, counting, and color overlay in electron microscopic images, *J. Struct. Biol.* **141**(3): 228–239.

Nagelhus, E. A., Horio, Y., Inanobe, A., Fujita, A., Haug, F. M., Nielsen, S., Kurachi, Y., and Ottersen, O. P., 1999, Immunogold evidence suggests that coupling of K$^+$ siphoning and water transport in rat retinal Muller cells is mediated by a coenrichment of Kir4.1 and AQP4 in specific membrane domains, *Glia* **26**(1):47–54.

Nagelhus, E. A., Mathiisen, T. M., Bateman, A. C., Haug, F. M., Ottersen, O. P., Grubb, J. H., Waheed, A., and Sly, W. S., 2005, Carbonic anhydrase XIV is enriched in specific membrane domains of retinal pigment epithelium, Muller cells, and astrocytes, *Proc. Natl. Acad. Sci. USA* **102**(22):8031–8035.

Nagelhus, E. A., Mathiisen, T. M., and Ottersen, O. P., 2004, Aquaporin-4 in the central nervous system: cellular and subcellular distribution and coexpression with KIR4.1, *Neuroscience* **129**(4):905–913.

Nagelhus, E. A., Veruki, M. L., Torp, R., Haug, F. M., Laake, J. H., Nielsen, S., Agre, P., and Ottersen, O. P., 1998, Aquaporin-4 water channel protein in the rat retina and optic nerve: polarized expression in Muller cells and fibrous astrocytes, *J. Neurosci.* **18**(7): 2506–2519.

Neely, J. D., Amiry-Moghaddam, M., Ottersen, O. P., Froehner, S. C., Agre, P., and Adams, M. E., 2001, Syntrophin-dependent expression and localization of aquaporin-4 water channel protein, *Proc. Natl. Acad. Sci. USA* **98**(24):14108–14113.

Nielsen, S., Nagelhus, E. A., Amiry-Moghaddam, M., Bourque, C., Agre, P., and Ottersen, O. P., 1997, Specialized membrane domains for water transport in glial cells: high-resolution immunogold cytochemistry of aquaporin-4 in rat brain, *J. Neurosci.* **17**(1):171–180.

Nusser, Z., Lujan, R., Laube, G., Roberts, J. D., Molnar, E., and Somogyi, P., 1998, Cell type and pathway dependence of synaptic AMPA receptor number and variability in the hippocampus, *Neuron* **21**(3):545–559.

Ottersen, O. P., 1987, Postembedding light- and electron microscopic immunocytochemistry of amino acids: description of a new model system allowing identical conditions for specificity testing and tissue processing, *Exp. Brain Res.* **69**(1):167–174.

Ottersen, O. P., 1989a, Quantitative electron microscopic immunocytochemistry of amino acids, *Anat. Embryol.* **180**:1–15.

Ottersen, O. P., 1989b, Postembedding immunogold labelling of fixed glutamate: an electron microscopic analysis of the relationship between gold particle density and antigen concentration, *J. Chem. Neuroanat.* **2**(1):57–66.

Ottersen, O. P., Zhang, N., and Walberg, F., 1992, Metabolic compartmentation of glutamate and glutamine: morphological evidence obtained by quantitative immunocytochemistry in rat cerebellum, *Neuroscience* **46**:519–534.

Panzanelli, P., Homanics, G. E., Ottersen, O. P., Fritschy, J. M., and Sassoe-Pognetto, M., 2004, Pre- and postsynaptic GABA receptors at reciprocal dendrodendritic synapses in the olfactory bulb, *Eur. J. Neurosci.* **20**(11):2945–2952.

Phend, K. D., Rustioni, A., and Weinberg, R. J., 1995, An osmium-free method of epon embedment that preserves both ultrastructure and antigenicity for post-embedding immunocytochemistry, *J. Histochem. Cytochem.* **43**(3):283–292.

Posthuma, G., Slot, J. W., Veenendaal, T., and Geuze, H. J., 1988, Immunogold determination of amylase concentrations in pancreatic subcellular compartments, *Eur. J. Cell Biol.* **46**(2):327–335.

Ragnarson, B., Ornung, G., Grant, G., Ottersen, O. P., and Ulfhake, B., 2003, Glutamate and AMPA receptor immunoreactivity in Ia synapses with motoneurons and neurons of the central cervical nucleus, *Exp. Brain Res.* **149**(4):447–457.

Ragnarson, B., Ornung, G., Ottersen, O. P., Grant, G., and Ulfhake, B., 1998, Ultrastructural detection of neuronally transported choleragenoid by postembedding immunocytochemistry in freeze-substituted Lowicryl HM20 embedded tissue, *J. Neurosci. Methods* **80**(2):129–136.

Rash, J. E., Davidson, K. G., Yasumura, T., and Furman, C. S., 2004, Freeze-fracture and immunogold analysis of aquaporin-4 (AQP4) square arrays, with models of AQP4 lattice assembly, *Neuroscience* **129**(4):915–934.

Rinvik, E., and Ottersen, O. P., 1993, Terminals of subthalamonigral fibres are enriched with glutamate-like immunoreactivity: an electron microscopic, immunogold analysis in the cat, *J. Chem. Neuroanat.* **6**(1):19–30.

Rossi, P., Sola, E., Taglietti, V., Borchardt, T., Steigerwald, F., Utvik, J. K, Ottersen, O. P., Kohr, G., and D'Angelo, E., 2002, NMDA receptor 2 (NR2) C-terminal control of NR open probability regulates synaptic transmission and plasticity at a cerebellar synapse, *J. Neurosci.* **22**(22):9687–9697.

Roth, J., 1996, The silver anniversary of gold: 25 years of the colloidal gold marker system for immunocytochemistry and histochemistry, *Histochem. Cell Biol.* **106**(1):1–8.

Ruud, H. K., and Blackstad, T. W., 1999, PALIREL, a computer program for analyzing particle-to-membrane relations, with emphasis on electron micrographs of immunocytochemical preparations and gold labeled molecules, *Comput. Biomed. Res.* **32**(2):93–122.

Salio, C., Lossi, L., Ferrini, F., Merighi, A., 2005, Ultrastructural evidence for a pre- and postsynaptic localization of full length trkB receptors in substantia gelatinosa (lamina II) of rat and mouse spinal cord, *Eur. J. Neurosci.* **22**(8): 1951–1966.

Shupliakov, O., Brodin, L., Cullheim, S., Ottersen, O. P., and Storm-Mathisen, J., 1992, Immunogold quantification of glutamate in two types of excitatory synapse with different firing patterns, *J. Neurosci.* **12**(10):3789–3803.

Somogyi, P., Halasy, K., Somogyi, J., Storm-Mathisen, J., and Ottersen, O. P., 1986, Quantification of immunogold labelling reveals enrichment of glutamate in mossy and parallel fibre terminals in cat cerebellum, *Neuroscience* **19**(4):1045–1050.

Somogyi, P., and Hodgson, A. J., 1985, Antisera to gamma-aminobutyric acid: III. Demonstration of GABA in Golgi-impregnated neurons and in conventional electron microscopic sections of cat striate cortex, *J. Histochem. Cytochem.* **33**(3):249–257.

Storm-Mathisen, J., Leknes, A. K., Bore, A. T., Vaaland, J. L., Edminson, P., Haug, F.-M. S., and Ottersen, O. P., 1983, First visualization of glutamate and GABA in neurones by immunocytochemistry, *Nature* **301**:517–520.

Storm-Mathisen, J., and Ottersen, O. P., 1990, Antibodies and fixatives for the immunocytochemical localization of glycine, In: Ottersen, O. P., and Storm-Mathisen, J. (eds.), *Glycine Neurotransmission*, Chichester: John Wiley & Sons, pp. 281–301.

Takumi, Y., Ramirez-Leon, V., Laake, P., Rinvik, E., and Ottersen, O. P., 1999, Different modes of expression of AMPA and NMDA receptors in hippocampal synapses, *Nat. Neurosci.* **2**(7):618–624.

Torp, R., Arvin, B., Le Peillet, E., Chapman, A. G., Ottersen, O. P., and Meldrum, B. S., 1993, Effect of ischemia and reperfusion on the extra- and intracellular distribution of glutamate, glutamine, aspartate, and GABA in the rat hippocampus, with a note on the effect of the sodium channel blocker BW1003C87, *Exp. Brain Res.* **96**:365–376.

Torp, R., Head, E., Milgram, N. W., Hahn, F., Ottersen, O. P., and Cotman, C. W., 2000, Ultrastructural evidence of fibrillar beta-amyloid associated with neuronal membranes in behaviorally characterized aged dog brains, *Neuroscience* **96**:495–506.

Torp, R., Ottersen, O. P., Cotman, C. W., and Head, E., 2003, Identification of neuronal plasma membrane microdomains that colocalize beta-amyloid and presenilin: implications for beta-amyloid precursor protein processing, *Neuroscience* **120**(2):291–300.

Valtschanoff, J. G., and Weinberg, R. J., 2001, Laminar organization of the NMDA receptor complex within the postsynaptic density, *J. Neurosci.* **21**(4):1211–1217.

van den Pol, A. N., 1989, Neuronal imaging with colloidal gold, *J. Microsc.* **155**(1):27–59.

van Lookeren Campagne, M., Oestreicher, A. B., van der Krift, T. P., Gispen, W. H., and Verkleij, A. J., 1991, Freeze-substitution and Lowicryl HM20 embedding of fixed rat brain: suitability for immunogold ultrastructural localization of neural antigens, *J. Histochem. Cytochem.* **39**(9):1267–1279.

Wang, B. L., and Larsson, L. I., 1985, Simultaneous demonstration of multiple antigens by indirect immunofluorescence or immunogold staining. Novel light and electron microscopical double and triple staining method employing primary antibodies from the same species, *Histochemistry* **83**(1):47–56.

4

Cell and Tissue Microdissection in Combination with Genomic and Proteomic Applications

STEPHEN D. GINSBERG, SCOTT E. HEMBY, ELLIOTT J. MUFSON, and LEE J. MARTIN

STEPHEN D. GINSBERG • Center for Dementia Research, Nathan Kline Institute, and Departments of Psychiatry and Physiology and Neuroscience, New York University School of Medicine, Orangeburg, NY 10962   SCOTT E. HEMBY • Department of Physiology and Pharmacology, Wake Forest University School of Medicine, Winston-Salem, NC 27157   ELLIOTT J. MUFSON • Department of Neurological Sciences, Rush University Medical Center, Chicago, IL 60612   LEE J. MARTIN • Division of Neuropathology, Departments of Pathology and Neuroscience, Johns Hopkins University School of Medicine, Baltimore, MD 21205

APPENDIX: DETAILED METHODOLOGY
  Acridine Orange Histofluorescence
  Microaspiration and LCM
  Tract-Tracing for Use with Microdissection
  aRNA Amplification
  TC RNA Amplification
  SELDI-TOF
  Supplies/Manufacturers
REFERENCES

Abstract: The combination of tissue microdissection protocols including discrete cell microaspiration and laser capture microdissection with high throughput gene expression profiling platforms such as cDNA microarrays and oligonucleotide microarrays enables the simultaneous assessment of many individual elements from a single cell or a population of homogeneous cells. This chapter outlines in detail the theoretical and practical background for selecting the appropriate tissues and conditions amenable to expression profiling. In addition, this report illustrates the usage of microdissection strategies and RNA amplification methodologies in concert with array technologies using tissues harvested from the central nervous system obtained from animal models of neurodegeneration and postmortem human brain tissues.

Keywords: brain, expression profiling, laser capture microdissection, microarray, molecular fingerprint, RNA amplification, SELDI-TOF

I. INTRODUCTION

The brain is a complex structure with heterogeneous neuronal (e.g., pyramidal neurons and interneurons) and nonneuronal (e.g., glial cells, epithelial cells, and vascular elements) cell populations. Advances in molecular biology provide the tools needed to sample gene expression from specific homogeneous cell populations within defined brain regions without potential contamination of adjacent neuronal subtypes and nonneuronal cells, and are an important goal of twenty-first century neuroscience. However, gene expression profiling of homogeneous populations of cells is a difficult task that demands a multidisciplinary approach including molecular biology, cell biology, neuroanatomy, and biomedical engineering. Individual cell types are likely to have unique patterns or a mosaic of gene and protein expression under normative conditions that is likely to be altered in pathological states. For example, distinct cortical and subcortical regions may serve entirely different functions and may be differentially affected in neurodegenerative diseases (Galvin, 2004; Ginsberg *et al.*, 1999b). Indeed, the molecular basis of why certain neuronal cell populations are vulnerable to neurodegeneration, often termed "selective vulnerability," can be elucidated by discrete cell analysis more readily than by utilizing regional and total brain preparations (Ginsberg and Che, 2005). Thus, the pattern of genomic and proteomic expression in a subpopulation of homogeneous cells or single cells is more likely to be informative than the pattern in a whole tissue homogenate, assuming the target population is well defined. With the advent of modern molecular and cellular techniques, it is now possible to isolate and study genomic DNA, RNA species, and proteins from microdissected tissue sources. At present, an optimal methodology is to evaluate single cells, identified either physiologically in living preparations (Eberwine *et al.*, 1992; Tkatch *et al.*, 2000) or by immunocytochemical or histochemical procedures in fixed cells in vitro or in vivo (Galvin and Ginsberg, 2004; Ginsberg and Che, 2004; Ginsberg *et al.*, 2004; Hemby *et al.*, 2003; Kamme *et al.*, 2003; Mufson *et al.*, 2002; Van Deerlin *et al.*, 2002). Unfortunately, the quantity of RNA harvested from a single cell, estimated to be $\sim$0.1–1.0 pg, is not sufficient for standard RNA extraction procedures (Phillips and Eberwine, 1996; Sambrook and Russell, 2001). Both exponential polymerase chain reaction (PCR) based analyses (Becker *et al.*, 1996; D'Amore *et al.*, 1997) and linear RNA amplification including amplified antisense RNA (aRNA) (Eberwine *et al.*, 1992, 2001; Ginsberg *et al.*, 1999a, 2000) and the newly developed terminal continuation (TC) RNA amplification (Che and Ginsberg, 2004, 2005; Ginsberg and Che, 2004; Ginsberg *et al.*, 2004) have been used in combination with single-cell microdissection procedures to enable the use of microarray analysis (Eberwine *et al.*, 2001; Ginsberg and Che, 2004). RNA amplification is a series of elaborate molecular-based methods used to amplify genomic signals in a linear fashion from minute quantities of starting materials for microarray analysis and other downstream genetic applications (Fig. 4.1). In this chapter, we illustrate the utility of combining discrete cell microdissection methodologies with RNA amplification for use in microarray analyses as well as pairing laser capture microdissection (LCM) with proteomic profiling for single cell and/or population cell resolution at the protein level. Utilization of tract-tracing methods in combination with gene expression analysis and proteomic profiling is also presented.

II. GENE EXPRESSION PROFILING USING FIXED TISSUES

A. Antemortem and Postmortem Variables

Assessment of single cell and homogeneous cell populations in optimally prepared, perfused fixed animal tissues as well as fixed postmortem human brain tissues is desirable due to the abundance of animal and human brain tissues that are archived within individual laboratories and brain banks and because of the use of relevant animal models to further understand disease mechanisms. At present, no consensus protocol exists for the fixation and/or extraction of brain tissues obtained from animals or from postmortem human tissues. Several laboratories have evaluated the effects of different fixation protocols on RNA quality, ease of tissue microdissection, and success of microarray analysis (Bahn *et al.*, 2001; Coombs *et al.*,

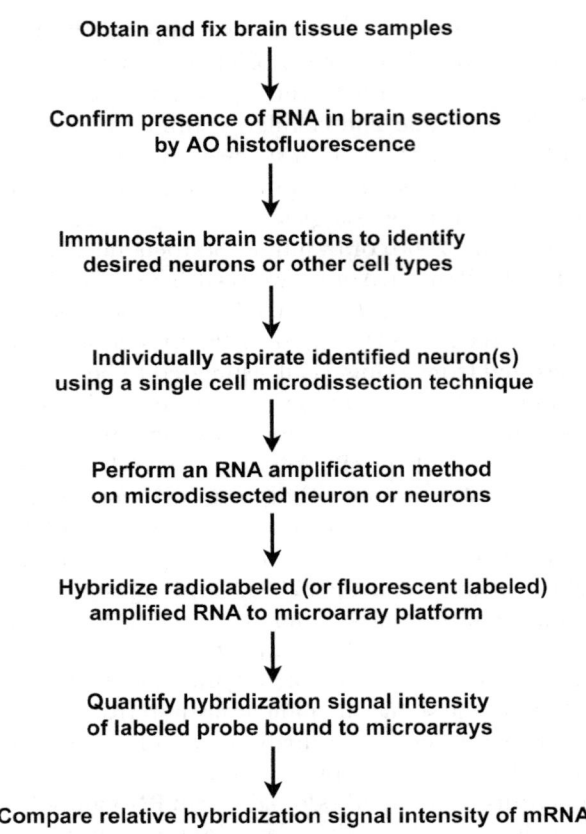

Figure 4.1. Schematic overview of the experimental design. Outline of the general procedures used to perform microdissection combined with high-throughput gene expression analysis using array platforms.

1999; Goldsworthy *et al.*, 1999; Van Deerlin *et al.*, 2000, 2002; Vincent *et al.*, 2002).

Despite potential advantages of discrete cell RNA amplification technology, several caveats must be considered when undertaking such studies in brain tissue. One factor is postmortem interval (PMI), or the time that elapses between time of death and preservation of the tissue sample. PMI is particularly relevant when obtaining postmortem human materials, as animal models can be fixed rapidly using perfusion techniques. Investigators must be cognizant of many factors including PMI and the time from dissection to tissue preservation that may affect the quality and quantity of recovered nucleic acids and proteins. Moreover, the choice of fixative for tissue preservation is an important factor affecting RNA stability. Fixatives include aldehydes (e.g., formalin, paraformaldehyde, and glutaraldehyde), alcohols (e.g., ethanol and methanol), oxidizing agents, and picrates. In general, fixatives either create cross-links or exert a precipitative effect that may alter the native structure of macromolecules. With regards to neuroscience, aldehydes and alcohols are the most commonly used fixatives. Aldehydes induce cross-linkage of lysine residues formed in proteins, and alcohols are protein denaturants. The means by which RNA is preserved is unknown but likely involves the inactivation of degradative enzymes. The choice of fixative must be balanced between optimizing tissue morphology and preserving nucleic acid integrity for evaluation. As reviewed by Van Deerlin *et al.* (2000, 2002), ethanol and depolymerized 4% paraformaldehyde-based fixatives provide optimal results for molecular-based studies. Another factor is the agonal state of the human cases examined and the presence of overlapping neurologic conditions. Agonal state refers to the nature and time period between the onset of the terminal phase of an illness and death. The agonal state of a patient prior to death can have profound effects on several parameters including tissue pH, RNA stability, and protein degradation (Bahn *et al.*, 2001; Leonard *et al.*, 1993; Van Deerlin *et al.*, 2000, 2002). For example, hypoxia, pneumonia, and protracted coma have been associated with alterations in RNA and protein levels (Barton *et al.*, 1993; Hynd *et al.*, 2003; Tomita *et al.*, 2004). Therefore, numerous variables, including antemortem characteristics, agonal state, duration of fixation, and length of storage, are relevant parameters that should be considered prior to the initiation of molecular studies that utilized human postmortem tissues.

B. Acridine Orange Histofluorescence and Bioanalysis

Of critical importance in discrete cell RNA assessment, as well as other molecular procedures, is the evaluation of RNA quality and quantity. A useful and relatively quick method for assessing RNA quality in tissue sections prior to performing expression profiling studies is the use of acridine orange (AO) histofluorescence. AO is a fluorescent dye that intercalates selectively into nucleic acids (Mikel and Becker, 1991; von Bertalanffy and Bickis, 1956) and has been used to detect RNA and DNA in brain tissues (Ginsberg and Che, 2004; Topaloglu and Sarnat, 1989; Vincent *et al.*, 2002; Zoccarato *et al.*, 1999). Upon excitation in the ultraviolet spectra, AO that intercalates into RNA emits an orange-red fluorescence, whereas AO that intercalates into DNA emits a yellowish-green fluorescence. AO can also be combined with immunocytochemistry within tissue sections to double label cytoplasmic RNAs and specific antigens of interest, and is compatible with confocal microscopy (Ginsberg *et al.*, 1997). In brain tissue sections, the pale background of white matter tracts that lack abundant nucleic acids contrasts AO-positive neurons. Nonneuronal cells tend to have less AO histofluorescence as compared to neurons and brain tumor cells (Sarnat *et al.*, 1987), suggesting that there is less overall RNA. It is important to note that individual RNA species (e.g., rRNA, tRNA, and mRNA) cannot be delineated by AO histofluorescence. Rather, this method provides a simple diagnostic test that can be performed on adjacent tissue sections to ensure the likelihood that an individual case has abundant RNA prior to performing expensive microdissection and microarray studies. A more definitive examination of RNA quality can be obtained via bioanalysis (e.g., 2100 Bioanalyzer, Agilent Technologies), which employs capillary gel electrophoretic methodologies to detect RNA quality and abundance (Che and Ginsberg, 2004, 2005; Ginsberg and Che, 2004). Bioanalysis enables visualization of results in an electropherogram and/or digital gel formats, and provides a means of RNA assessment at relatively high sensitivity. Investigators can also evaluate DNA and protein quality and abundance using bioanalysis platforms (Freeman and Hemby, 2004).

III. REGIONAL MICRODISSECTION METHODS

Microdissection of individual cells is performed to enable downstream gene expression profiling. Provided that procedures are performed on fresh, frozen, or well-fixed tissue sections and ribonuclease (RNase) free conditions are employed, both immunocytochemical and histochemical procedures can be utilized to identify specific cell(s) of interest (Ginsberg and Che, 2004, 2005). Several different methodologies have been used to aspirate individual cells or groups of cells including single-cell microaspiration and LCM techniques.

In addition to single-cell microdissection, regional dissections can also be performed, which may be useful to the investigator. Regional analysis is a powerful approach for the identification of transcripts that are enriched in a specific region, lamina, or nuclei that differ from adjacent or connected regions. Groups of related cells from discrete regions of brain or spinal cord can be readily dissected from paraffin-embedded tissue sections (e.g., 5–6 μm thick) or frozen tissue sections (e.g., 20–40 μm thick) by an experienced neuroanatomist. Unstained sections can be utilized, but optimal cellular resolution occurs using sections prepared for immunocytochemical or histochemical (e.g., Nissl stain) procedures. We have had success in scraping away areas of the tissue section that were not desired to reveal only the well-defined region of interest (Ginsberg and Che, 2002, 2004; Hemby *et al.*, 2002). Regional dissections can be performed on fresh, fixed, or thawed tissue blocks using a stereomicroscope along with a scalpel or micropunch. A caveat is that these approaches are highly operator dependent, and can be difficult to reproduce across samples. RNA is extracted from the resulting tissue for downstream applications, such as cDNA array analysis, quantitative real-time PCR (qPCR), serial analysis of gene expression (SAGE), and differential display, among others (Che and Ginsberg, 2005; Ginsberg and Che, 2002; Lein *et al.*, 2004; Zhao *et al.*, 2001). Regional dissections can also be used as the input source for protein in the case of fresh and/or frozen tissues for proteomics-based applications as well as for conventional neurochemical and immunoblotting procedures (Freeman and Hemby, 2004; Mouledous *et al.*, 2003; Palkovits, 1989). An advantage of regional analyses is that limited RNA amplification is necessary to generate significant hybridization signal intensity. For example, microdissection of the basal forebrain, hippocampal formation, midbrain, and nucleus accumbens has been performed to generate regional expression profiles in normal brains and in pathological conditions including Alzheimer's disease (AD) and cocaine self-administration (Backes and Hemby, 2003; Fasulo and Hemby, 2003; Ginsberg and Che, 2002, 2004; Tang *et al.*, 2003). A disadvantage of regional dissection procedures is the lack of single-cell resolution, as neurons, nonneuronal cells, vascular elements, and epithelial cells will be included in the dissection.

IV. SINGLE-CELL MICROASPIRATION METHODS

Discrimination and isolation of adjacent cell types from one another is critical because this enables the selection of relatively pure populations of individual cells and/or populations for subsequent analysis and avoids potential contamination from a variety of sources including glia, vascular epithelia, and other nonneuronal cells within the brain. One method of isolating individual cells or populations of homogeneous cells is termed single-cell microaspiration. Single-cell microaspiration entails visualizing an individual cell (or cells) using an inverted microscope connected to a micromanipulator, microcontrolled vacuum source, and an imaging workstation on an air table. Electrophysiology rigs can also be modified to aspirate cells from fixed tissue sections with minor modifications. Handheld and syringe-pump-driven vacuum sources can also be utilized; however, they are difficult to control and may cause inadvertent damage to the tissue section. Individual cells are carefully aspirated from the tissue section of interest, and placed in microfuge tubes for subsequent RNA amplification (Fig. 4.2). This methodology results in accurate dissection of the neurons of interest with minimal disruption of the surrounding neuropil (Ginsberg, 2001; Ginsberg *et al.*, 2004; Hemby *et al.*, 2002; Mufson *et al.*, 2002). An advantage of utilizing a single-cell microaspiration technique is the extremely high cellular (and potentially subcellular, compartmental, and/or dendritic) level of resolution for aspiration of single elements (Crino *et al.*, 1998; Ginsberg and Che, 2005; Hemby *et al.*, 2003). Disadvantages include the relative difficulty of performing the aspirating technique, experimenter error, and the lengthy time allotment necessary to perform microaspiration, especially if multiple cells are being acquired from different brain tissue sections. Moreover, investigators should be aware of the degree of heterogeneity of the cells of interest and the extent to which small numbers of cells may or may not be representative of a population of interest.

A key aspect of the success of single-cell and single-population gene expression analysis is that different cell types can be discriminated based

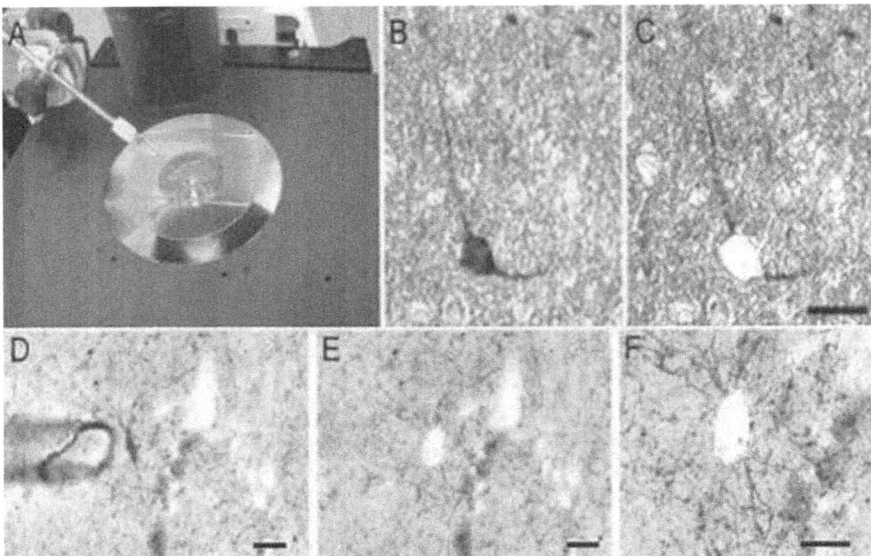

Figure 4.2. Microaspiration of single neurons. (A) Representative photomicrograph illustrating the placement of a human postmortem tissue section of the basal forebrain onto the microaspiration apparatus. (B) Section immunolabeled with an anti-neurofilament antibody depicting a representative layer II entorhinal cortex stellate cell obtained postmortem from a normal control human brain and the same section following microdissection of the immunostained neuron (C). Scale bar: 25 μm. (D) Human anterior nucleus basalis neuron visualized by dual immunolabeling for the cholinergic marker p75^{NTR} (brown cell bodies) and galanin (black punctate fibers). The microaspirating pipette can be visualized in the left plane of the field shown in (D). Photographs of the same tissue section are shown following microaspiration of the cholinergic neuron at low (E) and higher magnification (F). Scale bars in (D)–(F): 40 μm.

on their molecular fingerprint. For example, populations of neurons that express proteins selectively such as cholinergic basal forebrain neurons (Mufson *et al.*, 2002, 2003) or midbrain dopaminergic neurons (Fasulo and Hemby, 2003; Tang *et al.*, 2003) are amenable to single-cell RNA amplification and subsequent cDNA array analysis. Cells that lack a distinct or selective phenotypic signature can be analyzed using a variety of Nissl and immunocytochemical stains for downstream genetic applications (Ginsberg and Che, 2004, 2005; Kamme *et al.*, 2003). Although histological stains are typically not specific to an individual cell type or protein, much information can be gleaned by utilizing classical histological preparations in conjunction with contemporary protein (e.g., immunocytochemistry) and molecular biological methodologies. For example, hematoxylin and eosin (H&E) staining has been performed in combination with microdissection and PCR-based strategies as well as microarray platforms using RNA amplification methods (Becker *et al.*, 1996; Goldsworthy *et al.*, 1999; To *et al.*, 1998). Moreover, we have demonstrated that several Nissl stains including cresyl violet, H&E, and thionin perform as well as immunocytochemistry in terms of hybridization signal intensity detection when employing cDNA array analysis (Ginsberg and Che, 2004). In contrast, several dyes that bind to RNAs directly, such as AO and silver stain, do not perform well in combination with microdissection and subsequent cDNA array analysis (Ginsberg and Che, 2004).

V. LASER CAPTURE MICRODISSECTION

A. Introduction

The implementation of high-throughput microaspiration devices over the last few years has enabled rapid accession of single cells and homogeneous cellular populations for downstream genomic and proteomic analyses. Specifically, LCM is a strategy for acquiring histochemically and/or immunocytochemically labeled cells from in vivo and in vitro sources (Dolter and Braman, 2001; Ehrig *et al.*, 2001; Goldsworthy *et al.*, 1999; Lu *et al.*, 2004). LCM has become a widely used and reproducible technique that was developed originally at the NIH (Bonner *et al.*, 1997; Emmert-Buck *et al.*, 1996). There are two principal means of LCM: positive extraction and negative extraction.

B. Positive Extraction

Positive extraction (a method used by the PixCell IIe from Arcturus) employs a laser source directly on the cell(s) of interest for the purpose of microaspiration. There are four steps in positive extraction methods for capturing cells under direct visualization and recovering biomolecules. After locating the cells of interest in a tissue section, a small plastic cap (e.g., CapSure or CapSure HS LCM Cap) coated with a special thermoplastic film is placed over the area of tissue containing the cell targets. A nondestructive, low-power, near-infrared laser pulse is then directed through the cap at the target cell. The pulsed laser energy causes localized activation of the thermoplastic film that extends, embraces, and adheres to the target cell. Raising the thermoplastic cap separates targeted cells, now attached to the film, from surrounding undisturbed tissue (Fig. 4.3). Populations of cells attached to the cap are suitable for microscopic examination and downstream genetic analysis.

C. Negative Extraction

Negative extraction (or noncontact laser extraction) procedures employ a laser source to cut around the area of interest within a tissue section, and the microdissected material is catapulted into a microfuge tube (a method utilized by the PALM system, PALM Microlaser Technologies). A variety of conditions can modify the consistent success of cell capture, including tissue fixation. Tissues can be fresh, frozen, or fixed in alcohol or aldehydes

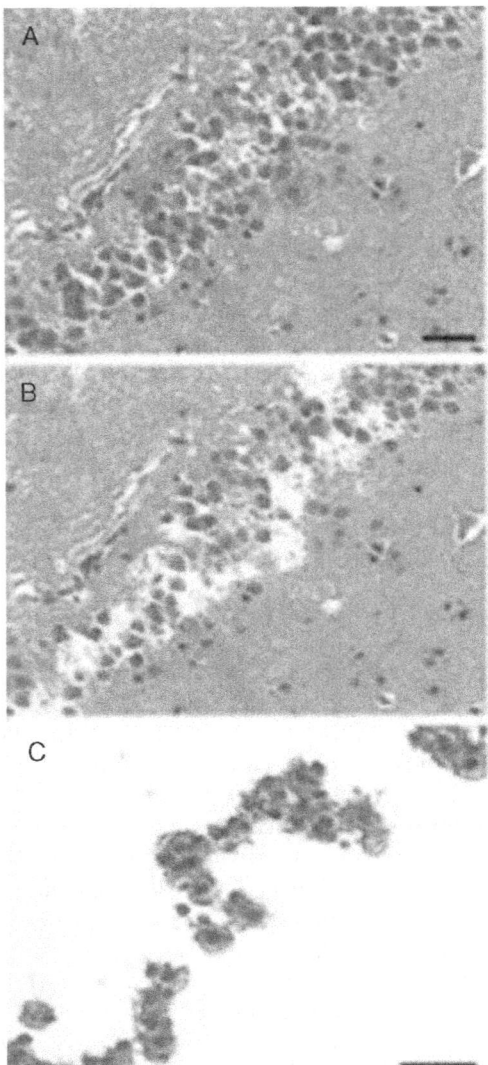

Figure 4.3. LCM of granule cells. Photomicrographs of a microaspiration of cresyl violet stained human hippocampal granule cells using LCM from a 6-μm thick ethanol fixed tissue section. (A) Section prior to LCM. (B) The cap is removed following laser pulses over desired cells, leaving spaces where microdissected granule cells originally resided. (C) Captured cells are visualized by placing the cap on a clean slide for contrast. Scale bar in (A) and (B): 25 μm; (C): 30 μm.

(Goldsworthy *et al.*, 1999; Su *et al.*, 2004). Other parameters include optimal section thickness (<10–14 μm) and the type of glass slide (e.g., uncoated, charged, poly-ʟ-lysine, and gelatin-coated) used to mount the tissue sections for subsequent microdissection. Both positive and negative extraction methods allow captured cells and their processes to be examined microscopically to confirm the identity and quality of isolated cell population(s). This quality control step ensures validity of the results obtained from downstream analysis. Single cells as well as dozens to hundreds of cells can be collected by LCM instrumentation. RNA, DNA, and protein can be extracted from microdissected cells and utilized as input sources for downstream applications such as microarray analysis, qPCR, as well as proteomics (Ehrig *et al.*, 2001; Fend *et al.*, 1999; Suarez-Quian *et al.*, 1999). We have utilized LCM to microdissect a variety of neuronal populations including hippocampal CA1 and CA3 neurons, dentate gyrus granule cells, and spinal motor neurons (see below) from mouse brains and postmortem human brains (Che and Ginsberg, 2004, 2005; Ginsberg and Che, 2002, 2004, 2005). LCM has been increasingly utilized to collect cells for downstream proteomic analyses including two-dimensional gel electrophoresis, tandem mass spectroscopy, and antibody-based protein chips (Craven *et al.*, 2002; Freeman and Hemby, 2004; Mouledous *et al.*, 2003; Simone *et al.*, 2000). In summary, the ability to access DNA, RNA, and protein from microdissected tissue samples via LCM-based technologies represents an exciting new avenue for studying homogeneous populations of brain cells.

VI. TRACT-TRACING COMBINED WITH DISCRETE CELL MICRODISSECTION

The combination of discrete cell dissection and RNA amplification allows the investigator to make specific assertions about disease- or drug-induced changes in gene and protein expression with unique certainty. Tract-tracing methodologies extend this capability by providing the means to identify and isolate specific processes and/or pathways of interest based on connectivity. Various tracers are employed to label neurons and neuronal processes in anterograde, retrograde, and bidirectional vectors. A discourse on the advantages and disadvantages of individual tracers is beyond the scope of this chapter. However, selection criteria depend on the experimental paradigm and the cell, region, or tissue type of interest (see these chapters in the current book for additional detail) (Lanciego and Wouterlood, 2006; Molnar *et al.*, 2006; Reiner and Honig, 2006).

Hemby and colleagues have undertaken a series of studies to evaluate the effects of psychotropic compounds on midbrain dopaminergic neurons as defined by their axonal targets (Backes and Hemby, 2003; Fasulo and Hemby, 2003). For example, in order to explore gene expression changes in tegmental-accumbal dopamine neurons following cocaine administration, the retrograde tracer Fluorogold (FG) was iontophoretically injected into the nucleus accumbens of rats. Following a 2-week period to allow for sufficient transport of the tracer, rats were sacrificed and the localization of injections was assessed by fluorescence microscopy. As depicted in Fig. 4.4A, a number of ventral midbrain neurons in the area corresponding

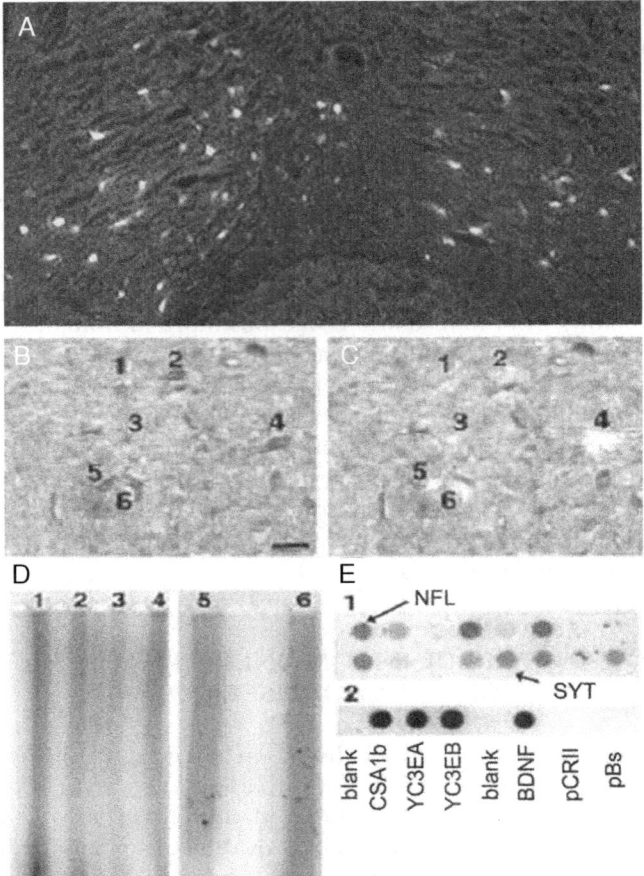

Figure 4.4. Tract-tracing in combination with microdissection and array analysis. (A) Representative section from rat midbrain for identification of FG-labeled cells. FG (4%) was iontophoresed into the nucleus accumbens of rats 2 weeks prior to sacrifice. Photomicrograph reveals significant midbrain FG-labeling within the ventral tegmental area of Tsai. (B) A representative section within the midbrain is shown following immunocytochemistry using an anti-FG antibody suitable for microaspiration. Scale bar: 20 μm. (C) FG-immunoreactive individual neurons were microdissected from the tissue section. (D) The six cells in B (1–6) were amplified by two rounds of aRNA and labeled with ^{32}P-CTP. aRNA was run on a 1% denaturing gel [numbers above each lane correspond to neurons in (B) and (C)]. (E) Radiolabeled aRNA from neuron #2 was used as a probe for identifying candidate cDNAs on a custom-designed array (E1) and subcloned differential display products (E2). Key (E1) top row; neurofilament-L (arrow); casein kinase II b; H67559; AA069725; T89891; AA076650; pulmonary surfactant associate protein (control); heme oxygenase 1 (control): bottom row; CG1 protein precursor; H89874; H70730; H89236; syntaxin (SYT; arrow); T92612; T90579; stathmin. Key (E2): blank; CSA1b; YC3EA; YC3EB; blank; brain derived neurotrophic factor (BDNF); vector (pCR II); vector (pBS). Accession numbers correspond to individual expressed sequence-tagged cDNAs (ESTs).

to the ventral tegmental area were FG-positive. Since not all of the projecting cells of the ventral midbrain are dopaminergic, the midbrain sections processed for tyrosine hydroxylase immunofluorescence to ensure assessment of dopamine-containing neurons using a mouse anti-tyrosine hydroxylase visualized with a Cy5 conjugated donkey anti-mouse secondary antibody. The procedure of dual labeling provides both certainty of anatomical connectivity and antigen specificity of the cells of interest. Alternatively, if cell-specific antigens are not available, antibodies directed against FG can be used to identify labeled neurons for microdissection and subsequent downstream genetic analyses (Fig. 4.4). When using a nonspecific antigen or histochemical stain to identify labeled neurons, it is imperative to further characterize the cell type post hoc using a validation technique such as qPCR. Multiple tracers can also be used within the same subject to identify different cell populations based on connectivity. Utility of employing multiple fluorescent tracers is dependent on the absorption/emission spectra. For example, we have used up to four tracers in rhesus monkeys to identify different populations of midbrain dopaminergic cells based on different projection paths (i.e., dorsolateral prefrontal cortex, orbitofrontal cortex, nucleus accumbens, and caudate/putamen) (Freeman and Hemby, 2004).

Investigators must be cognizant of various caveats when using tract-tracing methodologies. For example, the use of iontophoretic injections limits neuronal damage and the potential interpretational confound of labeling fibers en passant. The ability to iontophorese tracers may be limited by the chemical nature of the tracer and/or the vehicle required to solubilize the tracer. In addition, the influence of tracer uptake on neuronal function remains a relevant question. To date, equivocal data imply that tracers may damage RNA and/or protein integrity of cells in which the tracer is sequestered (Emsley *et al.*, 2001; Franklin and Druhan, 2000). Therefore, additional dose-response and toxicity studies are warranted to examine the extent to which various tracers may influence RNA and protein expression in neuronal populations.

VII. RNA AMPLIFICATION

A. aRNA

In order to generate a significant amount of RNA sufficient to perform microarray analysis and related high-throughput genetic readouts, an RNA amplification technique is often required when attempting expression profiling from single neurons, groups of neurons, or microdissected regions. PCR-based amplification methods are not optimal, as exponential amplification can skew the original quantitative relationships between genes from an initial population (Kacharmina *et al.*, 1999). Linear RNA amplification is another strategy that has been used successfully to generate enough input RNA for robust hybridization signal intensity on array platforms. The initial method of linear amplification termed aRNA amplification was developed by Eberwine and colleagues, and involves a T7 RNA polymerase-based amplification procedure that enables quantitation of the relative abundance of gene expression levels from identified single cells and populations (Eberwine *et al.*, 1992, 2001; Kacharmina *et al.*, 1999; Phillips and Eberwine, 1996). The resultant amplified aRNA maintains a proportional representation of the size and complexity of the initial input mRNAs (Eberwine *et al.*, 1992; VanGelder *et al.*, 1990). aRNA amplification entails the hybridization of a 66 basepair oligonucleotide primer consisting of 24 thymidine triphosphates (TTPs) and a T7 RNA polymerase promoter sequence [oligo d(T)T7] to mRNAs and conversion to an mRNA–cDNA hybrid by reverse transcriptase (Tecott *et al.*, 1988; VanGelder *et al.*, 1990) (Fig. 4.5). Upon conversion of the mRNA–cDNA hybrid to double-stranded cDNA, a functional

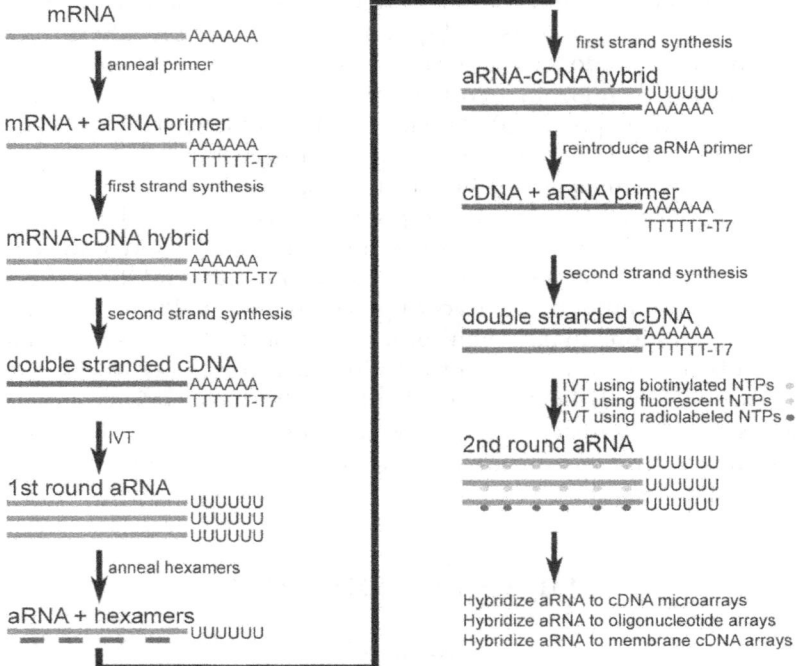

Figure 4.5. aRNA amplification scheme. An oligo d(T)T7 primer is hybridized to polyA+ mRNAs and a double-stranded mRNA–cDNA hybrid is formed by reverse transcription. The double-stranded mRNA-cDNA hybrid is then converted into double-stranded cDNA. Following the removal of tertiary structures and drop dialyzing the double-stranded cDNA against RNase-free water, the first round of aRNA synthesis occurs via in vitro transcription (IVT) using T7 RNA polymerase and NTPs. The second round of aRNA amplification begins by annealing random hexamers to the newly formed aRNA, and synthesizing a cDNA strand. The oligo (dT)T7 primer is then reintroduced and a double-stranded cDNA template is formed. aRNA probes are then generated with fluorescent, biotin, or radiolabeled second-round aRNA products.

T7 RNA polymerase promoter is formed. aRNA synthesis occurs with the addition of T7 RNA polymerase and nucleotide triphosphates (NTPs). Each round of aRNA results in an approximate 1000-fold amplification from the original amount of each polyadenylated [poly(A)+] mRNA in the sample (Eberwine *et al.*, 1992, 2001). Two rounds of aRNA are typically necessary to generate sufficient quantities of aRNA for subsequent downstream analyses. aRNA products are biased toward the 3′ end of the transcript due to the priming at the poly(A)+ RNA tail (Kacharmina *et al.*, 1999; Phillips and Eberwine, 1996). This 3′ bias exists for all amplified aRNA products and relative levels of gene expression can be compared (Che and Ginsberg, 2004; Madison and Robinson, 1998; Phillips and Eberwine, 1996). Moreover, amplified aRNA products tend not to be of full length (Ginsberg *et al.*, 1999a; Kacharmina *et al.*, 1999; Phillips and Eberwine, 1996). Although aRNA is a laborious and difficult procedure, we have generated successful results obtained from microaspirated cells from animal model and postmortem human brain tissues utilizing a wide variety of array platforms (Ginsberg *et al.*, 1999a, 2000; Hemby *et al.*, 2002, 2003; McClain *et al.*, 2005).

Several different strategies have been employed by independent laboratories to evaluate and improve linear RNA amplification efficiency (Iscove *et al.*, 2002; Klur *et al.*, 2004; Matz *et al.*, 1999; Wang *et al.*, 2000). The principal obstacle is the problematic second strand cDNA synthesis. This impediment is not specific to the aRNA protocol. Rather, this issue is endemic to all current RNA amplification methods. Key factors to improving RNA amplification include increasing the efficiency of second-strand cDNA synthesis and allowing for flexibility in the placement of bacteriophage transcriptional promoter sequences.

B. TC RNA Amplification

We have developed a new linear RNA amplification procedure that utilizes a method of terminal continuation. TC RNA essentially consists of synthesizing first-strand cDNA complementary to the RNA template, subsequently generating second-strand cDNA complementary to the first-strand cDNA, and finally IVT using the double-stranded cDNA as template (Che and Ginsberg, 2004, 2005) (Fig. 4.6). Synthesis of the first-strand cDNA complementary to template mRNA entails the use of two oligonucleotide primers: a poly d(T) primer and a TC primer. The poly d(T) primer is similar to conventional primers that exploit the poly(A)+ sequence present on most mRNAs. The TC primer consists of an oligonucleotide sequence at the 5′ terminus and a short span of three cytidine triphosphates (CTPs) or guanosine triphosphates (GTPs) at the 3′ terminus. In this manner, single-strand cDNA synthesis can be initiated by annealing a second oligonucleotide primer complementary to the attached oligonucleotide (Che and Ginsberg, 2004). By providing a known sequence at the 3′ region of first-strand cDNA and a primer complementary to it, hairpin loops will not form. Second-strand

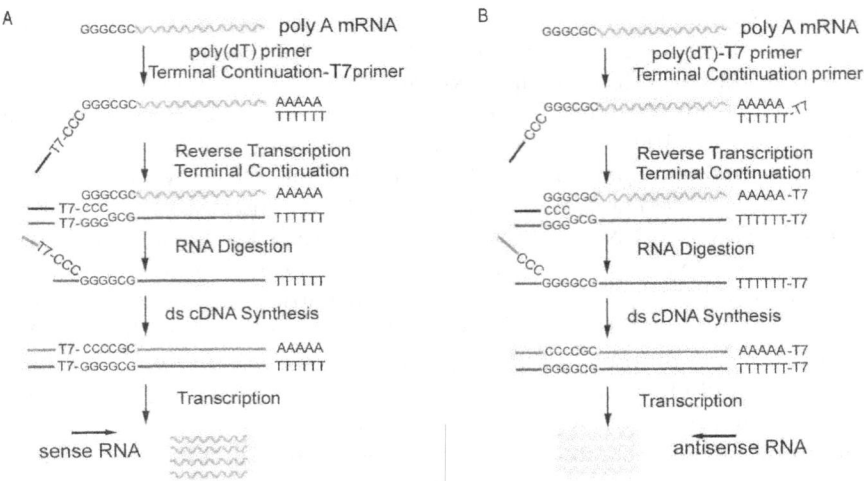

Figure 4.6. Overview of the TC RNA amplification method. (A) A TC primer (containing a bacteriophage promoter sequence for sense orientation) and a poly d(T) primer are added to the mRNA population to be amplified (green rippled line). First-strand (blue line) synthesis occurs as an mRNA–cDNA hybrid and is formed after reverse transcription and terminal continuation of the oligonucleotide primers. Following RNase H digestion to remove the original mRNA template strand, second-strand (red line) synthesis is performed using *Taq* polymerase. The resultant double-stranded product is utilized as template for IVT, yielding high-fidelity, linear RNA amplification of sense orientation (green rippled lines). (B) Schematic similar to (A), illustrating the TC RNA amplification procedure amplifying RNA in the antisense orientation (yellow rippled lines).

cDNA synthesis can be performed with robust DNA polymerases, such as *Taq*, and the TC reaction is highly efficient. One round of amplification is sufficient for downstream genetic analyses (Che and Ginsberg, 2004; Ginsberg and Che, 2004). Furthermore, TC RNA transcription can be driven using a promoter sequence attached to either the 3′ or the 5′ oligonucleotide primers. Therefore, transcript orientation can be in an antisense orientation (similar to conventional aRNA methods) when the bacteriophage promoter sequence is placed on the poly d(T) primer or in a sense orientation when the promoter sequence is attached to the TC primer, depending upon the design of the experimental paradigm (Fig. 4.6). TC RNA amplification offers high sensitivity, flexibility, and throughput capabilities for downstream genetic analyses. Following TC RNA amplification, a large proportion of genes can be assessed quantitatively as evidenced by bioanalysis and cDNA microarray analysis in mouse and human postmortem brain tissues (Che and Ginsberg, 2004; Ginsberg and Che, 2002, 2004, 2005; Mufson *et al.*, 2002). Robust linear amplification is consistently observed. Amplification efficiency of approximately 2500- to 3000-fold is demonstrated with commercially available purified mRNAs, and approximately 1000- to 1500-fold amplification is found after one round using biological samples of RNA extracted from a variety of brain sources (Che and Ginsberg, 2004). Results indicate a high degree of expression level similarity for high, moderate, and low expressed genes using the TC RNA amplification method. The threshold of detection of genes with low hybridization signal intensity is also greatly increased, as many genes that are at the limit of detection using conventional aRNA can be readily observed with the TC method (Che and Ginsberg, 2004). Importantly, increased sensitivity appears greatest for genes with relatively low abundance. Moreover, background hybridization is significantly attenuated when using TC RNA amplification (Ginsberg and Che, 2002, 2004; Mufson *et al.*, 2002).

VIII. MICROARRAY ANALYSIS OF MICRODISSECTED SAMPLES

Once an RNA amplification procedure is utilized to increase the input source of RNA species, biotinylated, fluorescent, or radiolabeled probes can be generated for subsequent hybridization to microarray platforms. Technical advances have fostered the development of high-density microarrays that allow for high-throughput analysis of hundreds to thousands of genes simultaneously. Synthesis of cDNA microarrays entails adhering cDNAs or ESTs to solid supports such as glass slides, plastic slides, or nylon membranes (Brown and Botstein, 1999; Eisen and Brown, 1999). A parallel technology uses photolithography to adhere oligonucleotides to array media (Lockhart *et al.*, 1996). Gene expression is assayed by harvesting total RNA or mRNA from sample tissues, labeling either by radioactive or by fluorescent methods, and hybridizing the labeled probes to arrays (Fig. 4.7). Arrays are washed to remove nonspecific background hybridization, and imaged using a laser scanner for biotinylated/fluorescently labeled probes and a phosphor imager for radioactively labeled probes. The specific signal intensity (minus background) of amplified RNA bound to each probe set (e.g., oligonucleotides or cDNAs/ESTs) is expressed as a ratio of the total hybridization signal intensity of the array, thereby minimizing variations due to differences in the specific activity of the probe and the absolute quantity of probe present. Gene expression data collected using single cells and/or homogeneous populations via RNA amplification and microarray analysis do not allow absolute quantitation of mRNA levels, but generate an expression profile of the relative changes in mRNA levels (Eberwine *et al.*, 2001; Galvin and Ginsberg, 2004; Ginsberg and Che, 2002; Ginsberg *et al.*, 2004; Hemby *et al.*, 2003; Madison and Robinson, 1998; Mufson *et al.*, 2002). Relative changes in individual mRNAs are analyzed by univariate statistics [e.g., analysis of variance (ANOVA) with post hoc Neumann–Keuls test] for individual comparisons (Ginsberg *et al.*, 1999a, 2000; Hemby *et al.*, 2002; Mufson *et al.*, 2002). Differential expression greater than approximately twofold is accepted conventionally as relevant for further examination (Freeman and Hemby, 2004; Galvin and Ginsberg, 2004; Ginsberg *et al.*, 2004; Hemby *et al.*, 2003; Mirnics *et al.*, 2000). Differentially expressed genes can be clustered

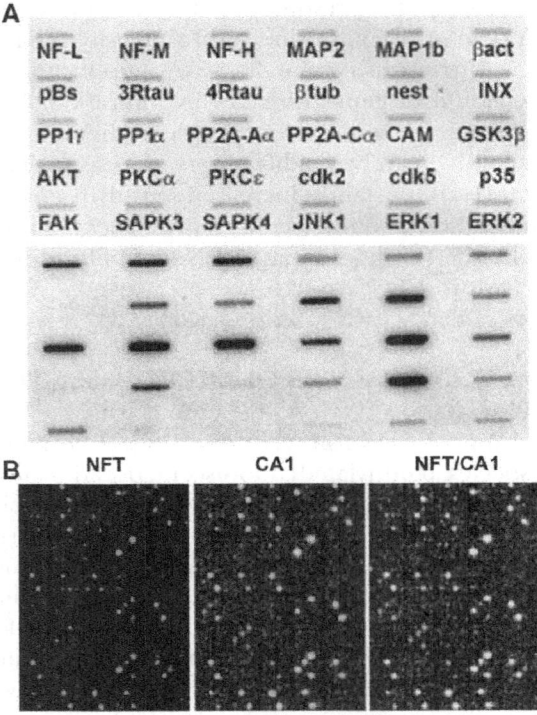

Figure 4.7. Representative array platforms. (A) A custom-designed cDNA array with 30 lanes is depicted. cDNAs are stained with bromophenol blue to show equal loading (top panel). The same array (A; bottom panel) is shown following hybridization with radiolabeled aRNA from a single CA1 neuron. Note the differential expression and abundance of cDNAs. (B) A portion of a high-density cDNA microarray, illustrating aRNA probes generated from neurofibrillary tangle (NFT)-bearing neurons (first panel; red), normal CA1 neurons (second panel; green), and an overlay of both (third panel). Yellow shows similar intensities for NFTs and normal neurons, green indicates a down regulation in NFTs relative to normal CA1 neurons, and red denotes an up regulation.

into functional protein categories for multivariate coordinate gene expression analyses (Freeman and Hemby, 2004; Ginsberg and Che, 2002; Kotlyar *et al.*, 2002). Computational analysis is critical for optimal use of microarrays due to the enormous volume of data that is generated from a single probe. Additionally, access to relational databases is desirable, especially when evaluating hundreds of ESTs that may or may not be linked to genes (and subsequent proteins) of known function.

IX. LCM IN COMBINATION WITH PROTEOMIC APPLICATIONS

Genomic and proteomic expression studies of tissues can be confounded easily because the cells of interest, for example, neurons, exist within a heterogeneous environment that contains many types of cells. LCM allows the isolation of neurons on a single-cell basis for cell-specific analysis (see section "Laser Capture Microdissection"). Once captured, these relatively pure cell populations can be analyzed using a variety of methods, including downstream proteomic applications. Specifically, intact proteins and mRNA can be recovered from LCM captured cells and analyzed quantitatively. For protein studies, the surface-enhanced laser desorption/ionization time-of-flight mass spectrometry (SELDI-TOF MS) approach is an excellent way to evaluate neuron proteomics using the ProteinChip Biology System (PBSII, Ciphergen Biosystems) (Issaq et al., 2002). The PBSII uses SELDI-TOF MS to retain proteins on a solid-phase chromatographic surface that are subsequently ionized and detected by TOF MS. Protein profiles can be generated from as few as 25–50 cells (Paweletz et al., 2001). The SELDI-TOF MS technology consists of three major components: the ProteinChip array, the chip reader apparatus, and the software. The ProteinChip array is a 10-mm-wide × 80-mm-long platform having 8 (or 16) 2-mm spots comprising a specific chromatographic surface. Each spot contains either a chemically (e.g., anionic, cationic, hydrophobic, and hydrophilic) or biochemically (e.g., antibody and receptor) treated surface for retaining entire classes of proteins or single target proteins, respectively. Chemically treated surfaces retain whole classes of proteins, while surfaces treated with biochemical agent (e.g., antibody or other type of affinity reagent) serve as bait and will interact with a specific target protein. Biochemically treated arrays are custom-made by the user. Sample (1–10 µl of protein extract from captured neurons) is applied to the surface, with protein specificity being achieved via the surface treatment and the application of solvents/buffers and washes. After an energy-absorbing molecule solution is added, the array is inserted into the ProteinChip reader to measure the molecular weight and relative amounts of bound proteins. The reader is a laser desorption ionization mass spectroscopy instrument equipped with a pulsed ultraviolet nitrogen laser. Laser activation of the sample causes its desorption/ionization and liberation of gaseous ions from the ProteinChip arrays. The ions enter the TOF MS module that measures the mass-to-charge ratio of each protein. Protein detection is displayed as a series of peaks. The readout generated by the TOF MS analysis is a trace showing the relative abundance versus the molecular weights of the detected proteins. The software converts the peak trace into a simulated one-dimensional gel electrophoresis display to identify differences in protein abundances between samples. This technology can be used to determine in normal and degenerating neurons at specific structural stages patterns of protein expression and the levels of specific proteins and specific posttranslationally modified proteins (e.g., cleaved, phosphorylated, and acetylated proteins).

A practical example illustrating the union of LCM and proteomics using postmortem human spinal cord is provided in Fig. 4.8. LCM can be used to isolate individual spinal motor neurons, yielding a pure cell preparation for downstream proteomic applications (Fig. 4.8A,B). Motor neurons are

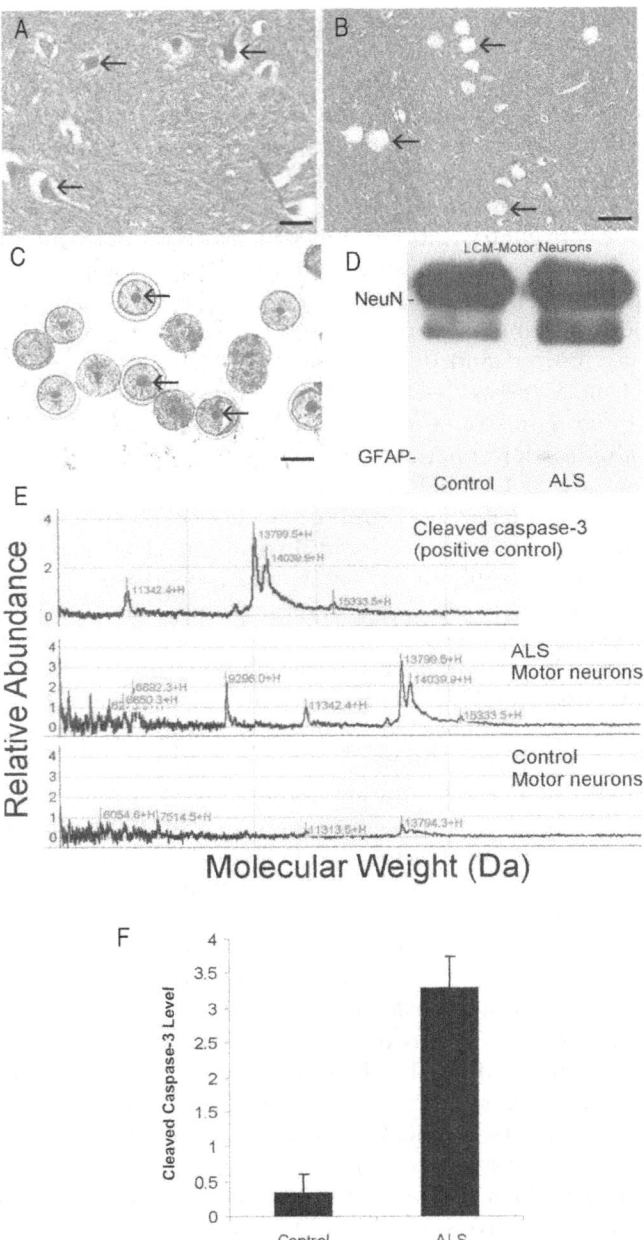

Figure 4.8. LCM and SELDI-TOF MS analysis of human ALS motor neurons. (A) Visualization of spinal cord motor neurons (arrows) in a Ponceau S stained cryostat section of human lumbar cord. Scale bar: 75 μm. (B) Human spinal cord section after harvesting motor neurons via LCM. The open empty circles in the section (arrows) show where the motor neurons were formerly located. Scale bar: 100 μm. (C) Confirmation of cell capture by direct visualization of caps with isolated motor

ideal for LCM because they are relatively large neurons with a low packing density. For example, we have used Ponceau S stained tissue sections to isolate target motor neurons from the surrounding neuropil for SELDI-TOF analysis. Captured cells can be viewed microscopically for confirmation (Fig. 4.8C). Moreover, Western blotting can be used to characterize the purity of human LCM samples. Astrocyte contamination, as assessed by glial fibrillary acidic protein (GFAP), is negligible. A high level of the neuronal nuclear protein NeuN and a very low level of GFAP in motor neuron cell lysates confirm the neuronal purity of the LCM samples (Fig. 4.8D). Even with long exposure times GFAP levels are barely detectable in motor neuron samples. Conversely, when cells with an astrocyte morphology are captured, the GFAP level is high and NeuN was not detectable. These pure motor neuron and astrocyte populations can be used for precise downstream molecular analysis of cell-specific events.

An example of the high resolution afforded by these types of applications is that the cell death protein, cleaved caspase-3, can be measured directly in human motor neurons obtained postmortem from normal control brains and subjects with amyotrophic lateral sclerosis (ALS). Approximately 14,000–15,000 motor neurons were isolated from fresh cryostat sections (stained with Ponceau S) from control lumbar spinal cords (three different cases for a total of ~45,000 motor neurons) and approximately 8000–10,000 motor neurons from ALS spinal cords (three different cases for a total of ~30,000 motor neurons) that were in the somatodendritic attritional stage of degeneration (Martin, 1999). Cleaved caspase-3 antibody (Cell Signaling Technology) was covalently bound to the surface of preactivated ProteinChip arrays (PS2 arrays, affinity capture surfaces). Covalently bound

---

Figure 4.8. (*Cont.*) neurons. The harvested motor neurons are surrounded by thermoplastic film. Scale bar: 225 μm. (D) Assessment of the purity of cell isolation by Western blotting of lysates of LCM acquired cells for NeuN and GFAP. Astrocyte contamination, as assessed by GFAP, is negligible. The high level of NeuN and very low level of GFAP in motor neuron cell lysates confirm the neuronal purity of the LCM samples. Even with long exposure times GFAP levels were only barely detectable in motor neuron samples from ALS cases. (E) Protein profiling in human ALS and control motor neurons by SELDI-TOF MS. PS2 ProteinChip arrays were used to isolate and quantify cleaved caspase-3. After sample preparation, the ProteinChip arrays were analyzed by laser desorption ionization TOF MS. For comparison purposes, the software of the SELDI Ciphergen system displays the data as a spectra view. Recombinant cleaved caspase-3 served as a positive control for identifying the molecular weights of the cleaved subunits. (F) Quantification of cleaved caspase-3 in human motor neurons. To identify differences in protein abundances between control and ALS cases, the software converts the peak trace into a simulated one-dimensional gel electrophoresis display to measure protein abundance. The values are mean ± standard deviation (SD). The measurements are normalized to parallel analyses of NeuN levels in the lysates. ALS motor neurons have significantly elevated ($p < 0.001$) levels of cleaved caspase-3 compared to age-matched controls (ANOVA with post hoc Neumann–Keuls test).

immunoglobulin (IgG) served as an antibody negative control. Crude cell lysates of captured motor neurons were applied to the ProteinChip with bound antibody, incubated, washed, and analyzed in a Ciphergen ProteinChip reader. Purified recombinant active caspase-3 (Medical and Biological Laboratories) was used as a positive control (Fig. 4.8E, upper retentate map) and it displayed prominent peaks at ~11.3, 13.8, and 14 kDa. Peptides corresponding to cleaved caspase-3 were found in ALS motor neurons (Fig. 4.8E, middle retentate map, peaks at ~11 and 14 kDa). No caspase-3 signal was observed in ALS or control cases with the nonspecific IgG bound to the chip. In age and postmortem delayed-matched control motor neurons, peaks of similar molecular weight were either not above background or were low (Fig. 4.8E, control motor neurons, lower retentate map). Quantification of cleaved caspase-3 (13.8 kDa protein) levels in control and ALS spinal motor neurons revealed highly significant ($p < 0.001$) increases in ALS motor neurons (Fig. 4.8F). The immunoassay results had remarkably low variability, likely due in part to the homogeneous population of motor neurons that was accrued via LCM.

Cell-based assays are critical for evaluating changes in protein cells undergoing degeneration. The use of tissue homogenate-based assays is suboptimal for this purpose because tissue homogenates cannot afford sufficient resolution. Thus, the interpretation of homogenate-based assays of tissues with a heterogeneous cellular composition is suspect. The strategy for measuring proteins in cells acquired via LCM represents a major step forward in the analysis of cell-specific degenerative events (Freeman and Hemby, 2004; Mouledous *et al.*, 2003). LCM and proteomic approaches are feasible and practical to apply, providing the availability of the equipment and service maintenance. The cellular resolution attained by LCM-based technologies for downstream proteomic applications is optimal for these types of in vivo investigations. LCM dramatically decreases the noise in the assays by minimizing contaminating cells. The integrity of the proteins and peptide fragments to be analyzed is maintained (Freeman and Hemby, 2004; Mouledous *et al.*, 2003). A pitfall of LCM-based technologies is that they are labor-intensive, and the number of captured cells required for quantitative signal detection is significant. However, the data gleaned by these types of studies represent definitive cell-specific events.

In summary, the analysis of human material as well as of appropriate animal models of neurodegeneration will provide direct results on the molecular events occurring within diseased cells. Employing LCM in combination with SELDI-TOF is ideal for dealing with asynchrony of neurodegeneration by providing structural–molecular correlations on neurons sampled at similar (and different) stages of degeneration. Human tissue experiments must be controlled at several levels with disease-specific controls that are matched for age, agonal state, and postmortem delay (Bahn *et al.*, 2001; Hynd *et al.*, 2003; Van Deerlin *et al.*, 2000, 2002). Moreover, interregional controls within the same case are necessary to rule out the possibility that observed changes are due to agonal state and tissue autolysis. Parallel studies of neurodegeneration in optimally prepared animal models can provide valuable side-by-side comparisons of relevant molecules. Thus, molecular profiles can be then brought into the context of the structural phenotype of the observed degeneration. Moreover, if captured cells (using regional, microaspiration, and/or LCM-based technologies) obtained from an animal model display expression profile(s) that differ vastly from the human condition it is designed to model, applicability comes into question. Ultimately, a particular animal model may be deemed to be inappropriate for further study within the context of single cell or homogeneous population cell analysis based on disparities in genomic and proteomic profiles from a human condition that they were designed to mimic.

X. ADVANTAGES/LIMITATIONS

A variety of tissues and cells can be used to extract mRNA for gene profiling experiments. When employing mRNA as a starting material, one cannot overemphasize the importance of the preservation of RNA integrity. RNA species are particularly sensitive to degradation by RNase. RNases are found in virtually every cell type, and they retain their activity over a broad pH range (Blumberg, 1987; Farrell, 1998). Thus, RNase-free precautions are essential for all microdissection-based studies. All biological samples require prompt handling, either through rapid RNA extraction, flash freezing, or through fixation to minimize degradation.

Reproducibility of single-cell expression profiling is a critical parameter that is improving. Advances at the level of tissue dissection, RNA amplification, microarray platforms, and developing powerful statistical methods will ultimately lead to greater utility and flexibility of these technologies. Recent advances include the utilization of pooled populations of individual cell types to reduce variability in expression levels yet maintain an expression profile for a single cell type. The likelihood of generating highly reproducible data is increased greatly by replicate array analysis of aliquots of the same amplified RNA sample. Validation of array results is important, and several independent alternative techniques are quite useful to reproduce changes seen on an array platform such as qPCR, SAGE, and/or in situ hybridization, among others.

When deciding whether or not to employ microaspiration and/or high-throughput array technologies, the most important aspect to consider is the question the researcher is interested in answering, and determining the method(s) that would be best suited to perform the experiment. Once a researcher has decided that an array experiment is appropriate, much consideration needs to go into sample size and preparation, tissue and/or cell quality, and importantly, input amount of RNA that will likely be generated. If the input source is a small sample of population of cells captured by LCM, then an RNA amplification method is requisite. A researcher then needs to calculate laboratory and technical effort, cost, and goals in order to determine the commitment level that will be needed to carry out array experiments. Sample preparation, RNA amplification, array hybridization, and array analysis usually require a long-term commitment, as many investigators have found out much to their dismay. A qPCR experiment would be more useful, for instance, if a researcher is trying to assess the regulation of a single gene product (or splice variants/isoforms of an individual gene family). An array experiment may yield the desired result, along with a plethora of potential data on dozens, hundreds, thousands of genes that may not be germane to central hypothesis. Quantitation of array platforms is typically relative, whereas qPCR can be more direct, using cycle threshold calculations as well as copy number (but this is difficult and not typically feasible for the casual qPCR user). qPCR can be reliable and cost effective, provided that the primer design is performed optimally. Alternatively, the solution hybridization afforded by RNase protection assays cannot be underestimated. RNase protection assays are especially useful when input sources of RNA are abundant, such as with in vitro paradigms. However, RNase protection assays in tissue sections, particularly fixed tissues, are not highly recommended. In summary, cDNA and oligonucleotide arrays are spectacular tools for high-throughput analyses within a myriad of paradigms and tissue sources. qPCR is a useful medium to low-throughput method that directly assays genes of specific interest. Our laboratory strategy is to combine the use of both assays, by defining expression profiling patterns on microarray platforms and validating individual gene level changes by independent qPCR analyses (Ginsberg and Che, 2004, 2005; Ginsberg *et al.*, 2004).

The combination of discrete cell microdissection procedures with microarray technologies allows for high-resolution, high-throughput expression profiling of dozens to hundreds to thousands of genes and proteins simultaneously from a single neuron or from a group of similar neurons. The next level of understanding of cellular and molecular mechanisms underlying normative function and pathological conditions lies in the ability to combine these aforementioned technologies with appropriate models to recapitulate the structure and connectivity of these systems in vivo and in vitro. Complex biological processes are not likely to be governed solely by the action of a single isolated gene. Rather, coordinate interactions of a multiplicity of genes may regulate normative function. When these gene programs or mosaics undergo increased or decreased expression during the lifespan, they may contribute to the mechanisms underlying disease pathogenesis. Single-cell and population-cell profiling techniques coupled with microarray platforms have the potential to quantify simultaneous expression levels of numerous genes and proteins in a given cell, thereby allowing for previously unobserved gene interactions, and ultimately protein interactions, to become more evident. Independent verification of individual gene level changes discovered by microarray analysis by alternate techniques is a critical component to a research program. Thus, a combination of multidisciplinary approaches is ideal for verification of gene expression level alterations, with the explicit knowledge that the sum of the evaluations may be more informative and reflect the actual biology of the system than an individual method.

APPENDIX: DETAILED METHODOLOGY

A. Acridine Orange Histofluorescence

This protocol was developed for AO histochemistry using animal model and human postmortem tissues embedded in paraffin (Ginsberg et al., 1997, 1998; Mikel and Becker, 1991). Briefly, tissue sections are deparaffinized in xylene, graded through a descending ethanol series, and placed in distilled water for 5 min. The sections are placed in a 0.2 M dibasic sodium phosphate/0.1 M citric acid (SC; pH 4.0) solution for 5 min prior to staining with AO (10 µg/ml; Sigma) in SC for 15 min. The sections are rinsed three times in the SC buffer, immersed in 50% ethanol in phosphate-buffered saline (PBS; 0.12 M; pH 7.4) for 2 min, cleared in xylene, and mounted with an antifading medium (Vectashield, Vector Laboratories). To reduce the intense autofluorescence of lipofuscin granules that are abundant in senescent human brain, selected tissue sections can be pretreated with either 0.05% potassium permanganate in PBS for 20 min followed by 0.2% potassium metabisulfite/0.2% oxalic acid in PBS for 30 s (Guntern et al., 1992) or 0.3% Sudan Black B (Sigma; w/vol in 70% ethanol) for 10 min (Yao et al., 2003) prior to AO histochemistry.

B. Microaspiration and LCM

Our laboratory utilizes a Nikon inverted microscope with MetaMorph 5.0 software (Universal Imaging Corporation), an Eppendorf micromanipulator, and an Eppendorf Transjector for single-cell microaspiration. Immunostained or histochemically stained tissue sections are not coverslipped or counterstained and are immersed in RNase-free 0.1 M Tris (pH 7.4) (Ginsberg et al., 1999a, 2000; Hemby et al., 2002, 2003). Individual cells are carefully aspirated from the tissue section and placed in microfuge tubes for subsequent TC RNA amplification. LCM is performed on immunostained or histochemically stained tissue sections that are dehydrated in an ascending series of ethanol and placed in fresh xylenes for a minimum of 15 min. LCM is performed using a PixCell IIe instrument (Arcturus). Caps containing desired captured cells are inverted into a microfuge tube containing Trizol reagent (Invitrogen) prior to initiating TC RNA amplification.

C. Tract-Tracing for Use with Microdissection

Fluorogold (4%; Fluorochrome Inc.) is dissolved in isotonic saline and backfilled into an autoclaved glass micropipette with the tip tapered to ~12–15 µm. FG is iontophoresed into brain regions via 5 µA of current with a 5-s on/off cycle for 10 min. Micropipettes should remain in place 5 min following infusion to allow proper diffusion into the neural tissue and prevent diffusion up the injection tract. FG is visualized using fluorescence microscopy with fluorescence excitation filter set at 340–380 nm and a barrier filter at 430 nm (Schmued and Heimer, 1990). Microaspiration and LCM are performed on uncoverslipped tissue sections as described above.

D. aRNA Amplification

For aRNA amplification, an oligo $d(T)T7$ primer (20 ng/μl) is hybridized directly to poly(A)+ mRNAs (Tecott *et al.*, 1988). A double-stranded mRNA–cDNA hybrid is formed by reverse-transcribing the primed mRNAs with dNTPs (1 mM) and 10 U reverse transcriptase (AMVRT, Sekigaku) for 3 h at $42°$C (VanGelder *et al.*, 1990). The double-stranded mRNA–cDNA hybrid is converted into double-stranded cDNA by heat denaturing for 5 min at $85°$C followed by the addition of dNTPs (1 mM) and 10 U T4 DNA polymerase (Invitrogen) and 10 U Klenow (Invitrogen), forming a functional T7 RNA polymerase promoter. Following the removal of tertiary structures and drop dialyzing the double-stranded cDNA against RNase-free water, the first round of aRNA synthesis occurs using 2000 U T7 RNA polymerase (Epicentre) and NTPs at $37°$C for 4 h (Eberwine *et al.*, 2001). The second round of aRNA amplification begins by annealing random hexamers to the newly formed aRNA, and synthesizing a cDNA strand. The oligo $d(T)T7$ primer is then reintroduced, which binds to the poly(A)+ sequence on the newly synthesized cDNA strand, and a double-stranded cDNA template is formed. aRNA is then tagged with fluorescent, biotinylated, or radiolabeled reagents to enable hybridization to the desired cDNA microarray, oligonucleotide platform, or membrane-based array.

E. TC RNA Amplification

TC RNA amplification consists of immersing microdissected cells or regional dissections in 250 μl of proteinase K solution (50 μg/ml; Ambion) for 12 h at $37°$C prior to extraction in Trizol reagent (Invitrogen). RNAs are reverse-transcribed in the presence of the poly $d(T)$ primer (10 ng/μl) and TC primer (10 ng/μl) in 1X first strand buffer (Invitrogen), 1 mM dNTPs, 5 mM dithiothreitol (DTT), 20 U of RNase inhibitor, and 5 U reverse transcriptase (Superscript III; Invitrogen) (Che and Ginsberg, 2004). The synthesized single-stranded cDNAs are converted into double-stranded cDNAs by adding into the reverse transcription reaction the following: 10 mM Tris (pH 8.3), 50 mM KCl, 1.5 mM $MgCl_2$, and 0.5 U RNase H (Invitrogen) in a total volume of 99 μl. Samples are placed in a thermal cycler and second-strand synthesis proceeds as follows: RNase H digestion step $37°$C, 10 min; denaturation step $95°$C, 3 min, annealing step $50°$C, 3 min; elongation step $75°$C, 30 min. 5 U (1 μl) *Taq* polymerase (PE Biosystems) is added to the reaction at the initiation of the denaturation step (i.e., hot start) (Che and Ginsberg, 2004). The reaction is terminated with 5 M ammonium acetate. The samples are extracted in phenol:chloroform:isoamyl alcohol (25:24:1) and ethanol precipitated with 5 μg of linear acrylamide (Ambion) as a carrier. The solution is centrifuged at 14,000 rpm and the pellet is washed once with 95% ethanol and air-dried. The cDNAs are resuspended in 20 μl of RNase-free H_2O and drop dialyzed on 0.025 μm filter membranes (Millipore) against 50 ml of 18.2 MΩ RNase-free H_2O for 2 h. The sample is collected off the dialysis membrane, and hybridization probes are synthesized by IVT using a fluorescent labeling kit (e.g., Cy3 and/or Cy5 labeling; Enzo Life Sciences) as per manufacturer's instructions. Alternatively, hybridization probes can be generated for membrane-based arrays using ^{33}P incorporation in 40 mM Tris (pH 7.5), 7 mM $MgCl_2$, 10 mM NaCl, 2 mM spermidine, 5 mM of DTT, 0.5 mM of ATP, GTP, and CTP, 10 μM of cold UTP, 20 U of RNase inhibitor, 1000 U T7 RNA polymerase (Epicentre), and 40 μCi of ^{33}P-UTP (GE Healthcare) for 4 h at 37°C (Che and Ginsberg, 2004).

F. SELDI-TOF

Patients were diagnosed with ALS by neurological examination using the El Escorial criteria (Brooks, 1994). Postmortem central nervous system tissues from these individuals were obtained from the Division of Neuropathology, Human Brain Resource Center, Johns Hopkins University School of Medicine. Neuropathological evaluation confirmed the clinical diagnosis of ALS (Martin, 1999). The cases studied were sporadic ALS. Postmortem samples of spinal cord from age-matched control individuals without neurological disease ($n = 3$) and patients with ALS ($n = 3$) were selected randomly for analysis.

At autopsy, spinal cord blocks (L5 segment) were dissected and snap-frozen in liquid nitrogen and stored at −80°C. Lumbar spinal cord blocks were sectioned at 12 μm in a cryostat, thaw-mounted onto Superfrost charged glass microscope slides, and stored at −80°C. For LCM, selected slides were stained briefly with Ponceau S prepared in a protease inhibitor cocktail and air-dried. A PixCell II LCM system was used for acquiring motor neurons using a laser spot size of 30 or 60 μm. Motor neuron isolates were lysed with cold 20 mM Tris-HCl (pH = 7.4) containing 10% (wt/vol) sucrose, 20 U/ml aprotinin (trasylol), 20 μg/ml leupeptin, 20 μg/ml antipain, 20 μg/ml pepstatin A, 20 μg/ml chymostatin, 0.1 mM phenylmethylsulfonyl fluoride, 10 mM benzamidine, 1 mM EDTA, and 5 mM EGTA. All of the protease inhibitors were purchased from Sigma. Protein concentration was determined by bicinchoninic acid protein assay kit (Pierce).

PS2 ProteinChip arrays (Ciphergen) were used for SELDI-TOF. PS2 arrays are recommended for use in covalent immobilization of biomolecules for the subsequent capture of target proteins from complex biological samples. PS2 arrays have spots that are preactivated with epoxide chemistry that covalently bind to free primary amine groups on the surface of biomolecules (e.g., antibodies) for immunoassays. The stably immobilized biomolecules capture proteins from biological samples through specific, noncovalent interactions. The PS2 surface is especially recommended when the aim is to include sensitive detection, low nonspecific binding, and target protein concentrations at less than 1% of total protein.

G. Supplies/Manufacturers

18.2 MΩ RNase-free H_2O (Nanopure Diamond, Barnstead, Dubuque, IA)

AMVRT (Sekigaku, Falmouth, MA)

ATP (Invitrogen, Carlsbad, CA)

Acridine Orange (Sigma, St. Louis, MO)

Ammonium acetate (Sigma)

Antipain (Sigma)

Aprotinin (Sigma)

Bioanalyzer (2100, Agilent Technologies, Palo Alto, CA)

Benzamidine (Sigma)

Chymostatin (Sigma)

Citric acid (Sigma)

Cleaved caspase-3 antibody (Cell Signaling Technology, Beverly, MA)

Caspase-3 active recombinant (Molecular and Biological Laboratories Wobum, MA)

Cresyl violet (Sigma)

dNTPs (Invitrogen)

DTT (Sigma)

EDTA (Sigma)

EGTA (Sigma)

Filter membranes (Millipore, Billerica, MA)

1X First strand buffer (Invitrogen)

Fluorescent labeling kit (Enzo Life Sciences, Farmingdale, NY)

Fluorogold (Fluorochrome Inc., Englewood, CO)

Klenow (Invitrogen)

Leupeptin (Sigma)

Linear acrylamide (Ambion)

MetaMorph 5.0 software (Universal Imaging Corp., Downingtown, PA)

Micromanipulator (Brinkmann-Eppendorf, Westbury, NY)

NTPs (Invitrogen)

Oxalic acid (Sigma)

PALM (PALM Microlaser Technologies, Bernried Germany)

P^{33}-UTP (GE Healthcare, Piscataway, NJ)

PBSII (Ciphergen Biosystems, Fremont, CA)

Pepsatin (Sigma)

Phenol:chloroform:isoamyl alcohol (Invitrogen)

Phenylmethsulfonyl fluoride (Sigma)

PixCell IIe LCM (Arcturus, Mountain View, CA)

Ponceau S (Sigma)

Potassium metabisulfate (Sigma)

Potassium permanganate (Sigma)
Protein assay kit (Pierce, Rockford, IL)
Proteinase K (Ambion, Austin, TX)
Purified mRNAs (Invitrogen)
Reverse transcriptase (Superscript III, Invitrogen)
RNase H (Invitrogen)
RNase inhibitor (Invitrogen)
Spermidine (Sigma)
Sudan Black B (Sigma)
Sucrose (Sigma)
Taq polymerase (PE Biosystems, Foster City, CA)
T4 DNA polymerase (Invitrogen)
T7 RNA polymerase (Epicentre, Madison, WI)
Tris (Sigma)
Trizol reagent (Invitrogen)
UTP (Invitrogen)
Vectasheild (Vector Laboratories, Burlingame, CA)

ACKNOWLEDGMENTS. We thank Shaoli Che, M.D., Ph.D., and Scott E. Counts, Ph.D., for their continued efforts on these projects. Support for these projects comes from the NINDS (NS34100, LJM; NS43939, SDG; NS48447, SD6), NIA (AG10668, EJM & SDG; AG14449, EJM & SDG; AG21661, EJM AG05146, LJM; AG16282, LJM), NCI (CA94520; SDG), NIDA (DA013772, SEH; DA013234, SEH), NIMH (MH074313, SEH) and Alzheimer's Association (SDG). We also express our appreciation to the families of the patients studied here, who made this research possible.

REFERENCES

Backes, E., and Hemby, S. E., 2003, Discrete cell gene profiling of ventral tegmental dopamine neurons after acute and chronic cocaine self-administration, *J. Pharmacol. Exp. Ther.* **307:** 450–459.

Bahn, S., Augood, S. J., Ryan, M., Standaert, D. G., Starkey, M., and Emson, P. C., 2001, Gene expression profiling in the post-mortem human brain—no cause for dismay, *J. Chem. Neuroanat.* **22:**79–94.

Barton, A. J. L., Pearson, R. C. A., Najlerahim, A., and Harrison, P. J., 1993, Pre- and postmortem influences on brain RNA, *J. Neurochem.* **61:**1–11.

Becker, I., Becker, K.-F., Röhrl, M. H., Minkus, G., Schütze, K., and Höfler, H., 1996, Single-cell mutation analysis of tumors from stained histologic slides, *Lab. Invest.* **75:**801–807.

Blumberg, D. D., 1987, Creating a ribonuclease-free environment, *Methods Enzymol.* **152:**20–24.

Bonner, R. F., Emmert-Buck, M., Cole, K., Pohida, T., Chuaqui, R., Goldstein, S., and Liotta, L. A., 1997, Laser capture microdissection: molecular analysis of tissue, *Science* **278:**1481–1483.

Brooks, B. R., 1994, El Escorial World Federation of Neurology criteria for the diagnosis of amyotrophic lateral sclerosis, *J. Neurol. Sci.* **124:**96–107.

Brown, P. O., and Botstein, D., 1999, Exploring the new world of the genome with DNA microarrays, *Nat. Genet.* **21**(Suppl):33–37.

Che, S., and Ginsberg, S. D., 2004, Amplification of transcripts using terminal continuation, *Lab. Invest.* **84:**131–137.

Che, S., and Ginsberg, S. D., 2006, RNA amplification methodologies, In: *Progress in RNA Research*, Nova Science Publishing, in press.

Coombs, N. J., Gough, A. C., and Primrose, J. N., 1999, Optimisation of DNA and RNA extraction from archival formalin-fixed tissue, *Nucleic Acids Res.* **27:**e12.

Craven, R. A., Totty, N., Harnden, P., Selby, P. J., and Banks, R. E., 2002, Laser capture microdissection and two-dimensional polyacrylamide gel electrophoresis: evaluation of tissue preparation and sample limitations, *Am. J. Pathol.* **160:**815–822.

Crino, P. B., Khodakhah, K., Becker, K., Ginsberg, S. D., Hemby, S., and Eberwine, J. H., 1998, Presence and phosphorylation of transcription factors in dendrites, *Proc. Natl. Acad. Sci. U.S.A.* **95:**2313–2318.

D'Amore, F., Stribley, J. A., Ohno, T., Wu, G., Wickert, R. S., Delabie, J., Hinrichs, S. H., and Chan, W. C., 1997, Molecular studies on single cells harvested by micromanipulation from archival tissue sections previously stained by immunohistochemistry or nonisotopic *in situ* hybridization, *Lab. Invest.* **76:**219–224.

Dolter, K. E., and Braman, J. C., 2001, Small-sample total RNA purification: laser capture microdissection and cultured cell applications, *Biotechniques* **30:**1358–1361.

Eberwine, J., Kacharmina, J. E., Andrews, C., Miyashiro, K., McIntosh, T., Becker, K., Barrett, T., Hinkle, D., Dent, G., and Marciano, P., 2001, mRNA expression analysis of tissue sections and single cells, *J. Neurosci.* **21:**8310–8314.

Eberwine, J., Yeh, H., Miyashiro, K., Cao, Y., Nair, S., Finnell, R., Zettel, M., and Coleman, P., 1992, Analysis of gene expression in single live neurons, *Proc. Natl. Acad. Sci. U.S.A.* **89:**3010–3014.

Ehrig, T., Abdulkadir, S. A., Dintzis, S. M., Milbrandt, J., and Watson, M. A., 2001, Quantitative amplification of genomic DNA from histological tissue sections after staining with nuclear dyes and laser capture microdissection, *J. Mol. Diagn.* **3:**22–25.

Eisen, M. B., and Brown, P. O., 1999, DNA arrays for analysis of gene expression, *Methods Enzymol.* **303:**179–205.

Emmert-Buck, M. R., Bonner, R. F., Smith, P. D., Chuaqui, R. F., Zhuang, Z., Goldstein, S. R., Weiss, R. A., and Liotta, L. A., 1996, Laser capture microdissection, *Science* **274:**998–1001.

Emsley, J. G., Lu, X., and Hagg, T., 2001, Retrograde tracing techniques influence reported death rates of adult rat nigrostriatal neurons, *Exp. Neurol.* **168:**425–433.

Farrell, R. E., Jr, 1998, *RNA Methodologies*, 2nd edition, San Diego: Academic Press.

Fasulo, W. H., and Hemby, S. E., 2003, Time-dependent changes in gene expression profiles of midbrain dopamine neurons following haloperidol administration, *J. Neurochem.* **87:**205–219.

Fend, F., Emmert-Buck, M. R., Chuaqui, R., Cole, K., Lee, J., Liotta, L. A., and Raffeld, M., 1999, Immuno-LCM: laser capture microdissection of immunostained frozen sections for mRNA analysis, *Am. J. Pathol.* **154:**61–66.

Franklin, T. R., and Druhan, J. P., 2000, The retrograde tracer fluoro-gold interferes with the expression of fos-related antigens, *J. Neurosci. Methods* **98:**1–8.

Freeman, W. M., and Hemby, S. E., 2004, Proteomics for protein expression profiling in neuroscience, *Neurochem. Res.* **29:**1065–1081.

Galvin, J. E., 2004, Neurodegenerative diseases: pathology and the advantage of single-cell profiling, *Neurochem. Res.* **29:**1041–1051.

Galvin, J. E., and Ginsberg, S. D., 2004, Expression profiling and pharmacotherapeutic development in the central nervous system, *Alzheimer Dis. Assoc. Disord.* **18:**264–269.

Ginsberg, S. D., 2001, Gene expression profiling using single cell microdissection combined with cDNA microarrays, In: Geschwind, D. H. (ed.), *DNA Microarrays: The New Frontier in Gene Discovery and Gene Expression Analysis*, Washington: Society for Neuroscience Press, pp. 61–70.

Ginsberg, S. D., and Che, S., 2002, RNA amplification in brain tissues, *Neurochem. Res.* **27:**981–992.

Ginsberg, S. D., and Che, S., 2004, Combined histochemical staining, RNA amplification, regional, and single cell analysis within the hippocampus, *Lab. Invest.* **84:**952–962.

Ginsberg, S. D., and Che, S., 2005, Expression profile analysis within the human hippocampus: comparison of CA1 and CA3 pyramidal neurons, *J. Comp. Neurol.* **487:**107–118.

Ginsberg, S. D., Crino, P. B., Hemby, S. E., Weingarten, J. A., Lee, V. M.-Y., Eberwine, J. H., and Trojanowski, J. Q., 1999a, Predominance of neuronal mRNAs in individual Alzheimer's disease senile plaques, *Ann. Neurol.* **45:**174–181.

Ginsberg, S. D., Crino, P. B., Lee, V. M.-Y., Eberwine, J. H., and Trojanowski, J. Q., 1997, Sequestration of RNA in Alzheimer's disease neurofibrillary tangles and senile plaques, *Ann. Neurol.* **41:**200–209.

Ginsberg, S. D., Elarova, I., Ruben, M., Tan, F., Counts, S. E., Eberwine, J. H., Trojanowski, J. Q., Hemby, S. E., Mufson, E. J., and Che, S., 2004, Single cell gene expression analysis: implications for neurodegenerative and neuropsychiatric disorders, *Neurochem. Res.* **29:**1054–1065.

Ginsberg, S. D., Galvin, J. E., Chiu, T.-S., Lee, V. M.-Y., Masliah, E., and Trojanowski, J. Q., 1998, RNA sequestration to pathological lesions of neurodegenerative disorders, *Acta Neuropathol.* **96:**487–494.

Ginsberg, S. D., Hemby, S. E., Lee, V. M.-Y., Eberwine, J. H., and Trojanowski, J. Q., 2000, Expression profile of transcripts in Alzheimer's disease tangle-bearing CA1 neurons, *Ann. Neurol.* **48:**77–87.

Ginsberg, S. D., Schmidt, M. L., Crino, P. B., Eberwine, J. H., Lee, V. M.-Y., and Trojanowski, J. Q., 1999b, Molecular pathology of Alzheimer's disease and related disorders, In: Peters, A., and Morrison, J. H. (eds.), *Cerebral Cortex, Vol. 14: Neurodegenerative and Age-Related Changes in Structure and Function of Cerebral Cortex*, New York: Kluwer Academic/Plenum, pp. 603–653.

Goldsworthy, S. M., Stockton, P. S., Trempus, C. S., Foley, J. F., and Maronpot, R. R., 1999, Effects of fixation on RNA extraction and amplification from laser capture microdissected tissue, *Mol. Carcinog.* **25:**86–91.

Guntern, R., Bouras, C., Hof, P. R., and Vallet, P. G., 1992, An improved thioflavine S method for staining neurofibrillary tangles and senile plaques in Alzheimer's disease, *Experientia* **48:**8–10.

Hemby, S. E., Ginsberg, S. D., Brunk, B., Arnold, S. E., Trojanowski, J. Q., and Eberwine, J. H., 2002, Gene expression profile for schizophrenia: discrete neuron transcription patterns in the entorhinal cortex, *Arch. Gen. Psychiatry* **59:**631–640.

Hemby, S. E., Trojanowski, J. Q., and Ginsberg, S. D., 2003, Neuron-specific age-related decreases in dopamine receptor subtype mRNAs, *J. Comp. Neurol.* **456:**176–183.

Hynd, M. R., Lewohl, J. M., Scott, H. L., and Dodd, P. R., 2003, Biochemical and molecular studies using human autopsy brain tissue, *J. Neurochem.* **85:**543–562.

Iscove, N. N., Barbara, M., Gu, M., Gibson, M., Modi, C., and Winegarden, N., 2002, Representation is faithfully preserved in global cDNA amplified exponentially from sub-picogram quantities of mRNA, *Nat. Biotechnol.* **20:**940–943.

Issaq, H. J., Veenstra, T. D., Conrads, T. P., and Felschow, D., 2002, The SELDI-TOF MS approach to proteomics: protein profiling and biomarker identification, *Biochem. Biophys. Res. Commun.* **292:**587–592.

Kacharmina, J. E., Crino, P. B., and Eberwine, J., 1999, Preparation of cDNA from single cells and subcellular regions, *Methods Enzymol.* **303:**3–18.

Kamme, F., Salunga, R., Yu, J., Tran, D. T., Zhu, J., Luo, L., Bittner, A., Guo, H. Q., Miller, N., Wan, J., and Erlander, M., 2003, Single-cell microarray analysis in hippocampus CA1: demonstration and validation of cellular heterogeneity, *J. Neurosci.* **23:**3607–3615.

Klur, S., Toy, K., Williams, M. P., and Certa, U., 2004, Evaluation of procedures for amplification of small-size samples for hybridization on microarrays, *Genomics* **83:**508–517.

Kotlyar, M., Fuhrman, S., Ableson, A., and Somogyi, R., 2002, Spearman correlation identifies statistically significant gene expression clusters in spinal cord development and injury, *Neurochem. Res.* **27:**1133–1140.

Lanciego, J. L., and Wouterlood, F. G., 2006, Multiple neuroanatomical tract-tracing, In: Zaborszky, L., Wouterlood, F., and Lanciego, J. L. (eds.), *Neuroanatomical Tract Tracing 3: Molecules-Neurons-Systems*, Springer New York, 336–365.

Lein, E. S., Zhao, X., and Gage, F. H., 2004, Defining a molecular atlas of the hippocampus using DNA microarrays and high-throughput in situ hybridization, *J. Neurosci.* **24**:3879–3889.

Leonard, S., Logel, J., Luthman, D., Casanova, M., Kirch, D., and Freedman, R., 1993, Biological stability of mRNA isolated from human postmortem brain collections, *Biol. Psychiatry.* **33**:456–466.

Lockhart, D. J., Dong, H., Byrne, M. C., Follettie, M. T., Gallo, M. V., Chee, M. S., Mittmann, M., Wang, C., Kobayashi, M., Horton, H., and Brown, E. L., 1996, Expression monitoring by hybridization to high-density oligonucleotide arrays, *Nat. Biotechnol.* **14**:1675–1680.

Lu, L., Neff, F., Dun, Z., Hemmer, B., Oertel, W. H., Schlegel, J., and Hartmann, A., 2004, Gene expression profiles derived from single cells in human postmortem brain, *Brain Res. Brain. Res. Protoc.* **13**:18–25.

Madison, R. D., and Robinson, G. A., 1998, IRNA internal standards quantify sensitivity and amplification efficiency of mammalian gene expression profiling, *Biotechniques* **25**:504–514.

Martin, L. J., 1999, Neuronal death in amyotrophic lateral sclerosis is apoptosis: possible contribution of a programmed cell death mechanism, *J. Neuropathol. Exp. Neurol.* **58**:459–471.

Matz, M., Shagin, D., Bogdanova, E., Britanova, O., Lukyanov, S., Diatchenko, L., and Chenchik, A., 1999, Amplification of cDNA ends based on template-switching effect and step-out PCR, *Nucleic Acids Res.* **27**:1558–1560.

McClain, K. L., Cai, Y.-H., Hicks, J., Peterson, L. E., Yan, X.-T., Che, S., and Ginsberg, S. D., 2005, Expression profiling using human tissues in combination with RNA amplification and microarray analysis: assessment of Langerhans cell histiocytosis, *Amino Acids,* **28**:279–290.

Mikel, U. V., and Becker, R. L., Jr., 1991, A comparative study of quantitative stains for DNA in image cytometry, *Anal. Quant. Cytol. Histol.* **13**:253-260.

Mirnics, K., Middleton, F. A., Marquez, A., Lewis, D. A., and Levitt, P., 2000, Molecular characterization of schizophrenia viewed by microarray analysis of gene expression in prefrontal cortex, *Neuron* **28**:53–67.

Molnar, Z., Blakey, D., Bystron, I., and Carney, R. S. E., 2006, Tract-tracing in developing systems and in post mortem human material using carbocyanine dyes, In: Zaborszky, L., Wouterlood, F., and Lanciego, J. L. (eds.), *Neuroanatomical Tract Tracing 3: Molecules-Neurons-Systems*, New York: Springer, 366–393.

Mouledous, L., Hunt, S., Harcourt, R., Harry, J. L., Williams, K. L., and Gutstein, H. B., 2003, Proteomic analysis of immunostained, laser-capture microdissected brain samples, *Electrophoresis* **24**:296–302.

Mufson, E. J., Counts, S. E., and Ginsberg, S. D., 2002, Single cell gene expression profiles of nucleus basalis cholinergic neurons in Alzheimer's disease, *Neurochem. Res.* **27**:1035–1048.

Mufson, E. J., Ginsberg, S. D., Ikonomovic, M. D., and DeKosky, S. T., 2003, Human cholinergic basal forebrain: chemoanatomy and neurologic dysfunction, *J. Chem. Neuroanat.* **26**:233–242.

Palkovits, M., 1989, Microdissection in combination with biochemical microassays as a tool in tract tracing, In: Heimer, L., and Zaborszky, L. (eds.), *Neuroanatomical Tract Tracing Methods 2: Recent Progress*, New York: Plenum Press, pp. 299–310.

Paweletz, C. P., Liotta, L. A., and Petricoin, E. F., III, 2001, New technologies for biomarker analysis of prostate cancer progression: laser capture microdissection and tissue proteomics, *Urology* **57**:160–163.

Phillips, J., and Eberwine, J. H., 1996, Antisense RNA amplification: a linear amplification method for analyzing the mRNA population from single living cells, *Methods Enzymol. Suppl.* **10**:283–288.

Reiner, A. J., and Honig, M. G., 2006, Dextran amines: versatile tools for anterograde and retrograde studies of nervous system connectivity, In: Zaborszky, L., Wouterlood, F., and Lanciego, J. L. (eds.), *Neuroanatomical Tract Tracing 3: Molecules-Neurons-Systems*, New York: Springer, 304–335.

Sambrook, J., and Russell, D.W., 2001, *Molecular Cloning: A Laboratory Manual, 3rd edition*, Cold Spring Harbor: Cold Spring Harbor Laboratory Press.

Sarnat, H. B., Curry, B., Rewcastle, N. B., and Trevenen, C. L., 1987, Gliosis and glioma distinguished by acridine orange, *Can. J. Neurol. Sci.* **14:**31–35.

Schmued, L. C., and Heimer, L., 1990, Iontophoretic injection of fluoro-gold and other fluorescent tracers, *J. Histochem. Cytochem.* **38:**721–723.

Simone, N. L., Remaley, A. T., Charboneau, L., Petricoin, E. F., 3rd, Glickman, J. W., Emmert-Buck, M. R., Fleisher, T. A., and Liotta, L. A., 2000, Sensitive immunoassay of tissue cell proteins procured by laser capture microdissection, *Am. J. Pathol.* **156:**445–452.

Su, J. M., Perlaky, L., Li, X. N., Leung, H. C., Antalffy, B., Armstrong, D., and Lau, C. C., 2004, Comparison of ethanol versus formalin fixation on preservation of histology and RNA in laser capture microdissected brain tissues, *Brain Pathol.* **14:**175–182.

Suarez-Quian, C. A., Goldstein, S. R., Pohida, T., Smith, P. D., Peterson, J. I., Wellner, E., Ghany, M., and Bonner, R. F., 1999, Laser capture microdissection of single cells from complex tissues, *Biotechniques* **26:**328–335.

Tang, W. X., Fasulo, W. H., Mash, D. C., and Hemby, S. E., 2003, Molecular profiling of midbrain dopamine regions in cocaine overdose victims, *J. Neurochem.* **85:**911–924.

Tecott, L. H., Barchas, J. D., and Eberwine, J. H., 1988, In situ transcription: specific synthesis of complementary DNA in fixed tissue sections, *Science* **240:**1661–1664.

Tkatch, T., Baranauskas, G., and Surmeier, D. J., 2000, Kv4.2 mRNA abundance and A-type K(+) current amplitude are linearly related in basal ganglia and basal forebrain neurons, *J. Neurosci.* **20:**579–588.

To, M. D., Done, S. J., Redston, M., and Andrulis, I. L., 1998, Analysis of mRNA from microdissected frozen tissue sections without RNA isolation, *Am. J. Pathol.* **153:**47–51.

Tomita, H., Vawter, M. P., Walsh, D. M., Evans, S. J., Choudary, P. V., Li, J., Overman, K. M., Atz, M. E., Myers, R. M., Jones, E. G., Watson, S. J., Akil, H., and Bunney W. E. Jr., 2004, Effect of agonal and postmortem factors on gene expression profile: quality control in microarray analyses of postmortem human brain, *Biol. Psychiatry.* **55:**346–352.

Topaloglu, H., and Sarnat, H. B., 1989, Acridine orange-RNA fluorescence maturing neurons in the perinatal rat brain, *Anat. Rec.* **224:**88–93.

Van Deerlin, V. M. D., Ginsberg, S. D., Lee, V. M.-Y., and Trojanowski, J. Q., 2000, Fixed post mortem brain tissue for mRNA expression analysis in neurodegenerative diseases, In: Geschwind, D. H. (ed.), *DNA Microarrays: The New Frontier in Gene Discovery and Gene Expression Analysis*, Washington, DC: Society for Neuroscience, pp. 118–128.

Van Deerlin, V. M. D., Ginsberg, S. D., Lee, V. M.-Y., and Trojanowski, J. Q., 2002, The use of fixed human post mortem brain tissue to study mRNA expression in neurodegenerative diseases: applications of microdissection and mRNA amplification, In: Geschwind, D. H., and Gregg, J. P. (eds.), *Microarrays for the Neurosciences: An Essential Guide*, Boston: MIT Press, pp. 201–235.

VanGelder, R., von Zastrow, M., Yool, A., Dement, W., Barchas, J., and Eberwine, J., 1990, Amplified RNA (aRNA) synthesized from limited quantities of heterogeneous cDNA, *Proc. Natl. Acad. Sci. U.S.A.* **87:**1663–1667.

Vincent, V. A., DeVoss, J. J., Ryan, H. S., and Murphy, G. M., Jr., 2002, Analysis of neuronal gene expression with laser capture microdissection, *J. Neurosci. Res.* **69:**578–586.

von Bertalanffy, L., and Bickis, I., 1956, Identification of cytoplasmic basophilia (ribonucleic acid) by fluorescence microscopy, *J. Histochem. Cytochem.* **4:**481–493.

Wang, E., Miller, L. D., Ohnmacht, G. A., Liu, E. T., and Marincola, F. M., 2000, High-fidelity mRNA amplification for gene profiling, *Nat. Biotechnol.* **18:**457–459.

Yao, P. J., O'Herron, T. M., and Coleman, P. D., 2003, Immunohistochemical characterization of clathrin assembly protein AP180 and synaptophysin in human brain, *Neurobiol. Aging* **24:**173–178.

Zhao, X., Lein, E. S., He, A., Smith, S. C., Aston, C., and Gage, F. H., 2001, Transcriptional profiling reveals strict boundaries between hippocampal subregions, *J. Comp. Neurol.* **441:**187–196.

Zoccarato, F., Cavallini, L., and Alexandre, A., 1999, The pH-sensitive dye acridine orange as a tool to monitor exocytosis/endocytosis in synaptosomes, *J. Neurochem.* **72:**625–633.

5

Molecules and Membrane Activity: Single-Cell RT-PCR and Patch-Clamp Recording from Central Neurons

WILLIAM H. GRIFFITH, SUN-HO HAN,
BRIAN A. McCOOL, and DAVID MURCHISON

WILLIAM H. GRIFFITH AND DAVID MURCHISON • Department of Medical Pharmacology and Toxicology, College of Medicine, Texas A&M University System Health Science Center, College Station, TX 77843–1114   SUN-HO HAN • Department of Pharmacology, University of California, Irvine, CA 92697–4625   BRIAN A. McCOOL • Department of Physiology and Pharmacology, Wake Forest University School of Medicine, Winston-Salem, NC 27157–1083

Abstract: This chapter summarizes methods for characterizing mRNA expression
and electrophysiological properties of central neurons using patch-clamp recording and single-cell reverse-transcription/polymerase chain reaction (scRT-PCR). A
simple scRT-PCR protocol can be used to identify neurons by the expression of
phenotypic marker mRNAs. The combination of these methods allows for the correlation of functional properties with molecular expression. Somewhat more complex
methods are available for quantitation of mRNA expression. Both traditional gel-
based PCR identification and real-time fluorescent PCR identification methods can
be employed. Advantages and requirements of various methods are discussed. Different types of tissue preparations are presented with emphasis on methods used
in our laboratories for acutely dissociated or cultured basal forebrain and amygdala
neurons. The basal forebrain contains a heterogeneous population of cholinergic
and GABAergic neurons, while the amygdala displays neurons with a complex receptor subunit composition. Investigation of neurons with this type of molecular
diversity benefits from techniques such as scRT-PCR for cell identification. We also
illustrate how these PCR methods can be combined with more complex experimental protocols, such as calcium buffering measurements using fluorescent dyes in
dissociated neurons from aged animals. The capacity to combine scRT-PCR with a
variety of experimental protocols allows the identification of unique cell types and
relationships between physiology and gene expression.

Keywords: aging, amygdala, basal forebrain, calcium-binding proteins, ChAT, fura-2
fluorescence, GAD, ion channel subunits, receptor subunits

I. INTRODUCTION

Single-cell reverse-transcription/polymerase chain reaction (scRT-PCR)
and antisense RNA amplification methodologies have generated a great
deal of excitement in the field of neuroscience because molecular expression profiles can now be determined in conjunction with a variety of experimental applications. Some of these approaches include combining electrophysiological recording and mRNA analysis (Dixon *et al.*, 2000; Eberwine
et al., 1992; Hinkle *et al.*, 2004; Lambolez *et al.*, 1992; Monyer and Lambolez,
1995; Sucher *et al.*, 2000; Surmeier *et al.*, 1996), extensive gene profiling of
single cells using macro- and microarrays (Eberwine *et al.*, 2001; Hinkle *et al.*,
2004), and analyzing changes in single-cell gene expression during disease
states, such as Alzheimer's disease (Chow *et al.*, 1998; Mufson *et al.*, 2002).
In addition, heterogenous cell types can be identified from seemingly homogeneous cell populations (Eberwine *et al.*, 2001; Elowitz *et al.*, 2002; Han

et al., 2002; Monyer and Markram, 2004; Pape *et al.*, 2001). Combinatorial approaches with molecular, genetic, and functional experimental paradigms are beginning to describe an almost unimaginable neuronal diversity and to reveal the challenges facing those who would like to understand the functional significance of that diversity.

Improved access to molecular methods has allowed electrophysiologists, among others, to greatly increase the resolving power of their experiments by correlational analysis of functional and molecular data from individual cells. In particular, scRT-PCR techniques have been used as a relatively easy and inexpensive way to examine the selected molecular expression patterns of neurons. For example, many early experiments correlated the functional properties of ion channels with specific mRNA expression patterns in the brain because, in most cases, the subunit compositions of native ligand- and voltage-gated ion channels were unknown. Some of these studies included glutamate receptors (Audinat *et al.*, 1996; Bochet *et al.*, 1994; Jonas *et al.*, 1994; Lambolez *et al.*, 1992), GABA receptors (Ruano *et al.*, 1997; Santi *et al.*, 1994), dopamine receptors (Surmeier *et al.*, 1996), and voltage-dependent potassium (Martina *et al.*, 1998; Song *et al.*, 1998) and calcium channels (Bargas *et al.*, 1994; Plant *et al.*, 1998). The adaptability of the scRT-PCR technique makes it amenable to a variety of experimental approaches, while providing a level of molecular information intermediate between histochemistry (see Stornetta and Guyenet, this volume) and expression array technologies (see Ginsberg *et al.*, this volume).

Because of the very small amount of genetic material in a single cell, some type of amplification is necessary in order to detect gene expression. In this review, we will discuss several methods, but we will focus on a simplified scRT-PCR protocol that we use in our laboratory to collect molecular data from acutely dissociated or cultured rat basal forebrain or amygdala neurons that have been electrophysiologically characterized by patch-clamp recording. Theoretically, the enzyme reverse transcriptase makes one or a few cDNA copies of each expressed mRNA in a cell. These cDNAs are less prone to degradation than are the original transcripts and can be amplified by PCR. Each cDNA of interest is targeted by a sequence-specific primer and is copied by a thermostable DNA polymerase. Temperature cycling produces a controlled exponential amplification of the cDNA, as each subsequent copy increases the number of templates. After a certain amount of amplification, the PCR reaction products become detectable. We will describe both traditional gel-based detection and the more recently developed real-time fluorescent detection, as well as quantitative considerations.

As with any investigative technique, there are limitations and assumptions associated with the interpretation of data from scRT-PCR experiments that must be taken into account. We will discuss these as they relate to our efforts to identify neuronal cell types and detect channel or receptor subunit composition in functionally characterized cells. Cell identification is particularly important in parts of the brain with diverse neuronal populations. For example, the basal forebrain contains both cholinergic and GABAergic neurons (Panula *et al.*, 1984; Rye *et al.*, 1984; Sarter and Bruno, 2002) that innervate the hippocampus, olfactory cortex, and cerebral cortex (Fibiger, 1982; Mesulam *et al.*, 1983; Zaborszky *et al.*, 1986). These cells have been implicated in cognitive processes, such as attention and some forms of memory (Bartus *et al.*, 1982; Olton *et al.*, 1991; Sarter and Bruno, 2000), and changes occur in these cells with age and Alzheimer's disease (Chow *et al.*, 1998; Coyle *et al.*, 1983; Decker, 1987; Fischer *et al.*, 1989; Mufson *et al.*, 2002). Basal forebrain neurons have proven to be a valuable model for investigating changes in ion channel function and calcium homeostasis during aging (Griffith *et al.*, 2000). In contrast, neurons in the lateral/basolateral amygdala are phenotypically more homogeneous and consist of principal glutamatergic neurons and GABAergic interneurons (McDonald, 1985; McDonald *et al.*, 1989). This brain region is intimately associated with the regulation of emotional behaviors like anxiety or fear (Fanselow and LeDoux, 1999; Killcross *et al.*, 1997) and plays a central role in drug-seeking behaviors (See *et al.*, 2003). Single-cell RT-PCR studies have revealed a surprisingly complex pattern of gene expression within individual glutamatergic or GABAergic lateral/basolateral amygdala neurons (Floyd *et al.*, 2003; McCool and Farroni, 2001). Most of the examples we use to illustrate this review were acquired from these systems.

II. METHODOLOGICAL CONSIDERATIONS

A. Types of RNA Amplification for Single-Cell PCR

Due to its low abundance, mRNA must be amplified in order to be detected in single neurons. Several strategies to detect mRNA expression in single cells have been devised, but all of them depend on reverse transcriptase to convert the original mRNA transcripts into cDNAs. The major differences in the techniques concern the amplification procedures, the types of primers, and the detection methods. The antisense RNA differs from others in that the amplification is accomplished by T7 RNA polymerase rather than by PCR. This method does not require any PCR for the detection of amplification products, but it can be used in conjunction. The advantages of this approach are that the amplification of the mRNA is a linear, and therefore readily quantifiable, process and that all of the transcript species present can be amplified simultaneously (Eberwine *et al.*, 1992, 2001; Hinkle *et al.*, 2004). Consult the chapter by Ginsberg *et al.* in this volume for a more thorough description of linear RNA amplification techniques. Other protocols are based on RT-PCR and feature exponential amplification of cDNA. These include 3′-end amplification followed by gene-specific PCR (Dixon *et al.*, 1998, 2000), or two-round PCR with nonspecific primers designed for conserved regions of closely related genes for first round, followed by second-round PCR with either specific primers or restriction enzyme detection (Audinat *et al.*, 1996; Lambolez *et al.*, 1992; Plant *et al.*, 1998). Another variation involves multiplex PCR where amplification is conducted with several specific primer pairs simultaneously (Edwards and Gibbs, 1994; Lindqvist *et al.*, 2002; Phillips and Lipski, 2000). Finally, Surmeier and colleagues (Surmeier *et al.*,

1996; Tkatch *et al.*, 1998; Yan and Surmeier, 1996) have used a method in which the cDNA yield from the RT is divided for separate amplification of each target sequence.

There are numerous possible modifications to the above methods, many of which have been reviewed for application to single cells (Audinat *et al.*, 1996; Dixon *et al.*, 2000; Eberwine *et al.*, 2001; Hinkle *et al.*, 2004; Monyer and Lambolez, 1995; Phillips and Lipski, 2000; Sucher *et al.*, 2000; Surmeier *et al.*, 1996). The important point is that various methods are amenable to a particular experimental priority or design. For example, if hundreds of genes are being investigated, then global RNA amplification using the non-PCR-based method of Eberwine *et al.* (1992) is desirable. However, this method is laborious and technically challenging and may not be practical for all gene expression studies. The different PCR protocols have the general advantage of amplification specificity and sensitivity. Problems can be encountered though, when attempting to optimize conditions for successful multiplex PCR with several primer pairs or when many cycles of amplification are employed. On the other hand, a straightforward RT-PCR protocol can permit the reliable detection of a small number of moderately abundant transcripts without requiring extra amplification steps. We use a relatively simple "one-round" single-cell RT-PCR protocol (modified from Surmeier *et al.*, 1996) to detect transcripts for phenotypic markers, receptor/channel subunits, and calcium-binding proteins (CaBP) in neurons of the basal forebrain and amygdala that have been functionally characterized. This allows the correlation of functional properties with molecular expression and the identification of different cell types.

Figure 5.1 gives an overview of our protocol for scRT-PCR of acutely dissociated or cultured neurons using random hexamer primers for the RT and specific forward and reverse primers for PCR of six target cDNAs. It should be noted that the specific conditions for optimizing the RT-PCR reactions were determined from work with known quantities of purified RNA or cloned DNA before being applied to single cells (discussed below, along with primer design). One of the target cDNAs is a positive control, which is usually chosen as an abundant transcript present in all neurons. We use the enzyme glyceraldehyde 3-phosphate dehydrogenase (GAPDH) as a positive control. For phenotypic markers, we use choline acetyltransferase (ChAT) and glutamic acid decarboxylase (GAD). This typically enables us to investigate the co-expression of three (four at most) other transcripts of interest. We have successfully detected expression of CaBP, voltage-gated calcium and sodium channel subunits, and NMDA and glycine receptor subunits. More detailed information for each step illustrated in Fig. 5.1 is provided in the following sections.

B. Harvesting Single Cells for Molecular Analysis

The first and most important aspect of an experiment designed to correlate physiological function with molecular expression in individual neurons

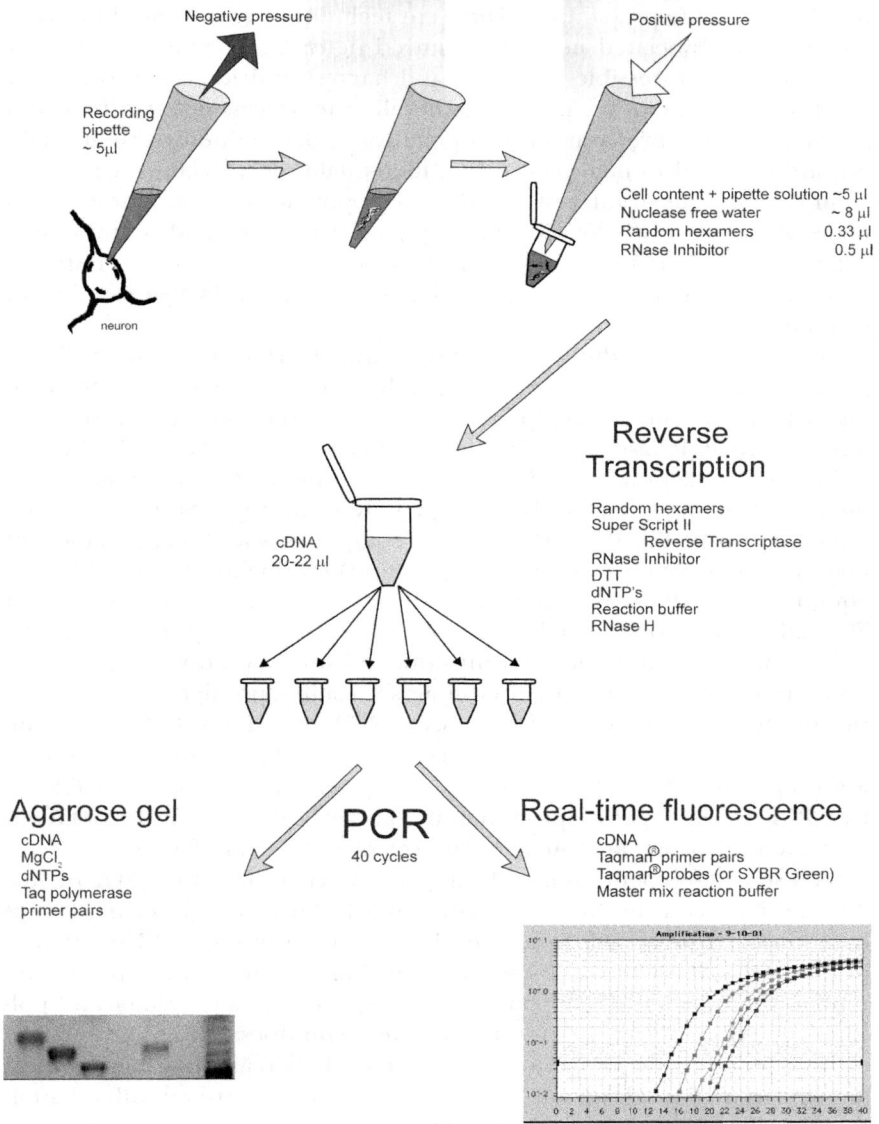

Figure 5.1. Diagram of the process for cell harvesting, reverse transcription (RT), and polymerase chain reaction (PCR). Top panel shows aspiration of a neuron into the recording pipette. RT was performed with random hexamers and Superscript® II reverse transcriptase. The cDNA yield was divided into six portions and subjected to 40 PCR cycles using a PTC-100TM Programmable Thermal Controller (MJ Research, Inc.) and agarose gel identification, or 40 PCR cycles via real-time PCR with an ABI Prism 7700 (Applied Biosystems) and fluorescent identification. Taq polymerase and specific primer pairs for the transcripts of interest were used for amplification (DTT, dithiothreitol; dNTP, deoxynucleotide triphosphate).

is to have a viable preparation. Different techniques are required to maintain acutely dissociated neurons, cultured neurons, or acute or cultured brain slices. It is possible to record and harvest individual neurons from all of these preparation types, and generally, the process of harvesting cells is similar for any preparation. It is practical to determine first what methods are required to make successful, high-quality patch-clamp recordings from the neurons of interest and then to apply the additional procedures necessary for RT-PCR. We will be focusing on dissociated and cultured neuronal preparations in the sections below, but a fine coverage of special methods for patch clamping in brain slices is provided by Petersen in this volume.

The most obvious difference between initial preparation for ordinary patch-clamp recording and recording to be followed by RT-PCR is the care taken to assure the patch-pipette, intracellular solution and RT reaction mixture are all RNase free. We use dedicated stocks of chemicals and materials that are kept separate from ordinary supplies. PCR-dedicated equipment, such as pipetters, are also kept separately. Latex gloves are worn when handling any of the RT-PCR-dedicated components and manipulation of these materials (such as solution preparation) is done inside a sterile hood equipped with UV lights. Gloves and all external surfaces are washed with 70% ethanol or commercially available RNase-inactivating solutions (e.g., RNaseZAP®, Ambion) prior to introduction into the hood. If gloves are powdered, some effort is required to wash residual powder from the exterior of the glove, especially if fluorescent PCR detection is to be used (the powder is usually highly fluorescent corn starch). Pipette and RT solutions are prepared with nuclease-free water and sterilized (autoclaved) RNase-free PCR tubes. Pipette tips containing an aerosol filter can be obtained sterile and nuclease- and nucleic acid-free from the manufacturer.

We use patch-pipettes pulled from glass (Garner Glass Co., 7052 or KG-33) that has been sterilized by baking (400°F, 2 h). This glass can be kept in a sterilized petri dish inside the UV hood. We prefer a filament type of glass (KG-33) due to the ease and reliability of filling the tip with the pipette solution without air bubble blockage. The pipette puller and polisher are washed with 70% ethanol-soaked Kimwipes® (Kimberly–Clark) without touching the actual heating elements. Pulled and polished pipettes are kept in dedicated storage jars that are periodically washed with ethanol. It is desirable to have the aperture of the patch-pipette tip as large as is consistent with obtaining stable recordings. Pipette tip size varies with cell type but is generally between 0.5 and 2.0 μm. Aliquots of intracellular pipette solution can be kept frozen in sterile nuclease-free tubes in a dedicated container until use. We have successfully used pipette solutions containing CsCl, cesium acetate, cesium methylsulfonate, or potassium gluconate as the principal salts. Other common ingredients, such as EGTA, TEA, NaCl, HEPES, etc., do not appear to affect the RT-PCR, and we have not had to alter the composition of our pipette solutions from what we use without RT-PCR. RNase inhibitor can be added to the pipette solution for added assurance against degradation of RNA. This precaution might be worthwhile in the cases of prolonged recording or low-abundance target transcripts.

A sterile plastic syringe (1 ml) can be fabricated as a sterile pipette filler by removing the plunger and melting the distal end over a small Bunsen burner and allowing the end to drop off under gravitational pull in a vertical orientation. When done correctly, this results in a narrowly tapered hollow tip that can be cut to a desired length with a sterile scalpel blade. The "trick" is to rotate the syringe for even heating directly in the flame (flame < 2 in. high and < 1 in. wide) until just before the plastic sags. These fillers are then disposable and inexpensive. In practice, it is not possible to have a sterile extracellular bath or recording chamber, because, of course, the preparation itself is not sterile. We do try, however, to keep the recording chamber, inflow lines, and glassware clean by washing them in 70% ethanol after each experiment. Efforts are made to keep dust to a minimum in the recording room. Extracellular solutions contain sugars and so should be made fresh daily, although we have successfully used solutions for several days if kept refrigerated overnight. Parts of the recording setup that can be safely wiped with ethanol (nonelectrical parts) are cleaned before each recording session. The investigator wears gloves throughout the process and keeps a squirt bottle of 70% ethanol handy for frequent hand rinsing.

The internal electrode wire should be rinsed or dipped in 10% bleach (sodium hypochlorite) and allowed to dry prior to each individual recording. Drying can be hastened by following the bleach with ethanol or by using compressed gas. A small positive pressure should be applied to the patch-pipette before lowering the tip below the bath surface and until the target cell is contacted. Avoid negative pressure except for seal formation, patch rupture, and cell aspiration.

Once the electrophysiological data have been gathered, the cell is harvested for RT by one of several means, depending on the experimental arrangement and preparation. We are most experienced with acutely dissociated or cultured basal forebrain neurons. Acutely dissociated basal forebrain neurons consist of the cell soma ($\sim$7–30-μm-long axis in rat) and the proximal processes (up to 60 μm long) that are superficially stuck to the bottom of the glass-recording chamber (Fig. 5.2A1). These cells can be lifted entirely from the bottom (Fig. 5.2A2), and thus away from other cells or cellular debris, while maintaining the patch seal. If the aperture of the pipette is large enough ($\sim$2.5 μm), strong negative pressure can be applied, and the entire cell can be carefully aspirated into the recording pipette (Fig. 5.2A3). As soon as the cell is inside, the pipette is quickly withdrawn from the bath. Often though, the pipette aperture is just small enough that the cell membrane and nucleolus combine to form an immovable plug. In this case, the pipette with attached cell is moved to an area of the bath that is visibly clear of cells and debris, and under fine manipulator control the tip is gently pushed against the bottom of the chamber to nudge the plug into the pipette. If this does not work, a slightly more vigorous push will partially break the tip and allow the cell to enter the pipette. Skillful breaking of

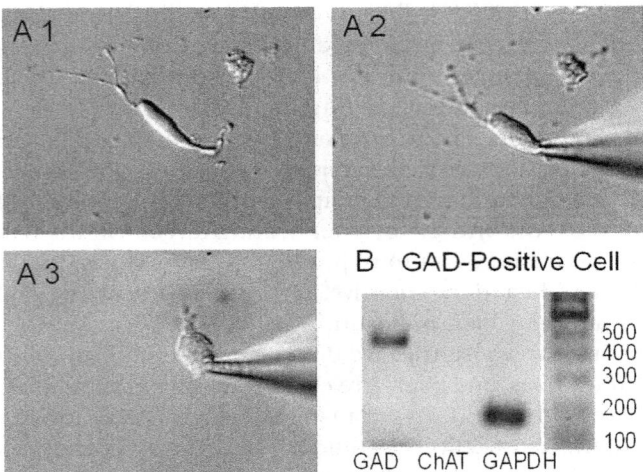

Figure 5.2. Single-cell RT-PCR from a GAD+ basal forebrain cell. (A1–A3) Aspiration of a basal forebrain neuron into a sterile patch-pipette is shown in sequence. Cell soma ~20μm on the long axis. Note that the cell was lifted off the bottom of the recording chamber before aspiration to ensure that only a single cell was collected. (B) Photomicrograph of the RT-PCR products from the same cell separated on an ethidium bromide-stained 2% agarose gel. The left lane shows a band of GAD67 mRNA, whereas no ChAT mRNA was detected. Glyceraldehyde-3-phosphate dehydrogenase (GAPDH) mRNA was used as a positive cell marker. Molecular weight markers are shown to the right (GAD, glutamate dehydrogenase; ChAT, choline acetyltransferase). (From Han *et al.*, 2002, with permission.)

the tip and coordinated application of negative pressure are necessary to prevent aspiration of bath solution or debris. If another cell, or even process is even partially aspirated, the pipette is discarded. We have aspirated both bath solution and noncellular debris to process as negative controls.

For cultured neurons, the concern for loose debris or cells is minimal because of the strong adhesion of the cells to the substrate. However, this makes it more difficult to aspirate an entire cell. This feature has been exploited by Eberwine and colleagues (Crino and Eberwine, 1996; Miyashiro *et al.*, 1994), who were able to aspirate portions of dendrites and show that there is some differential distribution of mRNAs between soma and processes. Therefore, to obtain an accurate profile of the transcripts present in a neuron, as much of the cell should be sampled as possible. In neuronal cultures that are moderately dense, it is not practical to try to aspirate a cell soma and neurites because of the numerous intercalated processes of neighboring cells. We have used a "vacuuming" method to aspirate the cell soma in an intact or slightly broken tip as shown in Fig. 5.3. Slightly moving the tip peels the soma off the substrate without the processes. While dissociated neurons are probably the easiest to harvest in relative completeness, Landfield and colleagues (Chen *et al.*, 2000) have shown that almost complete hippocampal CA1 neurons can be harvested from a "zipper" slice

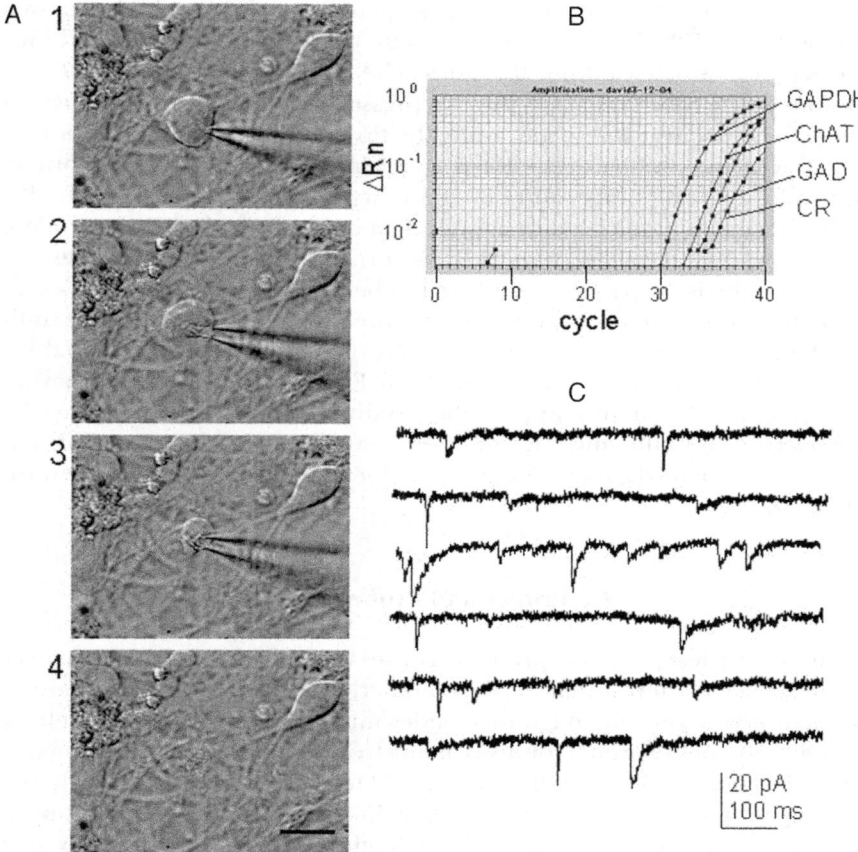

Figure 5.3. Combined patch-clamp recording of spontaneous synaptic currents and scRT-PCR from a ChAT+/GAD+ basal forebrain cell in culture. (A1–A4) Aspiration of the cell into the recording pipette is shown in sequence. Note that only the single cell is harvested without nearby debris. (B) Real-time RT-PCR amplification plot of transcript expression in the same cell. Expression of GAPDH, ChAT, GAD67, and the calcium-binding protein calretinin (CR) were detected. Calbindin and parvalbumin were not detected in this neuron. (C) Spontaneous miniature synaptic currents recorded from the same cell in the presence of TTX (0.5 µM). Scale bar: 20 µm.

preparation. In this case, the recording pipette is used to extract the cell from the slice before it is transferred to the mouth of a larger collection pipette that is used to aspirate the neuron. In other preparations though, it might not be possible to extract any of the cells reliably, and researchers must be content to aspirate a portion of the somatic cytoplasm (Pape *et al.*, 2001; Sucher and Deitcher, 1995).

Once the cell has been aspirated, we remove the pipette from the electrode holder and transfer it to an expelling apparatus, consisting of a holding platform for an RT tube and a syringe with a flexible plastic tube attached that has been washed with ethanol. The open end of the tube is placed over the butt-end of the pipette and the pipette tip is carefully introduced into the RT tube without hitting the sides. The tip is gently crushed into the bottom of the tube and slight positive pressure is applied to the syringe to expel the contents while withdrawing the tip from the initial RT solution so as not to create bubbles. The RT tube is immediately sealed, spun briefly in a low-speed centrifuge, and placed on dry ice. For our protocol, the RT tube contains 8 µl nuclease-free water, 1 µl dithiothreitol (DTT, 100 mM), 0.33 µl random hexamer primers (3 µg/µl), and 0.5 µl RNase inhibitor. A separate tube is prepared for each cell to be collected. These tubes can be prepared the day before and frozen, and then thawed and kept on ice until needed. Other arrangements are possible, such as including the initial RT ingredients in the pipette (Cao *et al.*, 1996; Eberwine *et al.*, 1992) or having the complete RT mixture in the tube (Audinat *et al.*, 1996). We freeze the harvested cells in the initial RT solution ($-80°C$), and conduct the RT and PCR within the next 3 days. Our protocol for RT is given in section "Reverse Transcription."

C. Primer and Probe Design

Successful design of PCR primers involves the identification of sequences in the gene of interest that fulfill several criteria. Among the most important considerations, the oligonucleotides must possess a favorable melting temperature of the primer for dissociating from its complimentary sequence to insure compatibility with the type of PCR to be performed. Generally, the melting temperature should be between 55 and $70°C$. This temperature is dependent on the content of G/C nucleotides, which should not exceed 55%. All primers that are to be used in the same PCR should have very similar melting temperatures so that the temperature parameters of the PCR can be optimized. Primers should be designed to avoid appreciable sequence-specific secondary structures like hairpin- or stem-loops. Additionally, extensive complimentary sequences between different primers can sequester these reagents away from more productive interactions with the template. Ultimately, both inter- and intraprimer interactions can interfere with the efficient PCR product formation that is essential for amplification of cDNAs derived from individual neurons. Fortunately, most software packages utilized for primer design automatically select primer sets using criteria that insure that each primer possesses optimal characteristics.

For single-cell RT-PCR, there are several additional concerns that should be addressed when designing PCR primers. First, one must consider that different means of analyzing PCR product formation will determine optimal primer selection. Single-cell PCR products that are analyzed using agarose gel electrophoresis must be large enough (> 100 bp) to distinguish from "primer dimers" that are common to single-cell approaches but should be small enough (typically < 500 bp) to insure that the target cDNA is well represented in the population of reverse transcription products. This is perhaps less of a concern when using RNA amplification approaches, but it still deserves some attention for low-abundance messages. Regardless, standard gel-based analysis often requires only forward and reverse primers, and these can be identified with publicly available software, like the San Diego Supercomputer Center's "Biology Workbench" (http://workbench.sdsc.edu/, Subramaniam, 1998). In contrast to gel-based methods of product detection, "real-time" analysis employs an additional fluorescent gene-specific oligonucleotide probe that, especially when used with the 5′-exonuclease assay (Whitcombe et al., 1998), provides exquisite sensitivity for low-abundance products. However, this approach has very specific requirements of the PCR primers, probes, and products. For example, the PCR product is typically ~100 bp since the annealing/extension phases are performed simultaneously at a temperature (60–65°C) that is suboptimal for most thermostable polymerases. In addition, the primer/probe melting temperatures and their sequence content have more specific requirements that make selection of appropriate target sequences in the gene of interest more demanding. For primer/probe design, we have had the most success employing commercial software specifically designed for a particular real-time PCR platform, such as Primer Express® (Applied Biosystems Inc.) for Taqman®-based probe detection.

Primer specificity is a topic worth special consideration. It must be determined that primers chosen for a given gene are highly specific for the product of interest. For example, it may be necessary that PCR products arising from cDNA be differentiated from those representing genomic DNA. For single neurons, genomic sources are unlikely to influence product formation (Johansen et al., 1995). However, in cases where message abundance is low and multiple rounds of PCR are necessary, primers can be designed to span multiple intron/exon boundaries. This does not necessarily require a direct knowledge of genomic structure in the specific species being used, given that, for the vast majority of genes (> 99%) the position of a given intron relative to coding region exons is highly conserved between humans, rats, and mice (Roy et al., 2003). Another specific concern, especially for neurotransmitter receptors, is that many mRNAs are highly similar to other gene products that arise either by transcriptional processing (e.g., splicing of alternative exons) or by gene duplication, giving rise to multiple, highly homologous subunits. In the former case, constraining at least one primer-binding site within an alternative exon can provide an efficient means of differentiating splice variants. Similarly, sequence comparisons between homologous subunits can identify highly conserved regions (> 85% identity/20 bp) that can be omitted from consideration when identifying promising target sequences. In many cases, nonhomologous regions between related subunits are often concentrated in 5′- and 3′-noncoding regions of the mRNA which are often more polymorphic than are coding regions. This may require precise knowledge of the target sequence in the particular species of interest. The sequence of a primer or probe can be checked for specificity after design by comparing it to the sequences in a gene database.

D. Gel-Based Identification Versus Real-Time
Fluorescent PCR Analysis

The simplified aspect of our RT-PCR protocol is that there is only a single round of PCR cycles for amplification. This limits the amount of amplification product available for detection, but reduces the possibility of a false or irrelevant positive amplification which increases with additional cycles of PCR. By using serial dilutions of known quantities of purified RNA, we were able to detect specific mRNAs from samples of as little as 2 pg/μl total RNA after 40 cycles of PCR by gel electrophoresis. Real-time fluorescence PCR was even more sensitive, as detailed below. It has been estimated that single neurons contain ~50 pg total RNA (Sucher *et al.*, 2000), so there should be no problem of starting with too little mRNA to detect. Detection was confirmed in gels by visual inspection and the separate identical scoring by two of three investigators. For real-time PCR detection, the amplification plot of the increase in fluorescence had to cross a threshold in the log-linear range and above the noise before the 39th cycle. These rather stringent requirements along with suitable controls (see section "Quantitation and Validation Experiments") insure that there is little likelihood of false positive amplification by this protocol. There is more chance of a false negative amplification with this protocol perhaps, than with others that involve more amplification, but some measures can be taken to increase the confidence in a negative result, as discussed below.

1. Agarose Gel PCR

Following RT, targeted cDNAs were amplified with specific primer pairs referenced in section "Listing of Primers and Probes." Forty cycles of PCR reactions generated products that were verified by gel electrophoresis. One drop of sterile mineral oil (Sigma) was added to each tube to cover the reaction mix and prevent evaporation and changes in salt concentrations during PCR. Once the tubes were placed in the PCR instrument (PTC-100TM Programmable Thermal Controller, MJ Research, Inc.), the temperature was elevated to 94°C for 2 min. During this period, Taq DNA polymerase 2.5 U (Promega) was added to enable a "hot start," which reduces the possibility of nonspecific amplification and primer dimers. The tubes then went through 40 cycles of three steps: 94°C (1 min), 50°C (1 min), and 72°C (1.5 min). After completion of all the cycles, the temperature was maintained at 4°C until gel electrophoresis.

The PCR product was extracted from the mineral oil by mixing with 100 μl of chloroform/isoamyl alcohol (24:1) and vortexing. The upper aqueous phase contained the PCR product and this was placed in a new tube mixed with the gel loading dye (Promega). The PCR products were loaded and ran for 2 h on a 1.5–2% agarose gel (Sigma) at 70 V (Horizontal Gel Electrophoresis System; Gibco BRL). The voltage and duration of the run are dependent on the size of the PCR products. Products were visualized by ethidium bromide staining and images were made using a digital camera system (MultiImage™ Light Cabinet, Alpha Innotech Corporation). Primer specificity was tested by RT-PCR of total RNA extracted from whole rat brain or basal forebrain tissue sections. Gels of the PCR products were seen to contain single bands at the expected molecular weights.

2. Semiquantitative Real-Time PCR (Taqman® Probes and SYBR® Green)

A relatively new method is available to quantitate PCR products using real-time changes in fluorescent intensity during each PCR cycle (Gibson *et al.*, 1996; Heid *et al.*, 1996). We currently use this method for our single-cell studies because of the increased throughput, enhanced sensitivity, and potential for quantitation. Because of the advantages, an emphasis will be given to this technique compared to gel-based PCR. Figure 5.4 shows an

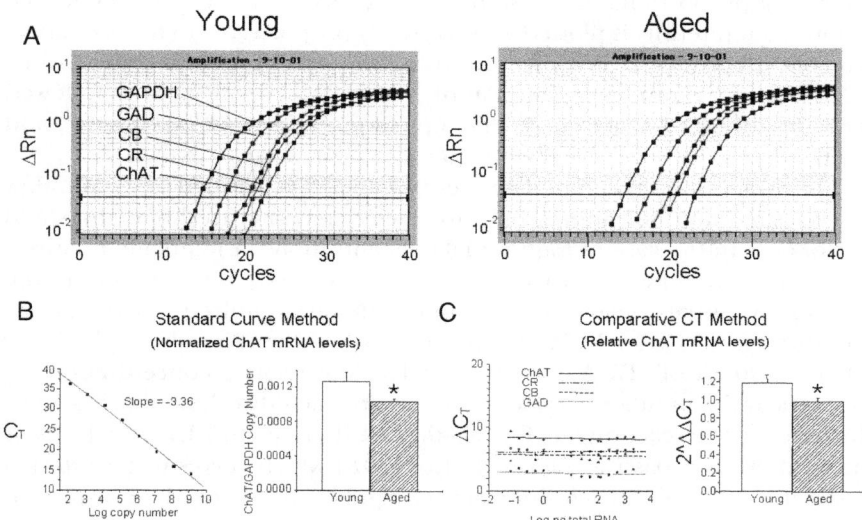

Figure 5.4. Semiquantitative RT-PCR. (*A*) Amplification plots of real-time PCR from young and aged total mRNA from rat basal forebrain; GAPDH, ChAT, GAD, CB, and CR expression are shown. Threshold cycle (C_T) for each transcript is shown as the intersection of the solid line with the fluorescent intensity plot. (*B*) Standard curve method for quantitation. A slope of −3.4 indicates a near 100% efficiency of the PCR reactions. ChAT values are normalized to GAPDH in both young ($n = 5$) and aged ($n = 5$) rats (* $p < 0.05$). (C) Comparative C_T method for quantitation. The validation experiment is shown to the left and plots the difference (ΔC_T) between the transcript and the normalizer (GAPDH). The slopes are < 0.1 and demonstrate that the efficiencies of the target and reference are approximately constant. The plot to the right shows the difference between the calibrator and the ΔC_T ($\Delta \Delta C_T$) for the same data shown in (A) (* $p < 0.05$).

amplification plot of the increase in fluorescence versus cycle number. In this example, PCR is performed with fluorescent-labeled probes. All the primers and probes were designed using Primer Express® software and their PCR products are approximately 50–150 bp, which are much smaller than products generated for regular gel-based PCR. Forward and reverse primers serve the same function as in gel-based detection, but an additional 25–30 bp Taqman® probe is utilized (see section "Listing of Primers and Probes"). The probe is designed to be complimentary to a sequence located between the forward and the reverse primers. Taqman® probes are constructed with a reporter dye (FAM, 6-carboxylfluorescein) at the 5′-end and the fluorescence quencher dye (TAMRA, 6-carboxytetramethyl-rhodamine) located at the 3′-end. The close contact between the two dyes is responsible for the fluorescence quenching. However, as Taq extends DNA synthesis from a PCR primer, it runs into the probe and displaces the probe's 5′-end. The resulting "Y" structure makes the 5′-end of the probe a substrate for the polymerase's innate 5′ to 3′ exonuclease activity which cleaves the probe, separating the reporter dye from the quencher dye and increasing the fluorescence. As amplification increases and more 5′-reporter dye is liberated, there is a proportional increase in fluorescence. For each PCR cycle, the fluorescent intensity is plotted and is directly proportional to the accumulation of PCR product. The PCR cycles using Taqman® probes included initial steps of 50°C/2 min for optimal AmpErase® UNG enzyme activity, followed by 95°C/10 min for activation of AmpliTaq® Gold DNA polymerase, and then 40 cycles of 95°C/15 s (melt), 60°C/1 min (anneal/extend).

The initial amount of template present can be determined by the number of cycles required for fluorescence to reach detection threshold (threshold cycle, C_T). In the case of high initial concentrations of template, relatively few cycles would be necessary to reach threshold (C_T is lower), but in the case of low initial concentrations, more cycles are required (C_T is higher). Because each PCR cycle (after the first two or three) results in doubling of the amount of cDNA, we can calculate the relative concentrations of samples by the number of cycles necessary to reach threshold fluorescence. Included in the reaction mix for real-time PCR (Taqman® Universal Master mix), there is a passive reference dye, ROX, which serves as an internal control to normalize each fluorescence signal. Therefore, the normalized reporter fluorescence is given by

$$Rn = \frac{\text{Emission intensity of reporter dye (FAM)}}{\text{Emission intensity of passive reference (ROX)}}.$$

This normalization corrects for extraneous fluorescence fluctuation. In addition, fluorescence intensities recorded in early cycles are considered background and are averaged and subtracted from fluorescence levels in later cycles. This is shown in the following equation:

$$\Delta Rn = (Rn+) - (Rn-).$$

Rn+ is the normalized fluorescent intensity in the first nonbackground cycle or beyond, and Rn− is the intensity of sample without template (negative control) or of background cycles. Finally, an amplification plot of ΔRn is graphed against each cycle, as described above.

In a similar manner, SYBER® Green also can be used as the fluorescent indicator. One advantage of SYBER® Green is the lower cost compared to Taqman® fluorescent probes. Amplification plots of PCR products using SYBER® Green can be generated by measuring the increase in fluorescence, resulting from the binding of SYBER® Green to double-stranded DNA. The passive reference molecule is also ROX, because the excitation–emission profile for SYBER® Green is similar to that of the FAM dye. When using SYBER® Green, an additional melting curve is performed to ensure that a single PCR product is produced. When SYBR® Green was used, the same primers for gel-based PCR were also utilized. The reaction mix contained the same ingredients for Taqman®-based detection, except 2× SYBR® Green PCR Master Mix (Promega) was included. The reaction parameters using the SYBR® Green detection protocol included 40 cycles of 95°C/1 min, 50°C/1 min, and 72°C/1.5 min.

Major advantages of real-time PCR include the high throughput and potential for quantitative analysis. We utilize the 96-well format of the ABI Prism 7700 sequence detection system to enable analysis of up to 14 cells (six sequence probes) along with appropriate controls all on one plate. This ability to analyze a relatively large number of cells is a definite advantage when combining both electrophysiology and gene analysis.

Several methods are available to quantitate results from real-time PCR experiments, including the standard curve method and the comparative C_T method. In the standard curve method, the initial concentration (copy number) for each sample is calculated. The comparative C_T method is similar to the standard curve method, except it uses a single point comparison and an arithmetic formula to achieve relative quantification. Figure 5.4 and the section below describe a comparison between these two methods in an experiment to quantitate tissue mRNA levels of the cholinergic marker, ChAT, from both young and aged rats. This example also illustrates a "validation" experiment required for using the comparative C_T method. The validation experiment confirms a linear amplification over an extended RNA concentration for each transcript, thus verifying the efficiency of the PCR. For the standard curve method, we utilized purified cloned DNA fragments that were cut by different restriction enzymes from cloned plasmid DNA. Serial dilutions of each cloned DNA fragment containing from 10^2 to 10^9 copy numbers were used to generate standard curves. The methods for restriction enzyme analysis are described in section "Special Methods for Validation Experiment".

E. Quantitation and Validation Experiments

Our primary use for real-time RT-PCR is to identify phenotypic marker mRNAs in single cells. Although we do not routinely quantitate our single-cell results, the following example summarizes a comparison of basal forebrain tissue mRNA levels between samples from young and aged rats (Fig. 5.4). We utilized both the standard curve and the comparative C_T methods for quantitation. Tissues from five young and five aged Fischer 344 rats were used for total RNA purification. The RNA was quantified by spectrometry after removing genomic DNA. Real-time RT-PCR was conducted on 10 ng samples and amplification plots for young and aged were constructed, as shown in Fig. 5.4A. In both young and aged samples, the GAPDH curve occurs at the earliest cycle number and GAD, calbindin (CB), calretinin (CR), and ChAT curves are detected later. The plateau fluorescence was reached around cycle 30, showing approximately the same level of fluorescence (ΔRn) regardless of sample concentrations. We set the threshold in the log-linear part of the curve where no limiting factors influenced amplification.

From DNA concentrations in each dilution and the molecular weight of PCR products we could calculate DNA copy number for each concentration. For example, cloned ChAT DNA was cut by the restriction enzyme *Pvu*II for a total 753 bp product that contained 323 bp ChAT PCR product sequences and 430 bp product of sections of plasmid DNA. The copy number of ChAT DNA molecules was calculated using the molecular weight of the 753 bp product and Avogadro's number (6.023×10^{23}). For ChAT, logarithmic copy number of initial concentration of DNA versus C_T was plotted as a standard curve (Fig. 5.4B, left panel). The curve for ChAT was fit by linear regression and had a slope of 3.36. According to the manufacturer's specifications, 100% efficiency of PCR reactions occurs with a slope of 3.4. These data suggest that our PCR efficiencies were close to 100%. Experiments were repeated in triplicate, and we used the same threshold fluorescence level that was utilized in standard curve preparation for calculating copy number from samples of young and aged rats. The copy numbers of ChAT were normalized by the copy number of GAPDH, and this ratio was compared in young and aged samples (Fig. 5.4B, right panel). An independent two-tailed *t*-test revealed that ChAT mRNA expression was significantly reduced in aged tissue ($p < 0.05$). Although this absolute standard curve method is a reasonable method for semiquantification analysis, there are several disadvantages, including cost and the number of samples for the standard curve that must be run for each target sequence. We have found this method to be less than ideal for single-cell analysis.

A second method of quantitation is available to overcome some of the disadvantages of the method discussed above: the comparative C_T method. This method utilizes sample normalization and calibration for relative comparisons. In the example above, the C_T value for ChAT is normalized by the C_T value of GAPDH. Samples are then normalized again by a "calibrator" to enable comparisons across reaction plates. We utilized 5 ng of purified whole brain RNA as the calibrator and the calibrator was subjected to the same RT-PCR as the samples on each plate. Because PCR results are an exponential function, normalization was achieved by subtracting C_T values. Normalized C_T (C_T of sample subtracted from C_T of GAPDH) is called the

ΔC_T, and it was further normalized by subtraction from the calibrator ΔC_T value, which produced a $\Delta\Delta C_T$ value. This $\Delta\Delta C_T$ was changed to a linear value by the conversion $2^\wedge \Delta\Delta C_T$ and then compared between each sample. In our example of quantification of tissue RNA, all of the young and aged samples were run on the same plate, and so ΔC_T values of one of the samples were utilized as calibrator and normalized to generate $\Delta\Delta C_T$ for the other samples. Results similar to above were obtained when the data were analyzed using the comparative C_T method. The comparative C_T method demonstrated a significant decrease in ChAT mRNA expression during aging (Fig. 5.4C, right panel, $p < 0.05$).

In order to perform valid comparisons using the comparative C_T method, a validation experiment must be performed. Because the normalizer (GAPDH) expression level may differ in each sample, the PCR amplification efficiency of GAPDH and other targets have to be consistent for different sample concentrations. This validation experiment is shown in Fig. 5.4C (right panel). A known quantity of total RNA purified from young basal forebrain was used for RT and different concentrations of cDNA corresponding to 0.06 ng to 2 μg starting RNA were subjected to PCR. All four lines were almost parallel, and more importantly, the slopes of each line ranged from 0.02 to 0.07. According to the manufacturer's specifications, if the slopes are less than 0.1, the validation test is successful and GAPDH can be used as normalizer. When semiquantitative analyses are desired, the comparative C_T method has the advantage that a single point (C_T value) can be used for comparisons, thus negating the need for a complete standard curve for each sample. This is particularly important for single-cell RT-PCR. Numerous reviews are available to discuss the general principles of quantitative real-time RT-PCR (Freeman et al., 1999; Medhurst et al., 2000; Stahlberg et al., 2004) and recently, a protocol for improved quantitative real-time RT-PCR from single cells has been developed (Liss, 2002).

Various controls should be used to confirm the specificity of amplification for both positive and negative results. Ideally, two positive and one negative control should be run with each PCR, and other controls can be run occasionally. The first positive control is the internal control to confirm the successful introduction of the cell into the RT tube and the subsequent completion of the RT-PCR reactions. As mentioned above, we use GAPDH transcript as the target for this control. This serves as the normalizer for quantitative methods also. The second positive control should contain mRNAs for all the targeted transcripts in quantities somewhat above the limits of detection. We have used whole tissue total RNA at concentrations of 1–5 ng/μl for this control that also serves as the calibrator (see above). Detection of the targeted cDNA in this control confirms the success and assesses the quality of the RT-PCR. Another control that can be used to confirm the specificity of the primers and success of the protocol is to test cells with well-known expression profiles. For example, cerebellar Purkinje neurons are GABAergic and we have detected GAD expression in them but not ChAT, as expected (unpublished observation). Cloning and sequencing of

PCR products is the ultimate control for primer specificity, and is discussed in more detail below (see section "Additional Methods for RT-PCR").

An important negative control is conducted by processing a cell normally, but omitting the reverse transcriptase. If amplification is detected, there could be contamination of the PCR by genomic or other cDNA. Some investigators control for amplification of genomic DNA by the use of a DNAase that can be inactivated by proteinase K and heat denaturation prior to RT (Sucher and Deitcher, 1995). However, this sort of amplification is not thought to be detectable in single cells undergoing only a single round of 40 PCR cycles. Careful primer design can minimize the potential for amplification of genomic DNA (see section "Primer and Probe Design"). For real-time PCR using the Taqman® system, any carryover cDNA contamination will contain "U" residues and is degraded prior to PCR by the enzyme Amperase® UNG (uracil-N-glycosylase). Other negative controls include running the RT-PCR without aspirating a cell, after aspirating some of the bath solution, and after aspirating noncellular debris. Ideally, the investigator conducting the PCR should be blind to the electrophysiological results to control for bias.

III. APPLICATIONS

A. Single-Cell Identification by mRNA Expression of Phenotypic Markers

For many experimental approaches, it is not necessary to quantitate the PCR products. Often, merely detecting expression of target transcripts is sufficient to correlate function with molecular identification. Neurons often can be identified by the expression of accepted phenotypic markers for metabolism or transport of neurotransmitters. The adaptibility of RT-PCR approaches to the identification of individual neurons makes it a useful technique for preparations such as dissociated neurons that cannot be subjected to other procedures commonly used to label phenotypic markers. Even in cases where a technique such as immunohistochemistry can be employed, RT-PCR might be employed because of the relative ease of processing, the potentially high-throughput capacity, and the ability to detect several molecular species for each cell.

Approximately 90% of the neurons of rat basal forebrain are considered to be cholinergic or GABAergic, and so we use primers designed to detect enzymes involved in the synthesis of acetylcholine (ChAT) and GABA (GAD). Other well-known phenotypic markers include tyrosine hydroxylase and vesicular glutamate transporters. A positive control marker should be targeted in each cell so that its expression can validate a detection failure of the phenotypic markers. Common positive controls include GAPDH, β-actin, 18S rRNAs, and neuron-specific enolase. It is a good practice to run a global positive control along with each PCR to further validate detection failures.

Thus, if a particular basal forebrain neuron expressed GAPDH, but ChAT or GAD was not detected, and the global positive control (purified whole basal forebrain tissue mRNA) showed expression of ChAT and GAD, then the confidence in the negative result is increased. A similar argument can be made for successful detection of a target transcript from a cell validating a negative detection of that transcript in other cells processed in the same PCR.

B. Single-Cell mRNA Expression and Patch-Clamp Recording

Combining physiological data with expression data can provide the strongest confirmation of a neuron's identity. This is particularly valuable in cases of ambiguous expressions of phenotypic markers. For example, GAD mRNA has been detected in some neurons that are clearly glutamatergic (Cao *et al.*, 1996) or that also express ChAT mRNA (Han *et al.*, 2002; Tkatch *et al.*, 1998). Combined patch-clamp and RT-PCR has allowed us to show that ChAT+/GAD+ neurons of the basal forebrain have physiological properties consistent with cholinergic, rather than with GABAergic cells (Han *et al.*, 2005). As shown in Fig. 5.5A, this acutely dissociated basal forebrain neuron

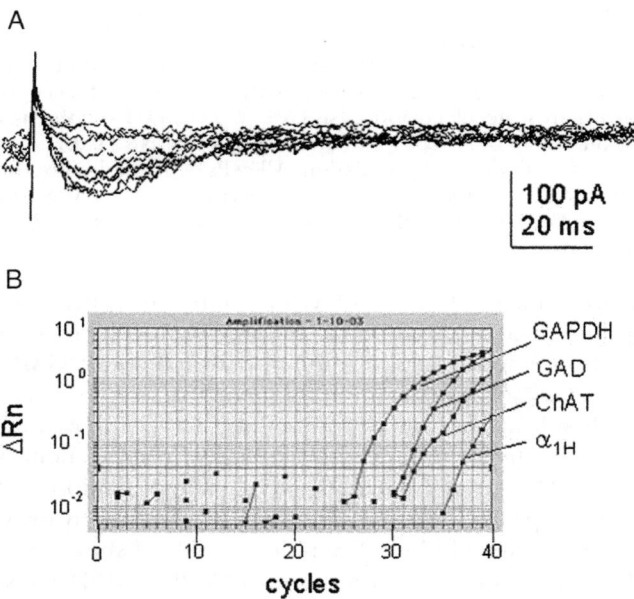

Figure 5.5. Combined patch-clamp recording of low-voltage-activated (LVA) calcium current and scRT-PCR from a basal forebrain neuron. (A) LVA currents are generated by an inactivation protocol, consisting of different prepulses from −110 to −50 mV, followed by a test pulse of −45 mV. The current in response to the test pulse is shown. (B) Real-time PCR amplification plot showing the increased fluorescence with each PCR cycle for specific transcripts for GAPDH, ChAT, GAD, and the LVA calcium channel $Ca_{v3.2}(\alpha_{1H})$ subunit. Transcripts for other LVA channel subunits ($Ca_{v3.1}$ and $Ca_{v3.3}$) were not detected.

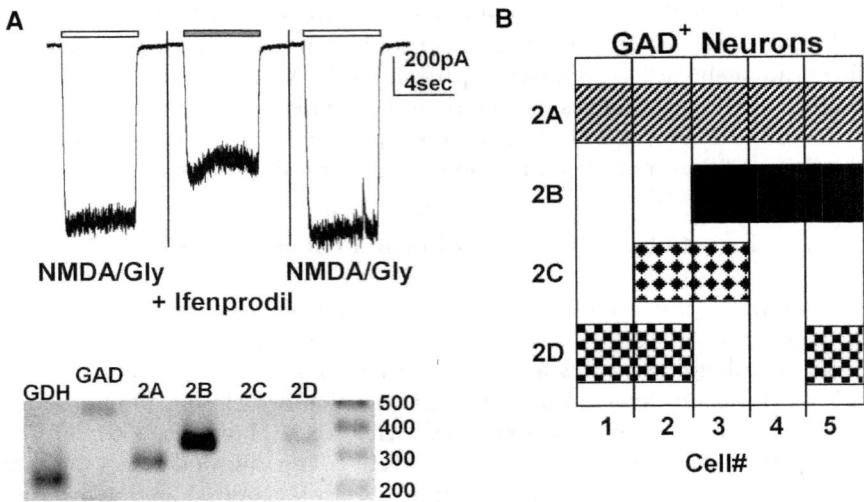

Figure 5.6. Combined patch-clamp recording of NMDA-gated currents and single-cell RT-PCR from a lateral/basolateral amygdala neuron. (A) NMDA (100 μM) and glycine (3 μM) were applied either with or without the NR2B-selective anatagonist ifenprodil (10 μM). The cellular contents of this neuron were then harvested and subjected to RT-PCR with primers for the ubiquitous gene glyceraldehyde phosphate dehydrogenase (GDH, bottom), for the 65 kD of the GABA synthetic enzyme glutamate decarboxylase (GAD), and for the four NR2 subunits. Products were visualized using standard agarose gel electrophoresis (size of products from the marker shown at right). (B) Summary of all GAD+ neurons shows that NR2 expression was quite heterogeneous in this neuronal population. All cells expressed more than one subunit. In addition, NR2C and NR2D subunit mRNAs were detected in several neurons, in contrast to the GAD− principal neurons that express primarily NR2A and NR2B (Floyd *et al.*, 2003).

displayed a low-voltage-activated (LVA) Ca^{2+} current typical of cholinergic cells in response to an inactivation voltage-step protocol. Figure 5.5B shows the real-time fluorescence amplification plot from that cell. GAPDH, ChAT, GAD, and LVA Ca^{2+} channel α subunit Ca_V 3.2 (α_{1H}) were detected.

Another valuable application of combined patch-clamp and RT-PCR is to identify and characterize the receptor subunits in native neurons. Figure 5.6A shows NMDA-activated currents and their modulation by the NR2B receptor subunit specific inhibitor, ifenprodil, in an acutely dissociated rat amygdala neuron. The ethidium bromide stained gel shown below detects GAPDH, GAD, and NMDA receptor subunits 2A, 2B, and 2D, consistent with the physiological result. The diversity of basolateral GAD+ amygdala neurons is represented in Fig. 5.6B. Note that each cell expresses multiple subunits.

C. Single-Cell mRNA Expression, Patch-Clamp Recording, and Fluorescent Calcium Measurements in Young and Aged Cells

Because the interface between scRT-PCR and patch clamping is the microelectrode, any application that can be combined with patch clamping

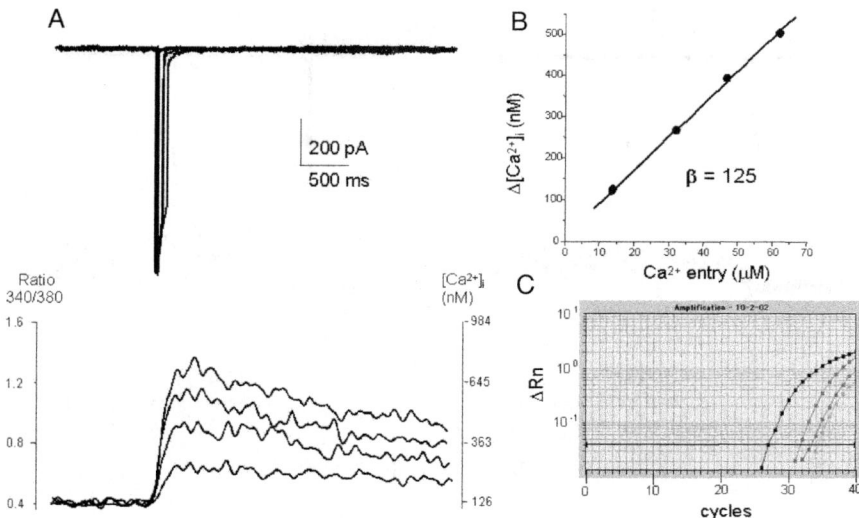

Figure 5.7. Ca^{2+} currents, fura-2 fluorescence ratio records, buffering curve, and PCR amplification plot for a young neuron expressing transcripts for ChAT, GAD, calretinin (CR), and GAPDH. (A) Ca^{2+} currents (top) and corresponding fura-2 fluorescent ratio records (bottom). (B) Buffering curve constructed from the data at left. The slope of the linear portion of the curve is essentially the reciprocal of the buffering value, 125. (C) Real-time PCR amplification plot showing the increased fluorescence with each reaction cycle for specific transcripts for the cell marker GAPDH, ChAT, GAD, and CR. Currents are generated by voltage steps $Vh = -60$ to 0 mV for different durations to create increasing Ca^{2+} influx. The ratio of fura-2 fluorescence at 340 and 380 nm excitation (340/380) increases with increasing intracellular Ca^{2+} concentrations.

can probably be combined with RT-PCR as well. We have taken advantage of this versatility by including scRT-PCR with patch clamping and fura-2-based Ca^{2+}-sensitive microfluorimetry to examine the Ca^{2+} buffering of rat basal forebrain neurons. We used standard methods for isolating neuronal high-voltage-activated Ca^{2+} currents from acutely dissociated neurons (Murchison and Griffith, 1996) with the addition of 50 μM fura-2 to the pipette solution and the application of a sterile RT-PCR protocol. The recordings shown in Fig. 5.7A were made with a Cs-acetate-based pipette solution, but the methods for recording and analyzing the fura-2 fluorescence signal are the same as previously published (Murchison and Griffith, 1998). Briefly, the Ca^{2+} currents are integrated to determine the amount of charge crossing the membrane and this is converted to an expected concentration of Ca^{2+} entry using a formula that estimates cell volume from the measured membrane capacitance. These values are graphed against the peaks of the calibrated fura fluorescence ratio signals. Fura-2 differentially changes the intensity of its fluorescent emission from excitation wavelengths of 340 and 380 nm in the presence of different Ca^{2+} concentrations such that an increasing ratio indicates increasing Ca^{2+} concentration. The resulting plot

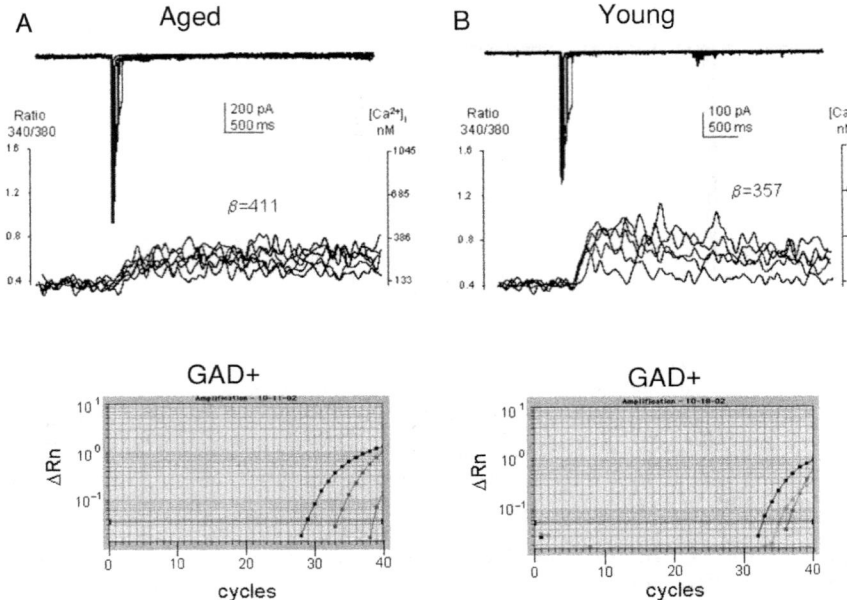

Figure 5.8. Ca^{2+} currents and corresponding fura-2 fluorescence ratio records in noncholinergic neurons identified by real-time RT-PCR from a 26-month rat (A) and a 4-month rat (B). Recordings are as described in Fig. 5.7. PCR amplification plots show the fluorescence increases associated with each reaction cycle for specific transcripts for the cell marker GAPDH, GAD, parvalbumin (PV), and calretinin (CR). The cell in "A" is identified as GAD+ and PV+, while the cell in "B" is identified as GAD+ and CR+. ChAT and CB were not detected in these neurons.

is called the buffering curve (Fig. 5.7B) and the reciprocal of the slope approximates the value β, which represents the number of buffered Ca^{2+} ions for every free ion. A larger buffering value indicates stronger cellular Ca^{2+} buffering. The real-time fluorescence amplification plot from this neuron is shown in Fig. 5.7C. GAPDH, ChAT, GAD, and the CaBP CR were detected in this cell, but the CaBPs calbindin (CB) and parvalbumin (PV) were not. Note that there is no interference of the fura-2 fluorescence with the fluorescent detection of the ABI Prism 7700 and Taqman® fluorescent probes because the excitation spectra do not overlap (fura-2 does not fluoresce at the excitation wavelengths of the probes).

The above combination of methods can be applied just as well to basal forebrain neurons from aged rats, as shown in Fig. 5.8. Young and aged GAD+ neurons are shown for comparison. Note that both GAD+ neurons display greater buffering than does the ChAT+/GAD+ neuron seen in Fig. 5.7, despite the detection of a CaBP in each of the cells. All of the target sequences detected in young basal forebrain neurons have been detected in aged neurons also (Han *et al.*, 2002, 2005).

IV. ADVANTAGES AND LIMITATIONS

Because scRT-PCR methods can be used with a number of experimental preparations, it is worthwhile to mention several advantages and disadvantages associated with various preparations. As shown in Figs. 5.2 and 5.4–5.8, acutely dissociated neurons are often used in a variety of experiments designed to study voltage- and ligand-gated currents. Visualization of an isolated neuron has obvious advantages for enhancing electrode placement and recording efficiency. Likewise, acutely dissociated cells are isolated and therefore easily harvested without contamination from other cells. A second important advantage of scRT-PCR analysis of acutely dissociated neurons is that these cells are very difficult to identify with more traditional methods, such as immunocytochemistry, because cells are not firmly attached to the glass surface and are not amenable to recovery after fixation. One disadvantage of acutely dissociated neurons is that they are removed from their surrounding milieu and may lack relevant external influences. This disadvantage is not unique to acutely dissociated neurons and may be raised for any in vitro preparation; thus this limitation does not apply particularly to RT-PCR experiments. Neurons in primary tissue culture (Fig. 5.3) have many of these same advantages for visualization and ease of electrical recording mentioned above. Additional advantages include functional synaptic circuits and cellular trophic interactions. One disadvantage may be an increased potential for aspiration of neighboring tissue into the harvesting pipette because individual neurons are not always isolated from one another. Another disadvantage of primary cultures is that neurons in culture do not necessarily represent adult expression patterns or cellular function. Cultured neurons have been identified both electrophysiologically and immunocytochemically for many years and so it will be interesting to determine how scRT-PCR cell identification compares with previous data. Finally, some of the most physiological preparations available to study cellular physiology are thin brain slices. In slices, cells can be visually identified and isolated for scRT-PCR (see references in earlier sections). Despite the advantage of intact tissue, a disadvantage of this preparation is the fact that the harvesting pipette must travel through surrounding tissue, and there is an increased likelihood of contamination. Nevertheless, the utility of scRT-PCR makes it worthwhile for application to a variety of experimental preparations.

The following advantages and limitations apply to RT-PCR protocols in general and to our scRT-PCR protocol as compared to other protocols.

A. Advantages

Utility: The protocol is amenable to a number of applications and is an excellent means of identifying neuronal phenotypes following functional characterization.

Sensitivity: If the efficiencies of the RT and PCR are close to optimal, a single copy of an mRNA target sequence should be amplified to reach detection threshold by 40 PCR cycles using real-time fluorescence.

Simplicity: Particularly if real-time detection is employed, the efficiency of data turnaround is higher and not as labor intensive as some protocols.

Reliability: Data are very reproducible with only a single round of PCR.

Quantifiability: By the use of real-time PCR, the relative abundance of a transcript can be estimated with some confidence.

Stability: Once amplified, cDNA is more stable than is labile RNA.

Precision: Expression profile data can be obtained from a discrete population of cells, avoiding the possible misinterpretation of tissue level profiling.

B. Limitations

Reliability: There is some potential for false negatives. If the efficiency of the RT or PCR is not close to optimal, then some transcripts that were present could go undetected.

Quantifiability: Different transcripts cannot be reliably compared quantitatively. Rare transcripts are potentially subject to more noise due to chance factors inherent in the first PCR cycles.

Selectivity: With our method, only a few targeted transcripts can be detected from any one cell. Only cDNAs targeted with specific primers are amplified.

Abundance: The small amount of RT product limits the number of targets that can be probed and the number of experimental replications possible.

Indirectness: mRNA expression is shown but protein expression is unknown.

Economy: Qualitative (gel-based) methods are the most economical but can be the least informative, while real-time approaches are more expensive.

V. APPENDIX: DETAILED METHODS

A. Neuronal Preparations

Adult male Fischer 344 or Sprague-Dawley rats or pregnant Sprague-Dawley females were purchased from Harlan (Indianapolis, IN). Animals took food and water ad libitum and were maintained on a 12 h light/dark cycle. Handling and care of the animals were in accordance with policies of Texas A&M University and the National Institute of Health.

Acutely dissociated basal forebrain neurons (medial septum and nucleus of the diagonal band) and lateral/basolateral amygdala neurons were obtained as described previously (McCool *et al.*, 2003; Murchison and Griffith, 1996). Briefly, isoflurane (Anaquest, Liberty Corner, NJ) anesthetized adult (2–27 mo) rats were decapitated. Coronal brain slices (400–450 μm) were microdissected to isolate the MS/nDB or amygdala and enzymatically treated with either trypsin (~0.7 mg/ml for basal forebrain; Sigma Type XI) or pronase (~0.5 mg/ml for amygdala, CalBiochem). After trituration of an individual hemislice, cells were dispersed onto the glass floor of a recording chamber and allowed to settle and stick to the glass for 7–8 min. The recording chamber is mounted on an inverted microscope and perfused at a rate of ~2 ml/min, resulting in a bath turnover of < 30 s. Experiments were performed at 20–21°C. Septal neurons were cultured from 20-day embryos using established methods (Hsiao *et al.*, 2004). Septal slices were obtained and treated briefly with trypsin. After mechanical dissociation, cells were centrifuged (1000 × g), resuspended (~4.5 × 10^6 cells/ml), and plated (~50 μl) on 10.5 × 22 mm glass coverslips (acid washed and coated with 50 μg/ml poly-D-lysine, Sigma) within 35-mm plastic petri dishes. Plated cells were maintained in a "culture media" containing equal volumes of D-MEM/F-12 and Neurobasal media (Gibco) with 2% B27 supplement (Gibco) and 1 μg/ml bovine serum albumin (Sigma). Up to eight plastic dishes were grouped in a large 150 × 20 mm covered glass petri dish and placed in a humidified 5% CO_2 incubator.

B. Reverse Transcription

The individual tubes containing the initial reaction mixture [8 μl nuclease-free water (Promega), 1 μl DTT (100 mM), 0.33 μl random hexamer primers (3 μg/μl), and 0.5 μl Rnase inhibitor (10 U/μl)] and each aspirated cell or negative control is removed from the −80°C freezer and allowed to warm on ice. They are then heated to 70°C for 10 min using a water bath or heating block. This step denatures various unwanted proteins. The RT tubes are then cooled on ice before adding the RT ingredients. Each tube receives 4 μl of 5× first strand buffer (250 mM Tris-HCl, 375 mM KCl, 15 mM $MgCl_2$), 1 μl DTT (0.1 M), 0.5 μl RNase inhibitor (10 U/μl), 0.4 μl mixed deoxynucleotide triphosphates (dNTPs 25 mM), and 1 μl Superscript® II reverse transcriptase (200 U/μl). This RT mixture is allowed to reach room temperature before a 50 min exposure to 42°C to synthesize the cDNA. The reaction is terminated by inactivating the reverse transcriptase with 70°C for 15 min. After cooling on ice, 0.5 μl RNase H is added to each tube and heated to 37°C for 20 min to remove the RNA from the RNA/DNA hybrid to yield single-stranded cDNA. The resulting cDNA can be used for PCR immediately or frozen for use later.

C. Solutions for PCR

Ingredients are from Invitrogen® unless indicated. For gel-based PCR, the reaction mixtures contained 2.0 mM $MgCl_2$, 0.5 mM of each of the dNTPs, 0.8–1.0 μM specific primers, 2.5 U Taq DNA polymerase (Promega), 5 μl 10× buffer (Promega), and 3–4 μl cDNA template from the single-cell RT reaction. For real-time PCR, each well contained cDNA template (3–5 μl), 2× Taqman® universal PCR master mix (14–16 μl, AmpliTaq® Gold DNA Polymerase, AmpErase® UNG, dNTPs with dUTP, passive reference dye ROX, optimized buffer component, Applied Biosystems), 0.3 μM specific forward and reverse primers, and 0.2 μM specific Taqman® probes.

D. Listing of Primers and Probes

1. Primers for Gel-Based PCR

The primers for ChAT, GAD, GAPDH, calbindin, calretinin, and parvalbumin that were used in gel-based PCR have been published previously (Han *et al.*, 2002). Likewise, primers for NMDA receptor subunits (Floyd *et al.*, 2003) and glycine receptor subunits (McCool and Farroni, 2001) have also been published.

2. Primers and Probes for Real-Time RT-PCR

	Forward and reverse primers	Taqman® probe
GAPDH	5′-CGCCCCTTCCGCTGAT-3′ 5′-TGACAATCTTGAGGGAGTTG TCA-3′	5′-CATGTTTGTGATGGG TGTGAACCACGAG-3′
ChAT	5′-CCGGTTTGTCCTCTCCACCAG-3′ 5′-GGGACCACGGGTCCATAACA-3′	5′-AGGTGCCCACAACC ATGGAGATG-3′
GAD	5′-ACGCCTTCGCCTGCAA-3′ 5′-GGACGCAGGTTGGTAGTAT TAGGA-3′	5′-TCCTCGAACGCGGG AGCGG-3′
Calbindin	5′-AATTGTAGAGTTGGCCCAT GTCTT-3′ 5′-TCAGTTGCTGGCATCGAAAG-3′	5′-CCCACCGAAGAGAAT TTCCTGCTGC-3′
Calretinin	5′-GCAGAGCTGGCGCAGATC-3′ 5′-CCCACGTGCTGCCTGAA-3′	5′-TGCCAACCGAAGAGA ATTTCCTTTTGTG-3′
Parvalbumin	5′-TCCTCAGATGCCAGAGACTT GTC-3′ 5′-CCGTCCCCGTCCTTGTC-3′	5′-AGGAAACAAAGACG CTGATGGCTGCT-3′
Ca$_v$3.2	5′-TGCCTCCGACTGGTTTGTAAC-3′ 5′-AGTCGTCGAAGGCCTCCAA-3′	5′-TGCCGCTCCGAACGT TGCAG-3′

E. Special Methods for Validation Experiment (Cloning, DNA Preparation, Restriction Enzymes, and Sequencing)

1. Cloning

For ligation, each PCR product was mixed with centrifuged pGEM-T vector (50 ng, Promega), 2× rapid ligation buffer, T4 DNA ligase (3 Weiss unit/μl, Promega), and nuclease-free water (Promega). This mixture was incubated at room temperature for 1 h or 15°C for a few hours. Transformation was performed using frozen JM109 high-efficiency component cells (Promega). Cells were thawed on ice just prior to use and mixed gently. Cells (50 μl) were transferred to a tube, containing 2 μl of the ligation reaction. Following gentle mixing and incubation for 20 min on ice, the tube was heat-shocked for 45–50 s in a 42°C water bath. Incubation on ice was performed for 2 min and then 950 μl SOS medium was added. After incubation for 1.5 h at 37°C with shaking (150 rpm), 100 μl of the transformation culture was spread onto agar SOC plates containing ampicillin (50 μg/ml), IPTG (isopropylthio-β-D-galactoside, Promega), and X-gal (chromogenic substrate 5-bromo-4-chloro-3-indolyl-β-D-galactoside, Promega) for double screening. Plates were incubated upside down overnight at 37°C. A white colony indicated insertion of foreign DNA (PCR product) into a polycloning site and these colonies were picked, plated, and incubated at 37°C for overnight again for single colony isolation. A white single colony was inoculated into LB broth with ampicillin (50 μg/ml) and cultured overnight for DNA preparation.

2. DNA Preparation

Large-scale DNA purification kits (Gibco BRL) were used for purifying DNA from cultured cells, containing cloned DNA of GAPDH, ChAT, GAD67, calbindin, calretinin, and parvalbumin. Overnight-cultured cells were harvested (4000–5000 rpm, 10 min) and were suspended in a cell suspension buffer (50 mM Tris-HCl, pH 8.0, 10 mM EDTA). A cell lysis solution containing NaOH (200 mM) and SDS (1%) was added followed by gentle tube inverting five times. Incubation occurred for 5 min with neutralization buffer (potassium acetate 3.1 M, pH 5.5) and then centrifugation was applied for 10 min ($15,000 \times g$ at room temperature). The supernatant was poured into equilibrated columns (equilibration buffer; 600 mM NaCl, 100 mM sodium acetate, pH 5.0, 0.15% Triton X-100) and flowed by gravity. The columns were washed twice with wash buffer (800 mM NaCl, 100 mM sodium acetate, pH 5.0) and elution buffer (1.25 M NaCl, 100 mM Tris–HCl, pH 8.5), and DNA was collected. 2-Propanol was used for precipitation and 70% ethanol rinsed the pellet. After 10 min of air-drying, the DNA pellet was dissolved with 200 μl TE buffer (10 mM Tris-HCl, pH 8.0, 0.1 mM EDTA). DNA insertion was verified by gel electrophoresis (1% agarose gel).

3. Restriction Enzyme Analysis and Sequencing

Purified DNA of cloned cultures from PCR products of GAPDH, ChAT, GAD67, calbindin, calretinin, and parvalbumin were digested by different restriction enzymes for verification of PCR product before sequencing and isolation of target fragments for real-time PCR standard curves.

The restriction enzyme *Pst*I for ChAT, *Nco*I for GAD, *Xmn*I for calbindin and parvalbumin, and *Sst*I for calretinin were used (*Pst*I, *Nco*I, *Sst*I from Gibco BRL; *Xmn*I from Promega). Each of the purified DNAs was mixed with the restriction enzyme (10 U/μl) and 10× buffers to reach final concentrations of 50 mM Tris-HCl (pH 8.0), 10 mM MgCl$_2$, and 50 mM NaCl for *Pst*I and *Sst*I, and 100 mM NaCl, 6 mM Tris-HCl (pH 7.5), 6 mM MgCl$_2$, 50 mM NaCl, and 1 mM DTT for *Nco*I and *Xmn*I. Overnight incubation at 37°C was performed and the size of the cut fragments was visualized by electrophoresis (1.2% agarose gel). The correct bands were confirmed by size, and the sequence of the inserted PCR product was performed using pairs of forward and reverse primers (DNA Technologies Lab, Department of Veterinary Pathobiology, Texas A&M University).

Different restriction enzymes were used to obtain purified cloned DNA of PCR products for GAPDH, ChAT, GAD67, CB, CR, and PV standard curves. *Eco*RI for GAPDH and *Pvu*II for ChAT, GAD, CB, CR, and PV were used to cut plasmid DNA. DNA fragments containing each PCR product sequence were visualized by gel electrophoresis on low-melting agarose gel. Bands of PCR products were sliced from the gel with a sterilized surgical blade (Bard-Parker), and Wizard® DNA Preps (Promega) were used for DNA purification. Agarose slices were incubated at 70°C until the agarose was completely melted. Resin (1 ml) was added and mixed thoroughly for 20 s. It was transferred to a minicolumn/syringe barrel assembled with a vacuum manifold and vacuum was applied. The column was washed by 2 ml of 80% 2-propanol and a vacuum was continued for 30 s to dry it. The column was transferred to a new tube and 50 μl of water or TE buffer was added. After 1 min incubation at room temperature, DNA was collected following centrifugation for 20 s at $10,000 \times g$. Eluted DNA was quantified by spectrometry and kept at −20°C until use.

F. Additional Methods for RT-PCR

1. Cloning and Sequencing of PCR Products

Even though PCR products displayed the correct size, nonspecific amplification or contamination with similar size products may occur. Therefore, we cloned each PCR product (ChAT, GAD, CB, CR, and PV) by ligation and transformation. pGEM plasmids containing each of the PCR products (3 µg total RNA purified from young rat basal forebrain) were separated by gel electrophoresis. They were distinguished by size, being heavier than pGEM alone but lighter than a pGEM+500 bp marker. All the plasmids came from single colonies and were purified by DNA purification kits obtained from Gibco BRL. Restriction enzyme analysis was performed to confirm the correct inserts of each product before sequencing. *Pst*I for ChAT, *Nco*I for GAD, *Xmn*I for CB and PV, and *Sst*I for CR were selected because they have only two cutting sites outside or inside of the inserts, thus producing only two fragments of known size. Depending on the direction of the inserts, there were two possible sets of fragment sizes generated after restriction enzyme treatment for each primer set. For further proof, plasmids with different inserts for ChAT, GAD, CB, CR, and PV were sequenced using forward and reverse primers and were shown to be a 100% match with their respective genomic sequence. The same verification experiment was repeated using total RNA from aged animals.

2. Optimization for Gel-Based PCR

Additional experiments were conducted to optimize the concentration of Mg^{2+} for different primers with PCR. Because salt concentrations play a critical role in amplification efficiency during PCR and different primers may have different optimal concentrations, we examined concentrations of 0.5, 1, 2, 3, and 4 mM Mg^{2+}. The strongest gel-based signals for each of the targets were detected with 2 mM Mg^{2+} concentration.

ACKNOWLEDGMENTS. This study was supported in part by NIH grant AG007805.

REFERENCES

Audinat, E., Lambolez, B., and Rossier, J., 1996, Functional and molecular analysis of glutamate-gated channels by patch-clamp and RT-PCR at the single-cell level, *Neurochem. Int.* **28:**119–136.

Bargas, J., Howe, A., Eberwine, J., Cao, Y., and Surmeier, D. J., 1994, Cellular and molecular characterization of Ca^{2+} currents in acutely isolated, adult rat neostriatal neurons, *J. Neurosci.* **14:**6667–6686.

Bartus, R. T., Dean, R. L., III, Beer, B., and Lippa, A. S., 1982, The cholinergic hypothesis of geriatric memory dysfunction, *Science* **217:**408–414.

Bochet, P., Audinat, E., Lambolez, B., Crepel, F., Rossier, J., Iino, M., Tsuzuki, K., and Ozawa, S., 1994, Subunit composition at the single-cell level explains functional properties of a glutamate-gated channel, *Neuron* **12:**383–388.

Cao, Y., Wilcox, K. S., Martin, C. E., Rachinsky, T. L., Eberwine, J., and Dichter, M. A., 1996, Presence of mRNA for glutamic acid decarboxylase in both excitatory and inhibitory neurons, *Proc. Natl. Acad. Sci. U.S.A.* **93:**9844–9849.

Chen, K. C., Blalock, E. M., Thibault, O., Kaminker, P., and Landfield, P. W., 2000, Expression of alpha 1D subunit mRNA is correlated with L-type Ca^{2+} channel activity in single neurons of hippocampal "zipper" slices, *Proc. Natl. Acad. Sci. U.S.A.* **97**:4357–4362.

Chow, N., Cox, C., Callahan, L. M., Weimer, J. M., Guo, L., and Coleman, P. D., 1998, Expression profiles of multiple genes in single neurons of Alzheimer's disease, *Proc. Natl. Acad. Sci. U.S.A.* **95**:9620–9625.

Coyle, J. T., Price, D. L., and DeLong, M. R., 1983, Alzheimer's disease: a disorder of cortical cholinergic innervation, *Science* **219**:1184–1190.

Crino, P., and Eberwine, J., 1996, Molecular characterization of the dendritic growth cone: regulated mRNA transport and local protein synthesis, *Neuron* **17**:1173–1187.

Decker, M. W., 1987, The effects of aging on hippocampal and cortical projections of the forebrain cholinergic system, *Brain Res. Rev.* **12**:423–438.

Dixon, A. K., Richardson, P. J., Lee, K., Carter, N. P., and Freeman, T. C., 1998, Expression profiling of single cells using 3 prime end amplification (TPEA) PCR, *Nucleic Acids Res.* **26**:4426–4431.

Dixon, A. K., Richardson, P. J., Pinnock, R. D., and Lee, K., 2000, Gene-expression analysis at the single-cell level, *Trends Pharmacol. Sci.* **21**:65–70.

Eberwine, J., Kacharmina, J. E., Andrews, C., Miyashiro, K., McIntosh, T., Becker, K., Barrett, T., Hinkle, D., Dent, G., and Marciano, P., 2001, mRNA expression analysis of tissue sections and single cells, *J. Neurosci.* **21**:8310–8314.

Eberwine, J., Yeh, H., Miyashiro, K., Cao, Y., Nair, S., Finnell, R., Zettel, M., and Coleman, P., 1992, Analysis of gene expression in single live neurons, *Proc. Natl. Acad. Sci. U.S.A.* **89**:3010–3014.

Edwards, M. C., and Gibbs, R. A., 1994, Multiplex PCR: advantages, development, and applications, *PCR Methods Appl.* **3**:S65–S75.

Elowitz, M. B., Levine, A. J., Siggia, E. D., and Swain, P. S., 2002, Stochastic gene expression in a single cell, *Science* **297**:1183–1186.

Fanselow, M. S., and LeDoux, J. E., 1999, Why we think plasticity underlying Pavlovian fear conditioning occurs in the basolateral amygdala, *Neuron* **23**:229–232.

Fibiger, H. C., 1982, The organization and some projections of cholinergic neurons of the mammalian forebrain, *Brain Res.* **257**:327–388.

Fischer, W., Gage, F. H., and Bjorklund, A., 1989, Degenerative changes in forebrain cholinergic nuclei correlate with cognitive impairments in aged rats. *Eur. J. Neurosci.* **1**:34–45.

Floyd, D. W., Jung, K. Y., and McCool, B. A., 2003, Chronic ethanol ingestion facilitates *N*-methyl-D-aspartate receptor function and expression in rat lateral/basolateral amygdala neurons, *J. Pharmacol. Exp. Ther.* **307**:1020–1029.

Freeman, W. M., Walker, S. J., and Vrana, K. E., 1999, Quantitative RT-PCR: pitfalls and potential, *Biotechniques* **26**:112–125.

Gibson, U. E., Heid, C. A., and Williams, P. M., 1996, A novel method for real time quantitative RT-PCR, *Genome Res.* **6**:995–1001.

Griffith, W. H., Jasek, M. C., Bain, S. H., and Murchison, D., 2000, Modification of ion channels and calcium homeostasis of basal forebrain neurons during aging, *Behav. Brain Res.* **115**:219–233.

Han, S-H., McCool, B. A., Murchison, D., Nahm, S. S., Parrish, A. R., and Griffith, W. H., 2002, Single-cell RT-PCR detects shifts in mRNA expression profiles of basal forebrain neurons during aging, *Mol. Brain Res.* **98**:67–80.

Han, S-H., Murchison, D., and Griffith, W. H., 2005, Low voltage-activated calcium and fast tetrodotoxin-resistant sodium currents define subtypes of cholinergic and noncholinergic neurons in rat basal forebrain, *Mol. Brain Res.* **134**:226–238.

Heid, C. A., Stevens, J., Livak, K. J., and Williams, P. M., 1996, Real time quantitative PCR, *Genome Res.* **6**:986–994.

Hinkle, D., Glanzer, J., Sarabi, A., Pajunen, T., Zielinski, J., Belt, B., Miyashiro, K., McIntosh, T., and Eberwine, J., 2004, Single neurons as experimental systems in molecular biology, *Prog. Neurobiol.* **72**:129–142.

Hsiao, S. H., DuBois, D. W., Miranda, R. C., and Frye, G. D., 2004, Critically timed ethanol exposure reduces GABA$_A$R function on septal neurons developing in vivo but not in vitro, *Brain Res.* **1008**:69–80.

Johansen, F. F., Lambolez, B., Audinat, E., Bochet, P., and Rossier, J., 1995, Single cell RT-PCR proceeds without the risk of genomic DNA amplification, *Neurochem. Int.* **26:**239–243.

Jonas, P., Racca, C., Sakmann, B., Seeburg, P. H., and Monyer, H., 1994, Differences in Ca^{2+} permeability of AMPA-type glutamate receptor channels in neocortical neurons caused by differential GluR-B subunit expression, *Neuron* **12:**1281–1289.

Killcross, S., Robbins, T. W., and Everitt, B. J., 1997, Different types of fear-conditioned behaviour mediated by separate nuclei within amygdala, *Nature* **388:**377–380.

Lambolez, B., Audinat, E., Bochet, P., Crepel, F., and Rossier, J., 1992, AMPA receptor subunits expressed by single Purkinje cells, *Neuron* **9:**247–258.

Lindqvist, N., Vidal-Sanz, M., and Hallbook, F., 2002, Single cell RT-PCR analysis of tyrosine kinase receptor expression in adult rat retinal ganglion cells isolated by retinal sandwiching, *Brain Res. Protoc.* **10:**75–83.

Liss, B., 2002, Improved quantitative real-time RT-PCR for expression profiling of individual cells, *Nucleic Acids Res.* **30:**e89.

Martina, M., Schultz, J. H., Ehmke, H., Monyer, H., and Jonas, P., 1998, Functional and molecular differences between voltage-gated K+ channels of fast-spiking interneurons and pyramidal neurons of rat hippocampus, *J. Neurosci.* **18:**8111–8125.

McCool, B. A., and Farroni, J. S., 2001, Subunit composition of strychnine-sensitive glycine receptors expressed by adult rat basolateral amygdala neurons, *Eur. J. Neurosci.* **14:**1082–1090.

McCool, B. A., Frye, G. D., Pulido, M. D., and Botting, S. K., 2003, Effects of chronic ethanol consumption on rat GABA(A) and strychnine-sensitive glycine receptors expressed by lateral/basolateral amygdala neurons, *Brain Res.* **963:**165–177.

McDonald, A. J., 1985, Immunohistochemical identification of gamma-aminobutyric acid-containing neurons in the rat basolateral amygdala, *Neurosci. Lett.* **53:**203–207.

McDonald, A. J., Beitz, A. J., Larson, A. A., Kuriyama, R., Sellitto, C., and Madl, J. E., 1989, Co-localization of glutamate and tubulin in putative excitatory neurons of the hippocampus and amygdala: an immunohistochemical study using monoclonal antibodies, *Neuroscience* **30:**405–421.

Medhurst, A. D., Harrison, D. C., Read, S. J., Campbell, C. A., Robbins, M. J., and Pangalos, M. N., 2000, The use of TaqMan RT-PCR assays for semiquantitative analysis of gene expression in CNS tissues and disease models, *J. Neurosci. Methods* **98:**9–20.

Mesulam, M. M., Mufson, E. J., Wainer, B. H., and Levey, A. I., 1983, Central cholinergic pathways in the rat: an overview based on an alternative nomenclature (Ch1–Ch6), *Neuroscience* **10:**1185–1201.

Miyashiro, K., Dichter, M., and Eberwine, J., 1994, On the nature and differential distribution of mRNAs in hippocampal neurites: implications for neuronal functioning, *Proc. Natl. Acad. Sci. U.S.A.* **91:**10800–10804.

Monyer, H., and Lambolez, B., 1995, Molecular biology and physiology at the single-cell level, *Curr. Opin. Neurobiol.* **5:**382–387.

Monyer, H., and Markram, H., 2004, Interneuron diversity series: molecular and genetic tools to study GABAergic interneuron diversity and function, *Trends Neurosci.* **27:**90–97.

Mufson, E. J., Counts, S. E., and Ginsberg, S. D., 2002, Gene expression profiles of cholinergic nucleus basalis neurons in Alzheimer's disease, *Neurochem. Res.* **27:**1035–1048.

Murchison, D., and Griffith, W. H., 1996, High-voltage activated calcium currents in basal forebrain neurons during aging, *J. Neurophysiol.* **76:**158–174.

Murchison, D., and Griffith, W. H., 1998, Increased calcium buffering in basal forebrain neurons during aging, *J. Neurophysiol.* **80:**350–364.

Olton, D. S., Wenk, G. L., and Markowska, A. M., 1991, Basal forebrain, memory and attention, In: Richardson, R. T. (ed.), *Activation to Acquisition: Functional Aspects of the Basal Forebrain Cholinergic System*, Boston: Birkhauser, pp. 247–262.

Panula, P., Revuelta, A. V., Cheney, D. L., Wu, J. Y., and Costa, E., 1984, An immunohistochemical study on the location of GABAergic neurons in rat septum, *J. Comp. Neurol.* **222:**69–80.

Pape, J. R., Skynner, M. J., Sim, J. A., and Herbison, A. E., 2001, Profiling gamma-aminobutyric acid (GABA(A)) receptor subunit mRNA expression in postnatal gonadotropin-releasing hormone (GnRH) neurons of the male mouse with single cell RT-PCR, *Neuroendocrinology* **74:**300–308.

Phillips, J. K., and Lipski, J., 2000, Single-cell RT-PCR as a tool to study gene expression in central and peripheral autonomic neurones, *Auton. Neurosci.* **86:**1–12.

Plant, T. D., Schirra, C., Katz, E., Uchitel, O. D., and Konnerth, A., 1998, Single-cell RT-PCR and functional characterization of Ca^{2+} channels in motoneurons of the rat facial nucleus, *J. Neurosci.* **18:**9573–9584.

Roy, S. W., Fedorov, A., and Gilbert, W., 2003, Large-scale comparison of intron positions in mammalian genes shows intron loss but no gain, *Proc. Natl. Acad. Sci. U.S.A.* **100:**7158–7162.

Ruano, D., Perrais, D., Rossier, J., and Ropert, N., 1997, Expression of GABA(A) receptor subunit mRNAs by layer V pyramidal cells of the rat primary visual cortex. *Eur. J. Neurosci.* **9:**857–862.

Rye, D. B., Wainer, B. H., Mesulam, M. M., Mufson, E. J., and Saper, C. B., 1984, Cortical projections arising from the basal forebrain: a study of cholinergic and noncholinergic components employing combined retrograde tracing and immunohistochemical localization of choline acetyltransferase, *Neuroscience* **13:**627–643.

Santi, M. R., Vicini, S., Eldadah, B., and Neale, J. H., 1994, Analysis by polymerase chain reaction of alpha 1 and alpha 6 GABAA receptor subunit mRNAs in individual cerebellar neurons after whole-cell recordings, *J. Neurochem.* **63:**2357–2360.

Sarter, M., and Bruno, J. P., 2000, Cortical cholinergic inputs mediating arousal, attentional processing and dreaming: differential afferent regulation of the basal forebrain by telencephalic and brainstem afferents, *Neuroscience* **95:**933–952.

Sarter, M., and Bruno, J. P., 2002, The neglected constituent of the basal forebrain corticopetal projection system: GABAergic projections, *Eur. J. Neurosci.* **15:**1867–1873.

See, R. E., Fuchs, R. A., Ledford, C. C., and McLaughlin, J., 2003, Drug addiction, relapse, and the amygdala, *Ann. NY. Acad. Sci.* **985:**294–307.

Song, W. J., Tkatch, T., Baranauskas, G., Ichinohe, N., Kitai, S. T., and Surmeier, D. J., 1998, Somatodendritic depolarization-activated potassium currents in rat neostriatal cholinergic interneurons are predominantly of the A type and attributable to coexpression of Kv4.2 and Kv4.1 subunits, *J. Neurosci.* **18:**3124–3137.

Stahlberg, A., Hakansson, J., Xian, X., Semb, H., and Kubista, M., 2004, Properties of the reverse transcription reaction in mRNA quantification, *Clin. Chem.* **50:**509–515.

Subramaniam, S., 1998, The biology workbench—a seamless database and analysis environment for the biologist, *Proteins* **32:**1–2.

Sucher, N. J., and Deitcher, D. L., 1995, PCR and patch-clamp analysis of single neurons, *Neuron* **14:**1095–1100.

Sucher, N. J., Deitcher, D. L., Baro, D. J., Warrick, R. M., and Guenther, E., 2000, Genes and channels: patch/voltage-clamp analysis and single-cell RT-PCR, *Cell Tissue Res.* **302:**295–307.

Surmeier, D. J., Song, W. J., and Yan, Z., 1996, Coordinated expression of dopamine receptors in neostriatal medium spiny neurons, *J. Neurosci.* **16:**6579–6591.

Tkatch, T., Baranauskas, G., and Surmeier, D. J., 1998, Basal forebrain neurons adjacent to the globus pallidus co-express GABAergic and cholinergic marker mRNAs, *Neuroreport* **9:**1935–1939.

Whitcombe, D., Brownie, J., Gillard, H. L., McKechnie, D., Theaker, J., Newton, C. R., and Little, S., 1998, A homogeneous fluorescence assay for PCR amplicons: its application to real-time, single-tube genotyping, *Clin. Chem.* **44:**918–923.

Yan, Z., and Surmeier, D. J., 1996, Muscarinic (m2/m4) receptors reduce N- and P-type Ca^{2+} currents in rat neostriatal cholinergic interneurons through a fast, membrane-delimited, G-protein pathway, *J. Neurosci.* **16:**2592–2604.

Zaborszky, L., Carlsen, J., Brashear, H. R., and Heimer, L., 1986, Cholinergic and GABAergic afferents to the olfactory bulb in the rat with special emphasis on the projection neurons in the nucleus of the horizontal limb of the diagonal band, *J. Comp. Neurol.* **243:**488–509.

6

Merging Structure and Function: Combination of In Vivo Extracellular and Intracellular Electrophysiological Recordings with Neuroanatomical Techniques

ATTILA SÍK

ATTILA SÍK • Centre de Recherche Université Laval Robert-Giffard, 2601 De la Canardiere, Québec, Québec G1J 2G3, Canada

Abstract: In order to understand the involvement of activity of single neurons in the context of the activity generated in small or larger neuronal networks, electrophysiological methods and morphological techniques need to be combined. In this chapter I describe a combination of methods designed to enable researchers to record the electrophysiological activity of single neurons in the intact brain, to analyze the interactions of these identified neurons with the surrounding neuronal network, and to investigate afterward the neuroanatomical characteristics of the recorded neurons.

Keywords: extracellular and intracellular recording, reconstruction, immunohistochemistry

I. INTRODUCTION

The deciphering of the function of neurons and neuronal networks in the intact brain is one of the major challenges in neuroscience. The electrophysiological properties of individual neurons and neuronal networks have been studied in great detail for many decades. However, to understand the role of various types of neurons in the genesis and/or the modulation of neural network activity as it is reflected in complex EEG patterns, it is necessary to record single-neuron activity simultaneously with the behavior of these single cells within the context of activity of the neuronal network in which they are embedded. In addition, the combination of intracellular and extracellular electrophysiological recordings with the subsequent morphological and neurochemical identification of the electrophysiologically characterized neurons provides additional information that is essential to understand the structural-functional relationship of neuronal networks.

Combined electrophysiological/network/morphological identification is necessary for the construction of realistic neuronal network simulation models. Such models need to take into account information about the complete physiological and anatomical properties: genomic and neurochemical content, receptor expression and other molecular information, knowledge about the patterns of dendrite and axon arborizations, information about the input (number, type, and origin of synapses terminating on the soma and on the dendritic trees), the output (number of axon terminals and the dendritic domain of postsynaptic targets), and finally information about the types and numbers of synaptically connected partners of different cell types. The degree of convergence and divergence of input/output among different neuronal populations is a crucial element in neuronal network simulations. Therefore, research has explored these questions, yet mostly in vitro. In spite of the obvious advantages of using slice preparations (e.g., controlled administration of pharmacological reagents, multiple intracellular recording, etc.), the in vitro approach has the substantial disadvantage that the neuronal network is artificially truncated. Measuring neuronal and network activity in the in vitro hippocampal slice preparation illustrates this point. In the intact hippocampus, the sizes, densities, and trajectories of the axonal arborizations of inhibitory and excitatory neurons are very different from each other (Li et al., 1993; Sik et al., 1993; Freund and Buzsaki, 1996). Since in a slice far more excitatory axons are truncated than inhibitory ones, the consequence of slicing is that the excitation–inhibition balance becomes unnaturally modified in the in vitro slice compared with the intact situation. It has been estimated that in the in vitro hippocampal slice preparation as much as 80–90% of excitatory terminals arising from an excitatory pyramidal cell may be truncated (Li et al., 1993; Sik et al., 1993), whereas inhibitory axon arborization remains relatively unaffected (e.g., "O-LM", basket cells, etc.; Sik et al., 1995). Several types of inhibitory neurons can be seriously truncated though [e.g., "backprojection" inhibitory neurons (Sik et al., 1994, 1995)]. Because of this artificial condition, neuronal network properties have to be investigated in vitro with great caution. To overcome this problem and to reliably investigate network properties of neurons, the most recent generation of studies has been designed to obtain information about the function of morphologically identified neurons in the intact brain. Several studies have been published in which intracellular recording of neurons was first obtained in anesthetized animals followed by labeling and subsequent neurochemical analysis of the same neurons (Sik et al., 1994, 1995, 1997).

Compared to the recording in in vitro slice preparations, the major challenge in the in vivo electrophysiological experiment is that the researcher has to perform the recordings without visual control. Furthermore, the brain in living animals is by no means as static as in slices, which makes it very difficult to conduct sustained intracellular recordings and to successfully fill the recorded cell and all its processes with a marker substance once the recordings have been completed. Therefore, it may not be surprising that the number of analyzed cells in research articles reporting in vivo intracellular recording is substantially lower than in articles using in vitro approach (Li et al., 1992; Sik et al., 1993, 1994, 1995, 1997).

In order to characterize the neuroanatomical and neurochemical properties of the electrophysiologically characterized cells, a combination of electrophysiology and various neuroanatomical techniques is required. The simplest technique used in the past to localize the position of the extracellular recording was to simply dip the electrode tip prior to recording in a dye such as methyl blue or fast green, which left in the brain tissue a mark of the position of the electrode (Grossman and Hampton, 1968; Simons, 1978; Takato and Goldring, 1979; Thomas and Wilson, 1965). The resolution of this method was obviously inadequate to determine the location of the electrode with great precision. When single-cell recording methods emerged, a reliable technique that allowed the researcher to successfully visualize the electrophysiologically recorded cell was highly sought after. At first, the new marker horseradish peroxidase (HRP), being at that time popular in neuroanatomical tract-tracing (Heimer and Robards, 1981), found a second "killer application" as an intracellular marker in neurophysiology. HRP clearly delineates the cell's geometry and it supplies as a bonus an electron-dense label that highlights the labeled neuron and its processes in ultrathin sections studied in the electron microscope (Chang *et al.*, 1981; Jankowska *et al.*, 1976; Kita and Kitai, 1986; Kitai *et al.*, 1976; Snow *et al.*, 1976; Tamamaki *et al.*, 1987; Tepper *et al.*, 1987). However, intracellular injection of HRP enables only a partial analysis of the axonal arborization of neurons because not all axons become filled in their entire trajectory. Fluorescent dyes like Procion yellow (Hassin, 1979; Kaneko, 1970; Kelly and Van Essen, 1974) and the highly fluorescent Lucifer yellow were also used with success (Stewart, 1978, 1981; Takato and Goldring, 1979). Although Lucifer yellow produces better details as a marker compared to HRP, it has a major disadvantage as well: as all fluorescent dyes it loses its light emission capability after a relatively short period of UV illumination (the so-called "bleaching" effect). Therefore, the full reconstruction of the axonal arborization of intracellularly labeled neurons has been difficult, impossible, or dependent on the ability of the researcher to apply the complicated technique of diaminobenzidine photoconversion (Sandell and Masland, 1988). Thus, complete reconstruction of a neuron requires the intracellular injection of a molecule with a low molecular weight, providing fast and complete diffusion into small neuronal appendages, combined with a visualization method that produces a stable, optical, and electron-dense end product. Sensitive intracellular recording followed by anatomical identification of the cell today is based on a widely used biotin-containing low-molecular-weight product called biocytin (Horikawa and Armstrong, 1988). Biocytin appeared to be superior to HRP in completely staining the dendritic trees and axonal arborizations in their finest details, both in vitro (Horikawa and Armstrong, 1988; Kawaguchi *et al.*, 1989) and in vivo (Kawaguchi *et al.*, 1990; Sik *et al.*, 1995, 1997). Recording of the electrophysiological activity of single neurons followed by injection of biocytin and completed with the analysis of the neuron's interactions with the neuronal network is used to study the complex question of neuronal function.

II. IN VIVO EXTRACELLULAR RECORDING

Various extracellular recording methods have been developed over the past decades with the purpose to monitor the electrical activity of neural networks. All these methods require that a low-resistance conductor should be placed in the region of interest. One of the simplest methods is to use a metal wire (Hubel, 1957). Because neurons of the same electrophysiological class generate similar action potentials, the only way to identify a given neuron from extracellularly recorded spikes is to move the electrode tip closer to its cell body (minimum distance is 20 μm in cortex) than to any other neurons. Neuron separation by this method is guaranteed by the differential proximity of the recording tip and one neuron, relative to the other neurons (see for details the chapter by Duque and Zaborszky in this volume). The substantially larger amplitude spikes, relative to the background "noise," guarantee neuron isolation. To record from another neuron, another electrode is needed. Because electrical recording from neurons is invasive, monitoring from larger numbers of neurons with the one-electrode–one-neuron approach inevitably increases tissue damage. Thus, improved methods are needed for the simultaneous recording of closely spaced neuronal populations with minimal damage to the hard wiring of the brain network.

The recent advent of localized multisite extracellular recording techniques has dramatically increased the yield of isolated neurons (Gray *et al.*, 1995; McNaughton *et al.*, 1983; Wilson and McNaughton, 1993). With only one recording site, signals from many neurons with similar size and orientation and which are at the same distance from the tip will provide the same magnitude signal, making single-cell isolation difficult. The use of two or more recording sites allows the triangulation of distances because the amplitude of the recorded spike is a function of the distance between the neuron and the electrode (Henze *et al.*, 2000; see for details the chapter by Nadasdy *et al.* in this volume). Wire tetrodes have numerous advantages over sharp-tip single electrodes, including larger yield of units, low-impedance recording tips, and mechanical stability. Because the recording tip needs not to be placed in the immediate vicinity of the neuron, long-term recordings in behaving animals are possible. Microelectromechanical system-based recording devices can reduce the technical limitations inherent in wire electrodes because with the same amount of tissue displacement the number of monitoring sites can be substantially increased (Bartho *et al.*, 2004; Buzsaki, 2004; Csicsvari *et al.*, 2003; Wise and Najafi, 1991). Whereas silicon probes have the advantages of tetrode recording principles, they are substantially smaller in size. Furthermore, multiple sites can be arranged over a longer distance, thus allowing the simultaneous recording of neuronal activity in various cortical layers (Buzsaki and Kandel, 1998). Currently available multishank probes can record from as many as hundred well-separated neurons.

Cortical pyramidal cells generate extracellular currents that flow mostly parallel with their somatodendritic axis. These extracellular features allow the separation of signals related to individual neurons. In practice, only a small fraction of all possible neurons can be reliably separated with the currently available probes and spike-sorting algorithms. Data processing and viewing algorithms are freely available. Neurophysiological and behavioral data can be explored by NeuroScope (http://neuroscope.sourceforge.net)

(Buzsaki *et al.*, 2004). Spike-sorting is performed in two steps, first automatically using KlustaKwik (http://klustakwik.sourceforge.net) (Harris *et al.*, 2000) and then manually using Klusters (http://klusters. sourceforge.net) (Hazan *et al.*, 2004). A further advantage of silicon probe monitoring of electrical activity is that the closely spaced recording sites have a known one or two dimensions. The multiple site approach allows the simultaneous monitoring of the extracellular flow of ions with high spatial resolution. From the measured voltages the current flow can be calculated, and the extracellular resistivity can be used to calculate the current density. Such current-source density measurements provide valuable information for identifying synaptic pathways and neuronal compartments responsible for generating the locally measured current. For example, if a spatial distribution of current-source density of a spontaneous field pattern matches that of the evoked currents by thalamic but not by callosal inputs, the firm conclusion can be drawn that the spontaneous pattern is generated by thalamic afferents (Bragin *et al.*, 1995; Buzsáki *et al.*, 2003; Nadasdy *et al.*, 1998).

A critical step in the reconstruction of a functional circuit is the identification of the anatomical nature of the recorded and spike-sorted units. This identification is possible only with a combined program comparing extracellular and intracellular spike shape and spike dynamics of morphologically identified neurons. The method requires several steps and dedicated experiments. The examples below are taken from such experiments in vivo in the hippocampus, but the method is compatible with neocortical areas or other structures as well, i.e., there seems to be no limit for the identification of extracellular spikes in behaving animals.

III. IN VIVO INTRACELLULAR RECORDING AND SINGLE-CELL LABELING

Extracellular recording of local fields and/or large numbers of neurons provides information about the cooperative activity of neuronal assemblies, a type of parameter that restricts the patterns of firing of individual cells. If one is interested in the intrinsic properties of individual cells, the inputs that an individual cell receives, the fluctuations of membrane potential, and so forth, then intracellular recording needs to be performed.

For in vivo intracellular recording, glass pipettes with very small tip diameters (<0.5 µm) called sharp electrodes are used most frequently. Recently, patch-pipettes have been used in vivo in order to record the electrical activity of single neurons (Ferster and Jagadeesh, 1992; Margrie *et al.*, 2002). More details on in vivo patch recording are presented in this volume in the chapter contributed by Petersen.

Sharp electrodes are manufactured using glass pipettes of different diameters with the aid of an electrode puller. Since in vivo recording often requires reaching deeper brain structures, the geometry of the electrode is different from the one used in vitro: the shanks of the pipettes must be sufficiently long. However, long shank electrodes may be too flexible. To reach deeper brain regions with a high degree of precision, it is necessary that the electrode does not drift from the planned path. This is difficult to achieve with flexible electrodes. Therefore, usually for in vivo intracellular recording, glass electrodes with rather thick glass walls are being used. An alternative is to use an even more rigid electrode made of quartz. The disadvantage using quartz is that this material requires a special and expensive electrode puller (e.g., Shutter P2000), where the glass is melted by laser light instead of a tungsten-heating element.

Once the electrode is pulled it is filled up with conductive electrolyte. If no special recording condition is required, usually 0.5 M potassium acetate solution is used for this purpose. Since the tip of the electrode is very small, the filling of the pipette is a process that has to be done with the necessary care and precaution. To support homogeneous filling, glass electrodes with inner filament are used, which, by the capillary effect, drastically increases the probability of a successful filling process of the pipette. Basic rule is that the recording pipette needs to be filled with electrolyte without any discontinuity (for example, air bubbles should be absent). This is not an easy task: simply placing the fluid inside the pipette from the back will certainly produce a certain amount of air bubbles, rendering the pipettes useless. To avoid the situation of ending up with a batch of sharp yet useless pipettes because of poor filling, the following steps are recommended: (1) place small droplets of conductive electrolyte at the open end of the pipette. Hold this still for a couple of minutes until the liquid fills the extreme tip of the pipette completely, (2) use a small diameter filling tube that is being inserted into the recording pipette down to its neck and then slowly and carefully fill the neck and shank up with the electrolyte, and (3) tap gently on the side of the pipette if bubbles are still present. This manipulation can eliminate discontinuities altogether; alternatively, one can use a very fine tungsten or other rigid wire to reach inside the pipette and try to remove any air bubble.

If the goal of the experiment is to anatomically identify the recorded neurons, some dye needs to be injected into the cells. For this purpose, a tracer needs to be incorporated in the electrolyte. This tracer can be a fluorescent dye like Lucifer yellow, or it can be a nonfluorescent substance like the widely used small biotinylated molecules (biocytin, neurobiotin), or even a mixture of several dyes. Dye injection can be achieved through the application of high pressure (Sik *et al.*, 1993; Tamamaki and Nojyo, 1993) or by the application of an electrical current (i.e., if the dye is polarized).

Most widely used intracellular labeling material is biocytin. This substance is a biotin–lysine complex of low molecular weight containing about 65% biotin, which retains a high affinity for avidin. Because of the high affinity of biotin to avidin, the conventional avidin–biotin complex method is conveniently used to reveal the recorded cell. The biotin-containing tracer can be injected either with positive current (i.e., biocytin) (Horikawa and Armstrong, 1988) or with both positive and negative current pulses (i.e., neurobiotin) (Kita and Armstrong, 1991).

IV. COMBINED TECHNIQUES

The combination of the described in vivo electrophysiological and anatomical methods allows both the functional and the structural characterization of neurons. First, the intrinsic electrophysiological properties of the neurons are investigated using intracellular recording methods (Fig. 6.1) in a condition where all the synaptic connections are intact and the extracellular environment is undisturbed. Second, the activity of each neuron is correlated with the neural network activity that is recorded by the extracellular electrode. Even though in many cases a rough classification of the recorded neuron can be achieved by analysis of the electrophysiological traces (excitatory, inhibitory neuron, etc.), the exact characterization requires labeling of the neuron (Fig. 6.2). Via the injection of a tracer the recorded cell can be visualized and identified neurochemically (Fig. 6.3). Because of its low molecular weight, biocytin diffuses easily into small structures. This characteristic makes the detection of complete axonal and dendritic arborization possible (Fig. 6.4). Using an additional immunoreaction and processing for electron microscopy, even the targets or afferents of the investigated neurons can be studied at the ultrastructural level (Sik *et al.*, 1995).

The power of the combined methods is demonstrated below in a hippocampal inhibitory (basket) cell, but naturally the same method can be adapted to any neuronal type. With the appreciation of the pivotal function of inhibitory cells in the orchestration of neural activity many questions need to be answered; for example, how inhibitory and excitatory inputs change the membrane potential of the neuron, how different types of inhibitory neurons fire, how different subtypes of inhibitory cells participate in network oscillations, how inhibitory cells are connected to excitatory versus inhibitory neurons, what is the size of an area that a single inhibitory neuron can innervate, how many excitatory and inhibitory neurons are innervated by a single inhibitory cell, etc. After analyzing the spontaneous activity of the basket cell during theta and non-theta network oscillations (not shown), we analyzed the response of the neuron to positive and negative current injections. The recorded inhibitory cell was firing rhythmically when theta activity was present in the extracellular field potential (Fig. 6.5B). The neuron showed firing frequency accommodation when positive current was injected intracellularly and found no sign of sag (I_h) current (Fig. 6.5A).

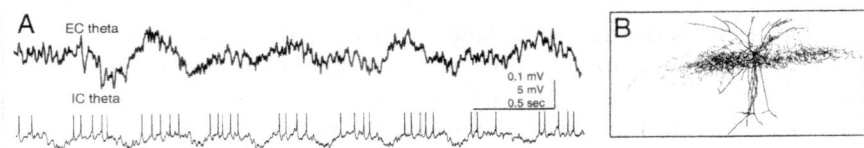

Figure 6.1. (A) Simultaneous recording of extracellular theta activity in the pyramidal cell layer of area CA1 of the rat hippocampus (EC theta) and intracellular activity of a basket neuron (IC theta). (B) Reconstruction of the dendritic and axonal arborization of the basket cell. (Reprinted with permission from Ylinen *et al.*, 1995.)

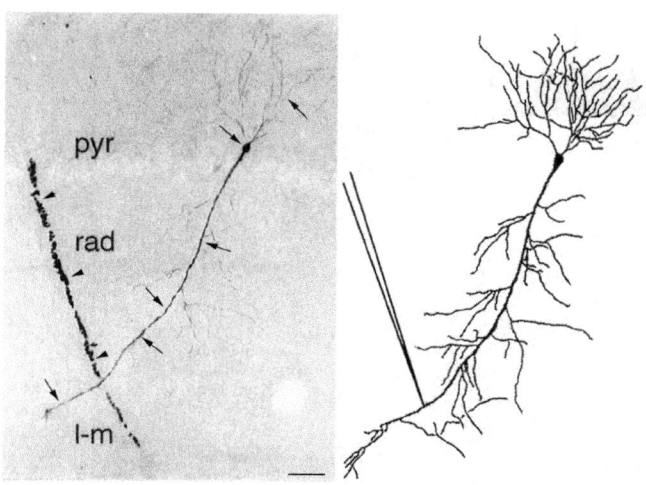

Figure 6.2. Photograph of the recording micropipette track (arrowheads) and the biocytin-filled neuron (arrows) and its camera lucida reconstruction. An intracellular recording was made from the apical shaft of a pyramidal cell at the border of stratum radiatum (*rad*) and lacunosum-moleculare (*l-m*) in area CA1 of the rat hippocampus. The electrode was moved beyond the dendrite during the experiment. The pipette track is filled with reaction product caused by peroxidase activity in red blood cells. *Abbreviation*: pyr, pyramidal layer. Scale bar: 50 μm. (Reprinted with permission from Kamondi *et al.*, 1998.)

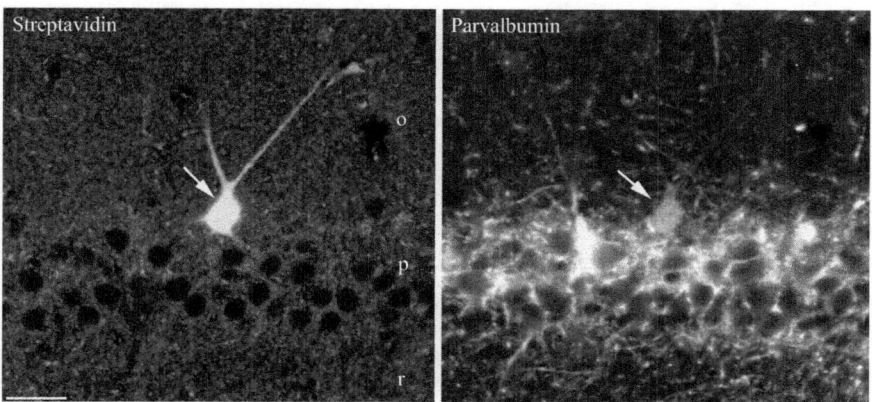

Figure 6.3. Immunohistochemical identification of the filled cell. The electrophysiological property of the neuron was recorded: the cell fired close to the top of theta oscillation and also during ripple activity (not shown). After recording and subsequent filling of the neuron, the visualization was achieved using fluorochrome-conjugated (Alexa 546) streptavidin (left panel). The picture was taken using a confocal laser microscope. The parvalbumin expression of the cell (arrow) was demonstrated by immunofluorescence using a different fluorochrome (Alexa 488) (right panel). The cell was later reconstructed and classified as basket cell (Dumont and Sik, 2006). *Abbreviations*: o, stratum oriens; p, stratum pyramidale; r, stratum radiatum. Scale bar: 25 μm.

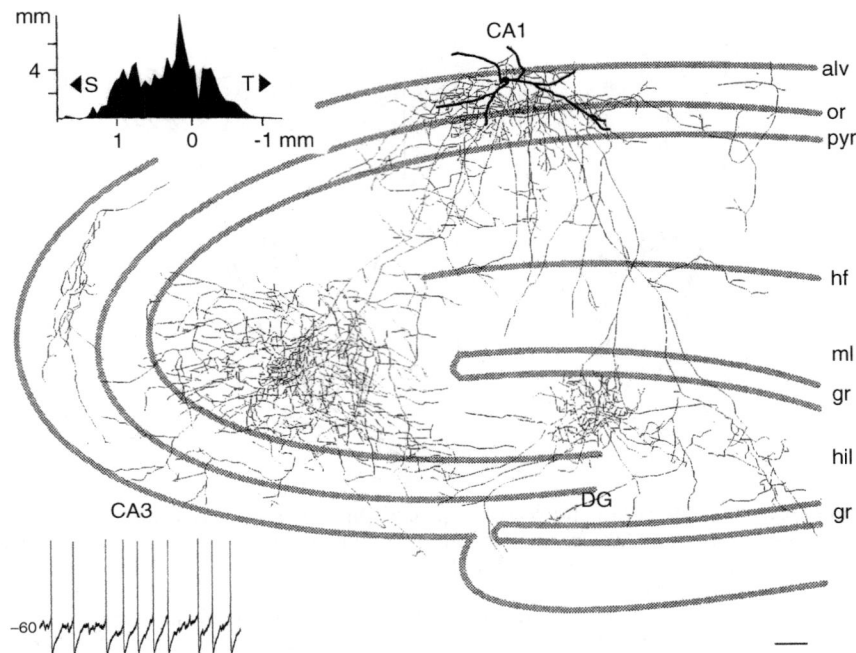

Figure 6.4. Reconstruction of an intracellularly recorded and subsequently filled feedback neuron in area CA1 of the rat hippocampus. After "developing" the filled neuron using the DAB–Ni method, the complete dendritic and axonal arborization of the neuron could be reconstructed. The cell body of the neuron is located in the alveus; axon collaterals are present in areas CA3, CA1, and in the hilus of dentate gyrus (DG). The upper left inset indicates the summated length of axon collateral along the septotemporal axis. The lower left inset shows the spontaneous activity of the neuron at resting membrane potential. *Abbreviations*: S, septal direction; T, temporal direction; alv, alveus; or, stratum oriens; pyr, stratum pyramidale; rad, stratum radiatum; hf, hippocampal fissure; ml, molecular layer; gr, granule cell layer; hil, hilus. Scale bar: 100 μm. (Reprinted from Sik *et al.*, 1994.)

——————————————————————————————————→

Figure 6.5. Complete electrophysiological, neurochemical, and neuroanatomical characterization of a hippocampal basket cell. (A) Response of the neuron to depolarizing and hyperpolarizing current injection. (B) Simultaneous recording of intracellular activity of the basket neuron and extracellular activity during theta oscillation. (C) Biocytin-labeled basket cell. (D) Fluorescent parvalbumin immuno-labeling of the same neurons (arrow). (E) Parvalbumin-containing target of the filled basket cell. In the inset, the white arrow indicates a putative synaptic contact between the biocytin-filled terminal and a parvalbumin-immunoreactive neuron. (F) Correlated electron microscopic analysis of the same bouton shows a symmetric synapse (arrow in the inset) on the cell body. (G) Partial reconstruction of the labeled basket cell indicating other parvalbumin-immunoreactive targets (large circles). Inset shows the position of the neuron in the CA1 region of the hippocampus. (H) Distribution of pyramidal cell and parvalbumin-immunoreactive inhibitory neuron targets of the intracellularly filled basket cell in the septotemporal direction. Overall, 99 boutons in contact with 64 parvalbumin-positive cells were counted. Graph in the

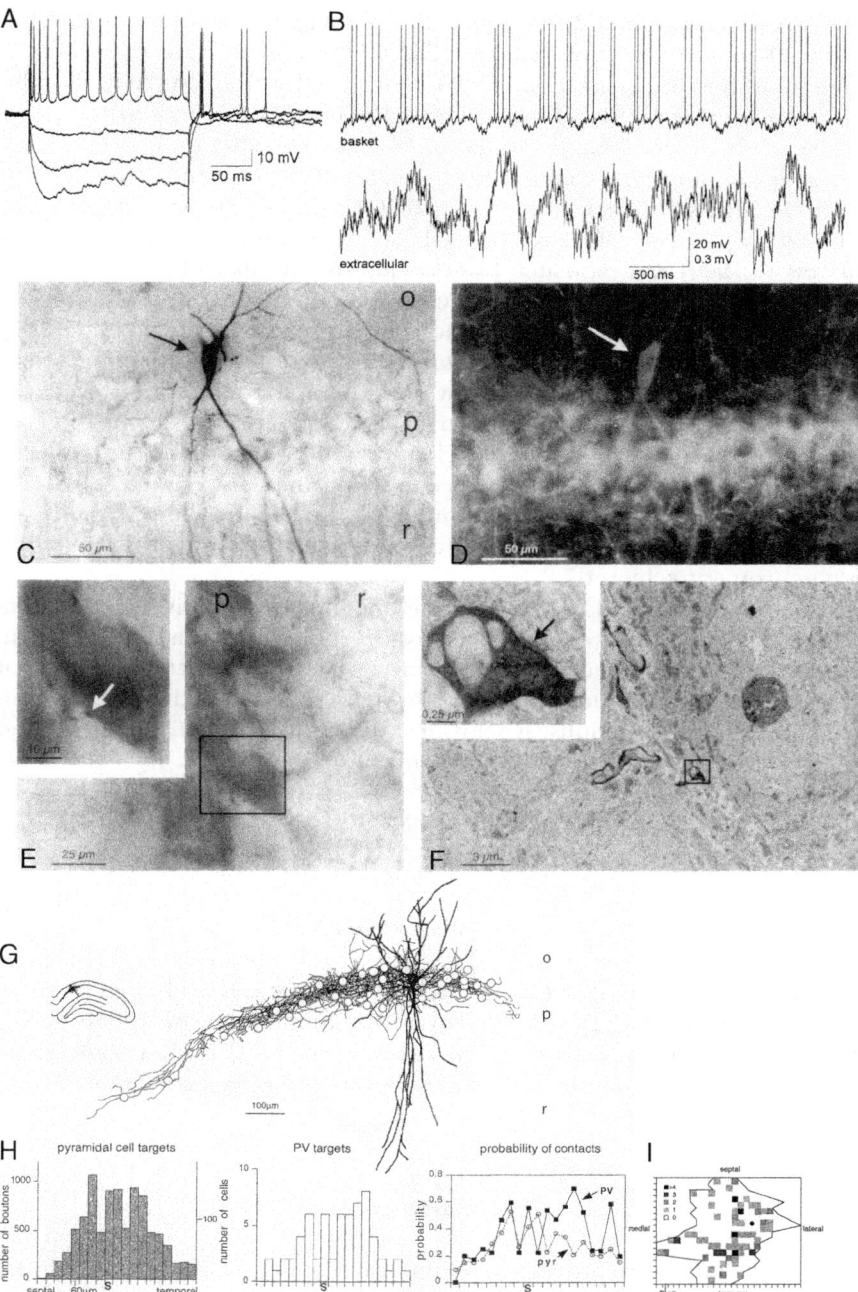

Figure 6.5. (*Cont.*) middle shows the probability of pyramidal and parvalbumin-immunoreactive inhibitory cells innervated by the filled neuron. (I) The 2D distribution of the interneuron–interneuron contacts is shown in H. *Abbreviations*: pyr, pyramidal; PV, parvalbumin; S, position of the soma. (Adapted, with permission, from Wang and Buzsaki, 1996, and Sik *et al.*, 1995.)

The filled cell contained parvalbumin (PV) (Fig. 6.5C,D) and innervated other PV-immunoreactive neurons (Fig. 6.5E). The putative synaptic contacts identified under the light microscope were further analyzed using the electron microscope (Fig. 6.5F). Indeed, the filled basket cells formed symmetric (inhibitory) synaptic contacts on other PV-containing neurons besides terminating on pyramidal cells. The number of synaptic contacts and synaptic targets were determined (Fig. 6.5G). Overall, 99 boutons in contact with 64 PV-positive cells were counted. The total number of pyramidal cell targets (~1500) was estimated by counting the number of boutons of the filled basket cell in each 60-μm-thick Vibratome® section, and assuming that a basket cell formed 9–10 boutons on a single pyramidal cell (Halasy *et al.*, 1996). The probability of contacts formed on PV-containing versus pyramidal cells was calculated by dividing the number of contacted PV or pyramidal cells by the total number of PV or pyramidal cells in the area innervated by the axon collaterals. The probability of postsynaptic contacts, however, decreased with the distance between the cell pairs (Fig. 6.5H,I). Thus, with the sequential application of the aforementioned methods the complete electrophysiological and neuroanatomical characterization of single neuron was achieved.

A computer simulation based on the obtained quantitative data demonstrates that inhibitory synaptic transmission could provide a suitable mechanism for synchronized oscillations in a sparsely connected network of inhibitory cells. This network can, through subthreshold oscillations in excitatory cell populations, synchronize discharges of spatially distributed excitatory neurons (Wang and Buzsaki, 1996).

V. SUMMARY OF ADVANTAGES AND LIMITATIONS

A. Advantages

Combination of in vivo intracellular recordings with morphological methods can provide crucial and detailed information about the structure and function of neurons. The quantitative data of connections of a single intact neuron are indispensable for further meaningful computer simulation of neural networks.

B. Limitations

In spite of the great advantages of the combination of in vivo intracellular or extracellular recording with neurochemical and morphological characterization of the neurons, several limitations still remain.

- Since the recording and subsequent filling of the recorded neurons are performed blindly, this is a very time-consuming process. Tedious work may result in only a handful of analyzed cells. Thus, if the experiment requires dozens of cells to be analyzed, it is advisable to use alternative methods such as the in vitro preparation or in vivo juxtacellular labeling.

- Anesthetics can alter the firing pattern of the neurons; therefore, careful judgment of the chosen drug is a prerequisite for reliable electrophysiological analysis. To overcome this problem, "juxtacellular" labeling was adapted to unanesthetized, drug-free animals by taking advantage of the head-restrained recording technique (Lee *et al.*, 2004, 2005).
- When the sharp electrode penetrates the cell, neurons often discharge artificially (the so-called "injury discharge"). If the membrane does not seal perfectly around the intercellular electrode, it can induce a higher activity rate than that in normal circumstances, resulting in erroneous identification of the neural activity.
- Precise application of pharmacological reagents into a small area is difficult or impossible.

Unless one asks specific questions that can be addressed only by intracellular recording (like membrane oscillations, EPSP, or IPSP measurement in an identified cell, etc.), this powerful but time-consuming approach can be substituted by juxtacellular labeling. Despite these limitations, we feel confident that the recording of electrophysiological activity of single neurons and analysis of their interactions with the neuronal network, in conjunction with the acquisition of accurate structural data on the synaptic architecture, provide sufficient data for realistic experimental modeling of neuronal function.

APPENDIX: DETAILED METHODS

A. Anesthetics

Different anesthetics can be used depending on the planned length of the recording session, whether the animal has to survive, etc. In acute terminal experiments (when the animal is sacrificed following the recording), urethane is the preferred anesthetic because the anesthetic effect lasts many hours. If the animal needs to survive longer than 12 h after the experiment, anesthesia by a cocktail of ketamine and xylazine is recommended.

- *Urethane*: Prepare urethane stock solution (5 g urethane in 10 ml of 0.9% NaCl) and inject i.p. (1.3–1.4 g/kg for rat);
- *Ketamine/xylazine*: Inject i.p. 75 mg/kg ketamine and 10 mg/kg xylazine (rat). If ketamine/xylazine is used in some cases, it has to be readministered during the experiment. Use ketamine only (20 mg/kg) whenever the animal shows signs of awakening. It can be injected i.p. or intravenously into the tail vein. This later provides more precision of the dosage but is harder to execute.

- *Combination of urethane and ketamine/xylazine*: Use 1.25 g/kg urethane and supplement the drug with 20 mg/kg ketamine and 2 mg/kg xylazine as needed.

B. Surgery and Implantation of Stimulating and Extracellular Recording Electrodes

- Shave off the hair from the area overlying the part of the brain studied, i.e., over the skull or the vertebrae.
- Place the animal into a sturdy stereotaxic apparatus. An antivibration table is highly recommended to make the intracellular recordings more stable and longer lasting.
- Use an animal thermoregulation device to keep the body temperature constant. A low body temperature dramatically decreases the activity of neurons.
- If the recordings will take prolonged periods, protect the animal's eyes with commercially available eye drops or with paraffin oil, etc. against dehydration of the cornea.
- Cut the skin with a scalpel and drill a small hole over the region where the recording will be performed. The hole should be as small as possible but should also provide sufficient space for manipulation. The ideal size is about 1 × 1 mm. If an extracellular recording electrode is placed into the same region through the same bone window, the size of the hole must be larger (1.5 × 1.0 mm).
- If a stimulating electrode is to be used, prepare a small bone window over the desired stimulation region.
- Implant the stimulating electrode and fix it with acrylic cement.
- Maneuver the extracellular recording electrode into position. The position of the electrode can be checked by driving the appropriate input pathway with the stimulating electrode. The extracellular recording electrode should be fixed by acrylic cement; thus, no extra electrode holder (which would occupy space) is necessary.
- Open the dura mater with the very sharp tip of a pointy surgical blade (for example, size 11) or by the tip of a small needle (like 27 g × 1/2″). Be careful not to cause any bleeding. If a large blood vessel is in the way, drill another hole or enlarge the original hole. Do not let the brain surface dry out, e.g., by putting a drop of 0.9% NaCl into the opening.

C. Intracellular Recording

After lowering the intracellular electrode into the area of interest, cells need to be impaled with the sharp electrode. There are differences in the membrane structure and intracellular ion content among cells that will result in differences of penetrability and survival of the cells following the penetration. As a general procedure, follow the next steps:

- Pull the electrode from 2.0-mm-diameter thick wall glass capillary.
- Fill the electrode with electrolyte (0.5 M potassium acetate) containing 1–3% biocytin.
- Keep the electrodes in a humid chamber (Petri dish with some moistened paper or cotton) to prevent clogging due to drying at the tip of the electrode.
- Insert the intracellular electrode into the region of interest, and cover the bone window with a mixture of paraffin and paraffin oil (1:1, kept warm before the application at ∼60°C) to prevent drying of the area and to reduce pulsation of the brain.
- Use quick, small steps (2 μm/step) to advance the intracellular electrode.
- When the recorded potential starts decreasing, it indicates that the electrode is pushed to a membrane. Use one of the following methods to penetrate into the cell: (a) "buzz" [small, short lasting (1–20 ms) electric current injection setting the electrode tip in motion causing small vibration of the tip. Many intracellular amplifiers have this button on their recording unit] and (b) mechanically, by very gently tapping the electrode holder or the motor. This will move the pipette a few micrometers, which may be sufficient to penetrate the cell membrane.
- After penetrating into the cell, apply a negative current to counteract the depolarization caused by ion leakage through the membrane opening through which the pipette tip has entered the cell. Hyperpolarize the cell till it stops firing and keep it at this state for a couple of minutes to allow the cell membrane to seal around the distal pipette shaft.
- Perform the electrophysiological recording. At the end of the session, use positive (or negative if it has beneficial effect) current pulses to inject the dye into the cell (300–500 ms at 1 Hz using 0.5–2.5 nA current). The necessary labeling time differs from cell to cell. A labeling time of 2–30 min is usually sufficient to obtain complete labeling of neurons. Cells with large axonal arborization, and/or long projection may require longer time with higher current.

D. Survival Time

The intracellularly injected dye spreads through the neuron and its processes by active transport mechanisms and by diffusion. If the reconstruction of complete axonal arborization is the goal, this requires a longer transport/diffusion period until the dye has filled all the thin axon collaterals down to their terminal arborizations. Neurons with axons projecting over longer distances require lengthier survival time. Typically, the survival time after the injection of the marker varies between 0 and 12 h. Longer survival is not recommended if biocytin or neurobiotin is used, because enzymes that might be activated due to the trauma may destroy these tracers. If a longer survival time is required, injection with biotinylated dextran amine is recommended instead of biocytin (BDA, MW 3000).

E. Fixation

Choosing the appropriate fixative is important if the intention is to determine the neurochemical features of the recorded neuron or other cells (like target or input cells). Most immunohistochemical staining procedures require fixation with a buffered solution of formaldehyde (depolymerized paraformaldehyde).

- Reanesthetize the animal if necessary.
- Open the thorax and insert a large diameter needle via the left ventricle into the ascending aorta.
- Place a clamp on the descending aorta, between the liver and the lungs.
- Open the right atrium to allow blood and perfusates to flow out.
- Flush the blood using 0.9% NaCl or 0.1 M PBS until the outflowing liquid from the heart is clear, typically 1–3 min.
- Switch the solution to the appropriate fixative (typically 4% formaldehyde in 0.1 M phosphate buffer, pH 7.4).
- Fix the brain for 30 min.
- Remove the brain from the skull. If the brain is too soft, postfixation may be carried out (1 h to overnight).

F. Visualization of the Intracellularly Filled Cells

Depending on the goal, the labeled cells can be visualized with either a fluorescent dye or a permanent marker [like 3,3′-diaminobenzidine (DAB) peroxidase reaction product]. We recommend a fluorescence-based visualization if the aim of the experiment is to further characterize the recorded cells immunohistochemically (see section "Immunolabeling"). If the reconstruction of the cell is the goal, the sections have to be kept in sequential order during the whole process.

- Section the brain using a vibrating microtome (section thickness 30–100 μm).
- Wash the fixative with PB (5×15 min).
- Treat the section with detergent if examination is planned other than electron microscopy (0.5% Triton X-100 in PB for 30 min).
- Wash with Tris-buffered saline (TBS) (3 × 15 min).
- React the sections with either HRP containing avidin–biotin complex (ABC; dilution 1:500 in TBS) or with fluorochrome-conjugated streptavidin (dilution 1:100–1:1000 in TBS depending on the type of the fluorescent dye) (1 h at room temperature for ABC; 5 h overnight at 4 h for fluorescence).
- Wash the reagent (3 × 15 min TBS).
- If fluorescent dye has been used, mount and coverslip with antifading medium and analyze the finding in a fluorescence microscope using the proper filtering. If ABC has been used, visualize the cell using the

following steps:

1. Prepare a DAB–Ni solution: measure 100 ml 0.01 M PB.
2. Add 30 mg 3,3′ diaminobenzidine (DAB) (☣ toxic and carcinogenic, decontaminate spoils and leftovers with chlorine bleach; ☼ light sensitive, cover with aluminum foil).
3. Add 40 mg NH_4Cl.
4. Add 5 ml 0.05 M $NiNH_4SO_4$ (Nickel ammonium sulfate; add dropwise under agitation).
5. Filter this DAB–Ni solution.
6. Remove the last wash, add 1 ml of DAB–Ni solution.
7. Allow the sections to be saturated with DAB–Ni (20 min room temperature, dark).
8. Add 10 µl of H_2O_2 solution (made up fresh by pipetting 10 µl of 30% H_2O_2 solution into 10 ml double distilled water) to start the reaction. A black reaction product will form in all structures that contain ABC (the reactive compound is the peroxidase). This reaction can take 5–40 min to fully develop. Inspect the section regularly in a microscope to monitor the progress of the reaction.

- When sufficient reaction product has formed, rinse the sections three times in TBS.
- Mount on glass slides and dry. Do not coverslip if the procedure is followed up with immunolabeling (see below).

G. Immunolabeling

In order to determine the neurochemical content of the recorded cell, an immunohistochemical reaction needs to be performed. Since the DAB or DAB–Ni reaction product masks the immunosignal of the cell, an immunofluorescence technique should be used in this case (Kawaguchi, 1993). Once the cell has been visualized by a streptavidin-conjugated fluorochrome (see above), a regular immunoreaction using fluorescence secondary antibody can be applied. The exact steps can be found in other books like *Immunohistochemistry* (Cuello, 1993). Briefly:

- Wash out the fixative (see above).
- Wash the sections by rinsing 3 × 10 min with 0.05 M TBS.
- Block aspecific immunosignal by incubating 45 min in blocking solution (TBS containing 5% normal goat serum (NGS) or other blocking serum; and 0.5% Triton X-100).
- Treat the sections with fluorochrome-conjugated streptavidin (dilution 1:100–1:1000 in TBS depending on the type of the fluorescent dye) (5h overnight at 4°C) in the dark.
- Wash out the reagent (3 × 15 min TBS).

- Use primary antibody diluted into TBS containing 0.5% NGS (or other serum), 0.01% sodium azide added. Incubate the sections in the incubation medium in the dark (overnight 2 days at 4°C) to allow antibodies to penetrate into the sections. Longer incubation results in better penetration of the antibody. During incubation, place the vials or well plates that contain the sections on a rocking plateau to ensure gentle agitation.
- Wash out the primary antibody (3 × 15 min TBS).
- Incubate the sections (in the dark) in the solution containing the fluorochrome-conjugated secondary antibody (dilution 1:100–1:1000 in TBS for 6 h to overnight). The excitation–emission spectrum of this fluorochrome should be of course different from that used to visualize the labeled cell.
- Wash 3 × 15 min with TBS.
- Mount on slides, allow sections to air-dry.
- Add antifading reagent like Mowiol (has to be at room temperature) and coverslip.
- Seal with nail polish.
- Analyze the neurochemical content of the cell using a fluorescence microscope equipped with the proper excitation–emission filters.
- Store at 4°C temporarily, or at −20°C for the long term.
- Then the permanent visualization of the filled cells using the ABC-DAB protocol is performed as described above.

H. Reconstruction

The entire dendritic and axonal arborization of the recorded neurons can be reconstructed after successful intracellular labeling. If sections are air-dried after mounting on slides, shrinkage in Z direction is substantial (about 80–90%). In the case when the real three-dimensional (3D) structure of the neuron is important, embedding of the section in plastic resin is necessary. The following protocol is an example using Durcupan, but other resins can also be used, such as Araldite, Spurr, and so forth.

- Dehydrate the sections in an ascending alcohol series (50%, 70%, 90%, 2 × 100% 10 min each).
- Change to intermediate solution (100% propylene oxide if Durcupan is used) (2 × 10 min). *Note*: Propylene oxide is volatile, toxic, and combustible. Use a fume hood and wear gloves.
- Place the section in pure Durcupan (overnight).
- Mount the sections, coverslip, and cure in an oven at 58°C for 24 h.

The cells can be reconstructed using a drawing tube or a microscope-computer equipped with Neurolucida software. The advantage of using the latter equipment is that the 3D information is preserved in each individual section. When a drawing tube is used, the reconstruction will be manufactured using the 2D projection of each section. Thus, with this procedure the 3D information in individual sections is lost (see further details in the chapters by Ascoli and Scorcioni and Duque and Zaborszky in this volume).

I. Equipment and Supplies (Some Recommended Equipment Is in Parenthesis, But Other Items Can Be Used)

1. Equipment for Surgery and Recording

Drill (NSK Emax)
Vibration isolation system (Newport VH Isostation)
Stereotaxic apparatus (Kopf Model 920)
Thermoregulator with heating pad (CWE TC 1000)
Operating microscope (Olympus SZ series)
Light source (WPI)
Inchworm motor system (Burleigh)

2. Equipment for Data Acquisition

Digital oscilloscope (Tektronix TDS 2014)
Noise reduction device (Hum bug) or traditional Faraday cage
Amplifier (Axon Multiclamp 700A computer-controlled microelectrode amplifier with Softpanel or Axoclamp-2B)
Analog-digital converter (Axon Digidata 1322A data acquisition system)
Data acquisition software (Axon pClamp 9.0 electrophysiology software)
Isolated pulse stimulator (A-M systems Model 2100)
Differential amplifier (A-M systems Model 3000) for recording EEG

3. Other Equipment

Micropipette puller (Sutter Instruments)

VI. CHEMICALS

Biocytin or Neurobiotin (Vector Laboratories)
Potassium acetate (Sigma)
Paraformaldehyde, glutaraldehyde, Durcupan (Electron Microscopy Sciences)

V. SOLUTIONS

- Phosphate buffer (PB) 0.2 M pH 7.4

    Stock solution A: 0.2 M NaH_2PO_4
    Stock solution B: 0.2 M Na_2HPO_4
    Add solution A to solution B in a 1:4 ratio until the pH reaches 7.4 to give 0.2 M PB.

- Tris-buffered saline (TBS) 0.05 M pH 7.4

  Trizma base 0.05 M
  Trizma acid 0.05 M
  0.9% NaCl

ACKNOWLEDGMENTS. This work was supported by a Canadian Institutes of Health Research (CIHR) grant and FRSQ. The author thanks Drs. Martin Deschênes and György Buzsáki for helpful discussion. Most of the work was performed in Dr. Buzsáki's laboratory at Rutgers University, NJ, USA.

REFERENCES

Bartho, P., Hirase, H., Monconduit, L., Zugaro, M., Harris, K. D., and Buzsaki, G., 2004, Characterization of neocortical principal cells and interneurons by network interactions and extracellular features, *J. Neurophysiol.* **92**:600–608.

Bragin, A., Jando, G., Nadasdy, Z., Hetke, J., Wise, K., and Buzsaki, G., 1995, Gamma (40–100 Hz) oscillation in the hippocampus of the behaving rat, *J. Neurosci.* **15**:47–60.

Buzsaki, G., 2004, Large-scale recording of neuronal ensembles, *Nat. Neurosci.* **7**:446–451.

Buzsaki, G., Hazan, L., Zugaro, M. B., and Csicsvari, J., 2004, Neuroscope: a viewer for neurophysiological data, *Soc. Neurosci.* (abstract):768.2.

Buzsaki, G., and Kandel, A., 1998, Somadendritic backpropagation of action potentials in cortical pyramidal cells of the awake rat, *J. Neurophysiol.* **79**:1587–1591.

Buzsáki, G., Traub, R., and Pedley, T., 2003, The cellular synaptic generation of EEG. In: Ebersole, J.S. and Pedley, T.A. (eds.), *Current Practice of Clinical Encephelography*, Philadelphia: Lippincott-Williams and Wilkins, (3rd ed.), pp. 1–11.

Chang, H. T., Wilson, C. J., and Kitai, S. T., 1981, Single neostriatal efferent axons in the globus pallidus: a light and electron microscopic study, *Science* **213**:915–918.

Csicsvari, J., Henze, D. A., Jamieson, B., Harris, K. D., Sirota, A., Bartho, P., Wise, K. D., and Buzsaki, G., 2003, Massively parallel recording of unit and local field potentials with silicon-based electrodes, *J. Neurophysiol.* **90**:1314–1323.

Cuello, A. C., 1993, Immunohistochemistry II, IBRO handbook series: Methods in the neurosciences. John Wiley & Sons, England.

Ferster, D., and Jagadeesh, B., 1992, EPSP–IPSP interactions in cat visual cortex studied with in vivo whole-cell patch recording, *J. Neurosci.* **12**:1262–1274.

Freund, T. F., and Buzsaki, G., 1996, Interneurons of the hippocampus, *Hippocampus* **6**:347–470.

Gray, C. M., Maldonado, P. E., Wilson, M., and McNaughton, B. L., 1995, Tetrodes markedly improve the reliability and yield of multiple single-unit recordings in cat striate cortex, *J. Neurosci. Methods* **63**:43–54.

Grossman, R. G., and Hampton, T., 1968, Depolarization of cortical glial cells during electrocortical activity, *Brain Res.* **11**:316–324.

Halasy, K., Buhl, E. H., Lorinczi, Z., Tamas, G., and Somogyi, P., 1996, Synaptic target selectivity and input of GABAergic basket and bistratified interneurons in the CA1 area of the rat hippocampus, *Hippocampus* **6**:306–329.

Harris, K. D., Henze, D. A., Csicsvari, J., Hirase, H., and Buzsaki, G., 2000, Accuracy of tetrode separation as determined by simultaneous intracellular and extracellular measurements, *J. Neurophysiol.* **84**:401–414.

Hassin, G., 1979, Pikeperch horizontal cells identified by intracellular staining, *J. Comp. Neurol.* **186**:529–540.

Hazan, L., Zugaro, M. B., Csicsvari, J., Creese, I., and Buzsáki, G., 2004, Klusters: a graphical application for spike sorting of extracellular units, *Soc. Neurosci.* (abstract):768.3.

Heimer, L., and Robards, M. J., 1981, *Neuroanatomical Tract Tracing Methods*, New York: Plenum Press.

Henze, D. A., Borhegyi, Z., Csicsvari, J., Mamiya, A., Harris, K. D., and Buzsaki, G., 2000, Intracellular features predicted by extracellular recordings in the hippocampus in vivo, *J. Neurophysiol.* **84**:390–400.

Horikawa, K., and Armstrong, W. E., 1988, A versatile means of intracellular labeling: injection of biocytin and its detection with avidin conjugates, *J. Neurosci. Methods* **25**:1–11.

Hubel, D. H., 1957, Tungsten microelectrodes for recording single units, *Science* **125**: 549–550.

Jankowska, E., Rastad, J., and Westman, J., 1976, Intracellular application of horseradish peroxidase and its light and electron microscopical appearance in spinocervical tract cells, *Brain Res.* **105**:557–562.

Kamondi, A., Acsady, L., and Buzsaki, G., 1998, Dendritic spikes are enhanced by cooperative network activity in the intact hippocampus, *J. Neurosci.* **18**:3919–3928.

Kaneko, A., 1970, Physiological and morphological identification of horizontal, bipolar and amacrine cells in goldfish retina, *J. Physiol.* **207**:623–633.

Kawaguchi, Y., 1993, Physiological, morphological, and histochemical characterization of three classes of interneurons in rat neostriatum, *J. Neurosci.* **13**:4908–4923.

Kawaguchi, Y., Wilson, C. J., and Emson, P. C., 1989, Intracellular recording of identified neostriatal patch and matrix spiny cells in a slice preparation preserving cortical inputs, *J. Neurophysiol.* **62**:1052–1068.

Kawaguchi, Y., Wilson, C. J., and Emson, P. C., 1990, Projection subtypes of rat neostriatal matrix cells revealed by intracellular injection of biocytin, *J. Neurosci.* **10**:3421–3438.

Kelly, J. P., and Van Essen, D. C., 1974, Cell structure and function in the visual cortex of the cat, *J. Physiol.* **238**:515–547.

Kita, H., and Armstrong, W., 1991, A biotin-containing compound N-(2-aminoethyl)biotinamide for intracellular labeling and neuronal tracing studies: comparison with biocytin, *J. Neurosci. Methods* **37**:141–150.

Kita, H., and Kitai, S. T., 1986, Electrophysiology of rat thalamo-cortical relay neurons: an in vivo intracellular recording and labeling study, *Brain Res.* **371**:80–89.

Kitai, S. T., Kocsis, J. D., Preston, R. J., and Sugimori, M., 1976, Monosynaptic inputs to caudate neurons identified by intracellular injection of horseradish peroxidase, *Brain Res.* **109**:601–606.

Lee, M. G., Hassani, O. K., Alonso, A., and Jones, B. E., 2005, Cholinergic basal forebrain neurons burst with theta during waking and paradoxical sleep, *J. Neurosci.* **25**:4365–4369.

Lee, M. G., Manns, I. D., Alonso, A., and Jones, B. E., 2004, Sleep-wake related discharge properties of basal forebrain neurons recorded with micropipettes in head-fixed rats, *J. Neurophysiol.* **92**:1182–1198.

Li, X.-G., Somogyi, P., Tepper, J. M., and Buzsaki, G., 1992, Axonal and dendritic arborization of an intracellularly labeled chandelier cell in the CA1 region of rat hippocampus, *Exp. Brain Res.* **90**:519–525.

Li, X. G., Somogyi, P., Ylinen, A., and Buzsaki, G., 1993, The hippocampal CA3 network: an in vivo intracellular labeling study, *J. Comp. Neurol.* **339**:181–208.

Margrie, T. W., Brecht, M., and Sakmann, B., 2002, In vivo, low-resistance, whole-cell recordings from neurons in the anaesthetized and awake mammalian brain, *Pflugers Arch.* **444**:491–498.

McNaughton, B. L., O'Keefe, J., and Barnes, C. A., 1983, The stereotrode: a new technique for simultaneous isolation of several single units in the central nervous system from multiple unit records, *J. Neurosci. Methods* **8**:391–397.

Nadasdy, Z., Csicsvari, J., Penttonen, M., and Buzsaki, G., 1998, Extracellular recording and analysis of electrical activity: from single cells to ensembles. In: Eichenbaum, H., and Davis, J. L. (eds.), *Neuronal Ensembles: Strategies for Recording and Decoding*, New York: Wiley-Liss, pp. 17–55.

Sandell, J. H., and Masland, R. H., 1988, Photoconversion of some fluorescent markers to a diaminobenzidine product, *J. Histochem. Cytochem.* **36**:555–559.

Sik, A., Penttonen, M., and Buzsaki, G., 1997, Interneurons in the hippocampal dentate gyrus: an in vivo intracellular study, *Eur. J. Neurosci.* **9:**573–588.

Sik, A., Penttonen, M., Ylinen, A., and Buzsaki, G., 1995, Hippocampal CA1 interneurons: an in vivo intracellular labeling study, *J. Neurosci.* **15:**6651–6665.

Sik, A., Tamamaki, N., and Freund, T. F., 1993, Complete axon arborization of a single CA3 pyramidal cell in the rat hippocampus, and its relationship with postsynaptic parvalbumin-containing interneurons, *Eur. J. Neurosci.* **5:**1719–1728.

Sik, A., Ylinen, A., Penttonen, M., and Buzsaki, G., 1994, Inhibitory CA1-CA3-hilar region feedback in the hippocampus, *Science* **265:**1722–1724.

Simons, D. J., 1978, Response properties of vibrissa units in rat SI somatosensory neocortex, *J. Neurophysiol.* **41:**798–820.

Snow, P. J., Rose, P. K., and Brown, A. G., 1976, Tracing axons and axon collaterals of spinal neurons using intracellular injection of horseradish peroxidase, *Science* **191:**312–313.

Stewart, W. W., 1978, Functional connections between cells as revealed by dye-coupling with a highly fluorescent naphthalimide tracer, *Cell* **14:**741–759.

Stewart, W. W., 1981, Lucifer dyes—highly fluorescent dyes for biological tracing, *Nature* **292:**17–21.

Takato, M., and Goldring, S., 1979, Intracellular marking with lucifer yellow CH and horseradish peroxidase of cells electrophysiologically characterized as glia in the cerebral cortex of the cat, *J. Comp. Neurol.* **186:**173–188.

Tamamaki, N., Abe, K., and Nojyo, Y., 1987, Columnar organization in the subiculum formed by axon branches originating from single CA1 pyramidal neurons in the rat hippocampus, *Brain Res.* **412:**156–160.

Tamamaki, N., and Nojyo, Y., 1993, Projection of the entorhinal layer-II neurons in the rat as revealed by intracellular pressure-injection of neurobiotin, *Hippocampus* **3:**471–480.

Tepper, J. M., Sawyer, S. F., and Groves, P. M., 1987, Electrophysiologically identified nigral dopaminergic neurons intracellularly labeled with HRP: light-microscopic analysis, *J. Neurosci.* **7:**2794–2806.

Thomas, R. C., and Wilson, V. J., 1965, Precise localization of Renshaw cells with a new marking technique, *Nature* **206:**211–213.

Wang, X. J., and Buzsaki, G., 1996, Gamma oscillation by synaptic inhibition in a hippocampal interneuronal network model, *J. Neurosci.* **16:**6402–6413.

Wilson, M. A., and McNaughton, B. L., 1993, Dynamics of the hippocampal ensemble code for space, *Science* **261:**1055–1058.

Wise, K. D., and Najafi, K., 1991, Microfabrication techniques for integrated sensors and microsystems, *Science* **254:**1335–1342.

Ylinen, A., Soltesz, I., Bragin, A., Penttonen, M., Sik, A., and Buzsaki, G., 1995, Intracellular correlates of hippocampal theta rhythm in identified pyramidal cells, granule cells, and basket cells, *Hippocampus* **5:**78–90.

7

Juxtacellular Labeling of Individual Neurons In Vivo: From Electrophysiology to Synaptology

ALVARO DUQUE and LASZLO ZABORSZKY

ALVARO DUQUE • Department of Neurobiology, Yale University School of Medicine, New Haven, CT 06510   LASZLO ZABORSZKY • Center for Molecular and Behavioral Neuroscience, Rutgers, The State University of New Jersey, Newark, NJ 07102

Animal Preparation for Electrophysiology
Electrophysiology and Labeling
Perfusion
Cutting and Pretreatment of Sections
Visualization of Biocytin-Filled Neuron and Digital Photography
Neurochemical Identification of Biocytin-Filled Neuron
Conversion of the Fluorescent Signal to DAB
Staining for a Second Antigen
Embedding for Electron Microscopy
3D Light Microscopy Reconstructions
Electron Microscopy and 3D Reconstruction from Ultrathin
    Sections
REFERENCES

Abstract: This chapter summarizes details pertaining to the extracellular recording and juxtacellular labeling method and its application to the characterization of basal forebrain neurons. Juxtacellular labeling is compared to other single-cell labeling techniques. Compatibility of the method with fluorescent retrograde/anterograde neural tracing, immunohistochemical, and electron microscopy techniques is also illustrated.

Keywords: basal forebrain, 3D reconstruction, electron microscopy, morphology, neurochemical identification

I. INTRODUCTION

Brain regions consist of elementary local circuits and cell assemblies specialized to carry out discrete computations that eventually give rise to behavior. These circuits are built from regionally specific combinations of limited types of individual neurons (see Nadasdy *et al.*, this volume). However, depending on the computational needs of the brain regions and the particular electrophysiological, neurochemical, and hodological characteristics of different neuronal populations, these circuits become increasingly complex across the neuraxis. Except for a few heuristic attempts by Cajal (1911) in the early part of the twentieth century, it was only in the 1970s when the imaginative cerebellar and cortical circuitry models of Szentágothai were published (Szentágothai, 1970, 1978). These general models were based on correlating (mostly indirectly) the 3D spatial architecture, derived from Golgi studies, with the synaptic pattern of putative circuits obtained from electron microscopical analysis (see also Arbib *et al.*, 1998, and partial list of Szentágothai's publications in Zaborszky *et al.*, 1992). To correctly interpret the complex entanglement of axons and dendrites under the electron microscope, however, it was necessary to combine the full cell visualization of the Golgi method (Golgi, 1883) with the power of electron microscopy to resolve individual synapses. After the initial attempt of Blackstad (1965), the development of gold toning of Golgi impregnated neurons (Fairen *et al.*, 1977) paved the way to identify the synaptic connections of individual neurons (Somogyi, 1977).

A further development in the analysis of neural circuits was the combination of intracellular electrophysiological recordings with horseradish peroxidase (HRP) labeling of the recorded cell (Jankowska *et al.*, 1976; Kitai *et al.*, 1976a, b; Light and Durkovic, 1976; Snow *et al.*, 1976; Cullheim and Kellerth, 1978). HRP, introduced as a neuroanatomical tracer in the early 1970s (Kristensson and Olsson; 1971; LaVail and LaVail, 1972), is an enzyme that catalyzes the reaction by which diaminobenzidine (DAB) precipitates and forms an electron-dense product (Graham and Karnovsky, 1966). The chapters of Somogyi and Freund in the previous edition of this series are excellent reviews on the study of the synaptic relationships of interconnected and chemically identified neurons, using a combination of electrophysiology, HRP filling, Golgi method (Golgi, 1883), and immunocytochemistry (Freund and Somogyi, 1989; Somogyi and Freund, 1989).

Intracellular recordings can also be combined with fluorescent dye labeling and histochemical or immunocytochemical methods to identify the transmitter of the recorded neurons (Aghajanian and Vandermaelen, 1982; Grace and Bunney, 1983a, b). However, morphological reconstructions of these neurons and study of their synaptology require conversion of the fluorescent signal to a permanent DAB end product (Buhl, 1993), a fact that makes this avenue much too cumbersome to be routinely applied.

In vivo technical advances, including the use of voltage-sensitive or calcium-sensitive dyes, for the study of individual synapses in the living animal as described in the chapter of Goldberg *et al.* (this volume) have the disadvantage of being applicable only to superficial layers of the cortex. Hence, they are not useful for investigating deep subcortical structures such as the basal forebrain or the basal ganglia. Intracellular or patch recordings are difficult to apply in vivo when the neuron to be recorded and labeled is in a deep subcortical structure or in an area densely packed with neuropil and fibers. The chances of breaking the fine pipette tips used for intracellular recordings increase dramatically as a function of the distance traveled by the micropipette electrode. Moreover, in vivo intracellular approaches are limited by the stability of the preparation. In particular, respiration and heartbeat produce movement, which is not reduced or only minimally reduced by the use of vibration-free tables. Movement in the preparation compromises the integrity of the micropipette tip, disrupts the recording, and usually quickly kills the cell under investigation. Therefore, most intracellular or patch recordings involving single-cell labeling in the basal forebrain or basal ganglia have taken place in vitro (Alonso *et al.*, 1996; Nambu and Llinas, 1997; Koos and Tepper, 1999, 2002). In vitro preparations are very stable and have, therefore, revealed much about intrinsic electrophysiological properties of these neurons, but the nature of the preparation usually does not allow full morphological reconstructions or chemical identification. Cutting afferents and efferents, however, and maintaining the neuron alive under typical in vitro conditions raises questions as to the validity and relevance of some of the results obtained. Moreover, behavioral correlates such as electroencephalograph (EEG) cannot be obtained in vitro.

In vivo extracellular recordings are usually easier to perform than intracellular recordings. However, for many years a major obstacle in their usefulness was the inability to label the recorded cell. A first approach to resolve this problem was to eject a dye, such as HRP, at the end of the recording session and label a small group of cells found in close proximity to the pipette tip. This effort, however, fell short of identifying "the recorded cell" although it did identify a small area where the recording took place. By decreasing the tip size (to 1–4 μm) of glass microelectrodes filled with 0.5–1.0% HRP in 2 M NaCl and ejecting HRP with 400–1000 nA negative current for 5–25 s after regular extracellular recordings, 30 years ago Lynch and colleagues (Lynch *et. al.*, 1974a, b) managed to label only 1–4 cells per attempt. However, in cases in which only one neuron was labeled, no evidence was provided that the stained neuron was actually the one recorded from. Twenty years later, working in the thalamic reticular nucleus, Pinault (1994) described a method for recording and labeling single neurons extracellularly and for the first time provided evidence that the labeled cell was the one recorded from. The usefulness of this methodology, called juxtacellular labeling, was finally established by Pinault in a 1996 study, which presented further evidence from different rat brain areas, including the neocortex, thalamus, basal ganglia, basal forebrain, and cerebellum, that the juxtacellularly labeled neuron was the one previously recorded from.

In this chapter we describe juxtacellular labeling of single neurons with details based on our own experience, using the method mostly in the basal forebrain of the rat. We will present our own control experiments providing further supporting evidence for the validity of the technique. We will elaborate on the usefulness of the technique by illustrating the compatibility of the labeling procedure with various tracer and immunocytochemical methods both at the light and at the electron microscopical levels to study the synaptology of chemically and electrophysiologically identified neurons, whose activity is also correlated with EEG recordings. The advantages, limitations, and drawbacks of the technique are also discussed.

II. GENERAL METHODOLOGY

Lesions or injections of anterograde and retrograde tracers may be applied prior to recording and labeling of single neurons (see section "Tracer Techniques in Combination with Electrophysiological Recording and Juxtacellular Labeling"). These extra steps can substantially increase the amount of information collected from a single neuron. These and all other procedures involving animals, and details pertaining to animal treatment should be in strict accordance with National Institutes of Health guidelines, which are readily available in publications such as the "Guide for the Care and Use of Laboratory Animals." Before starting any experiments, protocols need to be reviewed and approved by the respective Institutional Animal Care and Use Committee (IACUC). Figure 7.1 illustrates the general methodology.

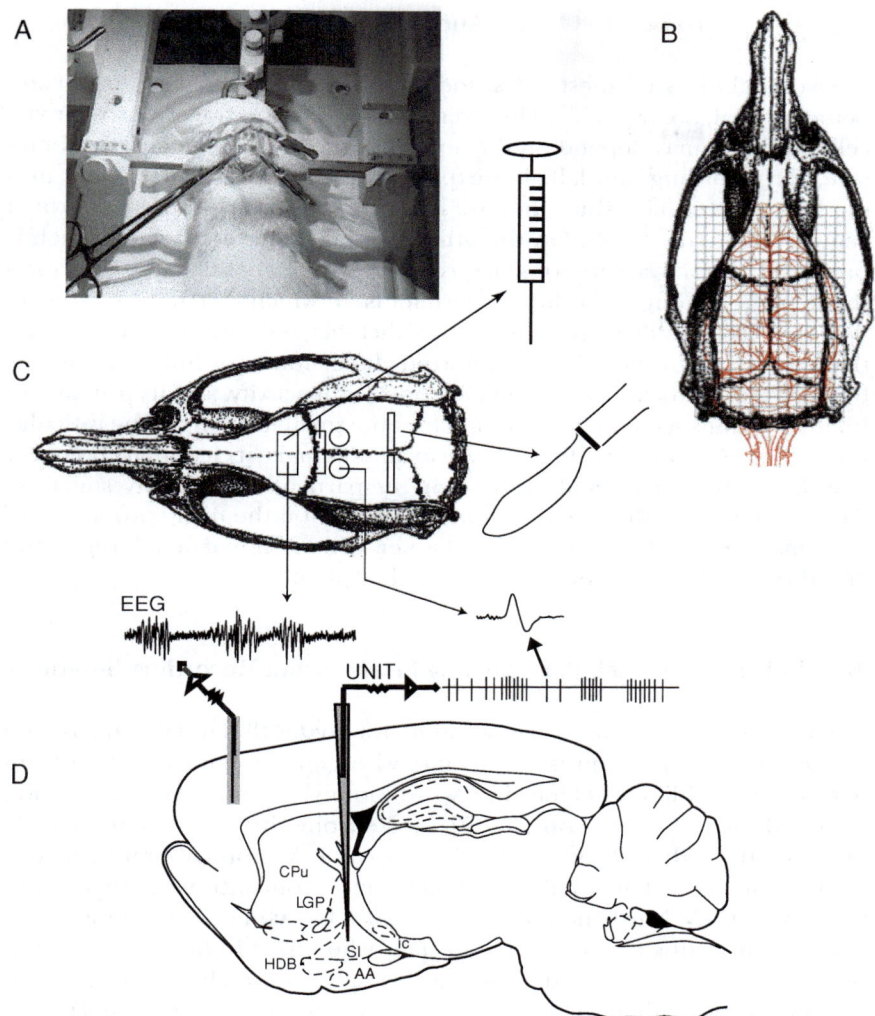

Figure 7.1. Schematic diagram of general methodology. (A) shows a digital photograph of the head of a rat fixed in a stereotaxic apparatus; four open holes on the cranium are visible. (B) shows a diagram of the rat skull with the superficial venous system superimposed on it. Special care is always taken to avoid rupture of the veins to minimize bleeding, and mediolateral measurements are taken from the midline of the superior sagittal sinus instead of from the midline of the bone fissure. (C) illustrate different procedures, some of which like the injection of tracers or other chemicals and the infliction of cuts or other lesions can be performed hours or days before recording and juxtacellular labeling of single neurons. (D) Single-cell extracellular recordings can be obtained concomitant with EEG recordings from one or several cortical and subcortical areas. *Abbreviations*: CPu, caudate putamen; LGP, lateral globus pallidus; SI, substantia innominata; HDB, horizontal limb of the diagonal band of Broca; AA, anterior amygdaloid area; ic, internal capsule.

A. Anesthesia

Several choices of anesthetics, including ketamine–xylazine, urethane, pentobarbital, are available. However, transport of the tracer during juxtacellular filling may depend on the activity of the cell being labeled. Therefore, for recording and labeling experiments barbiturate anesthetics may not be a good choice, since they are known to generally suppress neuronal activity (Richards, 1972). On the other hand, ketamine–xylazine, which is preferred for survival surgeries, may result in significant fluctuations in anesthetic plane in long recording experiments, in which variations in absorption may also result in relatively unpredictable redosing schedules. Urethane (ethyl carbamate) is our preferred choice for recording experiments because it appears to have little effect on neural activity, and its primary inhibitory actions seem to be restricted to some small neurons in the reticular formation (Rogers *et al.*, 1980). These examples make it clear that choices of anesthetic and other details depend on the particulars of the investigation. Discussion of the effects of different anesthetics on the firing properties of neurons is beyond the scope of this chapter. For additional details regarding anesthesia and surgical procedures, see theAppendix.

B. Choice of Juxtacellular Labeling Markers and Recording Solutions

The most logical choice of solution for juxtacellular labeling is one whose ionic composition is compatible with general extracellular recordings with the addition of biocytin [N_ϵ(+)-biotinyl-L-lysine, FW 372.5 g/mol; Sigma-Aldrich Co., St. Louis, MO] or Neurobiotin™ (*N*-(2-aminoethyl) biotinamide hydrochloride, FW 322.85 g/mol; Vector Laboratories Inc., Burlingame, CA). The usual recording solution concentration ranges from 0.5 to about 1 M NaCl and contains 0.5–5.0% biocytin or Neurobiotin.

Difficulty to dissolve biocytin, especially when used in higher concentrations, is sometimes reported. This may be easily resolved by warming up the solution, strong agitation, sonication, or any combination of these. Dissolving biocytin in water and then mixing it with the NaCl solution may help. In addition, we have performed recordings and juxtacellular labeling using typical K-based solutions at concentrations usually used in intracellular recordings. In these cases, we did not notice any particular advantage or disadvantage for the labeling procedure. In general, all these solutions can be stored for several weeks at 4°C or for several months at less than 0°C.

The use of compounds other than HRP, biocytin, or Neurobiotin for juxtacellular labeling is theoretically possible, but we are not aware of any particular studies trying other substances. We have tried using biotin dextran amine [BDA(s), Molecular Probes, Eugene, OR] compounds, but the results obtained so far are inconclusive. Therefore, the tracers of choice for juxtacellular labeling are biocytin and Neurobiotin. In our hands, both provide very similar results, a finding that is in agreement with studies comparing biocytin and Neurobiotin for intracellular labeling (Kita and Armstrong, 1991). According to Vector Labs, however, compared to biocytin and other neuronal labels, Neurobiotin is more soluble, iontophoreses better, and remains longer in cells. Neurobiotin also seems to be easier to dissolve than biocytin when used at higher concentrations.

Whether or not the solution needs to be filtered depends on how well the tracer is dissolved. If the solution appears completely crystal clear to the naked eye, it may not need to be filtered. Although filtering does not seem to have any adverse consequences. Because of the very small amounts of solution usually prepared (in the order of a few hundred microliters), it is advantageous to use very small filters, so that one loses the least amount of solution when filtering. Although any filter will do, the Cameo, 3 mm– 0.22 μm acetate syringe filters, or the like are very effective.

C. Electrodes and Recording Apparatus

1. Single Unit Extracellular and Labeling Electrode/Glass

Microelectrodes are pulled from glass capillaries. Usually, 1.0–2.0 mm capillaries containing a microfilament fused to the inner wall are convenient choices because of their strength and commercial availability. The microfilament is very useful because it facilitates filling of the electrode and makes the filling more uniform, which minimizes trapped air bubbles. If there are air bubbles trapped in the solution (they are easily seen under the microscope), it is best to remove them. This can be done by gently tapping on the electrode while holding it vertically. In the worse case, bubbles can be removed by introducing a very thin wire into the solution in the electrode so as to make the bubble(s) attach to it; once attached to the wire they can be pulled out. This is done under the microscope with the electrode placed horizontally, usually glued to a glass slide with a small piece of putty.

A tungsten wire with a small enough diameter can be prepared for this purpose by thinning it in a caustic solution. First, attach the positive pole, for instance that of a 9-V battery, to the wire and the negative pole to a carbon rod. Fill a small beaker with caustic solution, introduce the carbon rod into it, and then slowly introduce the tungsten wire into the solution. The current passed will "eat up" the wire, thinning it. After rinsing it in water, this wire can be used for removing bubbles.

One easy way to fill an electrode with solution is by first placing a drop of solution on its back and waiting for a couple of minutes for the tip to be filled by capillary action, and then the rest of the electrode can be filled (from the back) by using, for instance, a micropipette filling needle. Some convenient choices for micropipette filling needles are the MicroFil™ ones sold by WPI (World Precision Instruments, Sarasota, FL).

Of several pullers available, the Narishige PE-2 (Narishige, Tokyo, Japan) vertical puller has been a common choice, perhaps because it is an older model, which has been around for many years and is available in many laboratories. However, any puller able to handle the right size of glass capillaries will do. The tip of the electrode is broken under a microscope to a diameter in the range of 0.2–2.0 μm. The smaller tip, the higher the impedance of the electrode. High impedance results in the detection of fewer neurons, but has the advantage that they are more likely detected only at a closer range. Therefore, higher impedance electrodes are more selective as the possibility of detecting more than one cell is diminished. Higher impedance electrodes should be used for detecting smaller cells whose electrical potentials are also smaller and need to be detected at closer ranges. In our experience, pipettes filled with saline and 2% biocytin, with tips of about 1–1.2 μm, have very convenient impedances of approximately 20–40 MΩ measured in the brain.

2. Extracellular Electrode/MUA/EEG/LFP/Metal

Electroencephalographic (EEG) or multiunit activity (MUA) and local field potential (LFP) data can be collected at the time a single cell is recorded. The advantage is that one can then correlate the extracellular activity of a single cell with the activity of a network. However, one should keep in mind that the characteristics of the EEG or LFP activity are highly dependent on the type and depth of anesthesia. Furthermore, the signals are also different in different brain regions and are significantly related to the cortical layer where the electrode tip is placed. Hence, the interpretation of the strength of the relationship or correlation of the single-cell activity and the network activity should take these and other factors (such as the distance between the two electrodes) into account. EEG and LFP can be collected with a host of metal electrodes (i.e., stainless steel, tungsten, etc.), most of which are easy to make and are also readily commercially available in bipolar and monopolar choices. Good ready-to-use metal electrodes can be purchased from Frederick Haer & Co. (FHC, Bowdoinham, ME). If one prefers to make them, appropriate wire can be purchased from California Fine Wire (CFW, Grover Beach, CA). To make electrodes, the wire is cut to the desired length (usually 5–8 cm), taking care to keep the electrode straight. One or two millimeters of insulation is removed from one tip and a little bit more from the other tip, where an appropriate connector needs to be clipped or soldered. A full discussion of EEG and electrode construction is beyond the scope of this chapter and the reader should consult one of the many specialized papers on the subject.

3. Electrophysiological Recordings and Recording Apparatus

Both EEG and single unit signals can be simultaneously collected, amplified, filtered, and recorded using standard equipment. In our preparations, we have used a Neurodata IR-183 amplifier (Neurodata Instruments, New York, NY) to record and juxtacellularly label single cells. For convenience, both single unit and EEG tracers are usually displayed simultaneously in an oscilloscope or computer monitor and recorded directly to a computer hard drive via an interface. Recordings on magnetic tape for secondary storage and/or offline analysis are also common. Software and hardware combinations, such as Spike 2.0 from Cambridge Electronic Design Limited (CED, Science Park, Cambridge, England), are excellent choices for the acquisition of data via multiple channels. As computer power increases and cost decreases, direct storage of data into hard drives and quick online analysis with versatility of options become common practice.

D. Juxtacellular Labeling

After the collection of sufficient extracellular electrophysiological data to permit characterization of the recorded cell, one can proceed with juxtacellular labeling.

1. Getting Close to the Cell

Before labeling is attempted, the electrode needs to be in very close apposition to the cell being recorded; this is why the technique is called "juxtacellular" labeling. As the investigator advances the electrode closer to the cell, the amplitude of the unit's signal increases. The electrode should be advanced slowly in small steps of 1 or 2 μm. A sudden and substantial increase in baseline noise, sometimes accompanied by a transient jump in DC level, indicates a good position where to start experimenting with the entrainment of the cell.

2. Application of Pulses and Entrainment of the Cell

At the point when a juxtacellular position has been reached, anodal pulses in the range of 1–10 nA are usually used to eject the dye from the micropipette. These current pulses cause a vigorous response from the cell called "entrainment." An entrained cell fires action potentials at a higher frequency than normally observed both in response to and during the depolarizing phase of the current pulses being applied. Sometimes the neuron will also fire one or several spikes during the negative phase of the pulse. If the firing rate increases substantially during both phases of the pulse, the neuron is usually at risk of being killed. At that point the electrode should be moved back a few micrometers and/or the intensity of the pulses should be decreased. The experimenter can modulate entrainment to be on and off so as to avoid cellular damage. Uncontrolled firing at very high frequencies and during both phases of the pulse usually precedes irreparable cellular damage especially in slower firing neurons. The labeling procedure then requires that the experimenter pay attention to any electrophysiological change the neuron may undergo before and during labeling. Pulse intensity should be low at the start and slowly increased until the cell responds. If the cell does not respond, the electrode is moved closer and the procedure is repeated. Once the cell responds to the pulse, the position of the electrode and/or the intensity of the pulses may need to be adjusted to avoid cellular damage. If repeated attempts to engage a cellular response fail with anodal pulses, cathodal pulses can be attempted. Figure 7.2 illustrates the

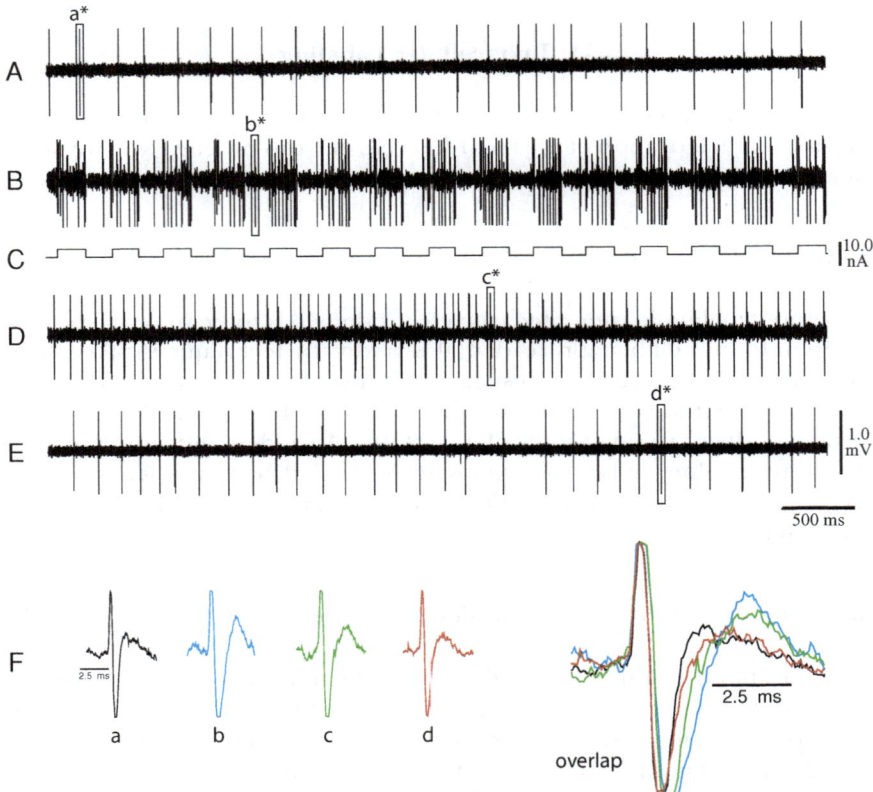

Figure 7.2. Extracellular recording and juxtacellular labeling protocol. Trace (A) shows the spontaneous firing of a neuron in the rat basal forebrain as recorded extracellularly. Trace (B) shows the entrainment of the neuron in response to the current pulses shown in trace (C). Notice that firing increases during the positive phase of the pulse but that the cell still fires spontaneously during some of the negative phases of the pulse (see b*). Also, notice that the base noise increased as compared to that in trace (A). Trace (D) shows that the spontaneous firing rate after cessation of the pulse is still higher than that before the entrainment, but with time (Trace E) it returns to preentrainment levels. Trace (F) shows selected extracellular action potentials before entrainment (a*), during entrainment (b*), and after entrainment (c*, d*). Their overlap illustrates how action potential width increases in a reversible manner during increased discharged frequency.

spontaneous firing of a neuron in the basal forebrain and its response to the current pulses.

The ability and the degree to which a neuron can increase firing rate probably depend on the neuron's molecular makeup and the state of the network where it resides. This is why some cells can entrain and fire at very high frequencies, while others cannot fire at such high frequencies in response to current pulses, although their firing rate still increases substantially, i.e., two- or threefold from their spontaneous rates. If the spontaneous firing rate is very low, say on the order of 1 Hz, then a sixfold increase should indeed be considered a very substantial change, despite the fact that a cell whose spontaneous firing rate increases just from 20 to 60 Hz appears to be more vigorously entrained. These vast differences that we have encountered in rat basal forebrain may be due to the diversity of cell types in this brain area. In particular, the compositions of their cellular membranes, ionic channels available, etc. in part dictate their very different intrinsic properties. Hence, it seems logical that they would respond in different ways to the same basic stimulus. Other important considerations arise, for instance, from geometrical constraints, the position of the electrode tip with respect to the soma, the shape and size of the soma, ephaptic relations, etc. In general, in our experience, faster firing neurons are much easier to label than slower firing ones, because they entrain easily and require less time for good quality labeling. Neurons whose firing rate cannot be modulated at all by the current pulse do not get labeled.

Generally, we applied pulses using an IR-183 Neurodata amplifier (Neurodata Instruments, New York, NY). Current pulses are 200 ms long (50% duty cycle). Entrainment for only a couple of minutes may be enough to label the soma. More complete staining of dendrites and axons requires in the order of at least 20 min.

3. Polarity of the Pulses

Both Neurobiotin and biocytin are zwitterions and should therefore be ejected from the micropipette with negative or positive current pulses. Pinault reported never to have seen labeling of neurons when applying iontophoretic negative (cathodal) pulses in the range of −50 nA to −1 μA (Pinault, 1996), an observation corroborated by others. However, in our own recordings and labeling of basal forebrain neurons, we have occasionally encountered cells that responded better, i.e., entrained better, in response to negative current pulses than to positive (anodal) current pulses. In these rare instances, the application of negative current pulses when using biocytin-filled electrodes did not result, as far as we could tell, in differences in the quality of labeling or required labeling time. Although these cells were not neurochemically identified, their morphology appeared similar to many other typical neurons of the basal forebrain. By chance, we did not attempt labeling of neurons by passing negative current pulses while using

Neurobiotin-filled electrodes. However, it is important to note that Kita and Armstrong (1991), although describing intracellular recordings and not juxtacellular labeling, reported that Neurobiotin is selectively ejected with positive current pulses. They also suggested that this property would be beneficial to electrophysiologists using hyperpolarizing currents to stabilize the membrane potential of neurons prior to recording.

4. Action Potential Shape and Cellular Response After Cessation of Pulse

To avoid false expectations of labeling, it is essential to make sure that the cell does indeed respond to the current pulses. If there is no response, the neuron will not get labeled. It is also important to corroborate that the neuron is still firing after cessation of the current pulse protocol and that it continues to fire as the electrode is slowly removed from its vicinity. The width of the action potential can be used to assess the health of a neuron and also serves to indicate that labeling has happened. In response to the current pulses, extracellular potentials can and usually become wider as shown in Fig. 7.2. Apparently this broadening indicates some micropuncture of the cellular membrane, a likely mechanism for the labeling (Pinault, 1996). After some time, extracellular action potential width does return to the level before injection, a possible indication of healing of the membrane. Also, it is likely that the firing rate of the neuron will be higher after cessation of the current pulse. However, over time it should return to baseline levels (Fig. 7.2).

E. Control Experiments: The Neuron Recorded is the One Labeled

In order to establish that the labeled neuron was indeed the neuron that was recorded, we designed several control experiments, some of them illustrated in Fig. 7.3. First, we applied current pulses 1–10 nA in intensity, for 15–30 min, without having first detected firing from a single neuron. This protocol resulted in no labeling of cells but if applied longer than approximately 20 min, it usually left a small accumulation of residual tracer in the tissue. Second, we applied current pulses of the same intensity for the same period of time as in the previous case, in the presence of a single unit, but without obtaining any response (entrainment) from the cell being detected. This protocol also resulted in no labeling of neurons. Third, we applied the same protocol while detecting a single unit and got the single unit to respond to the pulses. In this case a single-labeled neuron was later revealed and the morphological integrity of the neuron was preserved (Fig. 7.3A, B). Fourth, we labeled a neuron but killed it at the end of the labeling procedure by passing a high-voltage pulse. In these cases, when animals were perfused within approximately 1 h after the procedure, we indeed found a labeled neuron but usually its morphological integrity was

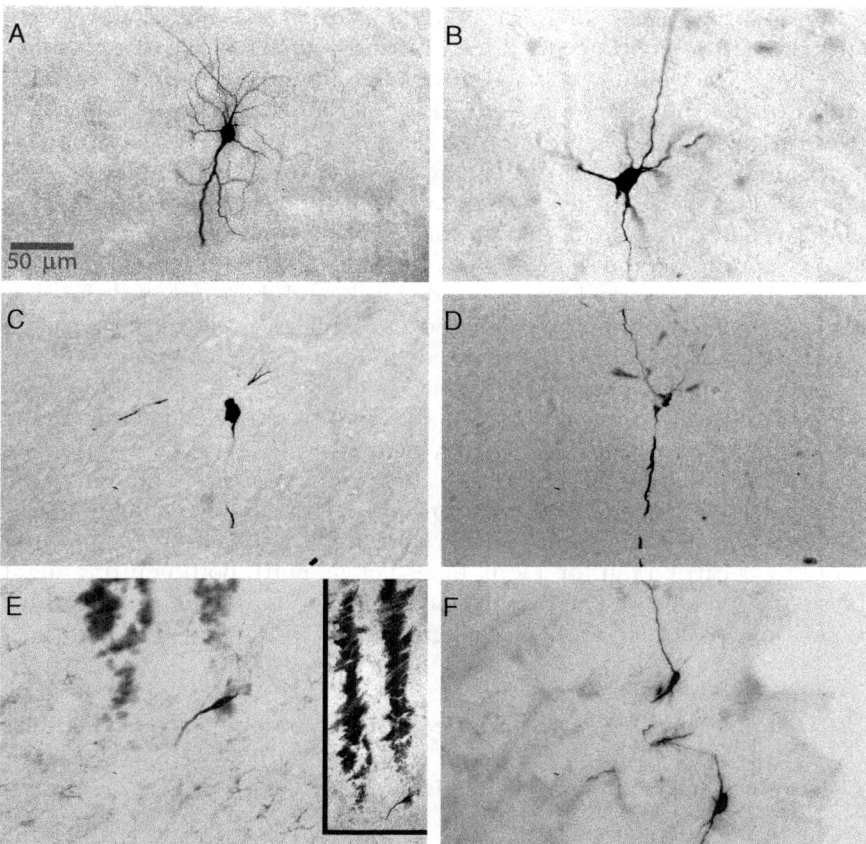

Figure 7.3. Control experiments: the neuron labeled is the one recorded. Delivery of current pulses without detection of a neuron or even when detecting a neuron that does not entrain results in no labeling of cells. (A) and (B) illustrate cases in which a single neuron was detected, recorded, and entrained (cells in two different animals). In each case, after juxtacellular labeling the electrode was retracted slowly while the cell was still firing. After proper histochemical procedures, a single-labeled cell was found. (C) and (D) illustrate cases in which after juxtacellular labeling of a single cell the neuron was "killed" by passing a high-voltage pulse. After histochemical procedures, a badly damaged single-labeled neuron was found in each case. (E) illustrates a case in which a single neuron was juxtacellularly labeled. The electrode was retracted dorsally until the cell was no longer responding to the pulses, yet these were passed for several minutes to create enough damage to mark the electrode track. Afterward, the electrode was totally removed from the brain and lowered again to the same depth at 75 μm lateral to the labeled cell, and pulses were passed again to mark a second electrode track (without detecting and entraining any other cells). After histochemical processing, one single-labeled cell was found at the end of a marked electrode track and 75 μm lateral to it, a second electrode track was found. (F) illustrates a case in which two neurons were detected and both responded (entrained) in response to current pulses. In that case, two labeled neurons were later found.

seriously compromised (Fig. 7.3C, D). If the animal was not perfused soon after the end of the procedure, the remains of the labeled cell sometimes were not found. Fifth, when we entrained a neuron for approximately 10 min. and then moved the electrode some micrometers away and proceeded to apply current pulses for another 20 or 30 min, without detecting and entraining a new cell or affecting the previously recorded neuron, we later found a single-labeled neuron and, most of the time, a single electrode track in the vicinity. Sixth, if by the same token, we first labeled a neuron for a few minutes, then retracted the electrode dorsally until the cell was no longer detected, marked the electrode track, and then removed the electrode completely out of the brain and moved it precisely, for instance, 75 μm laterally and returned the electrode to the same depth and proceeded to mark a second track, we later found a single track at the end of which we had a labeled neuron, together with a second track located laterally to the first 75 μm from it (Fig. 7.3E). Finally, if a single cell was entrained properly for labeling and then we still sporadically entrained a second neuron in the vicinity that was detected in the background, later we found a single strongly labeled neuron and a second weakly labeled neuron. However, if this procedure was carried out for a long time (i.e., 45 min), both neurons might be stained equally well, and then one could not determine with certainty which one was related to the main signal (Fig. 7.3F).

F. Histology and 3D Light and Electron Microscopy Reconstructions

1. Histology and Immunohistochemistry

Histological and immunohistochemical procedures for visualization and study of single juxtacellularly label neurons are identical to those followed for intracellularly labeled neurons. In short, animals are usually perfused with saline followed by a fixative such as acrolein, paraformaldehyde, glutaraldehyde, or a combination of these. Picric acid, glucose, and other chemicals are sometimes used to improve the outcome of a particular procedure. Protocols, including incubation times and amounts of chemicals used, are the choice of the investigator (see example in the Appendix), and there is a host of different recipes available in the literature.

2. 3D Light Microscopy Reconstructions

Modeling and experimental studies suggest that neuronal morphology (dendritic and axonal branching pattern) and class-specific connectivity play an important role in shaping network function (see chapters by Ascoli *et al.* and Nadasdy *et al.* in this volume). Two-dimensional (2D) camera lucida tracings of neurons have been performed following Cajal but all the focus-axis information is lost and the only view possible is that of the plane of sectioning. Another disadvantage of this technique is that the end product is a drawing; one cannot statistically summarize the drawing without somehow measuring it. Fortunately, recent technological advances allow the use of 3D computerized reconstructions to obtain precise numerical details about morphological characteristics and constraints for different neuronal populations. These quantitative methods are needed in order to be able to compare and contrast different types of neurons in a logical, mathematical form. Through mathematics, the biological constraints on the architectural nature of the brain and of its elements can be investigated. For instance, dendritic size, together with their extensions into specific spatial domains, can be an indirect measure of input density. Understanding at least some of the structural diversity of dendritic arbors, their orientation, and size can provide essential clues about their possible connections and hence, clues about the intricacies of dendritic function.

Axonal reconstructions of identified neurons, although more difficult than dendritic reconstructions, provide some clues as to whether the neuron is a projection neuron, a local interneuron, or both. En passant or terminal varicosities can be recorded during the light microscopical reconstructions. These data, together with available information on the number and type of neurons in the environment of the axonal arborizing space, can be used for deriving the connectional probability of the electrophysiologically identified neurons (Zaborszky *et al.*, 2002). It is likely that digital experiments will be increasingly used with the advance of 3D cellular databases equipped with proper warping tools for data integration (http://www.ratbrain.org; see also chapter by Nadasdy *et al.* inthis volume). The data from these "in silico" experiments can be used to generate hypotheses that in turn can be tested in actual experiments. For the state-of-the-art stereological assessment of neural connections, see the chapter by Avendano in this volume.

Typically, dendritic and axonal trees can be reconstructed from serial sections using computerized microscope systems. The most widely adopted commercial system is the Neurolucida® system (MicroBrightField Inc.; www.microbrightfield.com), which utilizes a computer-controlled stepping motor stage and a high-resolution miniature CRT monitor. The motorized stage allows data acquisition to extend distances significantly larger than a single microscope field of view. The miniature CRT is coupled with the camera lucida and superimposes a computer graphic display on the actual image of the microscopic specimen viewed through the oculars of the microscope. This permits the acquisition of microscopic data that require high resolution and clarity to visualize. Neurolucida® is operated on a PC equipped with the Microsoft Windows operating system. Digital reconstruction of single-labeled neurons as well as analysis of the morphological data using the Neurolucida software suite (Glaser and Glaser, 1990) as well as other software tools for extracting morphomertic measurements from the reconstructed neurons are described in detail in the chapter of Ascoli and Scorcioni (in this volume). Neurons in the Neurolucida® system are represented by the x, y, and z coordinates of manually traced points; thus morphological reconstruction is a very time-consuming process. There are other somewhat faster, but perhaps less precise, ways to combine traditional paper–pencil tracings with computer-assisted reconstructions of scanned images (see Ascoli and Scorcioni, this volume). Section "Applications" gives further details on reconstruction and morphological analysis of electrophysiologically and chemically identified neurons.

3. Electron Microscopy-3D Reconstruction from Ultrathin Sections

Electron microscopy of ultrathin sections produces high-resolution 2D images of cellular profiles and allows identification of synapses. However, most of the three-dimensional structural information is lost, although it can be recovered from reconstructing serial thin sections. The advantages of 3D reconstruction from ultrathin sections are several, including the ability to monitor plastic changes in the postsynaptic profiles (see chapter of Goldberg *et al.*, this volume), to better understand the complex organization of the neuropil, and to count synapses. Although the number of synapses can be deduced from 2D images using various correction factors or applying "unbiased" sampling strategies with optical dissectors as described in the chapter of Avendano in this volume, estimates of synapses can also be done from proper light microscopic reconstructions and application of an adequate electron microscopical sampling strategy. Although alignment of sections can be done manually (see Goldberg *et al.*, this volume), misalignment between sections due to mechanical and optical distortion often necessitates the use of computer programs. At present, the most popular program for alignment of EM images is the *Reconstruct* (V1.03.2) tool developed by Harris and Fiala (www.synapses.mcg.edu; http://synapses.bu.edu/tools/download.htm). This program computes the alignment from three points identified by the user in each image. A program under development that computes the fitting position using pixel density values results in more accurate alignment (Simon *et al.*, 2005).

G. Tracer Techniques in Combination with Electrophysiological Recording and Juxtacellular Labeling

1. Choice of Retrograde and Anterograde Tracers

Retrograde and anterograde tracers may be applied to the brain as an additional step prior to the juxtacellular labeling of single cells. Here, only some comments are presented with specific reference to juxtacellular labeling and the reader is referred to appropriate chapters in this volume for additional details about the tracers (Lanciego and Wouterlood, Reiner and Honig; Sesack *et al.*; and Wouterlood).

All tracer injections require animal survival to allow tracer transport from the injection site to the target area. It may be necessary to determine best concentrations and survival times empirically. There are many excellent fluorescent tracers and selecting one depends on what fluorescent tag will be used for visualization of the single cell. If, for instance, rhodamine (red) is used for the visualization of the single cell, then a yellow or blue (or both) fluorescent retrograde/anterograde tracer is a convenient choice. Fluorogold (Fluorochrome Inc., Denver, CO) and Fast Blue (Sigma, St. Louis, MO) are primarily retrograde tracers. The fluorescent dextrans are commonly used as anterograde or retrograde tracers (Reiner and Honig, in this volume), and these include rhodamine isothiocyanate, rhodamine B dextran, lysinated tetramethylrhodamine dextran (Fluoro-Ruby), and Fluoro Emerald. Their advantages seem to be less diffusion at the injection site and more permanent labeling than with the corresponding free dyes (Schmued *et al.*, 1990; Schmued and Heimer, 1990; Schmued, 1994). BDAs can also be used as retrograde and anterograde tracers and offer superior characteristics and versatility (Reiner *et al.*, 2000). Since BDAs contain biotin, their use in combination with biocytin or Neurobiotin labeling may be tricky and is therefore not recommended. Other more direct choices of anterograde and retrograde compounds for fluorescent detection include True Blue, Granular Blue, 4',6-diamidino-2-phenylindole, Diamidino Yellow, Nuclear Yellow, and Rhodamine Latex Beads, among others.

2. Sources of Artifact

There are several potential sources of artifact when tracers are injected into any part of the brain. Two particular problems are the size of the tracer uptake zone and the possibility of uptake of the tracer by fibers of passage at the injection site. To minimize the fibers of passage problem, the simplest rule is to minimize brain injury. To control the size of the tracer uptake zone, tracer injections need to be steady and slow (over 10–15 min), which in our experience creates less tissue damage and also minimizes the spread of the injection by avoiding stressing the tissue around the injection site. To avoid leakage of tracers along the pipette track, the exterior of the needle needs to be cleaned and dried before inserting into the brain and must be left in place for at least 5 min after the end of the injection. This technique appears to minimize damage to the tissue and also reduces the possible uptake of tracer by fibers of passage. However, the uptake of most tracers by fibers of passage cannot be entirely abolished. To minimize movement of the needle and improve reproducibility of injections, the syringe in use may be mounted on a Hamilton Chaney adapter. Also, many tracers may have a small component of transport in the opposite direction to the one intended, such as in the case of vestiges of retrograde transport for an otherwise mainly anterograde tracer injection or vice versa. In addition, because of the difficulty in limiting the injection exclusively to the area of interest, one strategy that can be used is to apply injections of different sizes (in different animals) so that larger injections can increase the chance of including the majority of the target area while smaller injections cover only portions of it.

III. APPLICATIONS

In this section we will illustrate the application of juxtacellular labeling of single neurons to the study of neural circuits of the basal forebrain of the rat. As outlined in Fig. 7.4 and illustrated in subsequent figures, we increase the level of complexity of the investigation of single basal forebrain neurons by increasing the number of procedures that are performed in addition to extracellular recording and juxtacellular labeling. These added procedures provide information about the transmitter, dendritic and axonal arborization pattern, and connectivity of single electrophysiologically identified neurons that could not otherwise be obtained, and which in the past has been only inferred from separate experiments. This multiplicity of methods is

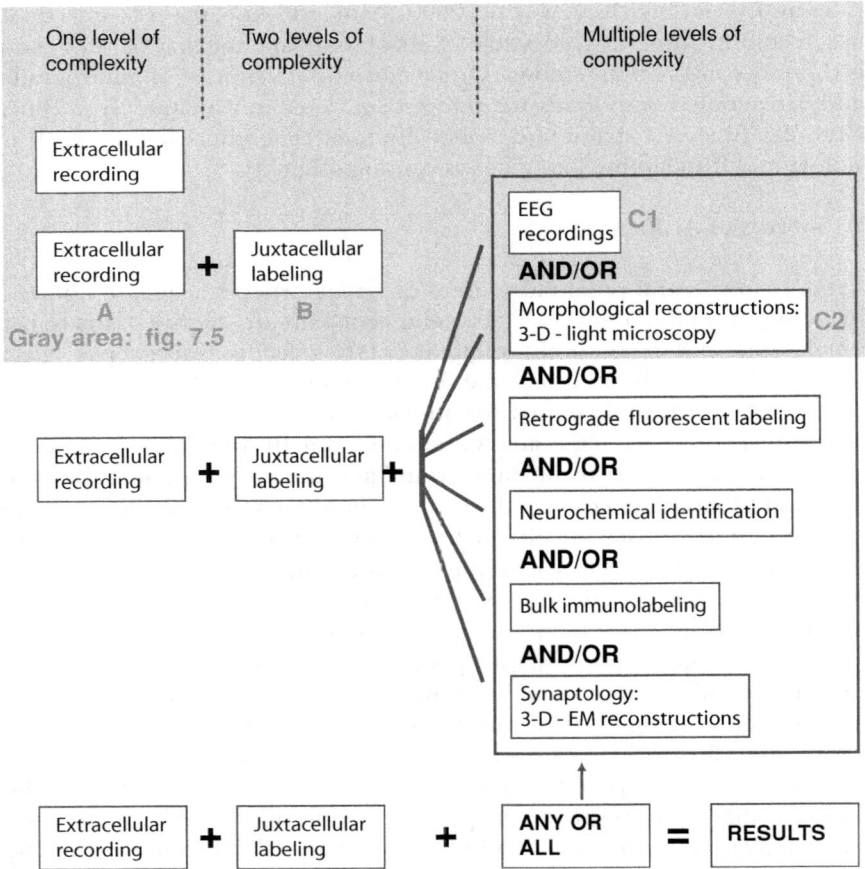

Figure 7.4. Schematic illustration of the general experimental strategy: extracellular recording and juxtacellular labeling of single neurons in combination with several methods that allow further cellular characterization. The added levels of complexity increase the amount of information collected about a single neuron. Letters A, B, C1, and C2 are illustrated in detail in Fig. 7.5.

necessary because simplified criteria, which in some brain areas have been successfully used to differentiate among different cell types, have failed in the basal forebrain due to the heterogeneity of electrophysiological, morphological, and neurochemical characteristics found in this region. This combination of techniques has allowed us to study the EEG correlation of the discharge properties of neurochemically and morphologically identified neurons in the basal forebrain (Duque *et al.*, 2000). Our studies, for instance, indicated that both cholinergic and parvalbumin-containing neurons increase firing during cortical low-voltage fast-electrical activity (LVFA), and therefore belong to the previously established category of "F" (fast) type basal forebrain neurons. On the other hand, neuropeptide Y positive (NPY) neurons that showed increased firing during cortical slow waves belong to the so-called S (slow) cell type. Whether or not this is the case for all basal forebrain cholinergic, parvalbumin, and NPY neurons is not known. Additional morphometric and synaptology studies in progress will allow us to build the "basic circuitries" of the basal forebrain that are critical to understand its functions.

A. Electrophysiological and Morphological Identification of Single Neurons

Figure 7.5 illustrates a typical experiment consisting of a single-cell extracellular recording with juxtacellular labeling and subsequent morphological reconstruction of the recorded neuron. This neuron was labeled for approximately 5 min by passing +8 nA current pulses.

From single-cell electrophysiological recordings, we classified this neuron as a bursty cell with a firing rate of 2.95 Hz. Its action potential shape is triphasic with a rise time of approximately 0.36 ms and total width of 2.88 ms. Labeling of the cell permitted us to locate the soma within the horizontal limb of the diagonal band of Broca. The soma measures 15×14 µm and its shape is round with six primary dendrites. The cell is a typical basal forebrain neuron according to a previous classification using the juxtacellular technique (Pang *et al.*, 1998). Panel C1 in Fig. 7.5 illustrates how one additional electrophysiological procedure, EEG, concomitant with the single-cell recording further allows the classification of this neuron into the "S" category of basal forebrain neurons. This categorization indicates that the firing rate increases during high-amplitude slow-cortical EEG activity and it is different from neurons that have a faster firing rate during fast cortical EEG activity, which were termed "F" cells (Detari and Vanderwolf, 1987; Dringenberg and Vanderwolf 1998; Detari, 2000). The simultaneous recording of EEG and single-cell electrophysiological data also revealed that there are neurons in the basal forebrain whose activities remain unchanged despite EEG cortical changes (Nunez, 1996).

Figure 7.5C2 illustrates how one additional procedure, 3D light microscopy reconstruction, further advances our knowledge of this single neuron. For instance, in addition to the soma shape and location, we now

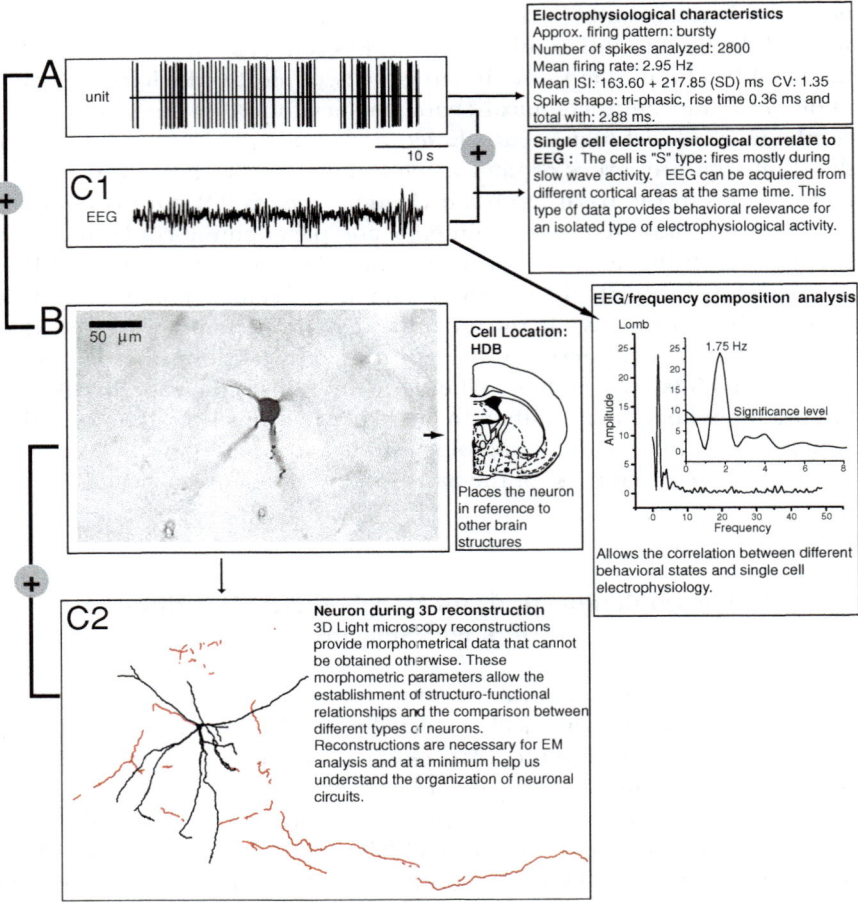

Figure 7.5. Example of extracellular recording and juxtacellular labeling in combination with EEG recording and morphological reconstruction. Arrows point to boxes containing examples of information that can be collected about the neuron from the different combined methods.

know the extent and direction of its dendritic tree. We know that the neuron has axon collaterals that communicate with nearby neurons and also a main projection axon that runs in the ventromedial direction. This extra information allows the formulation of hypotheses and the planning of further experiments. Do all "S" cells have axon collaterals? If so, what types of local neurons are contacted by "S" cells? Do they all have a projection axon?

B. Retrograde Labeling of Electrophysiologically Identified Neurons

Figure 7.6 illustrates the characterization of a basal forebrain neuron using an additional procedure to the ones illustrated in Fig. 7.5. In this case,

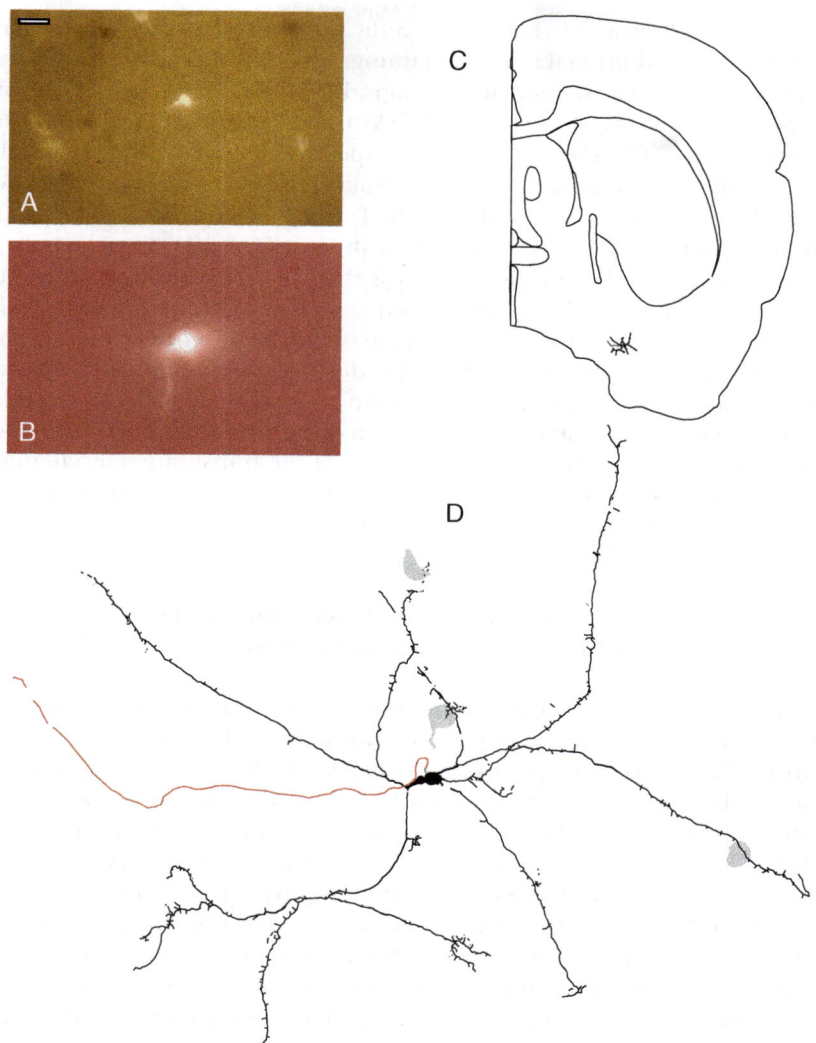

Figure 7.6. Juxtacellularly labeled ChAT negative, parvalbumin-negative basal fore-brain corticopetal neuron. This neuron was not tested for NPY. (A) The cell is retrogradely labeled with Fluorogold from the prefrontal cortex. (B) The single biocytin-filled neuron visualized with avidin-conjugated rhodamine. (C) Coronal section illustrating the position of this cell. (D) Partial neurolucida reconstruction showing that this neuron has only one projection axon (in red) and no axon collaterals in the neighborhood. The gray spots represent cholinergic neurons in close proximity to some dendritic appendages. Scale bar in (A) (applies to B also): 50 μm.

approximately a week prior to the single-cell recording and juxtacellular labeling, the animal received an injection of the retrograde tracer Fluorogold into the prefrontal cortex. The biocytin-labeled neuron was first visualized with avidin-conjugated rhodamine (Fig. 7.6B). This neuron was retrogradely labeled with Fluorogold (Fig. 7.6A), indicating that it projects to the prefrontal cortex. Additional immunostaining protocols revealed that this particular juxtacellularly stained and retrogradely labeled neuron was negative for both choline acetyltransferase (ChAT) and parvalbumin, markers for cholinergic and GABAergic neurons, respectively. After conversion of the rhodamine fluorescent signal (juxtacellular staining) to nickel-enhanced DAB, the tissue was processed for ChAT using DAB as end product. As the 3D light microscopy reconstruction shows (Fig 7.6D), there are three cholinergic cell bodies in the vicinity of this electrophysiologically identified and morphologically reconstructed cell. This identified neuron was F type and a "just" projection neuron, since only a single axon was found that lacked local collaterals. These extra procedures augmented the information collected about this single neuron. Now we know its single-cell electrophysiological characteristics and the correlation of these to the EEG (to the state of the animal). In addition, we know some of the transmitters or chemical markers for which it was negative, where it projects to, its morphology, and the important fact that it did not have axon collaterals.

C. Chemical Identification and Morphometry of Juxtacellularly Labeled Neurons

Figure 7.7 illustrates the addition of neurochemical identification to the characterization of a single basal forebrain neuron. In this case, the single neuron filled with biocytin was first visualized with avidin-conjugated rhodamine (Fig. 7.7C) and subsequently found to be positive for the calcium-binding protein parvalbumin, visualized with fluorescein isothiocyanate (FITC) (Fig. 7.7B). After conversion of the rhodamine fluorescent signal to nickel-enhanced DAB, the section containing the cell body of this electrophysiologically and chemically identified cell was immunolabeled for ChAT. As Fig. 7.7A shows, a cholinergic cell body (blue profile) was found in the vicinity of the reconstructed parvalbumin-positive neuron that seemed to be approached by axon collaterals of the electrophysiologically identified neuron.

Figure 7.8 displays some morphometric data regarding the dendritic tree of this parvalbumin-positive neuron, which had previously been electrophysiologically identified. Panel A in Fig. 7.8 is the dendrogram showing the dendritic tree of this neuron drawn in the same colors as used in the tracing of Fig. 7.7A. A dendrogram is a stylized drawing of a branched structure, e.g., axon or dendrite. The purpose of a dendrogram is to visualize the complexity of the 3D arborization pattern of the tree in a manner that can be easily compared among different neurons. This dendrogram shows that this particular parvalbumin neuron has eight dendritic trees, each dividing into several daughter branches with a total length of 8785 μm. The graph in Panel D, which displays the mean diameter of the dendritic branches, indicates that primary dendrites are thicker and that as dendritic order

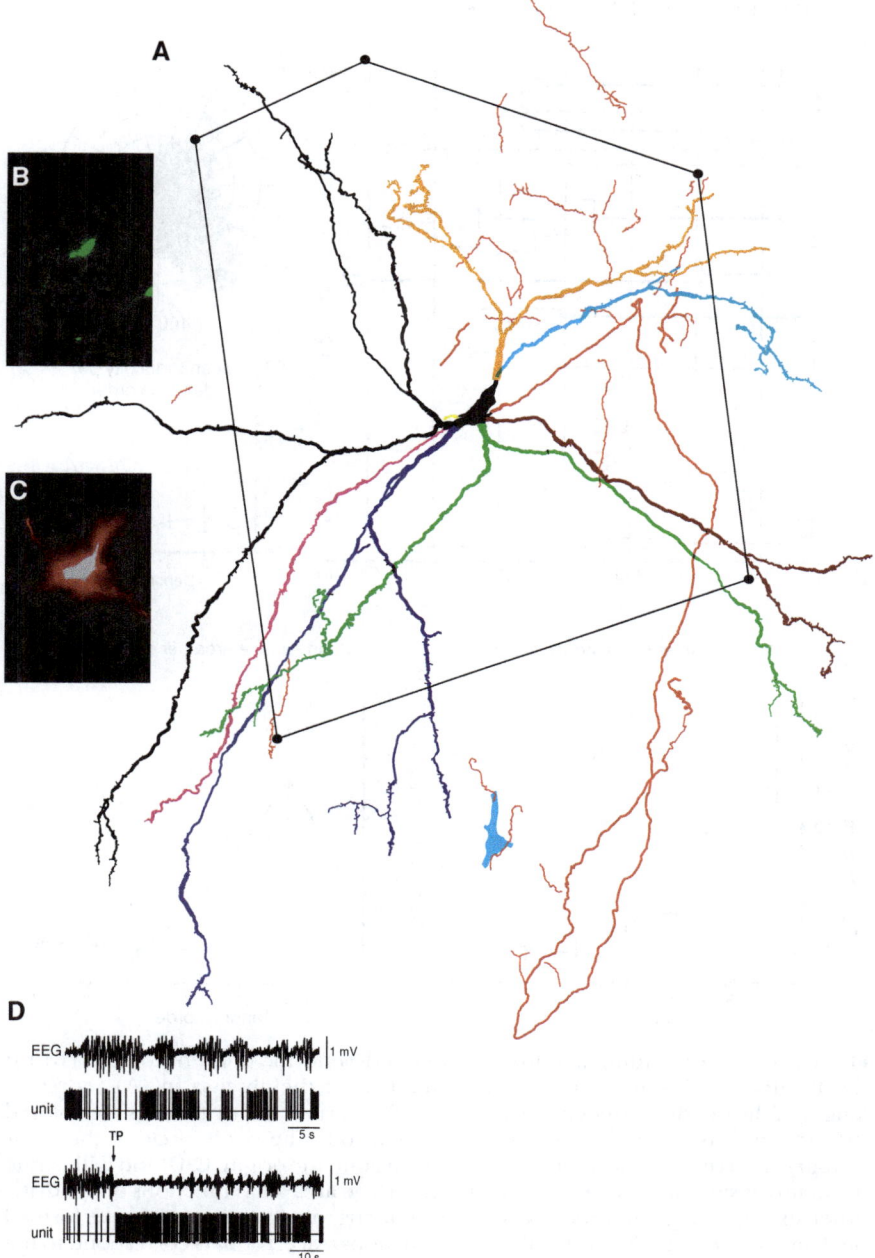

Figure 7.7. (A) Partial reconstruction of a juxtacellularly labeled basal forebrain parvalbumin-positive neuron. The axon is shown in red. The soma and one of the dendrites are shown in black and the other dendrites in different colors. Each dendrite matches its color in the dendrogram shown in Fig. 7.8A. The pentagon encloses pieces of the neuron that are found in one of the sections from the serial reconstruction and it corresponds to the block face shown in Fig. 7.9A. (B) The neuron stained for PV (FITC). (C) Biocytin-filled neuron visualized with rhodamine. (D) Concomitant unit and cortical EEG spontaneous and tail pinch (TP) induced activity.

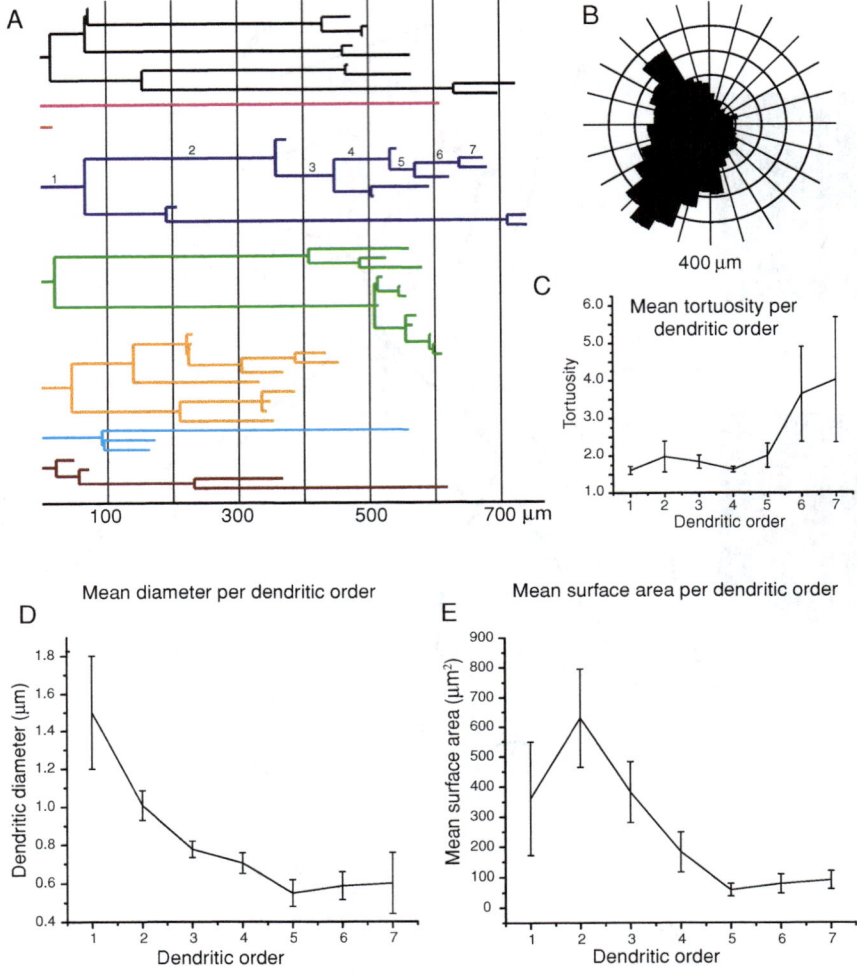

Figure 7.8. Corresponding dendrogram (A) and dendritic polar histogram (B) for the PV neuron shown in Fig. 7.7. The numbers at the abscissa in (A) represent length of the dendritic trees in micrometers. The various dendritic trees are colored the same way as in Fig. 7.7 to facilitate comparison. The outer circle of the polar histogram corresponds to a 400-μm diameter around the origin. C, D, and E illustrate mean tortuosity, mean diameter, and mean surface area all as functions of dendritic order, exemplifying different types of morphometric analyses that can be performed on data acquired with Neurolucida reconstructions. All error bars correspond to the SEM. For further explanation see text.

progresses dendritic branches become thinner, although the end portions of the dendrites have a tendency to be a bit thicker. This trend is similar to the situation observed in cholinergic neurons and is opposite to that in NPY neurons whose endings are very thin (Duque *et al.*, in preparation). The mean surface area, shown in Fig. 7.8E, is maximum in the case of secondary dendrites and about the same between primary and tertiary dendrites. Since tertiary dendrites, which are thinner than secondary and primary branches, have as much surface area as primary dendrites, it is reasonable to deduce that tertiary dendrites are longer than primary and secondary dendrites, which is indeed the case (not shown). Higher order dendrites have overall much less surface area. This type of analysis helps make predictions as to what to expect in terms of numbers of inputs, i.e., primary and tertiary dendrites could get roughly the same number of inputs, but second-order dendrites might get more inputs since they have on average more surface area. The polar histogram shown in Panel B gives a characterization of the directional distribution of dendritic growth projected onto the plane of sectioning, similar to the analysis described by McMullen and Glaser (1988). The algorithm for polar histograms breaks up the dendritic processes into line segments and determines the directions in which these line segments point and their corresponding lengths. The direction of the vector is calculated by projecting the line segment onto the plane of sectioning. The histogram represents total length by the distance from the origin and the angle that the vector makes with the x-axis plotted in the radial direction. Each sector in the polar histogram is the sum of all the dendritic growth in that particular range of angle. The dendrogram is another tool to compare quantitatively different neurons. The dendritic tortuosity displayed in Panel C gives a ratio between the actual length of a dendritic segment and the distance between its endpoints. Both the tortuosity and the dendritic orientation, as measured with the polar histogram, can determine spatial correlation between overlapping axonal and dendritic arbors, thus affecting the probability of connections (see also Zaborszky et al., 2002; Stepanyants et al., 2004). These analytical tools are provided in the Neurolucida software package. Additional analyses that would also permit elaborate comparison of dendritic morphometric parameters between various neuronal types (Scorcioni et al., 2004; Li et al., 2005) can be performed using L-Measure (Scorcioni and Ascoli, this volume), a JAVA program freely available both for download and for Web-based usage (http://www.krasnow.gmu.edu/L-Neuron).

D. Synaptology of Electrophysiologically and Chemically Identified Neurons

Embedding tissue sections containing electrophysiologically and chemically identified neurons into plastic allows the study of the different neurons' synaptic relationships. For example, we can study the input to specific dendritic compartments of an electrophysiologically and chemically identified neuron. Figure 7.9 shows the intermediary documentation, at the LM level, for a small dendritic segment of the parvalbumin neuron depicted in Fig. 7.7. Figures 7.10 and 7.11 show various boutons that impinge on the dendritic shaft of this parvalbumin neuron. Figure 7.12 shows a 3D rendering of the same piece of dendrite reconstructed from 66 ultrathin sections. As

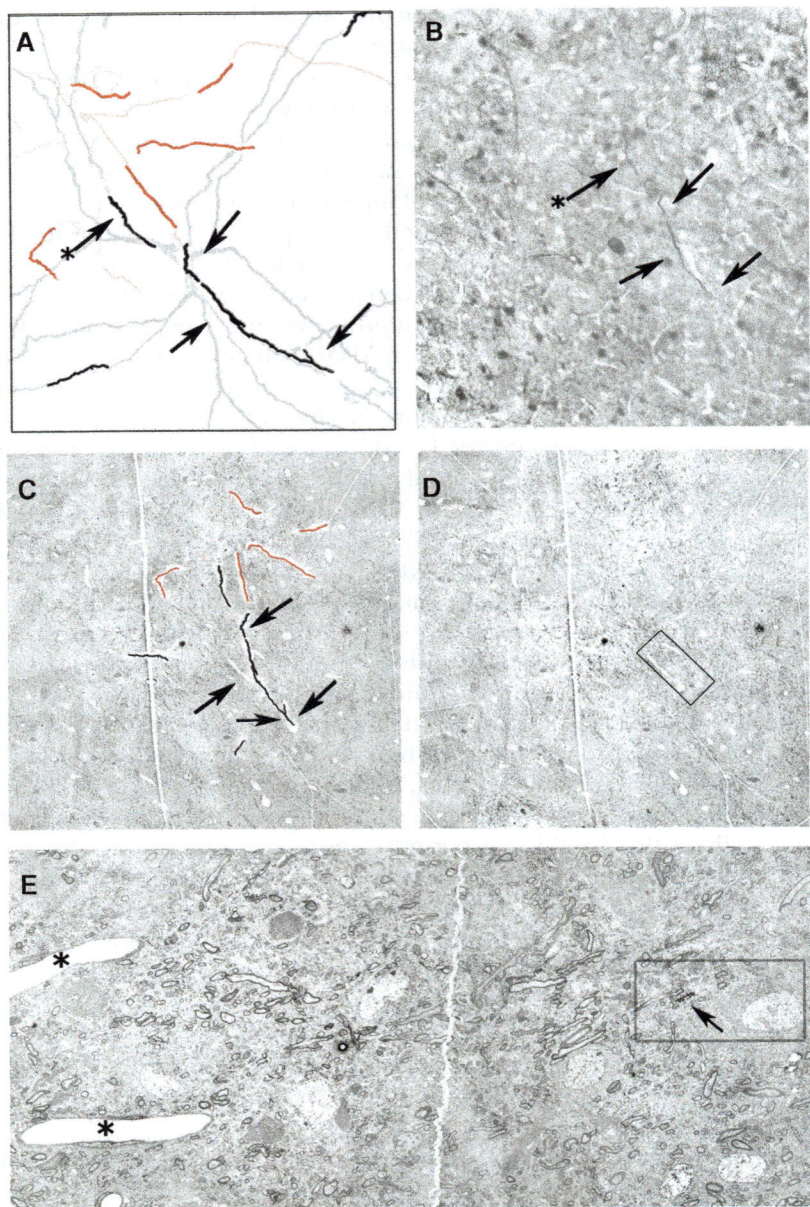

Figure 7.9. Correlated light–electron microscopy of the parvalbumin neuron shown in Fig. 7.7A. (A) Processes of this neuron contained in section 17 that was serially thin sectioned are shown against the full arborization of this neuron displayed as background. (Compare this panel with the enclosed processes in the pentagon of Fig. 7.7A). Axons are in red and dendritic processes are in black. (B) Blockface image of thick section 17. Three arrows point to a dendritic shaft that is also labeled in (A). Arrow with asterisk points to another dendritic shaft to facilitate comparison. (C, D) Low-magnification ($190\times$) montage of a thin section from this material with

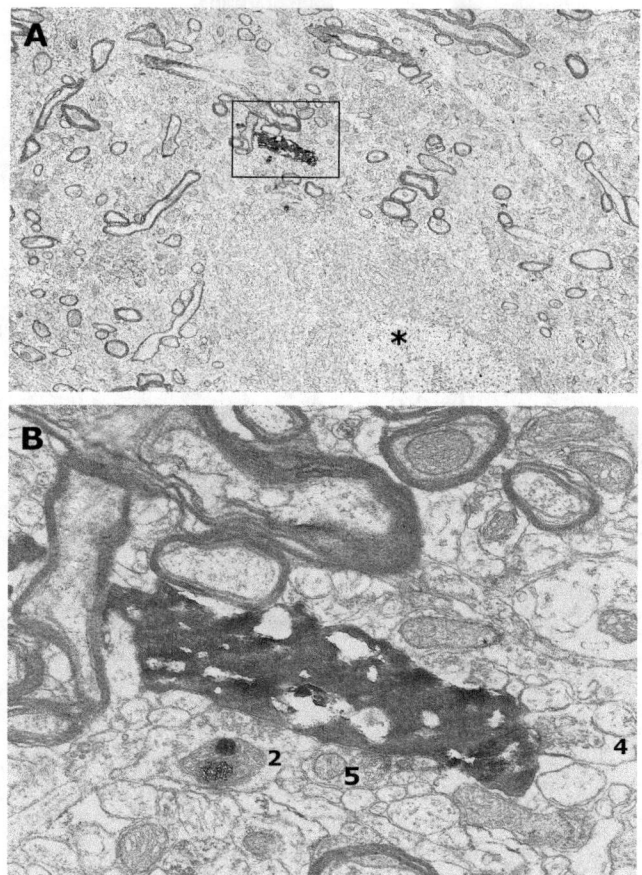

Figure 7.10. (A) Low-magnification electron micrograph (1400×) to show the boxed area from Fig. 7.9E. (B) High-power electron micrograph to illustrate the dendritic segment that has been 3D reconstructed from serial thin sections. Numbers denote different boutons.

the 3D model shows, this dendritic segment, which is about 3 μm in length, is surrounded by 11 boutons and most of them had synaptic appositions. Knowing that the total length of the dendrites of this parvalbumin neuron is 8785 μm, one can estimate that this neuron may receive as many as

←——

Figure 7.9. (*Cont.*) (C) and without (D) processes as seen in the Neurolucida file (A). The boxed area in (D) is enlarged in (E). (E) Low-power electron micrograph (440×) showing the boxed area from (D). The boxed area in (E) is shown at higher magnification in Fig. 7.10A. The two capillaries with asterisks are fiducial markers to compare (D) and (E). Arrow in the boxed area points to the small dendritic piece that is displayed in higher magnification in Figs. 7.10–7.11 and in the 3D model of Fig. 7.12.

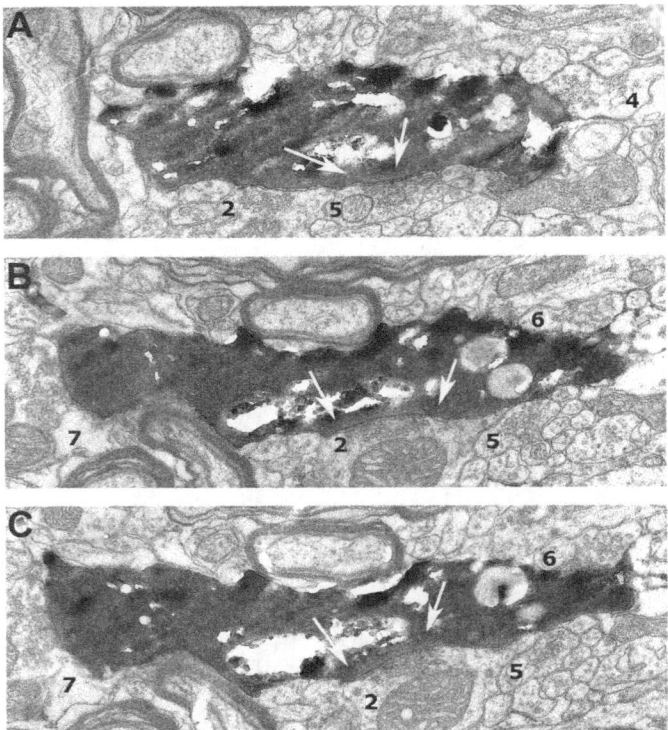

Figure 7.11. (A) Ultrathin section 23 in which boutons 2, 4, and 5 can be recognized (compare Fig. 7.10B). (B) and (C) show ultrathin sections 24 and 26 from the series of 66 sections with additional boutons 6 and 7. White arrows point to synaptic attachments.

32,000 synapses. However, since primary, secondary, etc. dendritic branches possess different diameters and may be contacted by different axons, for a more accurate estimation one has to select samples from each dendritic segment and the terminal branches of the dendritic trees. The dendrogram depicted in Fig. 7.8A helps to design the sampling strategy. As discussed under section "Chemical Identification and Morphometry of Juxtacellularly Labeled Neurons." dendritic length, diameter, and surface area values can affect the spatial arrangement of incoming boutons. In addition, analyses on electron micrographs (e.g., average bouton size) and data from tracing studies need to be taken into consideration if we want to arrive at a realistic estimation of the type and synaptic number than can impinge on single or statistically derived average neuron (Zaborszky *et al.*, 1975). Similar morphometric analyses on the axonal tree in conjunction with available data on the number of neurons that are in the space of axonal arborizations can help to estimate the number and postsynaptic targets of the local axon collaterals of electrophysiologically identified neurons (see Zaborszky and Duque, 2000).

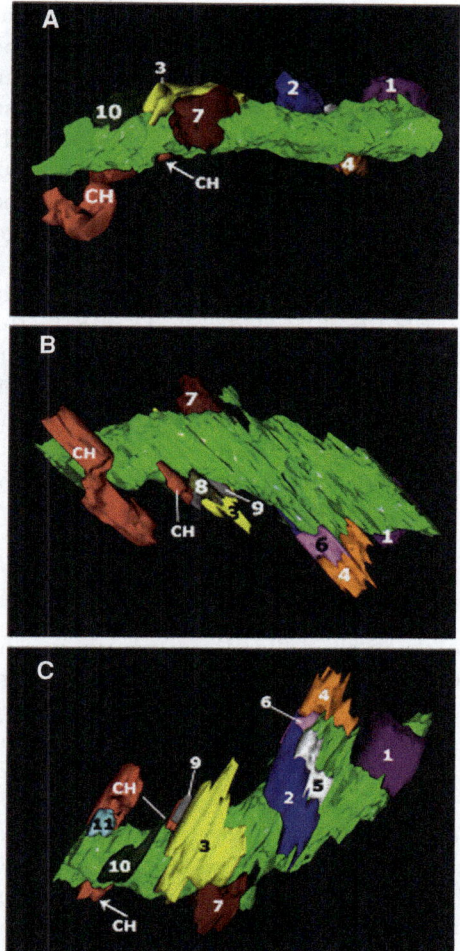

Figure 7.12. 3D partial reconstruction of the dendritic segment of the parvalbumin neuron illustrated in Fig. 7.7A. This 3D model consists of 66 ultrathin sections, rotated along the dendrite (green color). Numbered boutons are labeled with different color. Boutons labeled with numbers 2, 4, and 5 can also be seen in Figs. 7.9B and 7.10B. Boutons 2, 5, 6, and 7 can be identified on Fig. 7.11B, C. The red surface marked with CH corresponds to cholinergic dendritic profiles that are in close vicinity to the parvalbumin dendrite.

IV. SUMMARY OF ADVANTAGES AND LIMITATIONS

The fundamental advantage of any single-cell recording technique combined with the labeling of the recorded cell is that it allows the correlation of physiology and morphology at the cellular level. The advantage of juxtacellular labeling is its compatibility with a host of other techniques that, when combined, advance our ability to characterize single neurons in many different ways. By virtue of being extracellular, the recording can usually be maintained for prolong periods of time and it is less prone to disruptions due to movement. Hence, recording and labeling of neurons can be done both in superficial, i.e., cortical, as well as in subcortical structures that are several millimeters deep. Juxtacellular labeling allows for the staining of neurons in deep subcortical structures with a success rate of usually more than 80%. Neurons labeled with biocytin or Neurobiotin can be stained for light microscopy and/or electron microscopy using chromogens such as DAB. Juxtacellular labeling of single neurons is compatible with immunohistochemical techniques, making it possible to collect data about the chemical identity of the recorded and labeled cell. As seen in the previous examples, these advantages do not sacrifice the possibility of fine morphological evaluation at the light and/or the electron microscopy levels.

The staining obtained with juxtacellular labeling can reach Golgi-like quality in the soma and dendrites. It seems that myelinated axons are always filled. However, depending on labeling time and quality of entrainment, some axons are not fully labeled.

A. Advantages

1. The stability of the method allows long recordings, which in turn permit more complete electrophysiological characterizations.
2. The method allows for recording and successful labeling of neurons in various brain structures, particularly in very deep subcortical regions, where intracellular recordings are almost impossible to obtain.
3. At least in theory, the natural cellular membrane and intracellular environment is less disturbed since the cell is not being impaled with an electrode.
4. The method is species independent and compatible with immunohistochemistry and electron microscopy.

B. Limitations

1. Because the recording is extracellular, postsynaptic potentials cannot be recorded.
2. Quality of the "fill" depends on the length of time the cell is entrained and might not be as complete as the filling with intracellular techniques if the cell is not labeled long enough.
3. Axonal labeling may also be partial, but this again seems to depend on labeling time.
4. Occasionally, a glial cell in the vicinity of the labeled cell may also get labeled.
5. The combination of juxtacellular labeling with many other techniques, although powerful in providing exquisitely detailed characterization of single cells, can be limited in practice because it is labor-intensive and time-consuming.

APPENDIX

As illustrated in Fig. 7.4 and throughout examples in section "Applications", the advantage of juxtacellular labeling is its compatibility with many other techniques, which allows increasingly detailed characterization of a single cell. Figures 7.6–7.12 give an outline of possible steps for a sample experiment in which a neuron can be retrogradely labeled and then tested for several neurochemicals to be then converted to Ni-DAB, double labeled for ChAT bulk immunostaining, reconstructed, and finally analyzed at the EM level.

A. Animal Preparation Prior to Electrophysiological Recordings

Day 0

1. *Anesthesia*: Because these are survival surgeries, use a mixture of ketamine (85 mg/kg) and xylazine (15 mg/kg) i.p. Do not use urethane, as it affects gastrointestinal motility (Yuasa and Watanabe, 1994) and is toxic to the liver and lungs (Renuka and Dani, 1983). Do not use pentobarbital; it may affect retrograde and anterograde tracer transport (Rogers *et al.*, 1980). Gas anesthesia can also be used.
2. *Stereotaxis*: For stability and proper location of brain structures, anesthetized animals need to be placed correctly in an appropriate stereotaxic apparatus. Wound margins and points of contact between animal and stereotaxic apparatus are usually infiltrated with lidocaine solution (2%) and xylocaine ointment (5%), respectively. Aseptic conditions are necessary.
3. *Tracer injections*: Fluorogold is primarily a retrograde tracer. Prepare it at 2.5–4% in double distilled H_2O. Pressure inject 0.05–0.3 µl/ 10 min/per site using a 1 µl Hamilton syringe. Survival time: 2–14 days. Fast Blue (FB, Sigma): same as for FG.

B. Animal Preparation for Electrophysiology

Day 1

1. *Anesthesia*: Use urethane 1.3 g/kg i.p. Usually injected once. Caution: Urethane is highly toxic.
2. *Surgery*: Place animal in a stereotaxic apparatus as described before. Retract scalp and overlying fascia from the skull and puncture the atlanto-occipital membrane to allow drainage of some cerebral spinal fluid. Keep body temperature at 37–38°C preferably with a hot water circulating pad. Drill small burr holes in both hemispheres, over, for instance, the frontal cortex for EEG recordings [anteroposterior (AP) +1.6–1.8 mm, mediolateral (ML) ± 0.5–2.0, relative to bregma] and

over the basal forebrain (AP −0.3 to −1.0), L ±2.4–3.2 mm, relative to bregma) for single unit recordings.

C. Electrophysiology and Labeling

1. *Electrode fabrication:* Make recording microelectrodes from 2.0 mm outer diameter borosilicate glass capillaries (World Precision Instruments, Sarasota, FL) on a Narishige PE-2 vertical pipette puller. Break the tip of the electrode under visual guidance to approximately 1.0 μm in diameter. Fill the electrode with 0.5 M NaCl containing 4% biocytin. Measure in vitro impedance and use 10–30 MΩ electrodes.
2. *Set up EEG electrodes and obtain a signal:* Find and record a neuron long enough to allow statistical significance test to be performed for its single-unit electrophysiological characterization. Record EEG and single-cell electrophysiology at the same time, to allow correlation analysis.
3. *Labeling:* Get as close as possible to the cell. Entrain the cell by passing current pulses 1–10 nA in intensity. Monitor entrainment and move the electrode closer to or further away from the cell or increase or decrease pulse intensity as necessary to maintain entrainment without causing cellular damage. Entrain cell for at least 20 min to obtain good labeling.

D. Perfusion

Within a few hours after the termination of the labeling protocol, perfuse the animal transcardially.

Pass 100 ml normal saline followed by 200 ml of ice-cold fixative [4% paraformaldehyde, 15% saturated picric acid, and 0.05% glutaraldehyde in 0.15 M phosphate buffer (PB), pH 7.4], followed by 200 ml of the same fixative without glutaraldehyde.

Remove brain and postfix at 4°C overnight in the fixative without glutaraldehyde.

Optional: Cryoprotect in sucrose if the brain is to be cut frozen in a cryostat or on a freezing microtome.

E. Cutting and Pretreatment of Sections

Day 2

1. *Cutting:* Dissect the block of tissue containing the cell. The size of the block depends on what you expect and want to process. If long projection axons are expected, then cut the block so that it contains the area with the terminals of the axon. Notch one side of the brain so that you

can keep track of left and right hemispheres. Cut coronal or sagittal sections 50–60 μm thick with a Vibratome® or freezing microtome if only light microscopic processing is planned.

2. *Rinsing*: Select the sections to be processed and rinse them several times in cold PB until the yellowish color of the picric acid disappears. All rinses from here on should be done in cold 0.1 M PBS. Every rinse should take about 5 min, while gently agitating the sections in a shaker.

3. *Borohydride and peroxidase pretreatments*: Incubate sections in 1% sodium borohydride in PBS (to remove excess aldehydes) for 20 min. Rinse 3–5 times or until the bubbles are gone and the sections sink. Then incubate for 10 min in 1% hydrogen peroxide in cold PBS (to block peroxidases) and rinse again three times.

F. Visualization of Biocytin-Filled Neuron and Digital Photography

Days 3 and 4

1. Select 48 sections where you expect to find the soma of the juxtacellularly labeled neuron (so that the soma will be found within 2.4 mm of tissue, which is considered a very large margin of error). Store the rest of the sections in PBS at 4°C.

2. Incubate the sections overnight (4°C, in a shaker) in avidin-conjugated rhodamine (R) (1:500; Jackson ImmunoResearch Labs, West Grove, PA). This can be done in six scintillating vials, each containing 1 ml of solution and the corresponding sixth section of the series for a total of eight sections (make sure the sections are all submerged in the solution). This way, vial 1 will have sections 1, 7, 13, and so on; vial 2 will have sections 2, 8, 14, and so on.

3. Rinse the sections once or twice (to remove excess fluorescent particles).

4. Searching for the cell: Arrange the 48 sections individually in rostro-caudal order. (This can be done in two 24-well dishes.) Then mount the first section onto a glass slide and search for the cell under a epifluorescence microscope. Do not cover the section with a glass coverslip and do not use anything to enhance fluorescence. Just keep the section wet with cold PBS. These additional steps may substantially deteriorate the tissue and diminish the ability to do lengthy processing. If you do not find the cell, jump 150 μm and mount the fourth section and repeat the process. Instead of scanning sections in order, this speeds up the process of finding the cell; usually you will find dendritic processes and then you can just follow them very quickly (maybe through several sections) to the soma.

5. Document your finding by taking photographs, maybe at two or three different magnifications, usually 5×, 20×, and 40×. Low-magnification pictures can quickly determine the region where the cell is located.

More than 40× may be difficult because of the focusing over wet mounted tissue. Fluorescence may be very intense so that the soma may appear larger and blurry. Try to minimize this by illuminating at less than 100% and by using filters. At this time one can also determine if the cell is double labeled for instance with Fluorogold.

Note: We suggest using red fluorescent markers to label single cells because the normal human eye is more capable of detecting red than any other color.

G. Neurochemical Identification of Biocytin-Filled Neuron

Day 5

1. Incubate the section with the soma and a second control section selected from the ones in storage in a monoclonal rat anti-ChAT antibody (Rat anticholine acetyltranferase; 1:10; 2 days at 4°C; Boehringer Mannheim, Germany). Triton X can be added to help penetration of the antibodies and improve chances of positive immunotest, i.e., 0.02% Triton X in 0.1 M PB. This, however, may render the section useless for EM. Store the rest of the sections in 0.1 M PBS at 4°C or proceed to develop them (see conversion of fluorescent signal to DAB).
2. Rinse the section with the soma and the control section.
3. Incubate in a secondary antibody conjugated to, for instance, fluorescein isothiocyanate (FITC-conjugated goat anti-rat; 1:100–200; 4 h at room temperature; Jackson ImmunoResearch Labs, West Grove, PA). If the results of the incubation are poor, i.e., the fluorescent signal is not very good, the incubation can be done overnight at 4°C in a shaker. The problem is that in serial testing for different immunochemicals, this can add substantial time to the processing.

Day 6

Locate again the single-labeled cell with the "red" excitation/emission filter set of the fluorescence microscope and photograph it; change filters and determine if the same cell also emits green fluorescent light. If it does, then this is evidence that the juxtacellularly labeled cell expresses ChAT and it should be photographed again. If the identified cell is not ChAT positive, one can repeat the steps using a different primary antibody and a different control section. This procedure can be repeated two or three times. The secondary antibody can always be conjugated to FITC or some other "green" fluorochrome. This does not confound the results because one can observe how additional cells appear under "green" fluorescence in the field of interest, including or not the identified neuron. The latter neuron remains the only cell visible through the "red" fluorescence filter set.

H. Conversion of the Fluorescent Signal to DAB

1. Incubate the section containing the soma and all adjacent sections of interest (that were processed in avidin-conjugated rhodamine) in biotinylated peroxidase [1:200, "B" component of standard ABC (Avidin–Biotin Peroxidase complex) kit] (Vector Laboratories Inc., Burlingame, CA) for 2–4 h at room temperature (RT). The "A" or avidin component of the ABC kit is omitted because the single-stained neuron already contains the avidin (from the avidin-conjugated rhodamine).

2. Develop the neuron using 3,3′-diaminobenzidine tetrahydrochloride as a chromogen intensified with nickel (Ni) by incubating the sections for 10 min in a solution containing 0.05% DAB and 0.038% nickel ammonium sulfate and then adding hydrogen peroxide to a final concentration of 0.01%, while agitating for another 10–20 min. To determine the best timing, periodically wet mount the section with the soma or one with dendrites and see how dark they are using a regular transmitted light microscope. Try to balance the result between a very dark signal and a very low background. The overall darkness of the section containing the soma may be different because of the extra processing.

3. Rinse thoroughly to get rid of any Ni-DAB deposits outside the labeled neuron.

Note: The development of DAB can also be done by using 10% B-D-glucose and glucose oxidase instead of hydrogen peroxide. In short, after incubating the sections in the "B" component of the ABC kit as described above, incubate sections (15–25 min at RT) in a solution of 0.1 M PB containing 50 mg DAB, 40 mg ammonium chloride, 40 mg nickel ammonium sulfate, 0.4 mg glucose oxidase, and 200 mg B-D-glucose per 100 ml of solution. If one desires to filter the solution, this must be done before adding the glucose oxidase.

I. Staining for a Second Antigen

Days 7 and 8

1. *Select the section*: If the single cell was ChAT positive and one wants to double label the material, in order to find out if axon collaterals of the cholinergic neuron contact, for example, NPY neurons, then several sections around the soma of the cholinergic cell should be selected.

2. Incubate the sections in a primary antibody (Rabbit anti-NPY; 1:500; 2 days at 4°C; Peninsula Laboratories, Inc., Belmont, CA).

3. Rinse sections two or three times, then incubate in secondary antibody (Biotinylated Goat anti-Rabbit 1:200, 4 h RT; Jackson ImmunoResearch Labs, West Grove, PA).

4. Incubate in ABC as indicated in the ABC kit. Develop as described above, but using DAB only, without nickel enhancement. Develop long enough to make the signal light brown. Do not overdevelop because then it will be difficult to distinguish black from brown.

Note. If there are any concerns about again using avidin and biotin because of the possible detection of false signals beforehand, sections can be blocked using the avidin–biotin blocking agents sold by Vector (follow instructions in the package) or the signal can be developed by the peroxidase antiperoxidase (PAP) technique (see Pickel and Milner, 1989).

J. Embedding for Electron Microscopy

Days 9 and 10

1. Osmification: Sections to be investigated at the EM level should be osmicated in a solution containing 1% osmium tetroxide (Electron Microscopy Sciences, Fort Washington, PA) in phosphate-buffered saline (PBS), for approximately 30–40 min.
2. Dehydrate tissue in an ascending series of ethanols (30, 50, 70, 90, 100%). Do contrasting by treating the tissue with 1% uranyl acetate (Electron Microscopy Sciences, Fort Washington, PA) in 70% ethanol, for 30 min.
3. Dehydrate tissue with 1% propylene oxide (Electron Microscopy Sciences, Fort Washington, PA).
4. Infiltrate sections in durcupan (Fluka Chemie AG, Buchs, Switzerland) overnight and then flat embed them between liquid release agent-coated (Electron Microscopy Sciences, Fort Washington, PA) microscope glass slides and coverslips.

K. 3D Light Microscopy Reconstructions

Days 11–30

Neuron reconstructions at the light microscopy level can be carried out using several methods. Because 3D reconstruction offers more information than do 2D reconstructions obtained with typical camera lucida systems, it is advantageous to use a computerized 3D neuron tracing system such as the one offered by MicroBrightField Inc. (Williston, VT). Starting from the soma, the entire neuron is reconstructed from serial sections. The number of sections used in each case varies according to the cell. In the case of double-labeled material it is convenient to also plot other neurons, which are in close proximity to the axon of the juxtacellularly label cell. In the cases presented here the Neurolucida hardware system was interfaced with a Zeiss Axioscope microscope. Outlines of the sections, contours of structures, and fiducial markers were drawn with a 5× Plan-NEOFLUAR objective lens. Somata, dendritic, and axonal branches were traced with a 100×, oil immersion ACHROPLAN objective lens.

L. Electron Microscopy and 3D Reconstruction from Ultrathin Sections

Days 31–40

After photography and full serial reconstruction of the labeled neuron, coverslips can be removed and pieces of interest are dissected out of the tissue with a razor blade. Small pieces of tissue containing areas of interest (such as boutons) are mounted onto blank durcupan blocks and trimmed appropriately. Ribbons of ultrathin sections are cut on a Reichert Ultracut E ultramicrotome and picked up onto Formvar-coated single-slot grids (Electron Microscopy Sciences, Fort Washington, PA). In our examples, the ultrathin sections were analyzed on a Tecnai 12 transmission electron microscope and pictures were captured using either a Gatan Ultrascan digital camera US 4000 SP (11,000×) or conventional EM film. The films were developed and then digitized using a Microtek Scanmaker 4 scanner and further manipulated using Photoshop.

a. Ultrathin cutting

1. Set the block into the holder and using a blade trim it roughly.
2. Under the light microscope, determine the depth in which the area of interest is. With a glass knife, cut away the layers superficial to the area of interest.
3. Using a glass knife, trim the block.
4. Begin to cut ultrathin sections using a diamond knife with simultaneous light microscopic control.
5. Mount sections onto Formvar-coated single-slot grids (copper for general purposes or nickel if intending to do postembedding procedures).
6. Dry grids with filter paper and then place them into gridboxes.

Days 41–60

For quantifying synaptic contacts on electrophysiologically identified neurons, a certain portion of the dendritic tree or of the chemically and electrophysiologically identified neuron has to be reconstructed and the total number of synapses can be extrapolated knowing the total dendritic length and the number of synaptic boutons in the reconstructed volume. The small dendritic segment of the parvalbumin neuron shown in Fig. 7.12 was reconstructed from 66 ultrathin sections. The collection of digital images took about 20 EM h and about 10-h computer time was needed to reconstruct it in 3D.

ACKNOWLEDGMENTS. The research summarized in this review was supported by NIH grant NSO23945 to LZ and IR25 GM60826 to LZ and AD. The Tecnai 12 electron microscope was purchased from grant S10 RR13959 (LZ). Special thanks are due to Lennart Heimer for his comments on a previous version of this manuscript.

REFERENCES

Aghajanian, G. K., and Vandermaelen, C. P., 1982, Intracellular identification of central noradrenergic and serotonergic neurons by a new double labeling procedure, *J. Neurosci.* **2:**1786–1792.

Alonso, A., Khateb, A., Fort, P., Jones, B. E., and Muhlethaler, M., 1996, Differential oscillatory properties of cholinergic and noncholinergic nucleus basalis neurons in guinea pig brain slice, *Eur. J. Neurosci.* **8:**169–182.

Arbib, M. A., Erdi, P., and Szentágothai, J., 1998, *Neural Organization: Structure, Function, and Dynamics*, Cambridge: The MIT Press, p. 407.

Blackstad, T. W., 1965, Mapping of experimental axon degeneration by electron microscopy of Golgi preparations, *Z. Zellforsch. Mikrosk. Anat.* **67:**819–834.

Buhl, E. H., 1993, Intracellular injection in fixed slices in combination with neuroanatomical tracing techniques and electron microscopy to determine multisynaptic pathways in the brain, *Microsc. Res. Tech.* **24:**15–30.

Cajal, S. R., 1911, *Histologie du Système nerveux de l'Homme et des Vertébrés*, Paris: A Maloine.

Cullheim, S., and Kellerth, J. O., 1978, A morphological study of the axons and recurrent axon collaterals of cat sciatic alpha-motoneurons after intracellular staining with horseradish peroxidase, *J. Comp. Neurol.* **178:**537–557.

Detari, L., 2000, Tonic and phasic influence of basal forebrain unit activity on the cortical EEG, *Behav. Brain Res.* **115:**159–170.

Detari, L., and Vanderwolf, C. H., 1987, Activity of identified cortically projecting and other basal forebrain neurones during large slow waves and cortical activation in anaesthetized rats, *Brain Res.* **437:**1–8.

Dringenberg, H. C., and Vanderwolf, C. H., 1998, Involvement of direct and indirect pathways in electrocorticographic activation, *Neurosci. Biobehav. Rev.* **22:**243–257.

Duque, A., Balatoni, B., Detari, L., and Zaborszky, L., 2000, EEG correlation of the discharge properties of identified neurons in the basal forebrain, *J. Neurophysiol.* **84:**1627–1635.

Fairen, A., Peters, A., and Saldanha, J., 1977, A new procedure for examining Golgi impregnated neurons by light and electron microscopy, *J. Neurocytol.* **6:**311–337.

Freund, T. M., and Somogyi, P., 1989, Synaptic relationships of Golgi-impregnated neurons as identified by electrophysiological or immunocytochemical techniques, In: Heimer, L., and Zaborszky, L. (eds.), *Neuroanatomical Tract-Tracing Methods 2*, New York: Plenum Press, pp. 201–238.

Glaser, J. R., and Glaser, E. M., 1990, Neuron imaging with Neurolucida—a PC-based system for image combining microscopy, *Comput. Med. Imaging Graph.* **14:**307–317.

Golgi, C., 1883, Recherches sur l'histologie des centers nerveux, *Arch. Ital. Biol.* **3:**285–317.

Grace, A. A., and Bunney, B. S., 1983a, Intracellular and extracellular electrophysiology of nigral dopaminergic neurons—1. Identification and characterization, *Neuroscience* **10:**301–315.

Grace, A. A., and Bunney, B. S., 1983b, Intracellular and extracellular electrophysiology of nigral dopaminergic neurons—2. Action potential generating mechanisms and morphological correlates, *Neuroscience* **10**:317–331.

Graham, R. C., Jr., and Karnovsky, M. J., 1966, The early stages of absorption of injected horseradish peroxidase in the proximal tubules of mouse kidney: ultrastructural cytochemistry by a new technique, *J. Histochem. Cytochem.* **14**:291–302.

Jankowska, E., Rastad, J., and Westman, J., 1976, Intracellular application of horseradish peroxidase and its light and electron microscopical appearance in spinocervical tract cells, *Brain Res.* **105**:557–562.

Kita, H., and Armstrong, W., 1991, A biotin-containing compound N-(2-aminoethyl)biotinamide for intracellular labeling and neuronal tracing studies: comparison with biocytin, *J. Neurosci. Methods* **37**:141–150.

Kitai, S. T., Kocsis, J. D., Preston, R. J., and Sugimori, M., 1976a, Monosynaptic inputs to caudate neurons identified by intracellular injection of horseradish peroxidase, *Brain Res.* **109**:601–606.

Kitai, S. T., Kocsis, J. D., and Wood, J., 1976b, Origin and characteristics of the cortico-caudate afferents: an anatomical and electrophysiological study, *Brain Res.* **118**:137–141.

Koos, T., and Tepper, J. M., 1999, Inhibitory control of neostriatal projection neurons by GABAergic interneurons, *Nat. Neurosci.* **2**:467–472.

Koos, T., and Tepper, J. M., 2002, Dual cholinergic control of fast-spiking interneurons in the neostriatum, *J. Neurosci.* **22**:529–535.

Kristensson, K., and Olsson, Y., 1971, Retrograde axonal transport of protein, *Brain Res.* **29**:363–365.

LaVail, J. H., and LaVail, M. M., 1972, Retrograde axonal transport in the central nervous system, *Science* **176**:1416–1417.

Li, Y., Brewer, D., Burke, R. E., and Ascoli, G. A., 2005, Developmental changes in spinal motoneuron dendrites in neonatal mice, *J. Comp. Neurol.* **483**(3):304–317.

Light, A. R., and Durkovic, R. G., 1976, Horseradish peroxidase: an improvement in intracellular staining of single, electrophysiologically characterized neurons, *Exp. Neurol.* **53**:847–853.

Lynch, G., Deadwyler, S., and Gall, C., 1974a, Labeling of central nervous system neurons with extracellular recording microelectrodes, *Brain Res.* **66**:337–341.

Lynch, G., Gall, C., Mensah, P., and Cotman, C. W., 1974b, Horseradish peroxidase histochemistry: a new method for tracing efferent projections in the central nervous system, *Brain Res.* **65**:373–380.

McMullen, N. T., and Glaser, E. M., 1988, Auditory cortical responses to neonatal deafening: pyramidal neuron spine loss without changes in growth or orientation, *Exp. Brain Res.* **72**:195–200.

Nambu, A., and Llinas, R., 1997, Morphology of globus pallidus neurons: its correlation with electrophysiology in guinea pig brain slices, *J. Comp. Neurol.* **377**:85–94.

Nunez, A., 1996, Unit activity of rat basal forebrain neurons: relationship to cortical activity, *Neuroscience* **72**:757–766.

Pang, K., Tepper, J. M., and Zaborszky, L., 1998, Morphological and electrophysiological characteristics of noncholinergic basal forebrain neurons, *J. Comp. Neurol.* **394**:186–204.

Pickel, V., and Milner, T. A., 1989, Interchangeable uses of autoradiographic and peroxidase markers for electronmicroscopic detection of neuronal pathways and transmitter-related antigens in single sections, In: Heimer, L., and Zaborszky, L. (eds.), *Neuroanatomical Tract-Tracing Methods 2*, New York: Plenum Press, pp. 97–128.

Pinault, D., 1994, Golgi-like labeling of a single neuron recorded extracellularly, *Neurosci. Lett.* **170**:255–260.

Pinault, D., 1996, A novel single-cell staining procedure performed in vivo under electrophysiological control: morpho-functional features of juxtacellularly labeled thalamic cells and other central neurons with biocytin or Neurobiotin, *J. Neurosci. Methods* **65**: 113–136.

Reiner, A., Veenman, C. L., Medina, L., Jiao, Y., Del Mar, N., and Honig, M. G., 2000, Pathway tracing using biotinylated dextran amines, *J. Neurosci. Methods* **103**:23–37.

Renuka, and Dani, H. M., 1983, Effects of toxic doses of urethane on rat liver and lung microsomes, *Toxicol. Lett.* **15**:61–64.

Richards, C. D., 1972, On the mechanism of barbiturate anaesthesia, *J. Physiol.* **227**:749–767.

Rogers, R. C., Butcher, L. L., and Novin, D., 1980, Effects of urethane and pentobarbital anesthesia on the demonstration of retrograde and anterograde transport of horseradish peroxidase, *Brain Res.* **187**:197–200.

Schmued, L. C., 1994, Anterograde and retrograde neuroanatomical tract tracing with fluorescent compounds, *Neurosci. Protoc.* **50**:1–15.

Schmued, L. C., and Heimer, L., 1990, Iontophoretic injection of fluoro-gold and other fluorescent tracers, *J. Histochem. Cytochem.* **38**:721–723.

Schmued, L., Kyriakidis, K., and Heimer, L., 1990, In vivo anterograde and retrograde axonal transport of the fluorescent rhodamine-dextran-amine, Fluoro-Ruby, within the CNS, *Brain Res.* **526**:127–134.

Scorcioni, R., Lazarewicz, M. T., and Ascoli, G. A., 2004, Quantitative morphometry of hippocampal pyramidal cells: differences between anatomical classes and reconstructing laboratories, *J. Comp. Neurol.* **473**(2):177–193.

Simon, L., Noszek, A., Garab, S., and Zaborszky, L., 2005, Optimal alignment of EM serial sections. Program No. 458.113.2005. *Abstract Viewer/Itinerary Planner.* Washington, DC: Society for Neuroscience. Online.

Snow, P. J., Rose, P. K., and Brown, A. G., 1976, Tracing axons and axon collaterals of spinal neurons using intracellular injection of horseradish peroxidase, *Science* **191**:312–313.

Somogyi, P., 1977, A specific "axo-axonal" interneuron in the visual cortex of the rat, *Brain Res.* **136**:345–350.

Somogyi, P., and Freund, T. M., 1989, Immunocytochemistry and synaptic relationships of physiologically characterized HRP-filled neurons, In: Heimer, L., and Zaborszky, L. (eds.), *Neuroanatomical Tract-Tracing Methods 2*, New York: Plenum Press, pp. 239–264.

Stepanyants, A., Tamas, G., and Chklovskii, D. B., 2004, Class-specific features of neuronal wiring, *Neuron* **43**(2):251–259.

Szentágothai, J., 1970, Glomerular synapses, complex synaptic arrangements, and their operational significance, In: Schmitt, F. O. (ed.), *The Neurosciences: Second Study Program*, New York, NY: The Rockefeller University Press, pp. 427–443.

Szentágothai, J., 1978, The neuron network of the cerebral cortex: a functional interpretation, *Proc. R. Soc. Lond. B* **201**:219–248.

Yuasa, H., and Watanabe, J., 1994, Influence of urethane anesthesia and abdominal surgery on gastrointestinal motility in rats, *Biol. Pharm. Bull.* **17**:1309–1312.

Zaborszky, L., Csordas, A., Buhl, D., Duque, A., Somogyi, J., and Nadasdy, Z., 2002, Computational anatomical analysis of the basal forebrain corticopetal system, In: Ascoli, A. (ed.), *Computational Neuroanatomy: Principles and Methods*, NJ: Humana Press, pp. 171–197.

Zaborszky, L., and Duque, A., 2000, Local synaptic connections of basal forebrain neurons, *Behav. Brain Res.* **115**:143–158.

Zaborszky, L., Leranth, C., Makara, G. B., and Palkovits, M., 1975, Quantitative studies on the supraoptic nucleus in the rat: II. Afferent fiber connections, *Exp. Brain Res.* **22**:525–540.

Zaborszky, L., Palkovits, M., and Flerko, B., 1992, Janos Szentágothai: a life-time adventure with the brain. An appreciation on his eightieth birthday, *J. Comp. Neurol.* **326**:1–6.

8

Nonradioactive In Situ Hybridization in Combination with Tract-Tracing

RUTH L. STORNETTA and
PATRICE G. GUYENET

RUTH L. STORNETTA AND PATRICE G. GUYENET • Department of Pharmacology, University of Virginia, Charlottesville, VA 22908

Abstract: The use of in situ hybridization (ISH) for the detection of mRNAs in cell
bodies has greatly expanded our ability to detect cellular phenotypes in the central
nervous system. Riboprobes have been used in the past to identify neuropeptide
precursors, distribution of receptors, ion channels, and enzymes. More recently,
the discovery of unambiguous markers for the major ionotropic transmitters has
made possible the definitive identification of neurons involved in fast transmission. The advantages and disadvantages of different types of probes, including
DNA probes, oligonucleotides, and RNA probes for the detection of mRNAs are
described. Although in situ hybridization was pioneered with the use of radioactive
probes, nonradioactive alternatives are now readily available. The relative merits of
nonradioactive probes, specifically for combination with tract-tracing, are discussed.
This chapter focuses on in situ hybridization methods based on nonradioactive riboprobes and their use in combination with tract-tracing and immunocytochemistry.

Keywords: c-Fos, clone, digoxigenin, double in situ hybridization, Fluorogold, juxtacellular, plasmid, reverse transcription polymerase chain reaction, riboprobe

I. INTRODUCTION

The use of in situ hybridization (ISH) for the detection of mRNAs in cell
bodies has greatly expanded our ability to detect cellular phenotypes in the
central nervous system (CNS). The first descriptions of synthetic-labeled

RNA hybridizing to DNA in tissue (Gall and Pardue, 1969; John *et al.*, 1969) enabled the detection of specific DNA sequences not just on a gel or blot but in preserved intact specimens of the native tissue. Some of the first studies to use the ISH technique in brain tissues were performed in the 1980s and focused on the identification of various neuropeptides (Bloch *et al.*, 1986a, b; Hoefler *et al.*, 1986; Lanaud *et al.*, 1989; Pochet *et al.*, 1981; Shivers *et al.*, 1986; Siegel and Young, 1985; Terenghi *et al.*, 1987). Researchers have also been able to localize the mRNA coding for ion channels (Baldwin *et al.*, 1991; Brysch *et al.*, 1991; Hwang *et al.*, 1992; Lenz *et al.*, 1994; McKinnon, 1989; Perney *et al.*, 1992; Rudy *et al.*, 1992; Wang *et al.*, 1994), receptors (Goldman *et al.*, 1986; Malherbe *et al.*, 1990; Rogers *et al.*, 1991; Surmeier *et al.*, 1992; Wada *et al.*, 1988), and enzymes characteristic of certain neurotransmitters (Chesselet *et al.*, 1987; Julien *et al.*, 1987; Mezey, 1989; Seroogy *et al.*, 1989; Wuenschell *et al.*, 1986). More recently, the discovery of unambiguous markers for the major ionotropic transmitters γ-aminobutyric acid (GABA) (Esclapez *et al.*, 1994), glycine (Poyatos *et al.*, 1997), and glutamate (Fremeau *et al.*, 2004) has made possible the definitive identification of neurons involved in fast transmission. RNA detection in neuronal cell bodies is critical for the glutamate vesicular transporters 1 and 2 (VGlut1 and VGlut2) as well as for the glycine transporter (GlyT), since these proteins are present only in terminals and are not found in cell soma. GAD-65 and GAD-67 are also not as readily detected in cell bodies in the brainstem as the mRNA coding for these substances. Thus, the use of ISH has proved to be a critical technique for the identification of the cells involved in fast neurotransmission in the brain (Guyenet *et al.*, 2004).

The cytoplasmic localization of mRNA is an ideal technique when used in combination with retrograde tract-tracing, since the location and phenotypic identification of the projecting neurons is the goal of many tract-tracing studies. With the recent avalanche of sequence information now available, DNA templates may be generated for any known sequence and used for ISH. This technique is much faster and a positive outcome more likely than the generation of specific antibodies for a protein of interest. ISH also has the advantage of a somatic localization, unlike many proteins that are not present in large amounts in the soma, but transported to terminals or assembled into native form in the terminals. Of course, if one is interested in anterograde projections and determining the phenotype of terminal fields, ISH will not be useful.

The early studies describing ISH often relied on DNA probes. These had several difficulties including more labor-intensive cloning techniques. The probe itself was double-stranded and therefore could reanneal after denaturing and reduce the available hybridization sites (Lewis and Baldino, 1990). The engineering of a convenient vector incorporating the bacteriophage promoter next to a multiple cloning site featuring common restriction endonucleases allowed for the creation of a DNA template for the in vitro synthesis of RNA probes (Melton *et al.*, 1984). The RNA probe has several advantages. It is easier to procure the clone for the DNA template. The RNA probe is single-stranded and thus will not hybridize to itself. The

RNA–RNA hybrid formed in the tissue is stronger than a DNA–RNA hybrid and can withstand more stringent rinsing, resulting in lower background. The RNA–RNA hybrid will also resist the action of RNAse, another treatment that will substantially lower background. The use of oligonucleotide probes (Lewis *et al.*, 1985) is a further development in bringing the technique of ISH into laboratories with less familiarity with molecular biological techniques. Synthetic oligonucleotides are widely available, relatively inexpensive, and easy to label with either radioactive or nonradioactive methods. Radioactive-labeled oligonucleotides have been used successfully for many of the same mRNAs as the longer riboprobe counterparts including receptors and channels (Brysch *et al.*, 1991; Pelletier *et al.*, 1988; Wisden *et al.*, 1988). The major drawbacks with oligonucleotide probes are their relative lack of sensitivity for messages expressed at lower levels as well as the possibility for nonspecificity if the detected sequence has many identical regions to another sequence or splice variant.

Although ISH was pioneered with the use of radioactive probes, nonradioactive alternatives are now readily available. The caveat for nonradioactive probes is that they are generally less sensitive due to the lower incorporation of the digoxigenin-labeled nucleotide. However, this is not always the case, and in direct comparisons of some riboprobes, nonradioactive probes were reported to be equally sensitive to their radioactive counterparts (Clavel *et al.*, 1991; Kreft *et al.*, 1996; Mitchell *et al.*, 1993; Park *et al.*, 1991). Low expression levels are not a problem for the neuropeptide precursors, the cytoplasmic enzymes, or the vesicular transporters; however, it is a problem for messages that are expressed at a lower level, e.g., most of the receptors and ion channels. Some of this difficulty may be overcome by making the template longer (i.e., more base pairs), thus offering a greater amount of potential incorporation sites for the labeled nucleotide. On the positive side, the resolution of the nonradioactive method is higher since the signal is expressed directly within the cytoplasm and not as silver grains in an emulsion media overlying the cell. However, the problem of signal-to-noise ratio for low signal level must be realized. Also, if one is interested in quantitative analysis of message levels, the radioactive method is essential. However, for neuroanatomical studies, nonradioactive riboprobes and oligonucleotides can easily be combined with more traditional immunocytochemical methods for the detection of proteins as well as tract-tracing to discover connections of cells with specific phenotypes within the CNS (Johnson *et al.*, 2002; Stornetta *et al.*, 1999, 2002, 2003; Stornetta and Guyenet, 1999). The use of well-designed cDNA clones also provides a reliable source of material, free from the inconsistencies, availability, and specificity issues of polyclonal antibodies.

This chapter focuses on ISH methods based on nonradioactive riboprobes and their use in combination with tract-tracing and immunocytochemistry. For methods for oligonucleotides and/or radiolabeled probes, there are many other excellent references (see Chesselet, 1990; Darby, 2000; Valentino *et al.*, 1987; Wilkinson, 1998; Wisden and Morris, 1994; Young, 1990).

II. METHODOLOGICAL CONSIDERATIONS

A. Riboprobe Design

While it may be easy to obtain clones from colleagues, it is also relatively easy to design and produce a cDNA clone for a particular sequence of interest. This is often faster than waiting for the clone from a busy researcher, who may have problems shipping the clone overseas and has the distinct advantage of not having to sign material transfer agreements or involvement with other obligations. The NCBI Web site (http://www.ncbi.nlm.nih.gov/) with access to many different genetic databases is publicly available. Once the sequence of interest is found, the portion least likely to overlap with other closely related sequences can be determined by using the BLAST search (also found on the NCBI Web site) and noting where similar sequences align. Choose the least similar portion of the sequence for the cDNA template. We have had success with templates up to 3.3 kb in length.

B. Production of cDNA Clone

Design primers for reverse transcription polymerase chain reaction (RT-PCR) of RNA. Primer design is often available online from companies that offer primer synthesis. We have used the PrimerQuest tool from Integrated DNA Technologies (http://www.idtdna.com/SciTools/SciTools.aspx) by Steve Rozen and Helen Skaletsky with code available at http://www.genome.wi.mit.edu/genome_software/other/primer3.html. To create the template, use polyA+ selected RNA (in our case, the RNA is from whole brain or from medulla oblongata). Kits to extract oligo-dT isolated RNA are commercially available from many companies. We currently use a kit from Invitrogen (Carlsbad, CA; FastTrack 2.0 for isolation of mRNA from 0.4–1 g of fresh tissue). Prepared RNA is also commercially available. The next step is the RT-PCR to extract the particular sequence of interest from the sample RNA. We have used the Titan One Tube RT-PCR kit from Roche Applied Science (Indianapolis, IN) according to their directions with very good success. A single band of DNA of the correct length is necessary for use as a good template (Fig. 8.1, Step 1d). Performing a melting curve experiment on the RT-PCR by varying the annealing temperature of the PCR reaction will help in achieving this goal. Once the PCR product is obtained, subclone it into a vector usable for RNA in vitro transcription. We have had very good luck with the pCRII-TOPO vector from Invitrogen following the manufacturer's directions (Fig. 8.1, "RT-PCR," Steps 1a–d).

C. Transformation of cDNA Plasmid into Competent Cells (*E. coli*)

Whether the cDNA plasmid is obtained from outside the laboratory or within the laboratory as detailed in the previous steps, it must be transformed

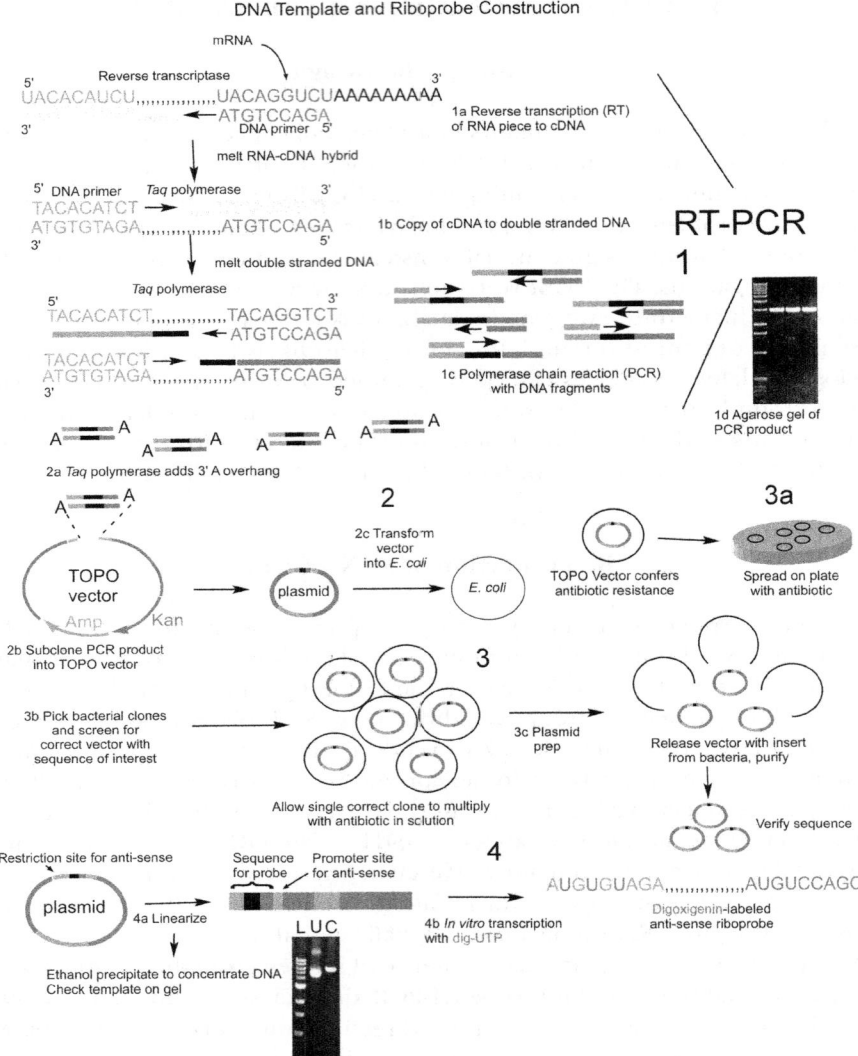

Figure 8.1. Overview of fabrication of DNA template and digoxigenin-labeled riboprobe. Step 1a: Annealing the DNA oligo that is 5′ to 3′ to the mRNA from a sample tissue extract of RNAs and creating a cDNA "copy" of the mRNA sequence of interest catalyzed by the reverse transcriptase enzyme. The RNA–cDNA hybrid melts (becomes single-stranded) by raising the temperature in the thermal cycler (PCR machine). Step 1b: Annealing the DNA oligo that is 3′ to 5′ to the cDNA (instead of the mRNA) and the copying of the cDNA to double-stranded DNA catalyzed by Taq polymerase. Once the copy is made, the cDNA–cDNA hybrid melts by raising the temperature of the PCR machine. Step 1c: The second annealing step is repeated 25–35 times to generate exponential quantities of cDNA. Step 1d: Agarose gel of DNA from PCR reaction—a single band of the correct length is produced. Step 2a: Taq polymerase adds a 3′ A overhang to the cDNA it copies. This property of the cDNA is exploited by the TOPO vector. Step 2b: The PCR product is subcloned into the

into competent cells and prepared in a reasonable quantity for further manipulation. This can be achieved with several commercially available kits. We currently use the One Shot Top 10 F' chemically competent *E. coli* from Invitrogen and the Wizard Plus Midipreps DNA purification system from Promega (Madison, WI) according to the manufacturer's instructions. After larger scale preparation of the DNA, the sequence should be verified before continuing, particularly in reference to the orientation of the clone in the vector. The concentration of the DNA can be determined with a spectrophotometer (Fig. 8.1, Steps 3a–c).

D. Production of Linear cDNA Template

Vectors useful for in vitro RNA transcription have multiple restriction sites on either end of the cloning site of the sequence of interest. Choose a restriction enzyme that cuts only once at the end of the sequence of interest (usually in the multiple cloning site) at the opposite end from the desired promoter site. The enzyme should leave a 5' overhang or blunt end (Fig. 8.1, Step 2a). (Note: Enzymes that leave a 3' overhang will result in the production of double-stranded RNA molecules, drastically reducing the yield of the in vitro transcription reaction.) The pCRII-TOPO vector has promoters on either side of the cloning site, thus allowing the production of either sense template or antisense template. We normally set up a large number of restriction enzyme reactions with about 40 μg of DNA total. Our experience is that the restriction enzymes work better in smaller volumes (about 20–30 μl per reaction). It is absolutely critical that the restriction

←───

Figure 8.1. (*Cont.*) TOPO vector (antibiotic resistance areas of the TOPO vector are indicated as "Kan" for kanamycin and "Amp" for ampicillin). The resulting circular piece of DNA (plasmid) is then mixed with chemically competent *E. coli*, incubated on ice for 5–30 min, the reaction heated to 42°C for 30 s, and then returned to the ice. Step 2c: The resulting transformation (the plasmid that is now incorporated into the *E. coli*) is spread on plates (previously prepared with a mix of antibiotic, agar, and appropriate growth medium). Step 3a: After sitting in a 37°C oven overnight, the plates will have small white dots (colonies of bacterial clones). Individual colonies are picked with a sterile toothpick or wire loop and placed into an aliquot of liquid growth media containing appropriate antibiotic. Step 3b: Screen a sample of the colonies to determine whether the colony contains the correct plasmid before growing the colony in large quantities. Step 3c: The "plasmid prep" is the procedure for growing large quantities of the correct bacterial colonies, and then releasing the plasmids from the bacteria and purifying the plasmid DNA. After this process, the plasmid sequence must be verified. Step 4a: The correct purified plasmid DNA is linearized with an appropriate endonuclease (restriction enzyme). This linearized DNA serves as the template and is concentrated with ethanol and checked on an agarose gel (L = ladder, U = uncut DNA plasmid, C = linear DNA template). Step 4b: The linear DNA is then transcribed into digoxigenin-labeled cRNA (riboprobe) in vitro using the appropriate RNA polymerase in a solution containing digoxigenin-labeled UTP.

enzyme cuts the DNA to completion. Any traces of circular plasmid DNA remaining will carry over to the in vitro transcription reaction, resulting in long stretches of noncoding plasmid sequence transcription and incorporation of much of the labeled nucleotide into this "garbage" sequence. Check the completion of the reaction by gel electrophoresis of 1 µl from each enzyme reaction as well as 1 µl of uncut ("supercoiled") plasmid DNA (Fig. 8.1, Step 4a). The uncut DNA will run at different lengths than the cut DNA. There should be only one clear band per restriction enzyme reaction, with no bands appearing like the uncut DNA. Combine all successful restriction reactions into one tube and perform a phenol–chloroform extraction to eliminate the enzyme and any other impurities from the now-linearized template. Concentrate the DNA template by ethanol precipitation. Measure the concentration by spectrophotometer (Fig. 8.1, Step 4a).

E. In Vitro Transcription of cRNA from Linear DNA Template

Assemble the components of the reaction mixture (with the exception of the enzyme) at room temperature and in the stated order to prevent the precipitation of DNA template by spermidine in the reaction buffer. Be aware of keeping everything as clean and RNAse-free as possible—use gloves and sterile tips and do the reaction assembly on a clean surface. Mix ingredients thoroughly in a sterile plastic Eppendorf tube by pipetting up and down after each addition (see Appendix section "In Vitro Transcription of cRNA from Linear DNA Template" for detailed recipes).

F. Test for Incorporation of Nonradioactive Label Using a Dot Blot on a Nytran Strip

This step is necessary to determine whether the riboprobe has been labeled with digoxigenin and also to determine the relative amount of digoxigenin that has been incorporated (see Appendix section "Test for Incorporation of Nonradioactive Label Using a Dot Blot on a Nytran Strip" for further details).

G. Tissue Preparation

Anesthetize rats with pentobarbital and perfuse transcardially with 100 ml phosphate-buffered saline (PBS; pH 7.4) followed by 500 ml of 4% paraformaldehyde in 0.1 M phosphate buffer (pH 7.4) at room temperature. Extract brains and postfix in the same fixative for up to 3 days at 4°C (postfixation does not appear to affect ISH; however, take care for postfixation time for any additional antibodies used). Section brains at 30 µm on a vibrating microtome at room temperature in 0.1 M phosphate buffer. (Brains may also be sunk in 25% RNAse-free sucrose and cut frozen.)

Collect sections in RNAse-free cryoprotectant 50 mM sterile phosphate buffer, 30% ethylene glycol (Sigma-Aldrich, St. Louis, MO), 20% glycerol (RNAse-free; Sigma-Aldrich) in sterile 24-well tissue culture plates. Sections may be kept in this solution at −20°C for up to 1 year.

H. In Situ Hybridization

1. Prehybridization

This step allows the tissues to adapt to the conditions of the hybridization buffer as well as serving as an important blocking step to prevent or at least decrease nonspecific hybrid formation. It is extremely important to make the prehybridization mixture with sterile, RNAse-free solutions, and with sterile dishes, pipette tips, etc. This is the time to be paranoid about RNAse! (see Appendix section "Prehybridization" for details).

2. Hybridization

In this step, the riboprobe is added directly to the "prehybridization" solution, and the conditions are optimized for hybrid formation. See Fig. 8.2 for a summary and Appendix section "Hybridization" for further details.

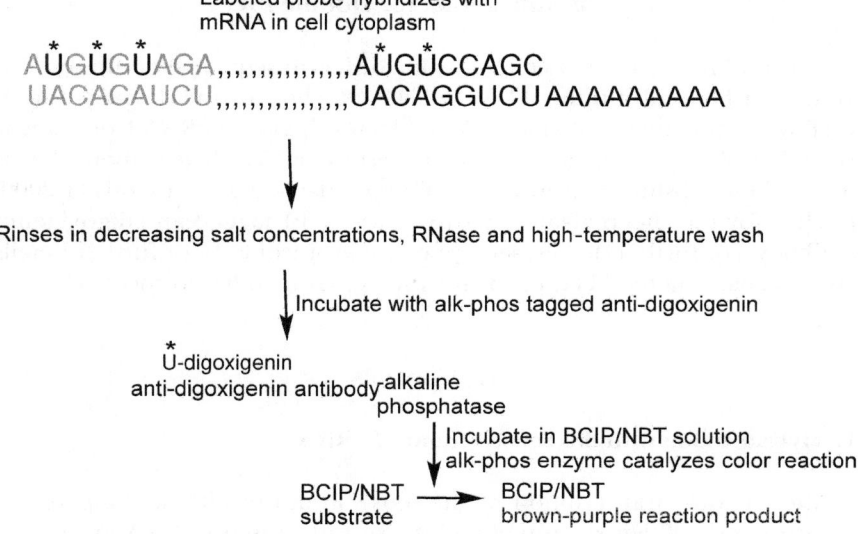

Figure 8.2. Overview of visualization of digoxigenin-labeled riboprobe.

3. Rinsing

The rinsing of the tissue after the hybrids are formed is necessary to decrease the background (nonspecific hybrids) as well as to destroy any riboprobe that is not hybridized. The basic idea is to take the tissue through solutions of decreasing salt concentration as the RNA–RNA hybrid is sensitive to salt concentration. Weaker bonds (i.e., nonspecific hybrids) will not survive the lower salt concentrations. One of the rinsing steps involves treatment with RNAse A, an enzyme that will destroy single-stranded RNA (e.g., nonhybridized riboprobe). The final step is a high-temperature rinse in the lowest concentration of salt solution, a condition in which only the strongest, most specific hybrids will survive. For a summary, see Fig. 8.2, and for details, see Appendix section "Rinsing Through Decreasing Salt Concentrations, RNAse A, and High Stringency Wash."

I. Immunocytochemistry for Revealing Digoxigenin and Other Proteins or Tract-Tracers of Interest

After the hybrids are formed and the tissue is rinsed, the hybrids are stable and proteins of interest may be revealed by using standard immunocytochemical protocols. The digoxigenin label as well as the Fluorogold (FG) tracer is revealed in this manner. See Fig. 8.2 for a summary and see Appendix section "Immunocytochemistry for Revealing Digoxigenin and Other Proteins or Tract-Tracers of Interest" for details.

J. Modifications for Double ISH

A second nonradioactive cRNA with a different sequence of interest can be transcribed, substituting FITC-12-UTP (Roche) for the digoxigenin-11-UTP (see Appendix protocol "In Vitro Transcription of cRNA from Linear DNA Template" for further details). To test the FITC-labeled riboprobe on the dot blot, substitute sheep anti-FITC peroxidase-tagged antibody (1:2000; Roche) for the sheep alkaline phosphatase (AP) tagged anti-digoxigenin antibody. For further details, see Appendix protocol G.1; for further details about visualizing the FITC probe for ISH, see Appendix protocol G.2.

K. Controls

1. Hybridization with the Sense Strand of cRNA

After in vitro transcription of the sense strand of cRNA, compare the incorporation of digoxigenin of both sense and antisense cRNAs by relative

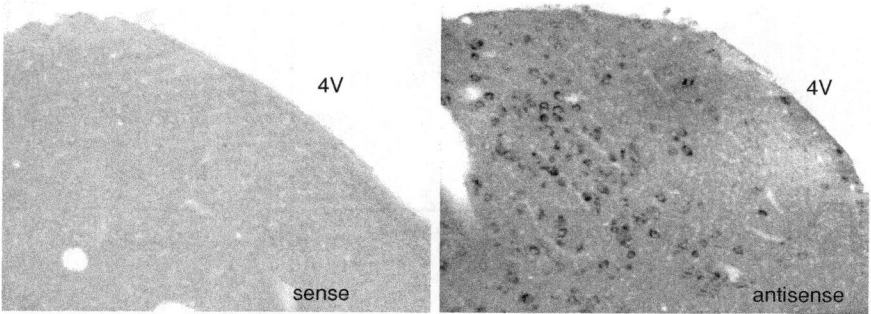

Figure 8.3. Sense/antisense control with cRNA for vesicular glutamate transporter-2 in medial vestibular nucleus of rat medulla oblongata.

appearance on the dot blot. Use a concentration about 2–3×, the equivalent amount of antisense cRNA, and do "side by side" ISH on tissue from the same brains with sense and antisense riboprobes. Allow the colorization procedure to progress for the same amount of time. There should be very strong signal in appropriate places for the antisense riboprobe-hybridized tissues and no signal in the sense-hybridized tissues (see Fig. 8.3). This is the standard control for specific hybridization signal.

2. Hybridization Signal in Expected Areas but No Signal Where No mRNA Should Be Found

We feel that the neuroanatomical consistency of the signal in areas that are known to express the mRNA versus no signal in areas where no mRNA is present is the best control (see Fig. 8.4). This is difficult in cases where the mRNA is ubiquitously or very broadly expressed. In these cases, one must rely on the sense/antisense control data. The internal consistency of the mRNA being expressed in certain types of neurons is also helpful (e.g., in the case of the VGlut2, we never saw the mRNA expressed in GABAergic neurons). One of the standards we have used for many mRNAs of interest is the lack of expression of mRNA in motor neurons (the larger motor neurons are notorious for expressing artifactual background in immunohistological procedures).

III. APPLICATIONS

A. Tract-Tracer Combined with ISH: VGlut1[specific] mRNA-Containing Neurons in Medulla Project to Cerebellum

After noting the striking distribution of VGlut1 mRNA in medulla oblongata in precerebellar nuclei and the fact that VGlut1 protein is expressed in

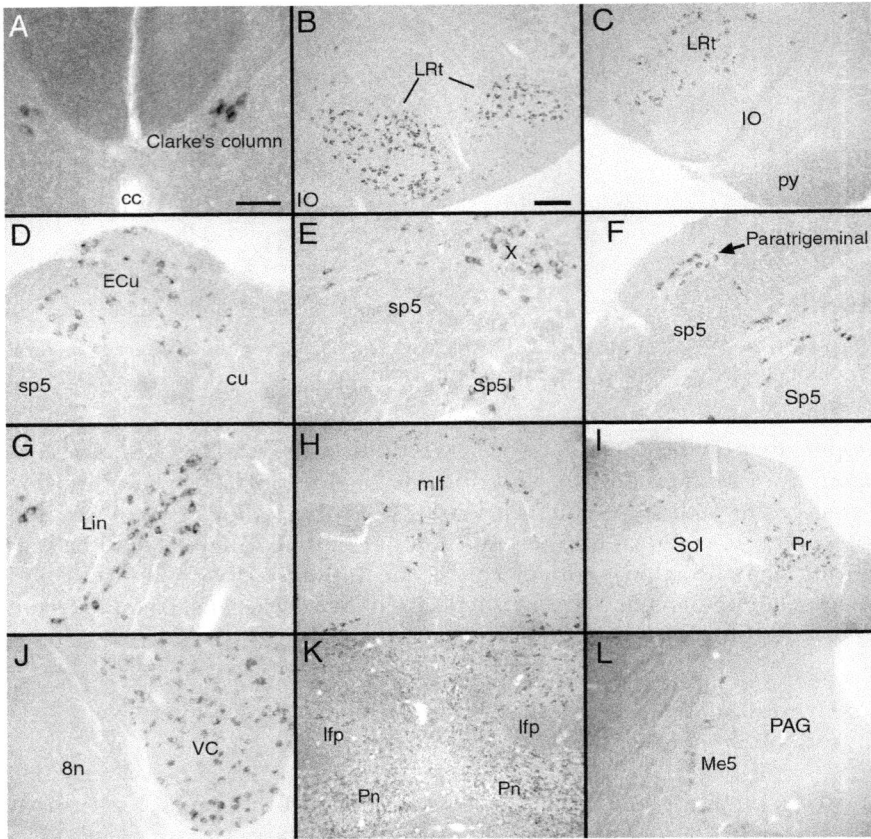

Figure 8.4. Discrete distribution of VGlut1 mRNA in rat brain and spinal cord. (A) Clarke's column of dorsal horn of spinal cord. (B) Lateral reticular nucleus (IO, inferior olive). (C) Lateral reticular nucleus [note the lack of label in the IO and the pyramidal tract (py)] (D) External cuneate nucleus (ECu) (sp5, spinal trigeminal tract; cu, cuneate fasciculus). (E) Spinal trigeminal tract and nucleus X. (F) Ectotrigeminal nucleus (E5) and paratrigeminal nucleus. (G) Linear nucleus (Lin). (H) Medial brainstem at Bregma level −12.80 mm (mlf, medial longitudinal fasciculus). (I) Prepositus nucleus (Pr) (Sol, nucleus of the solitary tract). (J) Ventral cochlear nucleus (VC) (8n, eighth nerve). (K) Pontine nucleus (Pn) (lfp, longitudinal fasciculus pons). (L) Mesencephalic trigeminal nucleus (Me5) (periaqueductal gray, PAG). Scale bar: 100 μm for A, D, E, G, J, and L. Scale bar:200 μm for B, C, F, H, I, and K.

mossy fibers in cerebellum (Bellocchio *et al.*, 1998; Hisano *et al.*, 2002), we tested the hypothesis that most cerebellar-projecting neurons in rat brainstem contain VGlut1 mRNA. Four 100-nl pressure injections of the retrograde marker FG (2% in sterile saline; Fluorochrome Inc., Englewood, CO; Schmued and Fallon, 1986) were placed 1–2 mm deep in various

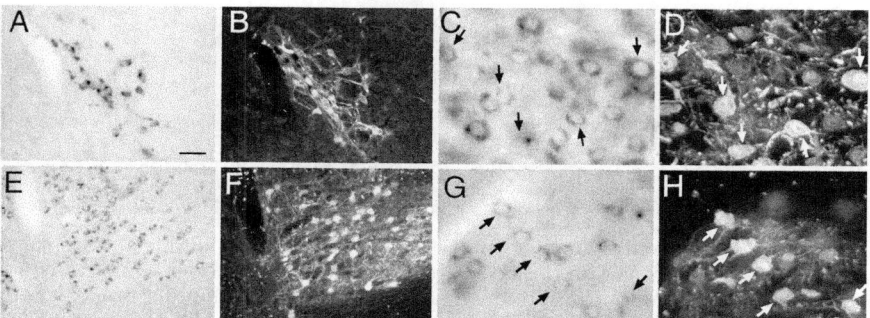

Figure 8.5. Examples of colocalization of VGlut1 mRNA and Fluorogold (FG) (from cerebellum). (A) VGlut1 mRNA in linear nucleus (brightfield). (B) FG immunoreactivity (Cy-3 revealed by fluorescence; same field as in A). Note that most of the cells are double labeled. (C) VGlut1 mRNA in pontine nucleus (brightfield). (D) FG immunoreactivity (Cy-3 revealed by fluorescence; same field as in C). Arrows point to a few of the many double-labeled cells. (E) and (G) VGlut2 mRNA in inferior olive (brightfield). (F) and (H) FG immunoreactivity (Cy-3 revealed by fluorescence; same field as in E). Arrows point to some of the double-labeled cells. Scale bar: 50 μm for A, B, E, and F. Scale bar: 20 μm for C, D, G, and H.

locations of the left cerebellar cortex. The brainstem was processed for ISH for VGlut1 mRNA in combination with FG immunocytochemistry. Immunocytochemical detection of FG was accomplished by incubating the tissue with a rabbit anti-FG antibody (1:10,000; Chemicon, Temeluca, CA), followed by a biotinylated anti-rabbit IgG (1:200; Vector, Burlingame, CA) and visualized with streptavidin Cy3 (1:1000; Molecular Probes, Eugene, OR). We found that the vast majority of FG-labeled neurons in pons and medulla (with the exception of the inferior olive) contained VGlut1 mRNA (Fig. 8.5).

B. Tract-Tracer Combined with ISH and Immunocytochemistry: Catecholaminergic Neurons in Rat Medulla Oblongata Containing VGlut2 mRNA Project to Spinal Cord

Fluorogold (2–3% in sterile saline) was pressure injected into the vicinity of the intermediolateral cell column (1 mm below the entry point of the dorsal roots) at the first and third thoracic segments bilaterally (50-nl injections, four injections per rat). The brainstem was processed for ISH with VGlut2 riboprobe in combination with antibodies for tyrosine hydroxylase (mouse monoclonal; Sigma) and FG (described above). We found many neurons immunoreactive for tyrosine hydroxylase that also contained VGlut2 mRNA and FG. Some of these neurons are illustrated in Fig. 8.6. This was the first study to demonstrate that brainstem presympathetic catecholaminergic neurons are glutamatergic.

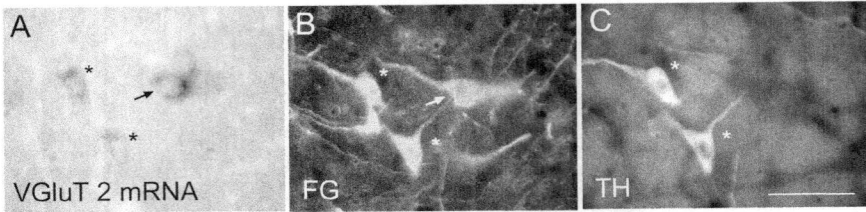

Figure 8.6. Rostral ventrolateral medulla (RVLM) catecholaminergic and noncate-cholaminergic neurons with projection to the thoracic spinal cord contain VGlut2 mRNA. (A–C) RVLM neurons in the same field photographed under brightfield in A show VGlut2 mRNA AP reaction product and under fluorescent light in B and C show FG immunoreactivity (ir) revealed with Cy-3 (B) and tyrosine hydroxylase (TH) ir revealed with Alexa 488 (C). Asterisks indicate two TH-ir bulbospinal cells that contain VGlut2 mRNA. The arrow points to a strongly VGlut2-positive bulbospinal cell that is devoid of TH-ir. Scale bar: 50 μm. (From Stornetta *et al.* 2002.)

C. Tract-Tracer Combined with Double ISH: Neurons Containing Both Preproenkephalin and VGlut2 mRNAs Project to Phrenic Motor Nucleus

FluoroGold was injected iontophoretically into the left ventral horn of spinal segment C4 after recording respiratory activity at the injection site. The rat medulla oblongata was processed for ISH with a VGlut2 digoxigenin-tagged riboprobe and a preproenkephalin (PPE) FITC-tagged riboprobe in combination with antibodies for FG (described above). Many bulbospinal neurons of the rostral ventral respiratory group in the caudal medulla oblongata contained both VGlut2 and PPE mRNAs. Examples of these neurons as well as the injection site are shown in Fig. 8.7. This study demonstrated that glutamatergic neurons controlling respiratory output could also use enkephalin as a neurotransmitter.

D. Tract-Tracer Combined with ISH and c-Fos: Baroactivated Neurons in Rostral Ventrolateral Medulla Contain PPE mRNA and Project to the Spinal Cord

FluoroGold was pressure injected into the first and third segments of thoracic spinal cord. Rats were cannulated for chronic blood pressure recording and intravenous injections. Injections of hydralazine into conscious, freely moving rats lowered blood pressure, and after 2 h animals were anesthetized and perfused. The brainstem was processed for ISH for PPE mRNA (detected by a digoxigenin-labeled riboprobe) in combination with antibodies for FG (visualized with Alexa 488) and c-Fos (goat polyclonal, Santa Cruz, visualized with streptavidin Cy3). Many bulbospinal neurons were c-Fos positive (baroactivated) and contained PPE mRNA. Examples of these neurons are shown in Fig. 8.8. This study demonstrates that c-Fos, a

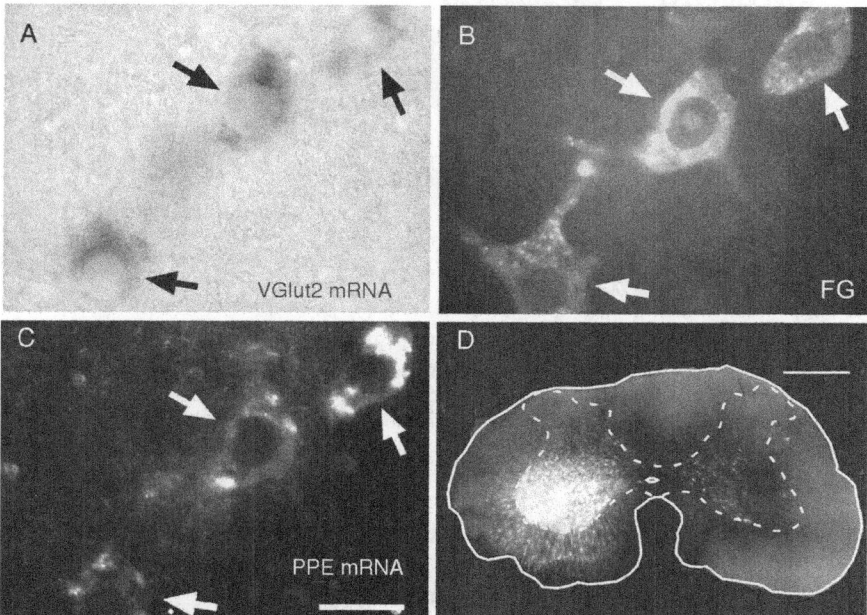

Figure 8.7. Colocalization of VGlut2 mRNA and PPE mRNA in rostral ventral respiratory group (rVRG) bulbospinal neurons. Cluster of rVRG neurons (arrows) containing (A) VGlut2 mRNA (BCIP/NBT reaction product; brightfield), (B) immunoreactivity to Fluorogold (Alexa 488, epifluorescence), and (C) PPE mRNA (Cy3; epifluorescence). Scale bar: 20 μm. (D) Fluorogold injection site into fourth cervical spinal cord segment. Composite photomicrograph of center of iontophoretic deposit. Scale bar: 500 μm. (From Stornetta *et al.*, 2003.)

useful marker of cell activation, can be colocalized with mRNAs of interest as well as with tract-tracers. In this particular example, this was the first demonstration of baroactivated presympathetic enkephalinergic neurons in brainstem.

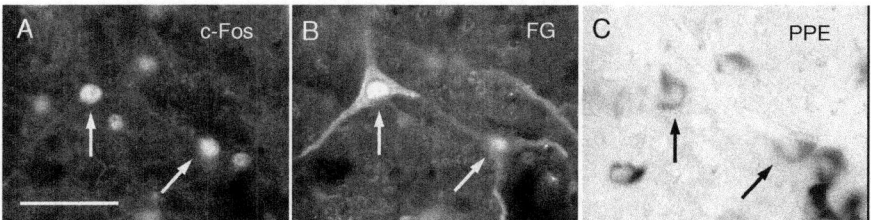

Figure 8.8. Preproenkephalin (PPE), c-Fos immunoreactive (ir) neurons in RVLM project to spinal cord. Cluster of RVLM neurons (arrows) containing (A) Fos-ir (Cy3; epifluorescence), (B) Fluorogold (FG)-ir (Alexa 488, epifluorescence), and (C) PPE mRNA (digoxigenin-labeled probe, brightfield). Scale bar: 50 μm. (From Stornetta *et al.*, 2001.)

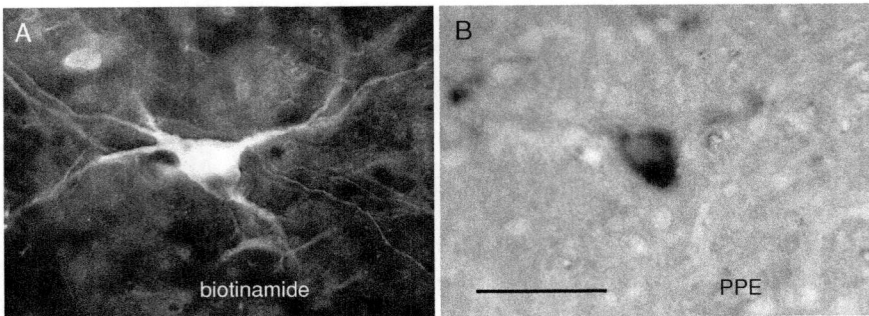

Figure 8.9. Blood pressure–sensitive neuron in RVLM antidromically activated from thoracic spinal cord contains PPE mRNA. (A) Neuron in RVLM recorded in vivo and juxtacellularly labeled with biotinamide (Alexa 488, epifluorescence). (B) Same neuron showing PPE mRNA (digoxigenin-labeled probe, brightfield). Scale bar: 50 μm. (From Stornetta *et al.*, 2001.)

E. ISH Combined with Juxtacellular Labeling: Barosensitive, Spinally Projecting Neurons of the Rostral Ventrolateral Medulla Contain PPE mRNA

Barosensitive vasomotor presympathetic cells were recorded in rostral ventrolateral medulla in vivo in halothane-anesthetized rats. Cells were labeled using the juxtacellular method (described in the chapter by Duque and Zaborszky, this volume) After transcardial perfusion, the brainstem was sectioned and processed for ISH and PPE mRNA (detected by a digoxigenin-labeled riboprobe) in combination with streptavidin Alexa 488 (Molecular Probes) for the detection of the biotinamide-labeled cell. An example of one of these PPE mRNA positive, spinally projecting (detected by antidromic activation from the spinal cord) barosensitive neurons is shown in Fig. 8.9. The ability to find the mRNA of physiologically identified neurons has allowed a major breakthrough in our ability to perform functional neuroanatomy. In this particular example, this was the first demonstration of enkephalinergic presympathetic neurons with firing patterns inversely correlated with blood pressure.

F. Viral Tracing Combined with Double ISH: Some Presympathetic Neurons of the Rostral Ventrolateral Medulla Contain Markers of GABA and Glycine

Rats were anesthetized and the adrenal gland was exposed and injected with pseudorabies virus (PRV; provided by L. Enquist, Princeton University, NJ). This technique is described in further detail in the chapter by Geerling *et al.* (this volume). After 3 days, animals were anesthetized and perfused transcardially. The rat medulla oblongata was processed for ISH with a GAD-67 digoxigenin-tagged riboprobe and a GlyT2 FITC-tagged riboprobe in

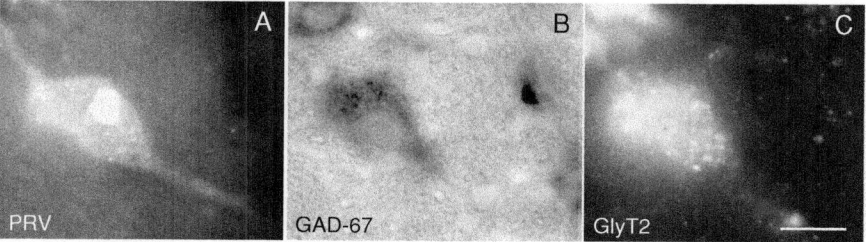

Figure 8.10. Presympathetic neuron in gigantocellular reticular nucleus ventral containing (A) pseudorabies virus (PRV)-ir (Alexa 488, epifluorescence), (B) GAD-67 mRNA (digoxigenin-labeled probe, brightfield), and (C) glycine transporter-2 (GlyT2) mRNA (FITC-labeled probe, Cy3, epifluorescence). Scale bar: 50 μm. (From Stornetta *et al.*, 2004.)

combination with antibodies for PRV (rabbit polyclonal, Enquist). An example of a neuron infected with PRV (resulting from injection in the adrenal gland) in an area of the gigantocellular nucleus in medulla that also contains GAD-67 and GlyT2 mRNAs is seen in Fig. 8.10. This demonstrates that some presympathetic neurons have the capacity for inhibitory control of sympathetic outflow. A major caveat with detection of mRNAs in cells infected with viruses is that viruses corrupt the RNA processing machinery and the native mRNAs are replaced by viral RNAs; thus, allowing only detection of native mRNA at very early time points in the cell's infection with virus (this issue is discussed in more detail in the chapter of Geerling *et al.*, this volume).

IV. ADVANTAGES AND LIMITATIONS

A. Advantages Over Radioactive Methods or Antibodies

1. Detection of nonradioactive riboprobes can be combined with the detection of common retrograde tract-tracers as well as many commercially available antibodies. Nonradioactive probes are particularly well suited for combination with multiple fluorescent tags.
2. Nonradioactive probes can be combined with other techniques such as pseudorabies viral tracing (Stornetta *et al.*, 2004) and juxtacellular labeling (Schreihofer *et al.*, 1999; Stornetta *et al.*, 1999, 2002, 2003).
3. Nonradioactive ISH reveals the transcripts in cell bodies making possible detection of substances normally transported to or solely present in terminals. This is essential for the identification of a particular cell phenotype in combination with retrograde tracers.
4. Nonradioactive ISH does not require special licensing/handling for radioactive materials.
5. Nonradioactive ISH is much faster than radioactive ISH and relative signal-to-noise ratio is easier to control.
6. Nonradioactive riboprobes can be produced in large amounts and kept frozen for at least 1 year before use.

7. Riboprobes can be generated for any known sequence fairly quickly and easily (within 1–2 weeks) compared to several months for antibody production.
8. Riboprobe specificity can be tested with sense controls as well as tissue specificity and can also be subjected to Northern blot analysis.

B. Disadvantages

1. Fluorescent retrograde tracers such as FG fade when exposed to ISH conditions and must be amplified with antibodies for best visualization after ISH.
2. Incorporation of nonradioactive label is less than for radioactive riboprobes and thus detection of low-level messages is more difficult and sometimes impossible with nonradioactive ISH.
3. Attention must be paid to sterile conditions as RNAse can wreak havoc with the procedure.
4. Extra equipment and reagents for molecular biology procedures, including centrifuges, gel electrophoretic apparatus, water baths, and enzymes, must be acquired/borrowed .
5. The protocol requires a few extra days as well as many extra steps over standard immunocytochemical methods.
6. The best available antibody to digoxigenin is AP-tagged and the reaction product of nitro blue tetrazolium (NBT)/5-bromo-4-chloro-3-indolyl phosphate, toluidine salt (BCIP) cannot be dehydrated or covered with "permanent" mounting media. The lack of dehydration results in the tissue looking a bit less clear than with peroxidase reaction products that can be dehydrated. This can be overcome by using a riboprobe transcribed with FITC-12-UTP (Roche) and visualized with a sheep anti-FITC antibody tagged with peroxidase and reacted with a traditional peroxidase substrate.

APPENDIX: DETAILED PROTOCOLS

A. In Vitro Transcription of cRNA from Linear DNA Template

A.1. In vitro transcription reaction mixture

Sterile H_2O (q.s. to 90 μl total volume)
1.5 μg DNA template (volume will depend on concentration)
9 μl 100 mM DTT (provided with enzyme)
18 μl ribonucleotides—UTP (ATP, CTP, GTP mix equal parts of 10 mM stock; Promega)
10.8 μl 2 mM UTP (we have tried varying the concentration of the unlabeled UTP~~this~~ concentration gives us the best results. Note that some unlabeled UTP is necessary for the reaction to work)

2.1 μl digoxigenin-11-UTP (Roche)
18 μl 5× transcription buffer
3 μl RNA polymerase (T3, T7, or SP6; Promega)

A.2. Incubate reaction in a 500 μl Eppendorf tube in a water bath for 2 h. For T3 or T7, incubate at 37°C. For SP6, incubate at 40°C.

A.3. Add 1.5 μl RQ1 DNase (Promega) and incubate for another 20 min at 37°C. (This will destroy the DNA template.)

A.4. Purify the probe with ProbeQuant G-50 micro columns (Amersham Biosciences, Piscataway, NJ) according to the manufacturer's instructions or by phenol–chloroform extraction and ethanol precipitation (we use the micro columns).

B. Test for Incorporation of Nonradioactive Label Using a Dot Blot on a Nytran Strip

B.1. Use a thin Nytran strip (Nytran Supercharge, SPC, pore size of 0.45 μm; Schleicher and Schuell, Keene, NH) razor cut (~0.5 cm ×6 cm) to fit in a 5-ml plastic test tube.

B.2. Wet the strip by immersion in 100% ethanol for 3 min.

B.3. Rinse in 2X saline-sodium citrate (SSC) (20×SSC stock: 3.0 M NaCl and 0.3 M sodium citrate in water, pH 7.0) for 3 min and air-dry on a piece of filter paper for about 10 min.

B.4. Spot 1-μl aliquots of dilutions of control digoxigenin-labeled RNA (Roche) along with 1-μl aliquots of dilutions of the probe to be tested on the strip (we use 1/100, 1/200, and 1/400).

B.5. Cross-link the RNA to the Nytran membrane with UV light [we use the Stratagene (La Jolla, CA) UV Stratalinker 1800 set on auto cross-link]

B.6. Immerse the strip in 2–3 ml 2×SSC in the test tube for 3-min shaking.

B.7. Immerse in 2–3 ml phosphate buffer-bovine serum albumin-triton (PBT) [0.1% bovine serum albumin (BSA)/0.2% Triton X-100 in PBS] and incubate for 15 min on shaker.

B.8. Incubate in anti-digoxigenin tagged with AP Fab fragments from sheep (1:1000; Roche) in PBT for 30 min on shaker.

B.9. Rinse 3 × 5 min in PBS.

B.10. Immerse in NMT (0.1 M NaCl/50 mM $MgCl_2$ in 0.1 M Tris, pH 9.5) for 5-min shaking.

B.11. Immerse in solution of NBT (4.5 μl/ml of NMT) and BCIP (3.5 μl/ml of NMT; NBT and BCIP from Roche). Spots should appear momentarily.

B.12. Stop the reaction by rinsing in TE 8.5 (Tris-EDTA). Use the color intensity relative to the control RNA to determine approximate concentration of probe. We find that, generally, the in vitro translated riboprobe should be about equal to the control RNA for best signal.

C. Prehybridization

C.1. Rinse sections in sterile PBS in sterile petri dishes. We use glass rods fashioned by flaming glass pipettes to seal the ends and make a hook for transferring sections between solutions. Rinse the glass hooks with RNaseZap (Ambion Inc., Austin, TX) prior to use.

C.2. Transfer sections to prehybridization mixture (300 μl per well in a sterile 24-well tissue culture dish). Four to eight sections will fit per well, depending on the size of the tissue. Incubate sections in this solution for 30–60 min shaking at room temperature and then at 37°C for 1 h.

C.3. Prehybridization mixture (in sterile H_2O)

0.60 M NaCl
0.10 M Tris-Cl (7.5)
0.01 M EDTA
0.05% sodium pyrophosphate
5% (w/v) dextran sulfate (must vortex and heat to 37°C to dissolve)
0.50 mg/ml yeast total RNA (Sigma R-7125)
0.05 mg/ml yeast tRNA (Roche 109495)
1×Denhardt's BSA [50X Denhardt's solution: 5 g Ficoll-70 (Sigma), 5 g polyvinylpyrrolidone, 5 g BSA (Fraction V; Sigma) q.s. with H_2O to 500 ml; may be kept in frozen aliquots]
50% deionized formamide (Sigma)
0.05 mg/ml poly A (Sigma P-9403)
10 μm of the four nucleoside triphosphates
10 mM dithiothreitol (DTT; Promega)

Make up prehybridization solution in larger batches and freeze in appropriate-size aliquots. Add 0.5 mg/ml herring sperm (Sigma D-6898-sodium salt of ribonucleic acids from herring testes) that has been boiled for 10 min to denature and quenched on ice just prior to use.

D. Hybridization

D.1. Either add riboprobe directly to wells or transfer the sections into a new solution of prehybridization mixture to which the riboprobe has been added. We found concentrations of 1–3 μl of riboprobe per well (resulting in 50–100 pg/μl) to be most effective for a good signal-to-noise ratio.

D.2. Incubate the sections with riboprobe at room temperature on the shaker for 15 min.

D.3. Incubate at 55–60°C overnight (shaking not necessary). The tissue culture dish cover comes in handy for this we have never had to worry about any extra precautions to stop evaporation.

E. Rinsing Through Decreasing Salt Concentrations, RNAse A, and High Stringency Wash

E.1. Transfer the sections to a mesh-well bottom dish (Nason Fabrications, Fort Bragg, CA) filled with 4 × SSC/10 mM sodium thiosulfate (NaTS).

E.2. Rinse 2 × 20 min @7 °C in this solution.

E.3. Using gloves and a work pad, change solution to 20 μg/ml RNAse A (Sigma) in RNAse buffer [0.5 M NaCl, 10 mM Tris (pH 8.0) 1 mM EDTA.]

E.4. Incubate @7 °C for 30 min.

E.5. Change solution to RNAse buffer and incubate @7 °C for 20 min.

E.6. Transfer to solution of 2 × SSC/10 mM NaTS @7 °C for 20 min.

E.7. Transfer to a solution of 0.5 × SSC @7 °C for 20 min.

E.8. Do a final "high-stringency" rinse in 0.1 × SSC at 50–55°C for 30–60 min. If background is a problem, this high-temperature rinse may be increased in temperature up to 60°C.

F. Immunocytochemistry for Revealing Digoxigenin and Other Proteins or Tract-Tracers of Interest

All rinses and incubations done at room temperature on shaker unless noted otherwise.

F.1. Rinse sections 3 × 5 min in TBS [0.1 M Tris (pH 7.4)/0.15 M NaCl.]

F.2. Incubate 30 min in 10% normal horse serum/0.1% Triton X-100 in TBS.

F.3. Incubate in sheep anti-digoxigenin tagged with AP (Roche) at 1:1000 in 10% normal horse serum/0.1% Triton X-100 in TBS first for 1 h at room temperature on the shaker and then overnight at 4°C shaking. Centrifuge the digoxigenin antibody prior to use, and use only the supernatant portion. Note: In our hands, the only antibody against digoxigenin that works well is the sheep AP-tagged antibody from Roche.

F.4. Add other antibodies of interest to this same mixture. Fluorescent tract-tracers such as FG (Fluorochrome Inc., Denver, CO) should be amplified by using an antibody against FG (e.g., rabbit anti-FG; Chemicon), since the ISH procedure causes fading of these markers.

F.5. Rinse sections 2 × 10 min in TBS.

F.6. Add direct-tagged fluorescent secondary antibody appropriate to FG or other antibodies at this time and incubate for 45–60 min.

F.7. Rinse 2 × 10 min in TBS.

F.8. Rinse 10 min in NMT (0.1 M NaCl/50 mM MgCl$_2$ in 0.1 M Tris, pH 9.5).

F.9. Transfer sections to a solution of NBT (4.5 μl/ml of NMT) and BCIP (3.5 μl/ml of NMT; 300 μl to 1 ml per well in a sterile 24-well tissue culture dish). Filter the NBT/BCIP solution before use (we use syringe microfilters of 0.22-μm pore size). It is important that the NBT/BCIP solution be in clean plastic or glassware. Any dust or dirt can cause the reaction to seed, and precipitate will form. Allow this reaction to proceed at room temperature protected from light on the shaker. Check the progress of the reaction after 30 min and then every 15 min until the dark reaction product is seen in cell bodies in appropriate areas and not in areas where the mRNA should not be present. This is critical to the success of the experiment. We use a dissecting microscope to inspect the tissue, while the reaction is progressing to ensure a good signal-to-noise ratio. The reaction should be stopped immediately if any background begins to appear.

F.10. Once there is dark reaction product and very low background (this may take between 45 min and 4 h, depending on the probe and the amount of mRNA expression), stop the reaction by transferring the sections back into the mesh-well dish in a solution of TE, pH 8.5.

F.11. Rinse 3 × 10 min in TE, pH 8.5.

F.12. Rinse 3 × 5 min in TBS and 1 × 5 min in 0.1 M phosphate buffer.

F.13. Mount from 0.1 M phosphate buffer. Air-dry and cover with Vectashield (Vector, Burlingame, CA) or another aqueous-based mounting media. Vectashield is good for protecting any other fluorophores in the reaction (e.g., fluorescent-tagged secondaries) from fading. Seal edges of coverslip with nail polish. Do not dehydrate the sections in alcohols or xylenes. Note: The NBT/BCIP reaction product is soluble in alcohols and organic solvents and is also light sensitive. The color will change slightly as the slides are exposed to light.

G. Modifications for Double ISH

Unfortunately, a commercially available FITC-tagged control RNA is not available to aid in determining exact concentrations of FITC-labeled cRNA from the dot blot test.

G.1. *Dot Blot Nytran Strip Test for FITC Probe.* Follow section "Test for Incorporation of Nonradioactive Label Using a Dot Blot on a Nytran Strip" for Steps B.1–B.7. Use sheep anti-FITC peroxidase-tagged antibody (1:2000; Roche) in place of the anti-digoxigenin antibody in Step B.8. Substitute a PBS rinse for the NMT rinse in Step B.10. To visualize the peroxidase tag, substitute Vector VIP (Vector Laboratories, one drop of each kit component per 3.3 ml of PBS) for the NBT/BCIP solution in Step B.11. Rinse in PBS rather than in TE as shown in Step B.12.

G.2. *Double ISH.* Perform the ISH exactly as described in sections "Prehybridization" and "Hybridization." Add the FITC-riboprobe with the digoxigenin–riboprobe in Step D.1. Add sheep anti-FITC peroxidase-tagged antibody (1:2000) along with any other primary antibodies in Step F.4. After Step F.5, insert the following steps:

   a. Incubate in biotin–tyramide at 1:75 in supplied diluent (Perkin-Elmer, Boston, MA) for 10 min in sterile 24-well tissue culture dish.

b. Transfer to mesh-well dish and rinse 3 × 5 min in TBS.
c. Incubate in streptavidin Cy3 at 1:200 (Jackson, West Grove, PA) plus any other secondaries as indicated in Step F.6.
d. Proceed with rest of protocol from Steps F.6 to F.13.

ACKNOWLEDGMENTS. This work was supported by grants HL 074011 and HL 28785 from the National Institutes of Health, Heart Lung and Blood Institute to P.G.G.

REFERENCES

Baldwin, T. J., Tsaur, M. L., Lopez, G. A., Jan, Y. N., and Jan, L. Y., 1991, Characterization of a mammalian cDNA for an inactivating voltage-sensitive K$^+$ channel, *Neuron* **7:**471–483.

Bellocchio, E. E., Hu, H., Pohorille, A., Chan, J., Pickel, V. M., and Edwards, R. H., 1998, The localization of the brain-specific inorganic phosphate transporter suggests a specific presynaptic role in glutamatergic transmission, *J. Neurosci.* **18:**8648–8659.

Bloch, B., Popovici, T., Chouham, S., and Kowalski, C., 1986a, Detection of the mRNA coding for enkephalin precursor in the rat brain and adrenal by using an "in situ" hybridization procedure, *Neurosci. Lett.* **64:**29–34.

Bloch, B., Popovici, T., Le Guellec, D., Normand, E., Chouham, S., Guitteny, A. F., and Bohlen, P., 1986b, In situ hybridization histochemistry for the analysis of gene expression in the endocrine and central nervous system tissues: a 3-year experience, *J. Neurosci. Res.* **16:**183–200.

Brysch, W., Creutzfeldt, O. D., Luno, K., Schlingensiepen, R., and Schlingensiepen, K. H., 1991, Regional and temporal expression of sodium channel messenger RNAs in the rat brain during development, *Exp. Brain Res.* **86:**562–567.

Chesselet, M. F., 1990, *In Situ Hybridization Histochemistry*, Boston: CRC Press.

Chesselet, M. F., Weiss, L., Wuenschell, C., Tobin, A. J., and Affolter, H. U., 1987, Comparative distribution of mRNAs for glutamic acid decarboxylase, tyrosine hydroxylase, and tachykinins in the basal ganglia: an in situ hybridization study in the rodent brain, *J. Comp. Neurol.* **262:**125–140.

Clavel, C., Binninger, I., Boutterin, M. C., Polette, M., and Birembaut, P., 1991, Comparison of four non-radioactive and 35S-based methods for the detection of human papillomavirus DNA by in situ hybridization, *J. Virol. Methods* **33:**253–266.

Darby, I. A., 2000, *In Situ Hybridization Protocols*, 2nd ed., Totowa, NJ: Humana Press.

Esclapez, M., Tillakaratne, N. J., Kaufman, D. L., Tobin, A. J., and Houser, C. R., 1994, Comparative localization of two forms of glutamic acid decarboxylase and their mRNAs in rat brain supports the concept of functional differences between the forms, *J. Neurosci.* **14:**1834–1855.

Fremeau, R. T., Voglmaier, S., Seal, R. P., and Edwards, R. H., 2004, VGLUTs define subsets of excitatory neurons and suggest novel roles for glutamate, *Trends Neurosci.* **27:**98–103.

Gall, J. G., and Pardue, M. L., 1969, Formation and detection of RNA-DNA hybrid molecules in cytological preparations, *Proc. Natl. Acad. Sci. U. S. A.* **63:**378–383.

Goldman, D., Simmons, D., Swanson, L. W., Patrick, J., and Heinemann, S., 1986, Mapping of brain areas expressing RNA homologous to two different acetylcholine receptor alpha-subunit cDNAs, *Proc. Natl. Acad. Sci. U.S.A.* **83:**4076–4080.

Guyenet, P. G., Stornetta, R. L., Weston, M. C., McQuiston, T., and Simmons, J. R., 2004, Detection of amino acid and peptide transmitters in physiologically identified brainstem cardiorespiratory neurons, *Auton. Neurosci.* **114:**1–10.

Hisano, S., Sawada, K., Kawano, M., Kanemoto, M., Xiong, G. X., Mogi, K., Sakata-Haga, H., Takeda, J., Fukui, Y., and Nogami, H., 2002, Expression of inorganic phosphate/vesicular glutamate transporters (BNPI/VGLUT1 and DNPI/VGLUT2) in the cerebellum and pre-cerebellar nuclei of the rat, *Mol. Brain. Res.* **107**:23–31.

Hoefler, H., Childers, H., Montminy, M. R., Lechan, R. M., Goodman, R. H., and Wolfe, H. J., 1986, In situ hybridization methods for the detection of somatostatin mRNA in tissue sections using antisense RNA probes, *Histochem. J.* **18**:597–604.

Hwang, P. M., Glatt, C. E., Bredt, D. S., Yellen, G., and Snyder, S. H., 1992, A novel K+ channel with unique localizations in mammalian brain: molecular cloning and characterization, *Neuron* **8**:473–481.

John, H. A., Birnstiel, M. L., and Jones, K. W., 1969, RNA-DNA hybrids at the cytological level, *Nature* **223**:582–587.

Johnson, A. D., Peoples, J., Stornetta, R. L., and Van Bockstaele, E. J., 2002, Opioid circuits originating from the nucleus paragigantocellularis and their potential role in opiate withdrawal, *Brain Res.* **955**:72–84.

Julien, J. F., Legay, F., Dumas, S., Tappaz, M., and Mallet, J., 1987, Molecular cloning, expression and in situ hybridization of rat brain glutamic acid decarboxylase messenger RNA, *Neurosci. Lett.* **73**:173–180.

Kreft, S., Zajc-Kreft, K., Zivin, M., Sket, D., and Grubic, Z., 1996, Application of the nonradioactive in situ hybridization for the localization of acetylcholinesterase mRNA in the central nervous system of the rat; comparison to the radioactive technique, *Pflugers Arch.* **431**:R309–R310.

Lanaud, P., Popovici, T., Normand, E., Lemoine, C., Bloch, B., and Roques, B. P., 1989, Distribution of CCK mRNA in particular regions (hippocampus, periaqueductal grey and thalamus) of the rat by in situ hybridization, *Neurosci. Lett.* **104**:38–42.

Lenz, S., Perney, T. M., Qin, Y., Robbins, E., and Chesselet, M. F., 1994, GABA-ergic interneurons of the striatum express the Shaw-like potassium channel Kv3.1, *Synapse* **18**:55–66.

Lewis, M. E., and Baldino, F., 1990, Probes for in situ hybridization histochemistry, In: Chesselet, M. F. (ed.), *In Situ Hybridization Histochemistry*, Boca Raton, FL: CRC Press, pp. 1–21.

Lewis, M. E., Sherman, T. G., and Watson, S. J., 1985, In situ hybridization histochemistry with synthetic oligonucleotides: strategies and methods, *Peptides* **6**(Suppl. 2):75–87.

Malherbe, P., Sigel, E., Baur, R., Persohn, E., Richards, J. G., and Mohler, H., 1990, Functional expression and sites of gene transcription of a novel alpha subunit of the GABAA receptor in rat brain, *FEBS Lett.* **260**:261–265.

McKinnon, D., 1989, Isolation of a cDNA clone coding for a putative second potassium channel indicates the existence of a gene family, *J. Biol. Chem.* **264**:8230–8236.

Melton, D. A., Krieg, P. A., Rebagliati, M. R., Maniatis, T., Zinn, K., and Green, M. R., 1984, Efficient in vitro synthesis of biologically active RNA and RNA hybridization probes from plasmids containing a bacteriophage SP6 promoter, *Nucleic Acids Res.* **12**:7035–7056.

Mezey, E., 1989, Phenylethanolamine N-methyltransferase-containing neurons in the limbic system of the young rat, *Proc. Natl. Acad. Sci. U. S. A.* **86**:347–351.

Mitchell, V., Gambiez, A., and Beauvillain, J. C., 1993, Fine-structural localization of proenkephalin mRNAs in the hypothalamic magnocellular dorsal nucleus of the guinea pig: a comparison of radioisotopic and enzymatic in situ hybridization methods at the light- and electron-microscopic levels, *Cell Tissue Res.* **274**:219–228.

Park, J. S., Kurman, R. J., Kessis, T. D., and Shah, K. V., 1991, Comparison of peroxidase-labeled DNA probes with radioactive RNA probes for detection of human papillomaviruses by in situ hybridization in paraffin sections, *Mod. Pathol.* **4**:81–85.

Pelletier, G., Liao, N., Follea, N., and Govindan, M. V., 1988, Distribution of estrogen receptors in the rat pituitary as studied by in situ hybridization, *Mol. Cell. Endocrinol.* **56**:29–33.

Perney, T. M., Marshall, J., Martin, K. A., Hockfield, S., and Kaczmarek, L. K., 1992, Expression of the mRNAs for the Kv3.1 potassium channel gene in the adult and developing rat brain, *J. Neurophysiol.* **68**:756–766.

Pochet, R., Brocas, H., Vassart, G., Toubeau, G., Seo, H., Refetoff, S., Dumont, J. E., and Pasteels, J. L., 1981, Radioautographic localization of prolactin messenger RNA on histological sections by in situ hybridization, *Brain Res.* **211**:433–438.

Poyatos, I., Ponce, J., Aragon, C., Gimenez, C., and Zafra, F., 1997, The glycine transporter GLYT2 is a reliable marker for glycine-immunoreactive neurons, *Mol. Brain Res.* **49:** 63–70.

Rogers, S. W., Hughes, T. E., Hollmann, M., Gasic, G. P., Deneris, E. S., and Heinemann, S., 1991, The characterization and localization of the glutamate receptor subunit GluR1 in the rat brain, *J. Neurosci.* **11:**2713–2724.

Rudy, B., Kentros, C., Weiser, M., Fruhling, D., Serodio, P., Vega-Saenz, D. M., Ellisman, M. H., Pollock, J. A., and Baker, H., 1992, Region-specific expression of a K+ channel gene in brain, *Proc. Natl. Acad. Sci. U. S. A.* **89:**4603–4607.

Schmued, L. C., and Fallon, J. H., 1986, Fluoro-gold: a new fluorescent retrograde axonal tracer with numerous unique properties, *Brain Res.* **377:**147–154.

Schreihofer, A.M., Stornetta, R.L., and Guyenet, P.G., 1999, Evidence for glycinergic respiratory neurons: Bötzinger neurons express mRNA for glycenergic transporter 2, *J. Comp. Neurol.* **407:**583–597.

Seroogy, K., Schalling, M., Brene, S., Dagerlind, A., Chai, S. Y., Hokfelt, T., Persson, H., Brownstein, M., Huan, R., Dixon, J., Filer, D., Schlessinger, D., and Goldstein, M., 1989, Cholecystokinin and tyrosine hydroxylase messenger RNAs in neurons of rat mesencephalon: peptide/monoamine coexistence studies using in situ hybridization combined with immunocytochemistry, *Exp. Brain Res.* **74:**149–162.

Shivers, B. D., Schachter, B. S., and Pfaff, D. W., 1986, In Situ hybridization for the study of gene expression in the brain, *Meth. Enzymol.* **124:** 497–510.

Siegel, R. E., and Young III, W. S., 1985, Detection of preprocholecystokinin and preproenkephalin A mRNAs in rat brain by hybridization histochemistry using complementary RNA probes, *Neuropeptides* **6:**573–580.

Stornetta, R. L., Akey, P. J., and Guyenet, P. G., 1999, Location and electrophysiological characterization of rostral medullary adrenergic neurons that contain neuropeptide Y mRNA in rat medulla, *J. Comp. Neurol.* **415:**482–500.

Stornetta, R. L., and Guyenet, P. G., 1999, Distribution of glutamic acid decarboxylase mRNA containing neurons in rat medulla projecting to thoracic spinal cord in relation to monoaminergic brainstem neurons, *J. Comp. Neurol.* **407:**367–380.

Stornetta, R. L., McQuiston, T. J., and Guyenet, P. G., 2004, GABAergic and glycinergic presympathetic neurons of rat medulla oblongata identified by retrograde transport of pseudorabies virus and in situ hybridization, *J. Comp. Neurol.* **479:**257–270.

Stornetta, R. L., Sevigny, C. P., and Guyenet, P. G., 2003, Inspiratory augmenting bulbospinal neurons express both glutamatergic and enkephalinergic phenotypes, *J. Comp. Neurol.* **455:**113–124.

Stornetta, R. L., Sevigny, C. P., Schreihofer, A. M., Rosin, D. L., and Guyenet, P. G., 2002, Vesicular glutamate transporter DNPI/GLUT2 is expressed by both C1 adrenergic and nonaminergic presympathetic vasomotor neurons of the rat medulla, *J. Comp. Neurol.* **444:**207–220.

Stornetta, R.L., Schreihofer, A.M., Pelaez, N.M., Sevigny, C.P., and Guyenet, P.G., 2001, Preproenkephalin mRNA is expressed by C1 and non-C1 barosensitive bulbospinal neurons in the rostral ventrolateral medulla of the rat, *J. Comp. Neurol.* **435:**111–126.

Surmeier, D. J., Eberwine, J., Wilson, C. J., Cao, Y., Stefani, A., and Kitai, S. T., 1992, Dopamine receptor subtypes colocalize in rat striatonigral neurons, *Proc. Natl. Acad. Sci. U. S. A.* **89:**10178–10182.

Terenghi, G., Polak, J. M., Hamid, Q., O'Brien, E., Denny, P., Legon, S., Dixon, J., Minth, C. D., Palay, S. L., Yasargil, G., and Chan-Palay, V., 1987, Localization of neuropeptide Y mRNA in neurons of human cerebral cortex by means of in situ hybridization with a complementary RNA probe, *Proc. Natl. Acad. Sci. U. S. A.* **84:**7315–7318.

Valentino, K. L., Eberwine, J. H., and Barchas, J. D., 1987, *In Situ Hybridization: Applications to Neurobiology*, New York: Oxford University Press.

Wada, K., Ballivet, M., Boulter, J., Connolly, J., Wada, E., Deneris, E. S., Swanson, L. W., Heinemann, S., and Patrick, J., 1988, Functional expression of a new pharmacological subtype of brain nicotinic acetylcholine receptor, *Science* **240:**330–334.

Wang, H., Kunkel, D. D., Schwartzkroin, P. A., and Tempel, B. L., 1994, Localization of Kv1.1 and Kv1.2, two K channel proteins, to synaptic terminals, somata, and dendrites in the mouse brain, *J. Neurosci.* **14:**4588–4599.

Wilkinson, D. G., 1998, *In Situ Hybridization: A Practical Approach*, 2nd ed., Oxford: Oxford University Press.

Wisden, W., and Morris, B. J., 1994, *In Situ Hybridization Protocols for the Brain*, London: Academic Press, Harcourt Brace &Co.

Wisden, W., Morris, B. J., Darlison, M. G., Hunt, S. P., and Barnard, E. A., 1988, Distinct GABAA receptor alpha subunit mRNAs show differential patterns of expression in bovine brain, *Neuron* **1:**937–947.

Wuenschell, C. W., Fisher, R. S., Kaufman, D. L., and Tobin, A. J., 1986, In situ hybridization to localize mRNA encoding the neurotransmitter synthetic enzyme glutamate decarboxylase in mouse cerebellum, *Proc. Natl. Acad. Sci. U. S. A.* **83:**6193–6197.

Young III, W. S., 1990, In situ hybridization histochemistry, In: Björklund, A., Hökfelt, T., Wouterlood, F. G., and Van den Pol, A. N. (eds.), *Analysis of Neuronal Microcircuits and Synaptic Interactions*, Amsterdam: Elsevier.

9

Viral Tracers for the Analysis of Neural Circuits

JOEL C. GEERLING, THOMAS C. METTENLEITER, and ARTHUR D. LOEWY

JOEL C. GEERLING AND ARTHUR D. LOEWY • Department of Anatomy and Neurobiology, P.O. Box 8108, Washington University School of Medicine, 660 S. Euclid Avenue, St. Louis, MO 63110   THOMAS C. METTENLEITER • Institute of Molecular Biology, Friedrich-Loeffler-Institut, D-17493

Abstract: Viral transneuronal tracing can be used to analyze neural circuits in the central nervous system (CNS). In particular, the pseudorabies virus (PRV) strain Bartha, an attenuated form of a pig alphaherpesvirus, is an excellent retrograde transneuronal tracer for labeling neural networks. This virus is transported from the axon terminal to the cell body of an infected neuron and enters the nucleus. There, it replicates, producing progeny virions that are distributed throughout the cytoplasm. These new viruses are then transferred into the axon terminals of second-order neurons that innervate the infected neuron, and the process is repeated. This technique has been used to analyze CNS networks involving chains of two or more functionally connected neurons. Due to the high sensitivity of viral transneuronal labeling, false-positive data can be generated, leading to potential pitfalls of interpretation—examples are discussed in this chapter. Protocols for growing PRV and viral tracing methodology are included.

Keywords: Bartha pseudorabies virus, herpes simplex virus, HSV1, PRV, rabies, transneuronal, transsynaptic

I. INTRODUCTION

The identification of neural networks is fundamental to understanding brain functions. In fact, the input and output connections of every part of the mammalian brain have been studied by neuroanatomical tracing methods that depend on the axonal transport of either proteins or fluorescent chemical markers. This approach, while providing important data, is limited to the analysis of single neurons, not circuits of functionally connected neurons. A clear operational guide regarding the mammalian brain is dependent on the knowledge of these circuits and the genetic expression patterns of individual neurons that form them.

Even though conventional tracing techniques label only single neurons, they can be used in combination to analyze multisynaptic circuits. A two-neuron circuit, for example, can be visualized by the combination of two tracer injections into the brain of the same animal. First, a group of neurons is retrogradely labeled following injection of a protein tracer, such as cholera toxin β-subunit (CTb), into its axon terminal field. Then, a second injection of an anterograde axonal tracer, such as the plant lectin *Phaseolus vulgaris* leucoagglutinin (PHA-L), is made at a central site that is known or predicted to project to the retrogradely labeled neurons. After several days, the brains from these animals are processed by a double immunohistochemical procedure that allows for the light microscopic identification of the so-called close contacts—PHA-L-labeled axon terminals abutting on CTb retrogradely labeled neurons. Definitive evidence is dependent on immunoelectron microscopy (see chapter by Sesack *et al.* in this volume). Since this technique is tedious and time-consuming, few reports have used this methodology, prompting a search for other approaches for the anatomical identification of functionally connected sets of neurons.

One of the first transneuronal tracers was identified in the early 1980s with the discovery that when the protein conjugate wheat germ agglutinin–horseradish peroxidase (WGA-HRP) was injected into the eye, it produced transneuronal labeling in second- and third-order visual relay neurons (Gerfen *et al.*, 1982; Itaya and van Hoesen, 1982). However, when this protein was injected into the brain, the volume of WGA-HRP injections had to be more limited—under these conditions, only vanishingly small amounts of WGA-HRP were transferred transsynaptically (Fig. 9.1). Other agents, such as the nontoxic fragment C of tetanus toxin, were also found to have limited utility as transneuronal tracers (Manning *et al.*, 1990). The approach was abandoned for more than a decade, but transgenic mice have been developed with neurons expressing plant lectin genes, such as WGA, and produce transneuronal labeling in defined neural circuits (Braz *et al.*, 2002; Horowitz *et al.*, 1999; Zou *et al.*, 2001).

Other attempts to develop anterograde transneuronal tracers included the use of radioactive amino acids (Wiesel *et al.*, 1974). These experiments required injections of large amounts of isotope into the target site, such as the eye. Then, after several days or weeks, the circuit could be identified in histological sections prepared for autoradiography. Since these studies require highly concentrated injections of expensive isotopes, this technique did not gain wide appeal. Even with these setbacks, the search continued for transneuronal tracers. In 1983, Martin and Dolivo made a key discovery when they demonstrated that viruses could be used to map central pathways (Martin and Dolivo, 1983).

Neurotropic herpesviruses (Fig. 9.2) have now proven extremely useful for detailed analysis of brain circuits. Briefly, viral retrograde transneuronal tracing occurs in the following manner. Live viruses are injected into a peripheral or central nervous system (CNS) target in a laboratory animal. After several hours, the viruses enter axon terminals innervating this structure. From here, the viruses are transported retrogradely to the parent cell bodies, undergo replication, and produce progeny virions, which become dispersed throughout the cytoplasm of each infected cell. These progeny virions are then transmitted to the incoming axon terminals that innervate the infected neurons, and the infectious process is repeated. Multiple

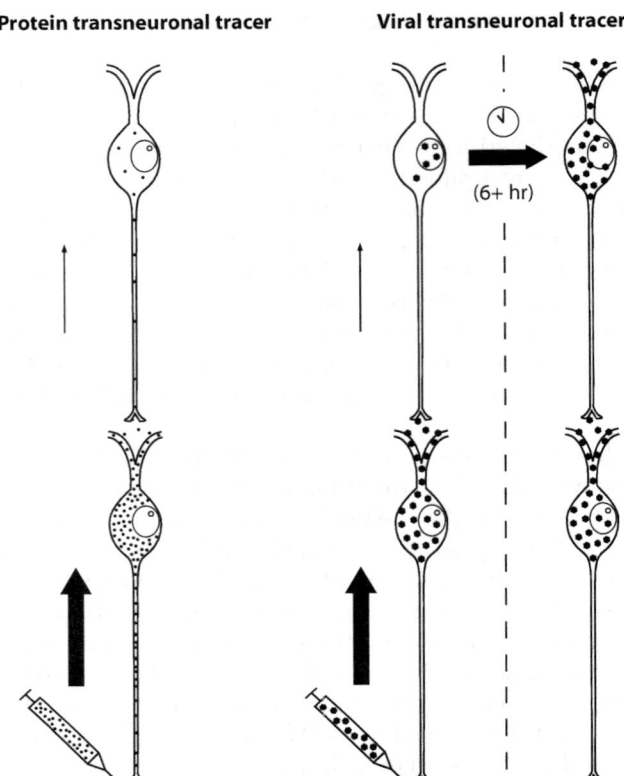

Figure 9.1. Comparison of protein and viral retrograde transneuronal tracers. Protein tracers, such as WGA-HRP (at left), can be used in transneuronal labeling studies, but only a small amount of tracer is transferred to second-order neurons. On the right, viral tracers, such as PRV, undergo replication within the second-order neurons and become self-amplifying transneuronal markers (Kuypers and Ugolini, 1990).

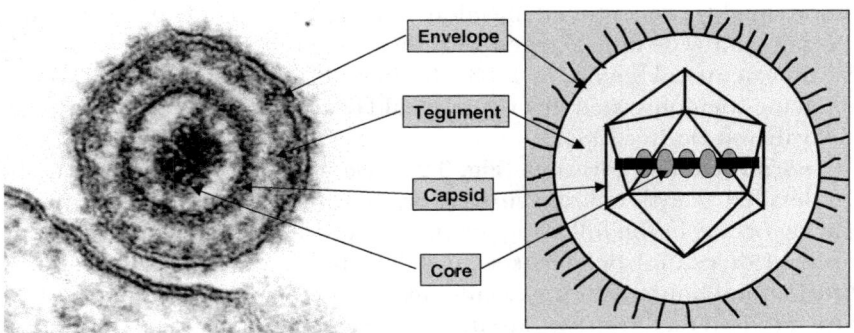

Figure 9.2. Morphology of a herpes virion. Herpesviruses, such as HSV1 and PRV, are composed of four structures. The core, a linear double-stranded DNA genome, is enclosed in an icosahedral capsid shell. The tegument, a layer of more than 15 different proteins, surrounds the capsid. The virus particle is enclosed in a lipid envelope in which viral glycoproteins are inserted (Mettenleiter, 2003).

rounds of replication and spread produce robust transneuronal labeling within neural circuits (Fig. 9.1).

II. HISTORICAL BACKGROUND OF VIRAL TRACING

One hundred years ago, the route of viral entry into the CNS was a matter of debate. The conventional view held that viruses spread locally, breaching epithelial boundaries, spaces, and fluids to enter the brain after establishing a foothold in the periphery. Another view proposed that viruses traveled the axonal processes of neurons as conduits into the CNS.

For herpesviruses, evidence subsequently accumulated in support of the concept of axonal transport. Goodpasture and Teague provided the earliest support for anterograde and retrograde axonal transport of viruses (Goodpasture, 1925; Goodpasture and Teague, 1923). In 1938, Albert Sabin, later famous for the development of polio vaccine, made the important observation that viruses enter the brain via preferential neural pathways (Sabin, 1938). For example, vesicular stomatitis and eastern equine encephalitis viruses ravaged the olfactory pathway to produce a lethal CNS infection, while pseudorabies virus (PRV), a herpes family virus, traveled in the sympathetic and trigeminal pathways without apparent olfactory infection.

The observations of Sabin were followed in the 1970s by clear demonstrations of specific axonal transport of herpesviruses to neuronal cell bodies and transneuronal spread in the CNS (Bak *et al.*, 1977; Cook and Stevens, 1973; Kristensson *et al.*, 1971, 1982). However, not until Martin and Dolivo (1983) published their study using PRV was it recognized that herpesviruses could be used as transneuronal tracers for defining neural circuits. In particular, they drew attention to the greatest advantage of using a virus as a transneuronal tracer—it replicates in each infected neuron. Thus, viruses can be viewed as self-amplifying markers, robustly labeling each hierarchical level of a neural circuit, in contrast to the diminishing transsynaptic diffusion of chemical tracers (Fig. 9.1). PRV tracing was used in two additional studies in the 1980s but, unfortunately, the investigators did not indicate the specific viral strain, source, and dose used in their experiments (Rouiller *et al.*, 1986, 1989). Attempted transsynaptic tracing with a wild-type form of PRV (Becker strain) was found to result in uncontrolled, nonspecific, and rapidly lethal infections, preventing its use as a specific transneuronal tracer (Strack *et al.*, 1989b).

Ugolini and colleagues injected herpes simplex virus type 1 (HSV1) into peripheral nerves and showed that it produced a transneuronal infection in rat brain (Kuypers and Ugolini, 1990; Ugolini *et al.*, 1987, 1989). However, the virus also spreads locally and nonspecifically to adjacent glial cells and neurons (Ugolini *et al.*, 1987). Subsequently, false-positive transneuronal labeling occurred in neurons connected to sites of nonspecific infection. This latter finding raised concern over whether viral infections could be contained within specific neural circuits, undermining enthusiasm for this method.

Because of these nonspecific infections, viral tracing was not widely exploited until the discovery that a less virulent derivative of PRV, Bartha PRV, produces highly specific retrograde transneuronal infections (Jansen *et al.*, 1993; Strack *et al.*, 1989b; Strack and Loewy, 1990). In contrast to the fulminant infections produced by wild-type PRV, Bartha PRV infections remain restricted mainly to synaptically linked chains of neurons and move only in the retrograde direction. Moreover, rats survive more than twice as long— up to 7 daysfollowing an injection of Bartha PRV into a peripheral target (Westerhaus and Loewy, 2001), whereas wild-type PRV kills rats within 3 days (Strack *et al.*, 1989b). This attribute allows transneuronal propagation to higher order neurons of a neural circuit (Enquist, 2002). Ensuing studies revealed specific genetic alterations responsible for the retrograde specificity and reduced infectivity of Bartha PRV (see below).

Bartha PRV is an effective and well-characterized retrograde transneuronal tracer that produces infections in most laboratory species (mouse, rat, gerbil, hamster, ferret, sheep, and chicken). Unlike other transneuronal viral tracers, such as HSV1 and rabies, PRV does not infect humans (Gustafson, 1975). Thus, Bartha PRV was quickly recognized as a safe and accessible tool for retrograde transneuronal tracing experiments. However, PRV does not cause infections in primates. Other neurotropic virusesHSV1 and rabieshave been tested and used for circuit analysis in monkeys. For further details, see section "Other Viruses Used for Transneuronal Tracing Studies."

Bartha PRV has been genetically modified in various ways to create new viral tools for neural circuit analysis. Advances in herpesvirus biology, including the availability of the complete PRV genome sequence (Klupp *et al.*, 2004), provide the opportunity to construct selective viral tracers, which should allow neuroscientists to unravel specific multisynaptic pathways in unprecedented detail.

III. BARTHA PRV AS A NEUROANATOMICAL TRACER

While Bartha PRV has primarily been used to identify chains of central neurons innervating peripheral targets, it has also been utilized for tracing circuits within the brain. In addition, viral tracing has been combined with various other well-established neuroanatomical techniques. Innovative methodologies continue to appear, making PRV an even more useful neurobiological tool. Genetically engineered PRV strains allow double transneuronal tracing experiments and the detection of transneuronally labeled neurons in living tissue, by fluorescent protein expression, for electrophysiological recording. Recombinant PRV has even been used in transgenic mice in an attempt to selectively label the inputs to specific neuronal phenotypes.

A. Phenotypic Characterization of PRV-Labeled Neurons

Bartha PRV was used to provide the first direct neuroanatomical identification of brainstem and hypothalamic neurons that regulate sympathetic outflow systems (Strack *et al.*, 1989a, b). Beginning with this work, it was demonstrated that peptide antigens could still be detected in infected neurons. This was a fortuitous discovery, since herpesviruses terminate protein synthesis in other cell types, via expression of the viral host shutoff protein gene, UL41 (Smiley, 2004). However, this viral host shutoff process does not occur in neurons (Nichol *et al.*, 1994).

Therefore, identification of the phenotype of infected neurons, by double-immunohistochemical labeling, can add important information about transneuronally labeled neurons (Fig. 9.3). In addition, investigators have successfully used in situ hybridization to demonstrate various mRNA transcripts in PRV-labeled cells (Boldogkoi *et al.*, 2002; Broussard *et al.*, 1996; Giles *et al.*, 2001; Song and Bartness, 2001; Stornetta *et al.*, 2004).

One caveat should be noted regarding the detection of marker molecules within PRV-infected neurons—expression of various peptides and enzymes within neurons infected with Bartha PRV tends to be reduced, often dramatically, relative to uninfected neighboring neurons of the same phenotype. Consequently, central injection of colchicine 24 h prior to killing an infected rat is sometimes used to boost the labeling of axonally transported peptides.

B. Use of PRV in Conjunction with Conventional Neural Tracers

The PRV transneuronal tracing technique can also be combined with conventional neural tracing methods. In one example, anterograde PHA-L tracing was combined with the PRV method to map the specific central regions targeted by the periaqueductal gray matter (PAG) to modulate sympathetic functions. In two series of rats, PHA-L injections were made into either the lateral PAG column, implicated in the fight-or-flight reactions, or the ventral PAG column, which mediates the opposite behavioral responses. Two days later, Bartha PRV was injected into the stellate sympathetic ganglion. After an additional 4 days, select regions of the hypothalamus and brainstem contained PHA-L terminals contacting PRV-labeled neurons. Specific sites, particularly the raphe magnus nucleus, through which PAG could modulate sympathetic activity, were thus identified (Farkas *et al.*, 1998). In this situation, PRV tracing alone was capable of demonstrating only that PAG neurons have multineuronal connections to the stellate ganglion—the specific presympathetic groups through which retrograde transneuronal PAG labeling had occurred were revealed by simultaneous anterograde tracing.

In a second example, a key hypothalamic relay nucleus was identified in a circuit implicated in circadian arousal functions—the pathway from the suprachiasmatic nucleus (SCN) to the locus coeruleus (Aston-Jones *et al.*,

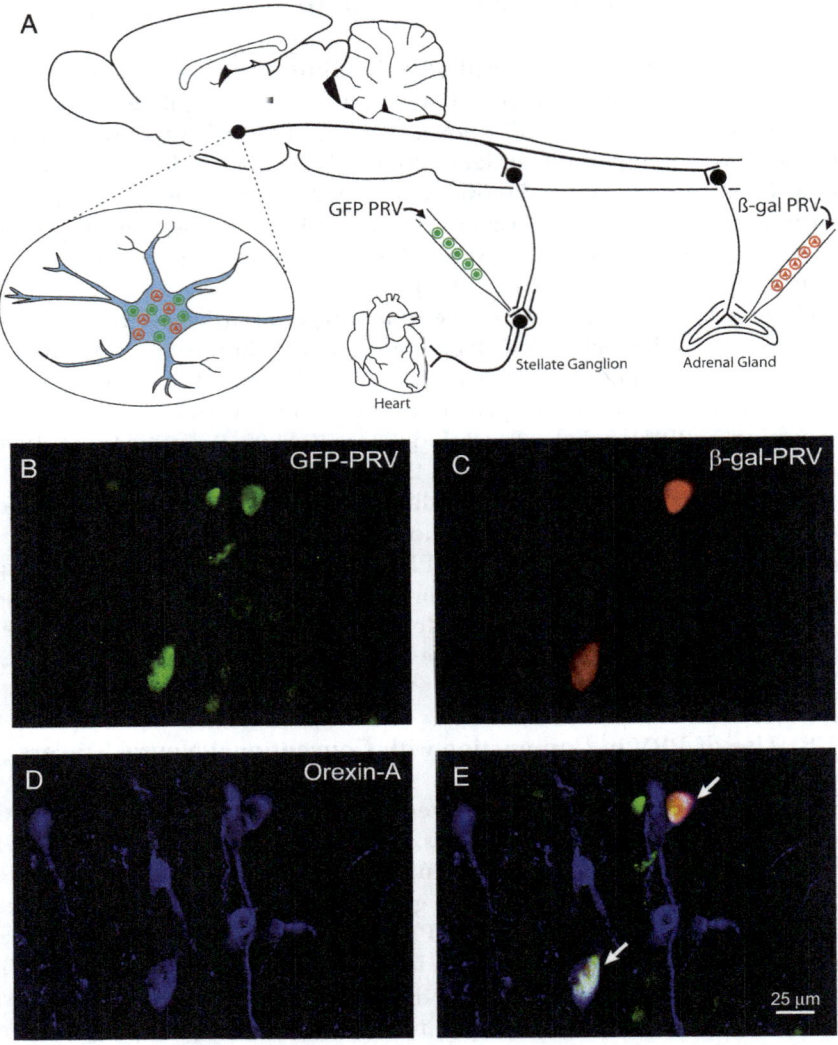

Figure 9.3. Double viral transneuronal tracing experiments can be used to identify specific neuronal phenotypes implicated as potential candidates regulating behavioral functions, such as the fight-or-flight response. In this example, orexin neurons (blue, panel D) in the lateral hypothalamus are transneuronally double labeled with Bartha PRV strains expressing two unique reporters: GFP-PRV from the stellate ganglion (green, B) and β-gal PRV from the adrenal gland (red, C). Triple-labeled cells appear white in panel E (Geerling *et al.*, 2003).

2001). The SCN was screened for neurons that project both transneuronally to the locus coeruleus and directly to the specific hypothalamic nuclei, such as the dorsomedial hypothalamic (DMH) nucleus. Bartha PRV was injected into locus coeruleus in animals with CTb injections into various hypothalamic nuclei. The highest number of double-labeled SCN neurons occurred in animals with CTb injections into the DMH, as opposed to other hypothalamic regions such as the paraventricular and lateral hypothalamic nuclei and the preoptic region. Since the DMH provides a direct input to locus coeruleus and DMH lesions blocked the circadian changes in locus coeruleus neural activity, a strong case was made for the existence of an SCN → DMH → locus coeruleus circuit (Aston-Jones *et al.*, 2001).

C. Double Retrograde Transneuronal Tracing with Two Isoforms of PRV

Neurons with branched axonal projections can be identified by double retrograde tracing techniques. For example, when two different retrograde tracers (e.g., Fluorogold and CTb) are injected into two central regions, it is possible to identify single neurons innervating both sites by the colocalization of the two tracers within the same cell.

Jansen *et al.* (1995a) demonstrated that this approach could be extended to transneuronal retrograde tracing by using two unique viruses. Two Bartha-derived PRV strains were separately injected into different sympathetic target tissues—adrenal medulla and stellate ganglion. The two recombinant viruses were uniquely identifiable. One contained a *lacZ* gene insertion within the nonessential gG gene locus. This virus could be identified by immunohistochemical staining for β-galactosidase (the protein product of the *lacZ* gene). In the other virus, the wild-type gC gene, which is mutated in Bartha PRV, was restored. Hence, this viral strain could be uniquely identified by its expression of the wild-type gC membrane protein, which is absent in the *lacZ* recombinant.

Double virus tracing can be a powerful method for detailing the exact central sites that regulate complex behavioral activities. The double-viral tracing technique was used to identify groups of putative central command neurons of the sympathetic nervous system, positioned to coordinate the activation of multiple different sympathetic target organs (Jansen *et al.*, 1995a). This study demonstrated the feasibility of double viral tracing. Later, other investigators applied this technique to demonstrate central neurons positioned to synchronize both somatomotor and sympathetic activations (Kerman *et al.*, 2003; Krout *et al.*, 2003), as well as brainstem neurons that could coactivate inspiratory and expiratory respiratory muscles that could discharge in parallel, which occurs during vomiting (Billig *et al.*, 2000). Double virus transneuronal tracing was used to demonstrate that vasopressin-containing SCN neurons (Ueyama *et al.*, 1999) and individual orexin neurons in the lateral hypothalamus (Geerling *et al.*, 2003) can coordinately regulate multiple sympathetic outflow systems. Double transneuronal labeling in orexin neurons is shown in Fig. 9.3. A similar strategy, using two uniquely identifiable strains of HSV1, has been used to compare the origins of sympathetic outflows within the spinal cord (Levatte *et al.*, 1998).

Since the introduction of the double virus labeling method, additional technical refinements have greatly increased its usefulness. The original study by Jansen and colleagues was extremely inefficient, requiring 256 rats in order to obtain eight animals with well-matched double infections. This 3% yield was partly due to the conservative viral dose used in this study, resulting in only a 20% success rate for single-virus studies (Strack et al., 1989b), but also to unmatched infectivity between the two viral tracing strains (Sams et al., 1995). These difficulties have been overcome with the addition of several new genetically engineered viral strains (see below).

One surprising aspect of this technique is that simultaneous infections can frequently be established by two separate viral strains, arriving from separate sites of origin, within the same neurons. This occurrence was not completely predictable, based on in vitro work demonstrating a cellular phenomenon called superinfection resistance. The herpesvirus membrane protein gD can inhibit the subsequent infection of the cell by additional virus (Campadelli-Fiume et al., 1988). Whether or not a similar effect is relevant for infections with Bartha PRV derivatives in vivo, a growing body of double viral tracing data unequivocally demonstrates that, in viral tracing paradigms, robust double infections can occur in many neurons.

Concern over superinfection resistance in double viral tracing was first raised when it was reported that infection by a Bartha PRV-derived strain with enhanced virulence, due to restoration of the virulence-enhancing wild-type gI gene (Whealy et al., 1993), was shown to greatly inhibit a second transneuronal infection in the same pathway by a β-galactosidase-expressing virus injected 24 h later (Kim et al., 1999). However, when the order was reversed, with the β-galactosidase strain injected first, it was unable to reciprocate this strong inhibition. Whether the first effect was due to a direct superinfection resistance mechanism in PRV-infected neurons, as opposed to indirect consequences of the greatly increased virulence of the first strain, was not resolved. For example, strong glial reaction at the injection site or rapid injury to the infected first-order neurons could have simply resulted in decreased entry and retrograde transport of virus injected a day later.

Banfield et al. (2003) found that when two isogenic Bartha recombinants were injected at the same time and site, they produced a double transneuronal infection in a large number of higher order neurons. More than 75% of the infected third-order paraventricular hypothalamic neurons expressed reporter genes from both viruses (green and red fluorescent proteins) after injection of a mixture of the two strains into one eye (Cano et al., 2003). Parallel experiments were performed on cultured dorsal root ganglion cells. These data showed that infection with one strain of PRV greatly reduced the susceptibility to infection by another PRV strain within 2 h and completely prevented superinfection after 4 h. Whether this in vitro time limit represents a similar constraint in vivo remains unknown. The robust double infections observed when the viral exposure times were matched exactly (by coinjecting them) combined with the indication that a small time window may exist for superinfection suggests that optimization of the double-virus technique requires matching viral rates of progression as closely as possible.

For this reason, an important aspect of double-viral tracing studies is the choice of viruses with similar transneuronal infection kinetics (Ter Horst, 2000). Optimal yields of double-infected neurons may be obtained when viral rates of transit are most closely matched, such that the two viruses arrive at an afferent site with minimum delay between strains. The first two double-viral tracing studies used two viruses with significantly different virulence characteristics (Jansen *et al.*, 1995a; Kim *et al.*, 1999). In each case, the infectivity of the *lacZ*/ β-galactosidase strain (a minimally altered version of Bartha PRV) was significantly reduced relative to the second strain (a version of Bartha to which a virulence-endowing membrane glycoprotein gene had been restored) (Kim *et al.*, 1999; Mettenleiter *et al.*, 1987, 1988; Sams *et al.*, 1995).

Following these original studies, double transneuronal tracing has benefited from the genetic engineering of isogenic viruses in which insertions of reporter genes have been targeted to the same genetic locus. For example, inserting the green fluorescent protein (GFP) gene within the same gG locus used for *lacZ* (Jons and Mettenleiter, 1997) has allowed more comparable double viral infections at equivalent doses and times (Geerling *et al.*, 2003; Krout *et al.*, 2003; Ueyama *et al.*, 1999). However, discordant rates of expression between the two reporter genes, within double-infected cells, can result in unequal detectability of the two strains, whether or not infectivity is equal. In some studies, two different promoters have been used to drive the expression of reporter genesthe intrinsic gG promoter and the human cytomegalovirus (CMV) immediate-early promoter. Different promoters could necessitate the use of two different viral doses to match the timing of reporter expression, even when the rates of viral spread may be comparable (Cano *et al.*, 2003). Differences in relative expression levels of reporter proteins can be largely overcome by the use of the CMV promoter in both strains, driving high levels of gene expression as early as possible (Banfield *et al.*, 2003). Nonetheless, any pair of viruses used in double tracing experiments should be compared to establish similar rates of transneuronal progression.

D. PRV Fills Neuronal Dendritic Trees

Card *et al.* (1993) noted that pseudorabies virions fill the entire dendritic tree of an infected neuron, out to the distal branches. This property can be useful for ultrastructural analysis (Carr *et al.*, 1999; Carr and Sesack, 2000). More interestingly, dendritic filling by PRV has been exploited to solve a long-standing problem in neuroanatomical tracingidentifying the synaptic afferents to the distal dendrites of a group of neurons.

When the dendrites of a particular group of neurons extend into adjacent cytoarchitectonic regionsoutside the boundaries of its parent cell group as defined in Nissl-stained sections~~it~~ can be difficult to determine, simply by retrograde tracer injections, which are true neural inputs to the particular cell group (Bourgeais *et al.*, 2003; Luppi *et al.*, 1995). In some instances, this problem can be overcome by injecting PRV within the center of a group~~after~~ replication in the soma, viral progeny spread throughout the neuron and can spread transsynaptically into even the most distal synaptic afferent terminals (Aston-Jones and Card, 2000). This approach has been used to demonstrate spinal lamina I afferents to the distal dendrites of amygdala-projecting neurons in the external lateral parabrachial nucleus (Jasmin *et al.*, 1997). More recently, Aston-Jones *et al.* (2004) injected PRV within the core of the locus coeruleus and found that the virus replicated and spread throughout these noradrenergic neurons, producing transneuronal infections in input neurons that contact the most distal dendritic branches.

E. Electrophysiological Recordings from Transneuronally Labeled Neurons

Various methods have been utilized to identify specific neurons for electrophysiological study in brain slices. In particular, fluorescent dyes such as Fluorogold have been used to identify retrogradely labeled neurons in living tissue (Kangrga and Loewy, 1995).

In 2000, Smith *et al.* first demonstrated that electrophysiological recordings could be targeted to identified neurons with multisynaptic connections to a specific target. These investigators created the Bartha-derivative PRV 152, designed to produce high, early expression of enhanced green fluorescent protein (EGFP, driven by a CMV promoter). They demonstrated that EGFP-expressing, PRV-infected neurons were easily identifiable in tissue slices. Most important, despite viral infection, the electrophysiological properties of these cells were comparable to uninfected neurons. This study was followed by a similar demonstration of electrophysiological recording from PRV-infected neurons, using a different GFP-expressing strain (Irnaten *et al.*, 2001).

These findings built confidence that recordings can be obtained from the neurons identified as multisynaptic afferents to a specific target. However, infected cells may show electrophysiological abnormalities (Fukuda *et al.*, 1983). Still, high-quality recordings have been made from visually identified GFP-PRV neurons and this method can be quite useful for studying neural circuits.

F. Genetically Engineered PRV for Highly Specific Tracing

Molecular biological tools have created opportunities for constructing viruses with improved properties as neural tracers (Boldogkoi *et al.*, 2004).

New viral tracers may allow more selective labeling within neuronal circuits. For example, the use of cell-specific conditional expression technology should allow the targeting of infection or viral reporter gene production to neurons of a particular phenotype. In this way, local CNS injections of specific viruses may produce transneuronal labeling restricted to the inputs of a functionally specific type of neuron.

The feasibility of this approach was demonstrated by DeFalco *et al.* (2001), who injected a genetically engineered Cre recombinase-dependent strain of PRV into a transgenic mouse that expresses Cre recombinase in only one neuronal phenotype. They began with Bartha PRV and removed its thymidine kinase (tk) gene, which is necessary for viral replication in vivo. The tk sequence was then reinserted, along with an EGFP reporter gene, in the nonessential gG locus (the same site used for the *lacZ* and GFP reporters described above), driven by a CMV promoter. A STOP sequence, flanked by loxP sites, was inserted upstream from the tk and EGFP sequences. This STOP sequence was positioned to prevent expression of these genes and, therefore, both viral replication and cellular expression of EGFP reporter.

However, the Cre recombinase enzyme can join the loxP sites, removing the intervening STOP sequence. Therefore, this replication-deficient virus, termed Ba2001, was injected into the brains of transgenic mice expressing Cre under the control of specific genetic promoters (neuropeptide Y or the leptin receptor). In these mice, Ba2001 could enter many types of neurons, but could replicate only in Cre-expressing neurons. In these specific neurons, infection with replication-competent virus was reported by concurrent EGFP expression. With these tools, the specific neural networks regulating NPY- or leptin receptor–expressing neuronal subpopulations within the hypothalamic arcuate nucleus could be selectively labeled.

This important methodological advancement raises the possibility that designer herpesviruses could become important tools for mapping neural circuits with unprecedented specificity. Unfortunately, in the 4-year period since publication, these findings have not been detailed or extended. Since certain CNS sites reported to be infected in the DeFalco report (DeFalco *et al.*, 2001), such as somatosensory cortex, seem incompatible with known inputs to the hypothalamus, the potential for spontaneous viral genetic mutation causing spurious labeling in vivo must be addressed.

The promise of custom-made viral tracers remains alluring (Boldogkoi *et al.*, 2004), but the enthusiasm surrounding this technology should not cause investigators to overlook the necessity of detailed neuroanatomical characterization of all new viral tracers. Any new viral strain should be carefully compared with a well-characterized strain, such as Bartha PRV, with respect to transsynaptic specificity, kinetics of infectious spread, tropism, and critical viral doses (Banfield *et al.*, 2003). Even minor alterations in what may appear to be insignificant regions of the PRV genome, such as the nonessential gG gene locus, can result in significant differences in infectivity (Cano *et al.*, 2003; Demmin *et al.*, 2001; Sams *et al.*, 1995). Finally, it is important that experimental results obtained with any viral strain be critically compared with existing neuroanatomical data.

IV. SPECIFIC RETROGRADE TRANSPORT OF BARTHA PRV

Since the first transneuronal tracing experiments with Bartha PRV, it was clear that this viral strain spreads preferentially, if not exclusively, in a retrograde direction (from the axon terminal to the cell body—the opposite direction of neural transmission). An early argument for retrograde specificity came from studies that showed that, when PRV was injected into skeletal muscle, it produced retrograde labeling in ventral horn motor neurons, but not in the dorsal root ganglia or central somatosensory sites (Rotto-Percelay *et al.*, 1992).

Patterns indicative of retrograde-only transport have also been observed after Bartha PRV injections within the CNS. After injection into the mediodorsal nucleus of the thalamus, O'Donnell *et al.* (1997) noted that infection within the cortex first occurred within deep layer neurons, consistent with retrograde transport from the thalamus. Card *et al.* (1998) showed that, unlike that of wild-type Becker PRV, injection of Bartha PRV in the prefrontal cortex did not produce an anterograde transneuronal infection in the striatum, a major efferent target. Chen *et al.* (1999) reported that, even when entering fibers of passage through an injection site, Bartha PRV did not produce anterograde transneuronal labeling.

Wild-type PRV clearly *does* spread anterogradely. This was known from Sabin's early observations of infection in the central trigeminal sensory pathways after olfactory instillation (Sabin, 1938). Later, it was observed that wild-type PRV injected into the eye produces a fulminant infection of all retinorecipient sites within the brain (Card *et al.*, 1991). Since there are relatively slight genetic differences between wild-type and Bartha PRVs, yet major differences in their transport properties, a search was initiated for the specific genes responsible for anterograde infectious spread.

At least three PRV genes appear necessary for anterograde spread: gE, gI, and Us9. All three of these genes are absent from Bartha PRV, due to a large deletion in the unique short (Us) region of the PRV genome (Lomniczi *et al.*, 1984). The deletion of any one of these genes from wild-type PRV eliminates anterograde viral transmission (Brideau *et al.*, 2000; Card *et al.*, 1992; Whealy *et al.*, 1993).

Both gE and gI are membrane glycoproteins and form a functional heterodimer (Mettenleiter *et al.*, 1988; Whealy *et al.*, 1993). These two genes had been previously characterized as encoding important PRV virulence-enhancing factors (Mettenleiter *et al.*, 1987, 1988). Deletion of either gene from wild-type PRV was shown to eliminate anterograde spread from the retina to the retinorecipient visual sites in the brain (Card *et al.*, 1992; Whealy *et al.*, 1993). Loss of anterograde spread in gE- and gI-null mutants has also been confirmed in the olfactory pathway (Babic *et al.*, 1996; Kritas *et al.*,

1994). Injection of a mixture of gE-null and gI-null PRVs within the eye, however, resulted in restoration of a wild-type anterograde infection pattern (Enquist *et al.*, 1994). This implies that both mutants infect retinal ganglion cells, but require the addition of their respective missing gene products—upon coinfection of the same cellfor productive anterograde spread. It is still unclear exactly how gE and gI allow anterograde transmission (Enquist *et al.*, 2002; Tomishima *et al.*, 2001).

In contrast, the mechanism by which the Us9 gene product influences anterograde transport is better characterized. As with gE and gI, absence of the Us9 gene inhibited the anterograde spread of wild-type PRV (Brideau *et al.*, 2000). Tomishima and Enquist (2001) further demonstrated in vitro that, without Us9, necessary membrane glycoproteins do not enter the axon of an infected neuron. While other viral proteins proceed normally into the axon, this lack of membrane protein trafficking prevents anterograde transmission of complete, infectious virions.

While the exact molecular mechanisms required for PRV anterograde infectious spread remain unknown, the studies cited above have highlighted some of the key factors. The identification of specific genetic mutations preventing anterograde transneuronal infections by Bartha PRV gradually cast doubt upon the only cited evidence that this strain could produce an anterograde infectionthe delayed infection of SCN after injection into the eye (Brideau *et al.*, 2000; Card, 2000; Card *et al.*, 1991, 1992; Enquist *et al.*, 1994; Husak *et al.*, 2000; Moore *et al.*, 1995; Smith *et al.*, 2000; Whealy *et al.*, 1993). Careful neuroanatomical analysis, however, revealed that this purportedly anterograde infection was actually produced by retrograde spread via multisynaptic autonomic outflows to the eye (see discussion under "Practical Considerations and Pitfalls"; Pickard *et al.*, 2002; Smeraski *et al.*, 2004).

In summary, a great deal of collective neuroanatomical experience with Bartha PRV indicates that this virus moves selectively in a retrograde direction. Three key PRV genes have been individually shown to be necessary for anterograde viral spreadBartha is deficient in each one. Together, these findings build a strong case that Bartha PRV is a retrograde neuronal tracer.

V. TRANSNEURONAL TRANSFER OF BARTHA PRV AT SYNAPTIC TERMINALS

The pattern of Bartha PRV transneuronal labeling is largely consistent with specific transfer at synapses without leakage to nearby neurons or local axons. For example, after PRV injection into any visceral tissue or autonomic ganglion, transneuronal labeling was consistently found in the parvocellular subdivision of the paraventricular hypothalamic nucleusan area known from earlier work to be a key site regulating autonomic functions. Importantly, nearby neurons lying in the intermingled magnocellular subdivision of this nucleus, which projects solely to the posterior pituitary, were not labeled (Strack *et al.*, 1989b). Such restricted labeling indicated that random cell-to-cell spread did not occur.

This consistent pattern led to the proposal that transneuronal spread of Bartha PRV occurred specifically through neuronal synapses (Strack and Loewy, 1990). However, beginning with the first use of Bartha PRV as a transneuronal tracer, it was observed that Bartha PRV can infect glia within infected neuronal sites (Rinaman *et al.*, 1993; Strack *et al.*, 1989b). Although only limited infections of astroglia occurred, this observation raised significant concern over the potential for false-positive labeling via not only local spread to unrelated neurons but also subsequent transneuronal propagation. Indeed, Ugolini *et al.* (1987) had reported that tracing with HSV1 resulted in formidable local spread to neurons within unrelated circuits. HSV1 injected into the mouse hypoglossal nerve spread from nerve roots in the ventral medulla to both glia and inferior olivary neurons and, via transneuronal transfer within only a few days, to neurons in the cerebellum. Although no such nonspecific infection had been reported for Bartha PRV, this potential roadblock was carefully examined in early experiments validating the virus as a transneuronal tracer.

In 1990, Strack and Loewy demonstrated that, after Bartha PRV was injected into the eye or the skin of the ear, retrograde labeling in the sympathetic superior cervical ganglion (SCG) was completely restricted to the subset of neurons afferent to the particular site of injection. Even after 4 days, infection did not spread locally within the SCG (Strack and Loewy, 1990). A similar result was reported for the CNS by Jansen *et al.* (1993). After injection of Bartha PRV into either the stellate ganglion or the adrenal medulla, coincident with CTb injection into the other sympathetic target, the percentage of spinal cord neurons labeled with both virus and CTb was not different from the double-labeled proportion found after injections of two conventional retrograde tracers (CTb and Fluorogold). This indicated that PRV infection within this first-order afferent site remains confined to specific sympathetic preganglionic neurons. When both experiments are considered together, a convincing case can be made against the likelihood that Bartha PRV produces lateral infections involving neighboring neurons (Scenario 3 in Fig. 9.4).

Consistent with these findings, diffusion of PRV through the neuropil may be hindered by the large size of infectious virions (200 nm) and by its binding to cell surface heparin sulfate moieties (Aston-Jones and Card, 2000). The spread of PRV to axons and local glia, but not to adjacent neuronal cell bodies, is also consistent with an earlier report indicating a greater herpesvirus-binding affinity for synaptic terminals and glial cells, relative to neuronal perikarya (Vahlne *et al.*, 1978).

However, besides local spread to adjacent neuronal cell bodies, one potential avenue of nonspecific PRV spread remainsthe leakage of virions into adjacent nonsynaptic axons (Scenario 2 in Fig. 9.4). Whereas the experiments cited above (Jansen *et al.*, 1993; Strack and Loewy, 1990) dispelled

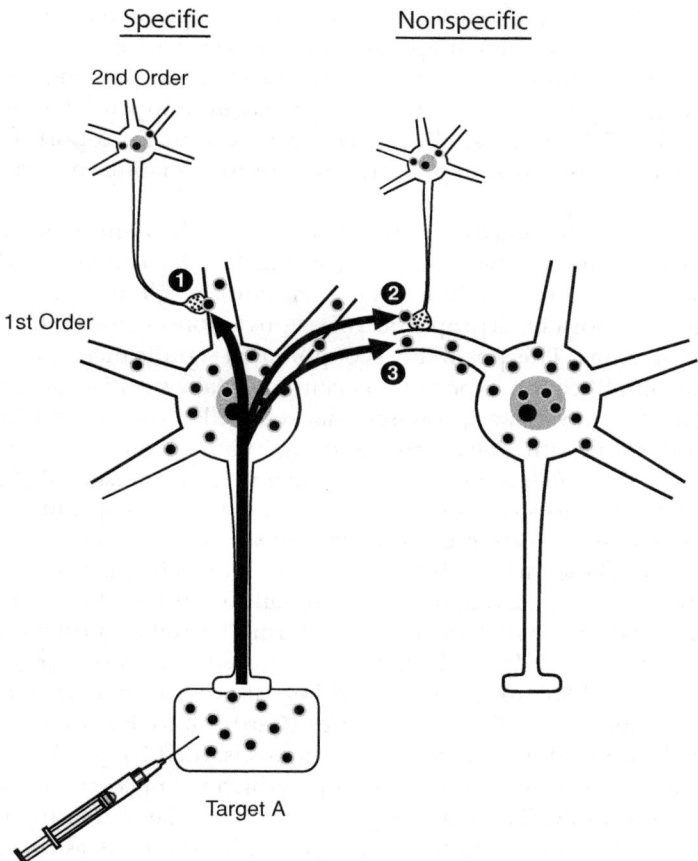

Figure 9.4. Viruses can produce both specific and nonspecific retrograde infections in the CNS. (1) The primary mode of transneuronal spread, for Bartha PRV and other viruses, is via direct transfer to the synaptic afferents of an infected neuron. (2) Spread of a viral tracer to adjacent axons and terminals that do not synapse upon the infected neuron may occur (see "Transneuronal Transfer of Bartha PRV at Synaptic Terminals"). (3) Lateral leakage of virus to neighboring neurons does not appear to occur with Bartha PRV (Jansen *et al.*, 1993), but may present a problem with other viral tracers, such as HSV1 (Ugolini *et al.*, 1987).

concern over the potential for local spread to neuronal cell bodies, only indirect tests dealt with the possibility that some of the PRV released from an infected neuron may infect adjacent axons or axon terminals.

First, when PRV was injected into the eye or into the skin of the ear, it produced second-order transneuronal labeling in the appropriate distribution of sympathetic preganglionic neurons in the spinal cord, as determined by prior electrophysiological data (Strack and Loewy, 1990). The eye- and ear-specific SCG neurons, through which transneuronal transport had occurred, are highly intermixed. Hence, this result indicated that transfer of PRV from infected first-order SCG neurons took place in a preferentially transsynaptic manner, not simply by transmission of virions to all nearby axon terminals. However, because the distributions of preganglionic neurons infected after injections into eye and ear overlapped between spinal levels T2 and T4, the possibility remained that a small proportion of the labeled neurons in this zone was the result of nonsynaptic viral transfer in the SCG.

This is the only tracing experiment to directly address the issue of Bartha PRV spread to the nearby axons. Clearly, Bartha PRV is *preferentially* transferred to synaptic afferents, but only circumstantial evidence exists concerning whether or not a small proportion of virions is nonspecifically transferred to adjacent axons. This possibility is of potential significance, given the exponential amplification expected to occur after false-positive labeling.

An attractive theory was proposed that reconciled the observation of astroglial infections with a lack of spread to adjacent neurons, and that offered a mechanism by which transsynaptic specificity may be preserved (Rinaman et al., 1993). Virions were observed, by electron microscopy, to be preferentially released from an infected neuron at sites of synaptic contact (Card et al., 1993). These virions did not appear to breach the synapses themselves; they spread parasynaptically and equally infiltrated the afferent axon terminal and the astroglial processes that form a barrier around the synaptic region (Card et al., 1993). Thus, it was proposed that astrocytic processes may absorb any PRV not incorporated into afferent terminals, preventing nonspecific spread to adjacent structures (Card, 1998). Furthermore, Card et al. (1993) noticed that, in contrast to neurons, the PRV produced within astrocytes did not acquire a viral envelope, which is a necessary component for infectious virions. These investigators proposed that, rather than serving as a source of PRV production and nonspecific local spread, astrocytes limit viral spread to nearby axons without producing normal infectious virions (Rinaman et al., 1993).

This appealing theory may explain the transsynaptic pattern of PRV spread and the lack of PRV spread to adjacent neurons (Jansen et al., 1993; Strack and Loewy, 1990), despite infection of adjacent glia (Rinaman et al., 1993; Strack et al., 1989b). However, the spread of some amount of virus to adjacent nonsynaptic axons remains an important possibility that cannot be addressed by circumstantial evidence or by inferential approaches. The issue of whether transneuronal PRV spread occurs exclusively via synaptic afferents remains unsettled. This possibility is an important consideration because the brain regions where nonspecific transfer presents the greatest obstacle to interpretation are those in which viral transneuronal tracing is most usefulsites such as the brainstem and hypothalamus, which contain spatially intermixed, yet functionally diverse, populations of neurons.

In summary, Bartha PRV is a retrograde transneuronal tracer that is preferentially taken up by synaptic terminals. Concomitant astroglial infections do not appear to lead to local nonspecific neuronal labeling and may even restrict PRV transfer to increase the probability that virus uptake occurs at synaptic sites. However, whether or not PRV is transferred *exclusively* to synaptic afferents remains unresolved.

VI. NEUROANATOMICAL TRACING WITH PRV—PRACTICAL CONSIDERATIONS AND PITFALLS

Viral tracing is a highly sensitive technique. When applied judiciously, the viral transneuronal labeling method can produce information regarding central neural circuits that is unattainable by other methods. Its high sensitivity is, however, inseparable from a significant potential for nonspecific labeling of unrelated neural circuits. The need for conservative interpretation of the patterns of central labeling is important because even attenuated viral tracers are capable of infecting many different cell types in the brain (neurons, astroglia, and ependymal cells) by various routes. An analysis of a series of sections throughout the brain should be performed for each PRV case in any given study to rule out the possibility that nonspecific viral labeling had occurred. For example, when PRV is injected into peripheral targets, such as an autonomic ganglion, inspection of the supraoptic and magnocellular paraventricular subnucleus can be used to determine whether a viremia had occurred, since labeling in these two sites would be the result of uptake from the vascular system. Similar screening scenarios are important for the evaluation of other types of experiments as well.

A. Viral Tracing in the CNS

The interpretation of data obtained from viral transneuronal infections within the CNS can be extremely difficult, compared with viral injections into peripheral structures. Viral entry into the CNS from the periphery can be isolated to a single neural channel, but the situation is not as straightforward as in the brain. Central neurons receive input from multiple CNS regions. Frequently, these regions are interconnected, greatly increasing the complexity of the potential routes of viral spread (Fig. 9.5).

In the late 1990s, it was demonstrated that Bartha PRV could be used to define central circuits (Jasmin *et al.*, 1997; Kaufman *et al.*, 1996; O'Donnell *et al.*, 1997), although earlier studies using HSV1 in monkeys had established the feasibility of this approach (Lynch *et al.*, 1994; Middleton and Strick, 1994). Subsequent evaluation of PRV tracing within the brain addressed significant concerns about this methodology, such as injection site analysis, nonspecific spread through the cerebrospinal fluid, and viral uptake by fibers of passage (Chen *et al.*, 1999).

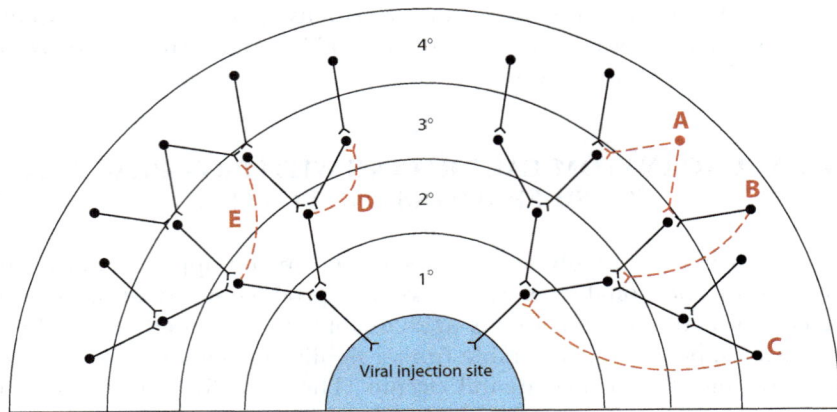

Figure 9.5. The brain's complex circuitry frequently offers multiple alternative hypotheses for the route by which a virus may have labeled a particular group of neurons. This complexity can complicate the interpretation of tracing data. In this diagram, each neuron symbolizes a neuroanatomical region with known axonal projections to a lower order site. Common types of neural connections that can complicate tracing data are shown as dashed lines. A simple time-course analysis of viral progression may differentiate between the potential routes of labeling in (C), in which two transneuronal steps separate the alternatives. However, this approach may not provide a clear answer for (B) and cannot resolve (A). The resolution of labeling in these situations may require additional neuroanatomical experiments. The reciprocally connected pair of neuronal groups depicted by (D) may lead to uncertainty as to whether a specific subset of neurons within the second-order group was labeled (1) directly from neurons in the first-order group or (2) from their target neurons in the infected third-order group, which had received virus from a different subset of second-order neurons in the same region. In (E), a similar hypothetical situation is depicted.

1. PRV Injection Site

Chen *et al.* (1999) directly addressed a number of potential pitfalls associated with Bartha PRV tracing within the brain. One of the critical issues is defining an injection site. This was problematic since after PRV was injected into brain parenchyma it rapidly entered local axons in which virions were transported away from the injection site. Several days later, when most PRV tracing experiments were terminated, immunohistochemical staining for PRV did not reveal the injection site. To avoid this problem, these investigators verified that a cocktail of PRV in a 0.05% CTb solution was useful for approximating the injection site (Chen *et al.*, 1999).

2. Bartha PRV in the Ventricular System

Another complicating issue associated with central PRV tracing studies is the possibility that the virus could enter the cerebrospinal fluid and cause nonspecific infections throughout the brain. Chen *et al.* (1999) injected Bartha PRV into the lateral ventricle of rats and found that, after 1 or 2 days, infections were confined to specific sites and not randomly distributed throughout the brain. Highly reproducible PRV labeling occurred, within a day after ventricular injection, in a specific subset of dorsal raphe neurons immediately beneath the cerebral aqueduct. These neurons are responsible for the serotonergic axonal plexus in the ependymal lining of the ventricular system (Chan-Palay, 1976), as demonstrated by rapid labeling of the same group of neurons, as well as their ependymal axonal plexus, by CTb (mixed with PRV for injection site localization) in the same animals. The presence of PRV and/or CTb within this specific ependymal-projecting subset of dorsal raphe neurons was proposed as practical marker for screening viral tracing cases after injection near a ventricle (Aston-Jones and Card, 2000; Chen *et al.*, 1999). After 2 or more days postinjection, PRV-labeled neurons were also found scattered in other regions, such as lateral septum and hippocampus. This labeling was hypothesized to have resulted from the uptake of PRV from infected ependymal cells that had lysed. Regardless of the cause for this labeling, these experiments showed that Bartha PRV injections into the brain ventricular system did not cause widespread infections.

3. Bartha PRV in the Vasculature

Vascular leakage of Bartha PRV is another problem, since central injections invariably cause disruption of some blood vessels. To date, six studies have shown that when Bartha PRV is injected into the venous system of rat in the doses used in central tracing studies it does not produce central infections (Westerhaus and Loewy, 1999). Inoculation of similar amounts of Bartha PRV directly into the arterial supply of the brain, however, has not been tested.

4. Uptake by Fibers of Passage

Another important issue is whether Bartha PRV is taken up by fibers of passage within the injection site. Peripheral nerves take up this virus when high doses are used (Dobbins and Feldman, 1994). Only one study has examined this issue for CNS injection sites. Bartha PRV was injected into the ventrolateral medulla of rats, in the region where crossed axons from the inferior olivary nucleus travel toward the inferior cerebellar peduncle. Viral injection here resulted in robust retrograde neuronal labeling in the contralateral inferior olivary nucleus (Chen *et al.*, 1999). Whether this uptake was due to entry into injured axons or whether uptake by intact fibers of passage is not certain, but these findings highlight a potential confound when PRV is used as a central tracer. Stereotaxic injections will cause a certain degree of damage to fibers passing near the target site—this complication should be considered when PRV is used as a central tracer.

5. Controlled Viral Tracing in the CNS

Some means of guarding against misinterpretation of nonspecific patterns of transneuronal infection include corroboration with previously characterized neuroanatomical connections, negative control lesion experiments, and positive control experiments excluding alternative pathways. Clearly, viral labeling should be viewed as specific only when underlying single neuronal connections can be verified by conventional neural tracing. If infected neurons are found in sites incompatible with a specific retrograde transsynaptic spread from the viral injection site, based on well-established neuroanatomical data, labeling should be considered nonspecific. If a hypothetical pathway cannot be convincingly constructed from the viral injection site to a particular infected group of neurons by piecing together well-characterized traditional tracing data, the specificity of labeling should be questioned.

When multiple different transneuronal routes could explain viral labeling data, combining single-neuron tracing with PRV injection can help distinguish the most likely pathway (Aston-Jones *et al.*, 2001; Farkas *et al.*, 1998). Also, lesion of a relay point in a proposed circuit should significantly decrease viral infection within a transneuronally labeled group of neurons. For example, in a study of the multineuronal circuit from the SCN to the medial prefrontal cortex, the paraventricular thalamic nucleus was hypothesized as the key relay point. Lesioning this cell group effectively blocked transneuronal labeling of the SCN after PRV injection within medial prefrontal cortex (Sylvester *et al.*, 2002). As shown in this study, lesion effects should be quantified and replicated in a sufficient number of cases to demonstrate a statistically significant reduction of viral labeling in lesioned animals versus nonlesioned animals after an identical postinjection survival time. Injection site analysis is necessary for every experiment to screen for the possibility that differences in labeling could be produced between lesioned and nonlesioned groups simply because of slight differences in the placements of viral injections. In addition, viral labeling in a positive control site should be quantified, in both lesioned and nonlesioned groups, to verify that the infection proceeded normally in alternative pathways that are not dependent upon the lesioned site as a relay.

A positive control lesion experiment, demonstrating that lesion of a relay within a different potential route does not reduce transneuronal labeling, can be helpful in verifying a hypothesized multineuronal pathway (Pickard *et al.*, 2002; Smeraski *et al.*, 2004). Lesions of alternative pathways can be critical for falsifying alternative hypotheses concerning the transneuronal route between injection site and labeled neurons (see "Accurate Interpretation of Viral Labeling Patterns"). However, the existence of multiple parallel pathways can still complicate data interpretation in some circumstances (Aston-Jones *et al.*, 2001; Farkas *et al.*, 1998).

The approaches described above are not always practical for all tracing objectives. Sometimes, control experiments will not guarantee a clear answer regarding the specific circuit being studied by viral transneuronal labeling. However, cases should always be screened for indications of nonspecific CNS labeling.

B. Viral Tracing in the Peripheral Nervous System

Viral tracing projects designed to study the central circuits controlling motor outflow systems are considerably easier to analyze than projects designed to study CNS circuits because the route of viral entry can be experimentally limited to a single outflow channel. For example, the central parasympathetic circuits regulating pancreas, airways, or heart were studied in isolation from the sympathetic system by performing PRV tracing experiments on rats with T1 spinal transactions. These lesions completely eliminated the possibility of viral entry into the brain via the sympathetic nervous system (Haxhiu *et al.*, 1993; Loewy and Haxhiu, 1993; Ter Horst *et al.*, 1996). Since Bartha PRV does not enter the CNS via afferent systems, the data generated in these particular studies clearly produced information regarding the organization of central parasympathetic motor systems.

PRV injections into sympathetic structures sometimes resulted in unexpected labeling of vagal motor neurons, which has been greatly reduced after bilateral subdiaphragmatic vagotomy (Geerling *et al.*, 2003). Even after bilateral vagotomy, occasional cases were generated with a relatively low number of PRV-infected neurons in the dorsal vagal nucleus (e.g., 10–20 bilaterally throughout a 1-in-5 series of the full extent of the nucleus). This residual labeling may have been due to the failure to transect all of the vagal fibers.

Practical considerations have required that various neuroanatomical criteria be set for what constitutes a specific viral tracing infection after peripheral PRV injection (Sams *et al.*, 1995; Strack *et al.*, 1989b). For example, after retrograde tracing from the stellate ganglion, Jansen *et al.* (1995b) discovered labeled neurons in the red nucleus. Since this nucleus was mainly considered a somatic premotor nucleus, the labeling was interpreted as nonspecific and used to screen individual cases for nonspecific viral spread. Although it is possible that labeling in the red nucleus resulted from specific transneuronal labeling of a previously unknown sympathetic outflow pathway, the admittedly subjective criterion served as a useful index for potential nonspecific labeling in this particular study.

Finally, the neurosecretory magnocellular neurons in the paraventricular and supraoptic nuclei should be carefully examined after any type of PRV injection into peripheral structures. Neurons in these two nuclei project exclusively to the capillary beds in the posterior pituitary. If any cell body labeling is found at these two sites, it indicates that false-positive labeling likely occurred due to uptake of PRV from the vascular system.

C. Accurate Interpretation of Viral Labeling Patterns

The data obtained from a PRV transneuronal tracing experiment can be very complex, even after only two or three rounds of retrograde transport and viral replication (Fig. 9.5). Exponential replication of the virus occurs, within a geometrically increasing population of infected afferent neurons, at every retrograde transmission. Large numbers of neurons in many sites can be labeled over several days.

Substantial increases in the number of infected neurons can lead to incorrect assumptions about the routes by which viral tracer has infected a particular group of neurons. The neuroanatomical material obtained from a PRV tracing experiment provides a complicated snapshot of all the neurons infected with PRV up to the time of death. The labeling pattern itself does not indicate the order in which particular neurons are infected. Determination of the multineuronal pathways indicated by such data is particularly problematic when an infected neuronal group is connected to numerous other infected sites. Frequently, more than one reasonable hypothesis can be generated to explain such complicated tracing data.

1. Determining the Route(s) of Transneuronal Labeling—Experimental Approaches

Multiple experimental approaches can test each hypothesis. First, a temporal estimate of viral progression can sometimes aid the generation of preliminary hypotheses about the hierarchical connections of an afferent pathway (Larsen *et al.*, 1998; Pickard *et al.*, 2002; Smeraski *et al.*, 2004). In this approach, a provisional timeline is created from the comparison of infected sites in animals killed at progressively longer postinjection time points.

Three problems, however, often prevent a simple temporal approach from discriminating between various alternative synaptic pathways. First, the rate of viral progression may be variable among experimental animals. Second, practical experience has revealed that retrograde infection does not occur in idealized waves with one stage of replication, and retrograde transfer occurring at one afferent level before infected neurons appear at a higher level. Rather, infections tend to progress in a continuous manner, with the number of neurons infected at one hierarchical level growing even after infected neurons have appeared in groups afferent to that site. Also, afferent neurons located farthest from an infected neuron require longer axonal transport times, resulting in a delay in labeling. This gradual spread becomes even more of a blur as the infection spreads through higher order afferents. When added to the variability in rate of viral progression between different animals, this problem can obstruct the presumed logical interpretation of a simple temporal approach. Third, synaptic connections in the brain frequently do not exist in an idealized hierarchical arrangement. Multiple retrograde avenues to an afferent site and complicated reciprocal connections between higher order afferents typify brain architecture (Fig. 9.5). When two or more potential routes of viral spread are possible, temporal analysis may be useful in distinguishing between possibilities, but corroboration by other approaches is often necessary.

Besides a time-course analysis of viral spread, combination with conventional neural tracing and lesion experiments (described above) can aid in differentiating between alternative pathways. When more than one hypothetical transneuronal route can explain the spread of virus to a labeled group of neurons, selective lesion studies of the different routes can help define the circuit (see above).

2. An Example: PRV Transneuronal Labeling in Retinorecipient Sites

One example plainly demonstrates the importance of this critical analytical approach to viral transneuronal tracing. When comparing the wild-type and attenuated Bartha strain of PRV, Card *et al.* (1991) observed two qualitatively different patterns of infection after injecting two different viruses into the vitreous body of the eye. Wild-type PRV produced a rapid infection in all the sites targeted by the retinal output, such as the lateral geniculate nucleus, superior colliculus, and SCN, consistent with the idea that this virus is transported in the anterograde direction. Bartha PRV, however, produced a greatly delayed infection in the SCN and did not produce an infection in the two main retinorecipient sites superior colliculus and lateral geniculate. Importantly, Bartha PRV infections within the SCN occurred only after long postinjection survival times (3–4 days). Given the high viral dose used 10^{6} plaque forming units (pfu), two orders of magnitude higher than the dose at which Bartha PRV was originally used for retrograde transneuronal tracing after 4 days of survival (Strack *et al.*, 1989a, b; Strack and Loewy, 1990) this time frame was consistent with the time required for retrograde transneuronal spread to the hypothalamus through a chain of multiple neurons.

Despite the extended time required for viral spread to the SCN, these observations were interpreted as evidence for anterograde transport of Bartha PRV. To reconcile this interpretation with the complete lack of anterograde transmission to the main retinal target sites, it was further assumed that only a subpopulation of retinal ganglion cells, which project to nonvisual sites such as the SCN, is vulnerable to a productive infection by Bartha PRV (Card, 2000; Card *et al.*, 1991). Despite a lack of evidence for anterograde transneuronal spread in other studies with Bartha PRV, this interpretation remained unchallenged.

The assumption of a very slow form of anterograde transneuronal spread, resulting in a selective infection of the retinorecipient neurons of the SCN, served as the basis for a number of subsequent viral tracing studies of this circuitry (Card, 2000; Hannibal *et al.*, 2001; Moore *et al.*, 1995; Smith *et al.*, 2000) and for investigations into the genetic properties conferring this unusual property upon Bartha PRV (Brideau *et al.*, 2000; Card *et al.*, 1992; Enquist *et al.*, 1994; Husak *et al.*, 2000; Tomishima and Enquist, 2001; Whealy *et al.*, 1993). The implication that Bartha PRV might be capable of anterograde transport casts doubt on several other investigations that provided evidence that this virus moves exclusively in a retrograde manner (Rotto-Percelay *et al.*, 1992; Strack *et al.*, 1989b).

It was not until over a decade later that an alternative explanation was tested. In the intervening years, Enquist *et al.* (1994), using a series of deletion mutants, evaluated the importance of individual genes to the anterograde spread of PRV. These important studies built a strong case for the necessity of three particular genes, deleted in Bartha PRV, for anterograde spread of PRV (see "Specific Retrograde Transport of Bartha PRV"). Further analysis of the retinal infections produced by viruses lacking two of these genes indicated that mutants deficient in anterograde transport can still infect all types of retinal ganglion cells (Enquist *et al.*, 1994; Husak *et al.*, 2000). This finding did not fit well with the proposal that Bartha PRV produces anterograde labeling in only a subset of retinorecipient nuclei by selectively infecting a small subpopulation of retinal ganglion cells (Card *et al.*, 1991).

Following these reports, Pickard *et al.* (2000) tested the possibility that the spread of Bartha PRV from the eye to the SCN and other sites might not be the result of slow anterograde spread from a specific subset of retinal ganglion cells, but, instead, due to retrograde transneuronal spread via the autonomic nerves innervating the eye. In the original tests of Bartha PRV transneuronal specificity, Strack and Loewy (1990) demonstrated that injection into the nearby anterior chamber of the eye produced robust retrograde transneuronal labeling of the sympathetic outflow to the eye. In addition, a prominent multisynaptic outflow from the SCN to the diverse sympathetic and parasympathetic targets had been demonstrated (Ueyama *et al.*, 1999). Accordingly, in both the hamster (Pickard *et al.*, 2002) and the rat (Smeraski *et al.*, 2004), it was shown that (1) enucleation of the eye 24 h after Bartha PRV injection (preventing anterograde spread of virus due to degeneration of the optic axons from the destroyed retinal ganglion cells) did not prevent later infection within the SCN and other retinorecipient sites, (2) PRV infection in autonomic preganglionic sites—the parasympathetic Edinger–Westphal nucleus and sympathetic ganglionic and preganglionic neurons—preceded the SCN infection, and (3) lesions of these autonomic sites prior to PRV injection virtually eliminated infection in the SCN. In the hamster, neurons in the retinorecipient portion of the SCN were not even the first to be infected. Rather, their target neurons in the subparaventricular zone, dorsal to the SCN, were infected before labeling occurred in the SCN (see also Card *et al.*, 1991). In addition, the first appearance of PRV within the rat SCN did not overlap the retinohypothalamic projection (identified by concurrent anterograde axonal labeling with CTb; Smeraski *et al.*, 2004). These findings clearly disproved claims that Bartha PRV could be used as an anterograde transneuronal tracer.

3. Thorough Neuroanatomical Hypothesis-Testing

A provocative pattern of transneuronal labeling can tempt assumptions about the nature of the underlying pathway from injection site to infected neurons. As the preceding example demonstrates, however, such assumptions should not prevent the rigorous testing of alternative hypotheses. Carefully analyzing viral tracing data before asserting confidence in a particular explanation can be both complicated and time-consuming. However, combining basic viral tracing with thorough and prudent neuroanatomical analysis can significantly advance our knowledge of complicated circuits within the CNS (Aston-Jones *et al.*, 2001; Krout *et al.*, 2003; Pickard *et al.*, 2002; Smeraski *et al.*, 2004; Sylvester *et al.*, 2002).

D. False-Negative Data After Viral Transneuronal Tracing

As with any neural tracing technique, the degree of uptake and subsequent labeling of afferents to an injection site is dependent, in part, upon the amount of tracer used. Hence, with small tracer injections, a lack of labeling can be observed in sites known to provide lighter innervation to an injection site. For PRV tracing, this was first noted by O'Donnell *et al.* (1997), when injections of Bartha PRV into the mediodorsal thalamic nucleus did not produce the retrograde labeling that was expected, based on prior retrograde tracing studies, within the basolateral amygdala, a light source of innervation. Despite substantial retrograde transneuronal infections via the dense pallidal afferents, the absence of basolateral amygdala labeling suggested that virions either selectively avoided particular afferent system or stochastically entered only a proportion of afferent terminals in a given site, based on the relative amount of virus and the density of axon terminals.

This latter possibility was tested by Card *et al.* (1999), who injected a range of different Bartha PRV concentrations (10^4–10^5 pfu) into the striatum. At 2 days postinjection, a clear dose dependency was observed for extent of viral transneuronal labeling in various sites afferent to the striatum.

Viral concentration and postinjection survival time are two critical variables that affect optimal transneuronal labeling. Since only a few papers have dealt with this subject, it is not possible to make generalizations at this time regarding the optimal conditions to label any given CNS circuit. Rather, these important experimental parameters need to be empirically determined, but a reasonable starting point for most experiments would involve injections of ~3000 virions of Bartha PRV and a survival range of 2–4 days.

VII. OTHER VIRUSES USED FOR TRANSNEURONAL TRACING STUDIES

Bartha PRV remains the only virus subjected to direct tests of its specificity as a retrograde transneuronal tracer (Card *et al.*, 1993; Chen *et al.*, 1999; Pickard *et al.*, 2002; Rinaman *et al.*, 1993; Rotto-Percelay *et al.*, 1992; Smeraski *et al.*, 2004; Strack *et al.*, 1989b; Strack and Loewy, 1990). However, various other viruses are also used for transneuronal tracing studies. Experiments with HSV1 and rabies have been used to produce transneuronal labeling with varying indications of specificity. Additional direct verifications of their directional and transsynaptic specificity could be highly useful, particularly since, unlike PRV strains, these can be used for tracing experiments in primates. In addition, restrictions on Bartha PRV usage in countries where PRV has been eradicated from most pig and cattle populations may leave these viruses as the only practical options for certain laboratories.

A. HSV1 as a Transneuronal Tracer

HSV1 has been used for transneuronal tracing in various species. Different HSV1 strains have been used for transneuronal studies in primates by Strick and colleagues (Clower *et al.*, 2001; Hoover and Strick, 1993, 1999; Lynch *et al.*, 1994; Middleton and Strick, 1994, 1996, 2001, 2002).

The transneuronal pattern of labeling produced by this virus is highly dependent on the specific strain used for tracing (Norgren and Lehman, 1998). The SC16 strain of HSV1, used in early studies by Ugolini, produced both retrograde and anterograde transneuronal labeling in the brainstem and cerebellum (Ugolini *et al.*, 1987). Another HSV1 strain, FMC, was used for retrograde transneuronal labeling of central neurons afferent to various autonomic targets in a series of studies by Blessing and colleagues (Blessing *et al.*, 1991; Ding *et al.*, 1993; Li *et al.*, 1992a, b, 1993; Wesselingh *et al.*, 1989).

The patterns of infection produced by the injection of different strains of HSV1 into monkey cortex indicated that the McIntyre-B strain preferentially caused a retrograde transneuronal pattern of labeling while the H129 strain produced an anterograde labeling pattern (Zemanick *et al.*, 1991). Subsequent analysis, however, revealed that neither virus is transported exclusively in one direction, despite a significant difference in directional preference. McIntyre HSV1 can produce transneuronal labeling in the anterograde direction (Norgren *et al.*, 1992). Also, H129 clearly produces a retrograde infection within first-order afferent neurons (Rinaman and Schwartz, 2004). This strain, unlike Bartha PRV, was not observed to spread transneuronally from retrogradely infected first-order afferent vagal motor neurons in rats with lesioned vagal afferent fibers, following injection into the stomach wall (Rinaman and Schwartz, 2004). However, the pattern of transneuronal infection produced by H129 in this study, after a presumed anterograde transneuronal infection within the nucleus of the solitary tract (NTS), may be more consistent with retrograde transneuronal labeling of neurons afferent to this site, rather than simply anterograde spread in NTS neurons to their efferent targets [e.g., the strong infection depicted in Fig. 2 of Rinaman and Schwartz, 2004, within a dorsal part of the bed nuclei, an NTS afferent site versus the dense NTS innervation in a more ventral region of the bed nuclei (Ricardo and Koh, 1978)]. Further testing of the directional specificity of transneuronal labeling produced by HSV1 strain H129 should reveal whether or not this virus will be useful as an anterograde transneuronal tracer.

One drawback of viral tracing with various strains of HSV1 is the lack of neuroanatomical experiments directly addressing the specificity of transneuronal labeling. In Ugolini's original HSV1 tracing study, a significant degree of nonspecific local spread of virus was reported (Ugolini *et al.*, 1987). This nonspecific spread resulted in false-positive anterograde transneuronal labeling. Further tracing work with this strain was then conducted without direct tests of the specificity of transneuronal labeling (Ugolini *et al.*, 1989). The potential for nonspecific labeling by various strains of HSV1 via both local and transneuronal routes limits the utility of this virus for many neural tracing objectives (Fig. 9.4).

B. Rabies as a Retrograde Transneuronal Tracer

Although the name "pseudorabies" may seem to imply a functional relationship between PRV and rabies, these two viruses are very different. Like HSV1, PRV is a member of the Alphaherpesvirinae family of neurotropic herpesviruses. It contains a double-stranded DNA genome, which is transcribed and replicated in the cell nucleus, and can cause lytic cell death shortly after infection or establish latency in vivo in neuronal tissue. The reason for the name "pseudorabies" was the CNS infection it produced in farm species at a time when few viruses (rabies being one of them) were known to invade the brain (Aujesky, 1902).

Rabies, in contrast, is a rhabdovirus a single-stranded, negative-sense RNA virus that replicates in the cytoplasm. Unlike herpesviruses, rabies infections of the CNS, while lethal, do not appear to cause widespread cell death. Hence, this virus has been used for retrograde transneuronal labeling in various paradigms, in both rodent and primate. The earliest tracing study with rabies demonstrated anterograde transneuronal infection within the brain after injection of a challenge virus strain (CVS) of rabies into mouse olfactory epithelium (Astic *et al.*, 1993). When CVS rabies was injected into the hypoglossal nerve, retrograde transneuronal labeling was produced in rats without obvious nonspecific spread (Ugolini, 1995).

In particular, no infected glial cells were observed and infection did not appear to spread locally, even several days after the onset of infection within primary infected neurons in the hypoglossal nucleus. This result stood in striking contrast to the nonspecific viral labeling originally observed with HSV1 (Ugolini *et al.*, 1987). Studies using rabies to produce transneuronal labeling have demonstrated a potential for its use in rodent and primate neural tracing experiments (Astic *et al.*, 1993; Graf *et al.*, 2002; Grantyn *et al.*, 2002; Kelly and Strick, 2003; Moschovakis *et al.*, 2004; Tang *et al.*, 1999; Ugolini, 1995). Rabies central transneuronal tracing methodology has been thoroughly reviewed by Kelly and Strick (2000).

C. Perspectives—HSV1 and Rabies

There are two major drawbacks associated with the use of HSV1 and rabies viruses as transneuronal tracers. First and foremost, these viruses infect humans, representing a potential hazard to laboratory personnel and requiring additional precautions, especially for rabies, which requires repeated vaccinations and strict precautions (Kelly and Strick, 2000). For transneuronal studies in nonprimate species, the use of Bartha PRV does not present this problem, since it does not infect humans (Gustafson, 1975). Second, information regarding the neuroanatomical specificity of labeling produced by these viruses is incomplete. Rabies and HSV1 have not yet been subjected to many of the experimental tests used to characterize Bartha PRV as a neuroanatomical tracer (Card *et al.*, 1993; Chen *et al.*, 1999; Pickard *et al.*, 2002; Rinaman *et al.*, 1993; Rotto-Percelay *et al.*, 1992; Smeraski *et al.*, 2004; Strack *et al.*, 1989b; Strack and Loewy, 1990).

Varying degrees of transneuronal and directional specificity have been inferred from the patterns of infection observed in various tracing paradigms (Ugolini, 1995; Ugolini *et al.*, 1987; Zemanick *et al.*, 1991). For HSV1, a problematic degree of nonspecific local spread, resulting in subsequent nonspecific transneuronal labeling, has been described (Ugolini *et al.*, 1987). In contrast, some strains of rabies may spread only in the retrograde direction in some paradigms (Kelly and Strick, 2000; Ugolini, 1995), although certain strains can clearly produce anterograde transneuronal labeling (Astic *et al.*, 1993). The transneuronal specificity of infection with rabies appears promising, especially in comparison with HSV1 (Ugolini, 1995), but has yet to be directly tested.

A high degree of transneuronal specificity may not be a prerequisite for transneuronal tracing in primate circuits involving massively parallel circuitry, such as primate cortical, basal ganglia, cerebellar, and thalamic pathways. So long as the bulk of viral transneuronal transfer occurs, stochastically, in a transsynaptic manner, it is possible that nonspecific viral spread is largely constrained to immediately adjacent portions of parallel pathways within the same circuits. In any case, rabies and HSV1 are currently the only viable options for transneuronal labeling experiments in primates.

VIII. CONCLUSION

Neurotropic viruses are extremely useful neural tracers for a variety of neuroanatomical objectives. Strains with well-characterized properties, such as Bartha PRV, can be used to gain valuable new data about mammalian neural circuits. When tracing studies are executed and interpreted within known technical limitations, they can provide information currently unattainable by other methods.

Combined advances in virology and molecular biology may allow the design of viruses that will provide selective information about particular neural networks, revealing CNS circuitry in unprecedented detail.

APPENDIX

A. Safety and Practical Issues

Bartha PRV has been successfully used to eradicate PRV from most pig and cattle populations in many countries, and consequently, various rules and restrictions have been developed concerning its use. In the United States, a BSL-2 laboratory is required for use of this agent.

B. Sources of Bartha PRV and Recombinant Strains

Bartha PRV and related recombinant strains can be obtained directly from individual researchers who work with this virus. Investigators can contact Drs. Arthur Loewy (USA) and Thomas Mettenleiter (Germany).

In addition, the Center for Neuroanatomy with Neurotropic Viruses was established at the University of Pittsburgh in the summer of 2004. Under the direction of Drs. J. Patrick Card (card@ns.pitt.edu) and Peter Strick (strickp@itt.edu), the Center will serve as a resource for investigators interested in obtaining various viral strains for tracing experiments.

C. Viral Growth, Aliquots, and Storage

The broad host range of PRV in vivo is reflected in a broad spectrum of cells that can be productively infected in vitro. One of the advantages of working with PRV is its ability to replicate to rather high titers in easily cultivable permanent cell lines. Primarily, kidney cell lines from rodents (rabbit RK-13 cells), ruminants (Madin-Darby bovine kidney, MDBK), or porcines (PSEK, PK-15) are used. However, other cell lines, e.g., monkey kidney cells (Vero), are also permissive for PRV infection. The highest titers of progeny virus are usually obtained on porcine cells, whereas RK-13 cells are preferentially used for transfection and for establishment of transgenic cells expressing viral genes for transcomplementation of respective viral deletion mutants. MDBK cells are ideal for plaque titration since they produce the clearest and most easily visible plaques. However, other cell lines can also be used.

All the three mentioned cell lines can be cultivated in Eagle's minimum essential medium (MEM) supplemented with either 5% (for MDBK, PSEK, and PK-15) or 10% fetal calf serum (for RK-13). They routinely grow well on disposable plastic tissue culture flasks and can be propagated by trypsinization (0.8-g NaCl, 0.4-g NaCl, 1-g dextrose, 0.58-g NaHCO$_3$, 0.5-g trypsin, 0.2-g EDTA, and 1l H$_2$O (pH 7.1–7.2), sterile filtered)from the culture flasks, dilution at a ratio between 1:3 and 1:10 with fresh medium supplemented with fetal calf serum, and reseeding.

To grow high-titered virus stocks, cells are infected with PRV at a multiplicity of infection between 0.01 and 0.1 pfu per cell. After 1 h of adsorption, the inoculum is removed and the cells are overlaid with fresh medium. After 2–3 days, a complete cytopathic effect develops with cells first exhibiting a rounded appearance, which, as viral infection progresses, leads to lysis of the cells. After lysis of the cell monolayer, the whole culture flask is frozen and thawed, supernatant and cell debris are collected in a plastic tube (e.g., Falcon 50-ml tube) and cellular debris is sedimented by low-speed centrifugation (e.g., Heraeus Minifuge, 10 min, 6000 rpm). All virus isolation steps should be performed on ice or in a refrigerated centrifuge (+4°C). The supernatant is removed and immediately aliquoted, routinely in between 0.5- and 1-ml aliquots, and frozen at −70°C or in liquid nitrogen. Storage at −20°C leads to rapid loss of infectivity.

For determination of the infectious titer, one aliquot is thawed on ice and serial 10-fold dilutions in medium are prepared. These are then plated onto monolayer cells preferably in 6- or 24-well tissue culture dishes. As mentioned above, MDBK cells are most suited for this purpose, although other cell types can also be used. After 1-h incubation at 37°C, the supernatant is removed and substituted by medium supplemented with 5% methylcellulose: 10-g methylcellulose, 3.76-g autoclavable MEM powder suspended in 390-ml H$_2$O are autoclaved (use magnetic stir bar that remains in a bottle). After cooling to room temperature, 200 mM L-glutamine and 880-mg NaHCO$_3$ (dissolved in 6-ml H$_2$O and sterile filtered) are then added. The stock solution is stored at 4°C and is diluted 1:4 in MEM/5% fetal calf serum for use.

Cells are then incubated in a 5% CO$_2$ atmosphere for 2–3 days. When plaques are clearly visible, the medium is removed, the monolayer is washed 3× with PBS, and cells are then fixed with 5% formaldehyde for 20 min, washed with PBS, and stained with 1% crystal violet in 50% ethanol for 5 min. Staining solution is removed and monolayers are washed extensively. Plaques are white on a blue background.

Routinely, titers up to, and sometimes in excess of, 10^7 pfu/ml can be obtained. If higher virus titers are required, the virus suspension can be concentrated by ultracentrifugation (e.g., for 1 h at 22,000 rpm in a Beckman TST-28 rotor). It is imperative that all steps are performed either on ice or at

+4°C. The virus pellet is gently resuspended in the desired volume of either TBSal (200 mM NaCl, 2.6 mM KCl, 10 mM Tris-HCl (pH 7.5), 20 mM MgCl $_2$, 1.8 mM CaCl$_2$) or MEM. Then, individual aliquots of 5 µl (or other desired amount) are made in plastic microcentrifuge tubes and stored at -70°C.

D. Dilution of PRV

A tube containing a single aliquot of PRV is allowed to thaw on crushed ice. For CNS injections, 2 µl of 0.1% cholera toxin β-subunit (CTb, product #03B, List Biologicals Inc., Campbell, CA) is added. The CTb is used as a marker for the injection site and can also provide limited single-neuron tracing data, in addition to viral transneuronal labeling, in the same animal. These proportions (adding 2 µl of 0.1% CTb solution to a 5-µl viral suspension) result in a final CTb concentration of about 0.03% slightly less than the 0.05% solution recommended by Chen *et al.* (1999). In situations where injection site determination is not required, a similar dilution can be made with sterile Dulbecco's Modified Eagle Medium.

It is important to note that addition of a 2-µl CTb solution dilutes the concentration of the viral suspension, a critical variable in interpreting viral tracing experiments (see discussions above). For example, adding 2 µl to a 5-µl suspension of 10^8 pfu/ml Bartha PRV will reduce the viral concentration to 7×10^7 pfu/ml. This change is relevant to the calculation of the amount of virus injected into experimental animals.

E. Injection of PRV

Injections are made with a glass micropipette that has been filled with the aid of an operating microscope. The pipette is secured to a micromanipulator attached to a stereotaxic frame, and then advanced into a specific brain target in a surgically prepared animal. The micropipette is attached by polyethylene plastic tubing to an air pressure regulator so that the virus can be ejected by applying pressure from a handheld 50-ml syringe. Commercial equipment, such as the Picospritzer (General Valve Corp., Fairfield, NJ), can be used as well. Alternatively, a glass micropipette can be glued onto a 1-µl Hamilton microsyringe (Fisher Scientific, Pittsburgh, PA) and used in a similar capacity. The advantage of using a glass micropipette is that a carefully measured volume of virus can be delivered under microscopic control. Generally, 40-nl injections of the solution described in section "Dilution of PRV" have produced good results. A 40-nl injection of PRV without added CTb (10^8 pfu/ml) contains about 4000 pfu. If diluted by the addition of CTb as described above, this same volume contains about 3000 pfu. Rats receiving PRV injections should be surveyed daily for signs and symptoms of viral infection, such as nasal inflammation, itching, and sneezing.

F. Preparation of Brain Sections for Immunohistochemical Staining

Survival periods usually range from 3 to 4 days, for CNS injections, up to as long as 7–8 days for peripheral injections, depending upon variables such as the distance of the injected target from the brain and the desired extent of retrograde transneuronal labeling. At the end of the survival period, brain sections are processed for immunohistochemical staining in the same manner as other neuroanatomical tracing methods.

Briefly, anesthetized animals are killed by perfusion through the heart with saline, followed by 4% paraformaldehyde made in 0.1 M sodium phosphate buffer (pH 7.4). The brain is removed, stored in 4% paraformaldehyde fixative for 2 days or more, and sectioned at 50 μm using a freezing microtome or cryostat. Histological sections are collected and stored in plastic tissue culture trays containing 0.1 M sodium phosphate buffer (pH 7.4) containing 0.1% sodium azide, which acts as an antibacterial agent. The quality of the histological staining tends to be reduced after longer storage periods.

G. Immunohistochemical Staining

Immunohistochemical staining follows standard protocols. For visualizing injection sites, CTb can be localized with a goat antibody supplied by List Biologicals (Campbell, CA), used at 1:40,000, and the avidin–biotin complex method (ABC, Vectastain kit, Vector Laboratories, Burlingame, CA). For visualizing PRV, monoclonal antibodies can be purchased from Chemicon (Temecula, CA). Detailed protocols for combined PRV and neuropeptide immunohistochemistry appear in several reports (Geerling *et al.*, 2003; Oldfield *et al.*, 2002; Sylvester *et al.*, 2002).

H. Disposal of PRV-Infected Material

Preparatory steps are taken before the animals are perfused plastic bags containing absorbent material (newspaper) are used to collect all fluids. After CNS removal, animal carcasses and these fluids are disposed in biohazard waste containers. Cages, water bottles, and bedding are sterilized and cleaned in a central veterinary facility. After completion of experiments, the tissue trays containing brain sections from PRV-infected animals are submerged in bleach until the tissue dissolves.

REFERENCES

Astic, L., Saucier, D., Coulon, P., Lafay, F., and Flamand, A., 1993, The CVS strain of rabies virus as transneuronal tracer in the olfactory system of mice, *Brain Res.* **619**(1–2):146–156.
Aston-Jones, G., and Card, J. P., 2000, Use of pseudorabies virus to delineate multisynaptic circuits in brain: opportunities and limitations, *J. Neurosci. Methods* **103**(1):51–61.

Aston-Jones, G., Chen, S., Zhu, Y., and Oshinsky, M. L., 2001, A neural circuit for circadian regulation of arousal, *Nat. Neurosci.* **4**(7):732–738.

Aston-Jones, G., Zhu, Y., and Card, J. P., 2004, Numerous GABAergic afferents to locus ceruleus in the pericerulear dendritic zone: possible interneuronal pool, *J. Neurosci.* **24**(9):2313–2321.

Aujesky, A., 1902, Not readily distinguishable from rabies, with unknown origin (in Hungarian), *Veteranarius* **25**:387–396.

Babic, N., Klupp, B., Brack, A., Mettenleiter, T. C., Ugolini, G., and Flamand, A., 1996, Deletion of glycoprotein gE reduces the propagation of pseudorabies virus in the nervous system of mice after intranasal inoculation, *Virology* **219**(1):279–284.

Bak, I. J., Markham, C. H., Cook, M. L., and Stevens, J. G., 1977, Intraaxonal transport of Herpes simplex virus in the rat central nervous system, *Brain Res.* **136**(3):415–429.

Banfield, B. W., Kaufman, J. D., Randall, J. A., and Pickard, G. E., 2003, Development of pseudorabies virus strains expressing red fluorescent proteins: new tools for multisynaptic labeling applications, *J. Virol.* **77**(18):10106–10112.

Billig, I., Foris, J. M., Enquist, L. W., Card, J. P., and Yates, B. J., 2000, Definition of neuronal circuitry controlling the activity of phrenic and abdominal motoneurons in the ferret using recombinant strains of pseudorabies virus, *J. Neurosci.* **20**(19):7446–7454.

Blessing, W. W., Li, Y. W., and Wesselingh, S. L., 1991, Transneuronal transport of herpes simplex virus from the cervical vagus to brain neurons with axonal inputs to central vagal sensory nuclei in the rat, *Neuroscience* **42**(1):261–274.

Boldogkoi, Z., Reichart, A., Toth, I. E., Sik, A., Erdelyi, F., Medveczky, I., Llorens-Cortes, C., Palkovits, M., and Lenkei, Z., 2002, Construction of recombinant pseudorabies viruses optimized for labeling and neurochemical characterization of neural circuitry, *Brain Res. Mol. Brain Res.* **109**(1–2):105–118.

Boldogkoi, Z., Sik, A., Denes, A., Reichart, A., Toldi, J., Gerendai, I., Kovacs, K. J., and Palkovits, M., 2004, Novel tracing paradigms—genetically engineered herpesviruses as tools for mapping functional circuits within the CNS: present status and future prospects, *Prog. Neurobiol.* **72**(6):417–445.

Bourgeais, L., Gauriau, C., Monconduit, L., Villanueva, L., and Bernard, J. F., 2003, Dendritic domains of nociceptive-responsive parabrachial neurons match terminal fields of lamina I neurons in the rat, *J. Comp. Neurol.* **464**(2):238–256.

Braz, J. M., Rico, B., and Basbaum, A. I., 2002, Transneuronal tracing of diverse CNS circuits by Cre-mediated induction of wheat germ agglutinin in transgenic mice, *Proc. Natl. Acad. Sci. U. S. A.* **99**(23):15148–15153.

Brideau, A. D., Card, J. P., and Enquist, L. W., 2000, Role of pseudorabies virus Us9, a type II membrane protein, in infection of tissue culture cells and the rat nervous system, *J. Virol.* **74**(2):834–845.

Broussard, D. L., Li, X., and Altschuler, S. M., 1996, Localization of GABAA alpha 1 mRNA subunit in the brainstem nuclei controlling esophageal peristalsis, *Brain Res. Mol. Brain Res.* **40**(1):143–147.

Campadelli-Fiume, G., Arsenakis, M., Farabegoli, F., and Roizman, B., 1988, Entry of herpes simplex virus 1 in BJ cells that constitutively express viral glycoprotein D is by endocytosis and results in degradation of the virus, *J. Virol.* **62**(1):159–167.

Cano, G., Passerin, A. M., Schiltz, J. C., Card, J. P., Morrison, S. F., and Sved, A. F., 2003, Anatomical substrates for the central control of sympathetic outflow to interscapular adipose tissue during cold exposure, *J. Comp. Neurol.* **460**(3):303–326.

Card, J. P., 1998, Practical considerations for the use of pseudorabies virus in transneuronal studies of neural circuitry, *Neurosci. Biobehav. Rev.* **22**(6):685–694.

Card, J. P., 2000, Pseudorabies virus and the functional architecture of the circadian timing system, *J. Biol. Rhythms* **15**(6):453–461.

Card, J. P., Enquist, L. W., and Moore, R. Y., 1999, Neuroinvasiveness of pseudorabies virus injected intracerebrally is dependent on viral concentration and terminal field density, *J. Comp. Neurol.* **407**(3):438–452.

Card, J. P., Levitt, P., and Enquist, L. W., 1998, Different patterns of neuronal infection after intracerebral injection of two strains of pseudorabies virus, *J. Virol.* **72**(5):4434–4441.

Card, J. P., Rinaman, L., Lynn, R. B., Lee, B. H., Meade, R. P., Miselis, R. R., and Enquist, L. W., 1993, Pseudorabies virus infection of the rat central nervous system: ultrastructural characterization of viral replication, transport, and pathogenesis, *J. Neurosci.* **13**(6):2515–2539.

Card, J. P., Whealy, M. E., Robbins, A. K., and Enquist, L. W., 1992, Pseudorabies virus envelope glycoprotein gI influences both neurotropism and virulence during infection of the rat visual system, *J. Virol.* **66**(5):3032–3041.

Card, J. P., Whealy, M. E., Robbins, A. K., Moore, R. Y., and Enquist, L. W., 1991, Two alpha-herpesvirus strains are transported differentially in the rodent visual system, *Neuron* **6**(6):957–969.

Carr, D. B., O'Donnell, P., Card, J. P., and Sesack, S. R., 1999, Dopamine terminals in the rat prefrontal cortex synapse on pyramidal cells that project to the nucleus accumbens, *J. Neurosci.* **19**(24):11049–11060.

Carr, D. B. and Sesack, S. R., 2000, Dopamine terminals synapse on callosal projection neurons in the rat prefrontal cortex, *J. Comp. Neurol.* **425**(2):275–283.

Chan-Palay, V., 1976, Serotonin axons in the supra- and subependymal plexuses and in the leptomeninges; their roles in local alterations of cerebrospinal fluid and vasomotor activity, *Brain Res.* **102**(1):103–130.

Chen, S., Yang, M., Miselis, R. R., and Aston-Jones, G., 1999, Characterization of transsynaptic tracing with central application of pseudorabies virus, *Brain Res.* **838**(1–2):171–183.

Clower, D. M., West, R. A., Lynch, J. C., and Strick, P. L., 2001, The inferior parietal lobule is the target of output from the superior colliculus, hippocampus, and cerebellum, *J. Neurosci.* **21**(16):6283–6291.

Cook, M. L. and Stevens, J. G., 1973, Pathogenesis of herpetic neuritis and ganglionitis in mice: evidence for intra-axonal transport of infection, *Infect. Immun.* **7**(2):272–288.

DeFalco, J., Tomishima, M., Liu, H., Zhao, C., Cai, X., Marth, J. D., Enquist, L., and Friedman, J. M., 2001, Virus-assisted mapping of neural inputs to a feeding center in the hypothalamus, *Science* **291**(5513):2608–2613.

Demmin, G. L., Clase, A. C., Randall, J. A., Enquist, L. W., and Banfield, B. W., 2001, Insertions in the gG gene of pseudorabies virus reduce expression of the upstream Us3 protein and inhibit cell-to-cell spread of virus infection, *J. Virol.* **75**(22):10856–10869.

Ding, Z. Q., Li, Y. W., Wesselingh, S. L., and Blessing, W. W., 1993, Transneuronal labelling of neurons in rabbit brain after injection of herpes simplex virus type 1 into the renal nerve, *J. Auton. Nerv. Syst.* **42**(1):23–31.

Dobbins, E. G., and Feldman, J. L., 1994, Brainstem network controlling descending drive to phrenic motoneurons in rat, *J. Comp. Neurol.* **347**(1):64–86.

Enquist, L. W., 2002, Exploiting circuit-specific spread of pseudorabies virus in the central nervous system: insights to pathogenesis and circuit tracers, *J. Infect. Dis.* **186**(Suppl. 2):S209–S214.

Enquist, L. W., Dubin, J., Whealy, M. E., and Card, J. P., 1994, Complementation analysis of pseudorabies virus gE and gI mutants in retinal ganglion cell neurotropism, *J. Virol.* **68**(8):5275–5279.

Enquist, L. W., Tomishima, M. J., Gross, S., and Smith, G. A., 2002, Directional spread of an alpha-herpesvirus in the nervous system, *Vet. Microbiol.* **86**(1–2):5–16.

Farkas, E., Jansen, A. S., and Loewy, A. D., 1998, Periaqueductal gray matter input to cardiac-related sympathetic premotor neurons, *Brain Res.* **792**(2):179–192.

Fukuda, J., Kurata, T., and Yamaguchi, K., 1983, Specific reduction in Na currents after infection with herpes simplex virus in cultured mammalian nerve cells, *Brain Res.* **268**(2): 367–371.

Geerling, J. C., Mettenleiter, T. C., and Loewy, A. D., 2003, Orexin neurons project to diverse sympathetic outflow systems, *Neuroscience* **122**(2):541–550.

Gerfen, C. R., O'Leary, D. D., and Cowan, W. M., 1982, A note on the transneuronal transport of wheat germ agglutinin-conjugated horseradish peroxidase in the avian and rodent visual systems, *Exp. Brain Res.* **48**(3):443–448.

Giles, M. E., Sly, D. J., McKinley, M. J., and Oldfield, B. J., 2001, A method for the identification of pseudorabies virus protein and angiotensin AT(1A) receptor mRNA expression in the same CNS neurons, *Brain Res. Brain Res. Protoc.* **8**(3):153–158.

Goodpasture, E. W., 1925, The axis-cylinders of peripheral nerves as portals of entry to the central nervous system for the virus of herpes simplex in experimentally infected rabbits, *Am. J. Pathol.* **1**:11–33.

Goodpasture, E. W., and Teague, O., 1923, Transmission of the virus of herpes fibrils along nerves in experimentally infected rabbits, *J. Med. Res.* **44**:139–184.

Graf, W., Gerrits, N., Yatim-Dhiba, N., and Ugolini, G., 2002, Mapping the oculomotor system: the power of transneuronal labelling with rabies virus, *Eur. J. Neurosci.* **15**(9): 1557–1562.

Grantyn, A., Brandi, A. M., Dubayle, D., Graf, W., Ugolini, G., Hadjidimitrakis, K., and Moschovakis, A., 2002, Density gradients of trans-synaptically labeled collicular neurons after injections of rabies virus in the lateral rectus muscle of the rhesus monkey, *J. Comp. Neurol.* **451**(4):346–361.

Gustafson, D. P., 1975, Pseudorabies, In: Dunne, H. W., and Leman, A. D. (eds.), *Diseases of Swine*, Ames, IA: Iowa State University Press, pp. 209–223.

Hannibal, J., Vrang, N., Card, J. P., and Fahrenkrug, J., 2001, Light-dependent induction of cFos during subjective day and night in PACAP-containing ganglion cells of the retinohypothalamic tract, *J. Biol. Rhythms* **16**(5):457–470.

Haxhiu, M. A., Jansen, A. S., Cherniack, N. S., and Loewy, A. D., 1993, CNS innervation of airway-related parasympathetic preganglionic neurons: a transneuronal labeling study using pseudorabies virus, *Brain Res.* **618**(1):115–134.

Hoover, J. E., and Strick, P. L., 1993, Multiple output channels in the basal ganglia, *Science* **259**(5096):819–821.

Hoover, J. E., and Strick, P. L., 1999, The organization of cerebellar and basal ganglia outputs to primary motor cortex as revealed by retrograde transneuronal transport of herpes simplex virus type 1, *J. Neurosci.* **19**(4):1446–1463.

Horowitz, L. F., Montmayeur, J. P., Echelard, Y., and Buck, L. B., 1999, A genetic approach to trace neural circuits, *Proc. Natl. Acad. Sci. U. S. A.* **96**(6):3194–3199.

Husak, P. J., Kuo, T., and Enquist, L. W., 2000, Pseudorabies virus membrane proteins gI and gE facilitate anterograde spread of infection in projection-specific neurons in the rat, *J. Virol.* **74**(23):10975–10983.

Irnaten, M., Neff, R. A., Wang, J., Loewy, A. D., Mettenleiter, T. C., and Mendelowitz, D., 2001, Activity of cardiorespiratory networks revealed by transsynaptic virus expressing GFP, *J. Neurophysiol.* **85**(1):435–438.

Itaya, S. K., and van Hoesen, G. W., 1982, WGA-HRP as a transneuronal marker in the visual pathways of monkey and rat, *Brain Res.* **236**(1):199–204.

Jansen, A. S., Farwell, D. G., and Loewy, A. D., 1993, Specificity of pseudorabies virus as a retrograde marker of sympathetic preganglionic neurons: implications for transneuronal labeling studies, *Brain Res.* **617**(1):103–112.

Jansen, A. S., Nguyen, X. V., Karpitskiy, V., Mettenleiter, T. C., and Loewy, A. D., 1995a, Central command neurons of the sympathetic nervous system: basis of the fight-or-flight response, *Science* **270**(5236):644–646.

Jansen, A. S., Wessendorf, M. W., and Loewy, A. D., 1995b, Transneuronal labeling of CNS neuropeptide and monoamine neurons after pseudorabies virus injections into the stellate ganglion, *Brain Res.* **683**(1):1–24.

Jasmin, L., Burkey, A. R., Card, J. P., and Basbaum, A. I., 1997, Transneuronal labeling of a nociceptive pathway, the spino-(trigemino-)parabrachio-amygdaloid, in the rat, *J. Neurosci.* **17**(10):3751–3765.

Jons, A., and Mettenleiter, T. C., 1997, Green fluorescent protein expressed by recombinant pseudorabies virus as an in vivo marker for viral replication, *J. Virol. Methods* **66**(2):283–292.

Kangrga, I. M., and Loewy, A. D., 1995, Whole-cell recordings from visualized C1 adrenergic bulbospinal neurons: ionic mechanisms underlying vasomotor tone, *Brain Res.* **670**(2):215–232.

Kaufman, G. D., Mustari, M. J., Miselis, R. R., and Perachio, A. A., 1996, Transneuronal pathways to the vestibulocerebellum, *J. Comp. Neurol.* **370**(4):501–523.

Kelly, R. M., and Strick, P. L., 2000, Rabies as a transneuronal tracer of circuits in the central nervous system, *J. Neurosci. Methods* **103**(1):63–71.

Kelly, R. M., and Strick, P. L., 2003, Cerebellar loops with motor cortex and prefrontal cortex of a nonhuman primate, *J. Neurosci.* **23**(23):8432–8444.

Kerman, I. A., Enquist, L. W., Watson, S. J., and Yates, B. J., 2003, Brainstem substrates of sympatho-motor circuitry identified using trans-synaptic tracing with pseudorabies virus recombinants, *J. Neurosci.* **23**(11):4657–4666.

Kim, J. S., Enquist, L. W., and Card, J. P., 1999, Circuit-specific coinfection of neurons in the rat central nervous system with two pseudorabies virus recombinants, *J. Virol.* **73**(11):9521–9531.

Klupp, B. G., Hengartner, C. J., Mettenleiter, T. C., and Enquist, L. W., 2004, Complete, annotated sequence of the pseudorabies virus genome, *J. Virol.* **78**(1):424–440.

Kristensson, K., Lycke, E., and Sjostrand, J., 1971, Spread of herpes simplex virus in peripheral nerves, *Acta Neuropathol. (Berl.)* **17**(1):44–53.

Kristensson, K., Nennesmo, L., Persson, L., and Lycke, E., 1982, Neuron to neuron transmission of herpes simplex virus. Transport of virus from skin to brainstem nuclei, *J. Neurol. Sci.* **54**(1):149–156.

Kritas, S. K., Pensaert, M. B., and Mettenleiter, T. C., 1994, Role of envelope glycoproteins gI, gp63 and gIII in the invasion and spread of Aujeszky's disease virus in the olfactory nervous pathway of the pig, *J. Gen. Virol.* **75**(Pt 9):2319–2327.

Krout, K. E., Mettenleiter, T. C., and Loewy, A. D., 2003, Single CNS neurons link both central motor and cardiosympathetic systems: a double-virus tracing study, *Neuroscience* **118**(3):853–866.

Kuypers, H. G., and Ugolini, G., 1990, Viruses as transneuronal tracers, *Trends Neurosci.* **13**(2):71–75.

Larsen, P. J., Enquist, L. W., and Card, J. P., 1998, Characterization of the multisynaptic neuronal control of the rat pineal gland using viral transneuronal tracing, *Eur. J. Neurosci.* **10**(1):128–145.

Levatte, M. A., Mabon, P. J., Weaver, L. C., and Dekaban, G. A., 1998, Simultaneous identification of two populations of sympathetic preganglionic neurons using recombinant herpes simplex virus type 1 expressing different reporter genes, *Neuroscience* **82**(4):1253–1267.

Li, Y. W., Ding, Z. Q., Wesselingh, S. L., and Blessing, W. W., 1992a, Renal and adrenal sympathetic preganglionic neurons in rabbit spinal cord: tracing with herpes simplex virus, *Brain Res.* **573**(1):147–152.

Li, Y. W., Ding, Z. Q., Wesselingh, S. L., and Blessing, W. W., 1993, Renal sympathetic preganglionic neurons demonstrated by herpes simplex virus transneuronal labelling in the rabbit: close apposition of neuropeptide Y-immunoreactive terminals, *Neuroscience* **53**(4):1143–1152.

Li, Y. W., Wesselingh, S. L., and Blessing, W. W., 1992b, Projections from rabbit caudal medulla to C1 and A5 sympathetic premotor neurons, demonstrated with phaseolus leucoagglutinin and herpes simplex virus, *J. Comp. Neurol.* **317**(4):379–395.

Loewy, A. D., and Haxhiu, M. A., 1993, CNS cell groups projecting to pancreatic parasympathetic preganglionic neurons, *Brain Res.* **620**(2):323–330.

Lomniczi, B., Blankenship, M. L., and Ben-Porat, T., 1984, Deletions in the genomes of pseudorabies virus vaccine strains and existence of four isomers of the genomes, *J. Virol.* **49**(3):970–979.

Luppi, P. H., Aston-Jones, G., Akaoka, H., Chouvet, G., and Jouvet, M., 1995, Afferent projections to the rat locus coeruleus demonstrated by retrograde and anterograde tracing with cholera-toxin B subunit and Phaseolus vulgaris leucoagglutinin, *Neuroscience* **65**(1):119–160.

Lynch, J. C., Hoover, J. E., and Strick, P. L., 1994, Input to the primate frontal eye field from the substantia nigra, superior colliculus, and dentate nucleus demonstrated by transneuronal transport, *Exp. Brain Res.* **100**(1):181–186.

Manning, K. A., Erichsen, J. T., and Evinger, C., 1990, Retrograde transneuronal transport properties of fragment C of tetanus toxin, *Neuroscience* **34**(1):251–263.

Martin, X., and Dolivo, M., 1983, Neuronal and transneuronal tracing in the trigeminal system of the rat using the herpes virus suis, *Brain Res.* **273**(2):253–276.

Mettenleiter, T. C., 2003, Pathogenesis of neurotropic herpesviruses: role of viral glycoproteins in neuroinvasion and transneuronal spread, *Virus Res.* **92**(2):197–206.

Mettenleiter, T. C., Schreurs, C., Zuckermann, F., Ben-Porat, T., and Kaplan, A. S., 1988, Role of glycoprotein gIII of pseudorabies virus in virulence, *J. Virol.* **62**(8):2712–2717.

Mettenleiter, T. C., Zsak, L., Kaplan, A. S., Ben-Porat, T., and Lomniczi, B., 1987, Role of a structural glycoprotein of pseudorabies in virus virulence, *J. Virol.* **61**(12):4030–4032.

Middleton, F. A., and Strick, P. L., 1994, Anatomical evidence for cerebellar and basal ganglia involvement in higher cognitive function, *Science* **266**(5184):458–461.

Middleton, F. A., and Strick, P. L., 1996, The temporal lobe is a target of output from the basal ganglia, *Proc. Natl. Acad. Sci. U.S.A.* **93**(16):8683–8687.

Middleton, F. A., and Strick, P. L., 2001, Cerebellar projections to the prefrontal cortex of the primate, *J. Neurosci.* **21**(2):700–712.

Middleton, F. A., and Strick, P. L., 2002, Basal-ganglia 'projections' to the prefrontal cortex of the primate, *Cereb. Cortex* **12**(9):926–935.

Moore, R. Y., Speh, J. C., and Card, J. P., 1995, The retinohypothalamic tract originates from a distinct subset of retinal ganglion cells, *J. Comp. Neurol.* **352**(3):351–366.

Moschovakis, A. K., Gregoriou, G. G., Ugolini, G., Doldan, M., Graf, W., Guldin, W., Hadjidimitrakis, K., and Savaki, H. E., 2004, Oculomotor areas of the primate frontal lobes: a transneuronal transfer of rabies virus and [4C]2-deoxyglucose functional imaging study, *J. Neurosci.* **24**(25):5726–5740.

Nichol, P. F., Chang, J. Y., Johnson, E. M., Jr., and Olivo, P. D., 1994, Infection of sympathetic and sensory neurones with herpes simplex virus does not elicit a shut-off of cellular protein synthesis: implications for viral latency and herpes vectors, *Neurobiol. Dis.* **1**(1–2): 83–94.

Norgren, R. B., Jr., and Lehman, M. N., 1998, Herpes simplex virus as a transneuronal tracer, *Neurosci. Biobehav. Rev.* **22**(6):695–708.

Norgren, R. B., Jr., McLean, J. H., Bubel, H. C., Wander, A., Bernstein, D. I., and Lehman, M. N., 1992, Anterograde transport of HSV-1 and HSV-2 in the visual system, *Brain Res. Bull.* **28**(3):393–399.

O'Donnell, P., Lavin, A., Enquist, L. W., Grace, A. A., and Card, J. P., 1997, Interconnected parallel circuits between rat nucleus accumbens and thalamus revealed by retrograde transsynaptic transport of pseudorabies virus, *J. Neurosci.* **17**(6):2143–2167.

Oldfield, B. J., Giles, M. E., Watson, A., Anderson, C., Colvill, L. M., and McKinley, M. J., 2002, The neurochemical characterisation of hypothalamic pathways projecting polysynaptically to brown adipose tissue in the rat, *Neuroscience* **110**(3):515–526.

Pickard, G. E., Smeraski, C. A., Tomlinson, C. C., Banfield, B. W., Kaufman, J., Wilcox, C. L., Enquist, L. W., and Sollars, P. J., 2002, Intravitreal injection of the attenuated pseudorabies virus PRV Bartha results in infection of the hamster suprachiasmatic nucleus only by retrograde transsynaptic transport via autonomic circuits, *J. Neurosci.* **22**(7): 2701–2710.

Ricardo, J. A., and Koh, E. T., 1978, Anatomical evidence of direct projections from the nucleus of the solitary tract to the hypothalamus, amygdala, and other forebrain structures in the rat, *Brain Res.* **153**(1):1–26.

Rinaman, L., Card, J. P., and Enquist, L. W., 1993, Spatiotemporal responses of astrocytes, ramified microglia, and brain macrophages to central neuronal infection with pseudorabies virus, *J. Neurosci.* **13**(2):685–702.

Rinaman, L., and Schwartz, G., 2004, Anterograde transneuronal viral tracing of central viscerosensory pathways in rats, *J. Neurosci.* **24**(11):2782–2786.

Rotto-Percelay, D. M., Wheeler, J. G., Osorio, F. A., Platt, K. B., and Loewy, A. D., 1992, Transneuronal labeling of spinal interneurons and sympathetic preganglionic neurons after pseudorabies virus injections in the rat medial gastrocnemius muscle, *Brain Res.* **574**(1–2):291–306.

Rouiller, E. M., Capt, M., Dolivo, M., and De Ribaupierre, F., 1986, Tensor tympani reflex pathways studied with retrograde horseradish peroxidase and transneuronal viral tracing techniques, *Neurosci. Lett.* **72**(3):247–252.

Rouiller, E. M., Capt, M., Dolivo, M., and De Ribaupierre, F., 1989, Neuronal organization of the stapedius reflex pathways in the rat: a retrograde HRP and viral transneuronal tracing study, *Brain Res.* **476**(1):21–28.

Sabin, A. B., 1938, Progression of different nasally instilled viruses along different nervous pathways in the same host, *Proc. Soc. Exp. Biol.* **38**:270–275.

Sams, J. M., Jansen, A. S., Mettenleiter, T. C., and Loewy, A. D., 1995, Pseudorabies virus mutants as transneuronal markers, *Brain Res.* **687**(1–2):182–190.

Smeraski, C. A., Sollars, P. J., Ogilvie, M. D., Enquist, L. W., and Pickard, G. E., 2004, Suprachiasmatic nucleus input to autonomic circuits identified by retrograde transsynaptic transport of pseudorabies virus from the eye, *J. Comp. Neurol.* **471**(3):298–313.

Smiley, J. R., 2004, Herpes simplex virus virion host shutoff protein: immune evasion mediated by a viral Rnase? *J. Virol.* **78**(3):1063–1068.

Smith, B. N., Banfield, B. W., Smeraski, C. A., Wilcox, C. L., Dudek, F. E., Enquist, L. W., and Pickard, G. E., 2000, Pseudorabies virus expressing enhanced green fluorescent protein: a tool for in vitro electrophysiological analysis of transsynaptically labeled neurons in identified central nervous system circuits, *Proc. Natl. Acad. Sci. U. S. A.* **97**(16):9264–9269.

Song, C. K., and Bartness, T. J., 2001, CNS sympathetic outflow neurons to white fat that express MEL receptors may mediate seasonal adiposity, *Am. J. Physiol. Regul. Integr. Comp. Physiol.* **281**(2):R666–R672.

Stornetta, R. L., McQuiston, T. J., and Guyenet, P. G., 2004, GABAergic and glycinergic presympathetic neurons of rat medulla oblongata identified by retrograde transport of pseudorabies virus and in situ hybridization, *J. Comp. Neurol.* **479**(3):257–270.

Strack, A. M., and Loewy, A. D., 1990, Pseudorabies virus: a highly specific transneuronal cell body marker in the sympathetic nervous system, *J. Neurosci.* **10**(7):2139–2147.

Strack, A. M., Sawyer, W. B., Hughes, J. H., Platt, K. B., and Loewy, A. D., 1989a, A general pattern of CNS innervation of the sympathetic outflow demonstrated by transneuronal pseudorabies viral infections, *Brain Res.* **491**(1):156–162.

Strack, A. M., Sawyer, W. B., Platt, K. B., and Loewy, A. D., 1989b, CNS cell groups regulating the sympathetic outflow to adrenal gland as revealed by transneuronal cell body labeling with pseudorabies virus, *Brain Res.* **491**(2):274–296.

Sylvester, C. M., Krout, K. E., and Loewy, A. D., 2002, Suprachiasmatic nucleus projection to the medial prefrontal cortex: a viral transneuronal tracing study, *Neuroscience* **114**(4):1071–1080.

Tang, Y., Rampin, O., Giuliano, F., and Ugolini, G., 1999, Spinal and brain circuits to motoneurons of the bulbospongiosus muscle: retrograde transneuronal tracing with rabies virus, *J. Comp. Neurol.* **414**(2):167–192.

Ter Horst, G. J., 2000, Transneuronal retrograde dual viral labelling of central autonomic circuitry: possibilities and pitfalls, *Auton. Neurosci.* **83**(3):134–139.

Ter Horst, G. J., Hautvast, R. W., De Jongste, M. J., and Korf, J., 1996, Neuroanatomy of cardiac activity-regulating circuitry: a transneuronal retrograde viral labelling study in the rat, *Eur. J. Neurosci.* **8**(10):2029–2041.

Tomishima, M. J., and Enquist, L. W., 2001, A conserved alpha-herpesvirus protein necessary for axonal localization of viral membrane proteins, *J. Cell Biol.* **154**(4):741–752.

Tomishima, M. J., Smith, G. A., and Enquist, L. W., 2001, Sorting and transport of alpha herpesviruses in axons, *Traffic* **2**(7):429–436.

Ueyama, T., Krout, K. E., Nguyen, X. V., Karpitskiy, V., Kollert, A., Mettenleiter, T. C., and Loewy, A. D., 1999, Suprachiasmatic nucleus: a central autonomic clock, *Nat. Neurosci.* **2**(12):1051–1053.

Ugolini, G., 1995, Specificity of rabies virus as a transneuronal tracer of motor networks: transfer from hypoglossal motoneurons to connected second-order and higher order central nervous system cell groups, *J. Comp. Neurol.* **356**(3):457–480.

Ugolini, G., Kuypers, H. G., and Simmons, A., 1987, Retrograde transneuronal transfer of herpes simplex virus type 1 (HSV 1) from motoneurones, *Brain Res.* **422**(2):242–256.

Ugolini, G., Kuypers, H. G., and Strick, P. L., 1989, Transneuronal transfer of herpes virus from peripheral nerves to cortex and brainstem, *Science* **243**(4887):89–91.

Vahlne, A., Nystrom, B., Sandberg, M., Hamberger, A., and Lycke, E., 1978, Attachment of herpes simplex virus to neurons and glial cells, *J. Gen. Virol.* **40**(2):359–371.

Wesselingh, S. L., Li, Y. W., and Blessing, W. W., 1989, PNMT-containing neurons in the rostral medulla oblongata (C1, C3 groups) are transneuronally labelled after injection of herpes simplex virus type 1 into the adrenal gland, *Neurosci. Lett.* **106**(1–2):99–104.

Westerhaus, M. J., and Loewy, A. D., 1999, Sympathetic-related neurons in the preoptic region of the rat identified by viral transneuronal labeling, *J. Comp. Neurol.* **414**(3):361–378.

Westerhaus, M. J., and Loewy, A. D., 2001, Central representation of the sympathetic nervous system in the cerebral cortex, *Brain Res.* **903**(1–2):117–127.

Whealy, M. E., Card, J. P., Robbins, A. K., Dubin, J. R., Rziha, H. J., and Enquist, L. W., 1993, Specific pseudorabies virus infection of the rat visual system requires both gI and gp63 glycoproteins, *J. Virol.* **67**(7):3786–3797.

Wiesel, T. N., Hubel, D. H., and Lam, D. M., 1974, Autoradiographic demonstration of ocular-dominance columns in the monkey striate cortex by means of transneuronal transport, *Brain Res.* **79**(2):273–279.

Zemanick, M. C., Strick, P. L., and Dix, R. D., 1991, Direction of transneuronal transport of herpes simplex virus 1 in the primate motor system is strain-dependent, *Proc. Natl. Acad. Sci. U.S.A.* **88**(18):8048–8051.

Zou, Z., Horowitz, L. F., Montmayeur, J. P., Snapper, S., and Buck, L. B., 2001, Genetic tracing reveals a stereotyped sensory map in the olfactory cortex, *Nature* **414**(6860):173–179.

Dextran Amines: Versatile Tools for Anterograde and Retrograde Studies of Nervous System Connectivity

ANTON REINER and MARCIA G. HONIG

BACKGROUND
METHODS USING DEXTRAN AMINES
  What Are the Dextran Amines?
  Anterograde Labeling with Dextran Amines
  Retrograde Labeling with Dextran Amines
  Retrograde Collateral Labeling with Dextran Amines
  Combining BDA10kDa Anterograde Labeling with Other
    Neuroanatomical Tracers
  Combining BDA3kDa Retrograde Labeling with Other
    Neuroanatomical Tracers
  Combining Dextran Amine Labeling with Immunohistochemistry
ADVANTAGES AND LIMITATIONS
  Merits of Dextran Amines Relative to Other Tracers
  Pitfalls and Solutions
APPENDIX
  Step-by-Step BDA Protocol for LM Single-Labeling Studies
  Step-by-Step BDA Protocol for LM Double-Labeling Studies
  Step-by-Step BDA Protocol for EM Studies
  Step-by-Step BDA Protocol for EM Double-Labeling Studies
  Expected Results
  REFERENCES

Abstract: Dextran amines are versatile and sensitive tools for anterograde and retrograde investigation of neural connectivity. Because of their tolerance of diverse

ANTON REINER AND MARCIA G. HONIG • Department of Anatomy and Neurobiology, University of Tennessee Health Science Center, 855 Monroe Avenue, Memphis, TN 38163 fixatives, they are ideal for various light and electron microscopic studies. They can be iontophoretically or pressure-injected, and then depending on the type of dextran amine used and the type of detection method, visualized by transmitted light microscopy, fluorescence microscopy, or electron microscopy. High-molecular-weight biotinylated dextran amines (BDAs; 10 kDa) yield sensitive and exquisitely detailed labeling of axons and terminals, while low-molecular-weight BDAs (3 kDa) yield sensitive and detailed Golgi-like retrograde labeling of neurons. Labeling with the BDAs can be visualized with an avidin-biotinylated horseradish peroxidase (ABC) procedure followed by a standard or a metal-enhanced diaminobenzidine (DAB) reaction, or with any of several fluorescent probes that bind to biotin. Fluorescent dextran amines can be directly visualized by fluorescence microscopy or rendered suitable for transmitted light or electron microscopic viewing by immunohistochemical detection of the given fluorophore. The variety of dextran amines and the methods for their visualization make them well-suited for multiple-label studies. The dextran amines can also be combined with other anterograde or retrograde tracers, or intracellular labeling, and the disparate markers separately visualized either by multicolor DAB or by DAB–VIP® labeling, or by multiple fluorescence viewing. In the same manner, pathway tracing with dextran amines and immunohistochemical labeling can be combined. The dextran amines are thus flexible and valuable pathway-tracing tools that have gained widespread popularity since being introduced.

Keywords: anterograde tracer, connectivity, dextran amines, double-label, neuroanatomical mapping, retrograde tracer

I. BACKGROUND

The selective silver impregnation of degenerating axons as a tool for anterograde tracing of neural pathways revolutionized neuroanatomy by making it possible to accurately and objectively delineate neural circuitry (Fink and Heimer, 1967; Nauta and Gygax, 1954). The anterograde tracing of neural pathways was further accelerated with the development of the autoradiographic detection of anterogradely transported radioactive amino acids (autoradiography, ARG) (Cowan *et al.*, 1972; Edwards, 1972; Hendrickson and Edwards, 1978), which proved to be a more reliable method than had been the silver stains of degenerating fibers. The limited anatomical resolution of axons and terminals provided by ARG-tracing methods was, however, a serious disadvantage. This limitation was overcome in the *Phaseolus vulgaris* leucoagglutinin (PHA-L) method, which provided superbly sensitive and selective anterograde labeling of axons and their terminals with limited uptake by fibers of passage at the injection site, with the added advantage that PHA-L–labeled axons and terminals could be visualized at both LM and EM levels (Gerfen *et al.*, 1989; Gerfen and Sawchenko, 1984; Sesack *et al.*, 2006, this volume; Zaborszky and Heimer, 1989). Two drawbacks of the PHA-L method, namely that iontophoretic injection methods are required and that PHA-L proved unreliable in the hands of many investigators, led to the quest for yet better pathway-tracing agents (Groenewegen and Wouterlood, 1990; Schmued *et al.*, 1990; Veenman *et al.*, 1992, 1995).

Shortly after the introduction of PHA-L, a number of investigators demonstrated the utility of fluorophore-bound dextran–lysine conjugates (called dextran amines) as pathway-tracing agents (Fritzsch and Wilm, 1990; Glover *et al.*, 1986; Nance and Burns, 1990; Schmued *et al.*, 1990). Fluorescent dextran amines rely on the ease with which dextran is transported by axons, and the fact that the lysine moiety makes it possible to fix the transported fluorophore-conjugated dextran in situ with standard aldehyde fixatives, to allow visualization of the cellular localization of the transported dextran amine by fluorescence microscopy. These tracers were found to combine high sensitivity and detailed resolution with simplicity and reliability in their delivery and uptake. While the fluorescent dextran amines were revealed to be in several respects superior to ARG and PHA-L, they were not well suited for studies involving detailed mapping of neuronal projection systems, primarily because fluorophores fade with viewing. This limitation was overcome by the development of the biotinylated dextran amines (BDAs), which possess the favorable transport properties of fluorescent dextran amines and can be visualized at the transmitted light level with a standard avidin-biotinylated horseradish peroxidase (HRP) (ABC) procedure (Brandt and Apkarian, 1992; Rajakumar *et al.*, 1993; Reiner *et al.*, 2000; Veenman *et al.*, 1992, 1995). Moreover, with the availability of antibodies against fluorophores such as rhodamine and fluorescein, it became possible to use peroxidase immunolabeling to also detect fluorophore-bound dextran amines (Reiner *et al.*, 2000, 2003). Further studies have revealed that depending on its molecular weight and the pH of the delivery vehicle, dextran amines could be used either as anterograde or as retrograde pathway-tracing agents (Fritzsch, 1993; Kaneko *et al.*, 1996; Medina *et al.*, 1997) in a wide variety of species (Albert *et al.*, 1999; Davila *et al.*, 2002; Fritzsch, 1993; Guirado *et al.*, 1999; Kenigfest *et al.*, 2000; Lanuza *et al.*, 1998; Lopes-Correa *et al.*, 1998; Scalia *et al.*, 1997; Striedter, 1994; Veenman *et al.*, 1992; Wang *et al.*, 2004). In addition, owing to its tolerance of diverse fixatives, BDA can be visualized at the LM or EM level, and it can be combined with various other pathway-tracing or immunohistochemical methods. The purpose of this chapter is to provide an overview of the uses of the dextran amines, and to provide simple protocols for their use.

II. METHODS USING DEXTRAN AMINES

A. What Are the Dextran Amines?

Dextrans are biologically inert hydrophilic polysaccharides that possess high water solubility, low toxicity, low immunogenicity, and resistance to cleavage by most endogenous glycosidases (Haugland, 1996). Dextrans of molecular weights ranging from 3 to 2000 kDa, conjugated to biotin or a wide variety of fluorophores, are commercially available from Molecular Probes (Eugene, OR), Sigma Chemical Company (St. Louis, MO), or Vector Laboratories (Burlingame, CA). For studies of nervous system connectivity, dextrans with covalently bound lysine residues are used, since the lysines allow the dextran to be conjugated to surrounding biomolecules by paraformaldehyde or glutaraldehyde fixation (Reiner *et al.*, 2000). Dextrans transport anterogradely and retrogradely in neurons and their axons, but unlike DiI and DiO are not transported in fixed nervous tissue (Fritzsch and Wilm, 1990). Dextrans of up to 70 kDa have been used in studies of nervous system connectivity, with smaller molecular weight dextrans transporting more quickly than larger. Moreover, 3-kDa dextran seems to transport more effectively in a retrograde direction than does 10-kDa dextran. The mechanism of BDA uptake appears to be pinocytotic for intact neurons (Jiang *et al.*, 1993). The dextran amines are, however, also taken up by damaged neurons and axons (Glover *et al.*, 1986; Lei *et al.*, 2004; Todorova and Rodziewicz, 1995; Veenman *et al.*, 1992). Dextran–lysine (called dextran amine) conjugated to biotin is referred to as BDA, while the fluorescent conjugates of dextran amines have been identified by either an abbreviated form of their name or a descriptive name. For example, 10-kDa tetramethylrhodamine-conjugated dextran amine, which we will, for clarity and consistency, here abbreviate as RDA10kDa, has also been called fluororuby (Schmued *et al.*, 1990). Similarly, 10-kDa fluorescein dextran amine (FDA10kDa) has been called fluoroemerald (Novikova *et al.*, 1997). Mixes of BDA10kDa and RDA10kDa, sometimes called miniruby (Liu *et al.*, 1993; Novikova *et al.*, 1997), or BDA3kDa and RDA3kDa, sometimes called microruby (Liu *et al.*, 1993; Novikova *et al.*, 1997), are also commercially available as are BDA10kDa–FDA10kDa mixes and BDA3kDa–FDA3kDa mixes (sometimes called mini- and microemerald) (Novikova *et al.*, 1997). Such mixtures of biotinylated and fluorophore-conjugated dextran amines allow fluorescence viewing to rapidly screen injection site and transport efficacy before proceeding to the steps needed for producing a permanent label suitable for transmitted light microscopic viewing of the BDA localization. Finally, in addition to different molecular weight FDA and RDA, Molecular Probes also offers dextran amines conjugated to a wide array of additional fluorophores, including the intensely fluorescent, fade-resistant Alexa fluorophores, which lend themselves to sensitive detection and differentiation of labeled structures, especially using confocal laser scanning microscopy (CLSM).

B. Anterograde Labeling with Dextran Amines

The 10,000 Da form of dextran amine is preferentially transported anterogradely, with yet higher weight forms transporting less effectively than 10 kDa (Fritzsch, 1993; Kaneko *et al.*, 1996; Medina *et al.*, 1997; Schmued *et al.*, 1990; Veenman *et al.*, 1992). A dextran amine of 10 kDa can be used in the same type of LM single-label anterograde pathway-tracing studies as

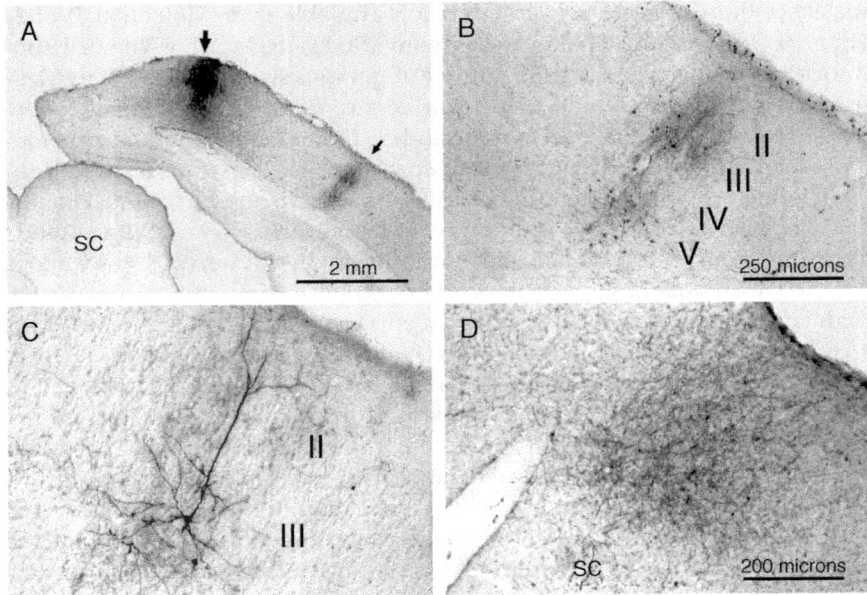

Figure 10.1. Iontophoretic BDA10kDa injection into primary visual cortex (area 17) of rat and resulting labeling in a series of transverse sections. Image A shows a low-power view of the BDA10kDa injection site in area 17 (large arrow), showing an anterograde projection to area 18 (small arrow). The superior colliculus (SC) is located in the lower left of the image. Image B shows a high-power view of the labeling in area 18 shown by the arrow in A. Note that labeled fibers in area 18 are located predominantly in layers II/III and V, as would be expected on the basis of the known projection of area 17 to 18 in rats (Peters, 1985). Some retrogradely labeled neurons are also evident in area 18, mainly in layers II/III and V, as would be expected on the basis of the known projection of area 18 to 17 (Peters, 1985). Although it is possible that the retrogradely labeled neurons contribute BDA10kDa-labeled collaterals to the fiber labeling in area 18, such labeled collaterals must be far fewer than the anterogradely labeled fibers since the distribution of labeled fibers and terminals in 18 conforms to that expected for anterograde labeling from 17 and not to that expected for labeled collaterals of layers II/III and V neurons (Peters, 1985). (C) Golgi-like retrograde labeling of a neuron in layer III of area 18 following the injection shown in image A. Image D shows labeled fibers and terminals in the superficial layers of the ipsilateral SC following the injection shown in image A. The magnification in D is the same as in C. (From Veenman *et al.*, 1992.)

PHA-L, biocytin, neurobiotin, CTb (cholera toxin B subunit), and WGA-HRP (HRP-conjugated wheat germ agglutinin), and it possesses several features that make it advantageous. First, a 10-kDa dextran amine reliably yields anterograde labeling after either iontophoretic or pressure injection into the nervous system (Figs. 10.1A,B,D and 10.2B). Second, it is easy to make small and well-defined injections of the dextran amine, thereby confining the injection to the region whose projections are being investigated and/or study the topographic order to the projection from the given region (Figs. 10.1A and 10.2A). Third, and as noted previously, the morphological detail

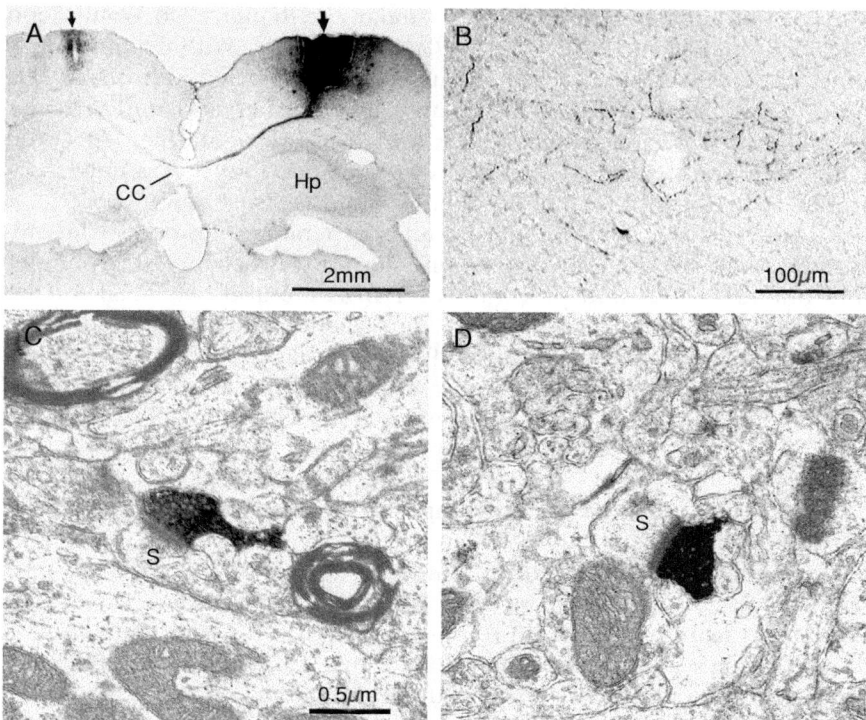

Figure 10.2. LM images of coronal sections showing iontophoretic BDA10kDa injection site in rat somatosensory cortex (A) and resulting anterograde labeling in ipsilateral striatum (B), and EM images of anterogradely labeled terminals in contralateral striatum after BDA10kDa injection into rat motor cortex (C, D). Image A shows a low-power view of an iontophoretic injection of BDA into rat somatosensory cortex (large arrow), showing anterogradely labeled fibers that cross in the corpus callosum (CC) above the hippocampus (Hp) and project to homotypic contralateral somatosensory cortex (small arrow). A blood vessel is present in the middle of the terminal field in the contralateral cortex. Image B shows a high-power view of labeled axons with varicosities in ipsilateral rat striatum following the iontophoretic injection of BDA into somatosensory cortex shown in A. Images C and D show EM views of anterogradely labeled terminals in striatum after injection of BDA10kDa into the contralateral motor cortex. Both terminals make asymmetric axospinous contact with the spines (S) of presumptive striatal projection neurons. The BDA+ terminals in C and D are densely labeled but clearly contain numerous unlabeled small clear vesicles. The tissue was prepared using freshly made 3.5% paraformaldehyde–0.6% glutaraldehyde–15% saturated picric acid as fixative. Both C and D are at the same magnification. From Veenman   *et al.*, 1992 (A, B) and Reiner *et al.*, 2003 (C, D).]

of the labeling is often exquisite (Figs. 10.1B,D and 10.2B). Fourth, the detailed labeling of terminals provided by dextran amine of 10 kDa, and the fact that BDA readily tolerates glutaraldehyde fixation with no significant attenuation of labeling makes BDA10kDa well-suited especially to characterizing the morphology and cellular targets of given brain regions at the LM and EM levels (Fig. 10.2C,D; Guirado *et al.*, 1999; Reiner *et al.*, 2003, 2004;

Van Haeften and Wouterlood, 2000; Veenman and Reiner, 1996; Wouterlood and Jorritsma-Byham, 1993; Wright *et al.*, 1999, 2001). With diaminobenzidine (DAB) visualization, terminals anterogradely labeled with BDA10kDa tend to have dense labeling of the axoplasm on the surface of vesicular membranes, so that synaptic vesicles are clearly recognizable. Moreover, BDA10kDa seems to have advantages over such other sensitive anterograde tracers as PHA-L and CTB in the simplicity of the visualization procedure, since both PHA-L and CTB require a multistep immunohistochemical procedure, while BDA requires only a one-step ABC procedure. The simpler and briefer visualization procedures for BDA are also beneficial for the preservation of ultrastructural detail (Van Haeften and Wouterlood, 2000; Wouterlood and Jorritsma-Byham, 1993). Note that BDA10kDa, however, is not exclusively an anterograde tracer and yields some retrograde labeling (Fig. 10.1B,C; Reiner *et al.*, 2000; Veenman *et al.*, 1992; Vercelli *et al.*, 2000).

C. Retrograde Labeling with Dextran Amines

Low-molecular-weight dextran amine (i.e., 3000 MW), when injected within an acidic vehicle (e.g., 10% BDA in 0.1 M sodium citrate–HCl, pH 3.0), is mainly transported retrogradely (Fritzsch, 1993; Kaneko *et al.*, 1996; Medina *et al.*, 1997; Veenman *et al.*, 1992). There is evidence that neuronal activity in terminals enhances their pinocytotic uptake of BDA3kDa, yielding more intense retrograde labeling (Jiang *et al.*, 1993). Since 3-kDa dextran amines are transported twice as rapidly as 10-kDa dextran amines (Fritzsch, 1993), shorter survival times are needed for BDA3kDa than that for BDA10kDa. Dextran amine of 3 kDa can be used in the same types of LM single-label retrograde pathway-tracing studies as HRP, Fluorogold, biocytin, neurobiotin, CTB, and WGA-HRP for routine mapping of the distribution of the neurons projecting to a given neural region and their topographic order. The labeling intensity and fixation tolerance that make BDA10kDa advantageous for anterograde labeling also make BDA3kDa advantageous for both LM and EM retrograde labeling studies (Figs. 10.2A–C and 10.3A,B). Moreover, because of the Golgi-like retrograde labeling it yields and its glutaraldehyde tolerance, BDA3kDa is especially useful for studies of the LM architecture of labeled neurons and for EM studies of their axodendritic and axospinous inputs (Fig. 10.4; Lei *et al.*, 2004). Note that 3-kDa dextran amine is, however, not exclusively a retrograde tracer and yields some anterograde labeling even when injected within an acidic vehicle (Jiao *et al.*, 2000; Medina and Reiner, 1997).

D. Retrograde Collateral Labeling with Dextran Amines

The sensitivity of BDA10kDa creates one of its major disadvantages as an anterograde tracer̶the retrograde labeling of neurons with BDA10kDa

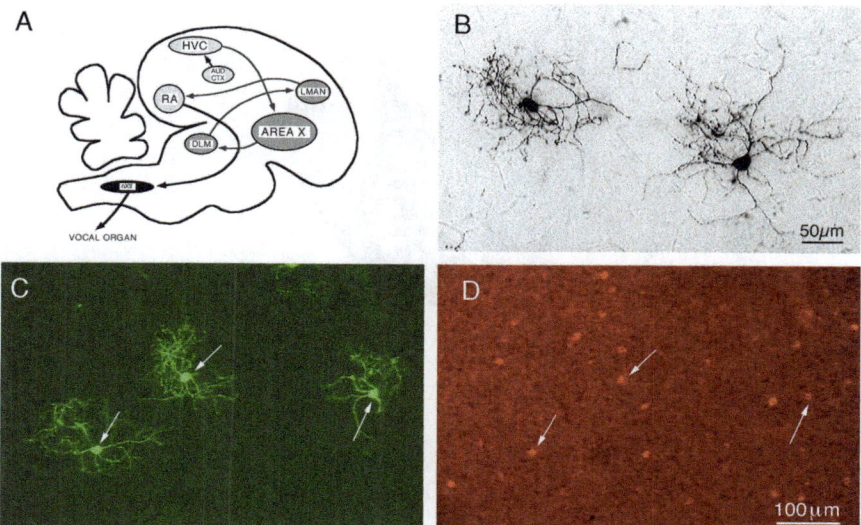

Figure 10.3. The schematic shown in A illustrates the circuit devoted to vocal learning in songbirds, as viewed in the sagittal plane. The serially connected components of this circuit are (1) the "cortical" area called HVC, which receives direct input from avian auditory cortex; (2) area X of the medial striatum (X); (3) the dorsolateral medial nucleus of the thalamus (DLM); (4) a "cortical" region termed the lateral magnocellular anterior nidopallium (LMAN); and (5) the "cortical" area called the robust nucleus of the arcopallium (RA), which projects to the vocal motoneurons of the brainstem. The images in B–D show fields of view within area X containing neurons retrogradely labeled from DLM with BDA3kDa. The image in B was captured using differential interference contrast microscopy and shows two such neurons in area X of male zebra finch visualized with an ABC/DAB procedure. These neurons have the morphological characteristics of pallidal neurons, which include being relatively large (12–14 μm) and possessing aspiny dendrites. The images shown in C and D are from the same individual field within area X, captured using CLSM, from a zebra finch that received a BDA3kDa injection into DLM. Image C shows DTAF-labeled BDA3kDa+ neurons, while D shows the TRITC-labeled LANT6+ neurons. All neurons that were labeled for BDA3kDa also contained LANT6, as indicated by the arrows. The results in C and D show that area X neurons that project to DLM also possess the pallidal trait of containing the neurotensin-related hexapeptide LANT6 (Lys^8-Asp^9-neurotensin$^{8-13}$). (From Reiner *et al.*, 2004.)

can, in some cases, be so complete as to label the axon collaterals of neurons retrogradely labeled by it (Chen and Aston-Jones, 1998). If those neurons have axon collaterals in the site to which the region injected with BDA10kDa projects (e.g., if the BDA-labeled neurons reside within the field of antero-gradely labeled BDA+ axons and terminals; Fig. 10.1A,B), then there may be no simple way to reliably distinguish between anterograde labeling and labeling of the collaterals of retrogradely labeled neurons. It would thus be uncertain whether a given labeled axon in the terminal field arises from the BDA injection site or is a collateral of a retrogradely labeled neuron.

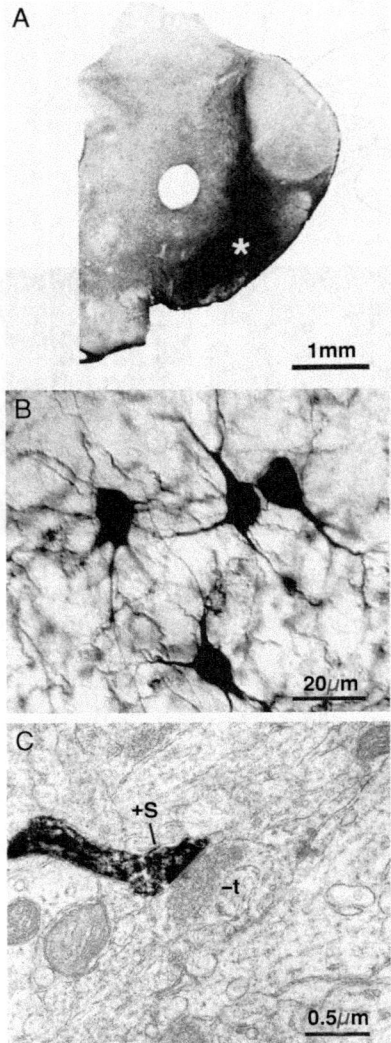

Figure 10.4. Example of a BDA3kDa injection into substantia nigra (A), and the LM level (B) and EM level (C) retrograde labeling of striatal neurons obtained. Image B shows perikarya and dendrites of striatonigral neurons labeled with BDA3kDa, while image C shows that this labeling included that of spines. A BDA3kDa-labeled spine (+s) is evident in C, and the labeled spine receives an asymmetric synaptic contact from an unlabeled terminal (−t) with the round vesicles characteristic of excitatory input from cortex. (From Lei *et al.*, 2004.)

One way to deal with this problem is to ascertain the projection targets of any population of neurons retrogradely labeled with BDA10kDa. In some cases, this information may be available from the literature, while in others it may be necessary to obtain it by anterograde labeling from the neuronal population in question. It may be the case that the retrogradely labeled neurons are known not to have collaterals in the region of the BDA anterograde labeling, in which case the confound potentially created by collateral labeling is obviated. In addition, the fact that BDA10kDa does not always yield extensive retrograde labeling also somewhat mitigates this problem (Brandt and Apkarian, 1992; Reiner *et al.*, 1993, 2000; Veenman *et al.*, 1992).

It is possible, however, to turn this disadvantage of BDA into a useful tool if one is interested in the collateral projections of the retrogradely labeled neurons (Chen and Aston-Jones, 1998). For example, we have shown that BDA3kDa injected into the pyramidal tract (PT) at pontine levels yields extensive Golgi-like labeling of pyramidal neurons in layer V of sensory and motor cortex in rats (Fig. 10.5A,B; Lei *et al.*, 2004; Reiner *et al.*, 2003). Accompanying this retrograde labeling is labeling of the collateral projections of these neurons to the striatum (Fig. 10.5C,D). Thus, this approach can be used to selectively label the corticostriatal projection of the PT-type cortical neurons. There are, in fact, at least two types of corticostriatal projection neurons in rats (Cowan and Wilson, 1994; Kincaid and Wilson, 1996; Wilson, 1987). One of these two types is the PT type, while the other type is termed the intratelencephalically projecting type (IT type), since it projects only within telencephalon. Because the laminar and regional distributions of the PT- and IT-type neurons in cortex are overlapping, it would be difficult to selectively label the input to ipsilateral striatum from only one of them by means of an injection of BDA10kDa directly into the cortex. Injection of BDA3kDa into the PT, however, provides a means for selectively labeling the PT-type neuron terminals in ipsilateral striatum and ascertaining their morphology and synaptic targets at the LM and EM levels (Lei *et al.*, 2004; Reiner *et al.*, 2003). It should be possible to use this same collateral labeling approach to selectively study the output of a particular neuronal population to one or more specific targets in circumstances in which that neuronal population (1) is intermingled with other neuronal populations within a field and whose projection is thus difficult to selectively label by anterograde labeling from the cell bodies of origin by direct injection of the region containing the cell bodies and (2) has a projection to at least one region that none of the surrounding other neuronal population in the region does. This unique projection then can be the target of a BDA3kDa injection to yield selective retrograde collateral labeling of the projection of that neuronal population. The approach for this collateral labeling with BDA3kDa is otherwise not different from that for conventional BDA anterograde or retrograde labeling.

E. Combining BDA10kDa Anterograde Labeling with Other Neuroanatomical Tracers

BDA10kDa can also be used in LM double-label studies in which one population of axons and terminals is labeled with BDA10kDa and (1) a

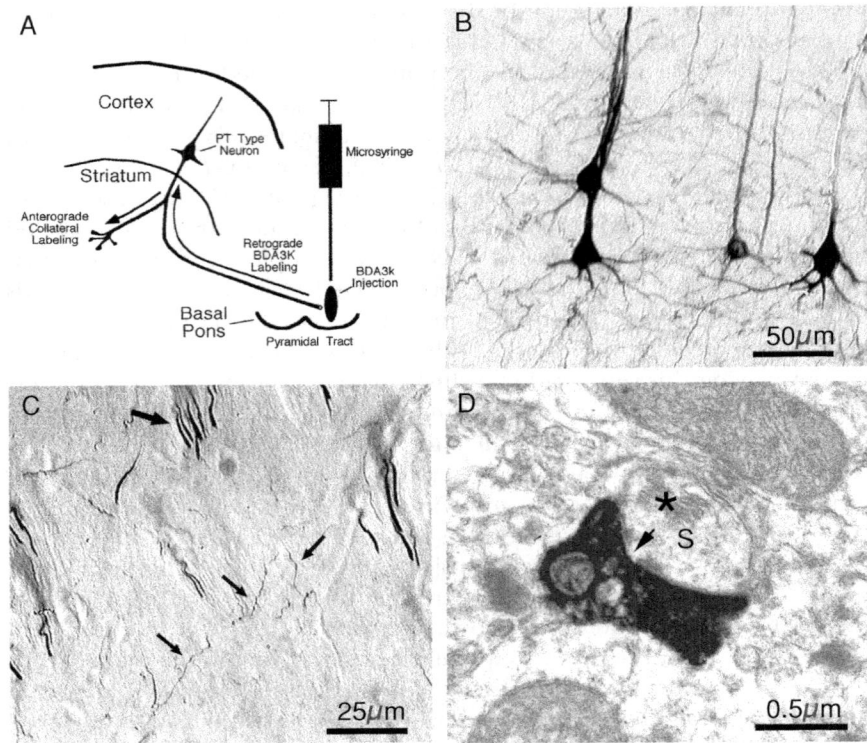

Figure 10.5. BDA3kDa injected into the axons of the pyramidal tract of rat at pontine levels, as shown schematically in A, yields extensive retrograde labeling of layer V cortical pyramidal neurons in ipsilateral sensory–motor cortex (B). The pyramidal neurons are labeled so thoroughly that their intrastriatal collaterals (small arrows in C), which arise from the pyramidal neuron axons as they pass through the striatum (large arrow in C), are extensively labeled. This allows selective visualization of the corticostriatal terminals of the pyramidal tract-type cortical neurons at the LM (C) and EM levels (D). The pyramidal tract neurons give rise to large terminals that make asymmetric synaptic contacts with the spines (S) of striatal neurons (D), which are identifiable by their size and the presence of spine apparatus (asterisk). The terminal shown is relatively large and associated with a perforated postsynaptic density (arrow). (From Reiner *et al.*, 2003.)

second population of axons is labeled with PHA-L, CTB, RDA10kDa, or FDA10kDa or (2) a population of neurons is retrogradely labeled with HRP, FluoroGold, CTB, RDA3kDa, FDA3kDa, or a viral tracer, such as pseudorabies, or labeled by intracellular filling (Alisky and Tolbert, 1994; Aston-Jones and Card, 2000; Dolleman-van der Weel *et al.*, 1994; Lanciego and Wouterlood, 1994; Luo *et al.*, 2001). The former type of study is useful, for example, to determine the topographic order in a projection system, with the two anterograde tracers injected side by side in the target of interest (Ojima and Takayanagi, 2001), while the latter studies serve to determine if input from a given source ends on neurons projecting to a particular target structure.

Such LM double-label studies require using two distinct methods for visualizing the two different populations of labeled structures. Compatible distinct pairs of labeling methods commonly used at the transmitted light level include two-color DAB (Antal *et al.*, 1990; Hancock, 1986; Hsu and Soban, 1982; Medina *et al.*, 1997; Veenman *et al.*, 1992), and DAB in combination with Vector Laboratories very intense purple (VIP®) (Gonzalo *et al.*, 2001; Lanciego *et al.*, 1997; Lanciego and Gimenez-Amaya, 1999). Since BDA can be detected using a fluorophore-conjugated antibiotin or a fluorophore-conjugated avidin, BDA10kDa and a second tracer can also be detected with pairs of distinct fluorophores (e.g., rhodamine and fluorescein), and viewed using conventional epi-illumination fluorescence microscopy or CLSM (Jiao *et al.*, 2000).

Alternatively, two fluorescent dextran amines such as RDA10kDa and FDA10kDa can be used for double anterograde labeling, or anterograde labeling with RDA10kDa or FDA10kDa can be combined with retrograde labeling with FDA3kDa or RDA3kDa, respectively. Such material can be directly viewed by fluorescence microscopy or CLSM, by separate permanent labels suitable for transmitted light viewing produced using antisera directed against rhodamine and fluorescein and two-color DAB, or by DAB and VIP (Kaneko *et al.*, 1996). A variant approach of injecting two tracers in one animal involves coinjecting an anterograde tracer such as BDA10kDa with a retrograde tracer such as CTB at the same site (Coolen *et al.*, 1999). This makes it possible to ascertain the inputs to (by retrograde CTB labeling) and the outputs of (by anterograde BDA10kDa labeling) a single site, with alternating sections labeled for each marker or different markers used to visualize the labeling in the same sections.

BDA10kDa, RDA10kDa, or FDA10kDa can also be used in EM double-label studies to determine the morphology of two defined types of axons with respect to each other or their targets (in combination with each other or PHA-L), or to determine the identity of the target structure of a defined type of axon (using a retrograde tracer to label the target structure and dextran amine to label the axons and terminals) (Alisky and Tolbert, 1994; Lanciego *et al.*, 1998a,b; Luo *et al.*, 2001; Van Haeften and Wouterlood, 2000). Such EM double-label studies usually (but not always) require using pairs of markers that are distinct at the EM level, for example, DAB and silver-intensified immunogold, DAB and benzidine dihydrochloride (BDHC), DAB and silver-intensified DAB, or DAB and VIP® (Anderson *et al.*, 1991, 1994; Groenewegen and Wouterlood, 1990; Van Haeften and Wouterlood, 2000; Wouterlood *et al.*, 1993; Zhou and Grofova, 1995). Note that the same EM marker (e.g., DAB) can be used for both structures to be visualized, if BDA-labeled input to a population of neurons retrogradely labeled with a second tracer is under study, and the retrogradely labeled neurons do not give rise to local collaterals with the same morphology as the BDA-labeled terminals.

F. Combining BDA3kDa Retrograde Labeling with Other Neuroanatomical Tracers

BDA3kDa can be used in LM double-label studies in which a set of perikarya is labeled with BDA3kDa and (1) a population of axons is labeled using PHA-L, CTB, RDA10kDa, or FDA10kDa or (2) a second population of neurons is retrogradely labeled, in this case with HRP, Fluorogold, CTB, RDA3kDa, FDA3kDa, or a viral tracer such as pseudorabies or by intracellular filling. Such LM double-label studies again require using two distinct methods for visualizing the two different populations of labeled structures, such as two-color DAB (Antal *et al.*, 1990; Hancock, 1986; Hsu and Soban, 1982; Medina *et al.*, 1997; Veenman *et al.*, 1992), DAB in combination with VIP (Gonzalo *et al.*, 2001), or two fluorophores (Jiao *et al.*, 2000). As true for fluorophore-conjugated 10-kDa dextran amines, fluorophore-conjugated 3-kDa dextran amines can be detected using antisera directed against the fluorophores (typically rhodamine or fluorescein), and BDA3kDa itself can be detected with an avidin-conjugated fluorophore or a fluorophore-conjugated antibiotin (Jiao *et al.*, 2000; Kaneko *et al.*, 1996). Again, a variant approach of injecting two tracers in one animal involves coinjecting a retrograde tracer such as BDA3kDa and an anterograde tracer such as PHA-L at the same site at the same time (Coolen *et al.*, 1999). This makes it possible to ascertain the inputs to (by retrograde BDA3kDa labeling) and the outputs of (by anterograde PHA-L labeling) a single site, with alternating sections labeled for each marker or different markers used to visualize the labeling in the same sections.

BDA3kDa can also be used in EM double-label studies of inputs to BDA3kDa-labeled neurons, and markers that are distinct at the EM level are then typically needed to distinguish the two classes of tracer-labeled structures (Alisky and Tolbert, 1994; Anderson *et al.*, 1991, 1994; Groenewegen and Wouterlood 1990; Lanciego *et al.*, 1998a,b; Van Den Pol and Decavel, 1990; Van Haeften and Wouterlood, 2000; Wouterlood *et al.*, 1993; Zaborszky and Heimer, 1989). Here too, using only one EM marker (e.g., DAB) for both categories of labeled structure is possible if the BDA3kDa retrogradely labeled neurons do not give rise to local collaterals with the same morphology as the anterogradely labeled input under study.

G. Combining Dextran Amine Labeling with Immunohistochemistry

BDA can also be used in LM double-label studies in which one population of axons and terminals is labeled with BDA10kDa or one population of neuronal perikarya is labeled with BDA3kDa, and a second population of axons/terminals, dendrites, and/or perikarya is labeled immunohistochemically for a neurotransmitter, neuropeptide, receptor, or any other molecule unique to its members (Figs. 10.3C,D and 10.6A–C; Jiao *et al.*, 2000; Lei *et al.*, 2004; Luo *et al.*, 2001; Reiner *et al.*, 2000, 2003, 2004; Veenman *et al.*, 1992).

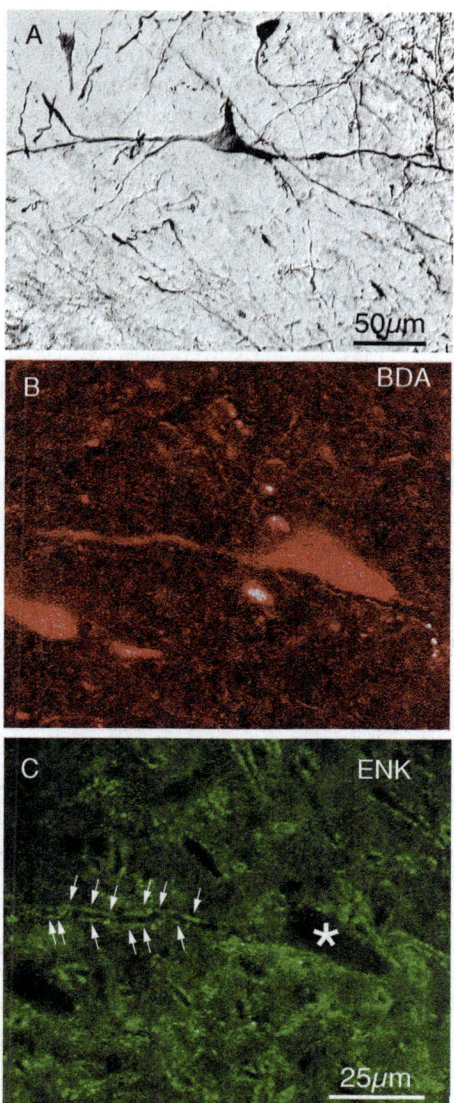

Figure 10.6. Injection of BDA10kDa into the subthalamic nucleus in pigeon yields retrograde labeling in globus pallidus. The labeled pallidosubthalamic projection neuron in image A was visualized using ABC/DAB and differential interference contrast microscopy. Note that it has a large perikaryon and long smooth dendrites. Images B and C are of a single field within globus pallidus, captured using CLSM. The BDA10kDa-labeled pallidal neuron was visualized with TRITC-conjugated antibiotin (B). In the same sections, ENK+ terminals were immunolabeled using a DTAF-conjugated secondary antiserum (C). The dendrite of the BDA-labeled pallidosubthalamic neuron in B is surrounded and contacted by ENK+ terminals (arrows) of striatal origin in C, as is characteristic of pallidal neurons that project to the subthalamic nucleus. The asterisk in C marks the site of the BDA-labeled neuron shown in B. (From Jiao *et al.*, 2000.)

Such studies are performed using two distinct markers for visualizing the two different populations of labeled structures, as in the case of combining BDA pathway tracing with another neuroanatomical tracer. Here too, it is possible to use fluorescent dextran amines rather than biotinylated if fluorescence viewing of labels is desired. Note that it may be necessary to colchicine-treat animals near the end of the dextran amine survival time to boost perikaryal immunoreactivity for the antigen of interest if it tends to be rapidly shipped out of the cell body via the axon (Anderson and Reiner, 1990).

BDA can also be used in EM double-label studies in which one population of axons and terminals is labeled with BDA10kDa or one population of neuronal perikarya is labeled with BDA3kDa, and a second population of axons/terminals, dendrites, and/or perikarya is labeled immunohistochemically for the distinctive neurotransmitter, neuropeptide, receptor, or other unique molecules it contains (Lei *et al.*, 2004). Such EM double-label studies are carried out in a manner similar to that for double-label studies involving BDA and a second neuroanatomical tracer. Here too, using only one EM marker (e.g., DAB) for both the BDA-labeled and the immunolabeled structures is possible if the BDA10kDa anterogradely labeled terminals do not have the same morphology as any of the immunolabeled terminals in the field under study, or if the BDA3kDa-labeled perikarya under study do not give rise to any terminals that resemble immunolabeled terminals in the field under study (Lei *et al.*, 2004).

III. ADVANTAGES AND LIMITATIONS

A. Merits of Dextran Amines Relative to Other Tracers

The efficacy of BDA as an anterograde or a retrograde tracer at the LM level has been confirmed for all major vertebrate groups, including monkeys (Brandt and Apkarian, 1992; Lanciego *et al.*, 1998a), cats (Richmond *et al.*, 1994), rats (Brandt and Apkarian, 1992; Veenman *et al.*, 1992; Wouterlood and Jorritsma-Byham, 1993), birds (Striedter, 1994; Tellegen *et al.*, 2001; Veenman *et al.*, 1992; Wang *et al.*, 2004), reptiles (Davila *et al.*, 2002; Guirado *et al.*, 1999; Kenigfest *et al.*, 2000; Lanuza *et al.*, 1998), amphibians (Fritzsch, 1993; Scalia *et al.*, 1997), bony fish (Albert *et al.*, 1999; Lopes-Correa *et al.*, 1998; Torres *et al.*, 2002; Xue *et al.*, 2001), and jawless fish (Pombal *et al.*, 1999). The efficacy of the dextran amines has also been shown in several different neural regions, in the intact CNS, in the PNS, in partially dissected preparations, in brain slices, and with both iontophoretic and pressure injection methods (Fritzsch, 1993; Novikov, 2001; Reiner *et al.*, 2000; Veenman *et al.*, 1992; Wang *et al.*, 2004). Iontophoretic BDA injections yield small well-defined injection sites and extensive and clear labeling of axonal projection systems over distances of at least several centimeters from the injection site (Figs. 10.1 and 10.2). A relatively wide range of pipette tip diameters

(10–50 μm) and current levels (2–5 μA) for iontophoretic injections and a wide range of survival times appear to be effective for transport of BDA. Pressure injection of BDA10kDa also yields extensive and detailed anterograde labeling, while pressure injection of acidic BDA3kDa yields extensive retrograde labeling. Finally, BDA labeling can readily be combined with other anterograde or retrograde labeling methods or with immunohistochemical labeling methods, and it can be used at the EM level, as noted above. BDA combines the advantages of amine-conjugated dextrans (sensitivity) and of biotinylated compounds (permanent label). For these reasons, the use of BDA10kDa for anterograde pathway tracing and BDA3kDa for retrograde pathway tracing has come to be widespread. The availability of fluorescent dextran amines or of mixes of fluorescent and BDAs adds further simplicity and versatility to dextran amines as pathway-tracing tools.

BDA10kDa appears similar to PHA-L and RDA10kDa in its sensitivity, although some have reported a superior sensitivity of BDA10kDa (Novikov, 2001). BDA10kDa has an advantage over PHA-L in that BDA10kDa requires only an ABC procedure followed by a DAB reaction to be visualized, whereas a multistep immunohistochemical procedure is required to visualize PHA-L (Van Haeften and Wouterlood, 2000; Veenman *et al.*, 1992; Wouterlood and Jorritsma-Byham, 1993). Further, some authors have reported that PHA-L labeling can be capricious (Groenewegen and Wouterlood, 1990; Schmued *et al.*, 1990), while we and others have found that BDA10kDa consistently yields excellent anterograde labeling (Lanciego *et al.*, 2000; Reiner *et al.*, 2000; Veenman *et al.*, 1992). Biocytin and neurobiotin also can be used for anterograde labeling and possess many of the advantages of BDA10kDa, but both biocytin and neurobiotin are quickly catabolized and therefore effective tracers over only short survival times (Izzo, 1991; King *et al.*, 1989; Kita and Armstrong, 1991; Lapper and Bolam, 1991; Novikov, 2001). Finally, BDA10kDa is also a more sensitive anterograde tracer than is WGA-HRP (Ferguson *et al.*, 2001).

BDA3kDa is more sensitive than HRP for retrograde labeling and is a more reliable and effective retrograde tracer than biocytin or neurobiotin (Kaneko *et al.*, 1996; Lapper and Bolam, 1991; Medina *et al.*, 1997). BDA appears to be more sensitive as both retrograde and anterograde tracers than CTB (Chen and Aston-Jones, 1998), although CTB is highly sensitive and useful (Reiner *et al.*, 1996; Shimizu *et al.*, 1994). For EM studies, BDA3kDa is advantageous over CTB due to the simplicity and brevity of the labeling procedure, which benefits EM ultrastructure, and due to the greater sensitivity and cellular detail provided by 3-kDa dextran amines (Van Haeften and Wouterlood, 2000; Wouterlood and Jorritsma-Byham, 1993). While Fluorogold may be as or more sensitive than BDA3kDa in terms of numbers of neurons retrogradely labeled from an injection of the same size, BDA3kDa yields more extensive dendritic labeling, and Fluorogold is toxic to both injected and retrogradely labeled neurons (Naumann *et al.*, 2000; Schmued *et al.*, 1993; Schmued and Fallon, 1986). BDA3kDa appears to be more robust than HRP or WGA-HRP as a retrograde neural tracer, except for retrograde labeling from peripheral muscle (Faulkner *et al.*, 1997; Richmond *et al.*, 1995; Todorova and Rodziewicz, 1995). In the case of retrograde labeling from muscle, dextran amines label fewer neurons than do other retrograde tracers, but do yield excellent dendritic labeling of the few motoneurons that are labeled.

The primary advantage of BDA over fluorescent dextran amines is that when visualized with DAB, BDA provides a permanent, electron-dense label rather than a transient, fluorescent one. Chang (1991) has, however, shown that Lucifer Yellow-conjugated dextran amine can be labeled immunohistochemically with a permanent, electron-dense label using antisera against Lucifer Yellow. Similarly, RDA can be visualized as a permanent label immunohistochemically with antisera against tetramethylrhodamine, while FDA can be visualized by antifluorescein immunolabeling (Kaneko *et al.*, 1996). Finally, while BDA10kDa appears to be a favorable anterograde tracer and BDA3kDa a favorable retrograde tracer, investigators must bear in mind that they need to optimize the procedures outlined here for their system.

B. Pitfalls and Solutions

1. Absence of Labeling for BDA After Peripheral Target Injection

While the 3-kDa dextran amines are excellent as retrograde tracers when injected into neural targets, they appear relatively poor for retrograde labeling from peripheral muscle (Faulkner *et al.*, 1997; Richmond *et al.*, 1993; Todorova and Rodziewicz, 1995). After muscle injection, dextran amines label fewer neurons than do many other retrograde tracers but do produce excellent dendritic labeling of the few motoneurons labeled. This attribute of dextran amines may reflect their tendency not to diffuse far from the injection and/or to a dependence on uptake by damaged nerve terminals.

2. Absence of Labeling for BDA After Nervous System Injection

There are three main reasons why BDA labeling might fail: a faulty ABC/DAB procedure for visualizing BDA, a failure to actually inject BDA, and too short a survival time. If there is no BDA labeling anywhere in the brain after ABC/DAB processing, including the injection target, then a defect in processing would be suspected. Carrying out tests of the ABC/DAB reagents by using immunohistochemistry to visualize some plentiful antigen (e.g., substance P) in sections in which the BDA labeling failed will reveal whether the fault lies with the ABC/DAB procedure. If the immunolabeling test yields no DAB-labeled structures, the investigator would need to systematically determine which ABC/DAB step or reagent is problematic. One potential source of problems in the ABC/DAB procedure is the use of sodium azide, which inhibits the peroxidase reaction, in the diluting solution. On the other hand, if the immunohistochemistry yields plentiful labeling and the track created by the microsyringe or micropipette used for the BDA injection is evident, then defective ejection of BDA or defective BDA would be suspected. To distinguish between these two possibilities, the investigator needs to consider several possibilities. Was the lot of BDA bad? Was it inappropriately stored? Was the pH of the BDA solution correct? Was the pipette or syringe clogged? Was the current polarity correct during the iontophoretic injection? Finally, if the injection site is labeled but transport is present only near the injection site, a larger injection or longer survival time is the likely solution.

3. Retrograde Labeling for BDA

The tendency of BDA10kDa, and similar-weight fluorescent dextran amines, to retrogradely label neurons, at least with some injections and for some brain regions, gives rise to the main limitation of BDA10kDa evident to us. The retrograde labeling with BDA10kDa seems greater after axonal damage and may be greater in amphibians and reptiles than in mammals. It is now clear that axon collaterals of such retrogradely labeled neurons can also be labeled. The presence of these labeled axon collaterals could then make it difficult to know whether labeled axons and terminals in a particular structure result from anterograde labeling of neurons at the injection site or from the retrogradely labeled neurons. A similar problem also exists for other sensitive anterograde tracers that result in some retrograde labeling, such as WGA-HRP, PHA-L, CTB, biocytin, and neurobiotin (Chen and Aston-Jones, 1998; Glover *et al.*, 1986; Groenewegen and Wouterlood, 1990; Izzo, 1991; King *et al.*, 1989; Kita and Armstrong, 1991; Nance and Burns, 1990; Schmued *et al.*, 1990; Shu and Peterson, 1988; Woodson *et al.*, 1991). In practice, this problem should not prevent a clear interpretation of the results of BDA injections, since such collateral labeling with dextran amine of 10 kDa should be minor in comparison with the anterograde labeling and since the retrogradely labeled neurons in most instances will not project to the same targets as the neurons at the injection site. Nonetheless, some caution must be exercised in interpreting the results of anterograde labeling using any of the above tracers.

4. Fibers of Passage

Another factor that can complicate the interpretation of the results obtained with various anterograde or retrograde tracers is the potential uptake of tracer by fibers of passage. This possibility has not been investigated extensively in the case of BDA, although it is clear that, like other tracers, BDA is taken up by axons that have been damaged by the injection (Brandt and Apkarian, 1992; Glover *et al.*, 1986; Lei *et al.*, 2004; Todorova and Rodziewicz, 1995; Veenman *et al.*, 1992). BDA also appears to be taken up pinocytotically by intact axon terminals at the injection site (Jiang *et al.*, 1993). Iontophoretic injection of tracer has been reported to minimize uptake by fibers of passage, in studies using the fluorescent retrograde tracer Fluorogold, since axonal damage is largely avoided by the thin glass micropipettes used for iontophoresis (Schmued and Heimer, 1990). It may thus be beneficial to use iontophoresis for dextran amine injection if the possibility of uptake by fibers of passage needs to be minimized.

APPENDIX

A. Step-by-Step BDA Protocol for LM Single-Labeling Studies

1. Injection

1. Intact vertebrates are deeply anesthetized and secured in a stereotaxic device or appropriate headholder, while semi-intact preparations (such as a partially dissected embryo or tissue slice) should be secured in an appropriate fashion. The region of interest is exposed, and a microsyringe or micropipette containing BDA is lowered into the target structure using stereotaxic methods or visual guidance.

2. For anterograde labeling using iontophoresis, micropipettes with tip diameters of 10–50 μm are filled with 10–15% BDA10kDa in 0.01 M sodium phosphate buffer (PB; buffers are at physiological pH, i.e., pH 7.2–7.4, unless otherwise noted). BDA10kDa can be injected iontophoretically using 2–5 μA positive current pulses (7 s on and 7 s off) for a period of 30–60 min. Smaller micropipette (2–4 μm tip diameters) and shorter iontophoretic sessions (2–5 min) favor smaller injections, which might be more suitable for small animals, small brain regions, or detailed studies of projection topography (Guirado *et al.*, 1999). During penetration and withdrawal of the pipette, the current is reversed to prevent leakage of tracer along the penetration track.

3. Pressure injections can be made with either a micropipette (20–50 μm tip diameter) or a microsyringe, using 2–15% BDA10kDa in 0.01 M PB for anterograde tracing and 10% BDA3kDa in 0.1 M sodium citrate–HCl (pH 3.0) for retrograde labeling. For pressure injections with a microsyringe, we use 0.01 μl steps per minute until the desired amount (typically 0.05–0.20 μl per injection site) has been injected. We and other investigators have typically pressure injected BDA3kDa (Kaneko *et al.*, 1996; Lei *et al.*, 2004; Medina *et al.*, 1997), but BDA3kDa can be injected iontophoretically as well.

4. Note that it is also possible to place a crystal of BDA directly in or on the desired site, or use fluorescent dextran amines (Nance and Burns, 1990; Pieribone and Aston-Jones, 1993; Schmued *et al.*, 1990).

5. The region overlying the injection site is then closed/sutured and the animal returned to normal housing. Semi-intact preparations are maintained using the standard procedures necessary for their viability (e.g., incubation in oxygenated saline solution or media) for a sufficient time to allow for transport (Veenman *et al.*, 1992).

2. Fixation

1. The postinjection survival period may be as brief as several hours if transport over only a small distance is needed (e.g., peripheral nerve labeling in a small specimen such as a chick embryo) (Veenman *et al.*, 1992) or 7–21 days for labeling over long CNS pathways in an adult vertebrate (Brandt and Apkarian, 1992; Veenman *et al.*, 1992).
2. For fixation of nervous tissue in intact animals for light microscopic studies, the chest cavity is opened after deep anesthesia, heparinized saline (12 mg heparin per ml physiological saline) injected into the heart to prevent blood clotting, and the animal perfused transcardially with 30–50 ml 6% dextran in 0.1 M sodium PB, followed by a fixative consisting of freshly made 4% paraformaldehyde in 0.1 M PB, to which we commonly add 0.1 M lysine and 0.01 M sodium periodate.
3. For small animals, tissues slices or partially dissected preparations, immersion fixation is adequate.
4. For EM grade fixation, freshly made 2% glutaraldehyde in 0.1 M PB, 1.25% glutaraldehyde and 1% paraformaldehyde in 0.1 M PB, 0.5% glutaraldehyde and 3% paraformaldehyde in 0.1 M PB, 0.6% glutaraldehyde, 3.5% paraformaldehyde, and 15% saturated picric acid in PB, or 4% paraformaldehyde in 0.1 M acetate buffer (pH 4.5) followed by 4% paraformaldehyde and 0.05% glutaraldehyde in 0.05 M borate buffer (pH 9.5) are all effective for visualizing BDA-labeled axons or perikarya (Brandt and Apkarian, 1992; Lei *et al.*, 2004; Reiner *et al.*, 2003; Veenman *et al.*, 1992). Perfused tissues can be postfixed.

3. Sectioning

1. The tissue of interest is dissected free as necessary and cryoprotected if it is to be sectioned frozen. We use cryoprotection in 0.1 M PB containing 20% sucrose, 10% glycerol, and 0.02% sodium azide for material to be sectioned frozen on a sliding microtome and 20% sucrose PB for tissue to be sectioned frozen with a cryostat. Tissue should be stored at 4°C until sectioned.
2. Sections are cut with either a sliding microtome, vibrating microtome (i.e., Vibratome®), or cryostat. Sliding microtome or vibrating microtome slices are collected as free-floating sections in 0.1 M PB and stored at 4°C until further processing. Cryostat sections are mounted onto gelatin-coated or Superfrost®/Plus slides, as they are cut and

stored at 4°C until processed on the slide. Gelatin-coated slides are prepared by dipping thoroughly washed slides in 2% gelatin in distilled water, draining them and then baking them at 50°C till dry.

3. Since BDA appears stable in fixed tissue (Brandt and Apkarian, 1992; Veenman *et al.*, 1992), sectioned or unsectioned tissue can be stored for months at 4°C in 0.02% sodium azide and PB prior to processing.

4. BDA Visualization

1. For visualizing BDA, we employ the ABC procedure (Hsu *et al.*, 1981), using the Vectastain ABC Elite kit (Vector Laboratories). With the Elite kit, one drop (50 µl) of avidin DH and one drop (50 µl) of biotinylated HRP are mixed in 2.5–10 ml 0.1 M PB, at least half an hour prior to use.

2. Sections are rinsed three times in 0.1 M PB and incubated in ABC solution in a vial on a rotator or orbital shaker for 30–60 min at room temperature, or overnight at 4°C (Brandt and Apkarian, 1992; Hsu and Soban, 1982; Veenman *et al.*, 1992; Wouterlood and Jorritsma-Byham, 1993).

3. Subsequently, the sections are rinsed several times in buffer, and then the labeling visualized using a brown DAB reaction (Anderson and Reiner, 1990; Hancock, 1986; Hsu and Soban, 1982; Zaborszky and Heimer, 1989) or a metal-intensified DAB procedure for a dark blue-black reaction product (Medina *et al.*, 1997; Veenman *et al.*, 1992), the latter of which tends to be more sensitive and provide better contrast for viewing the BDA labeling.

4. The metal-intensified procedure that we have used involves incubating tissue for 10–15 min in a solution containing 0.05% DAB tetrahydrochloride and 0.04% nickel ammonium sulfate in 0.1 M sodium PB (pH 7.2), followed by an additional 10–15 min of incubation with hydrogen peroxide added to a final concentration of 0.01% (Medina *et al.*, 1997).

5. For processing tissue fixed with a glutaraldehyde-containing fixative, a permeabilization step (e.g., 1 h at room temperature in 0.03% Triton X-100 in 0.1 M PB) carried out prior to the ABC procedure may help increase penetration of the ABC complex without damaging ultrastructure (Veenman *et al.*, 1992; Wouterlood and Jorritsma-Byham, 1993).

6. Investigators can also visualize BDA with a fluorophore-conjugated avidin or a fluorophore-conjugated antibiotin antiserum, although this obviates the advantage (i.e., permanence) of using BDA in single-labeling mapping studies. Finally, peroxidase immunolabeling can be used to detect RDA or FDA, using antibodies against rhodamine or fluorescein, respectively (Kaneko *et al.*, 1996; Reiner *et al.*, 2000, 2003).

5. Mounting and Viewing

1. Labeled free-floating and slide-mounted sections are rinsed, the free-floating sections mounted onto gelatin-coated or Superfrost/Plus slides, and slides then air-dried.
2. All DAB-labeled sections are next dehydrated through an ascending alcohol series to xylene, coverslipped with Permount and examined.
3. DAB-labeled sections can also be counterstained with neutral red, methyl green, or cresyl violet, depending on the color of the DAB reaction product used prior to being coverslipped.
4. In the case of BDA visualized by a fluorescent label or with the use of FDA or RDA, the sections should be rinsed in 0.1 M PB, mounted on gelatin-coated or Superfrost/Plus slides, and coverslipped with 9:1 glycerin:0.05 M carbonate buffer, 9:1 glycerol:PB saline containing *p*-phenylenediamine, or any of a number of commercially available fade-retarding coverslipping media (Anderson and Reiner, 1990; Jiao *et al.*, 2000).

B. Step-by-Step BDA Protocol for LM Double-Labeling Studies

1. Injection and Fixation

1. For multiple-label anterograde or retrograde pathway-tracing studies, all steps from the BDA injection up to the point of tissue processing are the same as for BDA single labeling.
2. It should be noted that if the transport times of the two tracers differ or if the distances from the projection target differ for the two injected substances, the injections need to be carried out at separate times. For example, assuming equal transport distances, markers transported more slowly than BDA (e.g., PHA-L) need to be injected well before the BDA, while markers transported more quickly than BDA (e.g., HRP) need to be injected after the BDA.
3. Tissue fixation and sectioning are as for tissue labeled with dextran amine alone, unless the second tracer constrains which fixative must be used. For example, HRP retrograde tracing requires high glutaraldehyde concentration in the fixative. Conversely, most antigens to be detected by immunolabeling call for little or no glutaraldehyde.

2. Visualization of BDA and Second Marker

1. For differentially visualizing the BDA and the second marker (e.g., a pathway-tracing agent or an antigen detected by immunolabeling) at the transmitted light level, two-color DAB procedures or DAB labeling

combined with VIP should be used (Alisky and Tolbert, 1994; Antal *et al.*, 1990; Hancock, 1986; Hsu and Soban, 1982; Lanciego *et al.*, 1997, 1998a, b, 2000; Lanciego and Gimenez-Amaya, 1999; Medina *et al.*, 1997; Reiner *et al.*, 1993; Veenman *et al.*, 1992).

2. For combining BDA with immunohistochemical labeling, the BDA detection can be carried out before or after the entire immunohistochemical procedure (Lei *et al.*, 2004; Reiner *et al.*, 2004). For example, immunohistochemical labeling of neurons can be performed using a brown DAB reaction, while BDA-containing axons and terminals can be labeled blue/black using a metal-intensified DAB reaction. Similarly, neural structures labeled with one tracer (such as HRP) can be visualized first with a blue/black metal–DAB reaction and BDA-labeled axons visualized second with a brown DAB reaction (Veenman *et al.*, 1992). Alternatively, both BDA and second neuroanatomical tracer can be visualized by immunofluorescence using separate fluorophores (or the dextran amines viewed directly if RDA or FDA is, e.g., used).

3. Other diverse pairs of markers are possible for distinct LM visualization of BDA and a second tracer/marker, such as DAB and silver-intensified immunogold (Anderson *et al.*, 1991; Chan *et al.*, 1990), DAB and BDHC (Anderson *et al.*, 1991; Levey *et al.*, 1986), DAB and a glucose oxidase reaction product (Piekut and Knigge, 1984), or DAB and an alkaline phosphatase reaction product (Falini *et al.*, 1982). Moreover, with the suitable combination of tracers and markers, LM triple and even quadruple labelings are possible (Anderson and Reiner, 1990; Kiss *et al.*, 1988; Lanciego *et al.*, 2000; Luo *et al.*, 2001; Staines *et al.*, 1988; Wessendorf *et al.*, 1990).

C. Step-by-Step BDA Protocol for EM Studies

1. Injection, Fixation, and Sectioning

1. Fibers and terminals that have been anterogradely labeled with BDA10kDa can be visualized at the EM level, as can neurons or their collaterals that have been retrogradely labeled with BDA3kDa (Guirado *et al.*, 1999; Lei *et al.*, 2004; Reiner *et al.*, 2003; Veenman and Reiner, 1996; Wouterlood and Jorritsma-Byham, 1993).

2. For the use of BDA10kDa or BDA3kDa in ultrastructural studies, the BDA injection and the procedures for visualization of BDA are the same as for LM studies. The major difference is that a fixative suitable for preservation of ultrastructural detail must be used. Wouterlood and Jorritsma-Byham (1993) have used freshly made 4% paraformaldehyde, 0.1% glutaraldehyde, and 0.02% picric acid in 0.1 M PB, and reported both good ultrastructural preservation and good BDA labeling. We have successfully used freshly made 0.6% glutaraldehyde, 3.5%

paraformaldehyde, and 15% saturated picric acid in 0.1 M sodium PB (pH 7.3) (Lei *et al.*, 2004; Reiner *et al.*, 2003; Veenman and Reiner, 1996). The upper limit of glutaraldehyde concentration that will be tolerated by BDA is uncertain, but available data indicate that up to 2.5% glutaraldehyde is consistent with good BDA labeling (Veenman *et al.*, 1992; Wouterlood and Jorritsma-Byham, 1993).

3. We routinely postfix our EM tissue overnight in the same fixative as used for perfusion but without glutaraldehyde, and then section it at 50 μm on a vibrating microtome (Lei *et al.*, 2004; Reiner *et al.*, 2003).

2. Visualization of BDA

Penetration of the various reagents during the ABC labeling procedure can be enhanced by freeze-thawing methods or brief 0.03% Triton X-100 treatment (Reiner *et al.*, 1993; Wouterlood and Jorritsma-Byham, 1993). The BDA-labeled tissue can then be osmicated, dehydrated, and plastic embedded by standard procedures.

D. Step-by-Step BDA Protocol for EM Double-Labeling Studies

When combining BDA with another pathway tracer or immunolabeling at the EM level, we recommend DAB labeling together with silver-intensified immunogold labeling, BDHC labeling, VIP labeling, or silver-intensified DAB labeling (Anderson *et al.*, 1991, 1994; Chan *et al.*, 1990; Groenewegen and Wouterlood, 1990; Lei *et al.*, 2004; Levey *et al.*, 1986; Van Haeften and Wouterlood, 2000). Generally, we carry out the BDA labeling, using a nickel-enhanced blue-black DAB reaction product followed by the second labeling (Lei *et al.*, 2004). Note that with the suitable combination of tracers and distinct markers, triple and even quadruple labelings at the EM level are possible (Anderson *et al.*, 1994; Anderson and Reiner, 1990; Groenewegen and Wouterlood, 1990; Kiss *et al.*, 1988; Staines *et al.*, 1988; Van Haeften and Wouterlood, 2000; Wessendorf *et al.*, 1990; Wouterlood *et al.*, 1993).

E. Expected Results

1. Injection Sites—LM

Iontophoretic injections of 10% BDA10kDa or BDA3kDa using micropipette with tip diameters of 20–50 μm yield small injection sites, typically ~500 μm in diameter (Figs. 10.1A and 10.2A), in which dense neuropil labeling and Golgi-like neuronal labeling are observed. The injection sites after pressure injections of 10–15% BDA are less well defined, with necrosis possibly present at the injection core. Use of 2% BDA for pressure injections may reduce such necrosis and yield better-defined injection sites (Naito and Kita, 1994). Finally, BDA can spread along the injection track (Veenman *et al.*, 1995; Wouterlood and Jorritsma-Byham, 1993). Slow delivery from a carefully cleaned pipette or syringe and slow withdrawal of the syringe or micropipette should mitigate this problem (Brandt and Apkarian, 1992; Veenman *et al.*, 1992).

2. Anterograde Labeling—LM

Injections of BDA10kDa, by either iontophoresis or pressure, yield extensive anterograde labeling of axons and terminals (Figs. 10.1 and 10.2). The BDA10kDa-labeled axons/terminals are labeled with great clarity and detail (Figs. 10.1B,D and 10.2B). BDA is transported at least 15–20 mm over a 1-week period. Longer survival times allow transport over even greater distances, and longer survival times may enhance labeling over shorter distances, due to increased cumulative buildup of BDA over time. It is clear that BDA transports relatively quickly, shows little degradation with time, and appears effective for a wide variety of projection systems, including those within the CNS as well as those from the CNS to such peripheral targets as muscle (Veenman *et al.*, 1992).

3. Retrograde Labeling—LM

Retrograde labeling is often seen with BDA10kDa and is routine with BDA3kDa (Figs. 10.1B,C, 10.3A–C, 10.4A,B, 10.5A,B, and 10.6A). For BDA10kDa, the numbers of neurons retrogradely labeled and the loci of the cell groups retrogradely labeled are variable even for injections of the same target nucleus, although a small variable fraction of the retrogradely labeled neurons shows a Golgi-like clarity and extent of labeling (Figs. 10.1C and 10.6A). In general, the use of micropipette with tip diameter less than 40 μm may minimize the amount of retrograde labeling from iontophoretic BDA10kDa injections, and retrograde labeling with pressure injections may be reduced by using a lower concentration of BDA10kDa (e.g., 2%), which still yields excellent anterograde labeling (Reiner *et al.*, 1993). Even when retrograde labeling occurs with BDA10kDa, retrogradely labeled neurons are not necessarily present in all known sources of input to the injected structure (Veenman *et al.*, 1992; Wouterlood and Jorritsma-Byham, 1993). BDA3kDa, by contrast, is a sensitive and dependable retrograde tracer.

4. Double Labeling—LM

Perikarya- or fibers/terminals-labeled brown with DAB using immunohistochemical methods can readily be distinguished from BDA-labeled axons stained blue-black with either nickel–DAB. Similarly, brown BDA-labeled axons can also be distinguished from blue-black immunohistochemically labeled perikarya and fibers in tissue double labeled for BDA and a second marker. In general, both sequences for carrying out the double labeling (i.e., immunohistochemistry then BDA, or BDA then immunohistochemistry) yield distinct two-color labeling, although carrying out the BDA/metal-intensified DAB procedure second can result in darkening of the brown DAB label. Similar results are obtained when combining BDA labeling with another pathway labeling method (e.g., HRP or PHA-L visualized immunohistochemically). In general, it appears critical to the success of the two-color procedure that the first DAB reaction be run to completion, to inactivate the peroxidase and thereby avoiding color mixing during the second DAB-based reaction. Lanciego and coworkers (Lanciego *et al.*, 1997, 1998a, b, 2000; Lanciego and Gimenez-Amaya, 1999) have also shown the efficacy of VIP in combination with DAB for two-color LM labeling.

5. Anterograde and Retrograde Labeling—EM

The characteristics of BDA labeling at the EM level depend on the marker used to visualize the BDA. When DAB is the chromogen used, the distribution and appearance of the DAB reaction product within BDA-labeled structures is indistinguishable from that seen when DAB is used to visualize PHA-L-labeled structures (Veenman and Reiner, 1996; Wouterlood and Jorritsma-Byham, 1993). BDA-labeled axons, terminals, perikarya, dendrites, and spines are uniformly filled with the flocculent DAB reaction product. If other EM markers such as silver-intensified immunogold, BDHC, or VIP are used to visualize the BDA or to visualize a second marker in combination with BDA, the appearance of the labeling will be as in previous studies using these markers (Anderson *et al.*, 1991, 1994; Chan *et al.*, 1990; Groenewegen and Wouterlood, 1990; Levey *et al.*, 1986; Van Haeften and Wouterlood, 2000).

ACKNOWLEDGMENTS. We thank Drs. C. L. Veenman, L. Medina, and Y. Jiao for their contributions to the development of dextran amines as pathway-tracing agents, and S. L. Cuthbertson for her excellent technical assistance in preparing some of the material presented here. Our research is supported by NS-19620 (AR), NS-28721 (AR), EY-05298 (AR), and NS-26386 (MGH).

REFERENCES

Albert, J. S., Yamamoto, N., Yoshimoto, M., Sawai, N., and Ito, H., 1999, Visual thalamotelencephalic pathways in the sturgeon *Acipenser*, a non-teleous actinopterygina fish, *Brain Behav. Evol.* **53**:156–172.

Alisky, J. M., and Tolbert, D. L., 1994, Differential labeling of converging afferent pathways using biotinylated dextran amine and cholera toxin subunit B, *J. Neurosci. Methods* **52**:143–148.

Anderson, K. D., Karle, E. J., and Reiner, A., 1991, Ultrastructural single- and double-label immunohistochemical studies of substance P-containing terminals and dopaminergic neurons in the substantia nigra in pigeons, *J. Comp. Neurol.* **309:**341–362.

Anderson, K. D., Karle, E. J., and Reiner, A., 1994, A pre-embedding triple-label electron microscopic immunohistochemical method as applied to the study of multiple inputs to defined nigral neurons, *J. Histochem. Cytochem.* **42:**49–56.

Anderson, K. D., and Reiner, A., 1990, The extensive co-occurrence of substance P and dynorphin in striatal projection neurons: an evolutionarily conserved feature of basal ganglia organization, *J. Comp. Neurol.* **295:**339–369.

Antal, M., Freund, T. F., Somogyi, P., and McIlhinney, R. A. J., 1990, Simultaneous anterograde labelling of two afferent pathways to the same target area with *Phaseolus vulgaris* leucoagglutinin and *Phaseolus vulgaris* leucoagglutinin conjugated to biotin or dinitrophenol, *J. Chem. Neuroanat.* **3:**1–9.

Aston-Jones, G., and Card, J. P., 2000, Use of pseudorabies virus to delineate multisynaptic circuits in brain: opportunities and limitations, *J. Neurosci. Methods* **103:**51–61.

Brandt, H. M., and Apkarian, A. V., 1992, Biotin-dextran: a sensitive anterograde tracer for neuroanatomic studies in rat and monkey, *J. Neurosci. Methods* **45:**35–40.

Chan, J., Aoki, C., and Pickel, V. M., 1990, Optimization of differential immunogold-silver and peroxidase labeling with maintenance of ultrastructure in brain sections before plastic embedding, *J. Neurosci. Methods* **33:**113–127.

Chang, H. T., 1991, Anterograde transport of Lucifer Yellow-dextran conjugate, *Brain Res. Bull.* **26:**813–816.

Chen, A., and Aston-Jones, G., 1998, Axonal collateral–collateral transport of tract tracers in brain neurons: false anterograde labeling and useful tool, *Neuroscience* **82:**1151–1163.

Coolen, L. M., Jansen, H. T., Goodman, R. L., Wood, R. I., and Lehman, M. N., 1999, A new method for simultaneous demonstration of anterograde and retrograde connections in the brain: co-injections of biotinylated dextran amine and the beta subunit of cholera toxin, *J. Neurosci. Methods* **91:**1–8.

Cowan, R. L., and Wilson, C. J., 1994, Spontaneous firing patterns and axonal projections of single corticostriatal neurons in the rat medial agranular cortex, *J. Neurophysiol.* **71:**17–32.

Cowan, W. M., Gottlieb, D. I., Hendrickson, A. E., Price, J. L., and Woolsey, T. A., 1972, The autoradiographic demonstration of axonal connections in the central nervous system, *Brain Res.* **37:**21–51.

Davila, J. C., Andreu, M. J., Real, M. J., Puelles, L., and Guirado, S., 2002, Mesencephalic and diencephalic afferent connections to the thalamic nucleus rotundus in the lizard, *Psammodromus algirus, Eur. J. Neurosci.* **16:**267–282.

Dolleman-van der Weel, M. J., Wouterlood, F. G., and Witter, M. P., 1994, Multiple anterograde tracing combining *Phaseolus vulgaris* leucoagglutinin with rhodamine- and biotin-conjugated dextran amine, *J. Neurosci. Methods* **51:**9–21.

Edwards, S. B., 1972, The ascending and descending projections of the red nucleus in the cat: an experimental study using an autoradiographic tracing method, *Brain Res.* **48:**45–63.

Falini, B., De Solas, I., Halverson, C., Parker, J. W., and Taylor, C. R., 1982, Double labeled-antigen method for demonstration of intracellular antigens in paraffin-embedded tissue, *J. Histochem. Cytochem.* **30:**21–26.

Faulkner, B., Brown, T. H., and Evinger, C., 1997, Identification and characterization of rat orbicularis oculi motoneurons using confocal laser scanning microscopy, *Exp. Brain Res.* **116:**10–19.

Ferguson, I. A., Xian, C., Barati, E., and Rush, R. A., 2001, Comparison of wheat germ agglutinin-horseradish peroxidase and biotinylated dextran for anterograde tracing of corticospinal tract following spinal cord injury, *J. Neurosci. Methods* **109:**81–89.

Fink, R. P., and Heimer, L., 1967, Two methods for selective silver impregnation of degenerating axons and their synaptic endings in the central nervous system, *Brain Res.* **4:**369–374.

Fritzsch, B., 1993, Fast axonal diffusion of 3000 molecular weight dextran amines, *J. Neurosci. Methods* **50**:95–103.

Fritzsch, B., and Wilm, C., 1990, Dextran amines in neuronal tracing, *Trends Neurosci.* **13**:14.

Gerfen, C. R., and Sawchenko, P. E., 1984, An anterograde neuroanatomical tracing method that shows the detailed morphology of neurons, their axons and terminals: immunohistochemical localization of an axonally transported plant lectin, *Phaseolus vulgaris* leucoagglutinin (PHA-L), *Brain Res.* **290**:219–238.

Gerfen, C. R., Sawchenko, P. E., and Carlsen, J., 1989, The PHA-L anterograde axonal tracing method, In: Heimer, L., and Zaborszky, L. (eds.), *Neuroanatomical Tract-Tracing Methods II, Recent Progress*, New York: Plenum Publishing Corporation, pp. 19–47.

Glover, J. C., Petursdottir, G., and Jansen, K. S., 1986, Fluorescent dextran-amines used as axonal tracers in the nervous system of the chicken embryo, *J. Neurosci. Methods* **18**:243–254.

Gonzalo, N., Moreno, A., Erdozain, M. A., Garcia, P., Vazquez, A., Castle, M., and Lanciego, J. L., 2001, A sequential protocol combining dual neuroanatomical tract-tracing with visualization of local circuit neurons within the striatum, *J. Neurosci. Methods* **111**:59–66.

Groenewegen, H. J., and Wouterlood, F. G., 1990, Light and electron microscopic tracing of neuronal connections with *Phaseolus vulgaris*-leucoagglutinin (PHA-L), and combinations with other neuroanatomical techniques, In: Bjorklund, A., Hokfelt, T., Wouterlood, F. G., and van den Pol, A. N. (eds.), *Handbook of Chemical Neuroanatomy. Volume 8. Analysis of Neuronal Microcircuits and Synaptic Interactions*, Amsterdam: Elsevier Science Publishers, pp. 47–124.

Guirado, S., Real, M. A., Davila, J. C., and Medina, L., 1999, The nucleus accumbens in the lizard *Psammodromus algirus*: chemoarchitecture and cortical afferent connections, *J. Comp. Neurol.* **405**:15–31.

Hancock, M. B., 1986, Two-color immunoperoxidase staining: visualization of anatomic relationships between immunoreactive neural elements, *Am. J. Anat.* **175**:343–352.

Haugland, R. P., 1996, *Handbook of Fluorescent Probes and Research Chemicals*, Eugene, OR: Molecular Probes Inc.

Hendrickson, A., and Edwards, S. B., 1978, The use of axonal transport for autoradiographic tracing of pathways in the central nervous system, In: Thompson, R. F., and Robertson, R. T. (eds.), *Methods in Physiological Psychology. Volume 2. Neuroanatomical Research Techniques*, New York: Academic Press, pp. 241–290.

Hsu, S. M., Raine, L., and Fanger, H., 1981, Use of avidin–biotin–peroxidase complex (ABC) in immunoperoxidase techniques. A comparison between ABC and unlabeled antibody (PAP) procedures, *J. Histochem. Cytochem.* **29**:577–580.

Hsu, S. M., and Soban, E., 1982, Color modification of diaminobenzidine (DAB) precipitation by metallic ions and its application for double immunohistochemistry, *J. Histochem. Cytochem.* **30**:1079–1082.

Izzo, P. N., 1991, A note on the use of biocytin in anterograde tracing studies in the central nervous system: applications at both light and electron microscopic level, *J. Neurosci. Methods* **36**:155–166.

Jiang, X., Johnson, R. R., and Burkhalter, A., 1993, Visualization of dendritic morphology of cortical projection neurons by retrograde axonal tracing, *J. Neurosci. Methods* **50**:45–60.

Jiao, Y., Medina, L., Veenman, C. L., Toledo, C., Puelles, L., and Reiner, A., 2000, Identification of the anterior nucleus of the ansa lenticularis in birds as the homologue of the mammalian subthalamic nucleus, *J. Neurosci.* **20**:6998–7010.

Kaneko, T., Saeki, K., Lee, T., and Mizuno, N., 1996, Improved retrograde axonal transport and subsequent visualization of tetramethylrhodamine (TMR)dextran amine by means of an acidic injection vehicle and antibodies against TMR, *J. Neurosci. Methods* **65**:157–165.

Kenigfest, N. B., Belekhova, M. G., Reperant, J., Rio, J. P., Vesselkin, N. P., and Ward, R., 2000, Pretectal connections in turtles with special reference to the visual thalamic centers: a hodological and gamma-aminobutyric acid-immunohistochemical study, *J. Comp. Neurol.* **426**:31–50.

Kincaid, A. E., and Wilson, C. J., 1996, Corticostriatal innervation of the patch and matrix in the rat neostriatum, *J. Comp. Neurol.* **374**:578–592.

King, M. A., Louis, P. M., Hunter, B. E., and Walker, D. W., 1989, Biocytin: a versatile anterograde neuroanatomical tract-tracing alternative, *Brain Res.* **497**:361–367.

Kiss, A., Palkovits, M., and Skirboll, L. R., 1988, Light microscopic triple-colored immunohistochemical staining on the same vibratome section using the avidin–biotin–peroxidase complex technique, *Histochemistry* **88**:353–356.

Kita, H., and Armstrong, W., 1991, A biocytin-containing compound *N*-(2-aminoethyl) biotinamide for intracellular labeling and neuronal tracing studies: comparison with biocytin, *J. Neurosci. Methods* **37**:141–150.

Lanciego, J. L., and Gimenez-Amaya, J. M., 1999, Notes on the combined use of V-VIP and DAB peroxidase substrates for the detection of colocalizating antigens, *Histochem. Cell Biol.* **111**:305–311.

Lanciego, J. L., Goede, P. H., Witter, M. P., and Wouterlood, F. G., 1997, Use of peroxidase substrate Vector® VIP for multiple staining in light microscopy, *J. Neurosci. Methods* **74**:1–7.

Lanciego, J. L., Luquin, M. R., Guillen, J., and Gimenez-Amaya, J. M., 1998a, Multiple neuroanatomical tracing in primates, *Brain Res. Protoc.* **2**:323–332.

Lanciego, J. L., and Wouterlood, F. G., 1994, Dual anterograde axonal tracing with *Phaseolus vulgaris* leucoagglutinin (PHA-L) and biotinylated dextran amine (BDA), *Neurosci. Protoc.* 94-050-06.

Lanciego, J. L., Wouterlood, F. G., Erro, E., Arribas, J., Gonzalo, N., Urra, X., Cervantes, X., and Gimenez-Amaya, J. M., 2000, Complex brain circuits studied via simultaneous and permanent detection of three transported neuroanatomical tracers in the same histological section, *J. Neurosci. Methods* **103**:127–135.

Lanciego, J. L., Wouterlood, F. G., Erro, E., and Gimenez-Amaya, J. M., 1998b, Multiple axonal tracing: simultaneous detection of three tracers in the same section, *Histochem. Cell Biol.* **110**:509–515.

Lanuza, E., Belekhova, M., Martinez-Marcos, A., Font, C., and Martinez-Garcia, F., 1998, Identification of the reptilian basolateral amygdala: an anatomical investigation of the afferents to the posterior dorsal venricular ridge of the lizard *Podarcis hispanica, Eur. J. Neurosci.* **20**:3517–3534.

Lapper, S. R., and Bolam, J. P., 1991, The anterograde and retrograde transport of neurobiotin in the central nervous system of the rat: comparison with biocytin, *J. Neurosci. Methods* **39**:163–174.

Lei, W. L., Jiao, Y., Del Mar, N., and Reiner, A., 2004, Evidence for differential cortical input to direct pathway versus indirect pathway striatal projection neurons in rats, *J. Neurosci.* **24**:8289–8299.

Levey, A. I., Bolam, J. P., Rye, D. B., Hallanger, A. E., Demuth, R. M., Mesulam, M. M., and Wainer, B. H., 1986, A light and electron microscopic procedure for sequential double antigen localization using diaminobenzidine and benzidine dihydrochloride, *J. Histochem. Cytochem.* **34**:1449–1457.

Liu, W. L., Behbehani, M. M., and Shipley, M. T., 1993, Intracellular filling in fixed brain slices using miniruby, a fluorescent biocytin compound, *Brain Res.* **608**:78–86.

Lopes-Correa, S. A., Grant, K., and Hoffmann, A., 1998, Afferent and efferent connections of the dorsolateral telencephalon in an electrosensory teleost, *Gymnotus carapo, Brain Behav. Evol.* **52**:81–98.

Luo, P., Haines, A., and Dessem, D., 2001, Elucidation of neuronal circuitry: protocol(s) combining intracellular labeling, neuroanatomical tracing and immunocytochemical methodologies, *Brain Res. Protoc.* **7**:222–234.

Medina, L., and Reiner, A., 1997, The efferent projections of the dorsal and ventral pallidal parts of the pigeon basal ganglia, studied with biotinylated dextran amine, *Neuroscience* **81**:773–802.

Medina, L., Veenman, C. L., and Reiner, A., 1997, Evidence for a possible avian dorsal thalamic region comparable to the mammalian ventral anterior, ventral lateral, and oral ventroposterolateral nuclei, *J. Comp. Neurol.* **384**:86–108.

Naito, A., and Kita, H., 1994, The cortico-pallidal projection in the rat: an anterograde tracing study with biotinylated dextran amine, *Brain Res.* **653**:251–257.

Nance, D. M., and Burns, J., 1990, Fluorescent dextrans as sensitive anterograde neuroanatomical tracers: applications and pitfalls, *Brain Res. Bull.* **25**:139–145.

Naumann, T., Härtig, W., and Frotscher, M., 2000, Retrograde tracing with Fluoro-Gold: different methods of tracer detection at ultrastructural level and neurodegenerative changes of back-filled neurons in long-term studies. *J. Neurosci. Methods* **103**:11–21.

Nauta, W. J. H., and Gygax, P. A., 1954, Silver impregnation of degenerating axons in the CNS. A modified technique, *Stain Technol.* **29**:91–93.

Novikov, L. N., 2001, Labeling of central projections of primary afferents in adult rats: a comparison between biotinylated dextran amine, neurobiotin® and *Phaseolus vulgaris*-leucoagglutinin, *J. Neurosci. Methods* **112**:145–154.

Novikova, L., Novikov, L., and Kellerth, J. O., 1997, Persistent neuronal labeling by retrograde fluorescent tracers: a comparison between Fast Blue, Fluoro-Gold and various dextran conjugates, *J. Neurosci. Methods* **74**:9–15.

Ojima, H., and Takayanagi, M., 2001, Use of two anterograde axon tracers to label distinct cortical populations located in close proximity, *J. Neurosci. Methods* **104**:177–182.

Peters, A., 1985, The visual cortex of the rat, In: Peters, A., and Jones, E. G. (eds.), *Cerebral Cortex. Volume 3. Visual Cortex,* New York: Plenum Publishing Corporation, pp. 19–80.

Piekut, D. T., and Knigge, K. M., 1984, Relationship of alpha MSH-specific neurons to the arcuate opiocortin neuronal system as determined by dual antigen immunocytochemical procedures, *Peptides* **5**:1089–1095.

Pieribone, V. A, and Aston-Jones, G., 1993, The iontophoretic application of Fluoro-Gold for the study of afferents to deep brain nuclei. *Brain Res.* **607**:47–53.

Pombal, M. A., Yanez, J., Marin, O., Gonzalez, A., and Anadon, R., 1999, Cholinergic and GABAergic neuronal elements in the pineal organ of lampreys, and tract-tracing observations of differential connections of pinealofugal neurons, *Cell Tissue Res.* **295**:215–223.

Rajakumar, N., Elisevich, K., and Flumerfelt, B. A., 1993, Biotinylated dextran: a versatile anterograde and rerograde neuronal tracer, *Brain Res.* **607**:47–53.

Reiner, A., Jiao, Y., Del Mar, N., Laverghetta, A. V., and Lei, W. L., 2003, Differential morphology of pyramidal-tract type and intratelencephalically-projecting type corticostriatal neurons and their intrastriatal terminals in rats, *J. Comp. Neurol.* **457**:420–440.

Reiner, A., Laverghetta, A. V., Meade, C. A., Cuthbertson, S. L., and Bottjer, S. W., 2004, An immunohistochemical and pathway tracing study on the striatopallidal organization of area X the in male zebra finch, *J. Comp. Neurol.* **469**:239–261.

Reiner, A., Veenman, C. L., and Honig, M. G., 1993, Anterograde tracing using biotinylated dextran amine, *Neurosci. Protoc.* 93-050-14.

Reiner, A., Veenman, C. L., Medina, L., Jiao, Y, Del Mar, N., and Honig, M. G., 2000, Pathway tracing using biotinylated dextran amines, *J. Neurosci. Methods* **103**:23–37.

Reiner, A., Zhang, D., and Eldred, W. D., 1996, Use of cholera toxin tracer reveals new details of the central retinal projections in turtles, *Brain Behav. Evol.* **48**:307–337.

Richmond, F. J, Gladdy, R., Creasy, J. L., Kitamura, S., Smits, E., and Thomson, D. B., 1995, Efficacy of seven retrograde tracers, compared in multiple-labeling studies of feline motoneurones, *J. Neurosci. Methods* **53**:35–46.

Scalia, F., Galoyan, S. M., Eisner, S., Haris, E., and Su, W., 1997, Biotinylated dextran amine and biocytin hydrochloride are useful tracers for the study of retinal projections in the frog, *J. Neurosci. Methods* **76**:167–175.

Schmued, L. C., Beltramino, C., and Slikker, W., Jr., 1993, Intracranial injection of Fluoro-Gold results in the degeneration of local but not retrogradely labeled neurons, *Brain Res.* **626**:71–77.

Schmued, L. C., and Fallon, J. H., 1986, Fluoro-Gold: a new fluorescent retrograde axonal tracer with numerous unique properties, *Brain Res.* **377**:147–154.

Schmued, L. C., and Heimer, L., 1990, Iontophoretic injection of fluoro-gold and other fluorescent tracers, *J. Histochem. Cytochem.* **38**:721–723.

Schmued, L., Kyriakidis, K., and Heimer, L., 1990, In vivo anterograde and retrograde axonal transport of the fluorescent rhohamine-dextran-amine, Fluoro-Ruby, within the CNS, *Brain Res.* **526:**127–134.

Sesack, S. R., Miner, L. H., and Omelchenko, N., 2006, Pre-embedding immunoelectron microscopy: applications for studies of the nervous system, In: Zaborszky, L., Wouterlood, F. G., and Lanciego, J. L. (eds.), *Neuroanatomical Tract-Tracing Methods 3: Molecules–Neurons–Systems*, New York: Springer/Kluwer/Plenum Publishers, pp. xxx–xxx..

Shimizu, T., Cox, K., Karten, H. J., and Britto, L. R. G., 1994, Cholera toxin mapping of retinal projections in pigeons (*Columba livia*), with emphasis on retinohypothalamic connections, *Vis. Neurosci.* **11:**441–446.

Shu, S. Y., and Peterson, G. M., 1988, Anterograde and retrograde axonal transport of *Phaseolus vulgaris* leucoagglutinin (PHA-L) from the globus pallidus to the striatum of the rat, *J. Neurosci. Methods* **25:**175–180.

Staines, W. A., Meister, B., Melander, T., Nagy, J. I., and Hokfelt, T., 1988, Three-color immunofluorescence histochemistry allowing triple labeling within a single section, *J. Histochem. Cytochem.* **36:**145–151.

Striedter, G. F., 1994, The vocal control pathways in budgerigars differ from those in songbirds, *J. Comp. Neurol.* **343:**35–56.

Tellegen, A. J, Arends, J. J., and Dubbeldam, J. L., 2001, The vestibular nuclei and vesitibuloreticular connections in the mallard (*Anas platyrhynchos* L.). An anterograde and retrograde tracing study, *Brain Behav. Evol.* **58:**205–217.

Todorova, N., and Rodziewicz, G. S., 1995, Biotin-dextran: fast retrograde tracing of sciatic nerve motoneurons, *J. Neurosci. Methods* **61:**145–150.

Torres, B., Perez-Perez, M. P., Herrero, L., Ligero, M., and Nunez-Abades, P. A., 2002, Neural substrata underlying tectal eye movement codification in goldfish, *Brain Res. Bull.* **57:**345–348.

Van Den Pol, A. N., and Decavel, C., 1990, Synaptic interactions between chemically defined neurons: dual ultrastructural immunocytochemical approaches, In: Bjorklund, A., Hokfelt, T., Wouterlood, F. G., and Van Den Pol, A. N. (eds.), *Handbook of Chemical Neuroanatomy. Volume 8. Analysis of Neuronal Microcircuits and Synaptic Interactions*, Amsterdam: Elsevier Science Publishers, pp. 199–271.

Van Haeften, T., and Wouterlood, F. G., 2000, Neuroanatomical tracing at high resolution, *J. Neurosci. Methods* **103:**107–116.

Veenman, C. L., and Reiner, A., 1996, Ultrastructural morphology of synapses formed by corticostriatal terminals in the avian striatum, *Brain Res.* **707:**1–12.

Veenman, C. L., Reiner, A., and Honig, M. G., 1992, Biotinylated dextran amine as an anterograde tracer for single- and double-label studies, *J. Neurosci. Methods* **41:**239–254.

Veenman, C. L., Wild, J. M., and Reiner, A., 1995, Organization of the avian "corticostriatal" projection system: a retrograde and anterograde pathway tracing study in pigeons, *J. Comp. Neurol.* **354:**87–126.

Vercelli, A., Repici, M., Garbossa, D., and Grimaldi, A., 2000, Recent techniques for tracing pathways in the central nervous system of developing and adult mammals, *Brain Res. Bull.* **51:**11–28.

Wang, Y., Major, D. E., and Karten, H. J., 2004, Morphology and connections of nucleus isthmi pars magnocellularis in chicks (*Gallus domesticus*), *J. Comp. Neurol.* **469:**275–297.

Wessendorf, M. W., Appel, N. M., Molitor, T. W., and Elde, R. P., 1990, A method for immunofluorescent demonstration of three co-existing neurotransmitters in rat brain and spinal cord, using the fluorophores fluorescein, lissamine rhodamine and 7-amino-4-methylcoumarin-3-acetic acid, *J. Histochem. Cytochem.* **38:**1859–1877.

Wilson, C. J., 1987, Morphology and synaptic connections of crossed corticostriatal neurons in the rat, *J. Comp. Neurol.* **263:**567–580.

Woodson, W., Reiner, A., Anderson, K. D., and Karten, H. J., 1991, The distribution, laminar location and morphology of tectal neurons projecting to the isthmo-optic nucleus and the nucleus isthmi, pars parvocellularis in the pigeon (*Columba livia*) and chick (*Gallus domesticus*): a retrograde labeling study, *J. Comp. Neurol.* **305:**470–488.

Wouterlood, F. G., and Jorritsma-Byham, B., 1993, The anterograde neuroanatomical tracer biotinylated dextran amine: comparison with the tracer *Phaseolus vulgaris* leucoagglutinin in preparations for electron microscopy, *J. Neurosci. Methods* **48**:75–87.

Wouterlood, F. G., Pattiselanno, A., Jorritsm-Byham, B., Arts, M. P. M., and Meredith, G. E., 1993, Connectional, immunocytochemical and ultrastructural characterization of neurons injected intracellularly in fixed brain tissue, In: Meredith, G. E., and Arbuthnott, G. W. (eds.), *Morphological Investigations of Single Neurons in Vitro*, New York: John Wiley and Sons, pp. 47–169.

Wright, A. K., Norrie, L., Ingham, C. A., Hutton, A. M., Arbuthnott, G. W., 1999, Double anterograde tracing of the outputs from adjacent "barrel columns" of rat somatosensory cortex neostriatal projection patterns and terminal ultrastructure, *Neuroscience* **88**:119–133.

Wright, A. K., Ramanthan, S., and Arbuthnott, G. W., 2001, Identification of the source of the bilateral projection system from cortex to somatosensory neostriatum and an exploration of its physiological actions, *Neuroscience* **103**:87–96.

Xue, H. G., Yamamoto, N., Yoshimoto, M., Yang, C. Y., and Ito, H., 2001, Fiber connections of the nucleus isthmian in the carp (*Cyprinus carpio*) and tilapia (*Oreochromis niloticus*), *Brain Behav. Evol.* **58**:185–204.

Zaborszky, L., and Heimer, L., 1989, Combination of tracer techniques, especially HRP and PHA-L, with transmitter identification for correlated light and electron microscopic studies, In: Heimer, L., and Zaborszky, L. (eds.), *Neuroanatomical Tract-Tracing Methods II, Recent Progress*, New York: Plenum Publishing Corporation, pp. 173–199.

Zhou, M., and Grofova, I., 1995, The use of peroxidase substrate Vector VIP in electron microscopic single and double antigen localization, *J. Neurosci. Methods* **62**:149–158.

11

Multiple Neuroanatomical Tract-Tracing: Approaches for Multiple Tract-Tracing

JOSÉ L. LANCIEGO and FLORIS G. WOUTERLOOD

JOSÉ L. LANCIEGO • Neuromorphology-Tracing Lab, Neurosciences Division, Center for Applied Medical Research (C.I.M.A.), University of Navarra Medical School, Pio XII Avenue, 31008 Pamplona, Spain   FLORIS G. WOUTERLOOD • Department of Anatomy, Vrije Universiteit Medical Center, Rm MF-G-136, P.O. Box 7057, 1007 MB Amsterdam, The Netherlands

Abstract: Experimental neuroanatomical tracing techniques are fundamental to the study of the structure of the central nervous system. In the last few decades, many new methods for axonal tracing and cell labeling have been introduced. Neuroanatomical tracing applied as an isolated method produces relatively straightforward answers, for instance, whether there is connectivity from compartment Y in nucleus A to layer X in area B. However, questions that deal with the intrinsic complexity of brain circuits require the application of multiple-tracing paradigms in which two or even three different tracers are combined in single histological sections. With such paradigms we can handle questions like "are the fibers arriving in layer X of area B in contact with neurons that project to compartment Z in nucleus C," "do these projection neurons receive as well innervation from area W," and "what is the neurochemical signature of these connectivity-identified neurons?" We illustrate this approach with examples from our studies on pallidonigral connectivity in association with nigrostriatal efferent neurons.

Analysis of the data acquired via a multiple-tracing approach provides more insight into the organization of the brain than does the analysis of data from single tracing, especially when it comes to network circuitry. Furthermore, by virtue of the simultaneous visualization of projections in the same section, these multiple techniques enable the precise determination of the degree of convergence or divergence of particular projections to a particular terminal zone or to particular neurons (the latter to be identified via retrograde tracing or via neurotransmitter immunocytochemistry). An additional advantage of multitracer methods is that the experimental animals can be most efficiently used and the number of used animals reduced.

In this chapter we will discuss in detail several existing protocols for the simultaneous detection of three different tracers, as well as methods in which we combine two tracers and the immunocytochemical detection of a neuroactive substance. Emphasis will be placed on providing a step-by-step account of each procedure. We will be dealing with peroxidase substrates and precipitates with different colors since these precipitates are persistent without specific storage measures and because at the end of the staining procedure the ensuing slides can be studied any time under any routine microscope.

Keywords: cholera toxin, dextran amines, Fluoro-Gold, immunocytochemistry, peroxidase, triple staining

I. TRACT-TRACING METHODS: HISTORICAL PERSPECTIVE

A. Retrograde and Anterograde Tracers

Modern neuroanatomical tracing is based on axonal flow, an inherent physiological transport mechanism in neurons first reported in 1948 by Weiss and Hiscoe. Depending on the direction of transport, neuroanatomical tracers are divided into two main groups, i.e., retrograde and anterograde tracers. Following injection into the brain, a retrograde tracer is taken up by axon terminals, incorporated in transport vesicles, and then transported back to the parent cell bodies. As a result of the delivery of a retrograde tracer to a given brain area, the somata of neurons that project to that area become labeled. However, in most cases the accumulation of retrogradely transported tracer is restricted to the cell somata and the main, thickest dendrites. Anterograde tracers, by contrast, are taken up by the cell somata or dendrites, and they are transported from the cell bodies into the axons, where they can finally be detected in the most distal arborizations and terminal boutons. The successful application of a sensitive anterograde tracer results in the staining of the cell somata and complete dendritic trees of the labeled cells, and in addition in the staining of the entire axonal configurations of the labeled neurons including their collateral branches and appendages, their terminal fields, arborizations, rosettes, and terminal boutons.

B. Previous and Current Generations of Tracing Methods

The previous generation of tracing methods, described in the first edition of *Neuroanatomical Tract-Tracing Methods*, included silver-degeneration methods (de Olmos *et al.*, 1981), Golgi silver impregnation (Blackstad, 1981; Millhouse, 1981), and the autoradiographic method capitalizing on the uptake (by neurons) of injected, radioactively labeled amino acids, their incorporation in proteins, and subsequent anterograde transport (Edwards and Hendrickson, 1981). As these methods were fully developed at the time when the first edition of *Neuroanatomical Tract-Tracing Methods* was published, we will not further discuss them here. The current generation of tracing methods includes the tracers discussed briefly below.

C. Horseradish Peroxidase and its Conjugates

The current generation of tracing methods has its origin in the publication by Kristensson and Olson in 1971, describing the uptake and retrograde axonal transport of the glycoprotein enzyme horseradish peroxidase (HRP). Transported HRP is histochemically visualized using the enzymatic electron transfer from the substrate diaminobenzidine (DAB; Sigma) to hydrogen peroxide through which a brown, insoluble DAB polymer is formed (Graham and Karnovsky, 1966; LaVail, 1975; see also Warr *et al.*, 1981). Improvements and refinements were added soon to the HRP transport technique (LaVail and LaVail, 1972; Mesulam, 1976, 1978, 1982). The enzyme was conjugated with the plant lectin wheat germ agglutinin (WGA; WGA-HRP; Gonatas *et al.*, 1979), generating a highly effective and sensitive bidirectional tracer (i.e., transported both anterogradely and retrogradely). Soon it was discovered that WGA-HRP could be transferred from one neuron to another neuron at the synaptic junction (the so-called transsynaptic transport).

D. Cholera Toxin Subunit B

A decade after the development of the HRP method, the β subunit of cholera toxin (CTB; List Biological Laboratories, Campbell, CA) was introduced as a substance transported both anterogradely and retrogradely (Trojanowski *et al.*, 1981). Like HRP, CTB works highly effectively both in the peripheral and in the central nervous systems (Stoeckel *et al.*, 1977). CTB has a general performance as a retrograde tracer that is superior to that of HRP (Trojanowski *et al.*, 1981, 1982; Wan *et al.*, 1982). Visualization of transported CTB can easily be combined with other immunocytochemical procedures, such as neuropeptide or neurotransmitter detection (Luppi *et al.*, 1987, 1990), and with other anterograde tracing techniques (Bruce and Grofova, 1992; Vetter *et al.*, 1993). Several procedures have been developed to further improve the performance of CTB in anterograde tracing experiments (Angelucci *et al.*, 1996).

E. Fluorescent Dyes

Parallel to the development of the HRP and CTB methods, retrograde tracing techniques were refined utilizing uptake and transport of fluorescent dyes: Evans Blue, Fast Blue, Diamidino Yellow, Propidium Iodide, True Blue, Nuclear Yellow, and Lucifer Yellow to mention just a few (for a review, see Akintunde and Buxton, 1992; Kuypers and Huisman, 1984). Soon the major advantages of fluorescence-based retrograde tracing were recognized by neuroscientists: fast, highly productive, compatibility with chemical characterization of neurons, and compatibility with anterograde and retrograde neuroanatomical tracing (Skirboll *et al.*, 1989). One of the advantages of fluorescence methods was also that the study of axonal collateralization became easier than with the silver methods. The rapid development in confocal laser scanning microscopy and in parallel that of new, bright, and stable fluorochromes have been instrumental for a renaissance in fluorescence tracing and chemical identification methods, beginning in the 1990s. Reliable detection of multiple markers in small structures, such as fibers and axon terminals, has been made possible by this instrument (Wouterlood, this volume).

The hydroxystilbamidine derivative known as Fluoro-Gold (FG; Fluorochrome) was first introduced as a retrograde tracer by Schmued and Fallon in 1986. This fluorescent dye has a peak excitation at 325 nm, with a peak emission at 440 nm. FG is widely used in tracing neuronal connections, not only due to its excellent stable fluorescence but also because of the simplicity of its application, its exclusive retrograde transport, the lack of uptake by intact or damaged fibers traversing the injection site (especially when performing iontophoretic delivery of the dye), and the elaborate filling of neuronal somata and main dendrites (Akintunde and Buxton, 1992; Divac and Mogensen, 1990; Novikova *et al.*, 1997; Pieribone and Aston-Jones, 1988; Schmued, 1994; Schmued and Heimer, 1990; Wessendorf, 1991). Since FG in itself is not electron dense, it is without histological processing not immediately suitable for ultrastructural visualization. Photoconversion of the fluorescence (Balercia *et al.*, 1992; Bentivoglio and Su, 1990), or an antibody against FG, is used to aggregate electron-dense material at loci where FG has been transported (Chang *et al.*, 1990; Van Bockstaele *et al.*, 1994). Hence, FG today has the added advantage of being visible both under the fluorescence microscope and under the confocal laser-scanning microscope, while it is compatible with biotinylated dextran amine (BDA; Molecular Probes) tracing and a host of other immunofluorescence methods. Finally, FG allows downstream processing for electron microscopic visualization (Köbbert *et al.*, 2000; Lanciego *et al.*, 1997, 1998a, b; Lanciego and Giménez-Amaya, 1999).

Until WGA-HRP tracing became available as a reliable anterograde tracer (Trojanowski *et al.*, 1981) and was recognized as being compatible with other tracing methods (Zaborszky *et al.*, 1984), combined anterograde–retrograde tracing methods hinged on anterograde degeneration as the only effective tracing method applicable in a common histological laboratory (immunocytochemistry combined with anterograde autoradiographic tracing being the province of a handful of highly specialized laboratories). Since silver-degeneration methods offer little compatibility with other light microscopic tracing methods, combined methods including anterograde tracing had been developed mainly for ultrastructural research (e.g., Blackstad, 1965, 1981; Somogyi *et al.*, 1979; Wouterlood *et al.*, 1985; Zaborszky and Cullinan, 1989).

F. *Phaseolus vulgaris* Leucoagglutinin and Biotinylated Dextran Amine

In the mid-1980s, the plant lectin *Phaseolus vulgaris* leucoagglutinin (PHA-L) was introduced as an anterograde neuroanatomical tracer (Gerfen and Sawchenko, 1984; Ter Horst *et al.*, 1984). The unique properties of this tracer include nearly exclusive anterograde transport and little uptake by fibers of passage. Furthermore, upon labeling all exquisite details of the labeled neurons are visible, and the tracer can be visualized in electron microscopy preparations (Groenewegen and Wouterlood, 1990; Wouterlood *et al.*, 1990; Wouterlood and Groenewegen, 1985, 1991). Finally, PHA-L tracing can rather easily be combined with a number of other methods. Gerfen and Sawchenko (1985) pioneered in this respect by combining PHA-L tracing with tyrosine hydroxylase (TH) immunofluorescence. Once set, this example was soon followed by others(Cullinan and Zaborszky, 1991; Gaykema *et al.*, 1991; Woolf *et al.*, 1986; Wouterlood *et al.*, 1987; Zaborszky and Cullinan, 1989). Innovations were also introduced, for example, by Antal *et al.* (1990) who demonstrated that conjugates of PHA-L with biotin and dinitrophenol are taken up like native PHA-L and transported anterogradely. Their work opened the avenue for the study of convergence or divergence to brain areas of axonal projections originating from different sources. PHA-L tracing combined with intracellular injection of Lucifer Yellow in neurons in sections of fixed brain labeled by retrograde transport of a fluorescent tracer was described by Wouterlood *et al.* (1992). Lanciego and Wouterlood (1994) showed the feasibility of PHA-L tracing in one preparation together with retrograde tracing, neurotransmitter immunocytochemistry, and with anterograde tracing using BDA. Biotinylated compounds and dextran amines had been introduced in the early 1990s (Fritzsch and Wilm, 1990; Schmued *et al.*, 1990), yet the breakthrough in the form of BDA arrived with the publication by Veenman *et al.* in 1992 (see chapter by Reiner and Honig in this volume). Through the years, BDA has become more popular than PHA-L as the tracer of choice, mainly because of simpler procedures, more stable results, and better properties, for instance, if one deals with electron microscopic issues.

G. Viruses

Viruses offer an instrument for tracing peripheral and central nervous connectivity based on retrograde transport, multiplication by overtaking metabolism, trans-synaptic transfer of new virions, transport, and so on. The unique aspect of using virus as a tracer is its built-in multiplication. In the chapter by Loewy *et al.* in this volume, the details of tracing using virus are fully discussed.

II. PARADIGMS AND SELECTION OF TRACERS FOR COMBINED USE

A. The Complexity of the Brain Requires Multiple Tracing

Complexity is an inherent feature of brain function. The anatomical study of complex brain circuits requires the thoughtful design of tracing strategies.

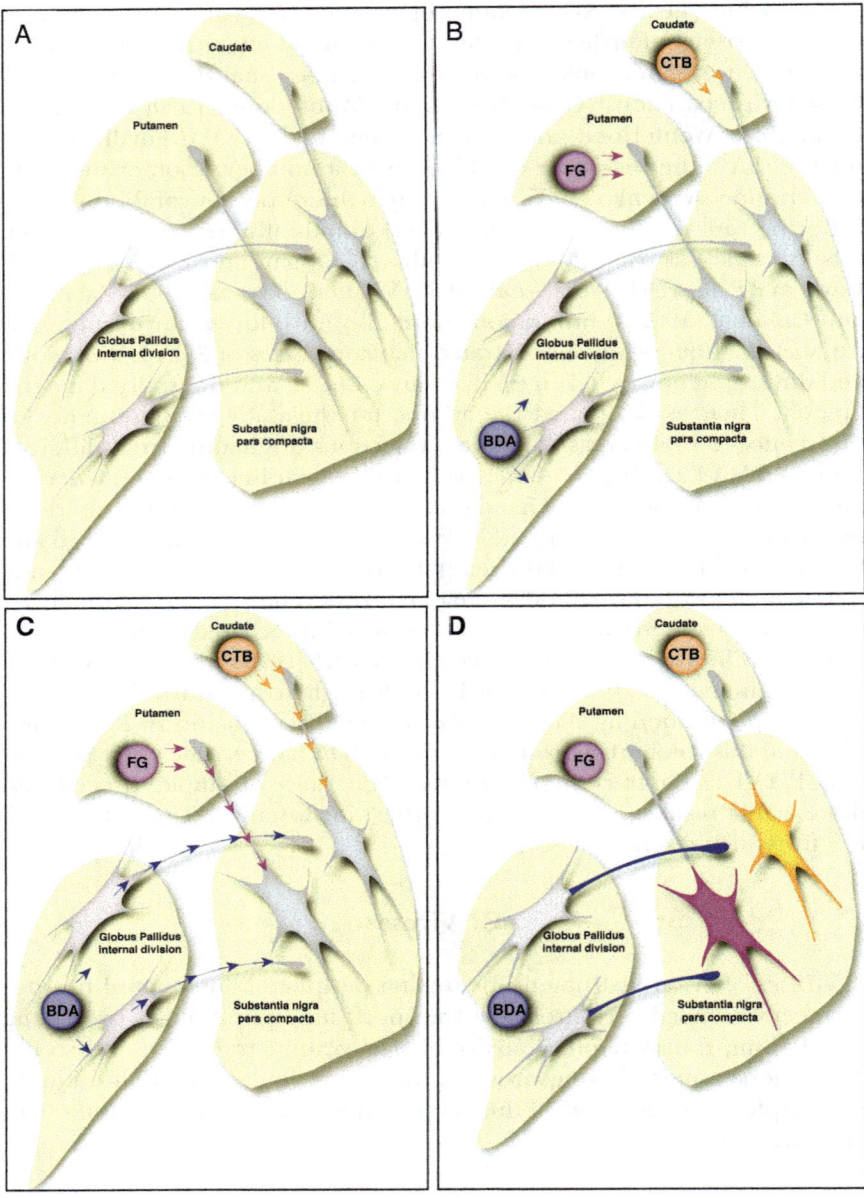

Figure 11.1. The study of the pallidal innervation of different subtypes of nigral dopaminergic neurons defined on the basis of their projection patterns as a typical example illustrating a problem to be approached by conducting a multitracer paradigm. (A) The goal: Analysis of the pallidal afferents to the substantia nigra pars compacta and their relationship with nigral neurons projecting either to the caudate nucleus or to the putamen. (B) The experimental paradigm: Injection of the tracer BDA in the internal division of the globus pallidus (GPi), followed by the delivery of the tracer CTB in the caudate nucleus and then the injection of FG in the

The scientific paradigm used here to highlight the multiple-tracing approach is basal ganglia connectivity (Figs. 11.1 and 11.2). Our working hypothesis is that the activity of nigral dopaminergic neurons is modulated by pallidal outflow. The anatomical substrate for such interference is connectivity between neurons located in the internal part of the globus pallidus and two subtypes of efferent neurons within the substantia nigra pars compacta, defined on the basis of their different target area within the primate caudate–putamen. One of the strategies to study these anatomical relationships is the simultaneous application of different neuroanatomical tracers at different loci in one and the same experiment (Fig. 11.1). Although single-tracer experiments usually provide valuable data on the organization of individual projections, they generally provide little information about potential convergence, divergence, or the degree of overlap between the circuits involved. A large number of such single-tracer experiments would be required just to show the topography of the connections, yet provide little information about the relationships of the incoming fibers with output neurons. By contrast, in a single animal the combination of various experiments as one multiple-tracing paradigm dramatically improves the view where fibers innervating an area end with respect to the precise location of neurons projecting from that area. Also, combination of several single-tracing experiments into one multiple-tracer project markedly reduces the number of required experimental animals. In a society in which animal experimentation is coming under ever increasing criticism, this is a premium. Yet, most of all, more and better information with respect to the relationships between the circuits of interest can be obtained with multiple-tracing experiments than can ever be achieved with single-tracing experiments.

←───

Figure 11.1. (*Cont.*) putamen. The delivery of all the injected tracers is performed in a single surgical session. (C) Tracer uptake and transport: BDA is taken up by dendrites of neurons located within the area of deposit and transported anterogradely to the striatum. Both CTB and FG are taken up by axon terminals arborizing within their respective injection sites and transported retrogradely to their parent cell bodies. (D) Expected results: Pallidal afferents to the substantia nigra pars compacta are labeled in blue-black by using DAB-Ni as a chromogen. Nigral neurons projecting to the caudate nucleus are labeled in brown with DAB substrate. Nigral neurons innervating the putamen are labeled in purple as a V-VIP precipitate. Injection sites became clearly defined within their respective targeted areas of deposit. The experimental design included BDA injection into the internal part of the GPi, followed first by CTB delivery in the ipsilateral caudate nucleus and then by injection of FG into the putamen. After 2 weeks of survival time, the animal was perfused, the brain extracted from the skull, and cryoprotected. Frozen coronal sections (40 μm thick) were obtained and processed for the simultaneous visualization of labeled structures (according to the step-by-step procedure delineated in the Appendix). At the end, within our area of study (the primate substantia nigra), labeled structures became apparent according to a three-color paradigm.

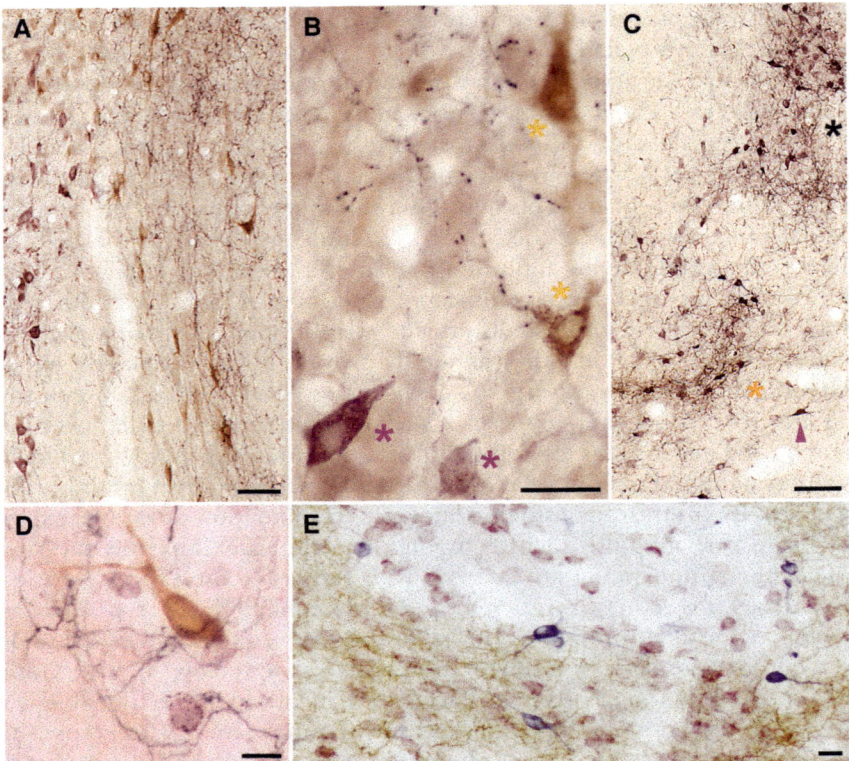

Figure 11.2. Examples illustrating different multitracer paradigms. (A, B) Dual retrograde tracing with FG and CTB, combined with BDA anterograde tracing. Photomicrographs of coronal sections through the primate substantia nigra. Projections from the internal part of the globus pallidus innervating the substantia nigra compacta are labeled with BDA (blue-black) and visualized by using the peroxidase substrate DAB-Ni. Nigral neurons projecting to the caudate nucleus are retrogradely labeled with CTB and visualized with DAB chromogen in brown (brown asterisk). Nigral neurons innervating the putamen are retrogradely labeled with FG and stained in purple by using V-VIP substrate (purple asterisk). Scale bar is 120 μm in (A) and 35 μm in (B). (C) Analysis of either convergence or divergence innervation arising from the substantia innominata (SI) or from the pedunculopontine nucleus (PPN) onto thalamic neurons giving rise to thalamostriatal projections. Photomicrograph is taken from a coronal section through the rodent thalamus at the level of the intralaminar nuclei. Projections from SI to the paracentral nucleus of the thalamus are labeled with PHA-L and stained in brown with DAB (brown asterisk). Projections from PPN to the central lateral nucleus of the thalamus are labeled with BDA and stained in dark blue-black with DAB-Ni (black asterisk). Neurons giving rise to thalamostriatal projections are retrogradely labeled after an FG deposit in the striatum. These neurons are stained in dark-light purple color by using V-VIP peroxidase substrate (purple arrowhead). Scale bar: 100 μm. (D) Combination of two tracers (BDA and FG stained with DAB-Ni and V-VIP, respectively) together with the immunocytochemical visualization of striatal giant cholinergic interneurons (brown-stained with DAB). Scale bar: 40 μm. (E) Combination of two tracers (BDA and FG stained with DAB and V-VIP, respectively) together with the histochemical visualization of NADPH diaphorase striatal interneurons (blue-stained with nitroblue tetrazolium). Scale bar: 45 μm.

TABLE 11.1. Best tracers to be selected for their combined use.

Tracer	Anterograde	Retrograde	Comments
Cholera toxin β subunit	+	+ + +	Second-choice retrograde tracer Commercial antibodies available Bidirectional transport[1] Pressure/iontophoretical delivery Survival time from 4 days to 4 weeks Taken up by fibers of passage
PHA-L	+ + +	–	Second-choice anterograde tracer Commercial antibodies available Capricious nature Iontophoretical delivery Survival time from 7 days to 3 months Lack of transport by fibers of passage
Fluoro-gold	–	+ + +	First-choice retrograde tracer Commercial antibody available Pressure/iontophoretical delivery Direct visualization by epifluorescence Survival time from 4 days to 1 year Taken up by fibers of passage[2]
Biotinylated dextran amine	+ + +	+	First-choice anterograde tracer No antibodies involved in visualization Pressure/iontophoretical delivery Survival time from 4 days to 2 months[3] Well suited to ultrastructural study Taken up by fibers of passage[4]

[1] A procedure improving the anterograde transport of CTB was reported by Angelucci *et al.* (1996).
[2] It is generally accepted that the iontophoretical delivery of FG minimizes the uptake by fibers of passage (Divac and Mogensen, 1990; Pieribone and Aston-Jones, 1988; Schmued and Fallon, 1986; Schmued and Heimer, 1990).
[3] Longer survival times were never tested by us. For more information, please see chapter by Reiner and Honig in this volume.
[4] See chapter by Reiner and Honig in this volume.

B. "Golden Rules": Requirements for a Tracer to Be Successful in a Multitracing Paradigm

Currently, we have at our disposal a wide variety of different tracers (Table 11.1), a huge array of antibodies to detect these tracers, and several possibilities to visualize them. Hence, in the process of designing multiple-tracing protocols the first question the investigator faces is that of the selection of the proper combination of tracers, i.e., which tracers are best suited for combined use, which order to maintain to successfully inject them, and how to inject them. In order to answer these difficult questions, one should keep in mind that the selected combination has to fulfill the following demands:

1. The first "golden rule" for multiple tracing is that the tracers involved should preferably be transported strictly unidirectionally (either anterogradely or retrogradely).

2. Only tracers of a similar nature can be combined together, i.e., nonfluo-rescent tracers cannot be combined with fluorescent tracers unless the latter can be appropriately detected in conventional light microscopy, for instance, after photoconversion or by detection via specific primary antibodies.

3. The tracers must have compatible survival times in order to avoid repet-itive surgical procedures, particularly given the increasing criticism of these procedures by ethical committees. In this regard, tracers such as HRP or biocytin are quickly transported and require only 1 or 2 days of survival time. However, such short range of survival times often impedes combination of these tracers with other commonly used trac-ers such as PHA-L, BDA, CTB, or FG, all of which require or tolerate longer survival times, from a few days to several months.

4. Commercially specific antibodies for the immunodetection of each individual tracer should be available, and these antibodies should not cross-react with each other or in any way interfere with the detection of any of the other markers under study.

5. It is desirable that the detection of the combined tracers can be per-formed by both immunoperoxidase and immunofluorescence meth-ods. When multiple immunoperoxidase detection is to be used, the individual chromogens should produce colored precipitates that can be unequivocally discriminated from each other (Table 11.2). Further-more, the resulting precipitates should preferably be permanent and compatible with existing cytoarchitectonic staining procedures. If im-munofluorescence is the chosen method to detect multiple tracers, dif-ferent fluorescence-coupled bridging antibodies should be available, each being labeled with a fluorescent dye with characteristic excitation and emission spectra clearly distinct from those of the other antibodies. Nevertheless, if this rule cannot be properly satisfied, modern confo-cal laser scanning microscopes are equipped with spectral detectors capable of distinguishing emission profiles overlapping up to 100% (one emission profile located within the other).

6. Finally, ultrastructural examination of labeled structures is always a desirable option, and thus all peroxidase substrates used to visualize tracers should be identifiable in the electron microscope by their own ultrastructural, electron-dense texture.

C. Winning Tracers

Although nearly all currently used tracers fulfill at least some of these de-manding criteria, some of these tracers more closely approach the ideal than others. According to our experience, BDA is undoubtedly the best tracer for anterograde tracing in multiple-labeling experiments. This is mainly due to its broad spectrum of survival time (from a few days to several weeks), its mostly anterograde transport especially when injected by iontophoresis,

TABLE 11.2. Commonly used peroxidase substrates.

Substrate	Color	Comments
Tetramethylbenzidine	Dark blue	Own electron-dense precipitate Difficult to combine with other chromogens Presumed carcinogenic
Diaminobenzidine	Brown	Incubation time of 10–40 min Own electron-dense texture Presumed carcinogenic Compatible with DAB-Ni, BHDC, V-VIP, and HGR
Nickel-enhanced DAB	Black	Strongest chromogen Incubation time of 5–10 min Compatible with DAB, BHDC, V-VIP, and HGR Presumed carcinogenic
Benzidine dihydrochloride	Blue	Own electron-dense texture Presumed carcinogenic Compatible with DAB-Ni and DAB
Vector very intense purple	Purple	Incubation time of 3–5 min Own electron-dense texture Presumed noncarcinogenic Soluble in ethanol Compatible with DAB-Ni and DAB
1-Naphthol/azur B	Blue-green	Not electron dense Compatible with DAB-Ni and DAB Fading over time
HistoGreen	Green	Incubation time of 1–5 min Presumed noncarcinogenic Soluble in ethanol So far, electron-dense texture not reported

Note. DAB, diaminobenzidine; BDHC, benzidine dihydrochloride; V-VIP, vector very intense purple; DAB-Ni, nickel-enhanced diaminobenzidine; HGR, HistoGreen.

straightforward histochemical detection using the avidin–biotin peroxidase (ABC) method without the need to use antibodies (this in turn simplifies the choice of antibodies to be used to detect the other markers), the availability of a broad range of fluorochrome-labeled streptavidin conjugates for the observation of transported BDA by confocal laser scanning, and sufficient tissue penetration of the reagents when used in ultrastructural analysis (Reiner and Honig, 2005, this volume; Wouterlood and Jorritsma-Byham, 1993). We consider PHA-L as the second option for anterograde tracing in combined paradigms. This tracer also offers a broad range of survival times (up to several months), and the axonal transport of this tracer is verified to be only in the anterograde direction. Both immunocytochemical and immunofluorescence detections have been used, and a good electron microscopy correlate is available. Nevertheless, many investigators report variable results using this tracer, and when the purpose is to conduct electron microscopy, the limited penetration of the anti-PHA-L antibodies into tissue sections may offer a histotechnical challenge (Wouterlood *et al.*, 1990).

For retrograde tracing, we feel that FG is undoubtedly the most efficient choice. The feasibility of using FG in multiple-tracer paradigms is impressive, given the advantages offered by several of its properties. The tracer can be delivered either by pressure or via iontophoresis, it has a broad range of survival times, and there is no anterograde transport. FG exquisitely fills the neuronal soma and main dendrites, and uptake by fibers of passage is minimal. Moreover, direct detection of transported FG is possible via fluorescence microscopy whereas indirect immunohistochemical or immunofluorescence detection can be combined with protocols for neurotransmitter immunodetection or other types of detection. FG can in addition be used in ultrastructural analysis. Our second choice for retrograde tracing is CTB, which shares several properties with FG, such as the wide range and flexibility of survival times, different detection methods (immunocytochemistry or immunofluorescence), and ultrastructural visualization. Although CTB is nicely transported in the retrograde direction, this tracer has several disadvantages when compared to FG, such as (1) larger tracer deposits are often required, (2) a moderate amount of anterograde transport is always present, (3) the transported tracer is often only located within the cell soma as a punctate deposit that does not enter the main dendrites, and (4) in our hands, the best results are obtained only with CTB applied through pressure injections. However, such injections always cover larger brain volumes than covered by iontophoretic injections.

D. Winning Combinations of Tracers

In the design of a multitracer paradigm centered around the purpose of defining the degree of convergence between projections arising from two different brain regions to the efferent neurons in another area, our currently favored choice is to combine dual anterograde tracing using BDA and PHA-L together with retrograde tract-tracing using FG (for more details, see Lanciego et al., 1998a; Fig. 11.2C). When the purpose of the design is to investigate whether two distinct subpopulations of efferent neurons located within a brain area are targeted by afferents arising from a different, third area, we feel that it is best to combine dual retrograde tracing with FG and CTB, together with anterograde tracing using BDA (Lanciego et al., 1998b, 2000; see also Fig. 11.2A,B). Additionally, several efficient triple-staining procedures that combine dual tracing using BDA and FG with the simultaneous detection of neuroactive substances are available (Gonzalo et al., 2001; Köbbert et al., 2000; Lanciego et al., 1997, 2000; see also Fig. 11.2C, D). In our experiments, we ultimately selected BDA as the anterograde tracer of choice for injecting the internal part of the globus pallidus (Lanciego et al., 2004; Fig. 11.2A,B), and FG and CTB as retrograde tracers. These two tracers were injected into the putamen and the caudate nucleus, respectively.

III. EVENTS AT THE INJECTION SITE

The injection site deserves special attention: how big is the injection spot, what is the effective injection spot, and do all neurons that lie within the demarcation of the injection spot internalize and transport the tracer?

The procedure by which a tracer is deposited at its desired locus is as follows:

1. opening of the skull and meninges, positioning of the tip of the pipette exactly on top of the pial surface of the brain, and reading its Z-coordinate;
2. lowering of the pipette to its final spatial position in the brain;
3. ejection of tracer substance from the pipette tip; and
4. retraction and subsequent closure of the wound.

Especially in Stage 3, tracer substance is ejected forcefully out of the micropipette or the injection needle, and it may take the way of the least mechanical resistance, i.e., some tracer may leak into the space between the pipette and the brain parenchyma, labeling cells here.

A. Uptake Mechanisms

Following its ejection from a pipette or a syringe, the tracer needs to pass the cell membrane in order to be transported by the neuron. The uptake mechanism of several tracers has been studied while the uptake of other tracers yet needs to be elucidated. There is evidence that glycoprotein tracers like HRP are taken up via a fluid-phase pinocytotic process (Gonatas *et al.*, 1979). The lectin conjugate of HRP binds to receptors on the external faces of cell membranes, thus enhancing the uptake (Trojanowski and Schmidt, 1984). PHA-L and BDA undergo receptor-mediated endocytosis (Fritzsch, 1993; Groenewegen and Wouterlood, 1990). By contrast, (electrically uncharged) FG may pass the cell membrane by simple diffusion and then become trapped via a pH gradient in endosomes (Wessendorf, 1991). The endosomes in turn are transported to the lysosome apparatus in the cell body in an attempt to degrade the ingested FG metabolically. A debate has been raging in the literature for the last 10 years as to whether cholera toxin enters cells via small pinocytotic-type fluid-phase vesicular carriers known as "caveolae" (Anderson and Edwards, 1993) or via clathrin-mediated endocytotic vesicles (Kirkham *et al.*, 2005). Whether the attenuated form of the toxin (i.e., CTB) uses either one internalization pathway is unknown.

B. BDA May Cause Track Labeling

During the insertion and retraction stages of the stereotaxic pipette or needle, tracer leakage may occur. BDA track labeling or uptake of the tracer by neurons located on the trajectory of the pipette when it was lowered to its final stereotaxic position in the brain or when it was retracted afterward have been reported by Dolleman-van der Weel *et al.* (1995). In our experience, track labeling can be suppressed by carefully cleaning the outer surface of the pipette before placement. Only when the positioning is successful and verified is the iontophoretic current applied to force the tracer out of the pipette. After iontophoresis, we leave the pipette in place for 15 min and then slowly retract, giving the brain parenchyma the chance to seal the injection area. An effective additional measure to suppress track labeling is to reverse the polarity of the iontophoretic current during the retraction of the pipette. If in spite of these measures track labeling is still a problem, the pipette or needle might be inserted into the brain at an angle so as to avoid track labeling vulnerable areas.

C. The Effective Injection Site Shape, Size, and Delivery of Tracer

Injection sites usually have a teardrop-like shape, with the diameter of the "teardrop" measured perpendicular to the axis of penetration of the application pipette or the needle. In all cases, the effective injection area is surrounded by an injection halo. The size and shape of an injection site depend in part on the cytoarchitectonics of the injection region, e.g., the presence of layers or bundles of myelinated fibers. In the case of BDA, diameters of the injection sites in the striatum (at injection time) vary between 200 and 300 μm (iontophoretical application with a pipette with an inner tip diameter of 20–30 μm) and up to 800 μm (single, large, pressure injection). Another issue that is directly related to the size of the injection site is the duration of the iontophoretical injection. As a rule of thumb, longer injection times result in larger deposits.

The effective site of injection of an anterograde tracer, e.g. BDA, is postulated here as that brain area located at the original injection coordinates that contains an aggregation of cells that stain above background with the detection method. Since this injection site is the spot visible after survival and histochemical staining, the border of this injection spot may not reflect the extent of diffusion of tracer away from the tip of the application pipette at or shortly after injection time. Yet, since we are appreciating labeling of fibers and terminals, the cell bodies to which these fibers belong must be labeled as well. There is little reason to assume that nonlabeled cell bodies give rise to labeled projections and, conversely, that label accumulates during the survival time in the distal axons and axon terminals, with no labeled cell bodies remaining in the original injection spots. In a longitudinal study of the fate of the tracer PHA-L, quite the contrary was seen. Label progressively disappeared from axon terminals, next from fibers, and finally from the cell bodies in injection sites (Wouterlood *et al.*, 1990).

D. PHA-L

One characteristic of PHA-L is that with pipette tip diameters larger than 30 μm there is no ejection of tracer when the iontophoretic current is applied. Consequently, the size of the injection site is always limited to a columnar or a teardrop-like space with a diameter of approximately 200 μm. If larger injections are required, for instance when the purpose is to cover an entire brain area, then multiple PHA-L injections have to be made in an array-like manner.

E. Cholera Toxin Subunit B

CTB can be delivered either by iontophoresis or by pressure. We prefer in our laboratory the pressure delivery method: 2% solution of CTB in sterile phosphate buffer (PB). Injection volume is typically in the range of 0.1–0.2 μl. The balance between the anterograde and the retrograde transport of the tracer is apparently related to the survival time. Short survival times (2–4 days) tend to increase the anterograde transport component, while survival times longer than a week have a tendency to enhance the retrograde transport component. The wide range of survival times for CTB, together with its feasibility for being delivered by using different methods, make CTB a very useful tracer for combined procedures as well as for coinjection protocols (see below).

F. Fluoro-Gold

In the last few years FG has become the tracer of choice for retrograde tracing purposes. This tracer is delivered either by pressure or by iontophoresis (if smaller injections are required). FG has the widest survival time spectrum, ranging from just a few days to 1 year (Akintunde and Buxton, 1992; Novikova *et al.*, 1997; Pieribone and Aston-Jones, 1988; Schmued, 1994; Schmued and Fallon, 1986; Schmued and Heimer, 1990). According to our experience, the most critical issue when dealing with FG tracing is the vehicle in which the tracer is dissolved. Best results using FG have been obtained by dissolving the tracer as a 2–3% solution in 0.1 M cacodylate buffer at neutral pH, as previously reported elsewhere (Schmued and Heimer, 1990). If the vehicle is not cacodylate buffer (e.g., distilled water, PBs, acetate buffer), then a nonhomogeneous solution easily forms that may precipitate and seriously compromise axonal uptake. In this regard, the use of FG for tract-tracing studies in primates is still somewhat controversial. It has been reported that FG may produce variable results when used in nonhuman primates (Schmued, 1994). Taking advantage of the superior solubility of FG in cacodylate buffer, we obtained satisfying retrograde labeling of both the nigrostriatal and the thalamostriatal pathways after pressure-injecting FG into the putamen of the primate *Macaca fascicularis* (Lanciego *et al.*, 1998b; see also Fig. 11.2A,B).

G. Survival Time

The survival time has to be adapted to the length of the pathways under scrutiny. According to the "golden rules" for multiple-tracing paradigms, the range of survival times of all the tracers involved should be sufficiently broad to facilitate their combined use. Hence, when carrying out experiments in rats, a survival time of 1 week is sufficient to produce high-quality staining of all the injected tracers. In our hands, BDA yields good results after survival times of up to 8 weeks in rodents (we have not tested longer survival times). In primates, good results were obtained over periods of 2 weeks, although Brandt and Apkarian (1992) have reported good BDA labeling in primates after survival times of up to 7 weeks.

H. Deposition of an Anterograde and a Retrograde Tracer in the Same Spot

If reciprocally connected networks are to be studied, it would be advantageous to use tract-tracing paradigms including the combined and simultaneous delivery of two different tracers (anterograde and retrograde) for disclosing the presence of reciprocal connections to a given brain area or nucleus. Two strategies may be followed to achieve coinjection: mixing tracers and injecting via two different pipettes or needles.

The injection of a mixture of FG and fluorescence-labeled dextrans reported by Nance and Burns (1990) was the first attempt to study reciprocal connections with one single injection (injection achieved by pressure). The only limitation of this elegant approach was the fading of signal over time, due to the fluorescent nature of both tracers. Later on, a different solution was provided by Risold and Swanson (1995) by coinjecting FG and PHA-L. In this approach, coinjections were achieved by using double-barreled glass pipettes. Since distally the two barrels had different tip diameters, the sizes of the injection sites were not comparable to each other. Probably the best circumvention of these drawbacks was provided by Coolen and Wood (1998) and by Coolen *et al.* (1999), by combining BDA and CTB. Both tracers share many features such as wide range of survival time, dilution parameters, etc. Coolen *et al.* (1999) coinjected BDA and CTB by using one pipette, which was lowered to the desired Z-coordinate. Ejection of the tracers from the pipette was achieved by pressure and iontophoresis (1:1 mixture of BDA and CTB, final concentration of 5 and 0.25%, respectively). This approach resulted in perfectly overlapping injection sites and in the availability of permanent detection through immunoperoxidase techniques. In our opinion this is the best choice when studying brain circuits in which the involved components exhibit reciprocal connections.

IV. VISUALIZATION METHODS: MULTIPLE IMMUNOPEROXIDASE FOR BRIGHT FIELD MICROSCOPY

It has long been a desire in modern neuroanatomy to be able to simultaneously and permanently visualize up to three transported markers in a single histological section in order to produce detailed mapping of brain circuits. The first attempts to achieve this were undertaken long ago, before the advent of strong fluorescent markers and fluorescent counterstains, and the natural approach was to combine these methods with classical counterstaining to determine the cytoarchitectonic context of the traced connectivity. Although technically extremely demanding, different approaches have been proposed and the possibilities seem almost unlimited, restricted only by the researcher's own imagination and technical skills.

The first successful attempts to simultaneously detect three different tracers were reported by Smith and Bolam (1991, 1992) using a protocol that combined retrograde transport of WGA-HRP with anterograde transport of PHA-L and biocytin. The tracers were detected with three different peroxidase substrates, namely tetramethylbenzidine (TMB) to visualize the transported WGA-HRP, nickel-enhanced DAB (DAB-Ni) to detect PHA-L, and finally, regular DAB to visualize biocytin. In its day, this technical tour de force was an imaginative solution, although the clear discrimination of the labeled elements was sometimes a difficult task. Subsequently, Dolleman-van der Weel *et al.* (1995) introduced a sequential light microscopy protocol combining anterograde tracing with PHA-L, rhodamine dextran amine (RDA), and BDA, stained respectively with a DAB-Ni (black precipitate), DAB (brown reaction product), and 1-naphthol/azur B (blue-green color) as the peroxidase substrates. Using this procedure, the labeled elements could be unequivocally distinguished, except in brain areas where the three tracers concurred "en massive." The only, although minor, drawback of this delicate procedure is that fading of 1-naphthol/azur B precipitate was noticed at room temperature, although the stain could be restored by repeating the incubation with the basic dye (Mauro *et al.*, 1985).

When trying to perform ultrastructural studies using three markers, the work published by Anderson *et al.* (1994) must be seen as a jewel in the tract-tracing literature. These authors designed an elegant preembedding triple-labeling method to clearly visualize the convergence of substance P- (SP-ir) and enkephalin-positive (ENK-ir) afferents onto tegmental dopaminergic neurons, the latter identified with a primary antibody against TH-ir. The antigens of interest were visualized with three distinct peroxidase substrates, resulting in three morphologically different electron-dense textures, namely DAB (flocculent) for the SP-ir terminals, silver-intensified immunogold (distinct particles) for the stain of ENK-ir terminals, and benzidine hydrochloride (needle-shaped crystalline material) for TH detection.

In general, in the process of the design of a triple-staining paradigm, a main limitation run into is the relatively small number of peroxidase substrates currently available (Table 11.2). This limitation does not always permit a straightforward color discrimination of the labeled elements. Furthermore, several peroxidase substrates used in light microscopical studies do not provide precipitates with a texture that can be easily differentiated in the electron microscope from "regular" DAB precipitate. In this regard, the introduction by Zhou and Grofova (1995) of the peroxidase substrate Vector® very intense purple (V-VIP) must be considered as an important advance in multiple-staining strategies. The peroxidase substrate V-VIP has two unique properties, notably a characteristic purple color that under the light microscope is distinguishable with ease from other commonly used chromogens (Fig. 11.2), and an electron-dense precipitate with a typical dense granular appearance. Initially developed as an alternative to DAB for double-labeling procedures, this new peroxidase substrate was rapidly incorporated into multiple-staining paradigms aimed at simultaneously detecting three different markers, specifically in those protocols in which three chromogens were combined, i.e., DAB-Ni (black), DAB (brown), and V-VIP (purple; Lanciego et al., 1997). The recent introduction of a new peroxidase substrate named HistoGreen (Thomas and Lemmer, 2005) has enriched the technical arsenal at our disposal for designing multiple colored detection of immunoperoxidase protocols.

A. Troubleshooting

When performing triple-staining procedures, the researcher is faced with three main problems that can reduce the quality of the final stain. The critical issues are (i) the choice of the most appropriate chromogen for each particular marker, (ii) background staining, and (iii) color mixing phenomena.

1. Chromogen Selection and Chromogen Sequence

We are dealing with three different peroxidase substrates (DAB-Ni, DAB, and V-VIP), each characterized by a different strength. The strongest chromogen is DAB-Ni, while V-VIP is the weakest. If we consider that the detection of BDA does not depend on antibodies, DAB-Ni is probably the best choice to visualize the BDA stain since this tracer stain fibers nicely with minimal background staining. As both DAB and V-VIP chromogens always produce some light background, we consider these two substrates as the most appropriate for cellular stains. Accordingly, we choose to use DAB as the substrate for CTB stain and V-VIP for the detection of FG.

Another important issue is the sequence in which the chromogens are applied during the staining procedure. By virtue of their differences in strength, the best results are obtained by using the strongest substrate first (DAB-Ni), then the DAB solution, while the weakest chromogen (V-VIP) should be used last. Different sequences were tested during the development of the three-color paradigm, always resulting in the appearance of intolerable background staining and color mixing phenomena.

2. Background Staining

Avoiding or suppressing background stain is critically important when attempting to obtain high-quality staining since background stain may seriously hamper the contrast between the different color reaction products of the chromogens. The appearance of an intolerable degree of nonspecific background staining often hinders the interpretation of the results since it may prevent the correct visualization of the finer structures labeled with a particular antigen (Groenewegen and Wouterlood, 1990). It is important to achieve a low level of background staining in triple-staining methods. Background staining is often seen when one antigen requires the prolonged incubation of its corresponding chromogen solution to obtain adequate staining. To avoid this pitfall, we recommend careful and continuous monitoring under the microscope of the progress of each reaction. In this way, the incubations with the different peroxidase substrates can be stopped in time before nonspecific background starts to build up. If such care is not taken, the background stain obtained in the first reaction will enhance the background stain obtained in the subsequent chromogen incubations. Nonspecific binding of the antibodies may also increase background stain, a drawback that can be minimized by adding 2% bovine serum albumin (BSA; Merck) to the incubation solution containing the primary antibodies.

3. Color Mixing

The tendency of a peroxidase substrate to adopt the color of other chromogens used previously is an undesired phenomenon that we refer to as "color mixing." Color mixing is often observed in circumstances where (i) the reaction product is at a high concentration, such as at the injection sites, (ii) cross-reactivity occurs between the antisera, (iii) prolonged incubations of the chromogen are necessary, and (iv) two or three of the antigens detected colocalize within the same structure. The best way to avoid color mixing consists of carefully selecting the antisera, preferably using antibodies raised in different animal species. Furthermore, monitoring the progress of the chromogen reactions at intervals by microscopy is indispensable. When the antigens under study colocalize within the same structure but are found in separate subcellular compartments, colocalization is not generally a major problem (Lanciego and Giménez-Amaya, 1999).

B. Results

   The reward of successful completion of a multitracing experiment is the full display in multiple colors of the complexity of neuronal connectivity. This is illustrated in Fig. 11.2, which is composed of photomicrographs of sections of monkey brain in which the specific triple-tracing paradigm included injection of BDA into the internal part of the GPi, CTB delivery in the caudate nucleus, and the FG deposit in the putamen. Thus, in our sections, we observed in the substantia nigra pars compacta a subpopulation of neurons retrogradely labeled with CTB (cell bodies containing brown precipitate). CTB-labeled neurons were distributed within cell clusters mainly composed of cells with similar projection patterns. Indeed, nigrocaudate-projecting neurons remained largely complementary to another cellular group stained in purple with V-VIP precipitate. These purple-stained cells are nigroputaminal neurons retrogradely labeled with FG. The cell bodies and dendrites of both types of retrogradely labeled nigrostriatal neurons were embedded in a dense terminal plexus of BDA-labeled fibers arising from GPi (stained black with DAB-Ni substrate). Cell bodies of both types of nigrostriatal retrogradely labeled neurons appeared within the terminal plexus of pallidonigral projections, indicating that both the nigrocaudate and the nigroputaminal projecting neurons receive pallidal innervation. The secondary observation of the complementary distribution of the different subtypes of nigrostriatal neurons and the distribution of pallidal afferents over both populations of projection neurons is the added value of triple-tracing paradigms.

APPENDIX

   In order to facilitate the use of this method by others, we provide a detailed description below that deals with the simultaneous detection of the anterograde transport of BDA in combination with the retrograde transport of CTB and FG. Only minor adjustments mainly related to the selection of antibodies are required when detecting different combinations of antigens.

A. Step-By-Step Procedure

1. Tracer Delivery, Survival Time, Perfusion, Cryoprotection, and Sectioning

   All injected tracers are delivered in a single surgical session. The anterograde tracer BDA is iontophoretically delivered as a 10% solution in 10 mM

PB, pH 7.25 (see section "Preparation of the BDA Solution for Injection") through a glass micropipette (inner tip diameter ranging from 20 to 35 μm), using a positive-pulsed direct current (7 s on/off) for 3–10 min (depending on the desired size of the injection site). Next, a 2% solution of FG in 100 mM cacodylate buffer, pH 7.3, is injected using the same iontophoretic parameters described above for BDA delivery. Finally, a total volume of 0.5–5 μl of a 2% solution of CTB in 0.1 PB, pH 6.0, is pressure-injected into the target area at a rate of 0.2 μl/min. For studies in primates, both BDA and FG were also pressure-delivered.

Regarding tract-tracing with CTB, survival times of 1 week for rodents and 2 weeks when dealing with primates always result in satisfactory bidirectional labeling. It has been proposed that the anterograde component of CTB transport is intensified with short survival times, while longer periods may enhance its retrograde transport (Angelucci *et al.*, 1996; Trojanowski *et al.*, 1981). However, in our hands, no significant variability of the anterograde/retrograde transport ratio of CTB was noticed when the survival times ranged from 2 days to 4 weeks.

Tissue fixation is always performed by trans-cardiac perfusion after deeply anesthetizing the animal with an overdose of 10% of chloral hydrate in distilled water. Once anesthetized, a saline Ringer's solution is used to perform a brief vascular rinse. Immediately afterward, the perfusion is continued with either 500 ml (rats) or 3000 ml (primates) of a solution containing 4% freshly depolymerized paraformaldehyde, 0.1% glutaraldehyde, and 0.2% of saturated picric acid solution in 125 mM PB, neutral pH. In the case of primates only, perfusion is continued with 1000 ml of a cryoprotective solution consisting of 10% glycerin and 1% dimethylsulfoxide (DMSO) in 125 mM PB, neutral pH. The brain is then extracted from the skull and stored in a cryoprotective solution composed of 20% glycerin and 2% DMSO in 125 mM PB, neutral pH (Rosene *et al.*, 1986). Finally, frozen tissue sections are obtained at a thickness of 40 μm on a sliding microtome and collected in 125 mM PB, neutral pH. Each series of sections is transferred to cryoprotectant and stored in a freezer at −40°C until they undergo further histological processing. In section "Cryoprotective Solutions" (see below) some advise regarding cryoprotection is offered.

2. Histological Processing

All procedures are carried out at room temperature, with gentle shaking on an orbital rotator, unless otherwise stated.

1. Inactivate the endogenous peroxidase activity, for 40 min (see section "Inactivation of the Endogenous Peroxidase Activity").
2. Rinse 3 × 10 min in 50 mM Tris-buffered saline–Triton X (TBS-Tx), pH 8.
3. Incubate in ABC solution for 90 min (see section "Preparation of the ABC Solution").

4. Rinse 2 × 10 min in 50 mM TBS-Tx, pH 8.
5. Rinse 2 × 10 min in 50 mM Tris-HCl, pH 8.
6. Stain with the first chromogen solution (DAB-Ni) for ~5 min (see section "Preparation of the DAB-Ni Solution").
7. Rinse 3 × 10 min in 50 mM Tris-HCl, pH 8.
8. Incubate for 60 h at 4°C in a cocktail solution of primary antibodies containing 1:2000 goat anti-CTB (List Biological) and 1:2000 rabbit anti-FG (Chemicon) in 50 mM TBS-Tx, pH 8 + 2% BSA.
9. Rinse 3 × 10 min in 50 mM TBS-Tx, pH 8.
10. Incubate for 2 h in a cocktail solution of bridge antibodies, containing 1:50 donkey anti-goat IgG (Nordic or Sigma) and 1:50 swine anti-rabbit IgG (Dako) in 50 mM TBS-Tx, pH 8.
11. Rinse 3 × 10 min in 50 mM TBS-Tx, pH 8.
12. Incubate for 90 min in 1:600 goat-PAP (Nordic or Sigma) in 50 mM TBS-Tx, pH 8.
13. Rinse 2 × 10 min in 50 mM TBS-Tx, pH 8.
14. Rinse 1 × 10 min in 50 mM Tris-HCl, pH 8.
15. Rinse 2 × 10 min in 50 mM Tris-HCl, pH 7.6.
16. Stain with the second chromogen solution (DAB) for approximately 10–30 min (see section "Preparation of the DAB Solution").
17. Rinse 3 × 10 min in 50 mM Tris-HCl, pH 7.6.
18. Rinse 1 × 10 min in 50 mM Tris-HCl, pH 8.
19. Rinse 2 × 10 min in 50 mM TBS-Tx, pH 8.
20. Incubate for 90 min in 1:600 rabbit-PAP (Dako) in 50 mM TBS-Tx, pH 8.
21. Rinse 2 × 10 min in 50 mM TBS-Tx, pH 8.
22. Rinse 1 × 10 min in 50 mM Tris/HCl, pH 8.
23. Rinse 2 × 10 min in 50 mM Tris/HCl, pH 7.6.
24. Stain with the third chromogen solution (V-VIP) for approximately 3–5 min (see section "Preparation of the V-VIP Solution").
25. Rinse 3 × 10 min in 50 mM Tris-HCl, pH 7.6.
26. Mount sections in 2% gelatin (Merck) in 50 mM Tris-HCl, pH 7.6.
27. Once dried thoroughly, quickly dehydrate the sections with 2 × 10 min rinses in 100% toluene (see section "On the Use of Toluene for Dehydration Purposes") and mount in DPX, Entellan, or similar mounting media.

B. Technical Tips

1. Preparation of the BDA Solution for Injection

BDA (biotin, dextran 10 KD, lysine fixable, catalog number D1956) is purchased from Molecular Probes Europe (Leiden, The Netherlands). For either iontophoretic or pressure injections, the tracer is dissolved as 2–15%

in 10 mM PB, pH 7.25 (Reiner *et al.*, 2000). Good results with BDA have also been reported by pressure-injecting either a 10% solution in saline (Brandt and Apkarian, 1992) or a 5% solution in distilled water (Rajakumar *et al.*, 1993).

2. Cryoprotective Solutions

Cryoprotection is achieved using a standard solution modified according to Rosene *et al.* (1986). This cryoprotectant solution consists of 20% glycerin and 2% DMSO in 125 mM PB at neutral pH. With this solution cryoprotection of a complete rat brain usually takes overnight at 4°C, or that for cat and monkey brain takes approximately 48 h at 4°C. Subsequently, it is possible to cut large brain sections without introducing freezing artifacts. In addition, this cryoprotective solution helps to preserve the ultrastructural details fairly well (Fig. 11.2). Another advantage when compared to other commonly used cryoprotectants, such as 30% sucrose in PB, is that the buffered combination of glycerin and DMSO produces a minimal amount of shrinkage. Finally, in our experience frozen tissue immersed in this cryoprotectant solution can be stored for up to 7 years at −85°C without showing any significant loss of immunoreactivity when compared to freshly processed tissue from the same animal.

3. Inactivation of the Endogenous Peroxidase Activity

Prior to immunocytochemical staining, we always inactivate the endogenous peroxidase activity in the tissue to minimize background staining. This procedure is especially advisable in cases of improperly perfused brains. The inactivation solution is a mixture of methanol and hydrogen peroxide (0.1 ml of H_2O_2 + 0.9 ml of H_2O in 15 ml of 100% methanol). Sections are incubated in the inactivation solution at a ratio of 1 min/μm section thickness (i.e., 40 min when dealing with 40-μm-thick sections).

4. Preparation of the ABC Solution

The ABC complex is supplied by Vector Labs (Burlingame, CA), catalog number Standard PK-4000. No differences in the final staining quality were observed when using the Standard ABC PK-4000 instead of the Vector ABC Elite kit. The ABC solution should be freshly prepared following the manufacturer's instructions. To prepare the incubation solution, add 8 μl of solution A to 1 ml of 50 mM TBS-Tx, pH 8, and shake gently. Then add 8 μl of solution B, shake gently, and wait for 30 min before using the solution.

5. Preparation of the DAB-Ni Solution

The DAB-Ni solution should be freshly prepared by dissolving 0.2 g of nickel ammonium sulfate (Carlo Erba) and 7.5 mg of DAB in 50 ml of 50 mM Tris/HCl, pH 8. Immediately before use, 10 µl of hydrogen peroxide must be added. Incubation usually takes 5–10 min, resulting in a black precipitate.

6. Preparation of the DAB Solution

The DAB solution is freshly prepared by dissolving 5 mg of DAB in 10 ml of 50 mM Tris-HCl, pH 7.6. Once dissolved, the resulting solution is filtered and 3.3 µl of hydrogen peroxide is added immediately prior to use. Incubation takes 20–40 min, resulting in a brown precipitate.

7. Preparation of the V-VIP Solution

The V-VIP solution is freshly prepared in 50 mM Tris-HCl, pH 7.6, with one drop from each vial (V-VIP substrate is presented as a kit containing four vials) to prepare 3.5 ml of the working solution. Incubation takes 5–10 min, resulting in a purple precipitate. Note that the final working solution of V-VIP presented here is a twofold dilution of the original recipe provided by the supplier. The use of a more diluted substrate results in a more manageable chromogen, making it easier to monitor the progress of the reaction, and therefore one can avoid typical drawbacks such as excessive background staining or color mixing phenomena.

8. On the Use of Toluene for Dehydration Purposes

V-VIP chromogen is soluble in ethanol. Accordingly, when using standard dehydration procedures in ascending series of ethanol followed by clearing in xylene, V-VIP stain may disappear or be diluted. To circumvent this problem, we perform the dehydration and clearing procedures by placing the slides with the thoroughly dried sections directly in 100% toluene (two rinses of 5 min each), which yields the best results.

ACKNOWLEDGMENTS. Most of the technical training received by José L. Lanciego was conducted under the expert guidance of his mentors, Prof. Francisco Collía at the University of Salamanca and Prof. Floris G. Wouterlood at the Amsterdam Vrije Universiteit Medical Center.

The data presented here are gathered from the work conducted by José L. Lanciego and his collaborators. Among others, we would like to express our gratitude to Drs. Elena Erro, José Arribas, Nancy Gonzalo, and María Castle. We also acknowledge the expert assistance provided by my technicians Ms. Elvira Roda and Ms. Ainhoa Moreno.

This work was partially funded by research grants from The Michael J. Fox Foundation, Fondo de Investigaciones Sanitarias ref. FIS 01/0237, Ministerio de Ciencia y Tecnología ref. BFI2003-02033, FEDER, Departamentos de Salud y Educación del Gobierno de Navarra, and by the "UTE Project FIMA."

REFERENCES

Akintunde, A., and Buxton, D. F., 1992, Quadruple labeling of brainstem neurons: a multiple retrograde fluorescent tracer study of axonal collateralization, *J. Neurosci. Methods* **45:**15–22.

Anderson, C. R., and Edwards, S. L., 1993, Subunit b of cholera toxin labels interstitial cells of Cajal in the gut of rat and mouse, *Histochemistry* **100:**457–464.

Anderson, K. D., Karle, E. J., and Reiner, A., 1994, A pre-embedding triple-label electron microscopic immunohistochemical method as applied to the study of multiple inputs to defined tegmental neurons, *J. Histochem. Cytochem.* **42:**49–56.

Angelucci, A., Clascá, F., and Sur, M., 1996, Anterograde axonal tracing with the subunit B of cholera toxin: a highly sensitive immunohistochemical protocol for revealing the fine axonal morphology in adult and neonatal brains, *J. Neurosci. Methods* **65:** 101–112.

Antal, M., Freund, T. F., Somogyi, P., and McIlhinney, R. A., 1990, Simultaneous anterograde labelling of two afferent pathways to the same target area with *Phaseolus vulgaris* leucoagglutinin and *Phaseolus vulgaris* leucoagglutinin conjugated to biotin or dinitrophenol, *J. Chem. Neuroanat.* **3:**1–9.

Balercia, G., Cheng, S., and Bentivoglio, M., 1992, Electron microscopic analysis of fluorescent neuronal labeling after photoconversion, *J. Neurosci. Methods* **45:**87–98.

Bentivoglio, M., and Su, H. S., 1990, Photoconversion of fluorescent retrograde tracers, *Neurosci. Lett.* **113:**45–87.

Blackstad, T. W., 1965, Mapping of experimental axon degeneration by electron microscopy of Golgi preparations, *Z. Zellforsch.* **67:**819–834.

Blackstad, T. W., 1981, Tract tracing by electron microscopy of Golgi preparations, In: Heimer, L., and RoBards, M. (eds.), *Neuroanatomical Tract-Tracing Methods,* New York: Plenum Press, pp. 407–440.

Brandt, H. M., and Apkarian, A. V., 1992, Biotin-dextran: a sensitive anterograde tracer for neuroanatomic studies in rat and monkey, *J. Neurosci. Methods* **45:**35–40.

Bruce, K., and Grofova, I., 1992, Notes on a light and electron microscopic double-labeling methods combining anterograde tracing with *Phaseolus vulgaris* leucoagglutinin and retrograde tracing with cholera toxin subunit β, *J. Neurosci. Methods* **45:**23–33.

Chang, H. T., Kuo, H., Whittaker, J. A., and Cooper, N. G. F., 1990, Light and electron microscopic analysis of projection neurons retrogradely labeled with Fluoro-Gold: notes on the application of antibodies to Fluoro-Gold, *J. Neurosci. Methods* **35:**31–37.

Coolen L. M., Jansen H. T., Goodman R. L., Wood R. I., and Lehman M. N., 1999, A new method for simultaneous demonstration of anterograde and retrograde connections in the brain: co-injections of biotinylated dextran amine and the beta subunit of cholera toxin, *J. Neurosci. Methods* **91:**1–8.

Coolen L. M., and Wood R. I., 1998, Reciprocal connections of the medial amygdaloid nucleus in the Syrian hamster brain: simultaneous anterograde and retrograde tract tracing, *J. Comp. Neurol.* **399:**189–209.

Cullinan, W. E., and Zaborszky, L., 1991, Organization of ascending hypothalamic projections to the rostral forebrain with special reference to the innervation of cholinergic projection neurons, *J. Comp. Neurol.* **306**:631–667.

de Olmos, J. S., Ebbesson, S. O. E., and Heimer, L., 1981, Silver methods for the impregnation of degenerating axoplasm, In: Heimer, L., and Robards, M. (eds.), *Neuroanatomical Tract-Tracing Methods*, New York: Plenum Press, pp. 117–170.

Divac, I., and Mogensen, J., 1990, Long-term retrograde labeling of neurons, *Brain Res.* **524**:339–341.

Dolleman-Van der Weel, M. J., Wouterlood, F. G., and Witter, M. P., 1995, Multiple antero-grade tracing, combining *Phaseolus vulgaris* leucoagglutinin with rhodamine- and biotin-conjugated dextran amine, *J. Neurosci. Methods* **51**:9–21.

Edwards, S. B., and Hendrickson, A., 1981, The autoradiographic tracing of axonal connections in the central nervous system, In: Heimer, L., and RoBards, M. (eds.), *Neuroanatomical Tract-Tracing Methods*, New York: Plenum Press, pp. 171–205.

Fritzsch, B., 1993, Fast axonal diffusion of 3000 molecular weight dextran amines, *J. Neurosci. Methods* **50**:95–103.

Fritzsch, B., and Wilm, C., 1990, Dextran amines in neuronal tracing, *Trends Neurosci.* **13**:14.

Gaykema, R. P., van Weeghel R., Hersh, L. B., and Luiten, P. G., 1991, Prefrontal cortical projections to the cholinergic neurons in the basal forebrain, *J. Comp. Neurol.* **303**:563–583.

Gerfen, C. R., and Sawchenko, P. E., 1984, An anterograde neuroanatomical tracing method that shows the detailed morphology of neurons, their axons and terminals: immunohis-tochemical localization of an axonally transported plant lectin, *Phaseolus vulgaris* leucoag-glutinin (PHA-L), *Brain Res.* **290**:219–238.

Gerfen, C. R., and Sawchenko, P. E., 1985, A method for anterograde axonal tracing of chemi-cally specified circuits in the central nervous system: combined *Phaseolus vulgaris* leucoag-glutinin (PHA-L) tract tracing and immunohistochemistry, *Brain Res.* **343**:144–150.

Gonatas, N. K., Harper, C., Mizutani, T., and Gonatas, J. O., 1979, Superior sensitivity of con-jugates of horseradish peroxidase with wheat germ agglutinin for studies of retrograde axonal transport, *J. Histochem. Cytochem.* **27**:728–734.

Gonzalo, N., Moreno, A., Erdozain, M. A., García, P., Vázquez, A., Castle, M., and Lanciego, J. L., 2001, A sequential protocol combining dual neuroanatomical tract-tracing with the visualization of local circuit neurons within the striatum, *J. Neurosci. Methods* **111**:59–66.

Graham, R. C., and Karnovsky, M. J., 1966, The early stages of absorption of injected horseradish peroxidase in the proximal tubules of mouse kidney: ultrastructural cytochemistry by a new technique, *J. Histochem. Cytochem.* **14**:291–302.

Groenewegen, H. J., and Wouterlood, F. G., 1990, Light and electron microscopic tracing of neuronal connections with *Phaseolus vulgaris*-leucoagglutinin (PHA-L) and combinations with other neuroanatomical techniques, In: Wouterlood, F. G., Van den Pol, A., Björklund, A., and Hökfelt, T. (eds.), *Handbook of Chemical Neuroanatomy*, Vol. 8, Amsterdam: Elsevier Science Publishers, pp. 47–124.

Kirkham, M., Fujita, A., Chadda, R., Nixon, S. J., Kurzchalia, T. V., Sharma, D. K., Pagano R. E., Hancock, J. F., Mayor, S., and Parton, R. G., 2005, Ultrastructural identification of uncoated caveolin-independent early endocytic vehicles, *J. Cell Biol.* **168**:465–476.

Köbbert, C., Apps, R., Bechmann, I., Lanciego, J. L., Mey, J., and Thanos, S., 2000, Current concepts in neuroanatomical tracing, *Prog. Neurobiol.* **62**:327–351.

Kristensson, K., and Olson, Y., 1971, Retrograde axonal transport of protein, *Brain Res.* **29**:363–365.

Kuypers, H. G. J. M., and Huisman, A. M., 1984, Fluorescent neuronal tracers, *Adv. Cell. Neuro-biol.* **5**:307–340.

Lanciego, J. L., and Giménez-Amaya, J. M., 1999, Notes on the combined use of V-VIP and DAB peroxidase substrates for the detection of colocalising antigens, *Histochem. Cell Biol.* **111**:305–311.

Lanciego, J. L., Goede, P. H., Witter, M. P., and Wouterlood, F. G., 1997, Use of peroxidase substrate Vector® VIP for multiple staining in light microscopy, *J. Neurosci. Methods* **74**:1–7.

Lanciego, J. L., Gonzalo, N., Castle, M., Sanchez-Escobar, C., Aymerich, M. S., and Obeso, J. A., 2004, Thalamic innervation of striatal and subthalamic neurons projecting to the rat entopeduncular nucleus, *Eur. J. Neurosci.* **19:**1267–1277.

Lanciego, J. L., Luquin, M. R., Guillén, J., and Giménez-Amaya, J. M., 1998b, Multiple neuroanatomical tracing in primates, *Brain Res. Protoc.* **2:**323–332.

Lanciego, J. L., and Wouterlood, F. G., 1994, Dual anterograde axonal tracing with *Phaseolus vulgaris* leucoagglutinin (PHA-L) and biotinylated dextran amine (BDA), *Neurosci. Protoc.* 94-050-06.

Lanciego, J. L., Wouterlood, F. G., Erro, E., Arribas, J., Gonzalo, N., Urra, X., Cervantes, S., and Giménez-Amaya, J. M., 2000, Complex brain circuits studied via simultaneous and permanent detection of three transported neuroanatomical tracers in the same histological section, *J. Neurosci. Methods* **103:**127–135.

Lanciego, J. L., Wouterlood, F. G., Erro, E., and Giménez-Amaya, J. M., 1998a, Multiple axonal tracing: simultaneous detection of three tracers in the same histological section, *Histochem. Cell Biol.* **110:**509–515.

LaVail, J. H. 1975, The retrograde transport method, *Fed. Proc.* **34:**1618–1624.

LaVail, J. H., and LaVail, M. M., 1972, Retrograde axonal transport in the central nervous system, *Science* **176:**1416–1417.

Luppi, P. H., Fort, P., and Jouvet, M., 1990, Iontophoretic application of unconjugated cholera toxin β subunit (CTB) combined with immunohistochemistry of neurochemical substances: a method for transmitter identification of retrogradely labeled neurons, *Brain Res.* **534:**209–224.

Luppi, P. H., Sakai, K., Salvert, D., Fort, P., and Jouvet, M., 1987, Peptidergic hypothalamic afferents to the cat nucleus raphe pallidus as revealed by double immunostaining technique using unconjugated cholera toxin as a retrograde tracer, *Brain Res.* **402:**339–345.

Mauro, A., Germano, I., Giaccone, G., Giordana, M. T., and Schiffer, D., 1985, 1-Naphthol basic dye (1-NBD), an alternative to diaminobenzidine (DAB) in immunoperoxidase techniques, *Histochemistry* **83:**97–102.

Mesulam, M. M., 1976, The blue reaction product in horseradish peroxidase neurohistochemistry, *J. Histochem. Cytochem.* **24:**1273–1280.

Mesulam, M. M., 1978, Tetramethylbenzidine for horseradish peroxidase neurohistochemistry: a non-carcinogenic blue reaction product with superior sensitivity for visualizing neural afferents and efferents, *J. Histochem. Cytochem.* **26:**106–117.

Mesulam, M. M., 1982, Principles of horseradish peroxidase neurochemistry and their applications for tracing neural pathways-axonal transport, enzyme histochemistry and light microscopic analysis, In: Mesulam, M. M. (ed.), *Tracing Neural Connections with Horseradish Peroxidase*, New York: Wiley: IBRO Handbook Series: Methods in the Neurosciences, pp. 1–551.

Millhouse, O. E., 1981, The Golgi methods, In: Heimer, L., and RoBards, M. (eds.), *Neuroanatomical Tract-Tracing Methods*, New York: Plenum Press, pp. 311–344.

Nance, D. M., and Burns, J., 1990, Fluorescent dextrans as sensitive anterograde neuroanatomical tracers: applications and pitfalls, *Brain Res. Bull.* **25:**139–145.

Novikova, L., Novikov, L., and Kellerth, J.-O., 1997, Persistent neuronal labeling by retrograde fluorescent tracers: a comparison between Fast Blue, Fluoro-Gold and various dextran conjugates, *J. Neurosci. Methods* **74:**9–15.

Pieribone, V. A., and Aston-Jones, G., 1988, The iontophoretic application of Fluoro-Gold for the study of afferents to deep brain nuclei, *Brain Res.* **475:**259–271.

Rajakumar, N., Elisevich, K., and Flumerfelt, B. A., 1993, Biotinylated dextran: a versatile anterograde and retrograde neuronal tracer, *Brain Res.* **607:**47–53.

Reiner, A., Veenman, C. L., Medina, L., Jiao, Y., Del Mar, N., and Honig, M. G., 2000, Pathway tracing using biotinylated dextran amines, *J. Neurosci. Methods* **103:**23–37.

Risold P. Y., and Swanson L. W., 1995, Evidence for a hypothalamothalamocortical circuit mediating pheromonal influences on eye and head movements, *Proc. Natl. Acad. Sci. U. S. A.* **92:**3902–3989.

Rosene, D. L., Roy, N. J., and Davis, B. J., 1986, A cryoprotection method that facilitates cutting frozen sections of whole monkey brain for histological and histochemical processing without freezing artifact, *J. Histochem. Cytochem.* **34**:1301–1316.

Schmued, L. C., 1994, Anterograde and retrograde neuroanatomical tract-tracing with fluorescent compounds, *Neurosci. Protoc.* 94-050-02.

Schmued, L. C., and Fallon, J. H., 1986, Fluoro-Gold: a new fluorescent tracer with numerous unique properties, *Brain Res.* **377**:147–154.

Schmued, L. C., and Heimer, L., 1990, Iontophoretic injection of fluorogold and other fluorescent tracers, *J. Histochem. Cytochem.* **38**:721–723.

Schmued, L. C., Kyriakidis, K., and Heimer, L., 1990, In vivo anterograde and retrograde axonal transport of the fluorescent rhodamine-dextran-amine, Fluoro-Ruby, within the CNS, *Brain Res.* **526**:127–134.

Skirboll, L. R., Thor, K., Helke, C., Hökfelt, T., Robertson, B., and Long, R., 1989, Use of retrograde fluorescent tracers in combination with immunohistochemical methods, In: Zaborszki, L., and Heimer, L., (eds.), *Neuroanatomical Tract-Tracing Methods 2*, New York: Plenum Press, pp. 5–18.

Smith, Y., and Bolam, J. P., 1991, Convergence of synaptic inputs from the striatum and the globus pallidus onto identified nigrocollicular cells in the rat: a double anterograde labeling study, *Neuroscience* **44**:45–73.

Smith, Y., and Bolam, J. P., 1992, Combined approaches to experimental neuroanatomy: combined tracing and immunocytochemical techniques for the study of neuronal microcircuits, In: Bolam, J. P. (ed.), *Experimental Neuroanatomy, A Practical Approach*, Oxford: Oxford University Press, pp. 239–266.

Somogyi, P., Hodgson, A. J., and Smith, A. D., 1979, An approach to tracing neuron networks in the cerebral cortex and basal ganglia. Combination of Golgi staining, retrograde transport of horseradish peroxidase and anterograde degeneration of synaptic boutons in the same material, *Neuroscience* **4**:1805–1852.

Stoeckel, K., Schwab, M. E., and Thoenen, H., 1977, Role of gangliosides in the uptake and retrograde transport of cholera and tetanus toxin as compared to nerve growth factor and wheat germ agglutinin, *Brain Res.* **132**:273–285.

Ter Horst, G. J., Groenewegen, H. J., Karst, H., and Luiten, P. G. M., 1984, *Phaseolus vulgaris* leucoagglutinin immunohistochemistry. A comparison between autoradiographic and lectin tracing of neuronal efferents, *Brain Res.* **307**:379–383.

Thomas M. A., and Lemmer B., 2005, HistoGreen: a new alternative to 3.3′-diaminobenzidine-tetrahydrochloride-dihydrate (DAB) as a peroxidase substrate in immunohistochemistry? *Brain Res. Protoc.* **14**:107–118.

Trojanowski, J. Q., Gonatas, J. O., and Gonatas, N. K., 1981, Conjugates of horseradish peroxidase (HRP) with cholera toxin and wheat germ agglutinin are superior to free HRP as orthograde transported markers, *Brain Res.* **223**:381–385.

Trojanowski, J. Q., Gonatas, J. O., Steiber, A., and Gonatas, N. K., 1982, Horseradish peroxidase (HRP) conjugates of cholera toxin and lectins are more sensitive retrograde transported markers than free HRP, *Brain Res.* **231**:33–50.

Trojanowski, J. Q., and Schmidt, M. L. 1984, Interneuronal transfer of axonally transported proteins: studies with HRP and HRP conjugates of wheat germ agglutinin, cholera toxin and the B subunit of cholera toxin, *Brain Res.* **311**:366–369.

Van Bockstaele, E. J., Wright, A. M., Cesari, D. M., and Pickel, V., 1994, Immunolabeling of retrogradely transported fluorogold. Sensitivity and application of ultrastructural analysis of transmitter-specific mesolimbic circuitry, *J. Neurosci. Methods* **55**:65–78.

Veenman, C. L., Reiner, A., and Honig, M. G., 1992, Biotinylated dextran amine as an anterograde tracer for single- and double-label studies, *J. Neurosci. Methods* **41**:239–254.

Vetter, D. E., Saldaña, E., and Mugnaini, E., 1993, Input from the inferior colliculus to medial olivocochlear neurons in the rat: a double label study with PHA-L and cholera toxin, *Hear. Res.* **70**:173–186.

Wan, X. C. S., Trojanowski, J. Q., and Gonatas, J. O., 1982, Cholera toxin and wheat germ agglutinin conjugates as neuroanatomical probes: their uptake and clearance, transganglionic and retrograde transport and sensitivity, *Brain Res.* **243**:215–224.

Warr, W. B., de Olmos, J. S., and Heimer, L., 1981, Horseradish peroxidase: the basic procedure, In: Heimer, L., and RoBards, M. (eds.), *Neuroanatomical Tract-Tracing Methods*, New York: Plenum Press, pp. 207–262.

Weiss, P., and Hiscoe, H. B., 1948, Experiments on the mechanism of nerve growth, *J. Exp. Zool.* **107:**315–396.

Wessendorf, M. W., 1991, Fluoro-Gold: composition and mechanism of uptake, *Brain Res.* **553:**135–148.

Woolf, N. J., Hernit, M. C., and Butcher, L. L., 1986, Cholinergic and noncholinergic projections from the rat basal forebrain revealed by combined choline acetyltransferase and *Phaseolus vulgaris*-leucoagglutinin immunohistochemistry, *Neurosci. Lett.* **66:**281–286.

Wouterlood, F. G., Bol, J. G. J. M., and Steinbusch, H. W. M., 1987, Double-label immunocytochemistry: combination of anterograde neuroanatomical tracing with *Phaseolus vulgaris*-leucoagglutinin and enzyme immunocytochemistry of target neurons, *J. Histochem. Cytochem.* **35:**817–823.

Wouterlood, F. G., Goede, P. H., Arts, M. P. M., and Groenewegen, H. J., 1992, Simultaneous characterization of efferent and afferent connectivity, neuroactive substances and morphology of neurons, *J. Histochem. Cytochem.* **40:**457–465.

Wouterlood, F. G., Goede, P. H., and Groenewegen, H. J., 1990, The in situ detectability of the neuroanatomical tracer *Phaseolus vulgaris*-leucoagglutinin, *J. Chem. Neuroanat.* **3:**11–18.

Wouterlood, F. G., and Groenewegen, H. J., 1985, Neuroanatomical tracing by use of *Phaseolus vulgaris*-leucoagglutinin (PHA-L): electron microscopy of PHA-L filled neuronal somata, dendrites, axons and axon terminals, *Brain Res.* **326:**188–191.

Wouterlood, F. G., and Groenewegen, H. J., 1991, The *Phaseolus vulgaris*-leucoagglutinin tracing technique for the study of neuronal connections, *Prog. Histochem. Cytochem.* **22:**1–78.

Wouterlood, F. G., and Jorritsma-Byham, B., 1993, The anterograde neuroanatomical tracer biotinylated dextran amine: comparison with the tracer PHA-L in preparations for electron microscopy, *J. Neurosci. Methods* **48:**75–87.

Wouterlood, F. G., Mugnaini, E., and Nederlof, J., 1985, Projection of olfactory bulb efferents to layer I GABA-ergic neurons in the entorhinal area. Combination of anterograde degeneration and immunoelectron microscopy in rat, *Brain Res.* **343:**283–296.

Zaborszky, L., and Cullinan, W. E. 1989, Hypothalamic axons terminate on forebrain cholinergic neurons: an ultrastructural double-labeling study using PHA-L tracing and ChAT immunocytochemistry, *Brain Res.* **479:**177–184.

Zaborszky, L., Leranth, C., and Heimer, L., 1984, Ultrastructural evidence of amygdalofugal axons terminating on cholinergic cells of the rostral forebrain, *Neurosci. Lett.* **21:**219–225.

Zhou, M., and Grofova, I., 1995, The use of peroxidase substrate Vector VIP in electron microscopic single and double antigen localization, *J. Neurosci. Methods* **62:**149–158.

12

Tract-Tracing in Developing Systems and in Postmortem Human Material Using Carbocyanine Dyes

ZOLTÁN MOLNÁR, DANIEL BLAKEY,
IRINA BYSTRON, and
ROSALIND S. E. CARNEY

ZOLTÁN MOLNÁR, DANIEL BLAKEY AND ROSALIND S. E. CARNEY
• Department of Human Anatomy and Genetics, University of Oxford, South Parks Road,
Oxford OX1 3QX, United Kingdom   IRINA BYSTRON • University Laboratory of
Physiology, University of Oxford, Parks Road, Oxford OX1 3PT, United Kingdom; Department
of Morphology, Institute of Experimental Medicine, 12 Acad. Pavlova, St Petersburg 197376,
Russia   ROSALIND S. E. CARNEY • Department of Neuroscience, Research Building
EP08, Georgetown University School of Medicine, Washington, DC 20057, USA

CONCLUDING REMARKS
APPENDIX
   Protocol for DiI Placement and Histological Processing (Modified
     from Molnár *et al.*, 1998)
   Protocol for Photoconversion (Modified from Lübke, 1993)
REFERENCES

Abstract: Rapid progress in neurobiology and genetics demands knowledge of fundamental aspects of brain development including the connectivity patterns within developing and adult brains. The primary focus of this chapter is on neuroanatomical tract-tracing using carbocyanine dyes which have several advantages over traditional tracing methods. First utilized for in vitro studies, a major breakthrough in the late 1980s was the demonstration that carbocyanine dyes act as anterograde and retrograde tracers in fixed tissue, eliminating the need for diffusion of tracers in vivo. Moreover, carbocyanine dyes are more efficacious than classical tracing methodologies especially during early stages of development, and consequently have been used to reveal the spatiotemporal patterns of axonal development in different species. Furthermore, the unique properties of the carbocyanine dye tracing method have opened up new avenues for tracing connections in human postmortem specimens. This is a key step in determining the precise connectivity of neural circuits in the human brain, and subsequently to relate this knowledge to pathological cases.

    The success of carbocyanine dyes as tracers, both in vitro and in fixed material, is reflected in the flurry of publications throughout the 1990s and into the present. However, there are relatively few systematic studies that have tested parameters to optimize their use or to give practical advice to enhance their efficacy. This chapter aims to bring together some of our experiences with the carbocyanine dye tracing method drawn from our studies in mammalian, reptilian, and human and nonhuman primate specimens.

Keywords: carbocyanine dyes, cortex, development, photoconversion, primate, thalamus

I. INTRODUCTION

    Topographic patterns of cortico-cortical and cortico-subcortical axonal projections have been elucidated in classical studies examining brain connectivity using techniques considered time-consuming, laborious, and expensive. Such techniques often require demanding surgical procedures to deliver tracers into living animals, which are subsequently sacrificed after sufficient transport time to reveal the tracer in the target tissue (for overview see Zaborszky and Heimer, 1989). These methods necessitate maintaining a surgical facility and appropriate housing of animal colonies. The application of carbocyanine dyes (Godement *et al.*, 1987; Honig and Hume, 1986) for neuroanatomical tract-tracing in fixed brains from embryonic and early postnatal animals provided an opportunity to simplify this process. The use of fixed brains gives the experimenter more precise temporal control, as fixation provides a snapshot of the status of axonal development during embryogenesis. Spatial control is achieved by having access to a particular region, various nuclei, or lamina under direct visual control in fixed brain specimens, in contrast to earlier methods requiring stereotactic injections. Next to the controllability of carbocyanine dye tracing, one of its biggest advantages is the opportunity it provides to trace in human postmortem brain tissue.

In this chapter, we discuss classical methods of neuroanatomical tract-tracing coupled with the emergence of carbocyanine dyes as a powerful tool in the field of developmental neurobiology. Whereas classical techniques provided fundamental knowledge of connectivity within the adult brain, the developmental timetable of these projections has been revised using carbocyanine dye tracing. Carbocyanine dye tracing continues to flourish in publications from investigative studies of rodent brain development and particularly in transgenic models. This has enabled developmental neurobiologists to form causal links between targeted gene ablation and abnormal axonal pathfinding.

II. OVERVIEW OF CLASSICAL AND MODERN TRACT-TRACING TECHNIQUES

A. Classical Neuroanatomical Tract-Tracing Techniques

The Golgi stain, used by neuroanatomists in the nineteenth century, provided a means of labeling neurons and was used to visualize their soma and processes. This technique was extensively used in embryonic and early postnatal animals and humans. However, the method was not conducive to the study of axonal tracts, as individual axons could not be traced from one brain region to the next. However, the eminent neuroanatomist Ramon y Cajal was able to perform some axon tracing analysis in embryonic brains using the Golgi staining techniques (Cajal, 1909). He also found that the stain gave better results in the nervous system of young and developing animals compared to the mature brain. Indeed, Golgi impregnation in fibers stops when the fiber acquires its myelin sheath (Wouterlood and Mugnaini, 1984). In the mid-twentieth century, axon tracing was accomplished by injuring neurons and analyzing their subsequent degeneration to follow the trajectory of the neurites (Nauta and Gygax, 1951). Modern techniques involve the injection of tracers into the brain, which are transported through axons anterogradely (e.g., autoradiography of labeled amino acids or *Phaseolus vulgaris*), and/or retrogradely (e.g., fluorescent dyes, rhodamine beads; see Zaborszky and Heimer, 1989). Transneuronal transport can also be achieved with wheat germ agglutinin–horseradish peroxidase.

B. Modern Techniques Using Carbocyanine Dyes

The neuroanatomical visualization of fiber projections has been markedly enhanced by the use of the lipophilic carbocyanine dye series

TABLE 12.1. Specifications of the most commonly used carbocyanine dyes (based on information provided by Molecular Probes).[1]

Common name	Chemical name	Absorption wavelength peak (nm)	Fluorescence emission peak (nm)	Molecular Probes catalog numbers
DiI	1,1′-Dioctadecyl-3,3,3′,3′-tetramethylindocarbocyanine perchlorate	551	569	D-282, D-383, D-384, D-3886, D-3911, N-22880
Fast DiI	1,1′-Dilinoleyl-3,3,3′,3′-tetramethylindocarbocyanine, 4-chlorobenzenesulfonate	551[2]	569[2]	D-3899, D-7756
DiA	4-(4-(Dihexadecylamino)styryl)-N-methylpyridinium iodide	456	590	D-3883, D-3897, D-7758
DiAsp	4-(4-(Didecylamino)styryl)-N-methylpyridinium iodide	456[2]	590[2]	D-291
DiO	3,3′-Dioctadecyloxacarbocyanine perchlorate	484	501	D-275, D-1125, D-3898, N-22881
DiD	1,1′-Dioctadecyl-3,3,3′,3′-tetramethylindodicarbocyanine, 4-chlorobenzenesulfonate salt	644	665	D-307, D-7757, N-22882

[1] http://probes.invitrogen.com/handbook/tables/0346.html.
[2] Spectral data stated represent estimated values based on similar compounds.

(Honig and Hume, 1986). Initially these dyes were used to study membrane fluidity by cell biologists. Subsequently, tracers such as DiI (1,1′-dioctadecyl-3,3,3′,3′-tetramethylindocarbocyanine perchlorate; all carbocyanine dyes listed are from Molecular Probes, Eugene, OR) and DiO (3,3′-dioctadecyloxacarbocyanine perchlorate), DiD (1,1′-dioctadecyl-3,3,3′,3′-tetramethylindodicarbocyanine, 4-chlorobenzenesulfonate salt), DiAsp (4-(4-(didecylamino)styryl)-N-methylpyridinium iodide), and DiA (4-(4-(dihexadecylamino)styryl)-N-methylpyridinium iodide) were used to label living, cultured neurons, enabling neuronal interactions and cell migration to be examined in vitro (Honig and Hume, 1986). See Table 12.1 for absorption and emission spectra for various carbocyanine dyes, listed with catalog numbers from Molecular Probes.

Additional applications of carbocyanine dye tracing in vitro have been studied incorporating time-lapse video microscopy to track axon growth or cell migration within developing systems (for review see Fishell *et al.*, 1995). A forebrain explant from murine cortex showed that DiI-labeled precursor cells can undergo lateral dispersion within the ventricular zone prior to radial migration so as to reach the developing cortical plate (Fishell *et al.*, 1993). Carbocyanine dye labeling of cells of the lateral ganglionic eminence in organotypic slice preparations was used in one of the first studies that showed tangential migration of interneurons to the dorsal cortex (De Carlos *et al.*, 1996). Time-lapse video microscopy was also used to describe growth cone morphology of thalamic axons in vitro in thalamocortical slices (Skaliora *et al.*, 2000) or cocultures (Yamamoto *et al.*, 1997).

The discovery by Godement *et al.* (1987) that the dye series is effective for the antero- and retrograde labeling of fiber populations in fixed tissue was a significant breakthrough. The dyes are thought to diffuse laterally through the plasma membrane of the cell at a rate of ~6 mm/day in living tissue (Honig and Hume, 1989), with a slower diffusion rate in fixed tissue of ~2 mm/day. The carbocyanine dye series contains several variants with different absorption/emission spectra permitting simultaneous use. DiI, which is excited in the green range and fluoresces in a red/orange color under a rhodamine filter, is often used in combination with DiA, which emits in a green wavelength, for two-color imaging (Honig and Hume, 1989). DiI and DiA can be detected with rhodamine and fluorescein optical filters, respectively. Such multicolor labeling is useful for visualizing multiple pathways concurrently, or for topographic labeling within a single axonal pathway (see Agmon *et al.*, 1996; Bicknese *et al.*, 1994; Miller *et al.*, 1993; Molnár *et al.*, 1998a). Figure 12.1 demonstrates several examples of the application of DiI tracing in various systems.

The mechanism of tracing with carbocyanine dyes is based on their lipid solubility. DiI becomes incorporated into the outer leaflet of the plasma membrane and can laterally diffuse with negligible transfer between intact cell membranes in living and fixed material (Godement *et al.*, 1987; Honig and Hume, 1986, 1989). There is some evidence that the dye molecules form detergent-like micelles which are sparingly soluble in the cytoplasm and can therefore "fill" the entire cell (Bruce *et al.*, 1997). The direction of diffusion can be either toward the cell body (retrograde) or toward the distal end of the axon (anterograde). Simultaneous retrograde and anterograde labeling can be either a desirable occurrence or an adverse occurrence, depending on the objective of the experiment. The ability to label bidirectionally can be useful to detect the source of fibers that project through the dye placement site. For example, carbocyanine crystal placement into the internal capsule of the embryonic forebrain reveals the origin of the early cortical

---→

Figure 12.1. Examples for the application of carbocyanine dye tracing in various systems. (A) Pyramidal neurons were backlabeled through their projections across the corpus callosum in a postnatal rat brain (P3). A DiI crystal was placed in the contralateral hemisphere and the brain was incubated for 3 weeks at 37°C. Numerous layer 5 and layers 2/3 cells were labeled. At this early stage, they all possess an apical dendrite with a terminal tuft reaching the marginal zone. (B) Two cerebral cortical slices were taken at P3 and cocultured for 2 weeks. Then, the cultures were fixed and DiI crystals were placed into one explant. This layer 5 pyramidal cell was backlabeled with DiI indicating in vivo-like connectivity between the explants. (C) Organized reciprocal connections cross the embryonic rat internal capsule in this horizontal section of an E20 rat brain counterstained with bisbenzimide. Multiple carbocyanine dye placements were made along an anteroposterior parasagittal line into the convexity of the cortex (DiA, DiI, and DiAsp). Since thalamic fibers have already arrived at the cortex prior to E16, each placement labeled a mixed bundle of thalamocortical and corticofugal axons. Six weeks incubation at room temperature

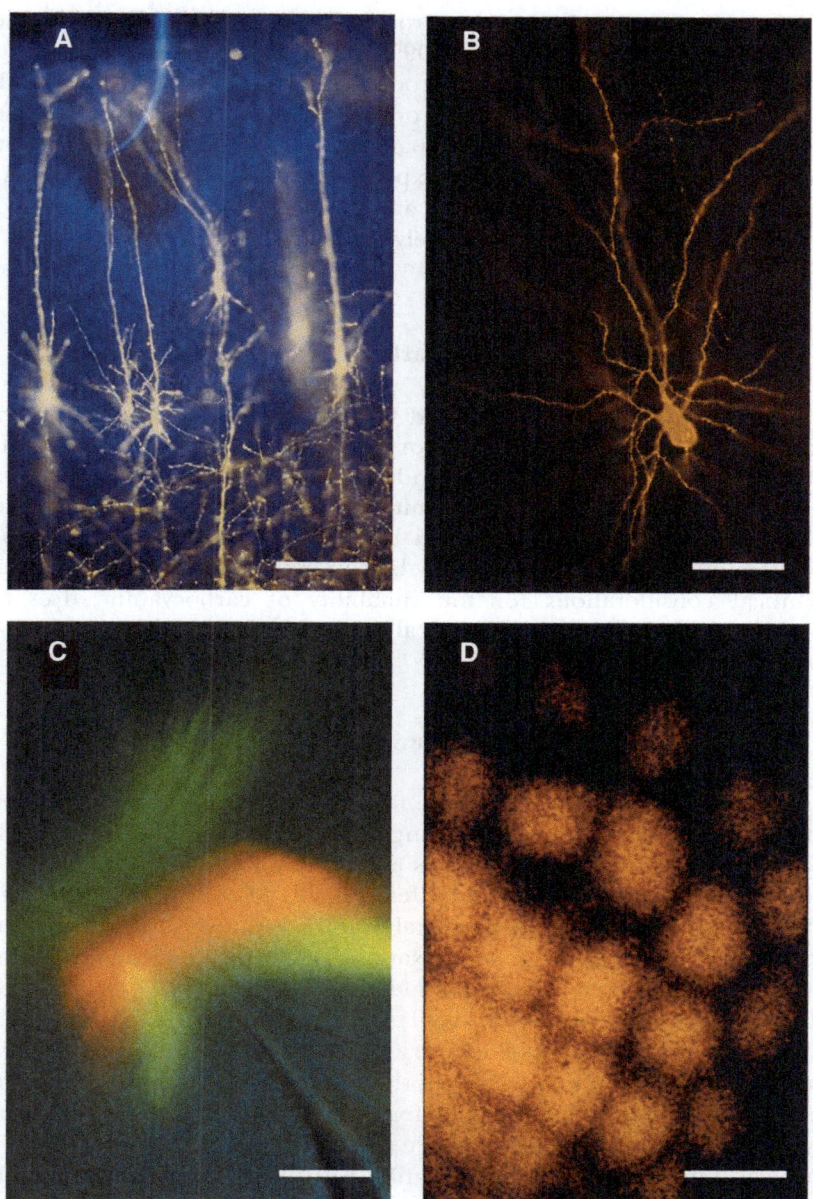

Figure 12.1. (*Cont.*) was used to enable full anterograde and retrograde diffusion. Three distinct bundles are clearly visible passing through the primitive internal capsule without substantial mixing or crossing. (D) Carbocyanine dye labeling (DiI) revealed the periphery-related patterning of thalamocortical axons in a tangential section through the barrel cortex in the primary somatosensory cortex of a P3 rat brain. A DiI crystal was placed into the ventrobasal complex and we used 4 weeks of incubation at 37°C. The nonlabeled darker areas represent the emerging septa. Scale bars: 50 μm (A); 20 μm (B); 100 μm (C); 250 μm (D).

and thalamic neurons that project through the internal capsule (McConnell *et al.*, 1989; Molnár and Cordery, 1999). As well as revealing the origin of axons by retrograde transport, anterograde diffusion of the dye reveals the target of the fibers of passage that incorporates the dye at the placement site. This can be useful for investigating patterns of connectivity that are poorly understood. According to our own experience, there is limited scope to manipulate the relative strength of the anterograde and retrograde transport by reducing or increasing, respectively, the incubation periods.

C. Methodology of Carbocyanine Dye Tracing

This section details the standard methodology used to apply carbo-cyanine dyes, either in solid form or as a solution, to fixed and in vitro preparations, tissue processing, and analysis. The section "Additional Techniques: Combination with Immunohistochemistry, Photoconversion of Carbocyanine Dye-Labeled Material, and Electron Microscopic Analysis" discusses other techniques that may be combined with dye tracing. Some technical considerations for the suitability of carbocyanine dyes are considered in the section "Technical Considerations and Limitations of Carbocyanine Dye Tracing."

1. Fixation and Storage of the Tissue

Chemical fixation using aldehyde fixatives is compatible with carbocya-nine dye tracing. Generally, in younger embryos [< E16 (embryonic day 16) in mice, E17 in rat], the heads are removed and fixed overnight in 4% formaldehyde in phosphate buffer (PB) 0.1 M pH 7.4 (4% formalin), whereas older embryos and postnatal animals require transcardial perfusion prior to overnight immersion fixation. Thereafter, tissue may be stored in phosphate-buffered saline with sodium azide (PB 0.1 M, 0.09% NaCl, 0.05% sodium azide, pH 7.4) to avoid microbial growth, and to prevent antigen masking if immunohistochemistry is desired. Additional glutaraldehyde during initial fixation may be used if the tissue is to be analyzed by electron microscopy; however, fluorescence from the glutaraldehyde may produce higher background and thus reduce the signal-to-noise ratio of the dye. Carbocyanine dye tracing is incompatible with alcohol fixation and paraffin embedding of tissue. It is also believed that freshly fixed brains produce better results, but we have observed successful labeling in material that has had extended storage in a fixative (4% formaldehyde). Nonetheless, a prolonged postfixation period is not recommended; the resultant high precipitation of membrane and cytoplasmic proteins can lead to slower diffusion of DiI (Bruce *et al.*, 1997), and polymerization of the formaldehyde releases alcohol moieties which may impair labeling. The carbocyanine dye diffusion speed has not been systematically analyzed as a function of the length or the strength of the fixation.

2. Tracer Delivery

There are several methodologies in use for the delivery of carbocyanine dyes in living and fixed tissue:

(A) *One of the most widely used methods is crystal placement.* Small individual crystals (~100 μm in size) are picked up and inserted into the tissue using platinum wires or insect pins. If the crystals are to be inserted through a membrane rather than just into nervous tissue, preboring a hole is usually desirable. The tissue, including cell membranes, is damaged at the implantation site, but this is believed to facilitate the dye uptake. This method is equally suitable for living and fixed tissues (see McConnell *et al.*, 1989; Molnár *et al.*, 1998a). Removing loose or sprinkled fragments of carbocyanine crystals, attached to the surface of the specimen with a damp tissue or by pipetting off with H_2O/buffer, avoids contamination of the label from the site of attachment. Otherwise, tracer diffusion from aberrant crystals may confound the interpretation of the experiment.

(B) *Solution injection.* Saturated solutions of the dyes may be made up in 95% ethanol, 5% DMSO. Picospritzers (Intracel, Royston, UK) are used to inject nanoliter volumes of dye into the tissue (see Agmon *et al.*, 1996; Métin and Godement, 1996). Close to the site of the injection, there is aspecific diffusion, which is partially due to the alcohol solution. This method of delivery is also suitable for fixed specimens in addition to living tissues in vivo and in vitro.

(C) *Solution soaking can be used to label culture explants or grafts in living tissue.* To label the entire surface of the tissue, the explant may be briefly immersed in a dye solution followed by rapid and extensive rinsing. "Microcrystals" of dye will precipitate onto the surface of the explant and neurite outgrowth can be studied within cocultured structures, for example, explants of thalamus and cortex (Molnár and Blakemore, 1991). This method is suitable for in vitro living tissue. During short culture periods (up to 4–6 days in vitro) this method labels neurites, but at later stages the dye becomes incorporated into the cytosol and the fiber labeling gradually disappears. This method will also reveal the movement of motile cells (such as macrophages) within the in vitro preparation as well as migratory cells that incorporated the label at the dye placement site.

(D) *Coating inert substrates.* Tungsten or gold particles are placed onto a slide and a drop of dye solution in methylene chloride is added. The solvent rapidly evaporates, leaving a coating of dye on the particles. These particles can then be delivered into the tissue using a "gene gun" (BioRad, Hercules CA). Single cells and low density cell populations can be labeled in this manner (Gan *et al.*, 2000). This method is suitable for both living and fixed tissues. Alternatively, nitrocellulose sheets may be soaked in the same manner, cut into 0.5-mm squares, and inserted into fixed tissue (Ma *et al.*, 2002).

3. Targeting the Tracer to Specific Anatomical Locations

Successful targeting of specific nuclei or anatomical structures is aided by using a high-magnification binocular light microscope. It is relatively easy in fixed material to dissect away overlying tissue to access nuclei or particular anatomical layers directly. For example, to perform a crystal placement in the lateral geniculate nucleus (LGN), the midbrain may be transected coronally, rostral to the superior colliculus to gain access to the posterior dorsal thalamus. Alternatively, to access the ventrobasal (VB) complex, the brain is bisected longitudinally to separate the hemispheres, and the medial part of the diencephalon is cut away (Molnár *et al.*, 1998a). The use of whole mount preparations, such as the intact brain, is advantageous for fixed tissue tracing as the whole axonal projection should be maintained. It is also possible to perform the tracer delivery in sectioned tissue (see sectioning in "Sectioning of the Tissue"); however, this is advisable only if the connections are already well understood, and can be preserved in the preparation. For example, thalamocortical slice preparations can maintain connectivity between thalamus and cortex for the somatosensory or the auditory pathways, yet a similar preparation is not possible for the visual pathway (Agmon and Connors, 1991).

4. Incubation Periods

The temperature and length of the incubation are two main parameters that can influence dye diffusion. These variables are specific to different carbocyanine dyes. Additional techniques of enhancing diffusion in fixed human specimens (Sparks *et al.*, 2000) are discussed in the section "Carbocyanine Dye Tracing in Primates, Including Normal and Pathological Postmortem Human Specimens" which reviews carbocyanine tracing in primates.

Incubation at 37°C is believed to enable a faster diffusion rate in fixed tissue (see Table 12.2 from Molnár *et al.*, 1998a), and carbocyanine dyes with shorter fatty acid chain diffuse faster than tracers with longer chains (see Fast

TABLE 12.2. Incubation periods (weeks) needed for carbocyanine dye (DiI) transport in embryonic and early postnatal rats to label thalamocortical projections (Molnár *et al.*, 1998b).

Age	Incubation period (weeks)	
	22°C	37°C
E13–E14.5	2	1
E14.5–E16	3–6	2–4
E18–P0	6	4
P0–P8	6–10	4–8

DiI, in section "Altering Experimental Variables to Attempt Enhancement of DiI Diffusion in Primate Brains"). For young embryonic specimens, room temperature is preferred because the incubation period is longer and thus easier to control, while background staining and transneuronal labeling (discussed below) are less apparent. Different carbocyanine dyes can vary in their optimal incubation periods in a given specimen. In mammalian fixed brain tissue, as aforementioned, DiI (18 carbons "C18") proves to be the most reliable tracer, yielding the highest intensity of fluorescent labeling and appropriate background staining. The incubation period required for DiA or DiAsp seems to be shorter than that used for DiI. Therefore, for an E16 tracing experiment, we routinely implant DiA crystals a week later than DiI crystals (see Molnár *et al.*, 1998a). This delayed implantation protocol optimizes the transport period for both dyes.

DiA, DiD, and DiAsp (see *Molecular Probes Handbook*) are also reasonable tracers in fixed embryonic and early postnatal rodent tissue, but the background staining tends to be higher, despite a shorter incubation period than that used for DiI. However, such parameters may strongly depend on several factors, such as the length of storage in fixative, the age of the specimen, and even the particular pathway studied. Table 12.2 gives a general idea on the incubation periods needed for DiI labeling of the thalamocortical pathway in embryonic and early postnatal rats (Molnár *et al.*, 1998b). DiD has also been used in combination with DiI in other studies (Catalano and Shatz, 1998), although we have not found these tracers to be as effective. While DiO is an excellent tracer in living tissue, its transport is not as impressive in fixed material.

5. Sectioning of the Tissue

Tracing with carbocyanine dyes limits the repertoire of the storage and sectioning methods generally used for tissue processing. The specimen should be stored without freezing in 4% formalin (0.1 M PBS with 0.05% NaN_3). Methods that require freezing (cryostat, sliding microtome) or wax embedding are not compatible, and thus Vibratome sectioning is most commonly used (Vibratome, Oxford Instruments or Vibroslicer, Leica VT1000S). Embryonic and early postnatal sections are fragile; therefore, it is advisable to embed the tissue in agarose (4% in PBS), which is heated and mixed until the agar is fully dissolved, typically in a microwave oven. The agarose must be allowed to cool prior to embedding the tissue; in most cases 50°C is appropriate. To further reduce any heat damage, we place the embedding molds on crushed ice and fill with heated agarose. Once the correct temperature has been reached, the brain is then placed into the agarose. For small embryonic brains low melting agar (e.g., Type VII, Sigma A-4018) should be used. After cutting it is not necessary to remove the agar surrounding the section, which in fact aids mounting the section onto the slide and maintain morphology.

6. Counterstaining, Mounting, and Analysis of Carbocyanine Dye-Labeled Material

This section deals with counterstaining and mounting of carbocyanine dye-labeled material for immediate analysis with epifluorescence and confocal microscopy. Further processing for photoconversion of DiI label into a permanent product, electron microscopy, and immunohistochemistry is addressed in the section "Additional Techniques: Combination with Immunohistochemistry, Photoconversion of Carbocyanine Dye-Labeled Material, and Electron Microscopic Analysis."

(A) *Counterstaining.* To reveal the major anatomical subdivisions and cytoarchitecture of various structures, fluorescent "Nissl" counterstains can be used. We routinely use Bisbenzimide (10 min in 2.5 µg/ml solution in PBS, protected from light), although acridine orange (10 µg/ml in PBS, 10 min) or DAPI or Neuro Trace 500/525 green or the Sytox counterstains are also excellent (see *Molecular Probes Handbook* for concentrations of other counterstains; www.probes.com). Obviously, the fluorescence signal of the counterstain must not conflict with that of the dye, and therefore the choice of counterstain is dependent on the carbocyanine dye used; e.g., DiA is not compatible with acridine orange.

(B) *Mounting media.* The recommended mounting media are PBS, various versions of PBS–glycerol solutions (1:1, 3:1, or 1:9 mixture), or Hydromount (National Diagnostics). Antiquenching agents (1% *p*-phenylenediamine, 10% PBS, and 90% glycerol) (e.g., Johnson *et al.*, 1982) can prolong the signal for numerous other methods, but we did not find this particularly beneficial for carbocyanine dyes (Molnár and Carney, unpublished observations, 2003). Since the label is robust, the major problem is the deterioration of the label at the cut surfaces and nonspecific signal. Degradation is inevitable after a few weeks, and therefore it is advisable that the material is analyzed shortly after sectioning (preferably within the first 3 days).

(C) *Coverslipping and storage.* During mounting, it is imperative to cover the sections before they dry out, otherwise the label becomes blurry and diffuse. Since the sections do not adhere to the slide firmly, lowering the coverslips should be done with extra care. The coverslips are sealed with Paraseal (Raymond A Lamb, London) or with nail varnish and are stored, light-protected at 4°C.

(D) *Epifluorescence analysis.* DiI and DiA labeling are photographed using TRTIC and FITC filters, respectively. For conventional photography, highly sensitive films (e.g., 400 or 1600 ASA Kodak Ectachrome for Slide) are recommended. For double and triple exposure, it is recommended that the bisbenzimide counterstain or DiA label is exposed for a shorter period than is DiI. Nowadays, digital cameras easily facilitate documentation. After single exposure, the individual frames can be adjusted for brightness, contrast, and can be superimposed on the counterstain label using image processing. Thus, combining dye labeling with nuclear counterstaining provides an accurate view of labeled axonal projections within the context of the anatomical structure. When only one carbocyanine dye is used, e.g., DiI, photographs taken under TRITC filter illumination can be converted to gray scale to create a high-contrast image (gray scale tone inversion may further enhance the image). Multicolor dye tracing provides information on the topographic relationship between different axonal tracts. In fiber systems that exhibit a high topographical organization, alternating dye combinations, such as DiA–DiI–DiA, have been used to reveal the trajectories of neighboring axonal tracts (Molnár *et al.*, 1998a).

(E) *Confocal microscopy.* Confocal microscopy can be used in selected regions not only to reveal precise cellular and subcellular colocalization, cell morphology, dendritic spines, and growth cone morphology of labeled axons, but also the precise topographic relationship between intimately associated axon tracts. Confocal microscopy of the primitive internal capsule following multiple DiI and DiA crystal placements in embryonic rat cortex and thalamus demonstrated the spatial relationship between the developing fiber systems (Molnár *et al.*, 1998a). Labeling from multiple sites of cortex also revealed that the spatial arrangement of adjacent corticofugal and thalamo-cortical axonal tracts was maintained in the *Snap25* knockout brain (Molnár *et al.*, 2002). The bright intensity of DiA labeling can lead to "bleedthrough" in filter settings detecting DiI labeling, but DiI label is not strong enough to cause the reverse effect. Therefore, it is recommended that DiI and DiA labeling is scanned separately during confocal microscopy to avoid false interpretation (see chapter by Wouterlood in this volume).

D. Additional Techniques: Combination with Immunohistochemistry, Photoconversion of Carbocyanine Dye-Labeled Material, and Electron Microscopic Analysis

Carbocyanine dye labeling is relatively robust, but like other fluorophores the signal will degrade over time and when subjected to certain requisite conditions such as illumination during epifluorescence and especially confocal microscopy. Furthermore, unspecific diffusion of the label from the cut surfaces after sectioning and coverslipping will increase the background staining. Hence, carbocyanine dye labeling should be documented as soon as possible with both epifluorescence and confocal microscopy. To avoid inclusion of regional photobleaching, start documentation at lower magnifications and increase as needed. Although tracing with carbocyanine dyes has limitations, it is possible to convert the label into a permanent product, which can be further utilized in combination with immunohistochemistry and in situ hybridization.

1. Carbocyanine Dye Tracing Combined with Immunohistochemistry

Carbocyanine dyes reveal elaborate neuronal morphology and patterns of connectivity, however, it is also desirable to determine the neurochemical properties of such cells. Standard immunohistochemical protocols use permeabilization agents (e.g., Triton X-100) to enable antibody penetration into this tissue, which would disrupt the cell membrane and cause leakage of the dye. Therefore, other axonal tracers such as fluorescent beads are often combined with tracing studies and immunohistochemistry (Voelker *et al.*, 2004) especially from postnatal ages upward. Antibody labeling in DiI-labeled specimens may be obtained by prolonged incubation with the primary antibody and avoiding the use of permeabilization agents, although only the uppermost ~20 μm should be used for analysis because of the limited penetration of the antibody. However, the upper ~5 μm of the tissue may also be unstable because of the leakage of carbocyanine dye from truncated axons; hence the thickness of the tissue with reliable double staining might be limited. The success of DiI tracing and immunofluorescence will largely depend on the antibody used. For example, DiI crystal placement in the internal capsule of E14.5 mice backlabeled cortical pioneer neurons, which expressed the Tbrain1 protein as shown by Tbr1 immunohistochemistry visualized with the Alexa 594 fluorochrome (Hevner *et al.*, 2002). Cytoplasmic markers are likely to be more successful in combination with carbocyanine dyes than nuclear markers, which would require deeper penetration of the antibody. Indeed, immunofluorescence for a nuclear mitotic marker, anti-phospho-histone H3 (1:500; Upstate) in murine cortex, with omission of Triton X-100 from the immunohistochemical protocol, leads to uneven staining with both strong- and weak-labeled mitotic cells (Carney, 2004). Since immunohistochemistry in carbocyanine dye-traced material is performed without the addition of detergent, a considerable amount of false negativity can be expected.

Simultaneous carbocyanine dye tracing and immunohistochemistry would be extremely advantageous to couple aspects of axonal connectivity and neurochemical properties. To this end, Molecular Probes (now owned by Invitrogen) developed a sulfonated DiI derivative (D-7777), supposedly compatible with permeabilization procedures. However, sulfonated DiI did not diffuse from the dorsal thalamus when implanted as a paste or as liquid (in DMSO) in fixed embryonic rat brains (Carney and Molnár, unpublished observations, 2004), and hence does not represent a viable alternative as a tracer at present.

Histological sections with carbocyanine labeling can be further used for permanent immunohistochemistry for robust antigens, such as TAG1, L1, calbindin, or calretinin, following initial documentation of the tracing (López-Bendito *et al.*, 2002).

2. Photoconversion of DiI Label to a Permanent Reaction Product

For the reasons outlined above, immunofluorescence is not always compatible with carbocyanine dyes. Therefore, protocols have been developed that convert the delicate fluorescent label to a permanent diaminobenzidine (DAB) product (Lübke, 1993; Sandell and Masland, 1988). These protocols permit subsequent tissue permeabilization using detergents, which dramatically increase the efficacy of many immunohistochemistry protocols.

Following sectioning of the tissue, the labeled material is impregnated with DAB and exposed to light through an epifluorescence microscope using appropriate filters for visualization of the dye, e.g., a rhodamine filter set for DiI. This initiates a redox reaction between the dye and the DAB, creating the characteristic brown precipitate in labeled cells and neurites. The material can subsequently be used for a plethora of techniques including permanent immunohistochemistry, immunofluorescence, and electron microscopy (Fujimori *et al.*, 1997; Liu *et al.*, 1999; Métin and Godement, 1996; Papadopoulos and Dori, 1993).

However, there are drawbacks to using this method. Due to the limited region illuminated by a medium/high power objective lens, larger areas must be photoconverted in a series of exposures. The technique is time-consuming, as each field may take up to 1 h (Blakey and Molnár, unpublished observations, 2005) to achieve acceptable DAB label density. Photoconversion of DiI-labeled thalamocortical axons in an E17 mouse brain is shown in Fig. 12.2. Photoconversion can be done with carbocyanine dyes

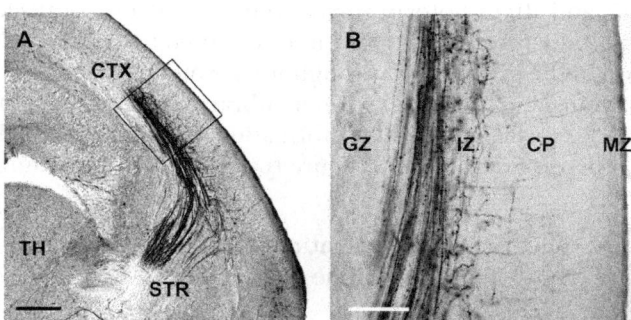

Figure 12.2. Photoconversion of DiI into a permanent DAB precipitate in labeled E17 mouse thalamocortical axons. An E17 mouse head was fixed by immersion in 4% paraformaldehyde overnight. The brain was dissected from the head and a single DiI crystal was inserted in the ventrobasal complex of the developing thalamus, thus filling a population of putative somatosensory axons. The brain was sectioned on a vibroslicer (Leica VT1000S) at a thickness of 75 μm. Slices were selected for axon labeling, and processed according to the protocol described above. As can be observed in panel A, only a limited proportion of the section can be photoconverted at any one time. However, individual fibers are filled and can be observed under higher magnification (Panel B). This material is now suitable for further processing. *Abbreviations:* CP, cortical plate; CTX, cerebral cortex; GZ, germinal zone; IZ, intermediate zone; MZ, marginal zone; STR, striatum; TH, thalamus. Scale bars: 300 μm (A); 100 μm (B).

other than DiI, using the appropriate filters. Note that to be successful, photoconversion should be performed while the staining intensity is still high, preferably as soon as possible after sectioning of the tissue. Some authors have also observed axon breakages following photoconversion, although this is often attributable to dehydrating procedures used subsequently to prepare the slides (Catalano *et al.*, 1996). Importantly, notice in the photoconversion protocol in the appendix of this chapter, that DAB is a suspected carcinogen; hence appropriate measures, such as use of a fume hood and decontamination procedures, must be employed.

3. Use of DiI-Labeled Material for Electron Microscopy

Ultrastructural investigation of labeled tissue is possible following photoconversion into a DAB precipitate (Lübke, 1993; Métin *et al.*, 2000). Time-lapse video microscopy to study in vitro growth cone movement has been combined with electron microscopy to gain knowledge about cell form, cytology, and cell–cell interactions (Fishell *et al.*, 1995). After photoconversion of the DiI label, sections are incubated in 1% osmium tetroxide in 100 mM PB for 30 min. Then the sections are rinsed with PB before uranyl magnesium acetate staining for 30 min (0.5% UrMgAc in 0.9% saline). Following this standard dehydration, blocking and cutting protocols are employed.

Photoconversion followed by electron microscopy can allow the differentiation of directly and transneuronally labeled cells (Bruce *et al.*, 1997). The two types of labeling exhibit different staining intensities under fluorescent illumination, and after photoconversion the amount and localization of the permanent product is also distinct. Directly stained tissue shows dense DAB precipitation in the cytosol and throughout the membranes; however, when it is labeled transneuronally only the membrane structures are stained. Although EM can reveal additional information about the labeled material, the electron-dense product can obscure the presynaptic structures.

E. Technical Considerations and Limitations of Carbocyanine Dye Tracing

Despite the many advantages that carbocyanine dye tracing conferred to investigative studies elucidating patterns of neuronal connectivity, cell migration, etc., there are some limitations and drawbacks to this technique.

1. Age and Size Considerations for Suitability of Carbocyanine Dye Tracing in Rodent Brains

It is believed that the age of the tissue alters the incubation period needed because of increased myelination and the greater axonal distances. It is noticeable that after the first postnatal week carbocyanine dye tracing becomes less reliable in rodent brains, and is not recommended after P9 (Molnár and Blakemore, 1995). Higashi *et al.* (2002) observed that another family of carbocyanine dyes (used as voltage-sensitive dyes) also had a better uptake at embryonic and early postnatal ages. However, Spires *et al.* (2004) achieved excellent labeling of callosal projections in rodent at P14, and some limited transport was even observed at adulthood. Axonal myelination might be the most obvious cause for the decreased carbocyanine dye transport in older brains, yet no systematic study has demonstrated the link directly. The limitations of DiI tracing in large brains are a more evident restriction in the primate brain. Strategies to overcome this problem are raised in the section "Carbocyanine Dye Tracing in Primates, Including Normal and Pathological Postmortem Human Specimens."

2. Transneuronal Labeling

DiI can diffuse transneuronally especially during incubation at high temperatures, and in the case of thalamocortical pathway may label radial glia (Godement *et al.*, 1987). This phenomenon is distinct from the transneuronal labeling achieved by wgHRP or tritiated proline [^{3}H], since these require active transport and uptake mechanisms. The mechanism of DiI transneuronal transport is not fully understood; however, one model proposes that where cells are in very close contact, dye molecules are able to passively diffuse from one membrane to the next (Hofmann and Bleckmann, 1999). However, the biological nature of this contact is not clear. Some studies have shown that removing calcium ions by the addition of EDTA to fixation and storage buffers can prevent this from occurring and improve the sharpness of the axon fills (Hofmann and Bleckmann, 1999).

We experienced transneuronal labeling of radial glia in some fixed embryonic rhesus monkey brains (see section "Carbocyanine Dye Tracing in Primates, Including Normal and Pathological Postmortem Human Specimens"), which had undergone extended incubation to encourage diffusion of DiI to the fullest extent. The transneuronal labeling that occurred after dorsal thalamic crystal placement, predominantly occurred at the corticostriatal boundary, labeling the radial glial pallisade of this region (Carney, 2004).

III. CARBOCYANINE DYE TRACING IN PRIMATES, INCLUDING NORMAL AND PATHOLOGICAL POSTMORTEM HUMAN SPECIMENS

A. Interspecies Comparison of Axonal Pathway Development

In this part of the review, we discuss previously employed methods for tracing in human brains and then give examples of our work, and that of others, in carbocyanine dye tracing studies in the primate brain.

1. Classical Tracing Methods in Human Specimens

Some of the classical studies aimed at determining axonal connectivity used the invasive technique of applying lesions to the brain, and mapping fibers which derived from, or transited through, the damaged region. This approach can also be applied in human specimens with focal lesions. After a period, the anterograde degeneration that followed lesioning was exploited to "map" the site of the lesion with the regions where degenerating axons were present, thus revealing topographic connections between brain regions. Degenerating axons could be visualized using silver stains by various protocols including the Bielschowsky, Nauta–Gygax, and Fink–Heimer methods, which are discussed next.

Silver impregnation of degenerating axons that has been attained in postmortem human tissues (Grafe and Leonard, 1980). Miklossy *et al.* (1991) further demonstrated the selective Nauta method applicable to tracing myelinated and unmyelinated axons in the central nervous system of human tissue, which was fixed up to 24 h after death, and after long-term storage (successful staining was achieved in one human brain stored in 10% formalin for 8 years). The time course of anterograde degeneration varies according to species, age, and fiber size. Miklossy *et al.* (1991) concluded that the optimal postsurvival time for use of the selective Nauta method following degeneration of nerve fibers was between 9 days and 5 months. A modification of the Nauta method was used in a study that described the pattern of callosal afferents in areas 17, 18, and 19 in human cortex (Clarke and Miklossy, 1990). Also, other methods that enabled visualization of myelin breakdown products have been used to identify anterograde degeneration of axon tracts in humans, 5–20 months after the onset of neurological symptoms, and in tissue with prolonged storage in formalin (Miklossy and Van der Loos, 1991). Beach and McGeer (1988) developed a method for using HRP in the human brain with minimal postmortem delay. HRP crystals were briefly applied to the white matter underlying the cortex, which was then washed off and the tissue was incubated for 1–2 days. Frozen sectioning and revelation of the HRP product using 3,3'-diaminobenzidine demonstrated retrogradely labeled pyramidal cells in layers 5 and 6 of the cortex (Beach and McGeer, 1988).

Dai *et al.* (1998a) utilized an in vitro anterograde tracing with neurobiotin to reveal the hypothalamic connections of the suprachiasmatic nucleus in postmortem human brains. It is suggested that this technique is particularly useful for tracing in the adult brain when myelination has precluded the use of carbocyanine dyes. This method has been used in a number of publications, regarding retinohypothalamic and intrahypothalamic projections in the human brain (Dai *et al.*, 1998b,c). Haber (1988) described fiber tracing in postmortem human brains using WGA-HRP.

B. Carbocyanine Dye Tracing in Postmortem Human Tissue

DiI has been effectively used to study prenatal development of human visual (FitzGibbon, 1997; Hevner, 2000; Qu *et al.*, 2002) and thalamocortical pathways (Bystron *et al.*, 2002, 2005), development of central vagal sensory and motor connections (Cheng *et al.*, 2004), connections of the nucleus of the solitary tract in the medulla oblongata (Zec and Kinney, 2003); retinal ganglion cells (Pavlidis *et al.*, 2003), and intrinsic connectivity of auditory areas in adult brain (Tardif and Clarke, 2001). DiI labeling of neural connectivity in the embryonic human diencephalon is shown in Fig. 12.3.

Developing connectivity of the hippocampus was investigated in fixed brain from fetal humans at 19–22 gestational weeks (19–22 GW). At 19 GW, DiI tracing revealed neurons from layers 2 and 3 of entorhinal cortex that project to the hippocampus and subiculum, and reciprocal connections to the entorhinal cortex originated from pyramidal neurons in CA1 and the subiculum (Hevner and Kinney, 1996). In addition to determining timetables of development of axonal pathways in the nonhuman and human primate brain, DiI tracing has been used to visualize cell types. During corticogenesis, early generated neurons accumulate subjacent to the pial surface and form the primordial plexiform layer (PPL) as shown in both cats and humans (Marin-Padilla, 1971, 1983). At 6 GW, the Cajal-Retzius (CR) cells appear in the PPL and later settle near the pial surface after the appearance of the cortical plate. The granule cells of the subpial granular layer (SGL) initially accumulate around the CR cells, and subsequently the two cell types become segregated while maintaining connectivity. The fate of polymorphic

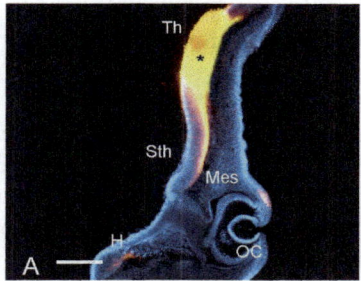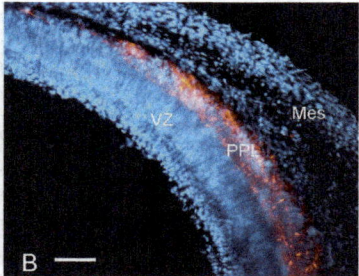

Figure 12.3. Carbocyanine dye tracing in postmortem embryonic human tissue at 6 gestational weeks. A single carbocyanine dye crystal (DiI) placed in the border between ventral and dorsal thalamus revealed early connectivity between the dorsal thalamus, subthalamus, and hypothalamus. (A) The crystal placement is indicated by the asterisk. (B) High-power view from a more posterior section of the same brain. The carbocyanine dye labeling is apparent in an extensive neural network in the dorsal thalamus. *Abbreviations:* Th, thalamus; STh, subthalamus; H, hypothalamus; OC, optic cup; Mes, mesenchyme; PPL, primordial plexiform layer; VZ, ventricular zone. Scale bars: 200 μm (A); 50 μm (B). The study was a part of collaborative project with Prof Blakemore and Prof Otellin.

CR and SGL cells is sealed around 24 GW when cell degeneration occurs, although a minority of CR cells exist throughout life (Meyer and Gonzalez-Hernandez, 1993).

DiI labeling from the pial surface in newborn ferret cortex visualized radial glia cells. Subsequently, DiI labeling was found in astrocytes identified with glial fibrillary acidic protein (GFAP) staining, thus demonstrating that radial glia transforms into astrocytes (Voigt, 1989). DiI/DiA labeling was used to recognize transitional forms of the radial glial cell transformation to astrocytes in the human brain (deAzevedo *et al.*, 2003).

1. Carbocyanine Dye-Tracing in Human Pathological Specimens

Notwithstanding the importance of determining axonal connectivity in fetal human subjects without developmental anomalies, carbocyanine dye tracing can be combined with standard histology to examine specific pathological specimens. Malformations of cortical development can result in a variety of clinical manifestations including epilepsy. DiI tracing was combined with immunohistochemistry in examining brain specimens of four human infants with various etiology in epilepsy, displaying subcortical or periventricular heterotopia, to assess the connectivity of the nodules (Hannan *et al.*, 1999). Crystal placement in the vicinity of nodules revealed short-distance diffusion, but the fibers did not penetrate them. DiI crystal placement within the nodules reveals interconnectivity between them in one case of heterotopia, suggesting that this altered connectivity may affect neuronal maturation, the balance of inhibition and excitation, and ultimate epileptogenic potential (Hannan *et al.*, 1999).

2. Large Brain Size in Primates May Limit the Applicability of Carbocyanine Dyes as Effective Tracers

Using DiI tracing in fixed embryonic rhesus monkey brains, we demonstrated that thalamocortical axons were present in the appropriate spatiotemporal position to exert the mitogenic influence on cortical precursors (Carney, 2004; Carney *et al.*, 2002, 2003), as suggested from previous in vivo and in vitro results in primates and rodents, respectively (Dehay *et al.*, 1989, 2001).

In our studies of embryonic rhesus monkey brains, although carbocyanine dye tracing was effective for examining early interactions between thalamus and cortex, there were some drawbacks that were sometimes encountered at later stages of development with large brain sizes. We placed DiI crystal in the dorsal thalamus, targeting the LGN at several embryonic ages. At E40, the onset of corticogenesis, the brain is relatively small and anterograde labeling of thalamocortical axons from DiI crystal placements revealed growth cones in the cortex (Carney *et al.*, 2002, 2003). However, at later ages, the lack of apparent growth cones on some axons indicated that the DiI had not diffused along the entire extent of the axon. In order to obtain the furthest diffusion possible we had incubated the brains at 37°C for up to 1 year following crystal placements of DiI in the LGN, to determine the timetable of development of geniculocortical axons to area 17. Unfortunately, the limited diffusion of DiI meant that only a minority of axons were labeled along the most distal part of the axons in area 17. A study of geniculocortical axon development in the human brain did not trace from the LGN but rather placed DiI crystals in the optic radiations of tissue blocks of human tissue at 20–22 GW (Hevner, 2000). On the other hand, as DiI diffuses both anterogradely and retrogradely, DiI placements in the optic radiation would be expected to retrogradely label corticofugal axons originating from area 17 in addition to geniculocortical axons from the LGN, and so this approach was not feasible in our study in monkey, which focused solely on thalamocortical projections.

3. Altering Experimental Variables to Attempt Enhancement of DiI Diffusion in Primate Brains

Undoubtedly the application of carbocyanine dye tracing in developmental neurobiology has yielded considerable advancements in the understanding of the development of axonal pathways in the embryonic brain. The importance of effective tracing in fetal human material including pathological specimens is clear, and modifications of the carbocyanine-dye-tracing technique have been attempted to increase diffusion from the injection site. Different incubation times were reported by several authors ranging from 6 to 15.5 months (Krassioukov *et al.*, 1998; Tardif and Clarke, 2001; Zec *et al.*, 1997). However, our observations and data of Lukas *et al.* (1998) suggest that incubation times of 5 months and more are unnecessarily long. Although neuronal tracing studies in embryonic and fetal human tissue did not focus on exact correlations between incubation times and tracing distance, such data were reported for postmortem material from adult donors. Carbocyanine crystals (DiI, DiA, DiO) were applied to the cervical spinal cord, sciatic nerve, and branchial plexus in humans and guinea pigs. This study was carried out to optimize tracing procedures and to reveal the validity of the combination of postmortem tracing with immunocytochemistry. Incubation in the dark at 37°C for 12–15 weeks proved optimal to achieve the longest tracing distances (28.9 ± 2.2 mm). Short postmortem times before fixation proved to be important. The concentration of tracers at the application site, longer incubation times, and incubation temperatures higher than 37°C did not result in longer tracing distances.

Using the Fast DiI (1,1-dioctadecy 1-3,3,3′,3′-tetramethylinsocarbocyanine perchlorate) version of the carbocyanine dye in human specimens still requires long incubation periods for dye diffusion (Mufson *et al.*, 1990; Zec *et al.*, 1997). Thus, neither prolonged incubation nor the use of a "Fast" carbocyanine dye could reach optimal diffusion in larger, primate brains (Carney and Molnár, unpublished observations, 2003), leaving the fixation method as an alternative parameter to be altered. Sparks *et al.* (2000) reasoned that aldehyde fixation that causes protein cross-linking may actually hamper DiI diffusion and therefore attempted a "delayed-fixation" approach. Human brain specimens of a short postmortem interval of 3 h or less were injected with Fast DiI and covered with paper towels soaked with phosphate buffer and incubated for 36 h at 4°C, prior to postfixation with 4% PFA. The authors report that such delayed fixation enabled a diffusion distance of 20–40 mm or more, which is at least three orders of magnitude greater than that achieved using conventional fixation protocols (Sparks *et al.*, 2000).

IV. EMERGING TECHNIQUES FOR NEUROANATOMICAL TRACT IMAGING

As scientific techniques progress, current popular techniques will be used less and less. Just as carbocyanine dye tracing relegated the need for classical axon tracing techniques, such as those using degeneration or electrophysiological stimulation at embryonic and early postnatal stages, the future may lie in magnetic resonance imaging (MRI) studies. White matter fiber tracts of the cerebrum contain densely packed myelinated axons along which the movement of water molecules occurs preferentially in directions that are perpendicular to the longitudinal axis of the axon. A major advantage of MRI studies is that it can be used in living animals and human subjects, repeatedly, hence permitting follow-up assessments, which are particularly information after brain injury. Pautler *et al.* (1998) showed that the olfactory and visual tracts in mice could be imaged in vivo using an MRI-visible contrast agent, manganese chloride ($MnCl_2$) solution, by topical application of $MnCl_2$ solution to the naris and intravitreal injection to target retinal ganglion cells.

Enhancement of the MRI technique, diffusion tensor MRI (DT-MRI) permits the visualization of white matter fiber tract in human subjects. Using probabilistic diffusion tractography (Behrens *et al.*, 2003b), the human thalamus could be segmented on the basis of its connectivity to the cortex (Behrens *et al.*, 2003a). The resulting subdivisions correspond to groups of nuclei and this approach has been used to produce a probabilistic atlas of the human thalamus based on its cortical connectivity (Johansen-Berg *et al.*, 2004). At present, however, carbocyanine dye tracing provides an unparalleled method to study developing brain circuitry. As MRI matures, its utility in research will become increasingly valuable.

V. CONCLUDING REMARKS

Carbocyanine dye tracing has several advantages over classical tracing techniques, although the drawbacks discussed above should be considered to determine the suitability of this method for the question at hand. Throughout this chapter, we refer to several examples where carbocyanine dye tracing has been used in the field of developmental neurobiology to enhance our understanding of the anatomical complexity of brain development. Undoubtedly the most enticing feature of carbocyanine dyes is the fact that they can be used in fixed tissue. This is especially important for neuronal tract-tracing of developing systems during the embryonic period. The use of fixed tissue eliminates the need for complex in utero surgery. Furthermore, direct knowledge of aspects of human nervous system development has come from carbocyanine dye tracing in fixed human tissue. This is particularly useful in pathological specimens. Simultaneous application of more than one carbocyanine dye has yielded significant advances in discovering the interactions between fiber systems and examining their precise topography in the developing brain. However, all methods of neuronal tract-tracing have their criticisms and carbocyanine dye tracing is no exception. Carbocyanine dyes label all axons and cells at the implantation site and cannot discriminate between anterograde and retrograde labeling. Another concern is that despite long incubations periods, the limited diffusion distance of carbocyanine dyes means that the entire axon may not be labeled in larger brains especially in later stages. Using this technique places a time constraint for data analysis as the fluorescence starts to fade a few days after the material is processed. Other methods such as photoconversion provide a means of turning carbocyanine dye fluorescence into a permanent product in small areas of tissue, but this technique introduces elements where safety precautions are needed. Nonetheless, standard carbocyanine dye tracing is a nonhazardous technique, which can be easily applied by researchers with no prior experience with this method.

We hope that this review puts carbocyanine dye tracing into the context of classical methodologies that were used to map axonal projections, but which proved impractical in developing systems or in human postmortem specimens. The advent of emerging imaging techniques, which have not reached sufficient resolution, may in turn dominate future experimental approaches. We hope that the true potential of the carbocyanine dye tracing method will be exploited by laboratories from various backgrounds.

APPENDIX

A. Protocol for DiI Placement and Histological Processing (Modified from Molnár *et al.*, 1998)

1. Materials

4% formaldehyde in phosphate buffer (PB) 0.1 M pH 7.4
PBS with sodium azide (PB 0.1 M, 0.09% NaCl, 0.05% sodium azide, pH 7.4)

Bisbenzimide (10 µg/ml in PBS)
Carbocyanine dye crystals (see Table 12.1 for details)
Platinum wires or insect pins
Fine forceps
Binocular dissecting microscope
Low melting agar (Type VII, Sigma A-4018)
Vibratome (Oxford Instruments) or Vibroslicer (Leica, VT1000S)
Slides, coverslips
Paraseal (Raymond A Lamb, London) or nail varnish
EPI fluorescence microscope with appropriate filters

2. Method

1. Fix brains with 4% formaldehyde.
2. Insert carbocyanine dye crystals with the help of an insect pin or platinum wire.
3. Keep brains at room temperature in PBS with sodium azide or at 4°C depending on the required incubation period (see Table 12.2 for examples).
4. Embed brains in agarose (4% in PBS) and prepare blocks according to the desired orientation for sectioning.
5. Cut sections (50–200 µm) with Vibratome (Oxford Instruments) or Vibroslicer (Leica, VT1000S). Collect serial sections in PBS solution.
6. Counterstain sections with bisbenzimide solution (10 min in 2.5 µg/ml solution in PBS), and then rinse them with PBS. *Safety note.* Bisbenzimide is a suspected carcinogen; hence gloves and appropriate precautions when handling or disposing of solutions must be used.
7. Transfer the sections to slides and coverslip with PBS solution without drying the sections. Seal coverslips with Paraseal or nail varnish.
8. Examine and document results within 3 days using the appropriate excitatory wavelength light (see details in Table 12.1 for filters). Store slides at 4°C.

B. Protocol for Photoconversion (Modified from Lübke, 1993)

1. Materials

Tris-HCl buffer 0.1 M pH 8.2 plus 0.9% NaCl (TBS)
0.05% 3,3′ diaminobenzidine tetrahydrochloride in TBS (TBS–DAB)
Cavity slides
Coverslips
EPI fluorescence microscope with appropriate filters depending on the carbocyanine dye used

2. Method

1. Rinse (2 × 10 min) sectioned material in TBS in the dark at room temperature.
2. Incubate with TBS–DAB for 30–60 min (dark, room temperature). *Safety note:* DAB is a suspected carcinogen, and so the microscope should be placed in a fume hood and decontamination procedures should be followed after photoconversion.
3. Transfer a section to the cavity slide, add a drop of TBS–DAB, and coverslip.
4. Expose to appropriate excitatory wavelength light through at least a 10× objective lens. (See details in Table 12.1 for filters.)
5. Change the TBS–DAB solution every 20 min.
6. Photoconvert until optimal staining is achieved, periodically stopping the reaction to check the progress of the conversion using bright-field illumination.
7. Rinse the section thoroughly with TBS.

The sections are then used for further protocols, such as immunohistochemistry, prepared for electron microscopy, or mounted onto slides, dehydrated through an alcohol series, and coverslipped under DePeX (BDH, Poole, UK).

ACKNOWLEDGMENTS. The study in our laboratory was supported by grants from the Medical Research Council UK; Human Frontiers Science Program (RGP0107/2001), The European Community (Grant QLRT-1999-30158), Wellcome Trust (Grant 063974/B/01/Z), and Swiss National Science Foundation (Grant 31-56032.98 to ZM). We thank Prof. Colin Blakemore and Ms. Patricia Cordery for their continued support. Dr. Colette Dehay and Dr. Henry Kennedy of INSERM U371, Lyon, France, kindly provided the monkey specimens used for tracing as part of a collaborative project. The DiI tracing study in the human brain was a part of collaborative project with Prof. Vladimir Otellin, Institute of Experimental Medicine, St Petersburg, Russia. R.S.E.C. was supported by an MRC studentship and the Goodger Scholarship from the Oxford Medical Science Division in the University of Oxford. D.B. was supported by a Wellcome Trust 4-year PhD scholarship. IB was sponsored, in part, by the Royal Society/Russian Academy of Science exchange program. We are grateful to Jamin De Proto for his help in editing and for useful discussions.

REFERENCES

Agmon, A., and Connors, B. W., 1991, Thalamocortical responses of mouse somatosensory (barrel) cortex in vitro, *Neuroscience* **41**:365–379.
Agmon, A., Hollrigel, G., and O'Dowd, D. K., 1996, Functional GABAergic synaptic connection in neonatal mouse barrel cortex, *J. Neurosci.* **16**:4684–4695.

Beach, T. G., and McGeer, E. G., 1988, Retrograde filling of pyramidal neurons in postmortem human cerebral cortex using horseradish peroxidase, *J. Neurosci. Methods* **23**:187–193.

Behrens, T. E., Johansen-Berg, H., Woolrich, M. W., Smith, S. M., Wheeler-Kingshott, C. A., Boulby, P. A., Barker, G. J., Sillery, E. L., Sheehan, K., Ciccarelli, O., Thompson, A. J., Brady, J. M., and Matthews, P. M., 2003a, Noninvasive mapping of connections between human thalamus and cortex using diffusion imaging, *Nat. Neurosci.* **6**:750–757.

Behrens, T. E., Woolrich, M. W., Jenkinson, M., Johansen-Berg, H., Nunes, R. G., Clare, S., Matthews, P. M., Brady, J. M., and Smith, S. M., 2003b, Characterization and propagation of uncertainty in diffusion-weighted MR imaging, *Magn. Reson. Med.* **50**:1077–1088.

Bicknese, A. R., Sheppard, A. M., O'Leary, D. D., and Pearlman, A. L., 1994, Thalamocortical axons extend along a chondroitin sulfate proteoglycan-enriched pathway coincident with the neocortical subplate and distinct from the efferent path, *J. Neurosci.* **14**: 3500–3510.

Bruce, L. L., Christensen, M. A., and Fritzsch, B., 1997, Electron microscopic differentiation of directly and transneuronally transported DiI and applications for studies of synaptogenesis, *J. Neurosci. Methods* **73**:107–112.

Bystron, I., Molnár, Z., Otellin, V., and Blakemore, C., 2005, Tangential networks of precocious neurons and early axonal outgrowth in the embryonic human forebrain, *J. Neurosci.* **25**:2781–2792.

Bystron, I., Otellin, V., Blakemore, C., and Molnár, Z., 2002, The early development of interconnections between thalamus and cortex in humans. FENS, Paris, France A039.3.

Cajal, S. R., 1909, Histologie du système nerveux de l'homme et des vertebres. Reprinted by Consejo Superior de Investigaciones Cientifias, Insituto Ramón y Cajal, 1952–1955.

Carney, R. S. E., 2004, *Thalamocortical development and cell proliferation in fetal primate and rodent cortex*, D. Phil Thesis, Department of Human Anatomy and Genetics, University of Oxford, UK.

Carney, R. S. E., Molnár, Z., Giroud, P., Cortay, V., Berland, M., Kennedy, H., and Dehay, C., 2002, Thalamocortical projections in the developing primate cortex. FENS, Paris, France A007.06.

Carney, R. S. E., Molnár, Z., Giroud, P., Cortay, V., Berland, M., Kennedy, H., and Dehay, C., 2003, *Novel Features of Thalamocortical Development in the Fetal Primate Cortex*, P3.09 ed., UK: British Neuroscience Association.

Catalano, S. M., Robertson, R. T., and Killackey, H. P., 1996, Individual axon morphology and thalamocortical topography in developing rat somatosensory cortex, *J. Comp. Neurol.* **367**:36–53.

Cheng, G., Zhou, X., Qu, J., Ashwell, K. W., and Paxinos, G., 2004, Central vagal sensory and motor connections: human embryonic and fetal development, *Auton. Neurosci.* **114**:83–96.

Clarke, S., and Miklossy, J., 1990, Occipital cortex in man: organization of callosal connections, related myelo- and cytoarchitecture, and putative boundaries of functional visual areas, *J. Comp. Neurol.* **298**:188–214.

Dai, J., Swaab, D. F., Van der Vliet, J., and Buijs, R. M., 1998a, Postmortem tracing reveals the organization of hypothalamic projections of the suprachiasmatic nucleus in the human brain, *J. Comp. Neurol.* **400**:87–102.

Dai, J., Van Der Vliet, J., Swaab, D. F., and Buijs, R. M., 1998b, Postmortem anterograde tracing of intrahypothalamic projections of the human dorsomedial nucleus of the hypothalamus, *J. Comp. Neurol.* **401**:16–33.

Dai, J., Van der Vliet, J., Swaab, D. F., and Buijs, R. M., 1998c, Human retinohypothalamic tract as revealed by in vitro postmortem tracing, *J. Comp. Neurol.* **397**:357–370.

deAzevedo, L. C., Fallet, C., Moura-Neto, V., Daumas-Duport, C., Hedin-Pereira, C., and Lent, R., 2003, Cortical radial glial cells in human fetuses: depth-correlated transformation into astrocytes, *J. Neurobiol.* **55**:288–298.

De Carlos, J. A., Lopez-Mascaraque, L., and Valverde, F., 1996, Dynamics of cell migration from the lateral ganglionic eminence in the rat, *J. Neurosci.* **16**:6146–6156.

Dehay, C., Horsburgh, G., Berland, M., Killackey, H., and Kennedy, H., 1989, Maturation and connectivity of the visual cortex in monkey is altered by prenatal removal of retinal input, *Nature* **337**:265–267.

Dehay, C., Savatier, P., Cortay, V., and Kennedy, H., 2001, Cell-cycle kinetics of neocortical precursors are influenced by embryonic thalamic axons, *J. Neurosci.* **21**:201–214.

Fishell, G., Blazeski, R., Godement, P., Rivas, R., Wang, L. C., and Mason, C. A., 1995, Optical microscopy: III. Tracking fluorescently labeled neurons in developing brain, *FASEB J.* **9**:324–334.

Fishell, G., Mason, C. A., and Hatten, M. E., 1993, Dispersion of neural progenitors within the germinal zones of the forebrain, *Nature* **362**:636–638.

FitzGibbon, T., 1997, The human fetal retinal nerve fiber layer and optic nerve head: a DiI and DiA tracing study, *Vis. Neurosci.* **14**:433–447.

Fujimori, K. E., Takauji, R., Yoshihara, Y., Tamada, A., Mori, K., and Tamamaki, N., 1997, A procedure for in situ hybridization combined with retrograde labeling of neurons: application to the study of cell adhesion molecule expression in DiI-labeled rat pyramidal neurons, *J. Histochem. Cytochem.* **45**:455–459.

Gan, W. B., Grutzendler, J., Wong, W. T., Wong, R. O., and Lichtman, J. W., 2000, Multi-color "DiOlistic" labeling of the nervous system using lipophilic dye combinations, *Neuron* **27**:219–225.

Godement, P., Vanselow, J., Thanos, S., and Bonhoeffer, F., 1987, A study in developing visual systems with a new method of staining neurones and their processes in fixed tissue, *Development* **101**:697–713.

Grafe, M. R., and Leonard, C. M., 1980, Successful silver impregnation of degenerating axons after long survivals in the human brain, *J. Neuropathol. Exp. Neurol.* **39**:555–574.

Haber, S., 1988, Tracing intrinsic fiber connections in postmortem human brain with WGA-HRP, *J. Neurosci. Methods* **23**:15–22.

Hannan, A. J., Servotte, S., Katsnelson, A., Sisodiya, S., Blakemore, C., Squier, M., and Molnár, Z., 1999, Characterization of nodular neuronal heterotopia in children. *Brain* **122**(Pt 2):219–238.

Hevner, R. F., 2000, Development of connections in the human visual system during fetal mid-gestation: a DiI-tracing study, *J. Neuropathol. Exp. Neurol.* **59**:385–392.

Hevner, R. F., and Kinney, H. C., 1996, Reciprocal entorhinal-hippocampal connections established by human fetal midgestation, *J. Comp. Neurol.* **372**:384–394.

Hevner, R. F., Miyashita-Lin, E., and Rubenstein, J. L., 2002, Cortical and thalamic axon pathfinding defects in Tbr1, Gbx2, and Pax6 mutant mice: evidence that cortical and thalamic axons interact and guide each other, *J. Comp. Neurol.* **447**:8–17.

Higashi, S., Molnár, Z., Kurotani, T., and Toyama, K., 2002, Prenatal development of neural excitation in rat thalamocortical projections studied by optical recording, *Neuroscience* **115**:1231–1246.

Hofmann, M. H., and Bleckmann, H., 1999, Effect of temperature and calcium on transneuronal diffusion of DiI in fixed brain preparations, *J. Neurosci. Methods* **88**:27–31.

Honig, M. G., and Hume, R. I., 1986, Fluorescent carbocyanine dyes allow living neurons of identified origin to be studied in long-term cultures, *J. Cell Biol.* **103**:171–187.

Honig, M. G., and Hume, R. I., 1989, DiI and diO: versatile fluorescent dyes for neuronal labelling and pathway tracing, *Trends Neurosci.* **12**:333–335, 340–331.

Johansen-Berg, H., Behrens, T. E., Robson, M. D., Drobnjak, I., Rushworth, M. F., Brady, J. M., Smith, S. M., Higham, D. J., and Matthews, P. M., 2004, Changes in connectivity profiles define functionally distinct regions in human medial frontal cortex, *Proc. Natl. Acad. Sci. USA* **101**:13335–13340.

Johnson, G. D., Davidson, R. S., McNamee, K. C., Russell, G., Goodwin, D., and Holborow, E. J., 1982, Fading of immunofluorescence during microscopy: a study of the phenomenon and its remedy, *J. Immunol. Methods* **55**:231–242.

Krassioukov, A. V., Bygrave, M. A., Puckett, W. R., Bunge, R. P., and Rogers, K. A., 1998, Human sympathetic preganglionic neurons and motoneurons retrogradely labelled with DiI, *J. Auton. Nerv. Syst.* **70**:123–128.

Liu, Q., Sanborn, K. L., Cobb, N., Raymond, P. A., and Marrs, J. A., 1999, R-cadherin expression in the developing and adult zebrafish visual system, *J. Comp. Neurol.* **410**:303–319.

López-Bendito, G., Chan, C. H., Mallamaci, A., Parnavelas, J., and Molnár, Z., 2002, Role of *Emx2* in the development of the reciprocal connectivity between cortex and thalamus, *J. Comp. Neurol.* **451:**153–169.

Lübke, J., 1993, Photoconversion of diaminobenzidine with different fluorescent neuronal markers into a light and electron microscopic dense reaction product, *Microsc. Res. Tech.* **24:**2–14.

Lukas, J. R., Aigner, M., Denk, M., Heinzl, H., Burian, M., and Mayr, R., 1998, Carbocyanine postmortem neuronal tracing. Influence of different parameters on tracing distance and combination with immunocytochemistry, *J. Histochem. Cytochem.* **46:**901–910.

Ma, L., Harada, T., Harada, C., Romero, M., Hebert, J. M., McConnell, S. K., and Parada, L. F., 2002, Neurotrophin-3 is required for appropriate establishment of thalamocortical connections, *Neuron* **36:**623–634.

Marin-Padilla, M., 1971, Early prenatal ontogenesis of the cerebral cortex (neocortex) of the cat (*Felis domestica*). A Golgi study: I. The primordial neocortical organization, *Z. Anat. Entwicklungsgesch* **134:**117–145.

Marin-Padilla, M., 1983, Structural organization of the human cerebral cortex prior to the appearance of the cortical plate, *Anat. Embryol.* **168:**21–40.

McConnell, S. K., Ghosh, A., and Shatz, C. J., 1989, Subplate neurons pioneer the first axon pathway from the cerebral cortex, *Science* **245:**978–982.

Métin, C., Denizot, J. P., and Ropert, N., 2000, Intermediate zone cells express calcium-permeable AMPA receptors and establish close contact with growing axons, *J. Neurosci.* **20:**696–708.

Métin, C., and Godement, P., 1996, The ganglionic eminence may be an intermediate target for corticofugal and thalamocortical axons, *J. Neurosci.* **16:**3219–3235.

Meyer, G., and Gonzalez-Hernandez, T., 1993, Developmental changes in layer 1 of the human neocortex during prenatal life: a DiI-tracing and AChE and NADPH-d histochemistry study, *J. Comp. Neurol.* **338:**317–336.

Miklossy, J., Clarke, S., and Van der Loos, H., 1991, The long distance effects of brain lesions: visualization of axonal pathways and their terminations in the human brain by the Nauta method, *J. Neuropathol. Exp. Neurol.* **50:**595–614.

Miklossy, J., and Van der Loos, H., 1991, The long-distance effects of brain lesions: visualization of myelinated pathways in the human brain using polarizing and fluorescence microscopy, *J. Neuropathol. Exp. Neurol.* **50:**1–15.

Miller, B., Chou, L., and Finlay, B. L., 1993, The early development of thalamocortical and corticothalamic projections, *J. Comp. Neurol.* **335:**16–41.

Molnár, Z., Adams, R., and Blakemore, C., 1998a, Mechanisms underlying the early establishment of thalamocortical connections in the rat, *J. Neurosci.* **18:**5723–5745.

Molnár, Z., Adams, R., Goffinet, A. M., and Blakemore, C., 1998b, The role of the first postmitotic cortical cells in the development of thalamocortical innervation in the reeler mouse, *J. Neurosci.* **18:**5746–5765.

Molnár, Z., and Blakemore, C., 1991, Lack of regional specificity for connections formed between thalamus and cortex in coculture, *Nature* **351:**475–477.

Molnár, Z., and Blakemore, C., 1995, How do thalamic axons find their way to the cortex? *Trends Neurosci.* **18:**389–397.

Molnár, Z., and Cordery, P., 1999, Connections between cells of the internal capsule, thalamus, and cerebral cortex in embryonic rat, *J. Comp. Neurol.* **413:**1–25.

Molnár, Z., López-Bendito, G., Small, J., Partridge, L. D., Blakemore, C., and Wilson, M. C., 2002, Normal development of embryonic thalamocortical connectivity in the absence of evoked synaptic activity, *J. Neurosci.* **22:**10313–10323.

Mufson, E. J., Brady, D. R., and Kordower, J. H., 1990, Tracing neuronal connections in postmortem human hippocampal complex with the carbocyanine dye DiI, *Neurobiol. Aging* **11:**649–653.

Nauta, W. J., and Gygax, P. A., 1951, Silver impregnation of degenerating axon terminals in the central nervous system: (1) technic. (2) chemical notes, *Stain Technol.* **26:**5–11.

Papadopoulos, G. C., and Dori, I., 1993, DiI labeling combined with conventional immunocytochemical techniques for correlated light and electron microscopic studies, *J. Neurosci. Methods* **46**:251–258.

Pautler, R. G., Silva, A. C., and Koretsky, A. P., 1998, In vivo neuronal tract tracing using manganese-enhanced magnetic resonance imaging, *Magn. Reson. Med.* **40**: 740–748.

Pavlidis, M., Stupp, T., Naskar, R., Cengiz, C., and Thanos, S., 2003, Retinal ganglion cells resistant to advanced glaucoma: a postmortem study of human retinas with the carbocyanine dye DiI, *Invest. Ophthalmol. Vis. Sci.* **44**:5196–5205.

Qu, J., Zhou, X., Zhang, L., Ni, H., Ashwell, K., and Lu, F., 2002, A preliminary study on development of human visual system in fetus by DiI-tracing, *Zhonghua Yan Ke Za Zhi* **38**:517–519.

Sandell, J. H., and Masland, R. H., 1988, Photoconversion of some fluorescent markers to a diaminobenzidine product, *J. Histochem. Cytochem.* **36**:555–559.

Skaliora, I., Adams, R., and Blakemore, C., 2000, Morphology and growth patterns of developing thalamocortical axons, *J. Neurosci.* **20**:3650–3662.

Sparks, D. L., Lue, L. F., Martin, T. A., and Rogers, J., 2000, Neural tract tracing using Di-I: a review and a new method to make fast Di-I faster in human brain, *J. Neurosci. Methods* **103**:3–10.

Spires, T. L., Molnár, Z., Kind, P. C., Cordery, P. M., Upton, A. L., Blakemore, C., and Hannan, A. J., 2005, Activity-dependent regulation of synapse and dendritic spine morphology in developing barrel cortex requires phospholipase C-β1 signalling, *Cereb. Cortex* **15**(4):385–393.

Tardif, E., and Clarke, S., 2001, Intrinsic connectivity of human auditory areas: a tracing study with DiI, *Eur. J. Neurosci.* **13**:1045–1050.

Voelker, C. C., Garin, N., Taylor, J. S., Gahwiler, B. H., Hornung, J. P., and Molnár, Z., 2004, Selective neurofilament (SMI-32, FNP-7 and N200) expression in subpopulations of layer 5 pyramidal neurons in vivo and in vitro, *Cereb. Cortex* **14**:1276–1286.

Voigt, T., 1989, Development of glial cells in the cerebral wall of ferrets: direct tracing of their transformation from radial glia into astrocytes, *J. Comp. Neurol.* **289**:74–88.

Wouterlood, F. G., and Mugnaini, E., 1984, Cartwheel neurons of the dorsal cochlear nucleus: a Golgi-electron microscopic study in rat, *J. Comp. Neurol.* **227**:136–157.

Yamamoto, N., Higashi, S., and Toyama, K., 1997, Stop and branch behaviors of geniculocortical axons: a time-lapse study in organotypic cocultures, *J. Neurosci.* **17**:3653–3663.

Zaborszky, L., and Heimer, L., 1989, *Combinatios of Tracer Techniques, Especially HRP and PHA-L, with Transmitter Identification for Correlated Light and Electron Microscopic Studies*, New York: Plenum Press.

Zec, N., Filiano, J. J., and Kinney, H. C., 1997, Anatomic relationships of the human arcuate nucleus of the medulla: a DiI-labeling study, *J. Neuropathol. Exp. Neurol.* **56**:509–522.

Zec, N., and Kinney, H. C., 2003, Anatomic relationships of the human nucleus of the solitary tract in the medulla oblongata: a DiI labeling study, *Auton. Neurosci.* **105**:131–144.

13

Combined Fluorescence Methods to Determine Synapses in the Light Microscope: Multilabel Confocal Laser Scanning Microscopy

FLORIS G. WOUTERLOOD

FLORIS G. WOUTERLOOD • Department of Anatomy, Vrije Universiteit Medical Center, Rm MF-G-136, P.O. Box 7057, 1007 MB Amsterdam, The Netherlands

Abstract: The dimensions of synapses are at or below the resolution limit of classical light microscopy. Under optimal conditions, one can appreciate processes of pre- and postsynaptic neurons that appose each other. Such appositions may be casual only and as such not functional in terms of synaptic communication. As a consequence, until quite recently, electron microscopy was the only means available to determine whether identified neurons synapse with each other. Technological developments, however, have created a middle ground between the strictly separated realms of light and electron microscopy. In this chapter I present a triple-fluorescence approach aimed at identifying the apposition of a presynaptic and a postsynaptic neuron, and simultaneously pinpointing a highly specific synapse-associated marker. This third marker identifies the presence of an active zone, necessary to distinguish casual appositions from functional synapses. Methods involved are neuroanatomical tracing, immunofluorescence, confocal laser scanning, and postacquisition computer processing followed by three-dimensional reconstruction and inspection. In my contribution, I will review the theory and practice involved in triple-labeling confocal fluorescence imaging. I begin by dealing with the dimensions of synapses and the structures involved, and relate the physical limitations of light microscopy to the problem of resolving synaptic structure. I then review the principles of image formation in fluorescence microscopy, and present the conditions that must be fulfilled in order to do sound multilabel confocal laser scanning: fluorochromes, lasers, channels, channel separation, and procedures to recognize and suppress unwanted phenomena such as crosstalk. In order to fully illustrate the points discussed, an actual triple visualization experiment will be described. Finally, I will emphasize several important aspects of "operator awareness", that is, the mind setting necessary to work with an advanced optoelectronic instrument like a confocal microscope and its sophisticated software. An aware user senses when some part of the complicated chain of processes is not producing what it is supposed to produce. If operator awareness is absent, strange results may be obtained.

Keywords: anterograde tracing, crosstalk, deconvolution, emission, excitation, fluorescence, neuron markers, synapses, three-dimensional reconstruction

I. INTRODUCTION

Synapses are at the very focus of neuronal functioning. While today the term synapse has a descriptive, morphological meaning, physiologists instead of neuroanatomists introduced the term long ago to underscore the concept of a functional juxtaposition of two neurons exchanging electrical nervous activity (Foster and Sherrington, 1897). In those days of the *belle époque*, neuroanatomists lacked instruments with sufficient resolution to study synapses, and the dispute between supporters of the novel neuron doctrine (Waldeyer, 1891) and those entertaining the earlier reticulum doctrine propagated by Gerlach (1858) lingered on for 50 years. The argument was finally settled in favor of the neuron doctrine after the invention of the electron microscope and the parallel development of appropriate preparative histologic techniques. In the early 1950s, the morphological correlate of Foster and Sherrington's functional synapse was revealed by Palade and Palay (1954, 1955). In the electron microscope, the ingredients of a typical central nervous system (CNS) synapse consist of a presynaptic axon terminal or bouton and a postsynaptic element that may be a dendritic spine, dendritic shaft, cell body, or even an axon hillock or axon terminal (Fig. 13.1A). Such a site where the outer membranes of two neurons are closely together will be referred to in this chapter as *juxtaposition* or *apposition*. It is evident that in an environment with a dense packing like that in the CNS, not all appositions can be synapses. Appositions involved in synapses display highly specialized areas with increased electron density: active zones. After the arrival of an action potential at a presynaptic bouton, synaptic vesicles docked at the active zone in this terminal fuse to the presynaptic membrane and release their neurotransmitter content into the synaptic cleft. The membrane postsynaptic to the active zone hosts postsynaptic receptors. Neurotransmitter molecules initiate, via their specific receptor, a chain of molecular events that finally generates an excitatory or inhibitory postsynaptic action potential. The point further exploited in this chapter is that the molecular machinery of the synapse includes unique proteins located pre- or postsynaptically. Excitatory and inhibitory events at synapses require completely different molecular machineries. As a consequence, if it could be possible to visualize a presynaptic axon terminal and its juxtaposed postsynaptic element, and to immunostain simultaneously some of the unique proteins belonging to either the excitatory or the inhibitory kind of molecular machinery (Fig. 13.1B), it might be possible to identify the presence of a synapse in the light microscope and to determine its neurochemical role at the same time. Translated into methodology terms, we need a triple-labeling experiment. We have successfully applied an antibody against ProSAP2/Shank3 as the "third marker" (Wouterlood et al., 2003). ProSAP2/Shank3 is a postsynaptic scaffolding protein involved in positioning the N-methyl-D-aspartate (NMDA) receptor at the postsynaptic density of excitatory synapses (Böckers et al., 1999, 2002). For inhibitory synapses, the protein gephyrin, i.e., a scaffolding protein for the gamma-aminobutyric acid A (GABA$_A$) receptor at the postsynaptic density, has been proposed as "third marker" (Sassoë-Pognetto and Fritschy, 2000).

The identification of synapses and their possible neurochemical role was until recently a scientific activity confined exclusively within the domain of the electron microscope (Sesack et al., this volume); however, electron microscopy requires fairly large investments in terms of resources, personnel, time, laboratory equipment, and instrument. Furthermore, as symbolized in the inset in Fig. 13.1A, the electron microscope is a sampling instrument

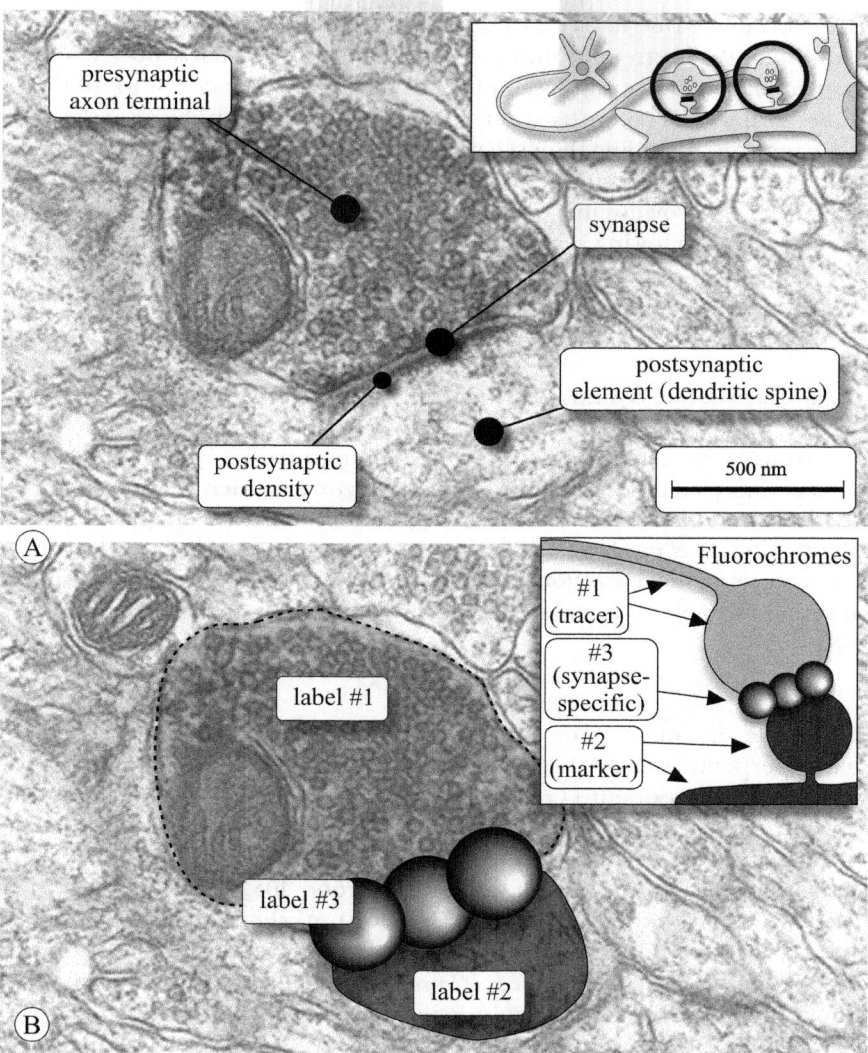

Figure 13.1. (A) Ingredients of a synapse: presynaptic axon terminal, pre- and postsynaptic membrane ("synapse"), postsynaptic element, in this case a dendritic spine. Synapses with marked asymmetry of the membrane specializations are thought to be excitatory. The postsynaptic density contains the molecular scaffolding machinery of the postsynaptic receptors. (B) Concept of a synapse in a light microscopical fluorescence paradigm: labeling of the presynaptic element (marker #1), labeling of the postsynaptic element (label #2, labels #1 and 2 may be neuroanatomical tracers or immunocytochemical markers). Labeling of a synapse-associated protein uniquely present in the postsynaptic density provides label #3. Inset: when fluorochromes are applied, a sandwich of fluorochromes 1, 2, and 3 will show up in the imaging system.

par excellence. Due to its enormous resolution and associated with this the requirement of extremely thin sections, the electron microscope is not the instrument of choice when the purpose of the investigation is to do three-dimensional (3D) reconstruction of large numbers of samples or to see complete neurons including their synapses. The modern confocal laser scanning microscope (CLSM), supplemented with image deconvolution and 3D reconstruction, provides just enough resolution to detect synapses, as will be argued in the following section.

II. THEORETICAL CONSIDERATIONS

A. Can We See Synapses? Practical Implications of Optics Theory

A typical CNS axon terminal is a three-dimensionally organized varicosity with a diameter of 0.5–1.0 μm. The active zone of such a bouton resembles a disk with a diameter of 0.2–0.3 μm and a thickness of approximately 50 nm (Peters *et al.*, 1991). Can we see such small structures in an optical microscope? The resolution of an optical system, or the smallest distance at which two points can be seen as separate points, is given by Ernst Abbe's equation

$$r = 0.61 \; \lambda/NA_{obj},$$

where the parameter r or lateral resolution distance is measured in the plane perpendicular to the optical axis (Inoué, 1995). Note that there is a wavelength component (λ) and a component related to the quality of the objective lens (NA, numeric aperture). This means that the wavelength of the light used is one of the factors that determine the resolution of the microscope. Based on this formula, with a good 40× dry objective (NA = 0.7), two points seen with blue light (λ = 450 nm) should be at least 392 nm apart in order to be seen as separate points. With red light (λ = 600 nm), the minimum distance becomes 522 nm, or 0.5 μm. Note that these are minimum theoretical distances between mathematical points. Such theoretical distances are always smaller than those practically attainable in tissue sections. With a high-quality oil immersion lens (NA = 1.30), the theoretical minimum distances under blue light and red light illumination become 211 and 281 nm, respectively. These numbers illustrate clearly that a synapse is a structure whose size lingers around and below the theoretical resolution limit of a normal light microscope. Because of this constraint, we cannot see under normal conditions with a light microscope whether single molecules in or around the synaptic junction belong to the presynaptic or the post-synaptic compartment. We may see clusters of molecules if such clusters are large enough or when we surround them with sufficient label to create aggregates of staining agent or precipitate in the order of 0.3–0.5 μm. Without doubt, these figures underscore the demand for high-quality objective lenses (high NA) if the aim of the microscope is to look (using bright or fluorescence light) at very small objects like the synapses between axon terminals and postsynaptic elements.

The above situation is further aggravated for the light microscopist by the fact that light is subject to diffraction. This is the way it is in nature. We have to accept that the projected image from a bright, one-dimensional point onto our eyes, a screen, or a detector is determined by laws of quantum physics and is seen by us and by our instruments as blurred.

Image formation in an optical system is as follows. The wave front of the light, or the photons if light is considered from a quantum physics point of view, distributes in a statistical fashion onto a screen or a detector, with a so-called primary projection maximum surrounded by primary projection minima, secondary projection maxima, secondary projection minima, and so on (Fig. 13.2A). This diffraction pattern is named an Airy distribution, or point spread function (PSF), if one likes to consider light as a stream of photons. The distance between the two primary projection minima in a two-dimensional plot equals the parameter r of Abbe's equation. Since the projection of a stream of photons on a screen is a spot, it is better to refer to the diameter of the disk whose center is the primary projection maximum while the edge is the primary projection minimum. This spot is called the *Airy disk*, and its diameter equals the parameter r of Abbe's equation. It is important to keep in mind that the shape of the diffraction pattern depends on the wavelength of the light involved. Green light ($\lambda = 500$ nm) has a sharper and narrower distribution curve compared with red light ($\lambda = 600$ nm) (Fig. 13.2B). As a rule of thumb, the higher the energy of the electromagnetic waves, the sharper the peak of the Airy distribution and the better the resolution. Blue light has a shorter wavelength and a higher energy than red light.

Considering two points close to each other, the outcome (r) of Abbe's equation in the previous section should be considered in terms of the distance between the peaks of two partially overlapping Airy distribution curves rather than an absolute distance between two mathematically defined, one-dimensional points. Two distributions of photons can still be distinguished from each other down to a minimum distance. This minimum distance is reached when the primary projection maximum of the first Airy distribution coincides with the first projection minimum of the second Airy distribution (Fig. 13.2C). This minimum distance, which equals the radius of the Airy disk ($1/2\ r$ of Abbe's equation), is called Rayleigh's criterion (named after Lord Rayleigh who published numerous papers on light theory, e.g., in 1891, on the behavior of light cast through a *pinhole*). The consequence of these physical laws is that a microscopist desperately trying to distinguish two blurred structures from each other by switching to a higher power lens finds that, beyond a certain magnification, this action does not further improve the image.

In classical optical and fluorescence microscopy with its inherent orthoscopic view, the microscopist typically deals with information present in one focal plane. Diffraction is likewise measured, and resolution is expressed

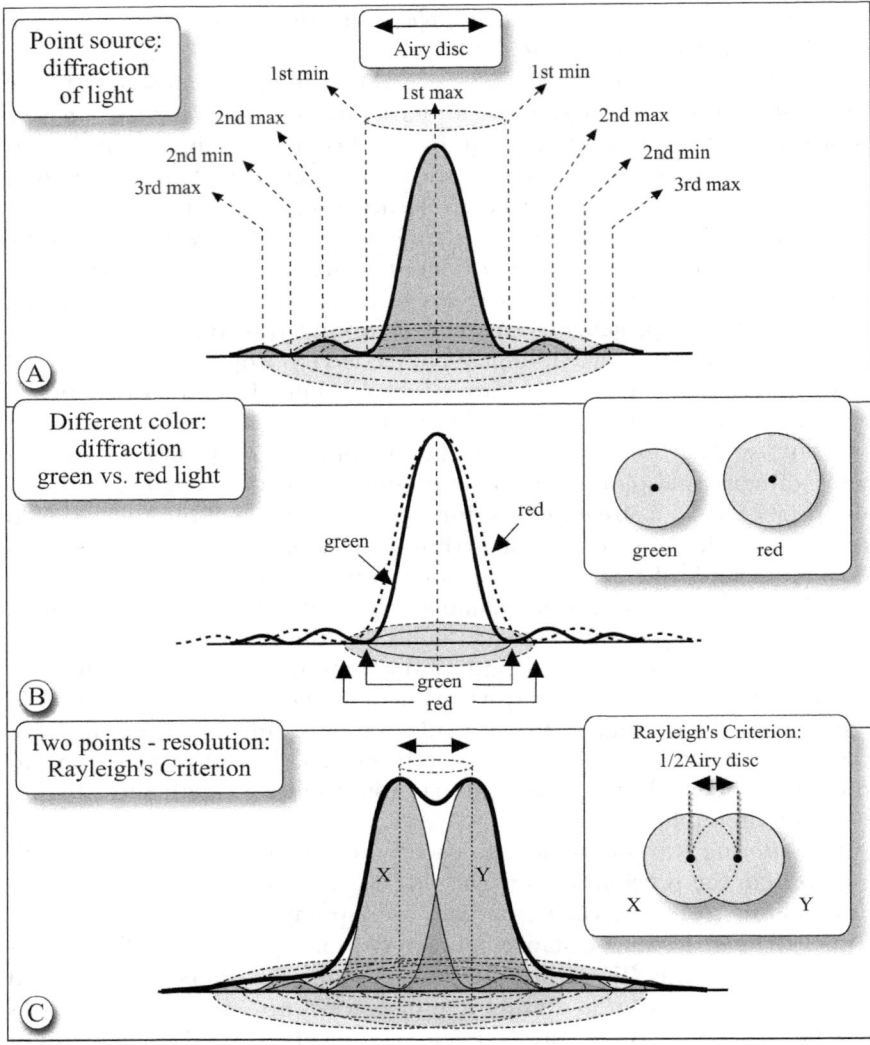

Figure 13.2. Basics of diffraction. (A) Light (photons) projected onto a screen distributes according to a diffraction pattern. The distance between the primary maximum and the first diffraction minimum is called one Airy disk radius. (B) Diffraction is wavelength dependent. The diffraction pattern of light with high energy (short wavelength, e.g., green light) shows a narrower peak than that of light with low energy (long wavelength, e.g., red light). A point light source using green light produces a smaller diffraction spot than that of a point light source using red light. An object "seen" with green light appears therefore smaller than the same object "seen" with red light. (C) Resolution according to Rayleigh's criterion: the smallest distance at which two separate points are still distinguishable as separate entities. The primary maximum of the diffraction pattern of point X coincides with the first diffraction minimum of point Y. This distance equals one Airy radius or half the diameter of the Airy disk. As can be inferred from Panel B, resolution depends on the wavelength of the used light.

in only one optical plane, the *XY* plane. This resolution is also referred to as the "radial resolution." In confocal laser scanning, one typically deals with the distribution of information in a 3D tissue volume. Accordingly, the microscopist has to take into account the axial resolution as well, that is, resolution measured along the optical axis or *Z* axis. This "axial resolution" is lower than that in the radial direction, since the mathematical expression for axial resolution is as follows:

$$r = 2\lambda\acute{\eta}/(\mathrm{NA_{obj}})^2,$$

where $\acute{\eta}$ is the refractive index of the mounting medium/immersion medium. With blue light (λ = 450 nm) and a high-quality oil immersion lens (NA = 1.30) and using oil immersion ($\acute{\eta}$ = 1.5), the theoretical minimum distance between two points in the *Z* direction at which these points are still distinguishable as points is 799 nm. Radial resolution in an optical system is 392 nm (see above), and therefore, it is approximately better than axial resolution by factor 2.

B. Pushing the Envelope: Improvements in Resolution and Image

Two improvements in optics have helped to push the limit of resolution a factor 1.4 down from the theoretically attainable values in a normal optical microscope (Inoué, 1995; Sheppard and Choudhurry, 1977). A third improvement has increased the detail seen by the observer's eyes and has made optical slicing possible. Note that these improvements belong to the category "optical and mathematical tricks" since the underlying fundamental quantum physics cannot be changed.

The first of these improvements is the use of monochromatic light, while the second improvement is postacquisition statistical processing of the signal, called deconvolution. Deconvolution can be considered a sort of reversing the statistics of an Airy distribution. There comes into spotlight the third improvement, which is the CLSM or the sublime instrument implementing these improvements. A laser provides a spot illumination of the object with monochromatic light. The confocal imaging system, whose centerpiece is a pinhole in front of its detectors, blocks haze and other out-of-focus information discomforting to the eye (Fig. 13.3). In-focus images generated by the laser scanning instrument are bitmaps stored on computer hard disk. Postacquisition statistical processing, i.e., deconvolution, "sharpens" the image further in a scientifically valid way.

1. Illumination with Monochromatic Light

The essence of white light is that it is a mixture of light of various wavelengths. As argued above, each wavelength has its own Airy distribution. Illumination of an object via a monochromatic illumination system produces a better image than illumination with white light, since a monochromatic

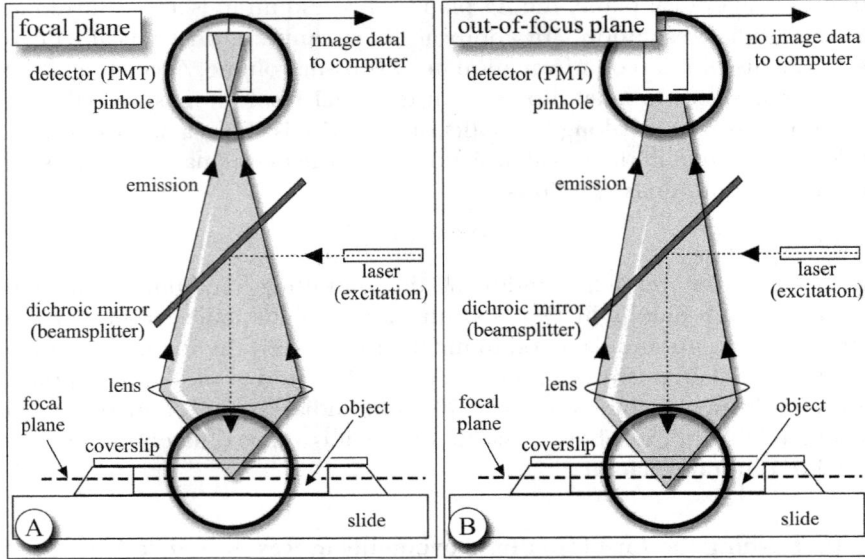

Figure 13.3. The essence of a confocal imaging system is a pinhole in front of the detector. (A) Fluorescence emitted from labeled structures located in the focal plane passes the pinhole and reaches the detector. (B) Emitted light from structures located in all planes other than the focal plane is rejected by the pinhole and does not reach the detector.

illumination system is dealing with only one Airy distribution instead of dealing with many. Furthermore, a lens refracts each wavelength in a slightly different way. The result is the color shift named "chromatic aberration."

Although lenses are usually color corrected to reduce chromatic aberration, the only way to reduce Airy-related blur would be to improve their numeric aperture. However, this parameter is bound by an absolute limit (NA = 1.4). The mixing of Airy distributions associated with different wavelengths combined with a touch of chromatic aberration results in increased blur of the details in the resulting image.

Excitation of a fluorochrome with a monochromatic illumination system avoids the conventional situation in which the object is illuminated with a cocktail of different Airy distributions (at least on the excitation side of the system). It also avoids chromatic aberration. The result is a markedly improved quality of the obtained image. Conventional fluorescence microscopes with their mercury or xenon lamps attempt to achieve via filtering of "excitation lines" from the lamp's light spectrum what a laser does by its very nature. Note that these mercury or xenon excitation lines are always narrow bands of wavelengths and by no means single discrete wavelengths. The fact that in addition to being monochromatic, laser light is also coherent (light waves are in sync) further contributes in a positive way to image formation. To put it simply, the truly monochromatic and coherent light of a laser produces a result superior to that obtained with a conventional fluorescence microscope. A laser illumination system also produces much better color separation in dual- or multifluorescence applications. The third advantage of a laser is the extremely small beam of high-intensity monochromatic light that can be used to scan a specimen with pulses of light. Note here that the light emitted by the fluorescent specimen is not monochromatic nor coherent. Filtering is necessary to narrow the bandwidth of the emitted light, especially in dual- or multiple-fluorescence applications.

2. Pinhole: Better Resolution at the Cost of Illumination

The most important optical improvement, however, offered by a confocal laser scanning instrument in comparison with a conventional fluorescence microscope is gained by the application of a pinhole in front of the detector. The pinhole is a device that allows only light emitted in a focal plane to pass, whereas emitted light originating from planes above and below the focal plane is rejected (Fig. 13.3; Minsky, 1957; Brakenhoff *et al.*, 1979; Inuoé, 1995). Thus, a confocal instrument possesses an intrinsic mechanism by which out-of-focus light (the major contributor to blur) does not reach the detector. The image looks as if it is sharper (which it is, since all information as well as blur over and under the focal plane is absent).

The diameter of the pinhole has its own effect on resolution, because Rayleigh's criterion also holds for projection apertures. In formula, Rayleigh's criterion applied to a pinhole is expressed as follows:

$$\theta = (1.22\lambda)/D,$$

where θ is the angular separation, λ the wavelength of the used light, and D the diameter of the pinhole. Most important in this respect is the *back-projected pinhole*, that is, the calculated diameter of the real pinhole projected back onto the fluorescence-emitting specimen. It is this back-projected pinhole that really matters and not the real size of the physical pinhole. Most manufacturers of confocal instruments refer in their documentation to this back-projected pinhole when they present data on their instrument's "pinhole." The formula implies that the smaller the pinhole, the better the angular separation, or resolution. Pinholes in general and especially small pinholes reject much light. In fact so much light is blocked by the pinhole (more than 99.99%) that the few photons that manage to pass the pinhole cannot be seen with the naked eye and have to be detected with an expensive and ultrasensitive electronic device: a photomultiplier. Since the emitted light from a fluorescent specimen is a fraction of the light used for excitation, it follows that a section with a fluorescent object needs to be literally flooded with high-intensity light in order to generate enormous number of photons of which only a fraction will ultimately reach the photomultipliers. A drastic measure like saturating a specimen with high-energy excitation light cannot be taken without dire consequences. Bleaching of the specimen is always a major source of concern. Apart from the application of antifading agents, a solution to bleaching is offered by the two-photon confocal microscope; however, the description of this complex instrument is outside the scope of the present chapter.

3. Intermediate Step: Pixelizing the Image

A photomultiplier is a photon counter and it generates an analogous signal. This signal is digitized and, in conjunction with the scanning movement of the laser beam, used to build up a bitmapped image of the structures of interest. These bitmaps can be further processed with a computer. The projection pattern of an image onto the photomultiplier detector is converted into discrete samples referred to as pixels. A pixel is a square area with a finite size and with a finite intensity level. The light intensity measured in this square area is a gray value, usually a number between 0 and 255 (8-bit intensity sampling) or between 0 and 4095 (12-bit intensity sampling). The size of the pixels compared with the size of the structures to be sampled is important. At this stage of image recording, the sampling rate according to Nyquist comes into the spotlight (Webb and Dorey, 1995). "Nyquist" provides a criterion inasmuch how dense sampling must be in a confocal instrument to satisfy Rayleigh's criterion. The Nyquist sampling rate applied to an Airy distribution implies that sampling must occur at a rate of at least twice the frequency of a distribution curve. In practice, four samples across the Airy disk of a projection diffraction spot originating from a single bright point is the minimum according to Nyquist. The publication by Webb and Dorey (1995) discusses the details of the process of converting a projection image into pixels.

By the application of a pinhole alone, resolution is not pushed beyond the theoretical limit (Inuoé, 1995). It is the contrast of the signal that is being improved and that provides the often-mentioned $\sqrt{2}$ better "resolution" (factor 1.4; Inuoé, 1995). In real-world terms, the theoretical minimum distance at 450 nm illumination (blue light) to distinguish two points in the radial direction is 280 nm, and 570 nm along the Z axis.

4. One Step Beyond Classical Resolution: Deconvolution

A real improvement of the resolution of the optical system is achieved via the combined use of a pinhole (see section "Intermediate Step: Pixelizing the Image") and an additional postacquisition data processing step called deconvolution.

Deconvolution (also called image restoration, deblurring) is, broadly speaking, the postacquisition computational processing of a blurry or noisy image with the purpose to obtain the very best resolution with the highest degree of statistical confidence (Bertero *et al.*, 1990; Holmes *et al.*, 1995; Snyder *et al.*, 1992; van der Voort and Strasters, 1995). As the generation of an Airy distribution of projected light is considered to be a "convolution" process, the reversal of this process is called "deconvolution." The statistical nature of the generation of an Airy pattern requires that "reversed" statistical calculations be applied in the deconvolution process. Although deconvolution can be performed on single images, this type of processing is in neuroscience mostly applied to Z series of confocal images. The reason is that most biological structures extend into three dimensions where, due to the very construction of the instrument, the X and Y directions have a resolution (radial resolution) different from that along the optical axis (axial resolution, Z direction). A Z series is a number of images taken in confocal mode. A stepping motor lifts the stage a small, controllable step between each subsequent image. A Z series is in fact a series of images each in the narrow focal plane, with the object moving stepwise along the Z axis through the focal plane. We will therefore deal with a limited and specialized application of deconvolution, notably the deconvolution of fluorescence images in Z series.

5. Three-Dimensional Shape of the PSF in a Laser Scanning Instrument

Basic to the theory behind deconvolution is the diffraction pattern of light as outlined earlier. Photons emitted by a point source (the light emitted by a molecule of fluorescent marker) distribute onto a plane or a detector according to a PSF, similar to the Airy distribution of the light wave front in a conventional microscope. The PSF for any given microscope is a compound PSF influenced by all optical components: the PSF_i. Even within one instrument, each objective lens–intermediate lenses-microscope combination has its own particular PSF_i, since the numeric aperture of the objective lens is paramount. It is important to realize that the PSF_i of a confocal instrument changes every time a different objective lens is selected.

In a routine light microscope, diffraction in the Z direction is neglected. By contrast, in 3D reconstructions from confocal images, knowledge of the Z component of the PSF_i is very important. One would expect that the axial component of the PSF_i of a confocal instrument is the same as the radial component. This is not the case, however, due to the factor-2 lower resolution in the axial direction versus that in the radial direction and, surprisingly, by the presence of the pinhole. The difference between the diffraction patterns in the XY and Z directions (and thus differences between the radial and the axial components of the PSF_i) can be understood intuitively as follows. The optical axis of a microscope is aligned with the pinhole. The optical axis "cuts through" the entire thickness of the section. As a consequence, all photons emitted by structures in the section along the path of the optical axis pass the pinhole, even the photons that have been generated in planes above and below the focal plane. The consequence of this photon behavior is that the "focal plane" in a confocal microscope is not a flat plane but a deformed plane "rippled" according to the diffraction pattern of the back-projected pinhole. In this plane, the very area where the objective lens performs best, the blur in the Z direction, unfortunately, is at its highest

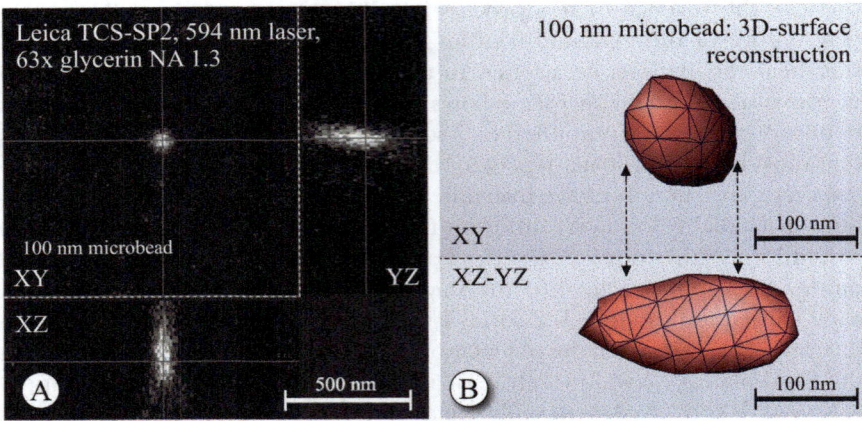

Figure 13.4. Image formation in a confocal microscope. Images of a 100-nm-diameter multifluorescent latex microsphere. (A) View of a confocal Z series in XY, XZ, and YZ directions. In the lateral direction (XY plane), the microsphere shows its true shape whereas in the axial direction (XZ and YZ) the microsphere appears elongated. (B) 3D reconstruction of this microsphere. The axial distortion of the sphere is caused by the different shape of the axial component of the point spread function (PSF_i) of a confocal microscope compared with the radial component.

and the confocality at its poorest. Fortunately, during scanning, the laser beam coincides only very shortly with the optical axis of the instrument. As a consequence of the presence of the pinhole, the 3D shape of the PSF_i of a confocal instrument resembles an ellipsoid rather than a sphere, with an axial or Z component definitely elongated compared with the radial or XY component (Hiesinger *et al.*, 2001; Shaw, 1995). This difference between radial and axial diffraction can easily be demonstrated by means of scanning very small fluorescent microspheres and by 3D reconstructing these spheres (Fig. 13.4, without postacquisition processing). In the XY plane, all microspheres appear spherical, while in the XZ and YZ planes they invariably look like mini rugby balls. The effect of the different shapes of the PSF_i measured radially versus axially is that the axial resolution of a confocal instrument is considerably lower (factor 2–2.5) than its radial resolution.

C. Key Instrument Parameters in Confocal Laser Scanning Microscopy

Since the theoretical optical considerations that hold for a conventional microscope are also valid for a confocal microscope, the key parameters to obtain high resolution in the confocal microscope are the wavelength of the light projecting through the optical system (that is, the emitted fluorescent light and not the incident light), the quality of the objective lenses (the higher the numeric aperture, the better), and the compound PSF of the optical system. Along with these factors, equipment like powerful and reliable lasers, good beamsplitters, excellent filter sets, and highly sensitive photomultipliers are a necessity. The opinions of the manufacturers differ with respect to the size and shape of the pinhole. The back-projected size of the pinhole should match the diameter of the Airy disk belonging to the fluorescence light emitted by the specimen. The main problem here is that the emitted fluorescence falls within a spectral emission band rather than being a fixed wavelength like the laser light used for excitation. Which emission wavelength to select? Conventional wisdom here is to use the wavelength at which peak emission intensity occurs.

Manufacturers are not specific about the physical shape and size of the pinholes implemented in their instruments. The shape of the pinhole is often determined by construction-related mechanical considerations. It may be a square rather than a circular aperture. Often the position of the pinhole is fixed in the Z direction (also a compromise) and, when the operator switches to an excitation laser with a different wavelength, the pinhole does not change its diameter. Theoretically, the diameter of the pinhole and the distance between the pinhole and the detector plane should vary according to the wavelength, yet this seems not to deter manufacturers from designing confocal instruments with fixed-position pinholes. Like many instruments, an actual confocal instrument is a compromise between theory and the practically attainable.

Since the PSF_i determines the amount of divergence of photons on their way from the fluorescent object to the detector, an advanced deconvolution program needs to know this PSF_i to do its job properly. Each confocal instrument has its own PSF_i. Although this PSF_i depends primarily on the objective lens, as argued above, construction factors play a role as well. The PSF_i at a particular magnification can be approximated via measurements on microspheres in the actual instrument and can be used in the computer program to calculate with high-statistical likelihood the origin of the photons.

The result of deconvolution calculations is a markedly improved image. Several deconvolution algorithms exist of which we use the Huygens II professional software (SVI, Hilversum, The Netherlands, http://www.svi.nl). Versions of Huygens II exist for Unix, Linux, Apple, and Windows platforms. Huygens II rapidly deconvolves Z series of images. According to Kano *et al.* (1996; images obtained in a two-photon confocal instrument), an improvement in resolution by two times in the XY plane can be obtained as well as an improvement by four times in the Z direction. This could amount to a resolving power with blue light (450 nm) of structures as small as 140 nm in the radial direction and 285 nm in the Z direction. Since the size of a CNS axon terminal is in the 0.5–1.0 µm range, the resolution of combined confocal scanning-deconvolution is therefore sufficient to study for instance colocalization of markers in nerve fibers and axon terminals. An active zone of a synapse (200–300 nm wide, 50 nm thick), if stained with a fluorescent marker that produces enough emission to hit the detector, will be rendered as a bright aggregate of fluorescence.

III. METHODOLOGY

A. Introductory

With the above physicooptical constraints and possibilities in mind, we use the confocal microscope to acquire images at high resolution. Through subsequent postacquisition processing, we improve the resolution via deconvolution.

We will now present the methodology used in carrying out our triple labeling experiment. Two rules of thumb apply throughout the entire process. First, overall performance depends on the weakest link in the chain. Second, preacquisition histology should be perfect. The ultimate goal is to identify the presynaptic terminal via neuroanatomical tracing (marker #1), to identify simultaneously the postsynaptic element (marker #2), and a protein uniquely associated with the synapse (marker #3). The outcome is a triple immunofluorescence protocol, which we will discuss in detail in this section. This protocol is illustrated with the projection in the rat from the presubiculum to parvalbumin interneurons in the entorhinal cortex (Wouterlood *et al.*, 2003).

In the introduction, we put forward that a synapse can be represented at the light microscopic level by a presynaptic axon terminal, a postsynaptic structure, and by molecules uniquely attached to a synapse as the intermediate marker. The challenge is to visualize this three-marker sandwich. Given the small dimensions of these sandwiches, we need to go beyond the classical limit of resolution to make them visible. We approach this suboptical resolution scale by using confocal laser scanning followed up with deconvolution and 3D reconstruction. Since "breaking the resolution barrier" occurs at the very end of a long and rather complicated chain of histochemical, physicooptical, and digital procedures, one should continuously keep in mind that the quality and reliability of the final 3D reconstructed image is completely dependent on the quality of every manipulation of the tissue sections in all the stages preceding the actual confocal laser scanning session and, of course, on the parameters applied during the laser scanning and postacquisition computer processing. The weakest link somewhere in the chain immediately affects the resolution at the end of the chain.

The multitude of factors influencing the end result of any confocal experiment can be grouped into four major clusters. As many factors as possible will be discussed while we proceed with the methodology:

1. Preacquisition histological procedures.
2. The confocal instrument itself: lenses, filters, detectors, and parameters.
3. Human factors like the skill, competence, and awareness of the person operating the confocal instrument.
4. Postacquisition image processing.

Sloppy histology, poor understanding of the basics, and incompetent operation of the instrument and computers can easily ruin the results and cannot be offset by the most sophisticated instruments, computers, and postacquisition computer processing.

The ultimate goal of multiple fluorescence is to observe sandwiches of aggregates of fluorochromes. This goal can be reached only when the brain is very well fixed such that proteins have had no chance before, during, or after fixation to diffuse out or move away from their original position. From this starting point, it follows that all membranes in general, and pre- and postsynaptic membranes in particular need to be in perfect shape. The same prerequisite of a very well fixed brain also holds for studies with the confocal instrument in which one wants to analyze colocalization of multiple markers in small cellular compartments such as axon terminals. On the other hand, fixation should not be too rigid because antibodies still must be able to penetrate into the sections in order to bind to their favorite epitopes. The fixation conditions are comparable to those described for preembedding electron microscopy by Leranth and Pickel (1989) in the previous issue of this book, except that in confocal laser scanning histochemistry, the use of detergents in the incubation media is allowed as a measure to enhance the penetration of the antibodies into the sections.

B. Anterograde Neuroanatomical Tracing and Follow-Up

The chapter by Lanciego in this book summarizes the pros and cons of various neuroanatomical tracers. Our experiment aims to visualize synapses, so we need to label the presynaptic axon terminal. This is best done via anterograde neuroanatomical tracing. A versatile anterograde tracer is biotinylated dextran amine (BDA; 10 kDa, Molecular Probes, Eugene, OR). This tracer is stable, its application relatively easy, it labels all the processes of neurons and their appendages throughout, and the detection is fairly straightforward with streptavidin conjugated to a fluorochrome of choice. BDA is also highly compatible with electron microscopy (Wouterlood and Jorritsma-Byham, 1993). An alternative anterograde tracer is the lectin *Phaseolus vulgaris* leucoagglutinin (Gerfen *et al.*, 1989; Gerfen and Sawchenko, 1984; Groenewegen and Wouterlood, 1990; Lanciego, 2005, this volume; Zaborszky and Cullinan, 1989; Zaborszky and Heimer, 1989). The procedural steps are listed in the Appendix.

C. Controls

Controls are extremely important in multilabel fluorescence staining. Each stage of the entire procedure requires its specific controls:

1. Immunofluorescence controls to test the specificity of the immunostaining.

2. Confocal instrument controls to check in a multichannel configuration the specificity of each channel and, separately, to check laser alignment. The laser alignment check is especially important if the purpose is to make 3D reconstruction of small structures presumed to lie very close to each other (markers #1 and #2 in our paradigm), or if the purpose is to study colocalization of markers in small structures (markers #2 and #3 in our paradigm).

1. Immunofluorescence Specificity Controls

Inherent in immunocytochemistry is the requirement to conduct sufficient control experiments to determine the specificity of the binding of the primary antibody to its epitope, the binding of the proper secondary antibody to the proper primary antibody, to exclude cross-reactivity, and to check whether nonspecific binding of antibodies to tissue components occurs (see Leranth and Pickel, 1989). In brief, control incubations should be designed to test that the primary antibodies react only with their specific epitopes and with no other tissue component, and to check that each of the secondary antibodies–fluorochrome conjugates reacts only with its corresponding primary antibody and not with the noncorresponding primary or secondary antibody. Primary antibodies can be tested with absorption controls (see Leranth and Pickel, 1989). Various controls with fluorochrome-tagged secondary antibodies are described by Wouterlood *et al.* (1998). It is important to know a priori that the immunocytochemical reactions have been successfully completed since otherwise no firm conclusions can be drawn from the images acquired in the confocal instrument. A thorough discussion of the application of fluorochromes in a multilabel experimental environment is provided by Wessendorf (1990), although this discussion stems from the preconfocal era. However, the principles of sound immunofluorescence practice as outlined by Wessendorf (1990) are still valid for confocal laser scanning configurations.

2. Confocal Instrument Controls

Central in confocal laser scanning is the concept of a *channel*. A channel is a specific configuration of the confocal instrument, which includes illumination with one of the available lasers and the corresponding specific filter, mirror, and detector settings such that the instrument is optimized to detect the associated fluorochrome and nothing else. This is especially important in multifluorescence confocal laser scanning. Each fluorochrome should be visible only in its own channel and not in channels set up to detect other fluorochromes. Signal detected in a channel associated with a different fluorochrome is considered *crosstalk* (also known as bleeding through).

Configuration of the channels of the instrument depends on the characteristics of the used fluorochromes. Conversely, the selection of proper fluorochromes is based on the laser wavelengths and filters available in the confocal instrument.

D. Fluorochromes and their Characteristics

The lasers used in confocal instruments produce spots of very high-intensity illumination of the section. Fluorochromes designed for use in such a harsh environment should therefore be particularly stable under high-intensity illumination. The classical fluorochrome fluorescein isothiocyanate is unsuitable in a laser illumination environment, since this dye is bleached away in a matter of seconds. Dyes with improved resistance to bleaching are the carbocyanine dyes (Amersham) and the Alexa Fluor™ dyes (Molecular Probes). Bleaching can be suppressed by the addition of an antifading agent to the mounting medium (see Longin *et al.*, 1993; Ono *et al.*, 2001).

In addition to resistance to bleaching, a fluorochrome for use in a CLSM should meet the following demands.

1. Excitation Spectrum

The shape of the excitation curve of a fluorochrome (excitation intensity plotted against wavelength) should be smooth, narrow, and steep, with an excitation maximum close to or coinciding with the wavelength of the assigned laser light of the confocal instrument. As an example, the fluorochrome Alexa Fluor™ 488 (excitation maximum at 491 nm; Table 13.1) will produce fluorescence with the highest intensity and quantum efficiency when illuminated with a 488 nm laser. Alexa Fluor™ 594 (excitation maximum of 590 nm) should be used in conjunction with a 594 nm laser. It

TABLE 13.1. Fluorochromes, their excitation peaks and the laser wavelength with which we excite these dyes in our confocal instruments, and potential of excitation crosstalk.

Fluorochrome	Excitation peak (nm)	Illumination with laser wavelength(s) (nm)	Excitation crosstalk with laser wavelength
Cy2™	489	488	—
Cy3™	554	543, 568	—
Cy5™	649	633, 647	—
Alexa Fluor™ 488	491	488	—
Alexa Fluor™ 546	556	543	—
Alexa Fluor™ 556	577	568	—
Alexa Fluor™ 594	590	594	—
Texas Red	595	594	568
Alexa Fluor™ 633	632 (shoulder at 580)	633, 647	568, 594
Alexa Fluor™ 647	650 (shoulder at 580)	633, 647	568, 594

Excitation peaks as provided by the manufacturers of the respective fluorochromes.

makes little sense to view fluorescence by, say, Alexa Fluor™ 594 through illumination with a laser that produces 488 nm or even 543 nm laser light. One may expect in those cases a low-quantum efficiency (a low intensity of the fluorescence related to the intensity of the excitation light, i.e., little bang for the buck and much bleaching), since the peak absorption of this Alexa dye is too far off the fixed excitation wavelengths supplied by the 488 and 543 nm laser light.

2. Emission Spectrum

Likewise, the curve of the fluorochrome showing the intensity of the emitted light plotted against the wavelength should be similarly shaped as its (ideal) excitation curve: smooth, narrow, and steep. Especially, the "tail" of the curve lingering toward the "red" end of the light spectrum should either be absent or else be as low and flat as possible. If the emission curve has an above-background spectral tail, it may cause emission crosstalk in double or multiple laser scanning, that is, inappropriate signal in channels configured toward the "red" end of the spectrum.

3. Resistance of Fluorochromes to Bleaching is Important for 3D Reconstruction

Bleaching of fluorescence signal (also called quenching or fading) may easily occur because of the very intense illumination of the fluorochrome by its assigned laser. 3D reconstructions are made on the basis of Z series of images. Imagine what happens if the fluorochrome offers little resistance to bleaching. The basic fact here is that the illumination part of most laser scanning microscopes is not confocal at all: the laser excites all fluorochrome molecules throughout the entire thickness of the section and all these excited molecules emit fluorescence; only *photons emitted from the focal plane* are detected. In a Z series, the region of interest (ROI) of the section will be exposed to a particular amount of laser light every time an image is acquired. If a Z series consists of n images, the ROI is exposed n times to the high-intensity laser light. In case of weak fluorescence, the operator may decide to scan each Z plane twice and average the result (the option *frame averaging* or *Kalman filtering*). In such a scenario, the ROI is exposed $2n$ times to the intense laser light. At the high magnification used in our experiments ($63\times$ immersion), all the laser light is concentrated by the objective lens onto a very small ROI. Thus, bleaching occurs faster at high magnification than at low magnification. In a Z series subject to bleaching, the first frame of the series will show a complete range of gray values. The second frame of the series may look less bright and crisp, while some low-intensity gray values will have disappeared. This reduction of brightness and contrast together with loss of low-gray values will progressively occur with the continuation of the Z series until at some point the acquired frame will be entirely black (no intensity left, everything bleached). When this occurs, completion of the Z sectioning becomes senseless and the subsequent 3D reconstruction becomes senseless as well. Thus, the operator of the confocal instrument always needs to be aware that bleaching is a potential danger. Countermeasures against bleaching are numerous. An experienced instrument operator knows to balance these countermeasures:

1. Addition of an antifading agent to the embedding medium prior to mounting (see Longin *et al.*, 1993; Ono *et al.*, 2001). There are several good antifading agents on the market, e.g., Vectashield™ Mounting Medium (Vector Laboratories, Burlingame, CA; www. vectorlabs.com) and ProLong (Molecular Probes). One can also apply one's own home-made additive (Platt and Michael, 1983).
2. Reduction of the laser intensity and increase in the gain (sensitivity) of the photomultipliers.
3. Reduction of the number of Z images.
4. Increase of the Z stepping increment.
5. Faster scanning (produces more noise, though).
6. Reduction of the image bitmap size.
7. No averaging of images (Kalman filtering off).
8. Increase in the scanning frequency.

E. Notorious: Crosstalk

One of the phenomena that may interfere in a negative way with the results in multilabel fluorescence studies is crosstalk. Crosstalk (also called "bleeding through") is, generally speaking, the observation of inappropriate fluorescence signal in a channel configured for imaging another fluorochrome. Crosstalk classically occurs always in a "higher" channel, that is, a channel configured around a laser–fluorochrome combination with longer wavelengths. We distinguish two types of crosstalk: emission crosstalk and excitation crosstalk.

1. Two Types of Crosstalk

Emission crosstalk is the excitation of a fluorochrome by its associated laser, and the inadvertent occurrence of some emission of this fluorochrome in the next, higher channel configured for a longer wavelength fluorochrome/laser combination. When emission crosstalk occurs, the appropriate channel shows a nice image while a faint copy of the image occurs in the inappropriate, "higher" channel. An example is excitation by a 488 nm laser of the fluorochrome Alexa Fluor™ 488, producing some emission in a channel configured around the combination of 543 nm laser/Alexa Fluor™ 546 (Figs. 13.5B and 13.6).

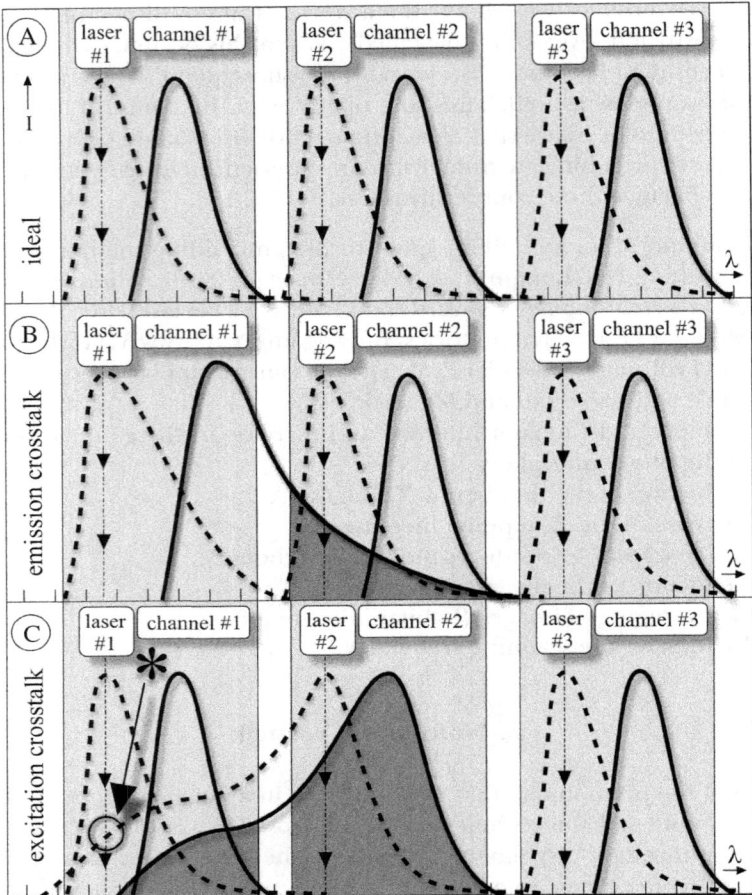

Figure 13.5. Diagram explaining channel separation and crosstalk in a confocal laser scanning instrument. Dashed line = excitation curve, solid line = emission curve. (A) Situation with ideally separated channels. For each channel, the laser excitation, fluorochrome excitation, and emission are strictly confined to the assigned wavelength frequency band. There is no interference between neighboring channels. (B) Emission crosstalk in channel #2 (shaded area): emission of the fluorochrome in channel #1 overflows in channel #2. This type of crosstalk can be avoided by sequential scanning. (C) Excitation crosstalk: the laser in channel #1 excites fluorochrome 1 but also fluorochrome 2 (emission of fluorochrome 2 in channels #1 and #2 is shown shaded), since the excitation curve of fluorochrome 2 extends into the wavelength frequency band of channel #1. This type of crosstalk cannot be avoided by sequential scanning. The signal produced in channel #1 by fluorochrome 2 has to be removed by postacquisition computer processing (so-called linear unmixing).

Excitation crosstalk is the effect in a double- (or triple-) fluorescence experiment that a particular laser excites next to "its own" associated fluorochrome also a second fluorochrome, e.g., one that belongs to the next, "higher" channel. Inadvertent signal of that next-channel fluorochrome is

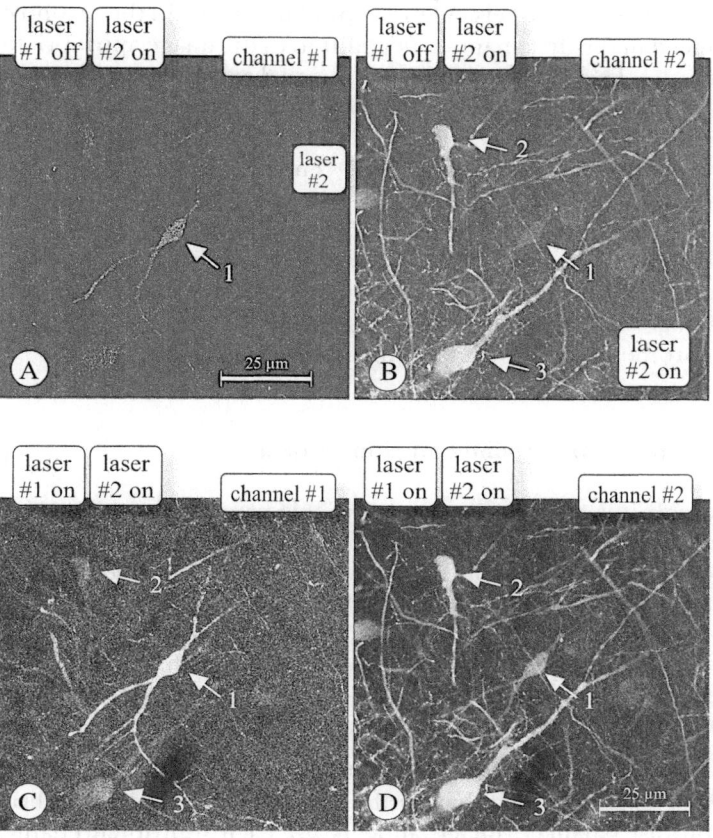

Figure 13.6. Practical examples of emission crosstalk, excitation crosstalk, and signal leakage. Section of rat hippocampal field CA1 immunostained with antibodies against calretinin (cell 1, Alexa Fluor 594) and parvalbumin (cells 2 and 3; Alexa Fluor 633). These markers were selected because CA1 cells express either calretinin or parvalbumin and never both markers. Channel #1: 594 nm laser, emission bandpass filter setting of 605–628 nm. Channel #2: 633 nm laser, emission longpass filter setting of 643–750 nm. The detector sensitivity for both channels had been optimized for its corresponding signal and was not further changed. Image pair A and B: Situation with only the laser in channel #2 switched on. In both channels, a ghost of the calretinin cell 1 is visible. In channel #1, this signal leakage effect is probably due to internal reflections or by incomplete cutoff by the bandpass filter assigned to channel #1. In channel #2, the ghost is caused by excitation and emission of the 594 fluorochrome by the 633 nm laser. The image pair in C and D was recorded with both lasers switched on. In C, ghost images of the parvalbumin cells 2 and 3 are visible caused by excitation crosstalk: the 594 nm laser excites the Alexa Fluor™ 633, and signal is picked up in channel #1. The ghost of the calretinin cell 1 in channel #2 is caused by emission crosstalk or by straightaway excitation of Alexa Fluor™ 594 in channel #2. All images at the same magnification.

produced in both channels. A faint copy of the image in the second channel is produced in the first channel, in addition to the image appropriate for the first channel. This "excitation" crosstalk signal resembles emission crosstalk, yet has a completely different cause (Fig. 13.6). Excitation crosstalk is much more difficult to recognize and avoid than emission crosstalk. An example is the excitation of Alexa Fluor™ 633 (emission in both 594 and 633 nm channels) by a 594 nm laser used to excite the fluorochrome Alexa Fluor™ 594 in the 594 nm channel (Figs. 13.5C and 13.6).

In addition to crosstalk phenomena, there may be internal reflections in the confocal instrument and incomplete cutoff of bandpass filters producing ghost images into a lower channel ("signal leakage"), e.g., signal of a "red" fluorochrome into a "green" channel (Fig. 13.6A).

2. Procedure to Determine Emission Crosstalk

Although it is possible to determine this type of crosstalk in double- or triple-stained sections, this check is best done with single-stained sections: for each channel, a section stained only with the fluorochrome assigned to that particular channel. In our laboratory, we have within reach these single-stained control sections for each of the laser excitation wavelengths with which our confocal instrument is equipped. In this check, we submit to the test three single-stained sections, each associated with its own channel of a three-channel setup.

1. Use the three single-stained sections to configure, one after another, three channels such that on the display screen a nice and brightness-contrast balanced image appears for each individual channel. Save these settings.
2. Insert the single-stained section associated with the first channel in the microscope.
3. Turn the laser intensities for the second and third channels back to zero but *do not change* the detector sensitivities for these channels.
4. Illuminate the section subsequently with the laser belonging to the first channel only (e.g., 488 nm).
5. Images remaining in the second and third channels represent emission crosstalk.
6. Repeat this procedure for the next combination of channels.

A countermeasure against emission crosstalk is to reduce the intensity of the laser in the first channel such that the crosstalk image in the second channel is no longer visible (of course, the laser for the second channel should be turned down temporarily to see the effect). Adjust, if necessary, the detector sensitivity for the first channel. Repeat the procedure for the next combination of channels: two and three (do not increase at this stage the sensitivity setting of channel 2, since the current setting of that channel has been determined in order to avoid crosstalk from channel 1). Next, continue multichannel laser scanning with these settings for laser intensities and detector sensitivities. If these measures do not help, then sequential scanning might offer a solution. Emission crosstalk can be avoided completely by resorting to a sequential scanning procedure.

In a multichannel configuration, crosstalk usually occurs in "higher" channels, for instance, crosstalk showing up in a 543 nm channel when the laser in the other channel is a 488 nm laser. This (emission-type) crosstalk occurs because the emission spectrum of any fluorochrome is always shifted to longer wavelengths compared with the excitation spectrum (so-called *Stokes shift*) and never to shorter wavelengths (which is impossible according to the second law of thermodynamics). The "sneaky" feature of excitation crosstalk is that it occurs in a channel configured around a shorter laser wavelength (a "lower" channel) than its appropriate channel. Given Stokes shift, this sounds paradoxical at first sight. There is, however, no conflict with the second law of thermodynamics at this point. The essence of this type of crosstalk is that the involved fluorochrome is excited by the laser belonging to the inappropriate, "lower" channel and that it produces its normal, Stokes-shifted fluorescence signal in both the inappropriate channel and the appropriate channels.

Excitation crosstalk is much harder to detect than emission crosstalk. Also, because most microscope operators trained as they are in recognizing Stokes shift and associated emission into a higher channel do not expect crosstalk in a "lower" channel, they are inclined to be aware of and test only for emission crosstalk and not for excitation crosstalk. Excitation crosstalk can be considerable when, for example, the fluorochrome Alexa Fluor™ 594 (excitation 594 nm laser) is combined with Alexa Fluor™ 633 (excitation 633 nm laser). The excitation curve of the latter fluorochrome possesses a shoulder that makes the dye sensitive to excitation at 594 nm. As a consequence, Alexa Fluor™ 633 produces signal simultaneously in both the 594 and 633 nm channels when the section is illuminated with 594 nm laser light (Figs. 13.5C and 13.6). Similarly, excitation crosstalk occurs for instance at 568 nm laser illumination with the combination Cy3™ (568 nm excitation maximum)–Alexa Fluor™ 633 (633 nm excitation maximum). It occurs also at 568 nm laser illumination with the combination Cy3™ (568 nm excitation maximum)–Alexa Fluor™ 647 (633 nm excitation maximum).

3. Procedure to Determine Excitation Crosstalk

This check is done with double- or triple-stained sections, with backup of single-stained sections. The test cannot be done with markers that are colocalized. Prior to testing, the single-stained sections should be tested to be sure that the individual fluorochromes do not produce emission crosstalk.

In this example, it is assumed that excitation crosstalk occurs in two adjacent channels:

1. Configure with a double-stained section two adjacent channels such that on the display screen a nice, brightness–contrast balanced image appears for each channel. Save these settings.
2. Replace the double-stained section with a single-stained section containing only the fluorochrome assigned to the *second* channel.
3. If now in the first channel an image is detected, this is due to excitation of the second channel's fluorochrome in the first channel. Increasing and decreasing the laser intensity for the first channel increases and decreases the amount of this excitation crosstalk, respectively.
4. Repeat these steps for the next combination of channels.

Note that this check is meant only to determine whether excitation crosstalk exists in the specimen. The elimination of excitation crosstalk is quite another story since, besides the fact that excitation crosstalk is hard to detect, one cannot filter to eliminate this type of crosstalk. Scanning in sequential mode offers no solution either since both fluorochromes are excited and produce emission whenever the first laser illuminates the specimen. The only way to get rid completely of excitation crosstalk is to replace the "offending" fluorochrome with a completely different fluorochrome. A way to intentionally reduce excitation crosstalk is via postacquisition image processing by a program called "dye separation" or "linear unmixing," provided that the intensity of the crosstalk signal is modest.

There exist several laser-fluorochromes combinations, which are relatively safe with respect to excitation crosstalk. Of course, these combinations should match the available lasers and filters of the confocal instrument. The bottomline is that with any combination of fluorochromes, the spectral excitation curves must be as much separated from each other as possible (as in Fig. 13.5A).

Examples of such "safe" combinations are the following:

1. Double-fluorescence labeling

1st laser (nm)	1st fluorochrome	Combined with	2nd laser (nm)	2nd fluorochrome
488	Cy2™		568	Cy3™
488	Alexa Fluor 488		568	Alexa Fluor™ 568
488	Cy2™		594	Alexa Fluor™ 594
488	Cy2™		594	Cy5™
488	Cy2™		647	Cy5™
488	Alexa Fluor™ 488		647	Alexa Fluor™ 594
488	Alexa Fluor™ 488		647	Alexa Fluor™ 647
488	Alexa Fluor™ 488		647	Alexa Fluor™ 647
488	Alexa Fluor™ 488		647	Cy5™

2. Triple-fluorescence labeling

1st laser (nm)	1st fluorochrome	Combined with	2nd laser (nm)	2nd fluorochrome	Combined with	3rd laser (nm)	3rd fluorochrome
488	Cy2™ or Alexa Fluor™ 488		543	Alexa Fluor 546		594	Alexa Fluor™ 594
488	Cy2™ or Alexa Fluor™ 488		543	Alexa Fluor 546		647	Alexa Fluor™ 633 or 647
488	Cy2™ or Alexa Fluor™ 488		543	Alexa Fluor 546		647	Cy5™

4. Practical Advice to Guard Against False Results Caused by Crosstalk

1. Always be aware of crosstalk.
2. Keep single-stained sections at hand.

When a particular set of scanning parameters for multifluorescence scanning has been determined such as laser intensities, detector sensitivities, filter selection, etc., first scan single-stained sections with these parameters and check whether an image is produced in the inappropriate "higher" or "lower" channels.

F. Operating the Confocal Instrument: "Operator Awareness"

A CLSM is a complicated optical and digital instrument, and a thorough understanding of what one is doing and what is happening (in physicooptical terms and in terms of instrument handling and software) is necessary. The reason is that by simply turning the controls of the instrument always some image can be produced in which the information contained may range from doubtful to completely worthless. The operator should always keep in mind that images obtained should represent the real world as close as possible and that the imaging should not be disturbed by some interference such as false positivity (e.g., channel crosstalk, sensitivity too high) or false negativity (e.g., out of focus, incorrect filter selection, sensitivity too low, fading, insufficient penetration of fluorescent marker). As the software in new confocal instruments is increasingly being equipped with all sorts of automatic functions, chances are on the rise that an unaware or inexperienced operator working "on autopilot" may program the detectors to acquire invalid or noninformation instead of a valid series of images.

In an instrument equipped with separate lasers, the perfect alignment of these lasers is a matter of concern. In a multiuser environment, instrument awareness of the operator is necessary at this point. Alignment and/or chromatic aberration can be tested with multifluorescent latex microspheres (e.g., the TetraSpeck® microspheres kit, Molecular Probes) or with sections containing small structures multilabeled on purpose (Wouterlood *et al.*, 1998). Also, for the purpose of calibration, we have control brain sections at hand containing axons labeled anterogradely with the neuroanatomical tracer BDA and incubated with a cocktail of streptavidin conjugates: Alexa Fluor™ 488, 546, and 633. By virtue of this triple-color cocktail, the same labeled fiber fluoresces under illumination with the 488, 543, or 633 nm lasers. In the confocal instrument, we scan in three channels and we overlay the acquired images. If misalignment or chromatic aberration occurs, this will reveal itself as "pixel shift," which is the slight deviation of the image in one channel compared with that acquired in a different channel. Note that in addition to radial pixel shift (in the XY plane) also axial pixel shift may occur (in the XZ and YZ planes). We use our triple-stained calibration sections to investigate both radial and axial pixel shifts (Fig. 13.7; see below for further details on pixel shift and image mismatch). The advantage of our calibration sections is that the labeled fibers are embedded in brain tissue. The brain parenchyma surrounding the fluorescence-marked structures is by no means isotropic. This anisotropism may cause distortion of the image. In this respect, our calibration sections are more realistic than slides containing multifluorescent latex microspheres embedded in an ultrahomogeneous, isotropic mounting medium.

G. Postacquisition Image Processing and 3D Reconstruction

Image processing software is available from a range of companies. Several manufacturers of confocal instruments offer a dye separation (linear unmixing) package to improve spectral separation if necessary. We prefer perfect signal separation at acquisition time, though, before relying on postacquisition dye separation. We consider the latter a measure of last resort.

Dye separation attempts, and often succeeds, in removing emission crosstalk from the acquired images. These programs use the spectral emission characteristics of the fluorochromes. As mentioned earlier, a nasty characteristic of the fluorochrome Alexa Fluor™ 633, and the same holds also for Cy5™, is that its excitation spectrum has a shoulder at 568 nm. This implies that these fluorochromes always produce fluorescence when a 568 nm laser is used (excitation crosstalk in the 568 nm channel, cf. Fig. 13.6) and that these fluorochromes under 568 nm laser illumination produce signal as well in the 633 nm channel, even when the 633 nm laser is switched off. When it is impossible in such and similar cases to use different fluorochrome combinations, then postacquisition dye separation is indicated as a helpful tool to remove the unwanted results of excitation crosstalk.

1. Deconvolution

As argued in the theoretical part of this chapter, the quantum physics of image formation predicts that any image recorded with an optical instrument is always blurred to some degree. The amount of blur depends on the

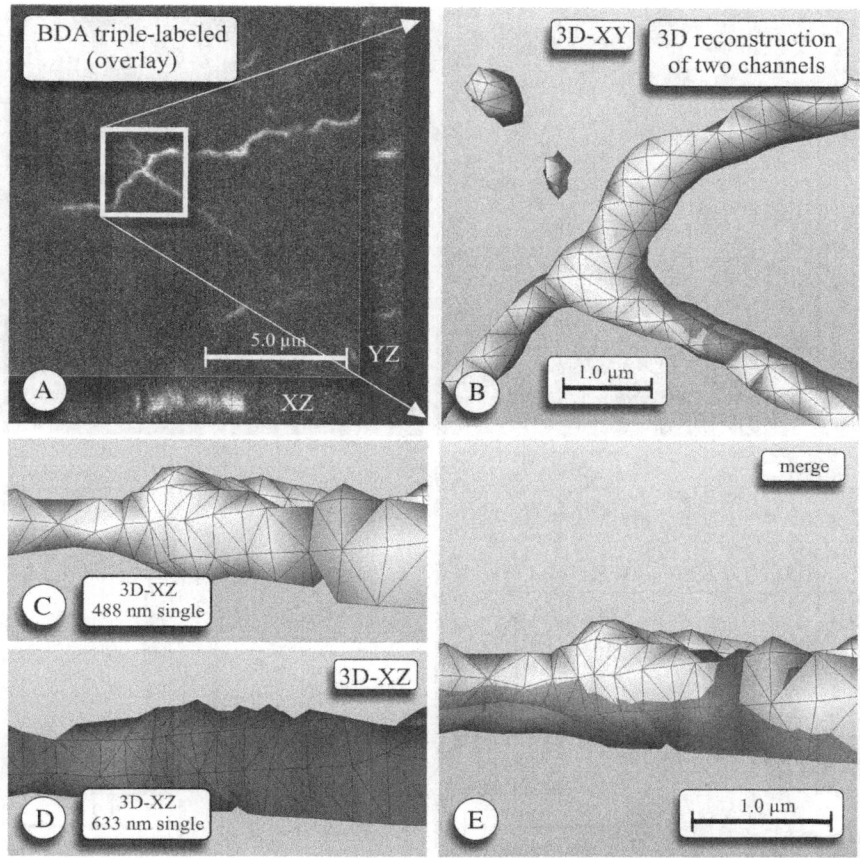

Figure 13.7. Pixel shift and image mismatch. Imaging of a BDA-labeled fiber stained with a cocktail of three fluorochromes. Image series not deconvoluted. (A) Composite *XY*, *XZ*, and *YZ* view of the image series. In the color image, shift of green and red pixels is seen in the axial direction; there is no shift in the lateral direction. (B) Enlarged portion of (A) 3D reconstructed (two of the three channels, *XY* view). There is no lateral shift of both images indicating that radial alignment of the lasers is perfect. (C, D) Single-channel 3D reconstructions of the image in the 488 and 633 nm channels (3D reconstruction turned 90°; *XZ* view). (E) Merge of C and D showing that in the axial direction, there is image mismatch. In this case, mismatch amounts to ~100 nm.

objective lense i.e. it is instrument specific information about how blur in a particular optical instrument is produced. This characteristics can be used to calculate from a blurred image via a statistical approach an image that resembles the original object with the highest degree of confidence. This reversal of the process of image formation is called deconvolution, "deblurring," or "image restoration" (Bertero *et al.*, 1990; Snyder *et al.*, 1992; van der Voort and Strasters, 1995). Deconvolution is the final step breaking the resolution limit barrier. Alternatively, deconvolution can be applied when an optical

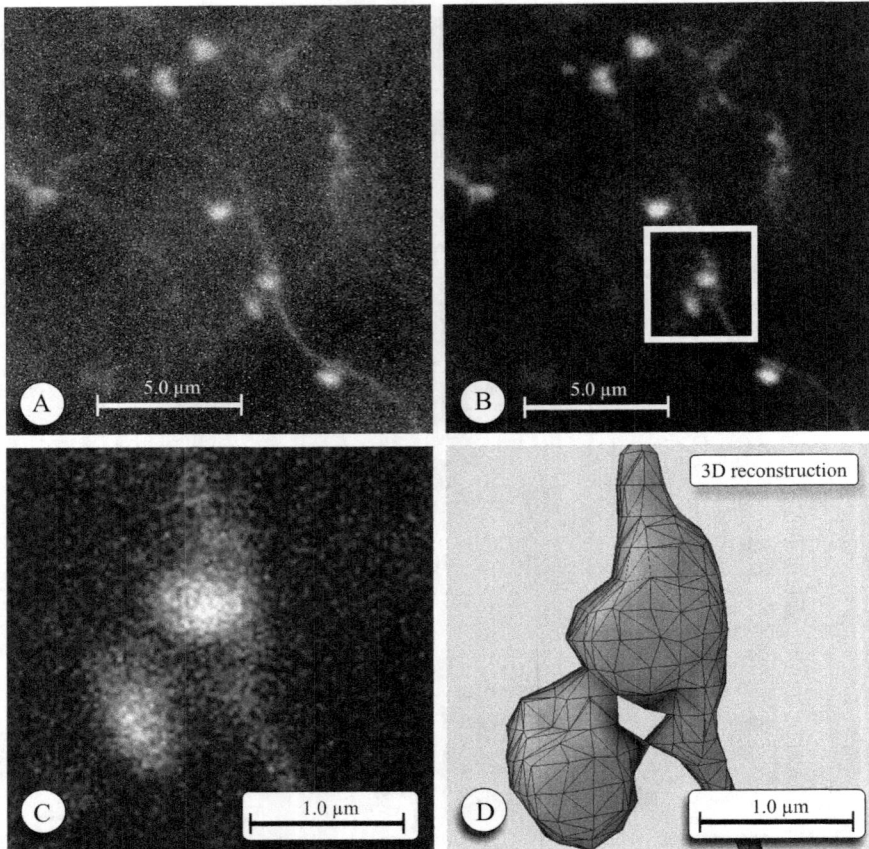

Figure 13.8. Deconvolution of a Z series of images of a BDA-labeled fiber. (A) Composite image of the Z series immediately after acquisition. (B) Same image series, deconvoluted with Huygens II software. (C) Detail of B. (D) 3D reconstruction of the detail in C.

signal is particularly blurry or noisy. Several deconvolution algorithms exist: blind deconvolution, iterative deconvolution according to Miller–Tikhonov, and iterative maximum likelihood estimate (MLE) deconvolution. MLE is specifically advised for the deconvolution of Z series of confocal images. The core of MLE consists of a statistical calculation that takes into account the PSF_i of the confocal instrument. The characteristics of the objective lenses and the other optical parts of the confocal instrument used for image acquisition, as well as the refractive indices of the tissue, immersion and embedding media, contribute to this PSF_i. The computer program uses the PSF_i to calculate, for each pixel of each acquired image in a Z series, the statistical likelihood of the exact origin of the photons emitted by the fluorescent specimen. The result is a statistically reliable, improved version of the Z image series (Fig. 13.8). The deconvolution program used in our laboratory can, however, only process the Z series of images belonging to one channel at a time, and so we have to run the program as many times as there are channels. After deconvolution, the Z series of the respective channels can be merged into a final multicolor image in which all immunostained structures are visible, color-coded according to their specific microscope channel. The increase in resolution can be as much as two times in the radial direction (XY) and four times in the axial direction (Z) (Kano *et al.*, 1996; this holds for two-photon confocal images).

Note that a multicolor 3D reconstruction is in fact an overlay of as many separate 3D reconstructions, one for every channel in the confocal instrument. For instance, a three-channel confocal imaging session produces three Z series of images. Each of these series is deconvoluted on its own and then merged into one final 3D reconstruction. The color code assigned to each single-channel reconstruction is conserved. Most operators of confocal instruments adhere to the convention to render the 3D reconstruction in the first channel in green, in the second channel in red, and that in the third channel in blue. The color of the fluorescence emitted by the fluorochrome is used as the associated color in the 3D reconstruction. This convention has of course nothing to do with the real colors since all confocal images are basically 8- or 12-bit gray scale bitmapped images.

2. Correction for Image Mismatch

Because in a multilabel experiment the images acquired in each channel are 3D, reconstructed independently from those acquired in the other channels, small deviations of the relative positions of reconstructed objects could go unnoticed. This so-called image mismatch is a source of both false positivity and false negativity when it comes to the detection of sandwiches of three (independently reconstructed) markers (Wouterlood *et al.*, 2002). Here, the intentionally triple-stained, single tracer containing sections comes back into focus (Fig. 13.7). The 3D reconstructions in each channel of the fibers in these sections should exactly match. If, for example, the 3D reconstructed image in one channel for some reason does not match with those in the two other channels, this is evidence that somewhere in the chain of image acquisition and processing something has gone wrong. The cause may be laser misalignment, deviation of the scanning mirrors, chromatic aberration, operator unawareness, and so on.

In each confocal image acquisition session we scan, for the purpose of instrument calibration, always a preparation containing intentionally triple-labeled fibers. The 3D reconstructed images of these fibers are used to detect image mismatch. If necessary, we can correct image mismatch by shifting the 3D reconstructions in each channel a few pixels in the appropriate X, Y, or Z direction. The result of this exercise is that in 3D reconstruction, the perfect overlay image of the triple-stained fibers appears like a structure painted with three layers of paint of different color. The amount of necessary shift is the correction factor that is next applied to the Z series of the images belonging to the scientific experiment. One assumption underlying this correction is that the parameters of the confocal instrument causing the misalignment are constant during the entire image recording session. Another assumption is that these parameters are similar for both the calibration section and the sections belonging to the scientific experiments.

Most confocal instruments are used in a multiuser environment. It is relevant in such an environment to conduct every now and then a check of the instruments to see that the alignment of the lasers and other components has not changed over time. It should also be noted that mismatch may occur in a confocal instrument just after switching on (a "cold" instrument), and then decrease until it becomes stable (in our instrument, a plateau is reached 30 min after powering on) (Wouterlood *et al.*, 2002).

3. Multichannel 3D Reconstruction

There are several competing software packages on the market capable of doing the job of calculating and rendering 3D reconstructions. In our laboratory, we have installed two packages: FluVR™ (SVI) and Amira™ (www.amiravis.com), which both run on the same Silicon Graphics workstation as the deconvolution software does. FluVR™ is a so-called volume-rendering program, whereas Amira™ is a surface-rendering program. Versions of both programs are available running under the Linux and Microsoft Windows® operating systems.

- *Volume rendering*: In this type of 3D rendering, a *Z* series of images is considered as a rectangular box filled with layers of cubes. Each layer corresponds with one frame of the image series. Each of the cubes (a voxel) has an assigned gray value. The program considers the gray value of each voxel as a measure for the absorption and emission of light by that voxel, and it starts a simulation in which it casts light from a virtual light source onto the space filled with voxels. Next, the program calculates the amount of light absorbed and the amount of fluorescence emitted by each voxel (depending on their gray value). A number of parameters such as distance from the light source, direction of the light, light intensity, transparency of voxels for excitation light, transparency of voxels for emission light, angle of inspection (camera position), and even reflection from the background can be modified interactively. The result of the calculation is a simulated 3D image of the fluorescence of all voxels of the *Z* series. Since all voxels including those that are "transparent" are involved, this type of rendering produces scientifically most valid 3D reconstructions. Since volume rendering includes calculations on all voxels of the entire *Z* series, it is time and memory resource consuming. Real-time rotation of 3D structures reconstructed via volume rendering is not possible on our computer. Multichannel volume rendering is possible (up to 32 channels, SVI; personal communication).
- *Surface rendering*: In this type of 3D rendering, a *Z* series of images is also considered as a rectangular box filled with layers of grayish cubes.

However, the operator selects a particular gray value (a threshold; in 8-bit images, there are 256 possible gray values). The program calculates lines connecting all voxels in the voxel space expressing the threshold gray value and draws on screen a wireframe consisting of these isolines (expressed as a grid of triangles). The software thus connects voxels in all the three directions. The surfaces of the wireframes are covered with a colored texture of choice. All further calculations are done with the coordinates of the corners of the triangles forming the wireframe. This saves a tremendous amount of processor resources. Surface rendering allows real-time rotation and zooming. These features are also available in the PC/Windows version of the program we use (Amira™, a high-end graphics board in the PC is recommended). Surface rendering has its analog in 3D computer games, where the objects or characters are based on wireframes clad with textures. Note that, apart from the use of RGB color images as textures, the number of colors that can be assigned to these animated characters is virtually infinite. The number of channels that can be rendered in surface rendering is therefore also virtually infinite.

IV. RESULTS

After surface 3D rendering of structures containing the three markers—presynaptic marker BDA (fluorochrome Alexa Fluor™ 546, laser 543 nm), postsynaptic marker parvalbumin (fluorochrome Alexa Fluor™ 594, laser 594 nm), and synapse-associated marker ProSAP2/Shank3 (fluorochrome Alexa Fluor™ 488, laser 488)—we noted consistent spatial separation of the reconstructed immunofluorescent material. Most reconstructed ProSAP2/Shank3 aggregates appeared exterior to BDA-labeled fibers. ProSAP2/Shank3 aggregates frequently resided in the interior of cell bodies and dendrites of parvalbumin-stained cell bodies and dendrites. In accord with the expectations (localization in the postsynaptic density), ProSAP2/Shank3 material mostly had a peripheral localization in parvalbumin-labeled structures, that is, just below the surface envelopes of parvalbumin-stained structures. There were no small ProSAP2/Shank3 aggregates seen deep in the interiors of parvalbumin-expressing structures or in BDA-labeled fibers and axon terminals.

In several cases where varicosities on BDA-labeled fibers appeared to be apposed to parvalbumin immunofluorescent dendrites, we noted ProSAP2/Shank3 immunofluorescent material sandwiched in between, immediately subjacent to the surface envelope of the target neuron, or immediately next to the apposition. We regard such a sandwich as the confocal analogon of a terminal bouton, forming a synapse with the parvalbumin-containing cell body. An example of such an apposition with aggregation of ProSAP2/Shank3 immunofluorescent material in a position inside a parvalbumin cell body facing a varicosity on a BDA-labeled fiber is shown in Fig. 13.9.

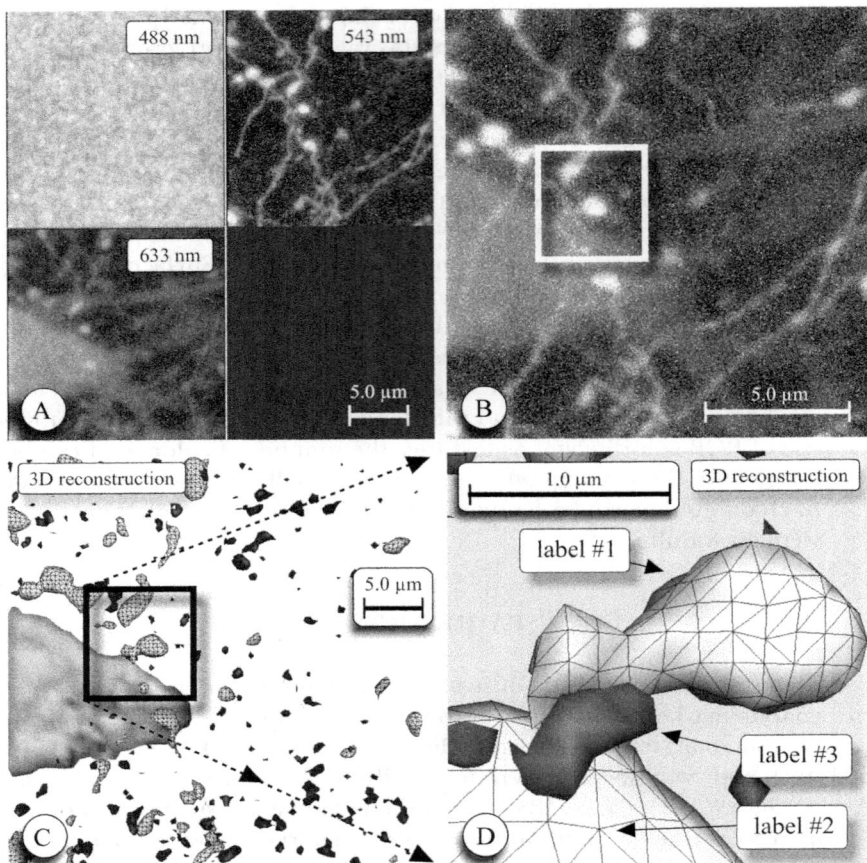

Figure 13.9. Result of triple channel confocal laser scanning, deconvolution, and 3D reconstruction. (A) Three-channel image generated by the confocal instrument of a Z series: ProSAP2/Shank3 (channel #1, 488 nm, label 3), BDA (channel #2, 543 nm, label 1), and parvalbumin (channel #3, 633 nm, label 2). (B) Channels #2 and #3 in overlay projection at higher magnification. BDA-labeled fibers stand out; the parvalbumin labeling is weak. (C) 3D reconstruction with Amira after deconvolution, channels merged. (D) Detail of the 3D reconstructions, showing a sandwich of the three markers indicative for a synaptic contact between the BDA-labeled fiber and the parvalbumin neuron. Label-2 structures (parvalbumin) rendered with a transparent texture.

V. CONCLUSIONS AND FUTURE

The above results of observable sandwiches of fluorescence material associated with an anterograde tracer, a postsynaptic marker, and a marker unique for a synapse lead to two conclusions. First, in this way at least appositions between processes of neurons can be made visible with great detail. Second, the application of the third marker provides the evidence in favor of the existence of a synapse. It is, in particular, the sandwiching that proves that a synapse occurs between the labeled presynaptic axon terminal on the one hand and the structure containing the postsynaptic marker on the other. Resolution indeed seems to be sufficient to distinguish structures with sizes in the order of 0.2–0.3 μm and to see whether there is apposition and/or colocalization at this edge of the theoretically possible resolution. In a separate series of experiments, we have conducted double staining, i.e., we labeled presynaptic fibers and axon terminals with BDA and we labeled presumed postsynaptic dendrites via filling through pericellular application of Neurobiotin™. We reconstructed numerous appositions of BDA-labeled axon terminals and dendrites containing Neurobiotin. In parallel electron microscopy experiments, we lesioned the source area in the brain and we studied the area with Neurobiotin-filled dendrites at high magnification under the electron microscope. We indeed noted synaptic contacts between degenerating axon terminals and containing dendrites (Wouterlood *et al.*, 2004).

At present available in the form of antibody against ProSAP2/Shank3 is only a marker for NMDA-regulated synapses, i.e., glutamatergic, excitatory synapses. Several antibodies have been developed against components of GABAergic, inhibitory synapses, of which the GABA$_A$ receptor scaffolding protein gephyrin (Sassoë-Pognetto and Fritschy, 2000) may be a candidate for labeling experiments similar to those presented in this chapter. In that case, we could in the future map on CNS neurons, the relative numbers of excitatory and inhibitory synapses with unprecedented speed.

VI. ADVANTAGES AND LIMITATIONS

The advantages and simultaneously the disadvantages of the confocal approach as outlined in this chapter are closely associated with the ability of the combination of confocal instrument and postacquisition processing to improve resolution of an optical system down to and slightly over its very limit as governed by the physical laws of optics and diffraction of light. Resolution offered by conventional double-label epifluorescence microscopy is limited due to the great depth of focus (fluorescence emitted by structures throughout an entire section reaches the eye) and, in association with this, the low resolving power. In fact, resolution offered by conventional fluorescence microscopy is barely sufficient to determine with confidence colocalization of markers in neuronal cell bodies. Determination of colocalization of markers in fibers is out of question. Resolving power is markedly improved in the CLSM. Combination of confocal microscopy with postacquisition deconvolution further improves the total resolution to a degree that colocalization of markers in fibers and axon terminals can be determined. Furthermore, with the addition of 3D reconstruction the observer can see rapidly and decisively whether axon terminals of a particular origin or chemical signature appose presumed postsynaptic structures, for instance processes belonging to a particular chemical category of neurons. The 3D computer reconstruction is essential here because it enables us to look at the apposing structures at any desired angle. By rotating and inspecting the 3D reconstruct, falsepositive apposition can be distinguished from true apposition because at angles of inspection different than the conventional, fixed orthoscopic look, spatial separation of the involved structures is rapidly detectable. In addition to this, the ProSAP2/Shank3 marker indicates whether there is a synaptic interface between the presynaptic terminal and the postsynaptic structure. Compared with double- and triple-labeling electron microscopic analysis, this means an enormous advantage in speed of analysis as well as in the number of synapses that can be inspected and counted in a given amount of time.

One disadvantage of the present approach is that we need more, better, and stable synapse markers, for instance a reliable marker associated with specific proteins present at the interface in inhibitory synapses. We assume that in the future, these markers will become available as the molecular structure of the postsynaptic density becomes better understood. A second and more fundamental disadvantage is that determination of synapses requires resolution at a level, which is at the edge of resolution attainable with optical systems. Note that the optical resolutions calculated in this chapter are ideal resolutions. In the extremely heterogeneous environment offered by brain tissue, the practical attainable resolution is always lower. Furthermore, the physical law of diffraction predicts that at the magnification necessary to pinpoint synapses, all stained "structures" will appear as blurry distributions of photons rather than the crisp images we are familiar with at low magnification. At a synapse, the pre- and postsynaptic markers are by nature extremely close. This, and the fluorescence associated with the synaptic marker sandwiched in between, causes overlapping distributions of photons in the detectors of the confocal microscope. This may cause confusion in many interpreters. If a crisp image is really necessary, then the electron microscope with its superior resolution (measured in tenths of nanometers instead of hundreds) is the instrument of choice. After all, electrons cannot be beaten as vehicles for the imaging of nanostructures.

APPENDIX

A. Surgery, Injection of BDA in the Rat

1. Anesthetize the rat deeply with an intraperitoneal injection of a mixture of four parts Ketaset (ketamine; 1% solution; Ket, Aesco, Boxtel, The Netherlands) mixed with three parts Rompun (xylazine; 2% solution, Bayer, Brussels, Belgium) (1 ml/kg body weight of this mixture).
2. Mount the animal in a stereotaxic frame.
3. Expose the skull and anesthetize the periost with lidocaine (10% spray; Astra Pharmaceutica BV, Zoetermeer, The Netherlands).

4. Drill an opening in the skull at the desired $X-Y$ coordinates and open the meninges.
5. Lower the tip of a borosilicate glass micropipette filled with tracer (BDA, 5% in 10 mM phosphate buffer, pH 7.25, pipette tip diameter 10–20 mm) to the desired rostrocaudal, lateral, and vertical coordinate.
6. Apply to the micropipette a positive pulsed 5 mA DC current (7 s on/7 s off) for 10–15 min. Leave the pipette in situ for 10 min after delivery of the tracer.
7. Retract the pipette, close the wound, and allow the animal to recover. The survival period postsurgery is usually 1 week.

B. Perfusion-Fixation, Sectioning, Storage

1. Inject an overdose of sodium pentobarbital (Nembutal, Ceva, Paris, France; intraperitoneally, 60 mg/kg body weight).
2. Perfuse transcardially, first with 100 ml of physiological saline solution of 38°C, pH 6.9, immediately followed by 1000 ml of 4% freshly de-polymerized paraformaldehyde and 0.1% glutaraldehyde in 125 mM phosphate buffer, pH 7.4 (room temperature). We use a perfusion system driven by compressed air, which delivers perfusion fluids at a constant, controllable hydrostatic pressure (Jonkers *et al.*, 1984). The thoracic aorta is clamped to ensure that all fixative is directed at the upper part of the body.
3. Immediately after perfusion carefully remove the brain from the skull. Cut 120-μm-thick slices with a vibrating microtome and collect these in chilled 125 mM phosphate buffer, pH 7.4 (vials kept on melting ice).
4. Infiltrate the slices with a cryoprotectant consisting of 20% glycerin and 2% DMSO in 125 mM phosphate buffer, pH 7.4 (Rosene *et al.*, 1986).
5. Transfer the slices to storage vials and place these in a freezer (−20 or −40°C) for later use.
6. Resection the slices into sections prior to immunohistochemistry for the purpose of obtaining better penetration of antibodies.

C. Resectioning Slices into Sections to Obtain Better Penetration of Antibodies

The third label in our paradigm (see Fig. 13.1) identifies a component uniquely associated with the synapse. We screened for this purpose many antibodies directed at protein components of receptor scaffolding molecules located in or at the postsynaptic density of the synapse. Although these antibodies work well in cell cultures, the vast majority of them suffer from insufficient penetration into brain sections of regular thickness (25–40 μm) or even worse, they simply do not penetrate at all. An antibody that works in our free-floating section incubation environment was raised against the

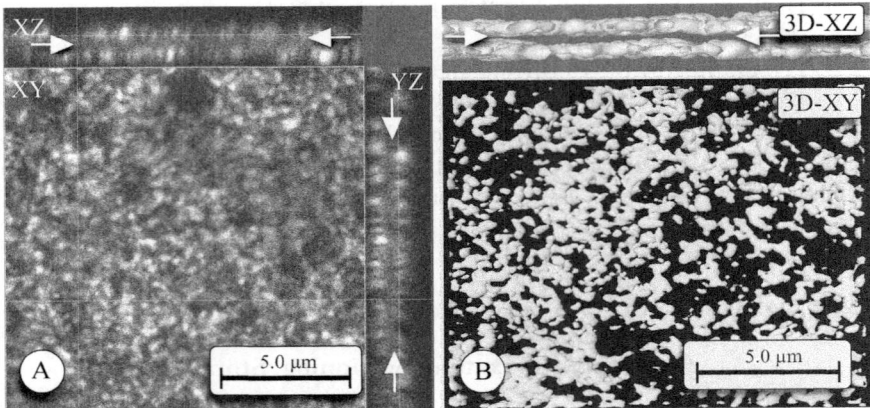

Figure 13.10. Insufficient penetration of antibodies into a section detected via Z confocal scanning. The marker is vesicular glutamate transporter 1, which is present in glutamatergic axon terminals. Image taken in stratum radiatum of CA1, hippocampus, 63× immersion lens NA 1.3, electronic zoom 8×. (A) Top and side views of a Z image stack (26 frames). Inspection of the series in simultaneous XY, XZ, and YZ rendering reveals intense staining of aggregates of immunofluorescence at the upper and lower surfaces of the section, while a band in its core is dark, with weak or low immunofluorescence (arrows). (B) 3D reconstruction of this Z series in XY and YZ view. The (artifactual) absence of staining in the core of the section (arrows) is even more dramatic after reconstruction.

protein PsoSAP2/Shank3 by Böckers *et al.* (1999). This protein is interpreted as an anchoring protein of the NMDA receptor at excitatory synapses, and it is located in the postsynaptic density (Böckers *et al.*, 2002). However, in pilot incubations, the penetration of this antibody appeared to be insufficient. The degree of penetration of an antibody can easily be measured in a confocal microscope by means of a Z scan of the complete section, from the upper to the lower surface (Fig. 13.10). A condition is that the marker is known to be distributed homogeneously throughout the area of interest. If in this Z scan, the distribution of the immunofluorescence occurs only at the outer surfaces of the section, or when the intensity of staining shows a gradient with good staining in a small superficial band at the upper and lower surfaces of the section and poor or no staining in between (as is visible in Fig. 13.10), then it is likely that penetration has been inadequate. Another phenomenon may lead to similar insufficient staining, notably a very high concentration of epitope in the section. The abundancy of epitope may lead to premature exhaustion of the antibody solution such that at a certain point no antibody is available to bind with epitope in the center of the section. Thus, a gradient type of immunostaining is produced similar to the poor-penetration type. The poor penetration phenomenon is different from bleaching caused by laser illumination since insufficient penetration shows up at both the upper and lower surfaces of a section, whereas bleaching produces a local loss of immunofluorescence throughout the section's thickness.

Several measures can be introduced to improve the penetration of antibodies or, for that matter, to attempt avoiding gradient types of immunostaining:

1. Prolong the incubation.
2. Incubate at a higher temperature.
3. Increase the concentration of antibody.
4. Reduce of the number of sections per incubation well.
5. Use less and/or smaller sections (if the epitope is present in such excess that the antiserum is fast exhausted, e.g., in case of neurotransmitters, transporters, postsynaptic density proteins, etc.).
6. Use thinner sections or sections cut on a cryostat.
7. Treat the sections with microwaves prior to or during incubation.
8. Add (excess) detergent.
9. Freeze-thaw the sections prior to incubation. This freeze-thawing is done by immersing the sections in a bath of isopentane that, in turn, is subsequently rapidly cooled down by liquid nitrogen (details in Wouterlood *et al.*, 1993).

As cryostat sections usually need on-slide incubation with antibodies, we resorted in the case of ProSAP2/Shank3 to the solution of cutting the thinnest sections possible. In order to do so, we started with 120-μm-thick slices. Slices with such a thickness (or thicker slices, e.g., 150 μm) are easy to cut with a vibrating microtome as well as easy to handle, manage, and store in a cryoprotection solution in a freezer. When needed, a 120-μm slice can be recovered from the freezer and resectioned according to the following procedure.

D. Preparation of Thin Sections for Free-Floating Incubation

1. Drip 30% sucrose on the cold stage on a freezing microtome until a mound is formed.
2. Flatten this mound by moving the knife of the microtome over it.
3. Take the 120 μm slice from its cryoprotectant and dip fast in 30% sucrose.
4. Place the 120 μm slice on a (gloved) finger and put it on the flat surface of the mound.
5. Allow to equilibrate, trim if necessary, and cut 10–15-μm-thick sections. Collect these sections in wells of a 24-well plate for further processing.

E. Triple-Fluorescence Staining Procedure

Continuous gentle agitation on a rocking plateau is always provided during the incubation to prevent the contents inside the wells from settling onto the bottom.

In between all incubation steps, the sections are thoroughly rinsed with incubation buffer: 50 mM Tris/HCl buffer with 0.875% sodium chloride, 0.5% Triton X-100, pH 8.0 (TBS-TX). We use excess of antibody solution. Steps are as follows:

1. Preincubate 1 h at room temperature with 5% normal goat serum.
2. Incubate for at least 48 h at 4°C with a cocktail of primary antibodies: mouse anti-parvalbumin (Sigma, St. Louis, MO; 1:500, marker #2) and guinea pig anti-ProSAP2/Shank3 (1:500; marker #3; antibody kindly supplied by Dr. Tobias Böckers, University of Freiburg, Germany).
3. Incubate for at least 24 h at 4°C with a cocktail consisting of streptavidin conjugated to the fluorochrome Alexa Fluor™ 546 (1:200, marker #1), goat anti-guinea pig IgG conjugated to Alexa Fluor™ 488 (1:100), and goat anti-mouse IgG conjugated to Alexa Fluor™ 594 (1:200).
4. Rinse with Tris buffer (6.06 g/l aqua dest, pH 7.4). Mount in Tris buffer with gelatin (0.2 g/100 ml Tris buffer, pH 7.4).
5. Dry and coverslip with DPX (Fluka Chemie AG, Buchs, Switzerland). After coverslipping, slides are always stored in a freezer at −20°C. This cold storage is intended to reduce fading of the fluorochromes over time. Sections containing fluorescence prepared as long as 8 years ago in our laboratory and stored at −20°C still contain sufficient fluorescence to be of good use.
6. Shrinkage of the tissue can be a problem if conservation of the 3D shape is paramount. Since a section adheres to a solid glass surface, drying will cause shrinkage mostly in the Z direction and to a lesser degree in the XY direction. Deformation of shape will be inevitable. In addition to shrinkage in the Z direction comes the reduced resolution in the axial direction in the confocal instrument. Shrinkage by drying and mounting in DPX can reduce the thickness of a section 60–75% compared with its original "wet" thickness. A measure to reduce shrinkage and deformation is to mount and embed directly in Aquamount™ (Gurr; BDH, Poole, UK), or to apply measures discussed by Bacallao *et al.* (1995).

F. Troubleshooting

Insufficient penetration, crosstalk, and bleaching are the biggest problems encountered in multichannel confocal laser scanning. Insufficient penetration can be solved by several measures as listed in section "Introductory." Emission and excitation crosstalk can be excluded with some combinations of laser excitation wavelengths and fluorochromes (see section "Fluorochromes and Their Characteristics"). If excitation crosstalk cannot be avoided, then postacquisition dye separation can be attempted. Bleaching can be suppressed by the application of antifading agents to the preparations and, at acquisition time, by being very conservative with the intensity setting of the lasers used for the excitation of the fluorochromes (see section "Controls"). Some fluorochromes resist bleaching much more than other fluorochromes. Remember that excessive bleaching in a Z series of images reveals itself as a one-way gradient over the frames of decreasing crispness, with structures becoming vague and finally merging with the background noise. The latter situation is devastating since surface-rendered 3D reconstruction is based on connecting voxels in the image series with corresponding gray levels. Thus, with bleached sections, 3D rendering is unreliable at least. Anticipating and preventing bleaching is therefore of vital importance. Several software solutions exist that recognize and allow correction for bleaching in Z series. The Huygens II software used by us is equipped with such an option.

ACKNOWLEDGMENTS. The know-how discussed in this chapter has been collected in several years of intense collaboration with a number of persons. I am much indebted to Peter Goede, Luciënne Baks-te Bulte, and Mariska Vonck for their extremely skilful immunohistochemical assistance. Various confocal instruments were at our disposal by the kind collaboration of several institutions and persons: Jan van Minnen, Faculty of Biology, Vrije University (Zeiss LSM 410), Jeroen Beliën, Department of Pathology, VUMC (Leica TCS-SP), and Wolfgang Härtig and Jens Grosche, Paul-Flechsig Institute of Neuroscience, Leipzig, Germany (Zeiss LSM-510). The continuous care by Nico Blijleven for the hardware and software with which we did the deconvolutions and 3D reconstructions is much appreciated. Several graduate students contributed as well to improvements in scanning methodology: Helen Pothuizen, Bas Jasperse, Ivo van den Elskamp, Michel van den Oever, Cathrin Canto, Malika Dahmaza, Robert Schuit, and Johanna Ramirez-Reatiga. I especially thank Amber Boekel for looking very critically at the procedures.

REFERENCES

Bacallao, R., Kiai, K., and Jesaites, L., 1995, Guiding principles of specimen preparation for confocal fluorescence microscopy, In: Pawley, J. B. (ed.), *Handbook of Biological Confocal Microscopy*, New York: Plenum Press, pp. 311–325.

Bertero, M., Boccacci, P., Brakenhoff, G. J., Malfanti, F., and van der Voort, H. T. M., 1990, Three-dimensional image restoration and super-resolution in fluorescence confocal microscopy, *J. Microsc.* **157**:3–20.

Böckers, T. M., Bockmann, J., Kreutz, M. R., and Gundelfinger, E. D., 2002, ProSAP/Shank proteins—a family of higher order organizing molecules of the postsynaptic density with an emerging role in human neurological disease, *J. Neurochem.* **81**:903–910.

Böckers, T. M., Kreutz, M. R., Winter, C., Zuschratter, W., Smalla, K. H., Sanmarti-Vila, L., Wex, H., Langnaese, K., Bockmann, J., Garner, C. C., and Gundelfinger, E. D., 1999, Proline-rich synapse-associated protein-1/cortactin binding protein 1 (ProSAP1/CortBP1) is a PDZ-domain protein highly enriched in the postsynaptic density, *J. Neurosci.* **9**:6506–6518.

Brakenhoff, G. J., Blom, P., and Barends, P., 1979, Confocal scanning light microscopy with high aperture immersion lenses. *J. Microsc.* **117**:219–232.

Foster, M., and Sherrington, C. S., 1897, *A Text Book of Physiology, Part III: The Central Nervous System*, 7th ed., London: Macmillan.

Gerfen, C. R., and Sawchenko, P. E., 1984, A method for anterograde axonal tracing of chemically specified circuits in the central nervous system: combined *Phaseolus vulgaris*-leucoagglutinin (PHA-L) tract tracing and immunohistochemistry, *Brain Res.* **343**:144–150.

Gerfen, C. R., Sawchenko, P. E., and Carlsen, J., 1989, The PHA-L anterograde axonal tracing method, In: Heimer, L., and Zaborszky, L. (eds.), *Neuroanatomical Tract-Tracing Methods 2, Recent Progress*, New York: Plenum Press, pp. 19–48.

Gerlach, J., 1858, *Mikroskopische Studien aus den Gebiete der menschlichen Morphologie*, Erlangen, Germany: Enke.

Groenewegen, H. J., and Wouterlood, F. G., 1990, Light and electron microscopic tracing of neuronal connections with *Phaseolus vulgaris*-leucoagglutinin (PHA-L), and combinations with other neuroanatomical techniques, In: Björklund, A., Hökfelt, T., Wouterlood, F. G., and van den Pol, A. N. (eds.), *Analysis of Neuronal Microcircuits and Synaptic Interactions. Handbook of Chemical Neuroanatomy*, Vol. 8, Amsterdam: Elsevier Biomedical Press, pp. 47–124.

Hiesinger, P. R., Scholz, M., Meinertzhagen, I. A., Fischbach, K.-F., and Obermayer, K., 2001, Visualization fo synaptic markers in the optic neuropil of Drosophila using a new constrained deconvolution method, *J. Comp. Neurol.* **429**:277–288.

Holmes, T. J., Bhattacharyya, S., Cooper, J. A., Hanzel, D., Szarowski, D. H., and Turner, J. N., 1995, Light microscopic images reconstructed by maximum likelihood deconvolution, In: Pawley, J. B. (ed.), *Handbook of Biological Confocal Microscopy*, New York: Plenum Press, pp. 389–402.

Inoué, S., 1995, Foundations of confocal scanned imaging in light microscopy. In: Pawley, J. B. (ed.), *Handbook of Biological Confocal Microscopy*, New York: Plenum Press, pp. 1–14.

Jonkers, B., Sterk, J., and Wouterlood, F. G., 1984, Transcardial perfusion fixation of the CNS by means of a compressed-air driven device, *J. Neurosci. Methods* **12**:141–149.

Kano, H., van der Voort, H. T. M., Schrader, M., van Kempen, G. M. P., and Hell, S. W., 1996, Avalanche photodiode detection with object scanning and image restoration provides 2–4 fold resolution increase in two-photon fluorescence microscopy, *Bioimaging* **4**:187–197.

Leranth, C., and Pickel, V. M., 1989, Electron microscopic preembedding double immunostaining methods, In: Heimer, L., and Zaborszky, L. (eds.), *Neuroanatomical Tract-Tracing Methods 2, Recent Progress*, New York: Plenum Press, pp. 129–172.

Longin, A., Souchier, C., Ffrench, M., and Bryon, P. A., 1993, Comparison of anti-fading agents used in fluorescence microscopy: image analysis and laser confocal microscopy study, *J. Histochem. Cytochem.* **41**:1833–1840.

Minsky, M., 1957, US patent no. 301467, Microscopy Apparatus.

Ono, M., Murakami, T., Kudo, A., Isshiki, M., Sawada, H., and Segawa, A., 2001, Quantitative comparison of anti-fading mounting media for confocal laser scanning microscopy, *J. Histochem. Cytochem.* **49**:305–311.

Palade, G. E., and Palay, S. L., 1954, Electron microscope observations of interneuronal and neuromuscular synapses, *Anat Rec.* **118**:335–336.

Palay, S. L., and Palade, G. E., 1955, The fine structure of neurons, *J. Biophys. Biochem. Cytol.* **1**:69–88.

Peters, A., Palay, S. L., and de Webster, F. H., 1991, *The Fine Structure of the Nervous System: Neurons and Their Supporting Cells*, 2nd ed, Oxford: Oxford University Press, 494 pp.

Platt, J. L., and Michael, A. F., 1983, Retardation of fading and enhancement of intensity of immunofluorescence by *p*-phenylendiamine, *J. Histochem. Cytochem.* **31**:840–842.

Rayleigh, L., and Strutt, J. W., 1891, On pin-hole photography, *Philos. Mag.* **11**:87–99.

Rosene, D. L., Roy, N. J., and Davis, B. J., 1986, A cryoprotection method that facilitates cutting frozen sections of whole monkey brains for histological and histochemical processing without freezing artifact, *J. Histochem. Cytochem.* **34**:1301–1315.

Sassoë-Pognetto, M., and Fritschy, J. -M., 2000, Gephyrin, a major postsynaptic protein of GABAergic synapses, *Eur. J. Neurosci.* **12**:2205–2210.

Shaw, P. J., 1995, Comparison of wide-field/deconvolution and confocal microscopy for 3D imaging, In: Pawley, J. B. (ed.), *Handbook of Biological Confocal Microscopy*, New York: Plenum Press, pp. 373–387.

Sheppard, C. J. R., and Choudhurry, A., 1977, Image formation in the scanning microscope, *Opt. Acta* **24**:1051–1073.

Snyder, D. L., Schulz, T. J., and O'Sullivan, J. A., 1992, Deblurring subject to nonnegativity constraints, *IEEE Trans. Sign. Proc.* **40**:1143–1150.

van der Voort, H. T. M., and Strasters, K. C., 1995, Restoration of confocal images for quantitative image analysis, *J. Microsc.* **158**:43–45.

Waldeyer, F., 1891, Über einige neuere Forschungen im Gebiete der Anatomie des Central-nervensystems, *Dtsch. Med. Wochenschr.* **17**:1213–1218, 1244–1246, 1267–1269, 1287–1289, 1331–1332, 1352–1356.

Webb, R. H., and Dorey, C. K., 1995, The pixilated image, In: Pawley, J. B. (ed.), *Handbook of Biological Confocal Microscopy*, New York: Plenum Press, pp. 55–67.

Wessendorf, M. W., 1990, Characterization and use of multi-color fluorescence microscopic techniques, In: Björklund, A., Hökfelt, T., Wouterlood, F. G., and van den Pol, A. N. (eds.), *Handbook of Chemical Neuroanatomy: Analysis of Neuronal Microcircuits and Synaptic Interactions*, Vol. 8. Amsterdam: Elsevier, pp. 1–46.

Wouterlood, F. G., Böckers, T., and Witter, M. P., 2003, Synaptic contacts between identified neurons visualized in the confocal laser scanning microscope. Neuroanatomical tracing combined with immunofluorescence detection of postsynaptic density proteins and target neuron-markers, *J. Neurosci. Methods* **128**:129–142.

Wouterlood, F. G., and Jorritsma-Byham, B., 1993, The anterograde tracer biotinylated dextran-amine: comparison with the tracer *Phaseolus vulgaris*-leucoagglutnin in preparations for electron microscopy, *J. Neurosci. Methods* **48**:75–88.

Wouterlood, F. G., Pattiselanno, A., Jorritsma-Byham, B., Arts M. P. M., and Meredith G. E., 1993, Connectional, immunocytochemical and ultrastructural characterization of neurons injected intracellularly in fixed brain tissue, In: Meredith, G. E., and Arbuthnott, G. W. (eds.), *Morphological Investigations of Single Neurons In Vitro. IBRO Handbook Series "Methods in the Neurosciences,"* No. 16, Chichester, UK: Wiley and Sons, pp. 47–74.

Wouterlood, F. G., van Denderen, J. C. M., Blijleven, N., van Minnen, J., and Härtig, W., 1998, Two-laser dual-immunofluorescence confocal laserscanning microscopy using Cy2- and Cy5-conjugated secondary antibodies: unequivocal detection of co-localization of neuronal markers, *Brain Res. Protoc.* **2**:149–159.

Wouterlood, F. G., van Haeften, T., Blijleven, N., Perez-Templado, P., and Perez-Templado, E., 2002, Double-label confocal laserscanning microscopy, image restoration and real-time 3D-reconstruction to study axons in the CNS and their contacts with target neurons, *Appl. Immunohistochem. Mol. Morphol.* **10**:85–102.

Wouterlood, F. G., van Haeften, T., Eijkhoudt, M., Baks-te-Bulte, L., Goede, P. H., and Witter, M. P., 2004, Input from the presubiculum to dendrites of layer-V neurons of the medial entorhinal cortex of the rat, *Brain Res.* **1013**:1–12.

Zaborszky, L., and Cullinan, W. E., 1989, Hypothalamic axons terminate on forebrain cholin-ergic neurons: an ultrastructural double-labeling study using PHA-L tracing and ChAT immunocytochemistry, *Brain Res.* **479**:177–184.

Zaborszky, L., and Heimer, L., 1989, Combination of tracer technqiues, especially HRP and PHA-L with transmitter identification for correlated light and electron microscopic studies, In: Heimer, L., and Zaborszky, L. (eds.), *Neuroanatomical Tract-Tracing Methods 2, Recent Progress*, New York: Plenum Press, pp. 49–96.

14

Advances in Understanding Cortical Function Through Combined Voltage-Sensitive Dye Imaging, Whole-Cell Recordings, and Analysis of Cellular Morphology

CARL C. H. PETERSEN

INTRODUCTION
VOLTAGE-SENSITIVE DYE IMAGING
WHOLE-CELL PATCH-CLAMP RECORDING
AN APPLICATION: ANALYZING THE SENSORY RESPONSE IN
 RODENT BARREL CORTEX
SUMMARY OF ADVANTAGES AND LIMITATIONS
APPENDIX
  Commercial Sources of Voltage-Sensitive Dye, Camera,
    and Imaging Software
  Staining Neocortex with Voltage-Sensitive Dye
  Voltage-Sensitive Dye Imaging
  Whole-Cell Recording
  Anatomical Analysis of Neuronal Structure and Position
REFERENCES

Abstract: Voltage-sensitive dyes can be used to image cortical network function with millisecond temporal resolution and with a horizontal spatial resolution of approximately 50 μm. This imaging technique can be combined with whole-cell patch-clamp

CARL C. H. PETERSEN • Laboratory of Sensory Processing, Brain Mind Institute, SV-BMI-LSENS AAB 105, Ecole Polytechnique Federale de Lausanne, CH-1015 Lausanne, Switzerland measurement of membrane potential followed by the anatomical analysis of neuronal morphology. Together, such experiments reveal the relationship of activity recorded in individual identified neurons with the spatiotemporally resolved ensemble dynamics of a cortical region. Application of these techniques to the rodent barrel cortex has advanced our understanding of the synaptic mechanisms underlying sensory responses to simple whisker stimuli.

Keywords: axonal and dendritic morphology, sensory processing, voltage-sensitive dye imaging, whole-cell patch-clamp recording

I. INTRODUCTION

Understanding the organizing principles and functions of the neocortex is likely to lead to insight concerning the mechanisms of sensory perception and behavior. Such complex mental processes are likely to result from the interactions of many neurons distributed throughout cortical and subcortical areas. A variety of different approaches have been taken to analyze large-scale brain activity. Electrophysiological techniques were among the first to be applied to analyze brain function. Recent technological advances in electrophysiology allow simultaneous extracellular recordings of action potential activity from hundreds of individual neurons located in multiple brain areas in awake behaving animals (Buzsaki, 2004; Nicolelis *et al.*, 2003). Such measurements are, however, limited to an analysis of suprathreshold spiking activity of neurons and give no information relating to the underlying mechanisms leading to the action potentials. More detailed measurements of neuronal function have been made using intracellular recordings either using sharp microelectrodes or by the patch-clamp technique (Hamill *et al.*, 1981; Neher, 1992; Sakmann, 1992). These intracellular electrophysiological techniques can record the neuronal membrane potential capturing both the subthreshold and the suprathreshold responses of neurons. However, rather few cells (usually only one or two cells) can be recorded simultaneously due to technical difficulties. Electrophysiological measurements offer excellent temporal resolution, but are limited in their ability to map the spatial extent of brain activity (even with over a hundred extracellular electrodes).

Imaging methods have been developed to give spatial information regarding brain activity. Spectacular images of brain function have come from functional magnetic resonance imaging; however, the temporal resolution of this technique is limited. Most current functional magnetic resonance imaging techniques focus on the blood oxygenation level-dependent signal and thus temporal resolution is limited not only by the measuring apparatus but also by the delayed coupling of neural activity to hemodynamic changes.

An obvious wish would be to directly image the electrical activity of neurons in the brain. Recent advances in both imaging technology and novel voltage-sensitive dyes now offer the opportunity for high temporal and spatial resolution recording of neocortical activity. Below, I describe a combination of a voltage-sensitive dye imaging technique with whole-cell recordings and postrecording anatomical analysis of synaptic circuits. The application of this combination to the rat somatosensory barrel cortex has advanced our understanding of cortical function.

II. VOLTAGE-SENSITIVE DYE IMAGING

Dye molecules that insert into the plasma membrane and change their optical absorption and/or emission properties, dependent upon the electrical field across the membrane, can be considered as voltage-sensitive dyes. The first optical measurements of action potentials were made on invertebrate preparations such as the squid giant axon and neurons of the leech (Salzberg *et al.*, 1973; Tasaki *et al.*, 1968). The first vertebrate in vivo voltage-sensitive dyes measurements visualized the spatiotemporal dynamics of sensory processing (Grinvald *et al.*, 1984). Since these pioneering steps, many further compounds have been successfully tested as voltage-sensitive dyes. Of equal importance, camera technology has advanced dramatically.

Voltage-sensitive dyes typically show a linear, approximately 10% change in fluorescence for 100 mV change in membrane potential. Despite the small amplitude of the signals, the spatiotemporal dynamics of the membrane potential of individual neurons can be imaged in vitro in individual mammalian neurons by loading voltage-sensitive dyes intracellularly with the whole-cell patch-clamp technique (Antic *et al.*, 1999; Djurisic *et al.*, 2004; Zecevic, 1996). Significant signal-to-noise improvements may be realized from excitation at the red spectral edge (Kuhn *et al.*, 2004).

In vivo imaging is hindered both by the small signal amplitude and by the heart-beat-pulsation-related artifacts, resulting primarily from changes in the blood oxygenation level. Shoham *et al.* (1999) made dramatic advances with in vivo imaging by using novel blue dyes, which are excited at long wavelengths where hemoglobin has little absorption. One of these dyes RH1691 (Fig. 14.1A) can be excited with 630-nm red light (Fig. 14.1B) at which wavelength it shows an increase in fluorescence (>665 nm) upon depolarization (Fig. 14.1C). RH1691 has proven extremely useful for high-resolution in vivo imaging of cortical function (Grinvald and Hildesheim, 2004) with low toxicity to the imaged neurons (Petersen *et al.*, 2003a).

III. WHOLE-CELL PATCH-CLAMP RECORDING

The patch-clamp technique developed by Bert Sakmann and Erwin Neher provided the first direct recordings of single ion channels (Neher and Sakmann, 1976). Further development of different recording configurations together with the remarkable mechanical stability and low noise

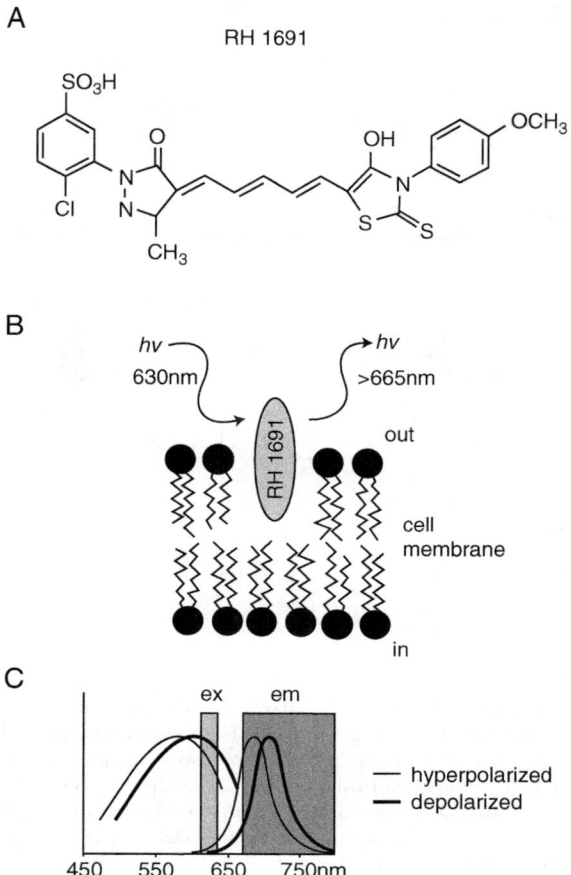

Figure 14.1. Part A shows the chemical structure of voltage-sensitive dye RH1691. After extracellular application, the dye molecules insert into the plasma membrane and upon excitation with 630-nm light they emit fluorescence that can be measured after 665-nm long-pass filters, as schematically indicated in Part B. A schematic drawing of the spectral shifts induced by changes in membrane potential is shown in Part C. During depolarization there is an increased absorption at 630 nm and an increase in fluorescence can be detected.

of the "gigaseal" (Hamill *et al.*, 1981) has extended the usefulness of this electrophysiological recording technique to virtually all biological preparations. The whole-cell patch-clamp configuration has proven particularly useful for recording both in acute brain slices and in vivo in the intact living animal (Fig. 14.2A,B). This is possible because the tip of the glass electrode patch-pipette can be kept clean by applying positive pressure to its inside with a resultant continuous outflow of intracellular solution. During the "search" for a cell to be recorded (Fig. 14.2C) this flux of liquid prevents debris from sticking to the patch-pipette and also "cleans" the cell membrane

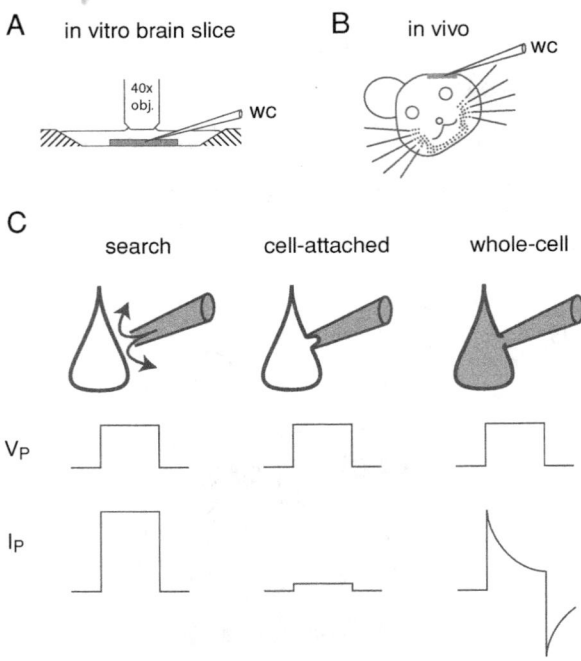

Figure 14.2. Whole-cell (WC) patch-clamp recordings can be made in vitro from neurons in brain slices (A) or in vivo from the intact living animal (B). Part C schematically shows various configurations of the patch-clamp technique. In the search mode, the pipette is under positive pressure, which keeps the tip of the electrode clean as it penetrates brain tissue. Large square shape currents (I_P) are evoked by square-shaped voltage command pulses (V_P) to the patch-clamp amplifier. When a cell is encountered current flow is decreased and gentle suction is applied. This allows formation of the gigaseal. This cell-attached configuration is characterized by very little current flow during voltage pulses, indicating a high resistance. Further gentle suction breaks down the membrane patch inside the tip of the electrode while maintaining the tight electrical seal of the membrane in contact with the glass pipette. This whole-cell configuration is characterized by large capacitative current transients in response to voltage pulses. The whole-cell configuration allows exchange of molecules from the patch-pipette with the inside of the cell.

priming it for the gigaseal. The cell to be recorded can be either visualized by various microscopy techniques (Margrie *et al.*, 2003; Stuart *et al.*, 1993) or electrically sensed by the increased resistance encountered as the pipette hits a cell (Blanton *et al.*, 1989). Upon encountering a target cell, the positive pressure is reversed by gentle sucking and an electrically tight gigaseal is formed. The gigaseal has resistance of >1 GΩ and has remarkable mechanical stability. This recording configuration is termed cell attached and can be used to record single channels in the patch membrane. The whole-cell recording configuration can be entered by further suction and can be monitored by the large increase in capacitance associated with the whole-cell membrane area.

The whole-cell configuration can remain stable for hours even in vivo despite the small movements of the brain (caused, for example, by heart-beat-pulsation), allowing high-quality measurements of membrane potential. The ease and stability with which whole-cell recordings can be obtained allow this experimental technique to be combined with other techniques such as voltage-sensitive dye imaging (Fig. 14.3).

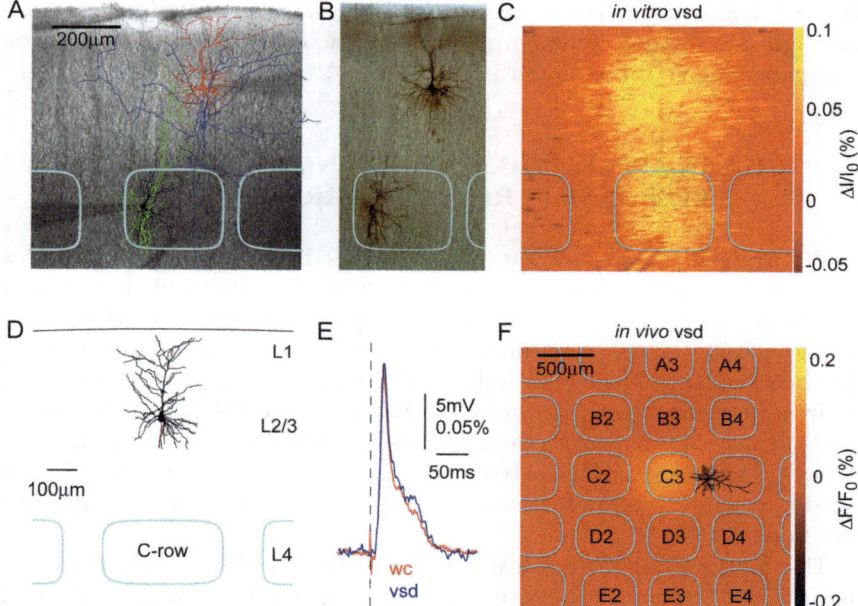

Figure 14.3. Part A shows a bright field photograph of a brain slice of the rat somatosensory cortex. The layer 4 barrels are outlined in cyan. One layer 4 neuron was recorded and later reconstructed with dendrites shown in black and axon in green. One layer 2/3 neuron was recorded and drawn with dendrites in red and axon in blue. A photograph of the stained neurons is shown in Part B. While the two neurons were recorded by the whole-cell patch-clamp technique, large-scale network activity was evoked by an extracellular stimulus delivered by a third electrode. This stimulation electrode was placed in layer 4 and the ensemble response was visualized with millisecond resolution with voltage-sensitive dye. Part C shows the voltage-sensitive dye image captured 12 ms after stimulation. The image shows columnar excitation. Part D shows a reconstruction of the somatodendritic compartment a layer 2/3 neuron recorded in vivo (viewed in a plane normal to the pial surface and along the row). During the whole-cell recording the cortical dynamics were imaged with voltage-sensitive dye. Part E shows the quantification of the C3 whisker-evoked voltage-sensitive dye response quantified over a 200 × 200 μm region centered on the soma of the recorded neuron. The time course of the optical response closely matches the time course of the changes in membrane potential of the recorded neuron. Part F indicated the lateral locations of the recorded neuron relative to the layer 4 barrels (cyan) and the evoked voltage-sensitive dye response recorded 15 ms after stimulation. (Reprinted in modified form with permission from Petersen and Sakmann, 2001 (© 2001 Society for Neuroscience), and Petersen *et al.*, 2003a (© 2003 Society for Neuroscience)).

In addition to measuring the electrophysiology of neurons, the whole-cell technique allows the introduction of small molecules into the cell cytoplasm. By including biocytin (biotinyl lysine) in the solution in the patch pipette, this label diffuses into intracellular milieu entering both axonal and dendritic compartments. After recording of the neuron, the brain can be fixed and sectioned. The location of the biocytin molecules can be revealed by the specific binding of the biotin motif (part of the biocytin molecule) to avidin conjugated to peroxidase, which in a reaction with diaminobenzidine forms a dark deposit. This allows visualization of the neuronal structure under light microscopy, and computer-aided tracing of dendritic and axonal compartments in three dimensions (Fig. 14.3).

IV. AN APPLICATION: ANALYZING THE SENSORY RESPONSE IN RODENT BARREL CORTEX

The rodent primary somatosensory cortex has aroused much interest in neurobiology because of its high degree of anatomical organization (Petersen, 2003). The mystacial vibrissae representation in this cortical area is segregated into discrete units termed barrels present in layer 4, which can be visualized in living brain slices (Petersen and Sakmann, 2000) as well as by numerous staining techniques. The layout of the barrels across the cortical map is identical to the layout of the whiskers on the snout of the rodent. This suggests that each of these barrels is intimately involved in processing the information from its corresponding whisker (Woolsey and Van der Loos, 1970).

The first level of neocortical processing begins with the layer 4 barrel neurons, which are directly connected to thalamic VPM neurons through glutamatergic synapses (Agmon and Connors, 1991). Both glutamatergic excitatory (spiny stellate and star pyramidal neurons) and diverse classes of GABAergic neurons in layer 4 receive direct VPM input (Bruno and Simons, 2002; Porter et al., 2001). Excitatory layer 4 neurons within the same barrel are strongly connected with approximately every third pair of neurons being synaptically connected, but there is very little synaptic connectivity between neighboring barrels (Feldmeyer et al., 1999; Petersen and Sakmann, 2000, 2001; Schubert et al., 2003; Shepherd et al., 2003). As seen in Fig. 14.4A,B, this pattern of physiologically measured synaptic connectivity likely results from the highly polarized dendritic and axonal arbors of these neurons, which rarely enter neighboring layer 4 barrels (Lübke et al., 2000; Petersen and Sakmann, 2000). Each layer 4 barrel, therefore, is an independent and irreducible unit consisting of a few thousand neurons which process information relating primarily to its isomorphic whisker. The ability to define the neuronal network in terms of both the number of participating neurons and the normal physiological input (since the location of the barrel in the sensory map can be established) makes the barrel cortex an ideal starting point for quantitative modeling of neocortical networks (Petersen, 2002).

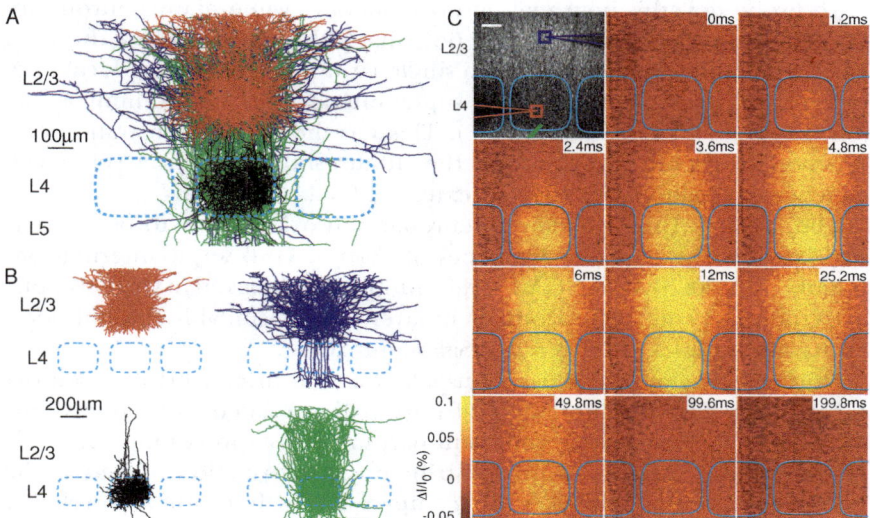

Figure 14.4. Parts A and B show the superposition of many excitatory neurons reconstructed from in vitro brain slice recordings and normalized according to the barrel width. Dendrites and cell bodies of layer 4 neurons shown in black are largely confined to the layer 4 barrel. The axons of the layer 4 neurons shown in green are laterally confined to the width of the layer 4 barrel but project heavily to both layers 2/3 and 4. The dendrites and cell bodies of the layer 2/3 pyramidal neurons are indicated in red and their axons in blue. The layer 2/3 axon spreads far laterally. Part C shows the functional activation of a barrel column evoked by extracellular stimulation of the layer 4 barrel and measured by voltage-sensitive dye imaging. In addition to the large stimulation electrode (green) in the layer 4 barrel, there is also a whole-cell recording pipette in layer 4 (red) and another in layer 2/3 (blue). The images demonstrate remarkably tight columnar activation throughout the duration of the response. (Reprinted in modified form with permission from Petersen and Sakmann, 2001 (© 2001 Society for Neuroscience)).

The excitatory layer 4 neurons project most densely into layer 2/3 with the horizontal axonal field spreading little wider than the underlying layer 4 barrel, thus defining anatomically a neocortical column (Fig. 14.4A, B). Excitatory synaptic connections from layer 4 to layer 2/3 pyramidal neurons occur frequently but have smaller efficacies and smaller NMDA receptor components than the layer 4 to layer 4 synapses (Feldmeyer *et al.*, 2002). The flow of excitation is strictly feed forward since there are no reciprocal excitatory connections from layer 2/3 to layer 4. Layer 2/3 pyramidal neurons synapse with their neighboring layer 2/3 pyramidal neurons, layer 5/6 pyramidal neurons (Reyes and Sakmann, 1999) and project to other cortical areas including contralateral somatosensory cortex, motor cortex, and secondary somatosensory cortex. Within the local circuits the axonal fields of layer 2/3 pyramidal neurons do not respect barrel column boundaries (Fig. 14.4A,B) extending far into the neighboring barrel columns.

To probe how this neuronal network operates when many neurons are excited, we imaged the membrane potential with voltage-sensitive dye (Fig. 14.4C). Stimuli were delivered to a single layer 4 barrel causing local excitation and spread of activity to the supragranular layer in a columnar fashion (Petersen and Sakmann, 2001). This was the first demonstration of a functional neocortical column at the subthreshold synaptic level, which matches the anatomically defined extent of the layer 4 axons.

The activity of the excitatory neuronal network is likely to be strongly regulated by the many diverse types of cortical GABAergic interneurons (Gupta *et al.*, 2000). When GABAergic inhibition is blocked in vitro, synaptic excitation can spread horizontally in layer 2/3 presumably through local excitatory synapses (Petersen and Sakmann, 2001).

During behavior, the whiskers usually operate in concert as a sensory organ. Therefore, the exchange of information related to the individual whiskers is likely to play a prominent part in cortical processing. One role for the barrel cortex is then to distribute the information related to the movement of a single whisker and compare this with information relating to movements of other whiskers. Such a process may occur in a defined spatial and temporal integrative process in the cortex.

The distributed nature of sensory signals originating from single brief sensory stimuli has been highlighted by combined in vivo voltage-sensitive dye imaging and whole-cell recordings (Fig. 14.5). This direct measurement of how cortical activity evoked by a single whisker is spatiotemporally distributed across the barrel cortex (Petersen *et al.*, 2003a) correlates well with measurements of receptive field properties of individual neurons analyzed by sequentially deflecting many whiskers (Armstrong-James *et al.*, 1992; Brecht *et al.*, 2003; Brecht and Sakmann, 2002; Moore and Nelson, 1998; Simons, 1978; Zhu and Connors, 1999). The earliest sensory response occurs ~8 ms following whisker deflection and is localized to the direct targets of the VPM input, the layer 4 barrel neurons and a fraction of neurons in mid-layer 5/6. In the next milliseconds, excitation propagates into layer 2/3 in a columnar fashion. Thus a functional neocortical column, bounded laterally by the layer 4 barrel structure, is depolarized 10–12 ms after whisker deflection (Fig. 14.5C). In the following milliseconds both infragranular neurons and neurons in neighboring barrel columns become excited, apparently mainly through local cortical synaptic circuits. Excitation spreads preferentially along the row orientation of the barrel cortex, for example deflection of the D2 whisker evokes first a response in the D2 barrel column and over the next milliseconds the largest responses are found in D1 and D3 neighboring barrel columns with smaller responses in the C2 or E2 columns. This oriented spread of excitation may serve a useful physiological function. The whisking behavior involves rapid whisker movements oriented largely in a plane along the rows. Thus during the forward motion of the whiskers, the D3 whisker will pass through a point in space a few milliseconds before the D2 whisker, which in turn will be followed by the D1 whisker moving through the identical spatial location. Thus whiskers lying in the same row will often

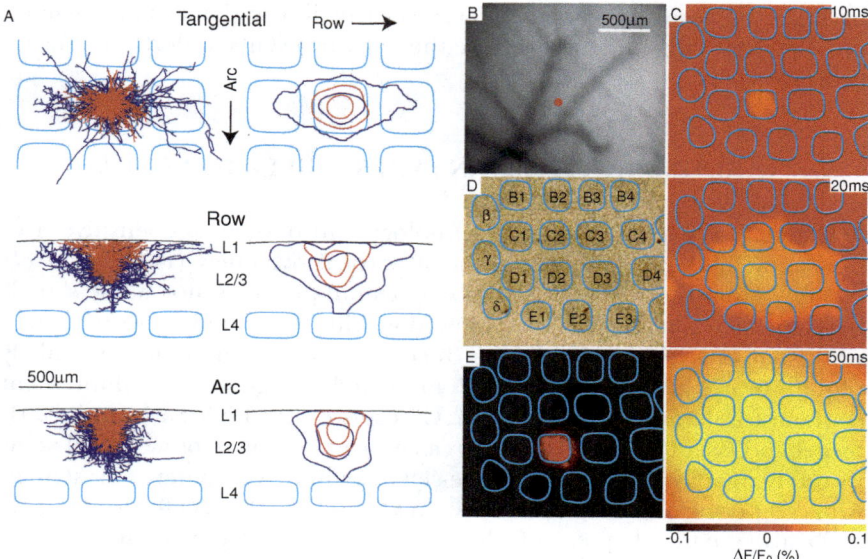

Figure 14.5. Part A shows the projection in three orthogonal directions of three dimensionally reconstructed axons (in blue) and dendrites (in red) of layer 2/3 pyramidal neurons recorded in vivo. The right-hand column shows the 10 and 50% contours of the length density of axon (blue) and dendrite (red) computed from the superimposed, gaussian smoothed, normalized computer-aided three-dimensional reconstructions. The axons extend preferentially in the row direction of the barrel cortex organization. Part B shows the blood vessels at the surface of the somatosensory cortex. Part C shows the voltage-sensitive dye signals recorded in response to deflection of the D2 whisker. The earliest signals occur ∼10 ms following whisker deflection and are localized to the homologous barrel column. In the next tens of milliseconds the signal propagates over a large cortical area preferentially in the row direction. After the functional imaging, DiI was injected into the location of the epicenter of the response. The DiI was allowed to diffuse and later the brain was sectioned tangentially to locate the layer 4 barrels (viewed under transillumination without staining in Part D). DiI fluorescence in layer 4 was found in the D2 barrel (E) in agreement with the location of the response to the D2 whisker deflection. (Reprinted in modified form with permission from Petersen *et al.*, 2003a (© 2003 Society for Neuroscience)).

sample the same point in space within milliseconds of each other. In order for the animal to process this information relating to individual whiskers distributed across the neocortical barrel field, it is likely to be important that this single-whisker-related information is rapidly exchanged along the rows of the barrel cortex. The rapid spread of the sensory response may mediate this integrative process, with propagation velocities along the row being twice as fast as along the orthogonal arc direction (Petersen *et al.*, 2003a). This spread of excitation may be mediated by local excitatory synaptic connections in layer 2/3 since their axons are preferentially oriented along the rows of the barrel cortex (Fig. 14.5A). The combined methodologies of single-cell recording, anatomical reconstruction, and ensemble imaging are therefore beginning to describe the synaptic events underlying sensory processing.

V. SUMMARY OF ADVANTAGES AND LIMITATIONS

Recent advances in imaging technology and new voltage-sensitive dyes now allow high-resolution imaging of cortical dynamics (Petersen *et al.*, 2003b). However, the current in vivo technique does not allow signals from individual neurons to be resolved within the stained neocortical network, but only ensemble activity. To overcome this limitation will likely require future generations of much improved voltage-sensitive fluorescent proteins (Ataka and Pieribone, 2002; Cacciatore *et al.*, 1999; Sakai *et al.*, 2001; Siegel and Isacoff, 1997). However, calcium as a measure of neuronal activity can be readily imaged with single-cell resolution within networks stained with cell permeant calcium-sensitive dyes (Peterlin *et al.*, 2000; Stosiek *et al.*, 2003; also see the chapter of Goldberg *et al.*, 2005, in this volume).

Since voltage-sensitive dye imaging is technically straightforward, this imaging technique can be readily combined with whole-cell patch-clamp recordings. Thus simultaneous measurements of single neuron and spatiotemporally resolved ensemble membrane potential measurements can be made. Together with anatomical analysis we can therefore begin to reconstruct the synaptic pathways, underlying simple sensory responses in the neocortex.

APPENDIX

A. Commercial Sources of Voltage-Sensitive Dye, Camera, and Imaging Software

1. Commercial Source of Voltage-Sensitive Dye RH1691

Optical Imaging Inc., PO Box 1262, Mountainside, NJ 07092-1262 (http://www.opt-imaging.com).

2. Commercial Sources of Cameras and Imaging Software

Imager 3001: Optical Imaging Inc., PO Box 1262, Mountainside, NJ 07092-1262 (http://www.opt-imaging.com).

NeuroCCD/NeuroPDA: RedShirtImaging LLC, 2 Stoneleigh Road, Fairfield, CT 06825 (http://www.redshirtimaging.com/).

MiCAM: SciMedia Ltd, 4 Executive Circle, Suite 170, Irvine, CA 92614 (http://www.scimedia.com).

B. Staining Neocortex with Voltage-Sensitive Dye

1. It is important to carry out animal experiments in accordance with local legislation. Anesthetize animal (e.g., juvenile 200 g Wistar rat injected intraperitoneally with urethane at 1.75 mg/g). Maintain the body temperature at 37°C.
2. Remove or reflect the skin covering the skull.
3. Carefully scrape the bone clean. Further cleaning of the bone can be performed, if necessary, with 1% hydrogen peroxide.
4. Apply a thin layer of cyanoacrylate glue to the surrounding region of the bone, away from where the craniotomy will be performed (this helps dental cement adhesion). Glue a metal head-plate to the skull with dental cement. The plate must be tangential to the region of cortex to be imaged and the hole in the plate must be positioned over this region.
5. When the dental cement has hardened, immobilize the skull and minimize movement of the brain by fixing the head-plate firmly between metal posts.
6. Perform a craniotomy of desired size by drilling within the hole of the head-plate. Be careful not to damage the underlying brain tissue.
7. Finally, remove the dura, leaving the pia of the underlying cortex exposed.
8. Dissolve the voltage-sensitive dye RH1691 (Shoham *et al.*, 1999) to 0.1–1 mg/ml in Ringer's solution: 135 mM NaCl, 5 mM KCl, 5 mM HEPES, 1.8 mM $CaCl_2$, and 1 mM $MgCl_2$.
9. Apply a small quantity (~250 μl) of this dye solution to the craniotomy. Seal this chamber with a glass coverslip and very slight pressure is applied to prevent brain edema.
10. Leave the cortex to stain for 1–2 h. During this period, the dye will diffuse into the superficial layers of the neocortex. At the end of the staining period, remove unbound dye by washing the cortex extensively with Ringer's solution. The cortex should now have a pale blue color.
11. Cover the craniotomy with 1% agarose, dissolved in Ringer's solution, and place a glass coverslip on top. The glass coverslip should not be just large enough to cover the width of the craniotomy, but little more. This will allow access for whole-cell recordings to be made.

C. Voltage-Sensitive Dye Imaging

1. Illuminate with ~530-nm green light and record the blood vessel pattern on the cortical surface using the camera.
2. Move the focal plane 300 μm into the cortex and excite the voltage-sensitive dye with epifluorescent light at 630 nm. Emitted light is long-pass filtered (>665 nm), forming the voltage-sensitive dye signal, which

should be recorded at frame rates faster than 100 Hz. Heart-beat-related signals form the largest artifacts. The timing of these known artifacts can be recorded via an electrocardiogram. The artifacts can then be removed by computer processing to improve the resolution of the collected signals.

D. Whole-Cell Recording

1. Fill whole-cell pipettes with intrapipette solution: 135 mM potassium gluconate; 4 mM KCl; 10 mM HEPES; 10 mM phosphocreatine; 4 mM MgATP; and 0.3 mM Na_3GTP; pH 7.2 adjusted with KOH and include 3 mg/ml of biocytin (to allow staining of the recorded neurons).
2. Pipettes should have a resistance of $\sim$5 MΩ. Monitor the tip resistance by applying brief voltage steps of 5 mV in the voltage-clamp mode while measuring the current flow on an oscilloscope. Apply a positive pressure of $\sim$200 mbar on the pipette.
3. Slowly advance the electrode through the agarose under the coverslip and into the cortex. When the tip is close to the chosen recording site, reduce the positive pressure to $\sim$30 mbar.
4. Advance the pipette in 2 μm steps until the tip resistance suddenly increases (indicating contact with a cell membrane). Release the pressure in the pipette and apply light suction until a gigaseal is formed.
5. Establish whole-cell recording configuration by rupturing the membrane in the pipette. This can be achieved by applying either brief suction pulses or slowly increasing the suction pressure.
6. After collecting data, slowly retract the whole-cell recording pipette while monitoring whole-cell capacitance transients with 5 mV voltage steps. During the retraction the excised patch configuration should be established. This insures that the neuron remains intact and viable for later anatomical staining.

E. Anatomical Analysis of Neuronal Structure and Position

1. Supplement anesthetic (e.g., by an additional intraperitoneal injection of 1 mg/g of urethane). Perfuse the animal transcardially with ice-cold 0.1 M phosphate buffer (pH $\sim$7.3) and then with $\sim$50 ml of 4% formaldehyde. Remove the brain from the skull and postfix overnight at 4°C.
2. Section the cortex tangentially with a vibratome at 100 μm and stain for biocytin using ABC kit (Vectastain Laboratories). The angle of mounting the brain for tangential sectioning is difficult to determine exactly, but is helped by the slightly blue color of the brain in the craniotomy remaining from the voltage-sensitive dye staining.

3. The blood vessels initially imaged with green illumination now provide the link between the location of the neuronal processes visualized by the anatomical stain and the functional voltage-sensitive dye images. The blood vessels, barrel patterns, and the axonal and dendritic processes can be traced in three dimensions, using an image-combining computerized microscopy system (e.g., Neurolucida, MicroBrightField, Inc. Williston, VT). For further details please refer in this volume to the chapter by Ascoli and Scorcioni and also the chapter by Bjaalie and Leergard.

4. In addition, other fluorescent labels can be introduced into the brain during the in vivo experiment for additional position information. Pipettes filled with DiI (1 mg/ml dissolved in dimethylformamide) can be inserted into the brain and DiI ejected through a brief pressure pulse. The DiI diffuses into the nearby tissue labeling axon and dendrites.

ACKNOWLEDGMENTS. I would like to thank the generous support of the Swiss National Science Foundation 3100A0-103832 "Synaptic mechanisms of sensory perception and associative learning" and the Leenaards Foundation "Neurobiology of addiction: interaction of barrel cortex and ventral tegmental area."

REFERENCES

Agmon, A., and Connors, B. W., 1991, Thalamocortical responses of mouse somatosensory (barrel) cortex in vitro, *Neuroscience* **41**:365–379.

Antic, S., Major, G., and Zecevic, D., 1999, Fast optical recordings of membrane potential changes from dendrites of pyramidal neurons, *J. Neurophys.* **82**:1615–1621.

Armstrong-James, M., Fox, K., and Das-Gupta, A., 1992, Flow of excitation within rat barrel cortex on striking a single vibrissa, *J. Neurophysiol.* **68**:1345–1358.

Ataka, K., and Pieribone, V. A., 2002, A genetically targetable fluorescent probe of channel gating with rapid kinetics, *Biophys. J.* **82**:509–516.

Blanton, M. G., Lo Turco, J. J., and Kriegstein, A. R., 1989, Whole-cell recording from neurons in slices of reptilian and mammalian cerebral cortex, *J. Neurosci. Methods* **30**:203–210.

Brecht, M., Roth, A., and Sakmann, B., 2003, Dynamic receptive fields of reconstructed pyramidal cells in layers 3 and 2 of rat somatosensory barrel cortex, *J. Physiol.* **553**:243–265.

Brecht, M., and Sakmann, B., 2002, Dynamic representation of whisker deflection by synaptic potentials in spiny stellate and pyramidal cells in the barrels and septa of layer 4 rat somatosensory cortex, *J. Physiol.* **543**:49–70.

Bruno, R. M., and Simons, D. J., 2002, Feedforward mechanisms of excitatory and inhibitory cortical receptive fields, *J. Neurosci.* **22**:10966–10975.

Buzsaki, G., 2004, Large-scale recording of neuronal ensembles, *Nat. Neurosci.* **7**:446–451.

Cacciatore, T. W., Brodfuehrer, P. D., Gonzalez, J. E., Jiang, T., Adams, S. R., Tsien, R. Y., Kristan, W. B., and Kleinfeld, D., 1999, Identification of neural circuits by imaging coherent electrical activity with FRET-based dyes, *Neuron* **23**:449–459.

Djurisic, M., Antic, S., Chen, W. R., and Zecevic, D., 2004, Voltage imaging from dendrites of mitral cells: EPSP attenuation and spike trigger zones, *J. Neurosci.* **24**:6703–6714.

Feldmeyer, D., Egger, V., Lubke, J., and Sakmann, B., 1999, Reliable synaptic connections between pairs of excitatory layer 4 neurons within a single "barrel" of developing rat somatosensory cortex, *J. Physiol.* **521**:169–190.

Feldmeyer, D., Silver, R. A., Lübke, J., and Sakmann, B., 2002, Synaptic connections between layer 4 spiny neurone-layer 2/3 pyramidal cell pairs in juvenile rat barrel cortex: physiology and anatomy of interlaminar signaling within a cortical column, *J. Physiol.* **538**: 803–822.

Grinvald, A., Anglister, L., Freeman, J. A., Hildesheim, R., and Manker, A., 1984, Real-time optical imaging of naturally evoked electrical activity in intact frog brain, *Nature* **308**:848–850.

Grinvald, A., and Hildesheim, R., 2004, VSDI: a new era in functional imaging of cortical dynamics, *Nat. Rev. Neurosci.* **5**:874–885.

Gupta, A., Wang, Y., and Markram, H., 2000, Organizing principles for a diversity of GABAergic interneurons and synapses in the neocortex, *Science* **287**:273–278.

Hamill, O. P., Marty, A., Neheri, E., Sakmann, B., and Sigworth, F. J., 1981, Improved patch-clamp techniques for high-resolution current recording from cells and cell-free membrane patches, *Pflügers Arch.* **391**:85–100.

Kuhn, B., Fromherz, P., and Denk, W., 2004, High sensitivity of stark-shift voltage-sensing dyes by one- or two-photon excitation near the red spectral edge, *Biophys. J.* **87**:631–639.

Lübke, J., Egger, V., Sakmann, B., and Feldmeyer, D., 2000, Columnar organization of dendrites and axons of single and synaptically coupled excitatory spiny neurons in layer 4 of the rat barrel cortex, *J. Neurosci.* **20**:5300–5311.

Margrie, T. W., Meyer, A. H., Caputi, A., Monyer, H., Hasan, M. T., Schaefer, A. T., Denk, W., and Brecht, M., 2003, Targeted whole-cell recordings in the mammalian brain in vivo, *Neuron* **39**:911–918.

Moore, C. I., and Nelson, S. B., 1998, Spatio-temporal subthreshold receptive fields in the vibrissa representation of rat primary somatosensory cortex, *J. Neurophysiol.* **80**: 2882–2892.

Neher, E., 1992, Ion channels for communication between and within cells, *Science* **256**:498–502.

Neher, E., and Sakmann, B., 1976, Single-channel currents recorded from membrane of denervated frog muscle fibres, *Nature* **260**:799–802.

Nicolelis, M. A., Dimitrov, D., Carmena, J. M., Crist, R., Lehew, G., Kralik, J. D., and Wise S. P., 2003, Chronic, multisite, multielectrode recordings in macaque monkeys, *Proc. Natl. Acad. Sci. U.S.A.* **100**:11041–11046.

Peterlin, Z. A., Kozloski, J., Mao, B. Q., Tsiola, A., and Yuste, R., 2000, Optical probing of neuronal circuits with calcium indicators, *Proc. Natl. Acad. Sci. U.S.A.* **97**:3619–3624.

Petersen, C. C. H., 2002, Short-term plasticity within the excitatory neuronal network of layer 4 barrel cortex, *J. Neurophysiol.* **87**:2904–2914.

Petersen, C. C. H., 2003, The barrel cortex—integrating molecular, cellular and systems physiology, *Pflügers Arch.* **447**:126–134.

Petersen, C. C. H., Grinvald, A., and Sakmann, B., 2003a, Spatiotemporal dynamics of sensory responses in layer 2/3 of rat barrel cortex measured in vivo by voltage-sensitive dye imaging combined with whole-cell voltage recordings and anatomical reconstructions, *J. Neurosci.* **23**:1298–1309.

Petersen, C. C. H., Hahn, T. T. G., Mehta, M., Grinvald, A., and Sakmann, B., 2003b, Interaction of sensory responses with spontaneous depolarization in layer 2/3 barrel cortex, *Proc. Natl. Acad. Sci. U.S.A.* **100**:13638–13643.

Petersen, C. C. H., and Sakmann, B., 2000, The excitatory neuronal network of layer 4 barrel cortex, *J. Neurosci.* **20**:7579–7586.

Petersen, C. C. H., and Sakmann, B., 2001, Functionally independent columns of rat somatosensory barrel cortex revealed with voltage-sensitive dye imaging, *J. Neurosci.* **21**: 8435–8446.

Porter, J. T., Johnson, C. K., and Agmon, A., 2001, Diverse types of interneurons generate thalamus-evoked feedforward inhibition in the mouse barrel cortex, *J. Neurosci.* **21**:2699–2710.

Reyes, A., and Sakmann, B., 1999, Developmental switch in the short-term modification of unitary EPSPs evoked in layer 2/3 and layer 5 pyramidal neurons of rat neocortex, *J. Neurosci.* **19:**3827–3835.

Sakai, R., Repunte-Canonigo, V., Raj, C. D., and Knopfel, T., 2001, Design and characterization of a DNA-encoded, voltage-sensitive fluorescent protein, *Eur. J. Neurosci.* **13:**2314–2318.

Sakmann, B., 1992, Elementary steps in synaptic transmission revealed by currents through single ion channels, *Science* **256:**503–512.

Salzberg, B. M., Davila, H. V., and Cohen, L. B., 1973, Optical recording of impulses in individual neurons of an invertebrate central nervous system, *Nature* **246:**508–509.

Schubert, D., Kotter, R., Zilles, K., Luhmann, H. J., and Staiger, J. F., 2003, Cell type-specific circuits of cortical layer 4 spiny neurons, *J. Neurosci.* **23:**2961–2970.

Shepherd, G. M., Pologruto, T. A., and Svoboda, K., 2003, Circuit analysis of experience-dependent plasticity in the developing rat barrel cortex, *Neuron* **38:**277–289.

Shoham, D., Glaser, D. E., Arieli, A., Kenet, T., Wijnbergen, C., Toledo, Y., Hildesheim, R., and Grinvald, A., 1999, Imaging cortical dynamics at high spatial and temporal resolution with novel blue voltage-sensitive dyes, *Neuron* **24:**791–802.

Siegel, M. S., and Isacoff, E. Y., 1997, A genetically encoded optical probe of membrane voltage, *Neuron* **19:**735–741.

Simons, D. J., 1978, Response properties of vibrissa units in rat SI somatosensory neocortex, *J. Neurophysiol.* **41:**798–820.

Stosiek, C., Garaschuk, O., Holthoff, K., and Konnerth, A., 2003, In vivo two-photon calcium imaging of neuronal networks, *Proc. Natl. Acad. Sci. U.S.A.* **100:**7319–7324.

Stuart, G. J., Dodt, H. U., and Sakmann, B., 1993, Patch-clamp recordings from the soma and dendrites of neurons in brain slices using infrared video microscopy, *Pflügers Arch.* **423:**511–518.

Tasaki, I., Watanabe, A., Sandlin, R., and Carnay, L., 1968, Changes in fluorescence, turbidity and birefringence associated with nerve excitation, *Proc. Natl. Acad. Sci. U.S.A.* **61:**883–888.

Woolsey, T. A., and Van der Loos, H., 1970, The structural organization of layer 4 in the somatosensory region (SI) of the mouse cerebral cortex: the description of a cortical field composed of discrete cytoarchitectonic units, *Brain Res.* **17:**205–242.

Zecevic, D., 1996, Multiple spike-initiation zones in single neurons revealed by voltage-sensitive dyes, *Nature* **381:**322–325.

Zhu, J. J., and Connors, B. W., 1999, Intrinsic firing patterns and whiskers-evoked synaptic responses of neurons in the rat barrel cortex, *J. Neurophysiol.* **81:**1171–1183.

15

From Dendrites to Networks: Optically Probing the Living Brain Slice and Using Principal Component Analysis to Characterize Neuronal Morphology

JESSE H. GOLDBERG, FARID HAMZEI-SICHANI,
JASON MacLEAN, GABOR TAMAS, ROCHELLE
URBAN, and RAFAEL YUSTE

JESSE H. GOLDBERG • McGovern Institute for Brain and Cognitive Sciences, Massachus-
etts Institute of Technology, Cambridge, MA   JASON MACLEAN AND ROCHELLE
URBAN • Department of Biological Sciences, Columbia University, New York, NY 10027
RAFAEL YUSTE • Howard Hughes Medical Institute, Department of Biological Sciences,
Columbia University New York, NY 10027   FARID HAMZEI-SICHANI • Department
of Physiology and Pharmacology, State University of New York, Downstate Medical Center,
Brooklyn, NY   GABOR TAMAS • Department of Comparative Physiology, University
of Szeged, Kozepfasor 52, Szeged H-6726, Hungary

Abstract: Recently, advances in optical imaging of the living brain slice preparation have permitted neuronal circuitry to be examined at multiple levels, ranging from individual synaptic contacts on dendrites to whole populations of neurons in a network. In this chapter, we describe three techniques that, together, enable a powerful dissection of neuronal circuits across multiple space scales. We describe methods for (1) combining whole-cell recording with two-photon calcium imaging and electron microscopic reconstruction to examine the functions of individual synapses and dendrites during synaptic stimulation, (2) imaging hundreds of neurons in the brain slice simultaneously to examine the spatiotemporal dynamics of activity in living neuronal networks, and (3) performing an unbiased, quantitative analysis of neuronal morphology that is increasingly necessary in light of the multiparametric structural diversity of distinct neuronal subclasses.

Keywords: cluster analysis, dendrite, imaging, microdomain, network, principal component analysis, two-photon calcium

I. INTRODUCTION

Brain slices have proven to be an excellent system to study the structure and function of cortical networks (Dingledine *et al.*, 1980). Early studies of pharmacologically disinhibited cortex showed that the in vitro circuit was functionally intact and capable of generating epileptiform activity (Prince and Connors, 1984), and plasticity studies with extracellular recording electrodes showed that synapses could undergo activity-dependent plasticity in vitro (Alger and Teyler, 1976; Sarvey *et al.*, 1989), with potentially great relevance to fundamental questions of learning and memory. Later, the advent of intracellular recording in brain slices revealed that distinct neuronal populations exhibited unique electrical behaviors (Connors and Gutnick, 1990), and, combined with post-hoc anatomical reconstructions and immunohistochemistry, have enabled the detailed study of hippocampal and neocortical networks, composed of excitatory pyramidal neurons as well as a great diversity of GABAergic interneurons (Freund and Buzsaki, 1996; Markram *et al.*, 2004; McBain and Fisahn, 2001; Somogyi *et al.*, 1998).

Further, the advent of simultaneous intracellular recordings from connected pairs of neurons revealed that connections between neurons are highly specific with respect to short-term synaptic dynamics, electrical coupling, and even the timing that governs synaptic plasticity (Gibson *et al.*, 1999; Gupta *et al.*, 2000; Markram *et al.*, 1997; Tamas *et al.*, 2000).

A major limitation of the electrophysiological approach to the brain slice preparation is inability to examine many spatially distinct components of cortical computation. This is important at the levels of both the individual neuron, where synaptic integration occurs in geometrically complex dendrites (Goldberg *et al.*, 2002; Poirazi and Mel, 2001; Yuste *et al.*, 1994), and the network at large, where ensembles of coactive neurons can be distributed unpredictably in space (Cossart *et al.*, 2003). Optical imaging techniques are ideal for approaching these issues. In this chapter, we describe three methods used to analyze the neuronal circuit at multiple levels. We describe (1) combining whole-cell recording with two-photon calcium imaging and electron microscopic reconstruction to examine the structure and function of dendrites, (2) imaging hundreds of neurons in the brain slice simultaneously to examine the spatiotemporal dynamics of activity in living neuronal networks, and (3) performing an unbiased, quantitative analysis of neuronal morphology that is increasingly necessary in light of the multiparametric diversity of neuronal subclasses.

II. TWO-PHOTON CALCIUM IMAGING IN DENDRITES

Two-photon calcium imaging has revolutionized the study of dendrites, allowing calcium signals to be examined in small structures such as dendritic spines (Yuste and Denk, 1995) and aspiny dendritic shafts (Goldberg *et al.*, 2003c). In this section, we will discuss combining intracellular recording with two-photon calcium imaging during synaptic activation. This combined approach is a powerful method for examining the two main functions of dendrites: chemical compartmentalization and synaptic integration. In these experiments, neurons from mouse neocortex are patched in whole-cell mode at physiologic temperature (37°C) and loaded via the patch pipette with a calcium-dependent fluorophore, i.e., a calcium indicator. Electrical synaptic activity is monitored through the recording electrode, and dendrites are visualized via two-photon excitation of the calcium indicator. Next, an electrode to be used for synaptic stimulation is placed immediately adjacent (ideally < 20 μm) to a specific dendritic region of interest (ROI). Electrical shocks delivered through this electrode excite local presynaptic axons that activate the dendrite of interest, and the resulting changes in intracellular calcium concentration are monitored by changes in the fluorescence of the calcium indicator.

To perform these experiments, we use a custom-made modified confocal microscope (described in detail in Majewska *et al.*, 2000; Nikolenko *et al.*, 2003), although turn-key commercial multiphoton microscopy systems are now readily available. We will discuss the three major considerations in performing these experiments: calcium indicator selection, targeting the stimulation electrode to the dendritic ROI, and synaptic activation of single synapses.

A. Selection of the Calcium Indicator

Calcium indicators are fluorophores whose fluorescence is dependent on calcium concentration. Important characteristics to consider when selecting an indicator are its resting fluorescence, its calcium-dependent change in fluorescence, and the kinetics of its chemical reaction with calcium. 1,2-Bis(o-aminophenoxy)ethane-N,N,N',N'-tetraacetic acid (BAPTA)-based high-affinity calcium indicators have fast Kon ($\sim10^8$ M/s) and are ideal for monitoring fast changes in calcium concentration, such as during synaptic activation and action potential backpropagation. In these experiments, the potassium salt form of the indicator is combined with standard internal recording solution and loaded into the cell during whole-cell recording. Typically, the dendritic tree is fully loaded approximately 20 min after gaining access to the cell.

Many different calcium indicators are available (Molecular Probes: http://www.probes.com), and are well suited to different experimental questions. We have used the potassium salt forms of the high-affinity indicators calcium green-1 (CG-1), Oregon green BAPTA-1 (OGB-1), and Fluo-4, and all three are excited by a mode-locked laser at 800 nm. A major advantage of CG-1 and OGB-1 is their high-resting fluorescence, which means that they can be used at lower concentrations (50–150 μM), reducing both exogenous buffer capacity and experimental misrepresentation of free calcium signals (see below). However, in part due to their high fluorescence at rest, their increase in intensity on binding calcium is compromised. Thus, Fluo-4 is better for detecting small or heavily buffered signals, such as calcium influx resulting from a single action potential or microdomains in GABAergic interneurons (Goldberg *et al.*, 2003c). The disadvantage of Fluo-4 is that it is dim at rest, necessitating higher concentrations (100–400 μM) of the dye to visualize the dendritic tree. This issue can be circumvented by loading the intracellular electrode with the calcium-insensitive dye Alexa594 in tandem with Fluo-4 during two-channel acquisition (Sabatini *et al.*, 2002).

In two-channel acquisition, two separate photomultiplier tubes (PMTs) are used to collect red and green light. In these experiments, both calcium-insensitive Alexa594 (25–100 μM) and calcium-sensitive indicator (see above) are loaded into the recording pipette. A dichroic mirror separates red and green emission light (e.g., 565 DCXR from Chroma Technology Corp., Brattleboro, VT) to separate the calcium-insensitive fluorescence from Alexa594, and the calcium-sensitive fluorescence from the Fluo-4. At an angle of 45°, this filter transmits ~80% of light between ~580 and 880 nm

(Red Channel: infrared light from laser to sample and red Alexa594 fluorescence back to internal PMT) and reflects light below 530 nm (Green Channel: emission from calcium indicator) to external PMT. We place an additional band pass 510/40 (transmits between 490 and 530 nm) filter in front of the external PMT to reduce scattered red light. Thus, morphology and calcium signal are represented on the red and green channels, respectively. The advantage of using a separate PMT for each signal is that low levels of calcium indicator can be used, reducing the experimental perturbation of calcium dynamics (see section "Summary of Advantages and Limitations").

B. Targeting the Stimulation Electrode to the Dendrite of Interest

Stimulating and detecting calcium signals in dendrites at individual synapses is in part a matter of chance, since the stimulation electrode must overlie an axon that happens to target the dendritic segment being imaged (Fig. 15.1). For this reason, the stimulation electrode has to be frequently moved to several locations even within the small vicinity of the dendrite of interest, but without compromising the quality of the electrical recording, i.e., without causing even slight movement of the brain slice. Thus, in order to reliably activate single or a limited number of synapses that converge onto a dendritic ROI, the stimulation electrode must be carefully constructed, controlled, and mounted on a stable micromanipulator. We use a borosilicate glass patch pipette (6–10 MΩ) bent at the tip by 60–80° with a microforge (Narishige, Japan). Bending the tip allows the electrode to enter the slice perpendicularly, reducing tissue deformity and allowing the electrode to

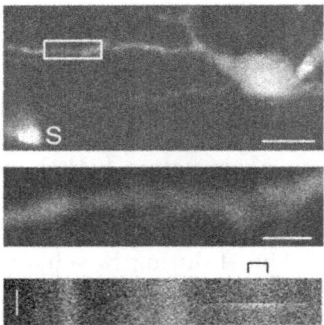

Figure 15.1. Two-photon imaging of calcium microdomains in aspiny dendrites. Top, interneuron from a slice of primary visual cortex filled with 100 μM Fluo-4. Stimulation electrode S was placed via visual guidance near dendrite of interest. Scale bar: 20 μm. Middle: dendritic segment as indicated by white box. Scale bar: 1 μm. Bottom: line scan placed in parallel through dendritic segment from above, and the calcium signal evoked by a single shock delivered through the stimulation electrode. Image is an average of four trials. Vertical scale: 400 ms; horizontal scale: 1 μm as above. (Reprinted, with permission, from Goldberg *et al.*, 2003c).

be repeatedly placed in the slice without compromising the quality of the whole-cell recording. Once constructed, the electrode is filled with a fluorescently labeled fluorophore [we use 20–200 µM Alexa488-dextran in standard artificial cerebrospinal fluid (ACSF)]. The dextran form of Alexa works well because it is not ejected from the electrode during electrical stimulation. This fluorescent labeling of the stimulation electrode is essential in order to visualize it during the experiment, allowing for precise control over where it is placed.

Before patching the neuron of interest, the stimulation electrode should be mounted on the micromanipulator, placed with its curved tip oriented toward the slice, adjacent to the patch electrode, to ensure that the objective can focus on the tip without physically touching its proximal segment.

Once the neuron is patched and the dendritic tree is filled with indicator, a region of the dendrite oriented in parallel to the position of the line scan is chosen (Fig. 15.1). With either spiny or aspiny dendrites, this approach maximizes the surface area of dendrite shaft (or number of spines) imaged and increases the likelihood of overlying an activated synapse.

With the dendritic segment of interest placed at the center of the imaging field, the objective is brought out of the slice until the fluorescently labeled tip of the stimulation electrode is in focus. The electrode is moved in the x- and y-axes until it is near the center, above the dendrite of interest, such that as it is gently lowered into the slice it will fall immediately next (< 15 µm) to it. Control of the stimulation electrode is perhaps the most crucial and experience-dependent part of the experiment.

C. Synaptic Activation of Single Synapses

The hallmark of a single synaptic calcium signal is the observation of successive failures and successes of a local calcium signal on repeated trials, reflecting the stochasticity of synaptic transmission (Goldberg *et al.,* 2003c; Yuste and Denk, 1995). Once the stimulation electrode is in position, adjacent to the dendrite of interest, it delivers single shocks (100–200 µs) in voltage mode (0.1–2 V) using a stimulus isolation unit (A.M.P.I Iso-Flex: http://www.ampi.co.il/isoflex.html). First, start with a small voltage setting (0.1 V) and monitor the excitatory postsynaptic potentials and currents (EPSP/C) response to observe if the neuron is being activated. Gradually ramp up the stimulus strength until calcium signals and subthreshold EPSP/Cs are observed. Frequently, no calcium signals will be observed until suprathreshold activation is achieved, reflecting strong activation of synapses off of the imaged dendrite of interest and global calcium accumulations secondary to action potential backpropagation.

If this occurs, or if repeated shocks fail to induce single synaptic calcium signals, the electrode must be lifted out of the slice, repositioned, and again dropped next to the dendrite of interest in its new position. This process may need to be repeated several times during an experiment.

Lastly, it is important to control direct activation of the dendrite by the stimulation electrode, as opposed to activation of axons that synapse on the dendritic segment under investigation. This may occur if the electrode is too close to the dendrite of interest (< 5 μm) and may result in global calcium signals and undershooting action potentials that may reflect the dendritic spiking. One way to control this artifact is by blocking synaptic transmission, such as with bath application of 20 μM DNQX, 50 μM d-APV, and 1 μM bicucculine, at the end of an experiment to ensure that the postsynaptic potential and the calcium signal are blocked. If a fast calcium signal persists, then either the stimulation strength was too high or the electrode was too close to the dendrite, and the data must be discarded.

III. ULTRASTRUCTURAL IDENTIFICATION OF IMAGED DENDRITIC SEGMENTS

During whole-cell recording, 0.3% biocytin is loaded into the recording pipette in addition to the calcium indicator. Standard procedures for the visualization of biocytin (Buhl *et al.*, 1994; Horikawa and Armstrong, 1988; Tamas *et al.*, 1997) can be used for the light and electron microscopic assessment of dendritic segments imaged previously with two-photon lasers (Fig. 15.2). Importantly, our experience shows that successful detection of biocytin is enhanced by restricting the number of scans (< 20) through a particular dendritic segment, presumably due to the photosensitivity of biocytin. The step-by-step protocol for electron microscopic reconstruction of imaged dendritic domains is presented in Appendix.

IV. IMAGING NEURONAL ENSEMBLES

The understanding of neuronal circuits has been, and will continue to be, greatly advanced by the simultaneous imaging of hundreds of neurons in the brain slice allowing for the examination of the spatiotemporal dynamics of activity in neuronal networks. In this section we describe the "bulk" loading of brain slices with acetoxy-methyl (AM) ester calcium indicators, in contrast to the calcium indicator salts described above, in order to image activity in large populations of neurons simultaneously. In our experience, voltage-sensitive dyes are still insufficient for the detection of activity in an individual neuron within a population of neurons, such as a cortical slice, due to their nonspecific staining pattern and poor spatial resolution (Yuste *et al.*, 1997). Calcium indicators (Tsien, 1989) that can be bulk-loaded into brain slices using their AM ester derivatives act as very good, albeit indirect, measures of action potential generation (Smetters *et al.*, 1999; Yuste and Katz, 1991) and provide single-cell resolution. Calcium entry via calcium channels resulting from the depolarization indicative of an action potential is sufficient to be imaged. These dyes still provide the best means of

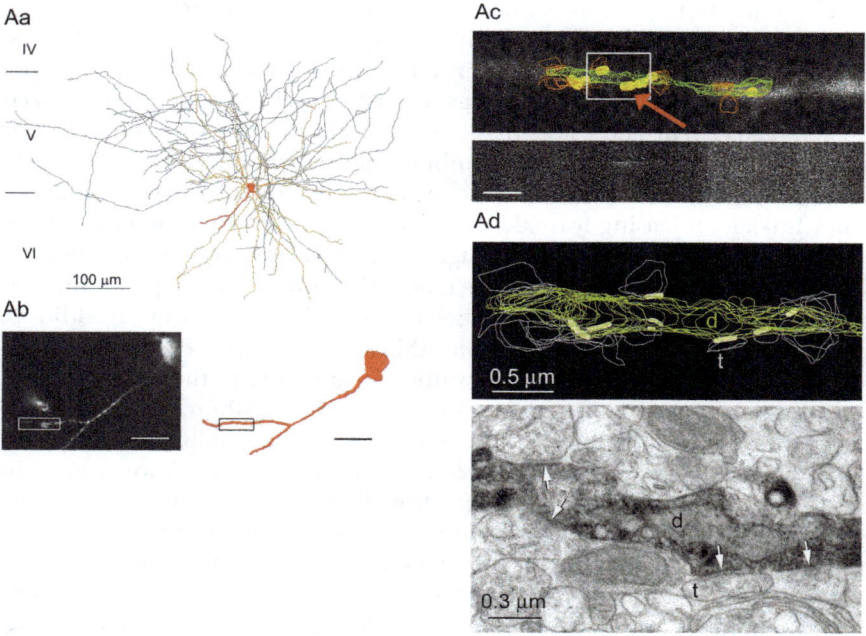

Figure 15.2. Ultrastructural reconstructions of imaged dendrites. (Aa) Reconstruction of biocytin-filled layer V fast spiking (FS) cell from a 300-μm-thick visual cortical mouse brain slice. Dendrites, orange; axon, black. Red dendritic segment represents ROI in (Ab). (Ab) Two-photon z projection of imaged ROI of an FS cell filled with 100-μM Fluo-4 (left) and corresponding region from the cell reconstruction (right). Boxes indicate the dendritic segment selected for line scan imaging and for expectation-maximization (EM) reconstruction. Scale bar: 20 μM. (Ac) Top: horizontal dendrite of interest with the cartoon of the serial EM reconstruction overlay at the precisely realigned section. Bottom: line scan through dendrite reveals the evoked single synaptic calcium signal. Note how its position appeared aligned to the synapse, as indicated by the red arrow in the cartoon. (Ad) Top: cartoon detail of the serial reconstruction. Dendrite, d, is labeled in green, and terminals, t, in white. The terminal labeled by "t" corresponds to the terminal of interest in Ac. Bottom: the electron micrograph focusing on the site aligned to the microdomain; arrows indicate synapses. (Reprinted, with permission, from Goldberg *et al.* 2003c). See Appendix for step-by-step protocol of how imaged dendrites were reconstructed.

imaging activity in large populations of neurons when single-cell resolution is desirable.

Since first used (Yuste and Katz, 1991), this method has been successfully applied throughout the central nervous system to study neuronal ensemble activity. Reports include studies on neocortex (Cossart *et al.*, 2003; Mao *et al.*, 2001), hippocampus (Tanaka *et al.*, 2002), cerebellum (Ghozland *et al.*, 2002), striatum (Mao and Wang, 2003), and spinal cord (Voitenko *et al.*, 1999), utilizing acute slices, cultured slices, or cultured dissociated neurons. Recently, the technique has also been applied in vivo to mouse, rat, and cat neocortex (Ohki *et al.*, 2005; Stosiek *et al.*, 2003).

The limitations of the technique are due to the properties of the dyes themselves. Calcium indicators, being charged molecules, do not easily cross the cell membrane and need, therefore, to be microinjected. To circumvent this problem, AM ester derivatives of the indicators were synthesized (Tsien, 1981). The AM esters mask negative charges, making the indicator molecules more lipophilic and membrane-permeant, thus allowing them to enter the cell. Once inside the cell, cytoplasmic esterases hydrolyze the acetyl ester linkage, releasing formaldehyde and free indicator, which then accumulates intracellularly as it is once again charged. However, the dependence on intracellular enzymatic cleavage makes this process cell dependent. This can result in differential loading efficiency in different neurons. In addition, the increased hydrophobicity of the AM ester derivatives of the indicators can cause problems in delivering sufficient amounts to their targets. This problem becomes significant in adult preparations, where the slice painting method appears to be the best loading strategy. Finally, while the time constant (i.e, rate) for the onset of quenching indicator fluorescence in neurons is rapid, the offset is proportionally slow due to saturation of the dye. Thus, while the calcium indicators are excellent measures for the onset of activity, they do not provide adequate temporal resolution to detect single action potentials during a burst of action potentials. This is in contrast to the voltage-sensitive dyes, which provide rapid onset and offset of signal. The trade-off between the two methods then is spatial versus temporal resolution, with the calcium indicators providing a far superior spatial resolution and the voltage-sensitive dyes providing greater temporal resolution, especially when examining the offset of activity.

A. "Bulk" Loading Acute Cortical Slices with Calcium Indicator

Different AM calcium indicators, under similar experimental conditions, have different loading efficacies. Further, indicators also vary in their effectiveness as detectors of action potentials, determined by the Kd and dynamic range of the dye, and importantly, in their affinity for neurons versus glia. The majority of experimental procedures in our laboratory use Fura-2, AM, which provides sufficient sensitivity to reliably detect, at minimum, two action potentials and is by far the best of the AM dyes at targeting neurons for loading. However, the methods described here can be used for any of the commercially available AM calcium indicators. Further, although the methodologies described below are for mouse brain slices, bulk loading of calcium indicators into neurons from rat and cat has been successfully applied by other groups (Ohki *et al.*, 2005).

Both juvenile and adult acute mouse cortical slices (PND10–PND30) can be loaded with the calcium indicators. Prepare the Fura-2, AM by dissolving 50 μg Fura-2, AM (Molecular Probes) in 15–48 μl DMSO and 2 μl of Pluronic F-127 (Molecular Probes) for a final concentration of 1–3.3 mM. The solubility of calcium AM indicators is rather poor. Thus, it is necessary to vortex the solution for 10–15 min prior to use. Place slices in loading chamber (a petri dish 35 × 10 mm) containing 2.5 ml of oxygenated ACSF. Pipette 5–10 μl of Fura-2, AM solution on top of each slice, resulting in a high initial concentration of Fura-2, AM. The concentration decreases as the Fura-2, AM diffuses away from the site of application, resulting in a final concentration in the entire chamber of approximately 10–20 μM. Alternatively, simply pipette the entire volume of the prepared Fura-2, AM at the far side of the loading chamber and allow the dye to passively diffuse throughout the chamber lading the slices. Once again this is the preferred method especially if the slices are from a younger animal. Load the slices in the dark for 20–30 min at 35–37°C with 95% O_2/5% CO_2 lightly ventilated into the chamber. As a rule of thumb, the loading time should be 10 min, plus as many minutes as the age of the animal in postnatal days. Remove slices from the loading chamber and place into incubation chamber containing oxygenated ACSF to allow them to rest (for more details see Sakmann and Neher, 1983). Imaging may begin after a 30-min recovery period.

B. Imaging Ensemble Activity

Two-photon imaging of calcium indicators allows highly sensitive detection of changes in neuronal calcium concentration with relatively little bleaching and photodamage. However, the simultaneous imaging of large populations of neurons is necessarily slow. Epifluorescent imaging of calcium indicators is sufficient to detect changes in neuronal calcium concentration and has the advantage that, using a fast camera, the detection of changes can be of higher temporal resolution than in the two-photon system. However, fluorescent imaging of bulk-loaded slices is subject to rapid photobleaching. Finally in our experience, spinning disk confocals, together with fast cameras, can be used to image thousands of neurons simultaneously, without significant photobleaching and with good signal-to-noise ratios, over long periods of time. In all three cases, cell activity, as indicated by fluorescence change, can be detected automatically, using custom written software (MatLab) allowing one to evaluate the spatiotemporal pattern of activity in the slice and in the neuronal ensembles. The software analyzes images of neurons, which are filtered and thresholded—round contiguous areas of brightness are recognized as neuronal somata. Activity is detected as a decrease (recall that fura *decreases* its fluorescence on binding calcium) in neuronal somata brightness that is greater than the surrounding area by some arbitrary number (determined using calibration experiments on each individual experimental setup) of standard deviations above the average (noise) fluctuation level.

For example, Fig. 15.3 illustrates the two-photon imaging of Fura-2, and the utilization of Fura-2, as an indicator of neuronal activity. Figure 15.3A illustrates thousands of loaded neurons in a neocortical slice imaged using a two-photon microscope. Figure 15.3B reveals the correlation between

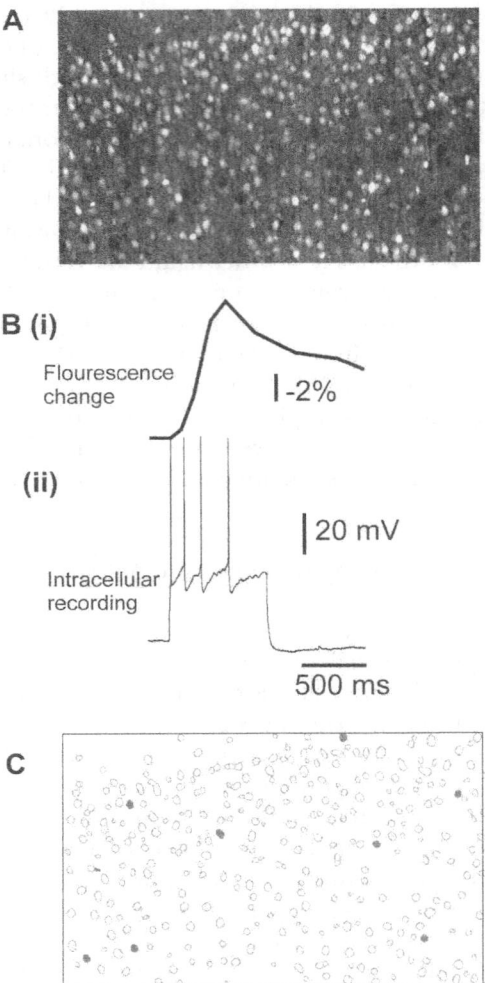

Figure 15.3. Calcium imaging of neuronal activity. (A) Two-photon image of 300-μm-thick transverse section of mouse somatosensory neocortex loaded with Fura-2, AM. Scale bar, 50 μm. Note how hundreds of neurons can be visualized. (Courtesy of Brendon O. Watson.) (B) Correspondence between action potentials and somatic fluorescence change. (i) Whole-cell recording of a burst of action potentials in response to a single intracellular depolarizing current step (150 pA, 200 ms). (ii) Normalized fluorescence change for the recorded neuron imaged with a cooled CCD camera (Micromax, Princeton Instruments). Recording pipette contained 25 μm of fura pentapotassium salt (comparable to the intracellular concentration of fura following bulk loading). While individual action potentials cannot be resolved at such a high firing frequency, onset of neuronal activity is accurately detected. (C) This fluorescent image can then be analyzed to detect spontaneously coactive neurons, as indicated by the filled contours.

neuronal activity and a change in fluorescence as detected by epifluorescence. The fluorescence of Fura-2, decreases in the presence of calcium, thus the negative fluorescent change in Fig. 15.3B (ii). As illustrated, even one-photon imaging of Fura-2, is sufficient to detect and resolve action potential generation in a single neuron within a field of hundreds or a thousand loaded cells. Because of the rapid time course for the onset of its response, Fura-2, lends itself to the elucidation of network dynamics when simultaneously imaging large populations of neurons. Figure 15.3C shows an example of automatically detected ensemble activity. The contours, corresponding to the somata of imaged neurons, which are filled indicate a group of neurons, which were simultaneously active.

V. QUANTITATIVE CLASSIFICATION OF NEURONAL MORPHOLOGY USING PRINCIPAL COMPONENT ANALYSIS

The goal in this section is to introduce principal component analysis (PCA) in simple terms and provide a practical guide for its use in the analysis of morphometric data. It is not intended to provide a rigorous introduction to PCA nor a comprehensive review of its many applications in data analysis. There are excellent publications addressing these two issues (Jolliffe, 2002).

The concept of PCA was introduced by Pearson (1901) and later independently developed by Hotelling (1933). PCA is a statistical method for transforming a set of data with presumably large number of correlated variables to a more parsimonious set, yet preserving as much original information as possible. Given a set of n cases with p measured variables (attributes), PCA is aimed at reconstructing a set of k uncorrelated variables, where $k \ll p$ (k is much smaller than p). These k uncorrelated variables are called principal components (PCs). PCs can be derived using the covariance or the correlation matrix of data. In this section, for reasons that will be mentioned later, the derivation of PCs is solely based on the correlation matrix.

PCA offers several advantages with respect to the analysis of high-dimensional data where numerous attributes need to be considered simultaneously. The aim in the analysis of the neuronal morphometric data using PCA is to describe the variance structure of the data in terms of a few morphological indices in order to facilitate detection of clusters and/or outliers.

A. Fundamental Concepts of Principal Component Analysis

If a large number of morphometric variables are measured for a set of n cases, and if these measured variables are correlated with each other, then it is possible to derive a limited number of uncorrelated variables named principal components that could faithfully represent the data. The mathematical operations underlying PCA ensure that PCs represent the maximum variance of the original data.

A PC is a linear combination of the original variables x_1, x_2, x_3, ..., x_p in the form $y_i = a_{i1} x_1 + a_{i2} x_2 + a_{i3} x_3 + \cdots + a_{ip} x_p$, where the coefficients a_{i1}, a_{i2}, a_{i3}, ..., a_{ip} are derived through the mathematical operations that maximize the variance of each PC y_i. Hence, the first PC accounts for the largest proportion of variance, the second PC for a smaller proportion, etc. Therefore, it is possible to represent a significant proportion of the original variance in terms of only a few PCs.

From a geometrical point of view, in a multidimensional variable space occupied by cases, a PC is the best fit line through the swarm of points. That is, the sum of squared distances from all the points to that line is at its minimum. In this sense, the first two PCs define the best fit 2D plane to the swarm of points in the original p-dimensional space. Similarly, the first three PCs define the best-fit 3D hyperplane.

Using this geometrical interpretation, one can imagine how PCA can reduce the dimensions necessary to represent the data. The upper bound on the number of PCs is the number of original variables. The number of PCs k can take any integer value from 1 to p; however, it is usually enough to choose k much smaller than p and still be confident that the structure of data remains intact. The ability to reduce dimensionality comes from the interrelatedness of the original variables. If the original variables are uncorrelated with each other, PCA will not be able to reduce the number of dimensions. Hence, one needs to examine cross correlations of variables before applying this method.

B. Example of Application

Let us apply the method of PCA to a set of neuronal morphometric data acquired through the NeuroExplorer software (MicroBrightField Inc., Williston, VT). The data contain 82 morphometric parameters (both raw and derived) from 67 neocortical neurons, filled with biocytin and reconstructed in three dimensions using Neurolucida. The morphological parameters include dendritic, somatic, and axonal variables. The correlation matrix was then computed, resulting in an 82 × 82 matrix (partly shown in Table 15.1). The correlation matrix shows both positive and negative values of correlation among variables. Thus, it is appropriate to apply PCA to this type of data hoping that one can reduce the 82-dimensional variable-space to a smaller and more manageable number of dimensions.

Table 15.2 shows some of the first eigenvectors and eigenvalues of the correlation matrix. The variance of each PC is indicated by its associated eigenvalues. For example, the third PC $y_3 = 0.030\times$ (somatic perimeter) $+(-0.023) \times$ (somatic area) $+ 0.240 \times$ (total axonal node) $+ \cdots + a_p x_p$. The third PC accounts for a variance of 6.740. The total variance accounted for by all the original variables is 82, the same as the number

TABLE 15.1. Correlation matrix of morphometric data (only the first 12 variables shown).

	SP	SA	ANT	TAL	TAA	ALTA	ND	TDN	TDL	ALD	TAD	DLTA
Somatic perimeter	1.000	0.790	0.149	0.328	0.295	0.057	0.164	0.118	0.420	0.323	0.455	-0.092
Somatic area		1.000	0.206	0.461	0.439	-0.026	0.283	0.128	0.446	0.290	0.611	-0.297
Total axonal node			1.000	0.803	0.770	-0.083	0.163	-0.010	-0.008	-0.110	0.176	-0.214
Total axonal length				1.000	0.943	-0.024	0.256	-0.087	0.193	-0.001	0.340	-0.148
Tile area of axon					1.000	-0.346	0.246	-0.086	0.131	-0.037	0.315	-0.229
Axonal length/tile area						1.000	-0.018	-0.001	0.139	0.081	-0.036	0.307
Number of dendrites							1.000	0.385	0.418	-0.173	0.376	-0.029
Total dendritic node								1.000	0.665	0.528	0.477	0.134
Total dendritic length									1.000	0.789	0.774	0.202
Average length of dendrites										1.000	0.600	0.207
Tile area of dendrites											1.000	-0.335
Dendritic length/tile area												1.000

TABLE 15.2. Eigenvectors and eigenvalues of the correlation matrix (only 12 shown).

	a_1	a_2	a_3	a_4	a_5	a_6	a_7	a_8	a_9	a_{10}	a_{11}	a_{12}
Somatic perimeter	-0.005	0.203	0.030	-0.053	-0.014	-0.167	-0.182	0.020	0.028	0.084	-0.056	-0.142
Somatic area	-0.050	0.202	-0.023	0.089	0.061	-0.058	-0.121	0.053	-0.036	0.103	-0.136	-0.053
Total axonal node	-0.155	0.044	0.240	0.106	0.027	-0.011	-0.081	0.082	-0.066	-0.005	0.075	-0.070
Total axonal length	-0.116	0.129	0.177	0.114	0.053	0.046	-0.190	-0.018	-0.093	0.180	-0.030	-0.030
Tile area of axon	-0.130	0.107	0.148	0.107	0.037	0.062	-0.236	-0.013	-0.079	0.195	0.008	0.060
Axonal length/tile area	0.078	0.048	0.060	-0.015	0.046	-0.044	0.179	-0.039	-0.025	-0.056	-0.097	-0.284
Number of dendrites	-0.029	0.051	-0.022	0.228	-0.036	-0.132	-0.029	-0.035	0.113	0.151	-0.093	-0.104
Total dendritic node	0.044	0.072	-0.029	0.215	-0.191	-0.207	0.124	0.067	0.026	-0.056	0.112	0.145
Total dendritic length	0.062	0.224	-0.001	0.160	-0.072	-0.182	0.069	-0.066	-0.046	0.055	0.025	0.152
Average length of dendrites	0.078	0.194	-0.002	0.055	-0.071	-0.124	0.078	-0.022	-0.098	-0.077	0.083	0.261
Tile area of dendrites	-0.028	0.247	-0.048	0.159	-0.063	-0.071	0.020	0.085	0.005	0.031	-0.081	0.067
Dendritic length/tile area	0.111	-0.067	0.076	0.010	0.005	-0.149	0.069	-0.250	0.001	0.103	0.117	0.095
Variance (eigenvalue)	15.173	9.348	6.740	5.842	4.741	4.292	3.461	3.228	2.846	2.535	2.167	1.971

of variables, since the variance of each standardized variable is equal to 1. One can then compute the quality of representation by assessing how much of the original variance is accounted by the selected PCs. For example, sum variance of the first 12 PCs can be computed by adding up the first 12 eigenvalues, $15.173 + 9.348 + 6.740 + 5.842 + 4.741 + 4.292 + 3.461 + 3.228 + 2.846 + 2.535 + 2.167 + 1.971 = 62.344$. Thus, the number of dimensions can be reduced from 82 to 12, yet keeping 62.344/82 or about 76% of the total variance of the data set.

Correlations of original variables with PCs are given in Table 15.3. These correlations are computed by multiplying each eigenvector column by the square root of its eigenvalue. These correlations are sometimes called loadings, a term borrowed from factor analysis (FA). Variables that highly correlate with a PC determine much of the "character" of that component and can be useful in interpreting PCs.

1. How Many Principal Components Should Be Retained?

There is no definite answer to this question. The number of PCs sufficient for an adequate description of the data depends on both the structure of the data and the goal of the analysis. However, guidelines, mostly based on intuition, have been suggested in choosing a subset of PCs. Since each individual variable in its standardized form contributes 1 to the total variance, it seems reasonable to exclude PCs that account for less than 1 unit of variance, i.e., have an eigenvalue less than 1 (Kaiser, 1960). However, Kaiser's criterion may not always hold and could lead to exclusion of some important PCs having low eigenvalue. In order to address this problem, Jolliffe (1972) has suggested a cutoff value of 0.7 for the variance of PCs. Based on simulation studies, PCs of sample data with variance greater than 0.7 correspond to population PCs with variance greater than 1 and thus should be retained. An alternative graphical method for choosing PCs is called a "scree" plot (Cattell, 1966). Variance of each PC is plotted against the ordinal number of each PC and those PCs that make the steeper part of the plot above the first inflection point are selected. The problem with the scree plot is that, depending on the data structure, there may not be any clear inflection point to use as a guide. If the sole purpose of PCA is to account for a significant portion of the total variance, one can arbitrarily choose a value such as 90% and include enough PCs to achieve this level of significance. All the above methods should be taken only as a guide, since there can be significant differences in the number of PCs retained by one method versus the others.

2. Reduction of Dimensionality by Choosing a Subset of Original Variables

For reasons that are outside the scope of this chapter, one may be interested in selecting a subset of the original variables rather than their linear

TABLE 15.3. Correlation of variables with individual PCs.

	PC1	PC2	PC3	PC4	PC5	PC6	PC7	PC8	PC9	PC10	PC11	PC12
Somatic perimeter	-0.021	0.621	0.078	-0.128	-0.031	-0.347	-0.340	0.036	0.048	0.133	-0.083	-0.199
Somatic area	-0.195	0.619	-0.059	0.215	0.132	-0.120	-0.225	0.095	-0.060	0.164	-0.201	-0.075
Total axonal node	-0.602	0.135	0.622	0.256	0.060	-0.023	-0.151	0.148	-0.111	-0.008	0.111	-0.098
Total axonal length	-0.453	0.394	0.461	0.276	0.115	0.095	-0.353	-0.032	-0.157	0.287	-0.045	-0.042
Tile area of axon	-0.508	0.328	0.384	0.259	0.081	0.128	-0.438	-0.023	-0.133	0.310	0.012	0.084
Axonal length/tile area	0.303	0.148	0.155	-0.036	0.099	-0.092	0.333	-0.069	-0.042	-0.089	-0.142	-0.399
Number of dendrites	-0.111	0.157	-0.058	0.551	-0.078	-0.274	-0.055	-0.063	0.190	0.240	-0.136	-0.147
Total dendritic node	0.171	0.219	-0.076	0.520	-0.416	-0.430	0.231	0.120	0.045	-0.089	0.165	0.203
Total dendritic length	0.241	0.684	-0.004	0.386	-0.157	-0.376	0.129	-0.119	-0.078	0.087	0.036	0.213
Average length of dendrites	0.303	0.593	-0.006	0.132	-0.155	-0.257	0.145	-0.040	-0.166	-0.123	0.122	0.367
Tile area of dendrites	-0.110	0.754	-0.124	0.384	-0.137	-0.146	0.038	0.153	0.008	0.049	-0.119	0.094
Dendritic length/tile area	0.431	-0.205	0.197	0.025	0.011	-0.308	0.128	-0.449	0.002	0.165	0.173	0.133

combination in the form of PCs (Jolliffe, 1972). One strategy is to retain one variable from each retained PC. The variable with the highest loading or correlation with the PC is retained. Conversely, one can exclude variables that are highly correlated with the excluded PCs and keep the remaining variables.

3. Detection of Clusters and Outliers

High dimensionality of the data precludes most of the common measures used for data visualization. PCA provides a parsimonious description of the data with a limited number of dimensions allowing questions to be asked regarding the existence of clusters, their topology, and the nature of outliers. This is an especially important issue in classification since methods of cluster analysis will always produce clusters whether or not these clusters "naturally" exist. Similarly, PCA facilitates detection of outlying and/or influential cases so that they can be eliminated before cluster analysis is performed. The choice of the clustering algorithm makes a significant difference in the composition of the clusters. Certain clustering algorithms (e.g., Ward's method) will detect mostly spherical clusters while others might be biased toward the detection of more oblong clusters (e.g., single linkage). PCA can verify the existence of clusters and provide clues about their topological features so that appropriate clustering algorithms can be used.

As a logical step after performing PCA, cluster analysis was used to define clusters of morphologically reconstructed neurons. Multiple clustering algorithms (K-Means, Hierarchical Agglomerative) produced similar cluster structures. An instance of cluster analysis utilizing Ward's method and squared Euclidean distances is shown in Fig. 15.4. The majority of neurons containing neuropeptide Y or paravalbumin are clustered together. The parameters used for clustering were strictly limited to morphological measurements from these neurons.

VI. SUMMARY OF ADVANTAGES AND LIMITATIONS

A. Two-Photon Imaging of Dendritic Calcium Microdomains

Advantages of two-photon excitation include reduced phototoxicity, higher spatial resolution, and deeper penetration into the sample (for more discussion see Denk and Svoboda, 1997; Denk *et al.*, 1990). These advantages are especially important for imaging small structures, such as dendrites and spines, which frequently lie deep in tissue and are sensitive to photodamage.

In addition to calcium-dependent fluorophores, voltage-sensitive dyes (Antic *et al.*, 1997) and sodium-dependent indicators (Rose *et al.*, 1999; Rose and Konnerth, 2001) have been used to functionally image dendrites.

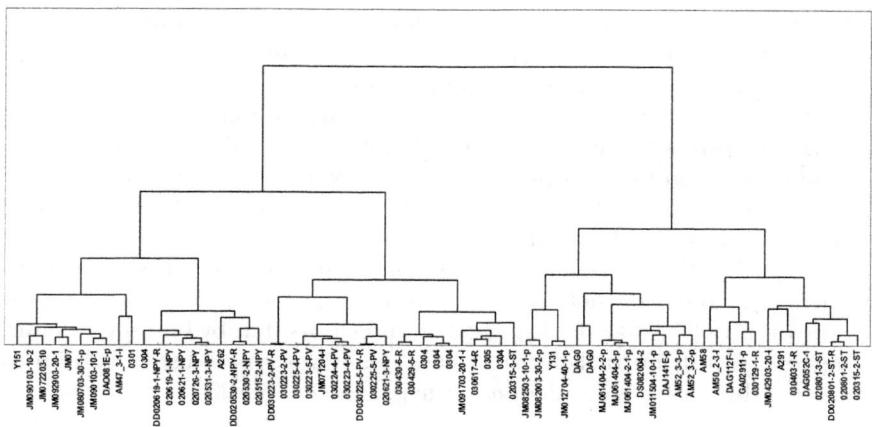

Figure 15.4. Dendrogram showing the five-cluster structure of the 67 morphologically reconstructed neurons. Neurons were assigned to clusters such that the sum of squared deviations from the mean values of parameters for each cluster is minimized. Squared Euclidean distances were used as a measure of dissimilarity among neurons and their clusters. Thus, the greater the similarity between neurons, the smaller the resulting Euclidian distance.

Calcium imaging offers several advantages. First, a large diversity of calcium indicators are available, and they offer good signal-to-noise ratio. Second, action potential backpropagation and subsequent opening of voltage-gated calcium channels appear to be a general function of dendrites. Thus, calcium imaging can be used as an indirect measure of dendritic voltage propagation (Goldberg *et al.*, 2003a; Spruston *et al.*, 1995; Waters *et al.*, 2005). Third, calcium is of particular interest because of its pluripotent role as a second messenger. As the major ion that both follows electrical potentials and binds proteins, it connects the electrical and organic worlds in excitable tissue. Thus, the spatial extent of a calcium signal offers key insights into the potential range of action of a given stimulus (Euler *et al.*, 2002; Goldberg and Yuste, 2005; Sabatini *et al.*, 2002). However, there are important limitations of calcium imaging in dendrites, and they must be understood to accurately interpret calcium fluorescent signals. Experimentally observed calcium fluorescence changes *do not reflect* native calcium signals; rather, they represent the chemical reaction between calcium and the indicator. Because high-affinity indicators bind virtually every ion that enters the cytoplasm, calcium extrusion, diffusion, and interactions with endogenous proteins are disturbed. Thus, fluorescent calcium signals differ significantly from "true" free calcium signals that would exist in the absence of indicator (for excellent discussion of these issues see Helmchen, 1999). First, because the unbinding process between calcium and its indicator is slow, calcium fluorescence transients are significantly slower than the free calcium signals they represent (Helmchen, 2002; Neher, 1998). Second, by binding calcium, indicators inhibit extrusion pumps, which in dendrites can act on fast time scales to control both the amplitude and the spatial range of calcium signals. Thus, slower calcium influxes, e.g., by *N*-methyl-D-asparate receptors, are overrepresented in amplitude, time, and space relative to faster influxes, such as through voltage-gated calcium channels during action potential backpropagation (Goldberg *et al.*, 2003a; Sabatini *et al.*, 2002). Third, calcium diffusion in dendrites is controlled by endogenous buffers, many of which are fixed (Allbritton *et al.*, 1992). The presence of highly mobile calcium indicator can thus distribute calcium over unphysiologically long ranges, such as through spine necks or along dendritic shafts. Lastly, calcium-dependent processes such as long-term potentiation (LTP) and long-term depression (LTD) may be disturbed by indicators that by binding calcium interfere with its normal signaling cascades. Thus, one must be careful to examine the *functions* of calcium in the presence of indicator.

B. Imaging Neuronal Ensembles

The major advantage of bulk loading calcium imaging is the excellent spatial resolution. In contrast to network imaging using voltage-sensitive dyes (Petersen and Sakmann, 2001), our technique has the ability to resolve single neurons. This is fundamental to understanding the spatial organization of coactive neural ensembles (Cossart *et al.*, 2003, 2005; Ikegaya *et al.*, 2004).

However, the disadvantage of this methodology is the relatively poor temporal resolution, particularly if the cessation of activity is an important variable. Calcium indicators have a relatively slow decay time constant and as a result the dyes are best utilized as indicators of the onset of activity rather than the offset. In this regard, voltage-sensitive dye imaging has improved temporal resolution and is better suited to questions focusing on the timing of network activity over the participation of individual cells (Petersen and Sakmann, 2001; see chapter by Carl C. H. Petersen in this volume).

C. Principal Component Analysis Versus Factor Analysis

Although PCA and FA can both reduce the dimensionality of a data set, they achieve this goal in ways that are fundamentally different. PCA assumes no explicit statistical model underlying the observed variance of the original data. The goal is maximal representation of the total variance by successive PCs. However, in FA, according to the underlying statistical model, the original variance is decomposed (partitioned) to common (shared among factors) and unique (accounted for by a single specific factor) components. Unique variances are not of interest in FA. In contrast, FA describes the common variance (buried in the off-diagonal elements of the correlation matrix) in terms of underlying factors. PCA and FA are techniques to explore different aspects of the correlation matrix, with PCA focusing on the diagonal elements of this matrix and FA focusing on the off-diagonal elements.

We will conclude this section with a word of caution. PCA, in the context presented in this chapter, provides a tool for descriptive data analysis of a sample. Therefore, much of the analysis deals with correlation matrices. Making inferences about a population, using PCA analysis of a population sample, requires analysis of covariance matrices and multivariate normality of the data. The former introduces significant problems in the analysis of morphometric data and the latter condition is usually not satisfied.

APPENDIX

A. Two-Photon Imaging of Dendritic Calcium Microdomains

1. Making the stimulation electrode is time consuming and should be done before brain slices are prepared. Pull a borosilicate glass pipette to 6–10 MΩ, and bend the tip 60–80° using a microforge. Make sure that the tip remains patent and that the segment distal to the curve is not longer than the working distance of the objective to be used in the experiment. Fill the pipette with standard ACSF and a fluorophore of choice (we use 20–200 µM Alexa488-dextran). Mount the electrode on the micromanipulator, and image the tip of the electrode to ensure that it can be visualized.

2. Prepare brain slices (we make slices 300 µm in mice aged 13–17 days) and intracellular electrodes (borosilicate glass pipettes pulled to 6–10 MΩ). Fill the intracellular electrode with standard internal pipette solution and a calcium-dependent fluorophore (we use Fluo-4 at a concentration of 50–400 µM), and bring both the intracellular electrode and stimulation electrode into the field, in focus just above the brain slice. Patch the neuron of interest with the intracellular electrode and hold the cell for 20 min to allow time for the fluorophore to diffuse into the dendritic tree.

3. Next, bring the stimulation electrode to the immediate vicinity of the ROI. Since moving the stimluation electrode in the slice can compromise the quality of the intracellular recording, all x- and y-axes positioning of the stimulation electrode should be done above the slice. To do this, visualize the dendritic tree, and zoom into a ROI. For imaging microdomains on aspiny dendrites, it is important to find a dendritic segment that runs parallel to the orientation of the line scanner. Maintain the x and y positions of the scanner at the ROI, zoom out, and come out of the slice in the z-axis until the fluorescently labeled tip of the stimulation electrode is in focus. Position the stimulation electrode, currently out of the slice, at the x and y coordinates corresponding to the ROI.

4. Once the stimulation electrode is in focus above the slice at x and y coordinates above the ROI, refocus on the dendrite and gently lower the stimulation electrode until it is in view, adjacent to dendrite to be imaged (see stimulation electrode, labeled S, in Fig. 15.1). It should

be clear that the advantage of taking the time to bend the electrode tip in Step 1 is that bringing the electrode into the slice does not deform the tissue or interfere with the intracellular recording.

5. Set the stimulation strength. Deliver brief (100 µs) single shocks through the stimulation electrode, and adjust the stimulation strength so that there are subthreshold EPSPs recorded by the intracellular electrode.

6. Next, image the ROI during these single shocks. Reliable stimulation of a single synapse on the imaged dendritic segment is difficult and in part a matter of chance. Activation of a single synapse on the imaged dendrite should result in all-or-none localized calcium signals on successive trials, reflecting the stochasticity of synaptic transmission. Achieving this result depends on perfectly adjusting the stimulation strength. If it is too low, there may be no calcium signal on the dendrite, even though EPSPs are recorded at the soma. This may occur if the axons triggered by the stimulation electrode do not synapse on the imaged dendritic segment. If this occurs, increase the stimulation intensity, staying within the subthreshold range, until calcium signals are observed. Alternatively, if stimulation strength is too high, nonlocalized signals may be observed along the entire dendritic region. This may occur if stimulation triggers multiple synapses converging onto the branch (Goldberg *et al.*, 2003b) and/or local dendritic spikes (Holthoff *et al.*, 2004; Schiller *et al.*, 2000).

7. If activation of a single synapse cannot be achieved with the electrode in its present position, it is advised to rapidly change the position of the stimulation electrode. Take the stimulation electrode out of the slice, readjust in x and y coordinates its position, and lower it back into the slice near the ROI, as in Step 4. In our experience, the stimulation electrode often had to be repositioned as many as five times during a single experiment to isolate single synapses and observe calcium microdomains.

B. Electron Microscopic Reconstruction of Imaged Dendritic Domains

1. All experiments were carried out according to the NIH Guide for Care and Use of Laboratory Animals (NIH publication no. 86-23, revised 1987). After the in situ phase of the experiment, the 300-µm-thick mouse brain slices are put between two Millipore filters to avoid deformations and fixed in 2.5% paraformaldehyde, 1.25% glutaraldehyde, and 15% saturated picric acid in 0.1 M phosphate buffer (PB, pH 7.4) for 12–36 h at 4°C.

2. After rinsing in 0.1 M PB (2 × 10 min at room temperature), slices are incubated in 10 and 20% sucrose in PB (30 min in each) and freeze-thawed in liquid nitrogen.

3. The slices are embedded in gelatine (10% in distilled water) and then resectioned at a thickness of 60 μm in 0.1 M PB to ensure proper transparency during light microscopic reconstruction.

4. The biocytin filled cells are visualized by the avidin-biotinylated horseradish peroxidase method with diaminobenzidine as chromogen.

5. Sections are postfixed with 1% (w/v) OsO_4 in 0.1 M PB and block stained in 1% (w/v) uranyl acetate in distilled water, dehydrated, and embedded into epoxy resin (Durcupan, Fluka) on glass slides.

6. Previously two-photon imaged areas were light microscopically photographed at serial (5 μm steps) focal depths on the permanent preparations.

7. Three-dimensional light microscopic reconstructions are carried out using Neurolucida and NeuroExplorer (MicroBrightField Inc.) with oil immersion objective at 1250× magnification.

8. A region containing the imaged cell is reembedded and a complete series of ultrathin sections is made from the soma and the imaged dendrite of the cell.

9. The imaged dendrite is identified protruding from the soma in the electron microscope and then the dendrite is followed on the serial ultrathin sections up to the imaged portion. The alignment of light microscopic and ultrastructural images is based on the distance from the soma, on the undulation/course of the dendrite, on the position of characteristic changes in dendritic diameter, or on the relative location of dendritic beads or spine-like structures. It is essential to observe all ultrathin sections in the area of interest by tilting the goniometer of the electron microscope ±75° for checking synapses cut at oblique planes.

10. Three-dimensional reconstructions of dendritic segments can be performed with Neurolucida and analyzed in NeuroExplorer software. Series of 50–120 ultrathin sections containing the ROI are photographed. Contours of the profiles of dendrites, synaptic terminals, and synaptic junctions are traced using a digitizing tablet on each photograph. Image alignment is based on minimally five points identifiable on neighboring sections using standard functions of the Neurolucida/NeuroExplorer software package.

C. Imaging Neuronal Ensembles

1. Prepare the Fura-2, AM by dissolving 50 μg Fura-2, AM (Molecular Probes) in 15–48 μl DMSO and 2 μl of Pluronic F-127 (Molecular Probes) for a final concentration of 1–3.3 mM.

2. Vortex the solution for 10–15 min prior to use.

3. Place slices in loading chamber (a petri dish 35 mm × 10 mm) containing 2.5 ml of oxygenated ACSF. Pipette 5–10 μl of Fura-2, AM solution on top of each slice, resulting in a high initial concentration of Fura-2, AM. Alternatively, pipette the entire volume of the prepared Fura-2, AM at the far side of the loading chamber and allow the dye to passively diffuse throughout the chamber lading the slices. Once again this is the preferred method especially if the slices are from a younger animal.

4. Load the slices (we use 300-μm-thick cortical slices from mice aged 13–30 days) in the dark for 20–30 min at 35–37°C with 95% O_2/5% CO_2 lightly ventilated into the chamber. As a rule of thumb, the loading time should be 10 min plus as many minutes as the age of the animal in postnatal days.

5. Remove slices from the loading chamber and place into incubation chamber containing oxygenated ACSF to allow them to rest.

6. Regardless of the method, imaging can begin 30 min following loading procedure.

REFERENCES

Alger, B. E., and Teyler, T. J., 1976, Long-term and short-term plasticity in the CA1, CA3, and dentate regions of the rat hippocampal slice, *Brain Res.* **110**:463–480.

Allbritton, N. L., Meyer, T., and Stryer, L., 1992, Range of messenger action of calcium ion and inositol 1,4,5-trisphosphate, *Science* **258**:1812–1815.

Antic, S., Major, G., Chen, W. R., Wuskel, J., Loew, L., and Zecevic, D., 1997, Fast voltage-sensitive dye recording of membrane potential changes at multiple sites on an individual nerve cell in the rat cortical slice, *Biol. Bull.* **193**:261.

Buhl, E. H., Halasy, K., and Somogyi, P., 1994, Diverse sources of hippocampal unitary inhibitory postsynaptic potentials and the number of synaptic release sites, *Nature* **368**:823–828.

Cattell, R., 1966, The scree test for the number of factors, *Multivariate Behav. Res.* **2**:245–276.

Connors, B. W., and Gutnick, M. J., 1990, Intrinsic firing patterns of diverse neocortical neurons, *Trends Neurosci.* **13**:99–104.

Cossart, R., Aronov, D., and Yuste, R., 2003, Attractor dynamics of network UP states in neocortex, *Nature* **423**:283–289.

Cossart, R., Ikegaya, Y., and Yuste, R., 2005, Calcium imaging of cortical networks dynamics, *Cell Calcium* **37**:451–457.

Denk, W., and Svoboda, K., 1997, Photon upmanship: why multiphoton imaging is more than a gimmick, *Neuron* **18**:351–357.

Denk, W., Strickler, J. H., and Webb, W. W., 1990, Two-photon laser scanning fluorescence microscopy, *Science* **248**:73–76.

Dingledine, R., Dodd, J., and Kelly, J. S., 1980, The in vitro brain slice as a useful neurophysiological preparation for intracellular recording, *J. Neurosci. Methods* **2**:323–362.

Euler, T., Detwiler, P. B., and Denk, W., 2002, Directionally selective calcium signals in dendrites of starburst amacrine cells, *Nature* **418**:845–852.

Freund, T. F., and Buzsaki, G., 1996, Interneurons of the hippocampus, *Hippocampus* **6**:347–470.

Ghozland, S., Aguado, F., Espinosa-Parrilla, J. F., Soriano, E., and Maldonado, R., 2002, Spontaneous network activity of cerebellar granule neurons: impairment by in vivo chronic cannabinoid administration, *Eur. J. Neurosci.* **16**:641–651.

Gibson, J. R., Beierlein, M., and Connors, B. W., 1999, Two networks of electrically coupled inhibitory neurons in neocortex, *Nature* **402**:75–79.

Goldberg, J., Holthoff, K., and Yuste, R., 2002, A problem with Hebb and local spikes, *Trends Neurosci.* **25**:433.

Goldberg, J. H., Tamas, G., Aronov, D., and Yuste, R., 2003c, Calcium microdomains in aspiny dendrites, *Neuron* **40**:807–821.

Goldberg, J. H., Tamas, G., and Yuste, R., 2003a, Ca2+ imaging of mouse neocortical interneurone dendrites: Ia-type K+ channels control action potential backpropagation, *J. Physiol.* **551**:49–65.

Goldberg, J. H., and Yuste, R., 2005, Space matters: local and global dendritic Ca2+ compartmentalization in cortical interneurons, *Trends Neurosci.* **28**:158–167.

Goldberg, J. H., Yuste, R., and Tamas, G., 2003b, Ca2+ imaging of mouse neocortical interneurone dendrites: contribution of Ca2+-permeable AMPA and NMDA receptors to subthreshold Ca2+ dynamics, *J. Physiol.* **551**:67–78.

Gupta, A., Wang, Y., and Markram, H., 2000, Organizing principles for a diversity of GABAergic interneurons and synapses in the neocortex, *Science* **287**:273–278.

Helmchen, F., 1999, Dendrites as biochemical compartments, In: Stuart, G., Spruston, N., and Hausser, M. (eds.), *Dendrites*, Oxford: Oxford University Press, pp. 161–192.

Helmchen, F., 2002, Raising the speed limit—fast Ca(2+) handling in dendritic spines, *Trends Neurosci.* **25**:438–441 (discussion 441).

Holthoff, K., Kovalchuk, Y., Yuste, R., and Konnerth, A., 2004, Single-shock LTD by local dendritic spikes in pyramidal neurons of mouse visual cortex, *J. Physiol.* **560**:27–36.

Horikawa, K., and Armstrong, W. E., 1988, A versatile means of intracellular labeling: injection of biocytin and its detection with avidin conjugates, *J. Neurosci. Methods* **25**:1–11.

Ikegaya, Y., Aaron, G., Cossart, R., Aronov, D., Lampl, I., Ferster, D., and Yuste, R., 2004, Synfire chains and cortical songs: temporal modules of cortical activity, *Science* **304**:559–564.

Jolliffe, I. T., 1972, Discarding variables in a principal component analyis. I: artificial data, *Appl. Stat.* **21**:160–173.

Jolliffe, I. T., 2002, Principal component analysis, *Springer Series in Statistics*, 2nd ed., New York: Springer-Verlag.

Kaiser, H. F., 1960, The application of electronic computers to factor analysis, *Educ. Psychol. Meas.* **20**:141–151.

Majewska, A., Yiu, G., and Yuste, R., 2000, A custom-made two-photon microscope and deconvolution system, *Pflügers Arch.—Eur. J. Physiol.* **441**:398–409.

Mao, B. Q., Hamzei-Sichani, F., Aronov, D., Froemke, R. C., and Yuste, R., 2001, Dynamics of spontaneous activity in neocortical slices, *Neuron* **32**:883–898.

Mao, L., and Wang, J. Q., 2003, Group I metabotropic glutamate receptor-mediated calcium signaling and immediate early gene expression in cultured rat striatal neurons, *Eur. J. Neurosci.* **17**:741–750.

Markram, H., Luebke, J., Frotscher, M., and Sakmann, B., 1997, Regulation of synaptic efficacy by coincidence of postsynaptic APs and EPSPs, *Science* **275**:213–215.

Markram, H., Toledo-Rodriguez, M., Wang, Y., Gupta, A., Silberberg, G., and Wu, C., 2004, Interneurons of the neocortical inhibitory system, *Nat. Rev. Neurosci.* **5**:793–807.

McBain, C. J., and Fisahn, A., 2001, Interneurons unbound, *Nat. Rev. Neurosci.* **2**:11–23.

Neher, E., 1998, Usefulness and limitations of linear approximations to the understanding of Ca++ signals, *Cell Calcium* **24**:345–375.

Nikolenko, V., Nemet, B., and Yuste, R., 2003, A two-photon and second-harmonic microscope, *Methods* **30**:3–15.

Ohki, K., Chung, S., Ch'ng, Y. H., Kara, P., and Reid, R. C., 2005, Functional imaging with cellular resolution reveals precise microarchitecture in visual cortex, *Nature* **433**:597–603.

Petersen, C. C., and Sakmann, B., 2001, Functionally independent columns of rat somatosensory barrel cortex revealed with voltage-sensitive dye imaging, *J. Neurosci.* **21**:8435–8446.

Poirazi, P., and Mel, B. W., 2001, Impact of active dendrites and structural plasticity on the memory capacity of neural tissue, *Neuron* **29**:779–796.

Prince, D. A., and Connors, B. W., 1984, Mechanisms of epileptogenesis in cortical structures, *Ann. Neurol.* **16**:S59–S64.

Rose, C. R., and Konnerth, A., 2001, NMDA receptor-mediated Na+ signals in spines and dendrites, *J. Neurosci.* **21**:4207–4214.

Rose, C., Kovalchuk, Y., Eilers, J., and Konnerth, A., 1999, Two-photon Na+ imaging in spines and fine dendrites of central neurons, *Pflugers Arch.* **439**:201–207.

Sabatini, B. L., Oertner, T. G., and Svoboda, K., 2002, The life cycle of Ca(2+) ions in dendritic spines. *Neuron* **33**:439–452.

Sakmann, B., and Neher, E., 1983, *Single Channel Recording*, New York: Plenum Press.

Sarvey, J. M., Burgard, E. C., and Decker, G., 1989, Long-term potentiation: studies in the hippocampal slice, *J. Neurosci. Methods* **28**:109–124.

Schiller, J., Major, G., Koester, H. J., and Schiller, Y., 2000, NMDA spikes in basal dendrites of cortical pyramidal neurons, *Nature* **404**:285–289.

Smetters, D. K., Majewska, A., and Yuste, R., 1999, Detecting action potentials in neuronal populations with calcium imaging, *Methods* **18**:215–221.

Somogyi, P., Tamas, G., Lujan, R., and Buhl, E., 1998, Salient features of synaptic organisation in the cerebral cortex, *Brain Res. Brain Res. Rev.* **26**:113–135.

Spruston, N., Schiller, Y., Stuart, G., and Sakmann, B., 1995, Activity-dependent action potential invasion and calcium influx into hippocampal CA1 dendrites, *Science* **286**:297–300.

Stosiek, C., Garaschuk, O., Holthoff, K., and Konnerth, A., 2003, In vivo two-photon calcium imaging of neuronal networks, *Proc. Natl. Acad. Sci. U. S. A.* **100**:7319–7324.

Tamas, G., Buhl, E. H., Lorincz, A., and Somogyi, P., 2000, Proximally targeted GABAergic synapses and gap junctions synchronize cortical interneurons, *Nat. Neurosci.* **3**:366–371.

Tamas, G., Buhl, E. H., and Somogyi, P., 1997, Massive autaptic self-innervation of GABAergic neurons in cat visual cortex, *J. Neurosci.* **17**:6352–6364.

Tanaka, E., Uchikado, H., Niiyama, S., Uematsu, K., and Higashi, H., 2002, Extrusion of intracellular calcium ion after in vitro ischemia in the rat hippocampal CA1 region, *J. Neurophysiol.* **88**:879–887.

Tsien, R. Y., 1981, A nondisruptive technique for loading calcium buffers and indicators into cells, *Nature* **290**:527–528.

Tsien, R. Y., 1989, Fluorescent probes of cell signaling, *Ann. Rev. Neurosci.* **12**:227–253.

Voitenko, N. V., Kostyuk, E. P., Kruglikov, I. A., and Kostyuk, P. G., 1999, Changes in calcium signaling in dorsal horn neurons in rats with streptozotocin-induced diabetes, *Neuroscience* **94**:887–890.

Waters, J., Schaefer, A., and Sakmann, B., 2005, Backpropagating action potentials in neurones: measurement, mechanisms, and potential functions, *Prog. Biophys. Mol. Biol.* **87**:145–170.

Yuste, R., and Denk, W., 1995, Dendritic spines as basic units of synaptic integration, *Nature* **375**:682–684.

Yuste, R., Gutnick, M. J., Saar, D., Delaney, K. D., and Tank, D. W., 1994, Calcium accumulations in dendrites from neocortical neurons: an apical band and evidence for functional compartments, *Neuron* **13**:23–43.

Yuste, R., and Katz, L. C., 1991, Control of postsynaptic Ca2+ influx in developing neocortex by excitatory and inhibitory neurotransmitters, *Neuron* **6**:333–344.

Yuste, R., Tank, D. W., and Kleinfeld, D., 1997, Functional study of the rat cortical microcircuitry with voltage-sensitive dye imaging of neocortical slices, *Cereb. Cortex* **6/7**:546–558.

16

Stereology of Neural Connections: An Overview

CARLOS AVENDAÑO

CARLOS AVENDAÑO • Department of Anatomy, Histology and Neuroscience, Autónoma University of Madrid, Medical School, 28029 Madrid, Spain

Abstract: Stereological methods have made a quiet revolution in the quantification of neural structures. By combining stochastic geometry and statistics, they have given neuroanatomists invaluable tools to estimate with accuracy and precision quantitative information in three dimensions, from data obtained on flat images, sections, and slabs of central and peripheral nervous tissues. This chapter reviews and discusses the application of a variety of stereological methods to quantify geometric parameters relevant for analyzing neural connections, such as neuron, axon, or synaptic number, axonal length, and volume of regions of interest. Special attention is given to histological and observation artifacts that may hamper the efficacy of otherwise unbiased and efficient sampling and measuring methods. Current strengths and future developments of stereology in this field are briefly reviewed as well. The chapter ends with a worked example of stereology applied to the quantitative analysis of connections in a connectional study using retrograde tracers.

Keywords: axon number, axonal length, morphometry, neuron number, synapse number

I. ASSESSING NEURAL CONNECTIONS

Cell-to-cell communication is a fundamental property of multicellular organisms. It takes place by the information-laden exchange of molecular signals between cells which abut, or are close to each other (autocrine and paracrine signals), and also by the effects of blood-borne molecules originating in distant cells (endocrine signals). A special kind of paracrine signaling in a broad sense (sometimes called *neurocrine*) is represented by the chemical neurotransmission in the nervous system. Neurons achieve local contacts with other neurons, which may be situated at remote places, by means of their long neurites (dendrites and axons) and a variety of synaptic specializations.

The generation of networks of local intercellular contacts at a distance from the cell body is a hallmark of the nervous system. Through extensive dendritic and axonal arborizations, individual neurons define ample domains of influence. The geometry of axons and dendrites, together with the distribution of ion channels, defines fundamental functional properties of the neuron. Basic geometric parameters, such as shaft caliber, tapering, or branching, initially shown to determine action potential propagation in axons (Goldstein and Rall, 1974), also shape the centrifugal and centripetal propagation of postsynaptic and action potentials in dendrites (Poirazi and Mel, 2001; Schaefer *et al.*, 2003; Spruston *et al.*, 1999). Spines and dendritic segments are also important in defining biochemical compartments in neurons, and these are dependent on their size, length, and shape (Helmchen, 1999; Korkotian and Segal, 2000). Different patterns in the density and spatial distribution of axonal or dendritic branches create a wide spectrum of wiring configurations, which, already described in early Golgi studies, are currently the subject of intensive experimental and theoretical research (Chklovskii *et al.*, 2002; Mitchison, 1992; Stepanyants *et al.*, 2004; Van Pelt *et al.*, 2001).

In organisms other than simple invertebrates, however, brainwork is mainly expressed at the population, rather than at the single-cell level. Neurons aggregate in spatially discrete sets (nuclei, modules, columns, and so on), which display common anatomical (in terms of form and connections) and functional (coherent activation and firing, substrate for representational mapping) properties. Moreover, if the geometric properties of single neurons and local axonal circuits are hard to tackle in quantitative terms, difficulties pile up when circuits established by neuronal populations are the target. So far, most of our current understanding of neural circuits rests largely on qualitative descriptions, and when quantitative morphological assessments are made, they are most often expressed in subjective and relative terms (heavy, dense, light, sparse, scattered, etc.), according to imprecise ordinal scales (Stevens, 1951). This situation, however, is changing for the better with the increasing awareness that sound quantitative neuroanatomy is a requisite to understanding the brain (Braitenberg and Schüz, 1998; Evans *et al.*, 2004).

II. TARGET PARAMETERS OF NEURAL CONNECTIONS

A. Structures Identified in Studies of Connections

Before the introduction of axoplasmic tracers and of methods to visualize specific molecules in neurons and neurites, most anatomical studies of neural connections had to rely on the Golgi and the myelin or axonal degeneration methods. Under optimal conditions, the Golgi method allows for the recognition of complete cell bodies and dendritic arbors with their spines and, at least, large stretches of unmyelinated axonal branches (Valverde, 1970). Camera lucida or digital imaging reconstructions provide data in two and three dimensions (2D and 3D), respectively, which can be used to parametrize and analyze geometrical and topological features of the various components of the neuron (Hillman, 1979; Uylings *et al.*, 1986a,b). Axonal degeneration studies, such as the Nauta and the Fink–Heimer methods, rely on the staining of abnormal (dying) and incomplete axonal profiles after focal lesions. While being very sensitive, only rough quantitative estimates of *en passant* and terminal connections can be obtained with these methods, which are more adequate for describing connections in qualitative and topographical terms.

The knowledge of neural connections made a dramatic leap forward with the introduction of tracing techniques, in the early 1970s, that exploited the axoplasmic transport of exogenous tracers, such as tritiated amino acids, horseradish peroxidase (HRP; free or conjugated to other molecules), lectins, and fluorescent molecules (references in Heimer and Zaborszky, 1989; see also chapters by Reiner and Lanciego, this volume). In addition to their biocompatibility, most of these tracers enable intense and, in some cases, fairly complete labeling of populations of cell bodies, dendrites, spines, and/or axonal trees and synaptic boutons to be obtained. Also, HRP or smaller molecules (such as Lucifer yellow, biocytin, or neurobiotin) can be administered to individual neurons, by intra- or juxtacellular injections, yielding in optimal conditions Golgi-like staining of the entire neuron. Additionally, different axoplasmic tracers can be combined among them, and/or with immunocytochemical labeling of specific molecules, to furnish informative connectional and neurochemical multiple labeling of neurons at single or population level (e.g., Vinkenoog *et al.*, 2005; see also Sesack *et al.*, this volume). Immunocytochemistry alone may provide excellent labeling of specific subsets of neurons or their neurites in selected regions, making them amenable to being measured and quantified. Finally, ultrastructural studies of neural connections are mostly used to obtain morphological and neurochemical data of synapses, spines, boutons, and other components of neurons labeled with tracers or immunocytochemistry prior to, or after embedding in resin (Fig. 16.1).

B. Parameters

Neural connections may be defined by a large number of parameters, and an appropriate choice among them should be made when designing a stereological (or, in general, any morphometric) study. On biological grounds, it is convenient to distinguish measures referred to individual neurons, as a whole (a cell soma, whole dendritic or axonal tree of a neuron, spines in a cell) or in part (length, surface area, number of spines, synapses, boutons, etc., referred to a dendritic or axonal subtree), from those referred to sets of structures belonging to neuronal populations (groups of neuron bodies, dendritic bundles, axonal networks, nerves or fiber tracts, synapses in a territory, etc.).

Different types of parameters may be defined on geometrical and statistical grounds: *First-order* parameters refer to primary geometrical measures, such as number, length, surface, and volume; they may also include relative measures, or ratios, such as numerical, length, surface, or volume densities. These parameters are direct descriptors of the target neurons or neurites, and will be the main target of this review (Table 16.1). *Second-order* measures, in a morphometric/stereological context, refer to statistical descriptions of the 3D spatial distribution of the first-order parameters. They inform about

$\longrightarrow$

Figure 16.1. Diagram illustrating the main neural components that may be defined by different methods of labeling single neurons and neuronal populations. The soma and dendritic arbor and, in optimal conditions, the full axonal tree can be labeled by the Golgi stain, and by targeting selected neurons with exogenous tracers or revealing specific molecules (top). Connections from neuron groups are usually studied by anterograde or retrograde labeling with tracers. And the global population of neural components in a region is also accessible to specific labeling and/or ultrastructural study (bottom).

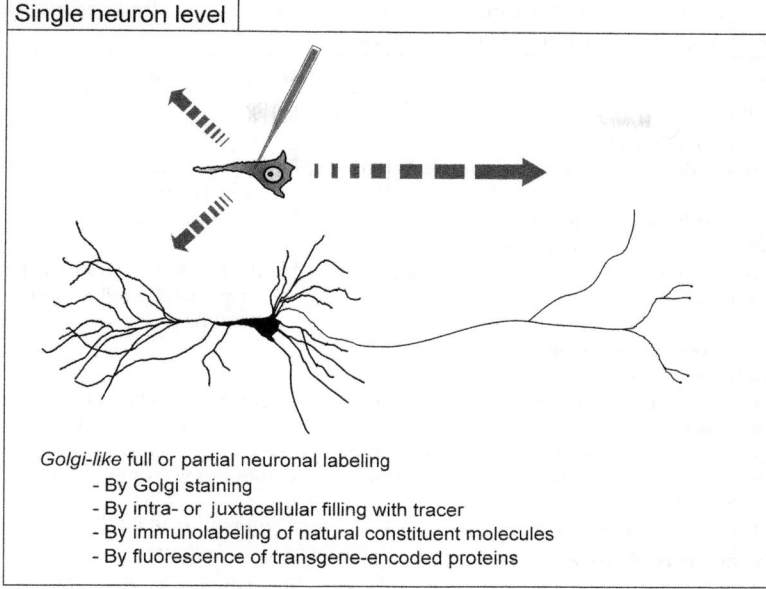

Single neuron level

Golgi-like full or partial neuronal labeling
- By Golgi staining
- By intra- or juxtacellular filling with tracer
- By immunolabeling of natural constituent molecules
- By fluorescence of transgene-encoded proteins

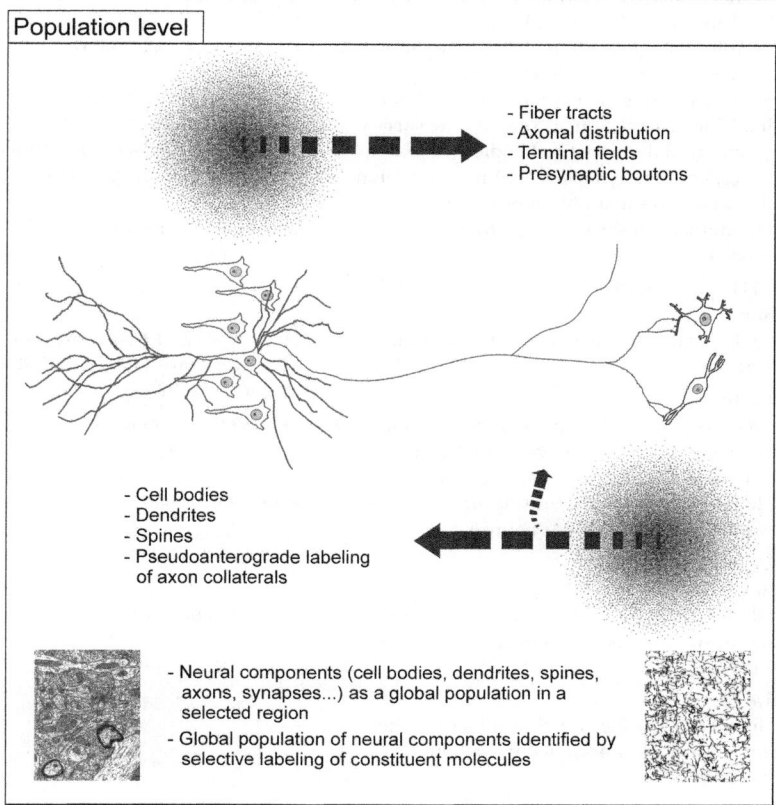

Population level

- Fiber tracts
- Axonal distribution
- Terminal fields
- Presynaptic boutons

- Cell bodies
- Dendrites
- Spines
- Pseudoanterograde labeling
  of axon collaterals

- Neural components (cell bodies, dendrites, spines, axons, synapses...) as a global population in a selected region
- Global population of neural components identified by selective labeling of constituent molecules

Figure 16.1.

TABLE 16.1. A selection of parameters of interest to stereologically assess neural connections, and some pertinent references

Parameter	Selected references
NUMBER	
Absolute numbers (N)	
Neuron bodies sending axons to a labeled target	Bermejo *et al.* (2003) and Negredo *et al.* (2004)
Neuron bodies labeled by a transneuronal anterograde marker, such as viruses	Glatzer *et al.* (2003)
Neuron bodies in a certain region, or expressing a certain molecular marker	West and Slomianka (1998) Donovan *et al.* (2002), and Avendoño *et al.* (2005)
Neuron bodies that effectively take up an anterograde tracer	—
Axons in a peripheral nerve, a central tract, or white matter	Larsen *et al.* (1998b) and Marner and Pakkenberg (2003)
Terminal or preterminal axons in a target territory	—
Presynaptic boutons (*en passant* or terminal) in a target territory	Calhoun *et al.* (1996)
Synapses in a target territory	Geinisman *et al.* (1996) and Tang *et al.* (2001)
Relative numbers (densities)	
N_L, number of spines, boutons, synapses, etc., per dendritic/axonal segment length	Rusakov and Stewart (1995)
N_A, number of cross-sectioned axons per unit area in a section of a nerve, tract, or territory	Partadiredja *et al.* (2003)
N_S, number of spines or synapses per unit area of dendritic/axonal/somatic membrane surface	Mayhew (1979b) and Rusakov *et al.* (1998)
N_V, numerical density of cell bodies (or axons, spines, boutons, synapses, etc.) per unit volume of a predetermined reference space	O'Malley *et al.* (2000) and Andersen and Pakkenberg (2003)
N_u, numerical density of spines, synapses, etc. per neuron	Turner and Greenough (1985) and Kleim *et al.* (1997)
LENGTH (path length)	
Absolute (L)	
Total dendritic or axonal length in a certain territory	Howard *et al.* (1992), Tang and Nyengaard (1997), and Mouton *et al.* (2002)
Relative (densities)	
L_V, length of neurite branches per unit volume of a predetermined reference space (nucleus, white matter sector, cortical layer, etc.)	Stocks et al. (1996) and Larsen *et al.* (2004)
L_u, length of axonal or dendritic arborizations (complete or partial) per neuron	Rønn *et al.* (2000)
SURFACE	
Absolute (S)	
Total membrane surface of dendrites, spines, axons, synaptic specializations, etc. in a given territory	Peterson and Jones (1993)
Relative (densities)	
S_S, membrane surface fraction (of a dendrite, spine, axon, etc.) occupied by synaptic specializations	—

TABLE 16.1. (*Cont.*)

Parameter	References
S_V, extension of membrane surface (of dendrites, spines, axons, synaptic specializations, etc.) per unit volume of a predetermined reference space (neuropil, dendrites, axons, etc.)	Machín-Cedrés (2004)
CROSS-SECTIONAL AREA	
Absolute (A) or relative (area fraction, A_A)	
Total or fractional area occupied by nerve or tract components (myelin, myelinated or unmyelinated axons, blood vessels, connective tissue) in a transverse section	Larsen (1998) and Schiønning *et al.* (1998)
Mean and distribution (a_n)	
For cross-sectional areas of a population of axons in a nerve or tract	Larsen (1998)
VOLUME	
Absolute (V)	
Volume of a nucleus, cortical layer, cell aggregate, injection site, projection territory, etc.	Blasco *et al.* (1999) and Bermejo *et al.* (2003)
Relative (volume fraction, V_V)	
Relative occupancy of a predetermined reference space by a defined neuronal (or vascular, glial, etc.) component	Jones (1999), Machín-Cedrés (2004)
Mean and distribution (v_n)	
For a population of cell bodies, presynaptic boutons, spines, etc.	Avendaño and Dykes (1996) and Lagares and Avendaño (1999)

the spatial architecture of the target neurons or neurites, either among themselves, or with respect to other structures, such as glial cells, blood vessels, etc. Second-order stereology is much less developed, and only a few comments will be made in section "Advantages, Limitations, and Perspectives." Neural shape may also be quantitated in topological terms, that is, by defining only the connectivity pattern of the (dendritic or axonal) tree architecture of a neuron. *Topological* measures include nodes or vertices (root, branch points, terminal tips) and segments (root, intermediate, terminal), and ignore metrical variables. Excellent reviews on the topological analysis of dendrites are available (Uylings and Van Pelt, 2002; Uylings *et al.*, 1989; Verwer *et al.*, 1992).

It is important to determine which is the minimum set of parameters that provide a satisfactory structural description of a neuron, seeking completeness, but avoiding redundancy. But it has to be realized that such a description only reaches its full neurobiological sense in a populational, and therefore statistical, sense (Ascoli, 2002). Seeking better estimators of shape and its variation and complexity within and between cell classes, metrical, topological, and/or spatial parameters have been combined into more sophisticated analytical methods. These provide measures which range from simple relations of the same parameter at various locations, as is the case of *taper* (ratio of a segment thickness between its end and its beginning) or *tortuosity* (ratio of trace length to Euclidean length for a given neurite segment), to fractal analysis, or to multivariate analysis of configurations of points in artificial *shape spaces*. No further comments will be made on these issues, but a growing body of literature is available to the interested reader (Adams *et al.*, 2004; Ascoli, 2002; Jelinek and Fernandez, 1998; Uylings and Van Pelt, 2002).

III. MORPHOMETRY AND STEREOLOGY

In its widest acceptance, morphometry has been defined as "Quantitative morphology; the measurement of structures by any method, including stereology" (Weibel, 1979). In practical terms, however, morphometric approaches in neuroscience have focused on the refinement of quantitative morphological descriptions and topological analyses of restricted structures, such as single neurons or limited axonal or dendritic fields. Stereology, in turn, has carved itself an expanding niche in the quantitation of large populations of cells and neurites.

Stereology combines stochastic geometry and statistics and is directly aimed at estimating values in a higher dimension from data obtained at a lower dimension on a properly sampled material (Cruz-Orive, 2002b). This is the case in most neurohistological and neuroimaging studies, which are based on sections or slices (physical or virtual) of parts of the central or peripheral nervous system. The need to apply correct sampling strategies due to the complexity and the large number of neural structures may be justified also for single cells. Figure 16.2 shows two examples of exceedingly complex axonal or dendritic arbors which, while amenable to costly individual complete reconstructions, would justify applying more efficient statistical sampling to estimate some morphometric parameters.

Heretofore, measurements of geometric parameters relevant for analyzing neural connections are carried out in most cases on sections of postmortem-processed tissue. Moreover, they are often performed on analogical or digital graphic reconstructions of images obtained from that tissue, in particular for dendritic and local axonal arborizations. Confocal images from labeled neurons in in vitro slices make an exception, but only in the case of small neurons with their whole axonal and dendritic trees in the slice. In this case the neurons, while partially deafferented, are otherwise intact; in any other case, we will be dealing with directly lesioned cells undergoing acute axotomy and dendrotomy to various degrees. A promising, but just incipient, development is the application of two-photon microscopy to the quantitative analysis of dendrites in vivo (Broser *et al.*, 2004).

Tissue processing introduces scores of structural deformations, which may be the cause of serious biases, if they are not properly controlled, or at

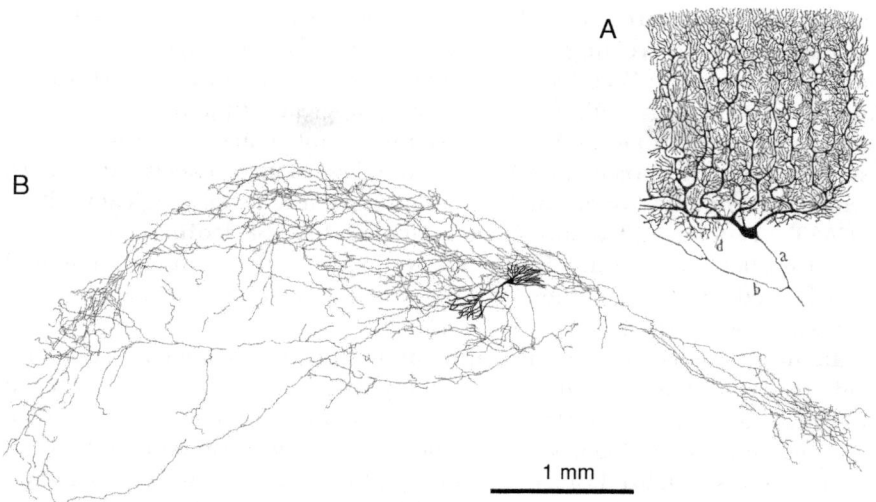

Figure 16.2. Examples of complex dendritic and axonal arbors of single neurons. (A) Profile drawn with camera lucida of a Golgi-stained human Purkinje cell (Cajal, 1904). (B) Pyramidal neuron in the rat hippocampus, filled intracellularly with HRP; the whole axonal arbor was stained and drawn with camera lucida. (Redrawn from Tamamaki *et al.*, 1984, with permission from Elsevier).

least taken into account (see section "Deformation Artifacts"). Structural deformations (or "morphological noise"; Horcholle-Bossavit *et al.*, 2000) also plague 3D reconstructions of dendritic or local axonal arbors made with the help of imaging software. They are due to histological, optical, and operator-linked distortions, worsened by the need to merge serial sections, which are unlikely to display identical distortions. Considerable differences in many parameters have been reported to occur between laboratories when reconstructing neurons of the same class, or, what is worse, between reconstructions of the same neuron carried out by different operators, or by the same operator working on two different systems (Scorcioni *et al.*, 2004).

Many stereological applications can be carried out entirely without any special equipment, other than a microscope with good immersion and dry optics, and a high-resolution *microcator*, or any alternative gadget destined to measure displacements of the stage in the z-axis (see section "Deformation Artifacts"). It cannot be denied, though, that stereology has already made an irreversible liaison with computer-assisted microscopy, arousing the interest of high-tech research-oriented companies to develop and improve interactive systems (Glaser and Glaser, 2000). This includes stereological tools, which simply could not be implemented without computer help (see section "Connections as Defined by Axonal Length"). These systems focus on providing software control for (1) video acquisition of live and still microscopic images, (2) creating and overlaying test grids on them, (3) moving the microscope stage, (4) collecting data on spreadsheets, and (5) performing some computations online (for local stereological estimators, such as the nucleator or the rotator). Yet, the systems do not "make decisions" on the sampling design, choice of probe(s), and recognition of objects and targets, which are left to the operator. The computer-assisted stereological toolbox (CAST; Visiopharm, Hørsholm, Denmark) and the StereoInvestigator (Microbrigthfield, Williston, VT) systems are probably the most complete and widely used, but several others[1] provide reasonable sets of the most popular stereological tools.

In the past years useful monographs on stereology have appeared that provide theoretical and practical information of great help for users, without delving too deeply into abstruse mathematical proofs (e.g., Evans *et al.*, 2004; Howard and Reed, 2005; Mouton, 2002). It is very recommendable, however, to add specialized practical training to reading. Intensive courses and workshops are available around the world, some of them sponsored by the International Society for Stereology (ISS; http://www.stereologysociety.org), which are very useful to boost beginners' acquaintance with the intricacies of stereology.

IV. ON BIASES AND ARTIFACTS

A. Biases in Counting and Measuring

An unbiased method is an accurate method, that is, a method that yields quantities whose mean coincides with the true value (see section "A Précis on Accuracy, Precision, and Efficiency"). Any systematic error (bias) in the method will likely lead to an estimation bias. But a clear distinction must be made between counting/measuring methods, and their application to real experimental material. The former have to prove only unbiasedness via statistical/geometrical arguments, and therefore their acceptance cannot await certification by "validation" against empirical "gold standards," or by showing "reproducibility" by various laboratories (Cruz-Orive, 1994, 2002a). Hence, possible sources of bias with respect to the method have to be sought in sampling design or mathematical errors, or in a wrongful choice of the geometric probes.

When an unbiased method is applied to anything closer to real life than a model data set, however, new possible sources of estimation bias can emerge.

---

[1] Stereologer (Systems Planning and Analysis, Inc., Alexandria, VA); Explora Nova Stereology (Explora Nova, La Rochelle, France); Digital Stereology (Kinetic Imaging, Liverpool, UK); MCID Basic Stereology System (Imaging Research Inc., St. Catharines, Ontario, Canada); Stereology Toolkit (BIOQUANT Image Analysis Corporation, Nashville, TN).

They are based on the ubiquitous presence of two types of artifacts. *Observation artifacts* are typically found in transmission and fluorescence microscopy on histological sections, but may also be present in electron microscopy (EM) images, as well as in virtual sections obtained by imaging techniques (such as PET, MR, or CT scans). *Deformation artifacts* originate in any dimension-changing deformation inflicted on the tissue by the various manipulations it suffers, from the processes of organ death and fixation to the staining and coverslipping of the sections. The following explanation of these artifacts is largely inspired on Cruz-Orive (1990) and Dorph-Petersen *et al.* (2001).

B. Observation Artifacts

Concerning observation, the general caveat is that objects to be counted or measured have to be unambiguously identified. This means, in first place, that they should be observed with sufficient resolving power, i.e., with a convenient spatial resolution of the imaging system. The use of inadequate magnifications, nonmatching refractive indexes of optical phases between the specimen and the objective, and objectives/condensers with low numerical aperture may fail to provide images above a crisp detection threshold.

However, even if the imaging system is adequate, further problems can derive from the nature of the sections and the staining of the target objects. Figure 16.3 summarizes these problems.

1. Signal Weakness

Lack of information due to incomplete staining in transmission microscopy, or to fading of light emission by a fluorophore. The former, in varying degrees, is a common result when immunolabeling thick sections, but is not restricted to it. Molecular size of the reagents, their solubility and ability to penetrate the tissue, and reaction time and temperature are factors that condition the completeness of the staining. Fading of fluorescence occurs naturally when a fluorophore is subjected to illumination with an appropriate excitation wavelength. An excellent review of the strengths and weaknesses of the use of fluorescent tracers for cell counting may be found in Peterson (2004); see also Negredo *et al.* (2004).

2. Overprojection

Excess of information in an image: when the section has a definite thickness, the objects of interest are opaque, and the surrounding tissue/embedding medium is translucent or transparent.

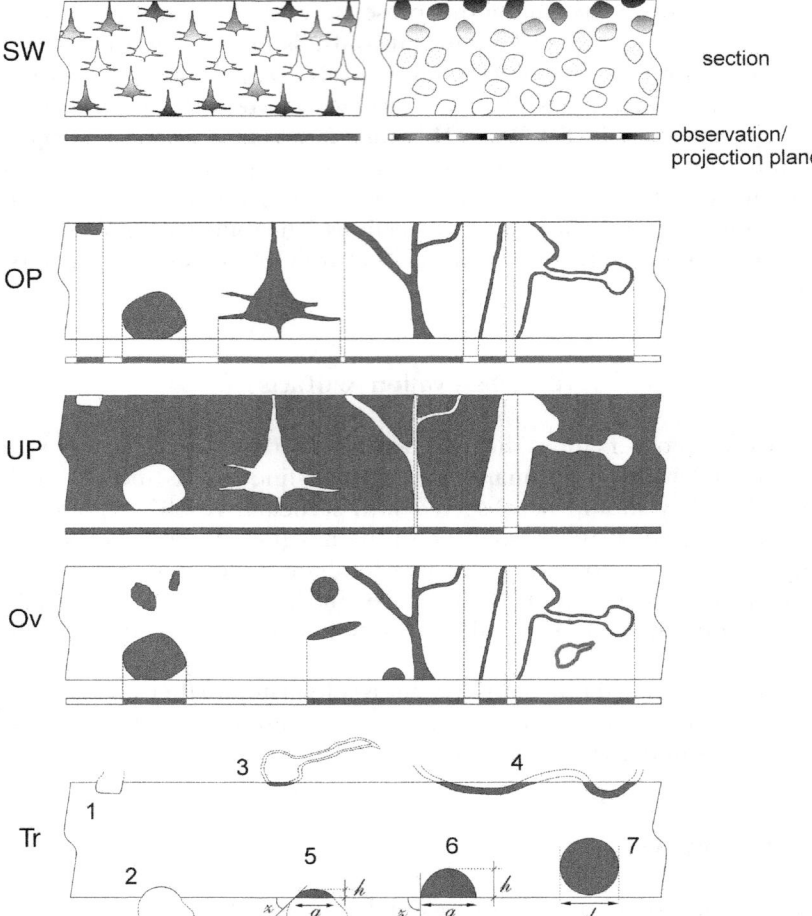

Figure 16.3. Observation artifacts in transmission microscopy on tissue sections. SW, signal weakness produced by poor stain penetration and/or fading in sections stained free floating (left) or after mounting (right). OP, overprojection, or *Holmes* effect. UP, underprojection. Ov, overlapping, Tr, truncation, physical/mechanical (with loss of material, 1–2), and optical/observational (3–5). Membranes or filaments cut tangentially are likely to be overlooked (3), or misidentified (e.g., as belonging to separate objects, 4). Fractions of spheroidal particles may be overlooked if cut at shallow angles (5). Larger particle fractions, on the other hand, may appear as entire particles (6, 7), because of overprojection. The four bottom diagrams are somewhat modified after Cruz-Orive (1990), with permission from the author.

3. Underprojection

Lack of information in an image: when the section has a definite thickness, the objects of interest are transparent, and the surrounding tissue/embedding medium is opaque.

4. Overlapping

Loss of information when objects overlap and mask each other in a section that is substantially thicker than the objects' size.

5. Truncation

Lack of information due to the physical loss of object fragments from the section surface (physical or mechanical truncation), or to the inability to detect object fragments present in the section, but which are cut at shallow angles (optical or observational truncation). Both are typical for particles; the latter is also common for membranes and flattened structures (particularly in ultrastructural studies of subcellular membranes and synapses).

Any of these artifacts, if moderate, can be handled by appropriate embedding, cutting, sampling, and choice of stereological procedures. Severe truncation or overlapping, and other causes of marked signal loss, however, cannot in general be corrected and should be avoided.

C. Deformation Artifacts

Stained tissue sections, the basic material on which measurements are to be made, are the end product of submitting a piece of previously living tissue to perfusion and fixation, dehydration, embedding, cutting, staining, mounting, and coverslipping. Obviously, a large number of interacting artifacts may be introduced by these procedures, and different types of deformation may result (Dorph-Petersen *et al.*, 2001; Fig. 16.4). Confidence that the estimates obtained represent real numbers and values would thus depend on the certainty that the dimensional distortions introduced by histological processing can be controlled.

1. From Fixation to Embedding

As soon as the tissue becomes anoxic, a cascade of autolytic effects sets out that leads to structural changes, first seen at the ultrastructural level. The rapid administration of fixatives is intended to preserve the structure (and the enzymatic reactivity, antigenicity, etc.), but it cannot entirely fulfill the objective. After an early intracellular swelling, the ensuing shrinkage/swelling balance is influenced by the osmolarity and ionic composition of the fixative solution, and its penetration and fixation rate. Moreover, not all tissue components, including membranes and subcellular organelles, display the same degree of deformation for a given osmolarity of the fixative. Hence, even though a good choice of the fixative and the fixation

A

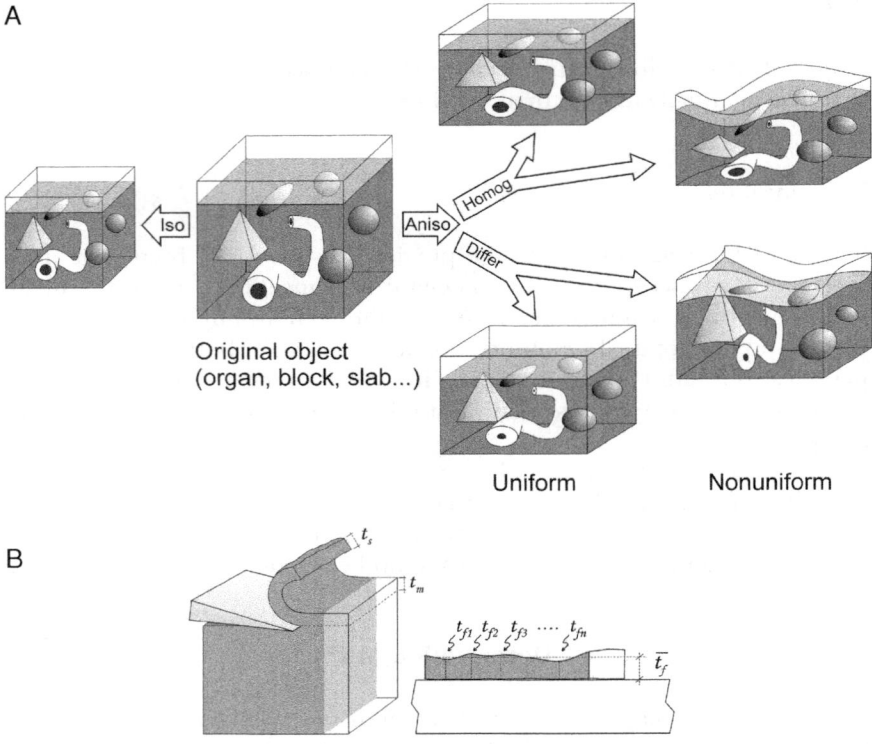

Original object
(organ, block, slab...)

Uniform   Nonuniform

B

Figure 16.4. (A) Different types of deformation of a mock block of brain tissue with two layers and some heterogeneous discrete components in one of them. To the left is shown an isotropic shrinkage (affecting all layers and components equally in 3D). To the right there are four examples of anisotropic shrinkage of the whole block, which affects the z-, but not the x- and y-axes. Shrinkage may affect homogeneously all components of the block in a uniform or nonuniform manner. Or it may be differential, affecting some components and not others. (B) Different meanings of thickness (t), shown on a diagram of the process of cutting and mounting a tissue section. t_m, thickness set at the microtome, or "block advance"; t_s, thickness of the section after cutting and in the collecting solution; t_f, final thickness measured on the completely processed section. This figure is inspired from Figs. 3 and 4 in Dorph-Petersen *et al.* (2001).

procedure may keep the overall volume of a brain or nerve block close to its premortem condition, it does not guarantee that cell nuclei and somata, dendrites, blood vessels, myelin, or extracellular space have not differentially shrunk or swelled. Specific details for different structures are lacking to a large extent, but good reviews are available that should help gauge the problem (Hayat, 1981, 2000).

Additional deformations are introduced by embedding procedures. When embedding entails dehydration and defatting, as in paraffin-, celloidin-, and epoxy resin-embedding, the net result is different degrees of shrinkage. For vibratome sectioning, fresh or fixed, but otherwise non-manipulated tissue is used, and therefore no further changes should be expected at this stage. In contrast, tissue processing for cryosectioning entails some additional shrinkage, because of the partial dehydration the tissue undergoes during cryoprotection in hyperosmolar sucrose or glycerol solutions. Water-soluble embedding media, such as glycol methacrylate, in general, yield moderate shrinkage, which is compensated by swelling at subsequent steps of processing of the sections (Gerrits et al., 1992). Again, whether deformation is homogeneous or differential (see below) is largely untested, but it is likely that the use of plastic, celloidin, or resin media results in a quite homogeneous matrix (Dorph-Petersen et al., 2001).

2. From Sectioning to Coverslipping

The knife advance produces an orthogonal (rotary and sliding microtomes) or zigzagging (vibratome) strain on the tissue block. This inevitably causes three kinds of mechanical effects on the resulting section: compression, creasing, and grazing. These effects can be substantially minimized when the chosen cutting thickness is the most adequate to the embedding medium, the knife's edge is sharp and unblemished, the clearance angle (that between the knife edge bevel and the tissue block) is set at the most effective minimum, and the cutting angle (angle of edge bevel) is shallow enough to minimize compression, but not too much, to avoid vibrating and causing chatter in the section (Ellis, 2000; Hayat, 2000).

Compression may be particularly severe when cutting very thin paraffin sections, so that the "height" of the section (parallel to the knife advance) may become over 50% shorter than the original "height" of the block (Dempster, 1943). Creasing consisting of regular undulations of the cutting surface is often found in vibratome sections. In general, irregularities of the surface are significant in vibratome and frozen sections, and less marked in sections from embedded material. Grazing, if not too coarse, is difficult to detect and is responsible for truncation of object fragments at the surface of the sections.

When sections are not mounted directly on glass slides (cryostat), they are collected in buffer (frozen, vibratome) or ethanol (celloidin), or floated on water (paraffin, resins) before mounting. As long as buffer osmolarity is controlled, no dimensional changes are likely to occur to frozen or vibratome sections. Celloidin sections are also stable in 70% ethanol. Paraffin sections and semi- and ultrathin resin sections, however, stretch on a warm water bath (or droplet), which compensates to a varying degree the compression and creasing underwent during cutting. Methacrylate sections display notable stretching on the water bath and in the mounting step, which may fully compensate the shrinkage suffered during the dehydration step of histoprocessing (Gerrits et al., 1987).

Mounted sections adhere firmly to the supporting slide and no longer change their planar (x–y axes) dimensions. Their height, or thickness (z-axis), however, may decrease over 50% for unembedded, hydrated sections that are allowed to dry on the slide (frozen, vibratome), while it remains essentially unchanged in celloidin- or methacrylate-embedded sections (Avendaño and Dykes, 1996; Bermejo *et al.*, 2003; Dorph-Petersen *et al.*, 2001). Differential vertical compression, resulting in changes in the particle distribution along the z-axis, has been recently reported (Gardella *et al.*, 2003). Overall, deformations along the z-axis are a source of considerable practical difficulties, and deserve specific consideration.

3. The Annoying t

Control of the reference space is a must for any direct stereological estimator of volume, as well as for most ratio estimators (number, length, and surface densities, as well as volume fractions). Since, as seen above, the tissue may undergo dimensional changes in any of the three axes: (1) volume estimations have to be corrected for any deformations taking place between the condition established as a reference baseline (living individual, fixed organ, etc.) and the processing stage at which the Cavalieri estimator of volume is applied (organ slabs, mounted sections, etc.) and (2) the estimation of the reference volume for ratio estimations, such as N_V or L_V, should ideally be made at the same sampling stage at which objects (cell bodies, axons, etc.) are counted with disectors; if V is estimated at an earlier stage, it is necessary to check whether additional deformations occurred between stages. When numbers are estimated using a fractionator sampling design (Gundersen *et al.*, 1988; Howard and Reed, 2005), the size and shape of the reference space are irrelevant. However, deformations along the z-axis (thickness) of the section are still relevant for this estimator, because the latest fraction that enters the equation in the estimator [see Eqs. (16.4) and (16.5) below] includes section thickness values.

When dealing with section thickness, it is important not to mix up concepts:

t_m: thickness set at the microtome. Equivalent to "block advance," it can be calibrated if doubts exist over the precision of the cutting device (Avendaño and Dykes, 1996; Dorph-Petersen *et al.*, 2001; Sterio, 1984).

t_s: thickness of the section in the collecting—or floating—solution. In addition to already showing local unevenness, it may differ from t_m because of compression during cutting, swelling by hydration or heating, etc.

t_f: final thickness measured at any specific location on the completely processed section. It is expected to vary in different positions because of the unevenness of the cutting surfaces.

$\bar{t}_f$: mean final thickness over individual measurements taken at locations properly sampled throughout the region of interest in the section.

The existence of many methods to estimate t_f, particularly for ultrathin sections (De Groot, 1988), indicates that none gathers sufficient simplicity, efficiency, accuracy, and general applicability to be proposed as a standard. For stereological measurements, however, ultrathin (below 100 nm) and semithin sections ($0.1 - 2.5$ µm) are generally used for applying physical disectors. Hence, it is t_m that matters, which is easier to gauge.

There is a compelling need to warrant unbiased and precise estimates of t_f on thick sections, however, given its importance in estimators based on the optical disector and fractionator. The difficulties it entails, and the frequent failure to adequately report on the mode and precision with which it was estimated, have served recently as a basis for some criticisms of stereological applications (Guillery, 2002). It is widely accepted that the most practical way to estimate t_f is by focusing through the section depth with high magnification immersion lenses of high numerical aperture (at least $60\times$ and 1.2 n.a.), and recording the positions of the upper and lower surfaces with a sensitive distance encoder (see Uylings *et al.*, 1986a,b for equations on optical depth due to both optics and accommodation). The optoelectronic microcators are excellent z-position recorders, with a resolution as good as ± 0.25 µm, but they are costly and not readily available to all laboratories. However, ingenious and inexpensive alternatives exist (Howard and Reed, 2005; Korkmaz and Tümkaya, 1997), with nearly as good a precision.

The introduction of observer biases when taking t_f readings is a real risk. The section surfaces may be irregular, fuzzy (in particular the bottom surface), indistinct, or just invisible. This is the case especially when working under dark field or fluorescence microscopy in spots without signal or, worse, where a strong signal arises from just the inner part of the section. All of these problems may be tackled by procedures that increase contrast at the surfaces (using semidark field illumination or differential interference contrast), but the specific application in each case takes learning and shared practice (Bermejo *et al.*, 2003; Dorph-Petersen *et al.*, 2001; Korkmaz and Tümkaya, 1997; Peterson, 2004). In an optical fractionator design, moreover, the ratio of the disector height (h) over $\bar{t}_f$ is the final sampling fraction. This is valid only when the variability of t_f within and between sections is very low, which seldom happens. If many objects are present in a few locations/sections that are thicker, and fewer in many locations/sections that are thinner, the mean fraction will tend to be larger and the object number estimated smaller than if the opposite situation is true. It has then been proposed that the contribution to $\bar{t}_f$ of the t_f measured at each spot, or at least that of the mean value of t_f for each section, be weighted by the relative contribution of objects counted (ΣQ^-) at the corresponding spot/section (Bermejo *et al.*, 2003; Dorph-Petersen *et al.*, 2001). The latter correction is given by

$$\bar{t}_{Q^-} = \frac{\sum_{s_1}^{s_n} \left(\bar{t}_i \sum q_i^- \right)}{\sum_{s_1}^{s_n} Q^-} \tag{16.1}$$

where $\bar{t}_{Q^-}$ is the mean value over the sections of the mean thickness per section, $\bar{t}_i$ weighted by the neuron count per section, q_i^-.

V. CONNECTIONS AS DEFINED BY NEURONAL NUMBERS

In studies of neural connections, neuron numbers usually refer to those cell bodies that are retrogradely labeled by axonal tracers, or are labeled through transneuronal anterograde transport by neurotropic viruses. Indirectly, a measure of connectivity is also given by the total number of projection neurons located in a regional domain whose target is known (e.g., granular cells of the dentate gyrus, parvocellular layers of the lateral geniculate nucleus, lateral cerebellar nucleus, etc.), or local circuit neurons in a defined territory (e.g., inhibitory interneurons in thalamic nuclei, granular cells of the olfactory bulb, bipolar cells of the retina, etc.). These neurons may have been labeled by a variety of histochemical or immunocytochemical procedures. Moreover, when using anterograde tracer injections, it is also desirable to grasp the magnitude of the cell population that took up the tracer by counting strongly labeled neuronal bodies.

A. Tools

1. Equality Guaranteed

Counting cells has to rely on a well-established stereological principle: all cells have to have the same probability of being counted. In order to comply with that principle, it is necessary to fulfill three requirements:

1. Counting in 3D demands that an unbiased 3D sampling probe is "inserted" in the target tissue. This can be accomplished in either of these two ways: by defining a known area in two registered serial sections of practical zero thickness separated a known distance, or by defining a known area at a certain focal plane in a thick section, and "moving" downward the focal plane a known distance along the z-axis. In both cases the 2D area is delineated by the already classical unbiased counting frame (Gundersen, 1977; Sterio, 1984), and the distance between the top and the bottom planes incorporates the third dimension (disector height h) to create a 3D unbiased "box" or "brick." If appropriate disector height and counting rules are respected, these probes are the basis of the *physical* and *optical disector* methods (Howard and Reed, 2005; Fig. 16.5).
2. A single feature of a cell has to be identified as the "counting unit," so that each cell is counted only once, regardless of its size or shape. The theory goes that "cell tops" (i.e., their first appearance as a recognizable target) fit the requirement, and so it is for nuclei or nucleoli

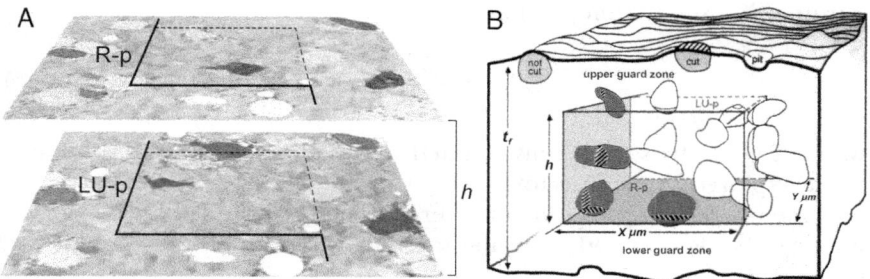

Figure 16.5. (A) The physical disector. An unbiased counting frame is overlaid on matching regions of two thin sections, separated by distance h, showing some stained and unstained cell body profiles. The frame consists of a solid line marking the exclusion territory and a dashed line bounding the acceptance area. A guard area is left around the frames so that all candidate profiles can be seen in their entirety. The upper disector is used as a reference (inclusion) plane (R-p) and the lower as a look-up (exclusion) plane (LU-p) when performing the disector counting from top to bottom; planes are reversed when performing the counting the other way. (B) The optical disector (slightly modified from Williams and Rakic, 1988, with permission from Wiley-Liss, Inc.). A virtual counting "box" is created inside a thick section by scanning a fraction of its thickness (h over t_f) with an unbiased frame, leaving guard zones around it. Darkened particles are excluded from the count for invading the forbidden space. As for the physical disector, the sampling volume is defined as $v(\text{dis}) = xyh$.

in a *physical disector* setting. The use of the cell body surface for this purpose, while possible, is often inconvenient, because of its irregularities and the ambiguous interface between body and dendritic bases. When using an *optical disector* it is better to rely on nucleolar or nuclear "optical equators" (planes where the nuclear membrane or the nucleolar surface is at its sharpest focus), because of the difficulty inherent to define membranes or surfaces cut at nearly tangential angles at the top (or bottom) of the body, nucleus, or nucleolus.

3. Since each disector explores only a tiny part of the target region, it is necessary to ensure that the whole region is surveyed with disectors distributed in a statistically correct manner. Whilst a simple or uniform random (UR) sampling may be used, it is more efficient to follow a systematic random sampling (SRS) regime (Gundersen and Jensen, 1987; Gundersen *et al.*, 1999).

2. Densities Vs. Total Numbers

It may occur that only a fragment of the whole target is accessible to study (brain biopsy, incomplete collection of a nucleus or region), or that the target boundaries are imprecise (some "reticular," transitional, or otherwise inconspicuously bounded cortical or subcortical areas). In such cases, it is possible to estimate only the real neuronal densities (number of cell bodies per unit of tissue volume, N_V) as

$$\hat{N}_V = \frac{\sum Q^-}{\sum v(\text{dis})} \tag{16.2}$$

where $\sum Q^-$ is the sum of cells counted, and $\sum v(\text{dis})$ is the total volume of tissue explored by the disectors.

If the whole target region (or reference space) is available, and its volume can be determined, then the total neuron number may be *indirectly* estimated:

$$\hat{N} = \hat{N}_V \cdot V_{\text{ref}} \tag{16.3}$$

It is evident that these estimates of N_V or N are sensitive to tissue deformations. N_V increases or decreases when net shrinkage or swelling occurs, respectively. And the estimate of N given by Eq. (16.3) may vary if V_{ref} is determined at a different stage of tissue processing. This may occur, for example, when only fragments of a region of interest are available to perform the number estimation, while volume estimates of the whole region were obtained in the living individual by imaging techniques. In this case, it is necessary to introduce a correction that accounts for any differential deformation that may have taken place (see section "Densities and Total Numbers").

When the whole region is accessible, it is possible to estimate directly the total number of cells by the *fractionator*, a procedure that is efficient, robust, and insensitive to tissue deformations (Gundersen, 1986). It can be associated with the optical disector method in what was called *optical fractionator* (West *et al.*, 1991), and has become a widespread tool to count neurons. Recent developments have improved its efficiency (Gundersen, 2002b). The rationale behind this procedure is simple: if a known fraction of the target region is adequately sampled, and the cell count in that fraction is $\sum Q^-$, then

$$\hat{N} = \sum Q^- \cdot f_T^{-1} \tag{16.4}$$

where f_T^{-1} is the inverse of the total, or final, fraction taken, which results from computing the fractions taken at the various steps of sampling. A typical paradigm consists of sampling a fraction of the serial sections, which contain the target territory (f_s), the areal fraction of the structure that is covered by the sampling frames in the sections (f_a), and the linear fraction of section thickness covered by the height of the disectors (f_h):

$$f_T^{-1} = \frac{1}{f_s}\frac{1}{f_a}\frac{1}{f_h} \tag{16.5}$$

The fractionator does not need any absolute measurements (of reference space, stage movement, magnification, etc.), and therefore is unaffected by dimensional changes of the tissue during processing. However, the last fraction f_h requires fair estimates of the final section thickness, and hence it is necessary to keep the annoying t on a tight rein (see section "Deformation Artifact").

B. Practical Tips

As a general recommendation, it is necessary to keep track of all data that are relevant to understand—and possibly replicate—the sampling protocol, and the values entered in the different estimators. Otherwise, errors may easily slip in, and skepticism may understandably grow in confused readers.

The whole target region should be defined unambiguously. If this is not possible (e.g., because of blurred borders or partial unavailability), the best possible sampling with disectors should be used instead of the fractionator.

A good design for the optical disector (or optical fractionator) must be efficient and not excessively labor-intensive. This is not always the case for the physical disector.

Vertical shrinkage may be considerable (up to 65%) when using vibratome or frozen sections. If these sections are obtained at a nominal thickness below 30 μm, it is possible that there is no room left for the optical disector height to be at least 10 μm. Thinner disectors risk being too "noisy," because of the limited focal resolution (never better than 0.5 μm).

The importance of measuring t with accuracy and a maximum precision cannot be sufficiently emphasized.

Staining must be complete, but, if too intense, it may create serious problems of overlapping.

A good selection of the counting units increases accuracy and efficiency. As commented above, cell bodies are not as good options as nucleoli or nuclei.

Fluorescent tracers of cell bodies and neurites are an excellent, and increasingly used option to label neurons and connections with chemical specificity. However, stereological applications to fluorescent material (and other dark-field applications) have particular difficulties, which cannot be ignored.

Confocal microscopy provides excellent focal resolution and noise elimination. However, the technique has its own constraints (apart from the costly equipment), and at present stereological applications in this field are just starting. While the future is certainly promising, conventional fluorescence optics with a good selection of filters together with Nomarski optics could yield nearly as good results (Peterson, 2004).

Stereology has been seldom applied to quantitate the neuron groups that are directly affected by the local deposits of tracer. However, such information could be of great value to assess the degree of divergence and convergence of afferent and efferent connections, evaluate the effective injection site (EIS), or study tracer clearance or fading in the injected area over time (see Appendix).

VI. CONNECTIONS AS DEFINED BY AXON NUMBERS

Two different scenarios pose quite different challenges to the task of counting axons. The easier situation is determining the number of axons in a peripheral nerve (or a central tract) cut more or less orthogonally to its major axis. The second, more complex, scenario is the estimation of the number of axons of a given kind that penetrates or distributes in a target territory.

A. Axon Numbers in 2D

Counting axons of peripheral nerves or central tracts is efficiently accomplished by applying a 2D fractionator sampling scheme to properly stained thin cross sections of the target structure (Mayhew, 1988, 1990). Thick sections could also in principle be used, but the fibers usually display wavy or slanted trajectories, which cause severe problems of overprojection and overlapping. What follows is summarized in Fig. 16.6, and borrows largely from the excellent review by Larsen (1998).

The procedure is particularly simple when only myelinated axons are to be counted. A 1-μm-thick section of the resin-embedded nerve or tract is stained with toluidine blue and coverslipped. The whole cross-sectional area containing fibers is sampled at high magnification, using a 100× oil-immersion lens. An unbiased counting frame of a known area a(frame) is systematically advanced in a meander path at preestablished dx and dy steps from a random starting point. If ΣQ is the number of myelinated axon cross sections included in all the frames, then the total number of axons in the nerve is estimated using the fractionator equation [Eq. (16.4)] $\hat{N} = \sum Q \cdot f_T^{-1}$, where f_T is now the only fraction performed, namely the ratio a(frame)$/dxdy$.

When unmyelinated axons are also a target, it is necessary to work on electron micrographs at a final magnification of at least 8000×. If the whole nerve or tract section fits the supporting metal grid, a complete sampling may be performed and the total N may be estimated by the fractionator equation as above (Fig. 16.6B). If the nerve is too large, however, it is possible that only a part of it is accessible. An alternative then exists, which is to estimate the density of axon profiles per area,

$$\hat{N}_A = \frac{\sum Q}{\sum a(\text{frame})} \tag{16.6}$$

from which the total axon number may be estimated, if the whole cross-sectional area of the nerve is available from semithin sections (and correcting for distortions in the ultrathin sections).

Most of these stereological analyses can be carried out capturing still video images (from semithin sections), or digitizing EM negatives, and merging them with customized test systems using any commercially available graphic

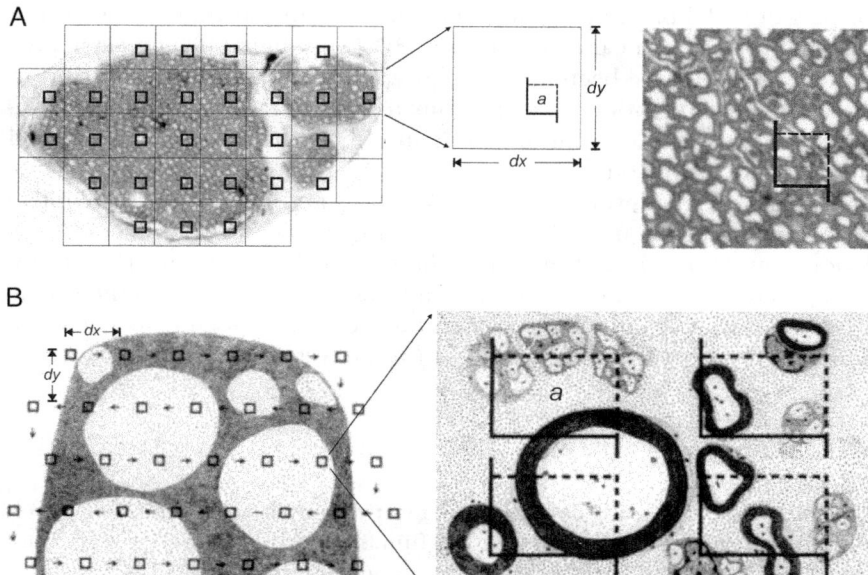

Figure 16.6. Counting axonal profiles in cross sections of a small nerve or fascicle by the 2D fractionator. (A) A virtual grid is overlaid on a toluidine blue-stained 1-μm-thick cross section of a distal branch of a cat median nerve. The grid may be created by a "meander" X-Y displacement of the microscope stage at $dxdy$ steps. Within the upper left grid square a counting frame is randomly located; the rest are systematically placed at $dxdy$ steps. At highest optical magnification practically all profiles of myelinated axons can be identified. (B) Unmyelinated axons can only be counted on EM images (slightly modified from Larsen, 1998, with permission from Elsevier). Sampling includes an additional intermediate step to locate small fields of vision where the EM pictures will be made. Each micrograph is examined with four sampling frames, to increase efficiency and reduce the impact of axon clusters on sample variance.

software. However, the efficiency improves greatly by using a stereological package (see section "Morphometry and Stereology") to control the X–Y stage displacements and generate the frames and other test systems. Sampling on the EM can also be expedited by using digital capture of the images instead of conventional photographic film.

B. Preterminal and Terminal Axonal Fields

A quantitative description of the innervation of a territory would seem awfully incomplete without estimating the number of incoming axons. However, this parameter has been largely neglected, probably because of the technical difficulties involved. A quantitative approach to axonal arbors has been practically restricted to metric and topological analyses of the whole or part of Golgi or intracellularly stained axons of individual cells, reconstructed in 2D (with camera lucida) or 3D (with digital imaging systems). Similar analyses have been more widely applied to quantitate dendritic arbors. Very recently, however, semiautomated vectorization for axonal reconstructions presents a promising alternative to axonal topology (see Ascoli and Scorcioni, this volume).

Stereology is a convenient, and often necessary, alternative to complete reconstructions, when dealing not only with axonal arbors from cell populations, but also with very complex arbors from single neurons (Fig. 16.2). Two parameters may be of particular interest: the number of somehow labeled preterminal and terminal axons in a defined target territory, and the number of varicosities (terminal or *en passant*) displayed by those axons.

1. Axons

Counting axons leads us away from metric analysis and takes us to topology. The number of axon segments is a function of the number of branching points (or *nodes*) and is independent of axonal length, thickness, or tortuosity. Most nodes correspond to axonal bifurcations, whereby three segments meet in the node (nodes with valence = 3); more rarely they mark trifurcations (valence = 4). Axon branches are not supposed to form closed loops, but rather give off collaterals in several intermediate steps until making terminal segments. Their analysis is then simpler than when fibrous or tubular structures arise and end in an interconnected meshwork. Such is the case of the microvessels in a brain region, which have been counted by a procedure that applies the disector method to a topologically defined microvessel unit (Løkkegaard *et al.*, 2001).

A simplified version of that procedure, suited to count axons in thick sections, would be given by the estimator

$$\hat{W}_V = \frac{\sum P_n(n_{\mathrm{val}} - 1)}{\sum v(\mathrm{dis})} \tag{16.7}$$

where W_V is the density of axon segments per unit volume of the target region, P_n is the number of nodes counted in the volume sampled with the disectors $[v(\mathrm{dis})]$, and n_{val} is the valence of each node counted. If tri- or multifurcations are very rare, the numerator may be simplified to $\sum P_n$. If the total volume of the target region is available or can be estimated (e.g., by the Cavalieri method), then the total number of axons in that region may be computed as

$$\hat{W} = \hat{W}_V \cdot V_{\mathrm{ref}} \tag{16.8}$$

When applying this procedure, some caveats are in order

1. Neither root segments, nor axon branches that divide no further within the disector space (Fig. 16.7) are included in the count. The

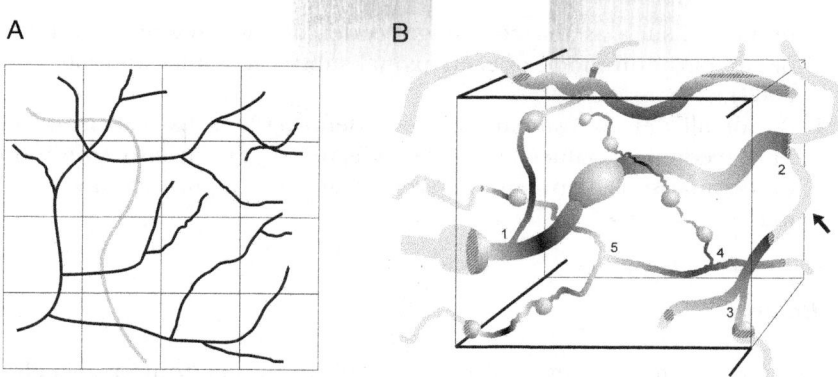

Figure 16.7. Diagrams to illustrate the stereological procedure to count axons. (A) In a simplified, 2D setting, two axonal profiles are shown. One is unbranched (gray), and the other (black) is highly ramified. The former has no nodes, and therefore is not counted at all. The second one shows 12 nodes, 10 bifurcations, and 2 trifurcations. If Eq. (16.7), with a simple adaptation to 2D, is applied, the estimation of axonal segments yields 26, which is the real number of segments, except for the root segment on the lower left corner. (B) A network of varicose axons of various types is sampled by an optical disector box. Axon portions inside the box are shown in dark shades; those outside are in light gray, and transects with the box walls are hatched. Five nodes (bifurcations; 1–5) are present within the inclusion volume of the sampling box, yielding a count of 10 segments [Eq. (16.7)]. In fact, there are 13 segments, but 3 of them correspond to (relative) "root" segments entering the box. Undercounting of root segments is only apparent: statistically they have been counted at their origin in another node, as shown by the fiber marked by an arrow. An additional axon winding on the upper part of the box does not have any node and therefore is not counted.

exclusion of root segments introduces a meaningless bias, since it is local branches what we look for. The omission of undivided branches is unbiasedly compensated by appropriate sampling within the target territory. As is usual in many stereological settings, about 200 events (nodes) counted in the same number of, or somewhat fewer, disectors systematically distributed should yield a satisfactory precision in the estimate.

2. Nodes have to be identified and defined unambiguously. This requires in first place that the staining has to be *reasonably* complete, that is, continuous along the full axon branches, or, if granular or discontinuous, it should leave no doubt as to the continuity of the stained segment fragments (Fig. 16.7B). Very thin axons can, in any case, fall below the optical resolution (about 0.3 μm). Moreover, the nodes have a volume, and therefore should be defined by their smallest possible single feature, such as the cusp of the smallest angle formed by the meeting segments.

3. Although the estimator is based on a topological feature of axons, it does not provide many interesting topological measures of axon

branches, such as the centrifugal order of each segment or node, or the partition mode of each axonal subtree (Uylings and Van Pelt, 2002).

4. As for all density estimators, tissue deformations have to be taken into account, if values are to be referred to the in vivo condition, or comparisons between subjects or groups of subjects are to be made.

2. Boutons

Terminal axon segments end in a swelling (terminal bouton), which is filled with synaptic vesicles, and display a synaptic membrane specialization. Similar swellings (varicosities, boutons *en passant*) often occur along the axon stem and collaterals, and may bear synaptic membrane specializations, depending on the cell of origin and the neurochemical content of the axon. Boutons vary in size from the barely visible smallest cholinergic varicosities in the cerebral cortex (0.3 μm in diameter; Chédotal *et al.*, 1994) to the several-micrometer-wide swellings in neurosecretory fibers in the hypothalamus, or the mossy fiber terminals in the hippocampus.

The density of axonal swellings in a target territory is a relevant measure of the local synaptic density. The total population of synaptic boutons can only be quantitated using EM, which gives enough resolution to identify all boutons, but introduces a number of difficulties (see section "Connections as Defined by Synapse Number"). When specific axonal populations are labeled, however, it is possible to estimate total numbers or densities by light microscopy, using the optical disector or the optical fractionator methods [see Eqs. (16.2)–(16.5) in section "Tools"]. Although still seldom applied, the stereological estimation of bouton numbers may provide a valuable, and anatomically precise, indirect estimation of synaptic contacts in a defined region (Calhoun *et al.*, 1996). Still, it is unjustified to presume that a one-to-one correlation exists between boutons and actual synaptic contacts (Mayhew, 1979b; see section "What is a Synapse").

VII. CONNECTIONS AS DEFINED BY AXONAL LENGTH

The length of a linear feature per unit volume of a reference space (length density, L_V) may be easily estimated by the classical stereological estimator

$$\hat{L}_V = 2 \cdot Q_A \tag{16.9}$$

where Q_A is the number of random intersections of the feature per unit area of a 2D probe (a bounded plane, virtual or physical) introduced in the space (Howard and Reed, 2005).

Despite its well-proven theoretical simplicity, the practical implementation of the estimator poses several difficulties. In first place, the encounters of the feature with the plane must fulfill not only translational (or positional) randomness, but also directional randomness, or isotropy. Secondly, real fibers (axons, dendrites, etc.) may be regarded as linear features of nonzero thickness, which are usually studied in sections of nonzero thickness. As we will see below, problems posed by length estimation differ when section thickness is ≪ than fiber thickness, as in transmission EM studies, and when it is ≥, typically for fibers in thick sections, and also applicable to very thin axons or dendrites in semithin sections.

A. Navigating the Isotropy Reefs

Axon branches may take any direction. However, in any given territory, not all directions are present with the same probability. That axons distribute anisotropically is a highly probable and prudent generalization. To guarantee randomness, therefore, it is necessary to proceed along either of two lines: (1) to generate sampling probes that are themselves isotropic, and to distribute them within the target space in a random pattern and (2) to randomize the tissue, by rotating tissue fragments along one or more axes. If, in order to increase efficiency, sampling is made systematic (or uniform), then the requisites for an isotropic uniform random (IUR) test system and sample, respectively, are fulfilled. These procedures, which will be summarized below, are clearly explained in recent stereological reviews (Calhoun and Mouton, 2001; Howard and Reed, 2005).

1. Isotropy in 2D and 3D

Isotropy in 2D consists in guaranteeing that all directions in the plane (represented by intersections of a line on a circumference) are equally probable. An IUR design in 2D using a set of parallel test lines spaced a distance d is easily achieved by randomly choosing an orientation angle ω such that $0° \leq \omega \leq 80°$ and positioning the set randomly on the target object. When multiple objects are measured, several orientations ($z = 3 - 4$ is usually appropriate) may be systematically applied: An orientation angle ω is randomly selected in the interval $0-(180/z)°$; the next orientations are set at $\omega + 180/z)°$, $\omega + 2(180/z)°$, etc. Alternatively, a prime number such as 37 may be chosen as the interval; the first orientation is randomly selected and then in the range $0-37°$. This approach has the advantage that, in practice, all orientations used are different.

Isotropy in 3D requires ensuring that all orientations in the 3D space (represented by intersections on the surface of a sphere of straight lines, or axes, rotating about its center) are equally likely to happen. It depends on the randomness of two angles: the longitude (the azimuthal coordinate ϕ, in the range $0-360°$, or $0-2\pi$), and the latitude or its complementary, the co-latitude (the polar coordinate θ, in the range $0-\pi/2$). While ϕ may be chosen with a standard random number table, θ can not. The reason is that, on the sphere's surface, spherical quadrilaterals of the same angular dimensions, $d\phi \times d\theta$, have larger areas the closer they are to the equator. Thence, if axes are directed to the angles of all quadrilaterals having the same angular dimensions, intersecting spots on the surface will crowd toward the poles. To avoid this bias, it is necessary to weigh the co-latitude with its sine; thus, the probability of a line (orientation) hitting any of the mentioned quadrilaterals is (Cruz-Orive, 2002b)

$$P = \frac{1}{2\pi} \cdot d\phi \cdot d\theta \cdot \sin\theta \qquad (16.10)$$

which means that the probability is 0 in an exact North Pole direction (when $\theta = 0$), it increases toward the equator, and then decreases to become again 0 in the South Pole direction.

2. Randomizing Tissue

There are two well-known procedures to generate IUR sections from a tissue block. The *orientator* (Mattfeldt *et al.*, 1990) is applicable when the block is large enough to permit that fragments or slabs of the block are subjected to successive stages of sectioning to ensure randomness in ϕ and θ. when the target object is divided into small fragments, they can be placed in separate spherical molds and embedded in resin. Once hardened, each resin ball is allowed to roll free and is then reembedded in a standard rectangular mold. This is the original *isector* method (Nyengaard and Gundersen, 1992), which more recently was adapted to preembed several tissue samples in spheres of agar, which in turn are embedded in methacrylate (Løkkegaard *et al.*, 2001).

In an IUR design, sections have to guarantee randomness in the three axes. It is possible, however, to ensure full isotropy if all sections are parallel to a fixed, arbitrary "vertical" axis. This vertical uniform random (VUR) design, developed by Baddeley *et al.* (1986; see also Howard and Reed, 2005), is of particular interest for research fields that, like neuroanatomy, rely heavily on serial sections. In this design all sections are made perpendicular to the "horizontal" plane and parallel to a randomly chosen orientation in that plane. It is important to realize that VUR sections are not isotropic, because they lack isotropy along the vertical axis (VA). As will be further commented below, the missing isotropy is provided by the *cycloids*, a special type of test lines whose length contains a continuous set of orientations that are proportional to $\sin\theta$.

Randomizing tissue samples, particularly in the IUR designs, has the inconvenience of having to break apart the original structure. This results in a marked loss of the anatomical references that are of great help in studies of neural connections. An exception may be found in the study of brain biopsies of cortical or subcortical territories, when overall data (e.g., the length of all axons in the white matter; Tang and Nyengaard, 1997) are sought. The problem also exists in VUR settings, because for length (but not number) estimations, serial sections cannot be made at an anatomically convenient, preestablished orientation. Rather, the plane of sectioning of each specimen has to be oriented randomly in a 0–360° interval to avoid bias. This has been a strong drive behind new developments of fully isotropic probes that can be used in any type of sections.

3. A Panoply of Isotropic Probes

Depending on the type of tissue and sections, density of target axons, and randomness design chosen, five different methods can be applied. As in the case of estimating number or volume, it is better to count dimensionless events (such as "hits" or intersections) than actually measure in 1D, 2D, or 3D. To ensure that such events are really dimensionless, the dimensions of the probe and the target feature must sum 3 (Howard and Reed, 2005; West, 1993). Therefore, all the probes used to estimate length are 2D (flat or curved bounded planes; Fig. 16.8). While some of these tests may be implemented manually on photomicrographs or live microscopic images, all benefit greatly from (and some actually require) the use of stereological software.

1. *Thin IUR sections*: an unbiased frame is applied in a UR pattern to the sections, using very high optical (or low EM) magnification (Fig. 16.8A). The length density of the axons is estimated as

$$\hat{L}_V = 2 \cdot \frac{\sum Q(ax)}{\sum a(\text{frame})} \qquad (16.11)$$

where $\sum Q(ax)$ is the total number of axon profiles counted in the sampling frames, and $\sum a(\text{frame})$ is the total area covered by the frames hitting the target region. The total axon length may be estimated if the volume of the target region is known. This method has been applied to estimate length density and total length of myelinated fibers in the white matter of human brains (Marner and Pakkenberg, 2003; Tang and Nyengaard, 1997).

2. *Thick VUR sections*: a cycloid, with its longer axis parallel to the VA of the section, is projected through the section depth. All intersections of target axons with the cycloid are recorded. For efficiency, an array of cycloids with associated points is recommended (Fig. 16.8B). Since axons are likely to span a number of sections, all sections must have the same thickness. The length density is estimated as (Gokhale, 1990; see also Batra *et al.*, 1995 and Stocks *et al.*, 1996)

$$\hat{L}_V = 2 \frac{\sum I_c}{\sum P_c l_c \bar{t}_s} \qquad (16.12)$$

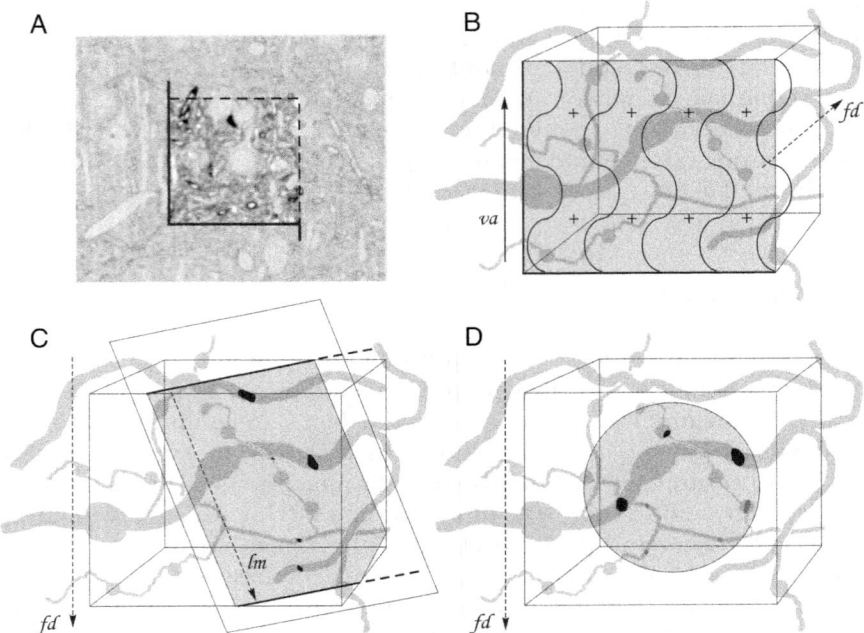

Figure 16.8. Stereological test systems for estimating axonal length. (A) Myelinated axons in systematically sampled IUR semithin sections of the rat cerebral cortex are counted with unbiased frames. Gray levels in the exclusion zones and an ample guard area have been dimmed. (B) An array of cycloids with points associated (one for four cycloids) is used to scan a "box" of tissue within a vertical thick section that contains a network of axons. The major axes of the cycloids are parallel to the vertical axis (va), which has to be identified in all sections. (C) Sampling of a thick section using a single isotropic virtual plane. The plane is computer generated at a random (isotropic in 3D) orientation. The line marking the intersection of the plane with the upper surface of the sampling box "moves" along the plane (lm), as the focal plane changes along the z-axis. Five intersections (black profiles) of axons with the virtual plane are counted. (D) Sampling a thick section with a virtual isotropic sphere. The sphere may be computer generated, or manually constructed as a series of concentric circumferences representing transects of the sphere surface at different focal depths inside the sampling box. Seven axon intersections are produced on the surface of the sphere. The focusing direction (fd) is indicated in (B–D).

where the numerator represents the total number of intersections of the cycloids with the projected axons, and the denominator is the total surface area of the cycloid used: ΣP_c is the number of cycloids (from the count of the associated points), l_c is the individual cycloid length, and t_s is the cycloid depth, represented by the mean section thickness.

3. *Total vertical projections* (TVP; Cruz-Orive and Howard, 1991): it is possible to estimate the total length of the branches when the whole arbor is present in a thick section by projecting it on a plane. The projection must be repeated a few times after UR rotations of the arbor. The total

length of the branches is a function of the number of intersections of the projected profiles with a grid of cycloids. Although applications to axonal arbors are lacking, this system was used to estimate the total dendritic length of Golgi-stained substantia nigra neurons (Howard *et al.*, 1992). Computer-based 3D reconstructions of axons or dendrites in confocal microscopy may be good candidates to apply TVP to estimate length in selected brain regions.

4. *Virtual isotropic planes*: parallel virtual planes at random orientations are generated by software and projected, under high magnification, inside thick sections obtained at any convenient arbitrary orientation. As the image is focused along the z-axis of the physical section, a "moving" line is visualized sliding along the virtual plane (Fig. 16.8C). In order to explore the same area with each random orientation, only the intersections of several equidistant parallel virtual planes within a "sampling box" of constant volume are used (Larsen *et al.*, 1998a). An unbiased estimator of length density is then

$$\hat{L}_V = 2 \cdot \frac{\sum I_{vp}}{\sum a(vp)} \qquad (16.13)$$

where $\sum I_{vp}$ is the total number of axonal transects with the virtual planes, and $\sum a(vp)$ is the sum of the sampling plane areas. The length of regenerating axons in spinal cord lesions has been recently estimated using this procedure (Larsen *et al.*, 2004). Virtual planes and the above estimator are implemented in the CAST software.

5. *Virtual isotropic sphere*: a translucent, computer-generated virtual sphere (or hemisphere) is UR placed across the target region in thick sections under high magnification. The sphere is contained within an unbiased box whose dimensions are greater than or equal to the sphere diameter (Fig. 16.8D). Since the surface of the sphere is an isotropic curved plane that contains all possible orientations, the tissue sections may have been cut at any arbitrary orientation. Length density is estimated as

$$\hat{L}_V = 2 \cdot (v/a) \cdot \frac{\sum I_{ss}}{\sum v(dis)} \qquad (16.14)$$

where (v/a) is the ratio of the sampling box volume to the sphere's surface area, $\sum I_{ss}$ is the total number of axonal transects with the sphere's surface, and $\sum v(dis)$ is the total volume of disectors used. As for the virtual plane estimator, the total length may be estimated here if the volume of the reference space is known, or a fractionator scheme is applied. Length density and total length of cholinergic fibers in the neocortex and the hippocampus have been estimated using this procedure (Calhoun and Mouton, 2001; Calhoun *et al.*, 2004). Although virtual spheres can be generated manually (Mouton *et al.*, 2002), the task is greatly facilitated by stereological software. Virtual spheres (or "space balls") have been recently implemented in the StereoInvestigator software.

B. A Caveat: Axons are Irregular Tubes, Not Lines

If isotropy conditions are properly met, simple linear features can be unbiasedly measured by any of the procedures described in the previous sections. When the features have a cross-sectional area, as for tubes, their axis (or *spine*) is hidden by the tube walls, and tube intersections with a plane become 2D profiles. This generates a number of problems, whose practical importance differs depending on whether the tubes have a stable thickness or present irregularities (swellings and constrictions), the degree of branching of the tubes, the frequency of free endings, and the type of section and probes used (Gundersen, 2002a; Larsen *et al.*, 2004).

1. When the arbor is located in a section of thickness much greater than the average tube (axon) diameter and is projected on a plane (see TVP design, previous section), in principle the number of intersections of the tube (axon) walls with the test lines is a close match of (twice) the number of intersections with the invisible axis (Fig. 16.9A). However, severe problems of overprojection and overlap may occur, leading to uncontrollable underestimations. Moreover, the thicker the section, the more difficult to combine full focal depth with enough magnification, both conditions being necessary to identify all profiles.
2. If, on the contrary, section thickness is less than or equal to the axon diameter, it is not possible to derive estimates of the real axon length from

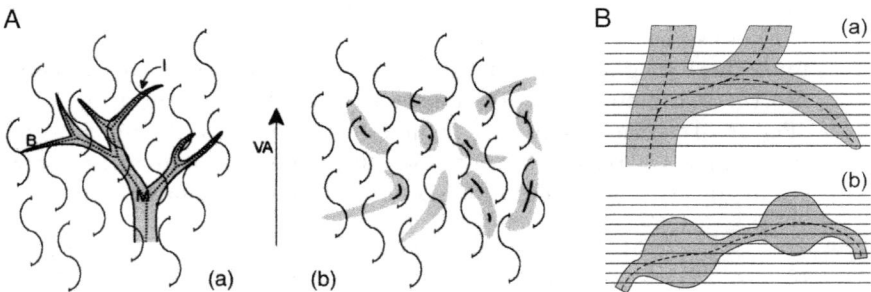

Figure 16.9. Problems in measuring length of tubular structures. (A) A tubular structure seen under TVP in a thick section (a) or in a vertical section of thickness less than or equal to the mean tubular diameter. In (a) the total length (*M*) is a function of the number of intersections (*I*) of the cycloids with the projected tube boundaries (B). In (b), however, there does not exist a simple relation between the tube profiles and their imaginary (invisible) axes, and therefore length estimations based on the intersections with the cycloids are biased (reproduced from Gundersen, 2002a, with permission from Blackwell Publ. Ltd.). (B) Thin sections of tubular structures yield a number of 2D profiles, which may unbiasedly represent intersections with the tube axis. For smooth tubes, the axial bending and curves are not a problem, because the number of random transects with the axis matches the number of whole profiles (a). Branching leads to underestimates, and terminal free endings to (negligible) overestimates (a). Nonsmooth tubes produce more severe degrees of overestimation (b). This is particularly the case for markedly beaded axons or dendrites. (Modified from Gundersen, 2002a and Mouton *et al.*, 2002).

the intersections observed between axon profiles with test lines. The reasons are multiple (Fig. 16.9B): curved trajectories are not a problem, but branching leads to underestimation, free ends to a slight overestimation, and irregularities in the fiber thickness lead to overestimation.

3. Scanning through thick sections at high magnification and with the finest possible depth of focus, using optical disectors (or confocal microscopy), eliminates several problems inherent to the TVP design. Its combination with appropriate test systems is, so far, the most reasonable approach to estimate length.

In sum, there is no simple solution to measure the length of tubular structures, such as axons, in a completely unbiased, efficient, and precise manner. The problem is still open to new theoretical and practical developments (Gundersen, 2002a; Mouton *et al.*, 2002).

VIII. CONNECTIONS AS DEFINED BY SYNAPSE NUMBER

The number of synapses that a neuron, or a system of neurons, makes in a target territory is undoubtedly a fundamental measure of neural connectivity. Since the late 1960s, transmission EM has been used to assess synapse numbers and density. But researchers soon realized that synapses were geometrically complex structures, and that inferences about numbers did not smoothly derive from linear and planar measures taken on EM images.

The simplest scenario would feature synapses as disk-shaped, flat or slightly convex, unambiguously bounded structures of similar (and not too small) size. In such a case, accurate number estimates could be given by classical formulas, such as the widely used $N_V = N_A / \bar{d}$, where $\bar{d}$ is the mean trace length of the synaptic profiles (Colonnier and Beaulieu, 1985; DeFelipe *et al.*, 1999). Yet, overprojection and truncation, which may occur even in this ideal situation (and be significantly greater in others), have led to the introduction of a variety of correction factors, with mixed results as to the improvement of the estimates (Calverley *et al.*, 1988; Colonnier and Beaulieu, 1985; Mayhew, 1979b; Zaborszky *et al.*, 1975).

Certainly, accepting assumption-based counting methods, which are hardly extendable to all conditions of the feature counted, does not help to meet the challenge posed by the practical problems inherent to synapse morphometry. In fact, the realization of these problems cannot but highlight the importance of using unbiased stereological methods, without ignoring the difficulties in their implementation.

A. What Is a Synapse?

Defining a synapse as a morphological entity that can be used as an unambiguous counting unit entails difficulties, which have been recognized some time ago (Mayhew, 1979a). For the chemical synapses, different components have been used for this purpose, each with specific advantages and limitations.

1. The Presynaptic Boutons Themselves

Vesicle-filled swellings along or at the end of axonal branches generally qualify as presynaptic entities, but, are they? Boutons identified by immunostaining of molecules associated with synaptic vesicles have been used as counting units with light microscopy (Calhoun *et al.*, 1996; Silver and Stryker, 1999). Yet, many vesicle-containing varicosities lack synaptic membrane specializations, particularly in cholinergic and bioaminergic axonal systems (Chédotal *et al.*, 1994; Descarries *et al.*, 1991). While representing other kind of (nonjunctional) synapse, they cannot be equated to classical (or junctional) synapses. Some would set stricter criteria for the latter, requiring that a few vesicles contact the axolemma. In addition, a significant number of boutons display paramembranous densities apposed to more than one postsynaptic element (Jones *et al.*, 1997; Sorra and Harris, 1993). Finally, synaptic boutons may fail to show presynaptic densities along a series of ultrathin sections, and this effect is size dependent.

2. The Synaptic Cleft

In principle, this space could be considered the anatomical kernel of a classical synapse, both symmetrical and asymmetrical. It presents, however, a major practical disadvantage because of its small transverse diameter (20–40 nm) and relative electron lucency. Since average ultrathin sections are at least 60 nm thick, and the flanking membranes are electron dense, the synaptic cleft is subject to severe problems of underprojection and truncation.

3. The Pre- and Postsynaptic Membrane Specializations

As pointed out by Mayhew (1979a), there was an early shift of emphasis toward accepting the whole apposition complex, or *synaptic plate* (the paramembranous densities plus the synaptic cleft), as the best unit for counting classical synapses. The membrane densities can be stained by the standard osmium-uranyl-lead procedure or the specific ethanolic phosphotungstic acid method; with either method they become electron dense. Moreover, the transverse thickness of the complex (or synaptic height) is larger than the average section thickness. Yet, recognition problems that arise from the size and morphology of the complex may significantly bias the estimates.

B. Unbiased Counting

Given the dimensions of the synapses, the physical disector (section "Connections as Defined by Neuronal Numbers") is the only stereological method capable of yielding unbiased estimates of synapse number, using the synaptic plate as the counting unit. Even so, counting synapses is no free ride, and faces a number of theoretical and practical problems, which have been recently discussed in detail by Tang *et al.* (2001).

1. Strict and educated criteria have to be applied to recognize synaptic profiles in 2D, so that they can be defined independently in each section of a disector pair (Coggeshall, 1999). The difficulties involved are substantial, as shown by comparing the results obtained with the disector and with a serial analysis of many sequential sections (stack analysis; De Groot and Bierman, 1983; Tang *et al.*, 2001). Any part of the synaptic plate included in the sampling area of the disector frame should qualify for the count. Observational truncation (section "Observation Artifacts") is a common problem, when tangential sections of the synaptic plate periphery offer limited information. If very rigorous criteria to recognize a synapse as such are set, truncation may account for nearly 25% underestimation (Tang *et al.*, 2001).

2. A few serial sections (3–7) are collected in a formvar- or pioloform-covered single-slot EM grid. Section thickness has to be measured on each section of every disector pair (t_1, t_2). Small's minimal fold method (De Groot, 1988; Small, 1968) is recommended for its relative easiness and not needing additional equipment. It is customary to use adjacent sections for a disector pair, but alternate sections may be used as well, on the condition that the thickness of the intermediate section (t_{int}) is also known. It is highly improbable that any synapse would fit entirely (and thus invalidate the procedure) in the missing section. Disector height h(dis) would equal $\bar{t} = (t_1 + t_2)/2$ if adjacent sections are used in one direction, and ($t_1 + t_2$) if (as recommended for efficiency) counting is done both ways. When alternate sections are used, and both ways are used to count, h(dis) $= t_1 + t_2 + 2t_{int}$.

3. Apart from substantial size differences, synapses may adopt complex shapes. Varying sizes or curvatures should not affect disector counting, unless so pronounced that a single synapse would appear as two or more disjoint profiles in two or more adjacent sections. But complexity usually refers to irregularities in the 2D geometry of the plate produced by simple and complex fenestrations in the paramembranous densities. Complete fenestration results in a *segmented* or double/multiple synapse. If marked changes in curvature are added, W-shaped and other bizarre 3D configurations can be obtained (Fig. 16.10). In order to avoid the overcounting bias due to mistaking a horseshoe-shaped or incompletely perforated for more than one synapse, Tang *et al.* (2001) incorporated a topological rule: if two disjoint plate profiles in one

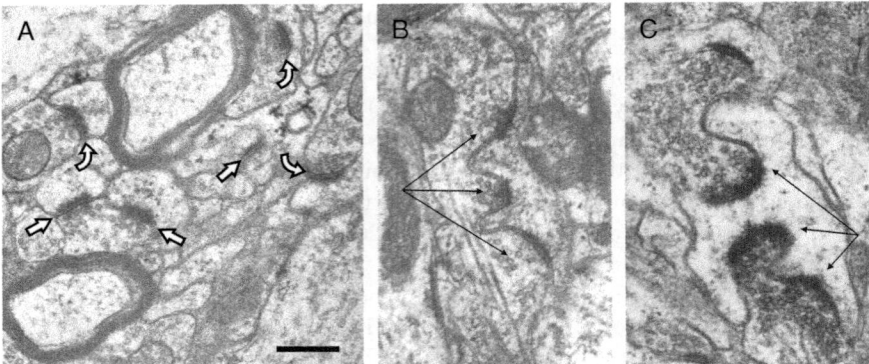

Figure 16.10. The shape of synaptic contacts in the same region may change significantly under different natural or experimental conditions. (A) shows a common field of the molecular layer of the dentate gyrus in a control rat. Most synaptic profiles correspond to flat (straight arrows) or convex (curved arrows) simple axospinous synapses. In (B) and (C) there appear two highly complex spines with fenestrated synaptic plates (thin arrows) from the same region, several weeks after partial denervation by transection of the perforant fibers. Calibration bar: 0.5 μm.

section were matched by a single plate in the paired section, it was a proof that the latter *bridged* the former; all were part of the same synapse, and therefore no synapse should be counted.

4. A new procedure has been proposed to avoid the limitations inherent to the physical disector. It consists in estimating N_V or N_L on unbiased 3D *bricks* projected within computer-assisted volume reconstructions of a test region, made from large stacks of serial ultrathin sections (Fiala, 2005; Fiala and Harris, 2001).

C. Of Pincushions and Haystacks: Sampling Designing

As a general rule, uniform (or systematic) random sampling is recommended for counting objects of any kind in stereology, and this includes synapses. However, the vast dimensional "distance" that exists between synapses and the reference structure (a whole brain, the striatum, a thalamic nucleus, etc.) makes a proper sampling an unusually demanding task. And it is more so when we want to sample, say, less than a trillionth ($<10^{-12}$) of the target region. This was the case in Tang *et al.*'s study of synapse number in human brain neocortex, but huge disproportions between sample size and reference space are likely to occur in most studies of regions larger than a very small brain nucleus. True, sample size would be of little relevance when the distribution of target objects is rather uniform. But since this is not usually the case for synapses, the problem remains, being a cause of concern (DeFelipe *et al.*, 1999). The following comments may help to ease that concern:

1. Using the cerebral cortex as an example, there is a much higher synapse density in the "thin" neuropil than in the remaining tissue, where cell bodies, blood vessels, large dendritic shafts, and myelinated axons concentrate (DeFelipe *et al.*, 1999). Yet, if the target objects are asymmetric synapses in general, some will be present in practically all sampled fields at a convenient magnification (30,000–40,000×), except in those fully hitting blood vessels or neuron bodies. Moreover, blood vessels and neuron bodies only account for a minimum of 3% in layer I and a maximum of 17% in layer IV of the rat parietal cortex (Machín-Cedrés, 2004); therefore, their impact on the efficiency of the procedure is small.

2. It is certainly harder to find pins in a haystack than in a pincushion; in fact, if there is only one pin and the haystack is bulky, any sampling design will very likely return 0 pins found. This does not invalidate an unbiased counting procedure; it will simply orient toward a more directed search, that is, to increase efficiency by sensibly narrowing the defined reference space. If data for specific types of synapse are sought, then specific practical solutions have to be applied. Good examples could be the symmetrical synapses around the initial segment of pyramidal cell axons, the axosomatic contacts, the synaptic contacts on a defined subpopulation of neurons, etc. These cases may show a very low overall density of events, with clustering in specific regions. In such cases, it is advisable to define—and count or measure—new convenient reference spaces, such as the number, volume, and/or membrane surface of initial segments, neuronal bodies, or labeled cells in a given region. They would be used as subindices for density estimations, and sampling would be directed only to the targeted space.

3. Two widespread misconceptions emerge from the true observation that the goal of many quantitative studies is just to make comparisons (Guillery and Herrup, 1997). One is that absolute numbers are of little interest. While difficult to disprove, it may be argued that the interest will rise exponentially when accurate data become available; moreover, real rather than artificial data greatly benefit biological simulations and modeling. The second is the assumption that biased or inaccurate procedures may not be too harmful, as long as the same are applied to the experimental conditions that are being compared. This is wrong, and potentially very harmful, because it does not take into account that the reference space or even the tissue deformation due to fixation may vary between experimental (or natural, such as aging) conditions (Haug *et al.*, 1984), or that the size and shape of the objects/synapses may vary substantially in the same region under different conditions (Fig. 16.10).

D. Densities and Total Numbers

While it is possible in principle to apply a fractionator sampling scheme to synapse counting, in practice it would be extremely inefficient to section exhaustively the (proportionately very large) sampled pieces, and keep track of the fraction used at every step (however, see Nyengaard *et al.*, 2005). Hence, it is more convenient to indirectly estimate the total number, from the estimation of the numerical density of synapses, and the volume of the reference space (section "Connections as Defined by Neuronal Numbers"; Geinisman *et al.*, 1996; Tang *et al.*, 2001). Since the latter has to be estimated independent of the former, it is important to control the dimensional changes that occur between the processing stages.

Volume shrinkage between the fixation step and the ultrathin sections may be determined by areal measurements in two steps:

1. From vibratome slices to resin-embedded blocks and semithin sections: the area of the target region, or a well-defined part of it, is estimated by point counting on the surface of the vibratome sections that will be used for further processing. Adjacent sections, mounted and stained, can be used instead, when no further X–Y dimensional changes are expected. Because shrinkage is not necessarily isotropic in 3D, it is recommended that vibratome sections are already isotropic (Tang *et al.*, 2001). When this is not possible (e.g., when the whole target region is serially cut in the vibratome), two tissue pieces from nearby regions are cut at two orthogonal angles, and a section from each is processed in parallel to the remaining sections. When the section or tissue block is embedded in resin, the area of the target region (or defined part) is again measured by point counting directly on the embedded piece or on the first semithin section taken from it. If doubts remain about the existence of additional areal shrinkage in the cut section, this may be easily estimated by comparing the total surface area of the section with the cut surface of the block.

2. From the semithin section to the ultrathin section: the area of an ultrathin section is estimated by point counting on a low-power EM printout of the section at a known magnification. Frequent calibration with a grating replica is recommended.

This procedure should be repeated for a few blocks or vibratome sections, especially when the tissue is markedly heterogeneous in terms of white matter, gray matter, and blood vessels content. The final *areal* shrinkage would be estimated as

$$\text{Shr}(a) = \frac{\sum A_{\text{block}}}{\sum A_{\text{vib}}} \frac{\sum A_{\text{ut}}}{\sum A_{\text{st}}} \qquad (16.15)$$

where areas, represented by point counts, refer to those obtained on vibratome sections (vib), the tissue block in the resin (block), and the semithin (st) and ultrathin (ut) sections. If heterogeneous samples of similar size are used, the estimation of $\mathrm{Shr}(a)$ must be averaged over them. If blocks differ in volume, the contribution of each block to estimate shrinkage must be weighted accordingly.

The final *volume* shrinkage, as a percentage, may then be estimated as (slightly modified from Tang *et al.*, 2001)

$$\mathrm{Shr}(\%\mathrm{vol}) = 100\left[\left(\sqrt{1 + \mathrm{Shr}(a)}\right)^3 - 1\right] \tag{16.16}$$

IX. A PRÉCIS ON ACCURACY, PRECISION, AND EFFICIENCY

An estimator of a parameter is *accurate* when the means of the estimates obtained concentrate about the true value, i.e., when its mean error (bias) is 0 for any fixed sample size. When the sampling and counting protocols are unbiased, estimates are accurate. If they are biased, the estimates will diverge consistently (systematic error) from the true value, regardless of the sample size. *Precision* is a measure of the degree of concentration of the individual estimates about their own mean. It will be high when the objects counted or measured are homogeneously distributed, and when sampling intensity is adequate. Since a very imprecise estimate, though unbiased, is of little value, the precision of an estimator must be determined, at least approximately. Finally, the *efficiency* of a method is a measure of the precision it offers per unit cost, computing as cost the animals used and the time (and money) spent in all procedures involved, from tissue processing to data collection and analysis. A practical glossary on these issues may be found in Cruz-Orive (2004).

The variability observed among individual estimates is a function of the natural, or biological, intersubject variance, increased in the imprecision, or "noise" inherent to the sampling/counting method (*error variance*). Variance can be expressed in relative terms as the squared coefficient of variation $[\mathrm{CV}^2 = (\mathrm{SD}/\mathrm{mean})^2]$ of the variable R studied (R = number, length, etc.), and a reasonable approximation to the error variance is given by the squared coefficient of error of the estimates of R for each subject (CE^2), averaged over all subjects. This may be expressed as (Gundersen, 1986; Howard and Reed, 2005)

$$\mathrm{OCV}^2(\hat{R}) = \mathrm{CV}^2(R) + \mathrm{mean}[CE^2(\hat{R})] \tag{16.17}$$

The CE is a good measure of the precision of individual estimates, and the evaluation of its contribution to the total, observed variance should help improve the sampling design. Since sampling in stereology is usually done at multiple levels (subjects, blocks, or sections containing the target region, areas sampled by sampling frames, disector volumes, and counting points), it may be interesting to identify which of these levels contributes more to the final variance. This can help design better strategies to increase precision, keeping efficiency high. However, from practical and theoretical reasons, it is generally accepted that the critical level is that just below the level that contributes most to the total variance. Certainly, we expect that the observed variance represents as faithfully as possible the natural or intersubject variance. Therefore, sampling should be increased at the block/section level until the error variance attributed to the sampling and counting procedure is reduced to levels considered acceptable. A popular recommendation is that the contribution of the error variance (expressed as CE^2) does not exceed 50% of the OCV^2 (Gundersen, 1986). This "rule" is, however, rather arbitrary and not backed by any theory. It is possible, on the other hand, to estimate at the outset a "tolerable" mean CE^2 due to the stereological method from Eq. (16.17), by fixing the error levels α and β and the true relative difference of means between groups we want to detect, and making a guess about the number of subjects that may be reasonably used to detect that difference, as well as of the expected biological variance of the target quantity among subjects (see Cruz-Orive, 2004 for a complete explanation).

In measuring error variance, however, there is another catch. This error cannot be determined exactly under systematic sampling—the one commonly used in stereology—as it is done for independent sampling ($CE = SEM/mean$). The values of geometrical properties (areal size, number of object profiles present) obtained on adjacent sections collected in a systematic series are not independent, and their variance is a function of two independent factors, the variance of the points or intersections counted for each section, and the variance due to the systematic sampling of sections, which depends on how—not on how much—the individual counts vary from one section to the next. The estimation of CE becomes even trickier when using ratio estimators, such as N_V or L_V, and the derived two-stage estimators, like $\hat{N} = \hat{N}_V \cdot \hat{V}_{ref}$. In such cases the overall CE is proportional to the sum of the individual CEs, decreased by a factor proportional to the covariance between them (Howard and Reed, 2005). On the other hand, there are statistical limitations that forbid such computations, under certain sampling conditions (Pakkenberg and Gundersen, 1997).

An update on the methods to estimate the error variance in different stereological designs may be found in recent papers by Gundersen, Cruz-Orive, and their colleagues (Cruz-Orive, 1999, 2004; Cruz-Orive *et al.*, 2004; García-Fiñana and Cruz-Orive, 2000; Geinisman *et al.*, 1996; Gundersen *et al.*, 1999). It must be remarked that making a good sampling design greatly benefits from analyzing variance and evaluating precision in a pilot study. This is an onerous, but inescapable part of the stereological practice for greenhorn quantitative neuroanatomists—another good reason for acquiring direct experience in some of the practical courses mentioned at the beginning of the chapter.

X. ADVANTAGES, LIMITATIONS, AND PERSPECTIVES

Shortly after the introduction of the disector (Sterio, 1984), the neural tissue became a cherished target for the application of the "new" stereological tools (Braendgaard and Gundersen, 1986; Pakkenberg and Gundersen, 1988). In less than two decades, stereology has become firmly established as a required set of methods in neuroscience. Its solid theoretical foundations opened up a new period of high expectations concerning the use of unbiased counting and measuring methods. In some notable cases, these expectations were formulated as requirements (Saper, 1996). However, and more importantly, "neurostereology" is smoothly spreading mainly because growing numbers of neuroscientists are convinced of its value. Whatever reluctant attitudes may remain, they are likely to fade with the dissemination of and the delving into the advantages and limitations of the methods proposed (Cruz-Orive, 2002a).

Before proceeding, it is convenient to stress that counting and measuring represent an added value in neuroanatomy, but can never replace observation and description. As a particular advice for students, the qualitative analysis of the tissue is not only necessary to evaluate the results of the staining method used, but is invaluable to design and refine whatever quantitative method is to be applied.

Regional volume measurements and cell counting were, and probably will be for a long time, the first objectives of "neurostereology." The field of application is boundless, from basic numerical descriptions of neural structures in different species, to development and aging, neuropathology, and a variety of neural changes secondary to physiological, pharmacological, or traumatic manipulations. Cell counting, however, was carried out by a variety of nonstereological methods since the late 19th century. Understandably, it is around cell counting where concern has grown among researchers and journal editors alike over which are the "best" methods, and how they should be reported and discussed in a publication. If properly applied by trained operators, the stereological methods fulfill the first requirement for any "good" method: accuracy, regardless of the type of counting objects. But the conditional above points to the importance of a good design, a good execution, and a good control of artifacts. Also for neural quantification, the devil is in the details.

Length and surface measurements of brain structures were essentially neglected before the advent of the new, design-based, stereological methods. While barely starting today, their study will likely expand considerably in the next few years.

New developments are expected in the implementation of stereological applications to confocal microscopy (Howell et al., 2002; Kubínova et al., 2004), neuroimaging in experimental and clinical settings (Roberts et al., 2000), and, in general, all computer-assisted procedures directed to grab, digitize, and manipulate images (see Bjaalie and Leergaard and Nadasdy et al., this volume). It is advisable, however, to not be enticed by the automatic analysis trap. At present, any quantification of neural tissue needs a considerable amount of direct personal interaction with images. In fact, computer assistance in stereology boosts efficiency mainly by quickly bringing to the operator's attention those and only those spots on which he or she must make decisions.

The multiple spatial relationships among various cell or subcellular populations are currently the focus of a variety of computational approaches (Nadasdy and Zaborszky, 2001; Zaborszky *et al.*, 2002). Second-order stereology, whose main objective is the quantitative analysis of the spatial arrangements of objects (Cruz-Orive, 1989), deserves therefore a separate mention. Neural components are not scrambled into a chaotic meshwork. Rather, they display exceedingly complex and plastic spatial configurations, which reveal genetically conditioned developmental events, and give support to brain functions. Certainly, the possibility to tackle in quantitative terms some of the architectonic intricacies of the brain is appealing. Second-order methods, however, are theoretically complex and have not been fully developed for 3D distributions. Not surprisingly, they have not been applied yet to the nervous system. The interested reader may enjoy getting updated on this promising field by consulting Mayhew (1999), Mattfeldt and Stoyan (2000), and Krasnoperov and Stoyan (2004).

The stereological estimators of geometrical parameters are fairly well developed, and their unbiasedness is guaranteed when sampling is correct. However, the estimators proposed to predict variance and error are only approximations, good enough under some conditions, but still under development in others (Cruz-Orive, 1999). Further refinements are foreseeable. Finally, the very existence of powerful and unbiased methods to count and measure is renewing the interest in investigating and controlling the artifacts derived from histological processing and observation. The possibility of making direct and accurate quantitative inferences from processed tissue to in vivo conditions will then no longer be a bridge too far.

APPENDIX: A CASE STUDY

A single deposit of a nonfluorescent retrograde tracer (e.g., HRP or biotinylated dextran amine) has been made in a thalamic nucleus (TN) of a rat. The main quantity of interest is the total number of retrogradely labeled neurons in the neocortex (Cx). As additional measures of convergence, the volume of and the neuron number in the region receiving the tracer will also be estimated.

A. Design and Equipment

A single or separate brain blocks containing the entire extent of TN and Cx are serially cut at 50 μm in the freezing microtome. Alternate sections are processed for, at least, revealing the tracer and Nissl staining. Eventually, additional series are collected for other staining purposes. This, however, is contingent upon the spatial extension of the labeled targets in TN and Cx along the cutting direction. A minimum of seven sections (and not more than 10–12) containing the target(s) is required.

The material is well suited to apply the Cavalieri estimator (to estimate the volume of tissue affected by the injection) and the optical fractionator (to estimate total neuronal number). The former may be applied without any special equipment: images of the injection site may be projected (on paper, with a table projector, or a camera lucida; or on a video/computer screen, with a video camera coupled to the microscope) or photographed. The test system consists of a quadratic point grid drawn either on paper or on a transparent film, depending on whether the image is to be projected on the grid, or the grid is to be laid over the image.

The optical fractionator requires, at the very least, a high numerical aperture, high-power ($60\times$–$100\times$) immersion lens, and a high-resolution microcator (see section "Deformation Artifacts"). Manual control of the X–Y stage displacements can also be greatly helped by digital encoders (e.g., MD2 or MD3 Microscope Digitizers, AccuStage, Shoreview, MN) at a moderate cost. It is very recommendable, however, to use an integral stereological system (see section "Morphometry and Stereology"; Footnote 1). The test system consists of an unbiased sampling frame (Fig. 16.5).

B. Procedure

A quick glance of the material shows that the injection site spans less than 1 mm along the direction of the block advance, while labeled neurons are found along an 8 mm long stretch of Cx. Therefore, sampling intensities should vary for TN and Cx: every second section will be used for volume and number estimation in TN, and every 20th will be used for number estimation in Cx.

1. Total Number of Labeled Neurons in Cx

The cell nucleus is chosen as the counting unit. The first section that will be used for counting is randomly chosen among those reacted for the tracer within the first 1000 μm of the Cx block. Subsequent sections will be picked orderly at 1000 μm intervals. A pilot evaluation of the area sampling intensity may then be performed on a section containing roughly an intermediate number of labeled neurons, compared to those sections holding the largest numbers and the fewest.

a. Pilot Study

At low power (using a $2.5\times$ or $4\times$ dry lens), the Cx area that includes labeled cells is quickly outlined, taking care not to leave outside any labeled cell, but sparing obviously unlabeled regions. A fraction of that area is then sampled by systematically displacing the stage at a fixed X, Y distance, using a 60× or 100× oil-immersion lens. A sampling frame that leaves sufficient guard space around it (at least equal to the largest nuclear diameter) is laid over the center of every field sampled. Expecting at least 55% vertical shrinkage, and therefore a final mean section thickness of around 22 μm, the disector height is set at 14 μm, so that 3 μm is left for the upper and the rest for the lower guard areas. All labeled neurons whose nucleus' equatorial plane comes into focus within the permitted frame boundaries are counted, and readings of section thickness are taken at least every second field. Once finished, we have to focus on the following items, in order to tailor adequately the design for the full study:

1. Ideally, $\Sigma Q^-_{(pilot)} \approx 25$ so that $\Sigma Q^-_{(full)} \approx 200$ (given that about eight sections will be sampled). If notably fewer or more neurons are counted, sampling intensity should be either increased or decreased, by modifying the X–Y sampling intervals or the size of the sampling frame, or both.
2. A certain variability of tissue thickness is expected both within and between sections. If moderate (namely, a relative variability CV < 15%), the frequency of thickness reading may be reduced (say, to one every 4–5 fields), and the contribution of the mean thickness of each section to the final mean thickness is weighted according to Eq. (16.1) (section "The Annoying t"). Large variations in thickness very likely point to nonoptimal quality of the sections, making it mandatory to measure thickness at every sampled field and weigh individually each measurement according to Dorph-Petersen *et al.* (2001).
3. For practical purposes, it is recommendable that the average number of cell counts per sampling field be in the range 0.5–5. Lower mean densities are unavoidable, however, when few neurons are labeled, or when the distribution of the labeled population is very uneven. Higher densities should be avoided by decreasing the frame size.

b. Full Study

Once the final sampling design is established, it is applied to all systematically sampled sections. The most complete stereological systems incorporate all data into datasheets, from which all needed computations are easily performed. If no such systems are available, data have to be fed manually to a custom-made spreadsheet. An example of an ad hoc organized datasheet is shown in Table 16.2. Procedures to estimate precision by computing the CE of the estimates are also included in the table.

2. Volume of TN Affected by the Injected Tracer

The determination of the "EIS," that is, the territory from which the tracer has been taken up by neurons, which eventually are recognized as

TABLE 16.2. Example of datasheet organization with itemized data

Case no.	Section no.	Labeled neurons in Cx		Volume of injection site in TN	Neuron number in injected spot in TN	
		$\Sigma Q^-_{(\text{section})}$	Mean t_{section}	$\Sigma P_{i(\text{section})}$	$\Sigma Q^-_{(\text{section})}$	Mean t_{section}
	1					
	2					
	3					
	..					
	..					
	n					
		$\Sigma Q^-_{(\text{Cx})} =$	Mean t_{Q^-}	$\Sigma V =$	$\Sigma Q^-_{(\text{TN})} =$	Mean t_{Q^-}
		$A =$		$A =$	$A = 4$	
		$B =$		$B =$	$B =$	
		$C =$		$C =$	$C =$	
		$\text{Nug}_{(Q^-)} =$		$\text{Nug}_{(V)} =$	$\text{Nug}_{(Q^-)} ==$	
		$\text{Var}(N)_{\text{SRS}} \approx$		$\text{Var}(V)_{\text{SRS}} \approx$	$\text{Var}(N)_{\text{SRS}} \approx$	
		$\text{CE}_{(Q^-)} \approx$		$\text{CE}_{(V)} \approx$	$\text{CE}_{(Q^-)} \approx$	
		$\text{est} N =$		$\text{est} V =$	$\text{est} N =$	

For number estimates,

$$A = \Sigma Q^-_i \, \Sigma Q^-_i \quad (i = 1, \ldots, n)$$
$$B = \Sigma Q^-_i \, \Sigma Q^-_{i+1}$$
$$C = \Sigma Q^-_i \, \Sigma Q^-_{i+2}$$

For volume estimates, substitute point counts (ΣP_i) for ΣQ^-_i

$$\text{Nug}_{(Q^-)} = \Sigma Q^-$$
$$\text{Nug}(V) = 0.0724 (ba^{-1/2})(n\Sigma P_i)^{-1/2}$$

where b is the mean perimeter length of injected spot profiles and a is the mean cross-sectional area of injected spot.

[For approximately spheroidal injections, the "shape" factor $(ba^{-1/2})$ is around 4.5.]

$$\text{Var}_{\text{SRS}} \approx \alpha [3(A - \text{Nug}) + C - 4B]$$

The value of α ranges between $1/240$ and $1/12$, depending on whether the values (point counts, cell numbers) are uniform or change smoothly over consecutive sections, or display marked irregularities (cf. García-Fiñana and Cruz-Orive, 2000).

$$\text{CE}(Q^-) \approx [(\text{Nug} + \text{Var SRS})]^{1/2} (\Sigma Q^-)^{-1}$$

Mean t_{Q^-}: compute using Eq. (16.1); see section "The annoying t."

retrogradely labeled, is a complex problem. The EIS varies for different tracers and survival times (Ahlsen, 1981; Conde, 1987; Gerfen and Sawchenko, 1984; Mesulam, 1982; Shook *et al.*, 1984) and may not correspond to the tracer deposit, which is apparent on microscopic examination, or "virtual injection site" (VIS; Mesulam, 1982). It is therefore convenient to indicate the criteria used to define the injection site. A commonly accepted criterion is the presence of clear-cut, uniform deposits of reaction product in the neuropil and cell bodies, but this can be extended to the surrounding territory displaying an overall heavy staining of cell bodies within a more weakly and irregularly labeled neuropil.

The cross-sectional area of the VIS is determined by point counting on consecutive sections (when deposits are small) or a systematic sample of sections. In general, between 5 and 10 sections should suffice to yield a precise volume estimate applying the Cavalieri estimator (Table 16.2).

3. Number of TN Neurons Present in the Injected Area

Small tracer deposits usually allow distinguishing cell bodies within it. If this is the case, their number may be estimated by applying the optical fractionator to the same sections used to determine the injection size. However, regardless of the injection size, several problems may hamper such an approach: (1) heavily stained cell bodies (and neuropil) usually produce noticeable, and often severe, overlapping artifacts in thick sections; (2) heavy overall labeling may also make it hard to distinguish between glial cells and small neurons; and (3) tissue damage often occurs in the injection core, leading to a complete local loss of signal.

An alternative strategy consists in estimating the number of neurons in a mirror "VIS" of the contralateral hemisphere, generated by overlaying mirror images of the cross-sectional profiles of the injection site on roughly symmetrical spots of the contralateral TN. While natural side differences, or asymmetries due to oblique sectioning may result in a tolerable degree of inaccuracy, notable tissue deformations due to local hemorrhage or tracer toxicity may, however, rule out this strategy.

Once the entire VIS is defined, the total neuron number enclosed in it is estimated by the optical fractionator method, following a similar procedure to that shown above (section "Total Number of Labeled Neurons in Cx"; Table 16.2).

ACKNOWLEDGMENTS. I thank Prof. Luis Cruz-Orive for reading the final draft of this chapter and contributing valuable suggestions. This work was supported in part by Project IST-2001-34892 (ROSANA) from the E.U.

REFERENCES

Adams, D. C., Slice, D. E., and Rohlf, F. J., 2004, Geometric morphometrics: ten years of progress following the 'revolution', *Ital. J. Zool.* **71**:5–16.

Ahlsen, G., 1981, Retrograde labelling of retinogeniculate neurones in the cat by HRP uptake from the diffuse injection zone, *Brain Res.* **223**:374–380.

Andersen, B. B., and Pakkenberg, B., 2003, Stereological quantitation in cerebella from people with schizophrenia, *Br. J. Psychiatry* **182**:354–361.

Ascoli, G. A., 2002, Neuroanatomical algorithms for dendritic modelling, *Network* **13**:247–260.

Avendaño, C., and Dykes, R. W., 1996, Quantitative analysis of anatomical changes in the cuneate nucleus following forelimb denervation: a stereological morphometric study in adult cats, *J. Comp. Neurol.* **370:**491–500.

Avendaño, C., Machin, R., Bermejo, P. E., and Logares, A., 2005, Neuron numbers in the sensory trigeminal nuclei of the rat: A GABA- and glycine-immunocytochemical and stereological analysis, *J. Comp. Neurol.* **493:**538–553.

Baddeley, A. J., Gundersen, H. J. G., and Cruz-Orive, L. M., 1986, Estimation of surface area from vertical sections, *J. Microsc.* **142:**259–276.

Batra, S., König, M. F., and Cruz-Orive, L. M., 1995, Unbiased estimation of capillary length from vertical slices, *J. Microsc.* **178:**152–159.

Bermejo, P. E., Jiménez, C. E., Torres, C. V., and Avendaño, C., 2003, Quantitative stereological evaluation of the gracile and cuneate nuclei and their projection neurons in the rat, *J. Comp. Neurol.* **463:**419–433.

Blasco, B., Avendaño, C., and Cavada, C., 1999, A stereological analysis of the lateral geniculate nucleus in adult *Macaca nemestrina* monkeys, *Vis. Neurosci.* **16:**933–941.

Braendgaard, H., and Gundersen, H. J. G., 1986, The impact of recent stereological advances on quantitative studies of the nervous system, *J. Neurosci. Methods* **18:**39–78.

Braitenberg, V., and Schüz, A., 1998, *Cortex: Statistics and Geometry of Neuronal Connectivity*, Berlin: Springer-Verlag.

Broser, P. J., Schulte, R., Lang, S., Roth, A., Helmchen, F., Waters, J., Sakmann, B., and Wittum, G., 2004, Nonlinear anisotropic diffusion filtering of three-dimensional image data from two-photon microscopy, *J. Biomed. Opt.* **9:**1253–1264.

Cajal, S. R. 1904, *Textura del Sistema Nervioso del Hombre y los Vertebrados*, Madrid: N. Moya.

Calhoun, M. E., Jucker, M., Martin, L. J., Thinakaran, G., Price, D. L., and Mouton, P. R., 1996, Comparative evaluation of synaptophysin-based methods for quantification of synapses, *J. Neurocytol.* **25:**821–828.

Calhoun, M. E., Mao, Y., Roberts, J. A., and Rapp, P. R., 2004, Reduction in hippocampal cholinergic innervation is unrelated to recognition memory impairment in aged rhesus monkeys, *J. Comp. Neurol.* **475:**238–246.

Calhoun, M. E., and Mouton, P. R., 2001, Length measurement: new developments in neurostereology and 3D imagery: erratum, *J. Chem. Neuroanat.* **21:**257–265.

Calverley, R. K., Bedi, K. S., and Jones, D. G., 1988, Estimation of the numerical density of synapses in rat neocortex. Comparison of the "disector" with an "unfolding" method, *J. Neurosci. Methods* **23:**195–205.

Chédotal, A., Umbriaco, D., Descarries, L., Hartman, B. K., and Hamel, E., 1994, Light and electron microscopic immunocytochemical analysis of the neurovascular relationships of choline acetyltransferase and vasoactive intestinal polypeptide nerve terminals in the rat cerebral cortex, *J. Comp. Neurol.* **343:**57–71.

Chklovskii, D. B., Schikorski, T., and Stevens, C. F., 2002, Wiring optimization in cortical circuits, *Neuron* **34:**341–347.

Coggeshall, R. E., 1999, Assaying structural changes after nerve damage, an essay on quantitative morphology, *Pain* (Suppl. 6)**:**S21–S25.

Colonnier, M., and Beaulieu, C., 1985, An empirical assessment of stereological formulae applied to the counting of synaptic disks in the cerebral cortex, *J. Comp. Neurol.* **231:**175–179.

Conde, F., 1987, Further studies on the use of the fluorescent tracers fast blue and diamidino yellow: effective uptake area and cellular storage sites, *J. Neurosci. Methods* **21:**31–43.

Cruz-Orive, L. M., 1989, Second-order stereology: estimation of second moment volume measures, *Acta Stereol.* **8:**641–646.

Cruz-Orive, L. M., 1990, Observation artifacts in stereology, *Notes for the Stereology Course of the ISS at GB-Martin Mere* (Unpublished).

Cruz-Orive, L. M., 1994, Toward a more objective biology, *Neurobiol. Aging* **15:**377–378.

Cruz-Orive, L. M., 1999, Precision of Cavalieri sections and slices with local errors, *J. Microsc.* **193:**182–198.

Cruz-Orive, L. M., 2002a, *Land mensuration without pontification: a tale of Ancient Greece*, Unpublished Work.

Cruz-Orive, L. M., 2002b, Stereology: meeting point of integral geometry, probability, and statistics. In memory of Professor Luis A. Santaló (1911–2001), *Math. Notae* **41**:49–98.

Cruz-Orive, L. M., 2004, Precision of the fractionator from Cavalieri designs, *J. Microsc.* **213**:205–211.

Cruz-Orive, L. M., and Howard, C. V., 1991, Estimating the length of a bounded curve in three dimensions using total vertical projections, *J. Microsc.* **163**:101–113.

Cruz-Orive, L. M., Insausti, A. M., Insausti, R., and Crespo, D., 2004, A case study from neuroscience involving stereology and multivariate analysis, In: Evans, S. M., Janson, A. M., and Nyengaard, J. R. (eds.), *Quantitative Methods in Neuroscience*, Oxford, UK: Oxford University Press.

DeFelipe, J., Marco, P., Busturia, I., and Merchán-Pérez, A., 1999, Estimation of the number of synapses in the cerebral cortex: methodological considerations, *Cereb. Cortex* **9**:722–732.

De Groot, D. M. G., 1988, Comparison of methods for the estimation of the thickness of ultrathin tissue sections, *J. Microsc.* **151**:23–42.

De Groot, D. M., and Bierman, E. P., 1983, The complex-shaped "perforated" synapse, a problem in quantitative stereology of the brain, *J. Microsc.* **131**:355–360.

Dempster, W. T., 1943, Paraffin compression due to the rotary microtome, *Stain Technol.* **18**:13–26.

Descarries, L., Séguéla, P., and Watkins, K. C., 1991, Nonjunctional relationships of monoamine axon terminals in the cerebral cortex of adult rat, In: Fuxe, K., and Agnati, L. F. (eds.), *Volume Transmission in the Brain: Novel Mechanisms for Neural Transmission*, New York: Raven Press, pp. 53–62.

Donovan, S. L., Mamounas, L. A., Andrews, A. M., Blue, M. E., and McCasland, J. S., 2002, GAP-43 is critical for normal development of the serotonergic innervation in forebrain, *J. Neurosci.* **22**:3543–3552.

Dorph-Petersen, K. A., Nyengaard, J. R., and Gundersen, H. J. G., 2001, Tissue shrinkage and unbiased stereological estimation of particle number and size, *J. Microsc.* **204**:232–246.

Ellis, R. C., 2000, The microtome: function and design, http://home.primus.com.au/royellis/microt/microt.htm, 1–13 (+15 Figures and 2 Tables).

Evans, S. M., Janson, A. M., and Nyengaard, J. R., 2004, *Quantitative Methods in Neuroscience*, Oxford, UK: Oxford University Press.

Fiala, J. C., 2005, Reconstruct: a free editor for serial section microscopy, *J. Microsc.* **218**: 52–61.

Fiala, J. C., and Harris, K. M., 2001, Extending unbiased stereology of brain ultrastructure to three-dimensional volumes, *J. Am. Med. Inf. Assoc.* **8**:1–16.

García-Fiñana, M., and Cruz-Orive, L. M., 2000, Fractional trend of the variance in Cavalieri sampling, *Image Anal. Stereol.* **19**:71–79.

Gardella, D., Hatton, W. J., Rind, H. B., Glenn, G. D., and Von Bartheld, C. S., 2003, Differential tissue shrinkage and compression in the z-axis: implications for optical disector counting in vibratome-, plastic- and cryosections, *J. Neurosci. Methods* **124**:45–59.

Geinisman, Y., Gundersen, H. J. G., Van der Zee, E., and West, M. J., 1996, Unbiased stereological estimation of the total number of synapses in a brain region, *J. Neurocytol.* **25**:805–819.

Gerfen, C. R., and Sawchenko, P. E., 1984, An anterograde neuroanatomical tracing method that shows the detailed morphology of neurons, their axons and terminals: immunohistochemical localization of an axonally transported plant lectin, *Phaseolus vulgaris* leucoagglutinin (PHA-L), *Brain Res.* **290**:219–238.

Gerrits, P. O., Horobin, R. W., and Stokroos, I., 1992, The effects of glycol methacrylate as a dehydrating agent on the dimensional changes of liver tissue, *J. Microsc.* **165**:273–280.

Gerrits, P. O., van Leeuwen, M. B. M., Boon, M. E., and Kok, L. P., 1987, Floating on a water bath and mounting glycol methacrylate and hydroxypropyl methacrylate sections influence final dimensions, *J. Microsc.* **145**:107–113.

Glaser, J. R., and Glaser, E. M., 2000, Stereology, morphometry, and mapping: the whole is greater than the sum of its parts, *J. Chem. Neuroanat.* **20**:115–126.

Glatzer, N. R., Hasney, C. P., Bhaskaran, M. D., and Smith, B. N., 2003, Synaptic and morphologic properties in vitro of premotor rat nucleus tractus solitarius neurons labeled transneuronally from the stomach, *J. Comp. Neurol.* **464:**525–539.

Gokhale, A. M., 1990, Unbiased estimation of curve length in 3D using vertical slices, *J. Microsc.* **159:**133–141.

Goldstein, S. S., and Rall, W., 1974, Changes of action potential shape and velocity for changing core conductor geometry, *Biophys. J.* **14:**731–757.

Guillery, R. W., 2002, On counting and counting errors, *J. Comp. Neurol.* **447:**1–7.

Guillery, R. W., and Herrup, K., 1997, Quantification without pontification: choosing a method for counting objects in sectioned tissues, *J. Comp. Neurol.* **386:**2–7.

Gundersen, H. J. G., 1977, Notes on the estimation of the numerical density of arbitrary profiles. The edge effect, *J. Microsc.* **111:**219–222.

Gundersen, H. J., 1986, Stereology of arbitrary particles. A review of unbiased number and size estimators and the presentation of some new ones, in memory of William R. Thompson, *J. Microsc.* **143:**3–45.

Gundersen, H. J., 2002a, Stereological estimation of tubular length, *J. Microsc.* **207:**155–160.

Gundersen, H. J. G., 2002b, The smooth fractionator, *J. Microsc.* **207:**191–210.

Gundersen, H. J. G., Bagger, P., Bendtsen, T. F., Evans, S. M., Korbo, L., Marcussen, N., Møller, A., Nielsen, K., Nyengaard, J. R., Pakkenberg, B., Sørensen, F. B., Vesterby, A., and West, M. J., 1988, The new stereological tools: disector, fractionator, nucleator and point sampled intercepts and their use in pathological research and diagnosis, *APMIS* **96:** 857–881.

Gundersen, H. J. G., and Jensen, E. B., 1987, The efficiency of systematic sampling in stereology and its prediction, *J. Microsc.* **147:**229–263.

Gundersen, H. J. G., Jensen, E. B. V., Kiêu, K., and Nielsen, J., 1999, The efficiency of systematic sampling in stereology—reconsidered, *J. Microsc.* **193:**199–211.

Haug, H., Kuhl, S., Mecke, E., Sass, N. L., and Wasner, K., 1984, The significance of morphometric procedures in the investigation of age changes in cytoarchitectonic structures of human brain, *J. Hirnforsch.* **25:**353–374.

Hayat, M. A., 1981, *Fixation for Electron Microscopy*, New York: Academic Press.

Hayat, M. A., 2000, *Principles and Techniques of Electron Microscopy. Biological Applications*, Cambridge: Cambridge University Press.

Heimer, L., and Zaborszky, L., 1989, *Neuroanatomical Tract-Tracing Methods 2. Recent Progress*, New York: Plenum Press.

Helmchen, F., 1999, Dendrites as biochemical compartments, In: Stuart, G., Spruston, N., and Häusser, M. (eds.), *Dendrites*, Oxford: Oxford University Press, pp. 161–192.

Hillman, D. E., 1979, Neuronal shape parameters and substructures as a basis of neuronal form, In: Schmitt, F. O., and Worden, F. G. (eds.), *The Neurosciences: Fourth Study Program*, Cambridge, MA: MIT Press, pp. 477–498.

Horcholle-Bossavit, G., Gogan, P., Ivanov, Y., Korogod, S., and Tyc-Dumont, S., 2000, The problem of the morphological noise in reconstructed dendritic arborizations, *J. Neurosci. Methods* **95:**83–93.

Howard, C. V., Cruz-Orive, L. M., and Yaegashi, H., 1992, Estimating neuron dendritic length in 3D from total vertical projections and from vertical slices, *Acta Neurol. Scand. Suppl.* **137:**14–19.

Howard, C. V., and Reed, M. G., 2005, *Unbiased Stereology*. 2nd ed., Oxford: BIOS Scientific Publishers.

Howell, K., Hopkins, N., and Mcloughlin, P., 2002, Combined confocal microscopy and stereology: a highly efficient and unbiased approach to quantitative structural measurement in tissues, *Exp. Physiol.* **87:**747–756.

Jelinek, H. F., and Fernandez, E., 1998, Neurons and fractals: how reliable and useful are calculations of fractal dimensions? *J. Neurosci. Methods* **81:**9–18.

Jones, T. A., 1999, Multiple synapse formation in the motor cortex opposite unilateral sensorimotor cortex lesions in adult rats, *J. Comp. Neurol.* **414:**57–66.

Jones, T. A., Klintsova, A. Y., Kilman, V. L., Sirevaag, A. M., and Greenough, W. T., 1997, Induction of multiple synapses by experience in the visual cortex of adult rats, *Neurobiol. Learn. Mem.* **68**:13–20.

Kleim, J. A., Vij, K., Ballard, D. H., and Greenough, W. T., 1997, Learning-dependent synaptic modifications in the cerebellar cortex of the adult rat persist for at least four weeks, *J. Neurosci.* **17**:717–721.

Korkmaz, A., and Tümkaya, L., 1997, Estimation of the section thickness and optical disector height with a simple calibration method, *J. Microsc.* **187**:104–109.

Korkotian, E., and Segal, M., 2000, Structure–function relations in dendritic spines: is size important? *Hippocampus* **10**:587–595.

Krasnoperov, R. A., and Stoyan, D., 2004, Second-order stereology of spatial fibre systems, *J. Microsc.* **216**:156–164.

Kubínova, L., Janacek, J., Karen, P., Radochova, B., Difato, F., and Krekule, I., 2004, Confocal stereology and image analysis: methods for estimating geometrical characteristics of cells and tissues from three-dimensional confocal images, *Physiol. Res.* **53**:S47–S55.

Lagares, A., and Avendaño, C., 1999, An efficient method to estimate cell number and volume in multiple dorsal root ganglia, *Acta Stereol.* **18**:185–195.

Larsen, J. O., 1998, Stereology of nerve cross sections, *J. Neurosci. Methods* **85**:107–118.

Larsen, J. O., Gundersen, H. J., and Nielsen, J., 1998a, Global spatial sampling with isotropic virtual planes: estimators of length density and total length in thick, arbitrarily oriented sections, *J. Microsc.* **191**:238–248.

Larsen, J. O., Thomsen, M., Haugland, M., and Sinkjaer, T., 1998b, Degeneration and regeneration in rabbit peripheral nerve with long-term nerve cuff electrode implant: a stereological study of myelinated and unmyelinated axons, *Acta Neuropathol.* **96**:365–378.

Larsen, J. O., Von Euler, M., and Janson, A. M., 2004, Virtual test systems for estimation of orientation-dependent parameters in thick, arbitrarily orientated sections exemplified by length quantification of regenerating axons in spinal cord lesions using isotropic, virtual planes, In: Evans, S. M., Janson, A. M., and Nyengaard, J. R. (eds.), *Quantitative Methods in Neuroscience*, Oxford: Oxford University Press, pp. 264–284.

Løkkegaard, A., Nyengaard, J. R., and West, M. J., 2001, Stereological estimates of number and length of capillaries in subdivisions of the human hippocampal region, *Hippocampus* **11**:726–740.

Machín-Cedrés, R., 2004, *Respuesta de la corteza sensorial a cambios crónicos en la estimulación funcional: Un estudio estereológico ultraestructural en la rata*, Doctoral Thesis, Autonoma University of Madrid.

Marner, L., and Pakkenberg, B., 2003, Total length of nerve fibers in prefrontal and global white matter of chronic schizophrenics, *J. Psychiatr. Res.* **37**:539–547.

Mattfeldt, T., Mall, G., Gharehbaghi, H., and Møller, P., 1990, Estimation of surface area and length with the orientator, *J. Microsc.* **159**:301–317.

Mattfeldt, T., and Stoyan, D., 2000, Improved estimation of the pair correlation function of random sets, *J. Microsc.* **200**:158–173.

Mayhew, T. M., 1979a, Basic stereological relationships for quantitative microscopical anatomy—a simple systematic approach, *J. Anat.* **129**:95–105.

Mayhew, T. M., 1979b, Stereological approach to the study of synapse morphometry with particular regard to estimating number in a volume and on a surface, *J. Neurocytol.* **8**: 121–138.

Mayhew, T. M., 1988, An efficient sampling scheme for estimating fibre number from nerve cross sections: the fractionator, *J. Anat.* **157**:127–134.

Mayhew, T. M., 1990, Efficient and unbiased sampling of nerve fibers for estimating fiber number and size, In: Conn, P. M. (ed.), *Quantitative and Qualitative Microscopy—Methods in Neurosciences*, Vol. 3, New York: Academic Press, pp. 172–187.

Mayhew, T. M., 1999, Second-order stereology and ultrastructural examination of the spatial arrangements of tissue compartments within glomeruli of normal and diabetic kidneys, *J. Microsc.* **195**:87–95.

Mesulam, M.-M., 1982, Principles of horseradish peroxidase neurohistochemistry and their applications for tracing neural pathways, In: Mesulam, M.-M. (ed.), *Tracing Neural Connections with Horseradish Peroxidase*, New York: John Wiley and Sons, pp. 1–151.

Mitchison, G., 1992, Axonal trees and cortical architecture, *Trends Neurosci.* **15:**122–126.

Mouton, P. R., 2002, *Principles and Practices of Unbiased Stereology: An Introduction for Bioscientists*, Baltimore, MD: Johns Hopkins University Press.

Mouton, P. R., Gokhale, A. M., Ward, N. L., and West, M. J., 2002, Stereological length estimation using spherical probes, *J. Microsc.* **206:**54–64.

Nadasdy, Z., and Zaborszky, L., 2001, Visualization of density relations in large-scale neural networks, *Anat. Embryol.* **204:**303–317.

Negredo, P., Castro, J., Lago, N., Navarro, X., and Avendaño, C., 2004, Differential growth of axons from sensory and motor neurons through a regenerative electrode: a stereological, retrograde tracer, and functional study in the rat, *Neuroscience* **128:**605–615.

Nyengaard, J. R., and Gundersen, H. J. G., 1992, The isector: a simple and direct method for generating isotropic, uniform random sections from small specimens, *J. Microsc.* **165:**427–431.

Nyengaard, J. R., Witgen, B. M., Grady, S. M., Anderson, B. B., and Gunderson, H. J. G., 2005, Estimation of synapse number using electron microscopy and a new fractionator principle with varying sampling fractions, *Abstract/Viewer Intinerary planner, Soc. Neurosci. Prog.* no. 454.14.

O'Malley, A., O'Connell, C., Murphy, K. J., and Regan, C. M., 2000, Transient spine density increases in the mid-molecular layer of hippocampal dentate gyrus accompany consolidation of a spatial learning task in the rodent, *Neuroscience* **99:**229–232.

Pakkenberg, B., and Gundersen, H. J., 1988, Total number of neurons and glial cells in human brain nuclei estimated by the disector and the fractionator, *J. Microsc.* **150:**1–20.

Pakkenberg, B., and Gundersen, H. J. G., 1997, Neocortical neuron number in humans: effect of sex and age, *J. Comp. Neurol.* **384:**312–320.

Partadiredja, G., Miller, R., and Oorschot, D. E., 2003, The number, size, and type of axons in rat subcortical white matter on left and right sides: a stereological, ultrastructural study, *J. Neurocytol.* **32:**1165–1179.

Peterson, D. A., 2004, The use of fluorescent probes in cell-counting procedures, In: Evans, S. M., Janson, A. M., and Nyengaard, J. R. (eds.), *Quantitative Methods in Neuroscience*, New York: Oxford University Press, pp. 85–114.

Peterson, D. A., and Jones, D. G., 1993, Determination of neuronal number and process surface area in organotypic cultures: a stereological approach, *J. Neurosci. Methods* **46:**107–120.

Poirazi, P., and Mel, B. W., 2001, Impact of active dendrites and structural plasticity on the memory capacity of neural tissue, *Neuron* **29:**779–796.

Roberts, N., Puddephat, M. J., and McNulty, V., 2000, The benefit of stereology for quantitative radiology, *Br. J. Radiol.* **73:**679–697.

Rønn, L. C., Ralets, I., Hartz, B. P., Bech, M., Berezin, A., Berezin, V., Møller, A., and Bock, E., 2000, A simple procedure for quantification of neurite outgrowth based on stereological principles, *J. Neurosci. Methods* **100:**25–32.

Rusakov, D. A., Harrison, E., and Stewart, M. G., 1998, Synapses in hippocampus occupy only 1–2% of cell membranes and are spaced less than half-micron apart: a quantitative ultrastructural analysis with discussion of physiological implications, *Neuropharmacology* **37:**513–521.

Rusakov, D. A., and Stewart, M. G., 1995, Quantification of dendritic spine populations using image analysis and a tilting disector, *J. Neurosci. Methods* **60:**11–21.

Saper, C. B., 1996, Any way you cut it: a new journal policy for the use of unbiased counting methods, *J. Comp. Neurol.* **364:**5.

Schaefer, A. T., Larkum, M. E., Sakmann, B., and Roth, A., 2003, Coincidence detection in pyramidal neurons is tuned by their dendritic branching pattern, *J. Neurophysiol.* **89:**3143–3154.

Schiønning, J. D., Larsen, J. O., Tandrup, T., and Braendgaard, H., 1998, Selective degeneration of dorsal root ganglia and dorsal nerve roots in methyl mercury-intoxicated rats: a stereological study, *Acta Neuropathol. (Berl.)* **96:**191–201.

Scorcioni, R., Lazarewicz, M. T., and Ascoli, G. A., 2004, Quantitative morphometry of hippocampal pyramidal cells: differences between anatomical classes and reconstructing laboratories, *J. Comp. Neurol.* **473:**177–193.

Shook, B. L., Abramson, B. P., and Chalupa, L. M., 1984, An analysis of the transport of WGA-HRP in the cat's visual system, *J. Neurosci. Methods* **11:**65–77.

Silver, M. A., and Stryker, M. P., 1999, Synaptic density in geniculocortical afferents remains constant after monocular deprivation in the cat, *J. Neurosci.* **19:**10829–10842.

Small, J. V., 1968, Measurement of section thickness, In: *4th European Regional Conference on Electron Microscopy*, Rome, pp. 609–610 (Abstract).

Sorra, K. E., and Harris, K. M., 1993, Occurrence and three-dimensional structure of multiple synapses between individual radiatum axons and their target pyramidal cells in hippocampal area CA1, *J. Neurosci.* **13:**3736–3748.

Spruston, N., Stuart, G., and Häusser, M., 1999, Dendritic integration, In: Stuart, G., Spruston, N., and Häusser, M. (eds.), *Dendrites*, New York: Oxford University Press, pp. 231–260.

Stepanyants, A., Tamas, G., and Chklovskii, D. B., 2004, Class-specific features of neuronal wiring, *Neuron* **43:**251–259.

Sterio, D. C., 1984, The unbiased estimation of number and sizes of arbitrary particles using the disector, *J. Microsc.* **134:**127–136.

Stevens, S. S., 1951, Mathematics, measurement, and psychophysics, In: Stevens, S. S. (ed.), *Handbook of Experimental Psychology*, New York: John Wiley.

Stocks, E. A., McArthur, J. C., Griffen, J. W., and Mouton, P. R., 1996, An unbiased method for estimation of total epidermal nerve fibre length, *J. Neurocytol.* **25:**637–644.

Tamamaki, N., Watanabe, K., and Nojyo, Y., 1984, A whole image of the hippocampal pyramidal neuron revealed by intracellular pressure-injection of horseradish peroxidase, *Brain Res.* **307:**336–340.

Tang, Y., and Nyengaard, J. R., 1997, A stereological method for estimating the total length and size of myelin fibers in human brain white matter, *J. Neurosci. Methods* **73:**193–200.

Tang, Y., Nyengaard, J. R., De-Groot, D. M. G., and Gundersen, H. J. G., 2001, Total regional and global number of synapses in the human brain neocortex, *Synapse* **41:**258–273.

Turner, A. M., and Greenough, W. T., 1985, Differential rearing effects on rat visual cortex synapses: I. Synaptic and neuronal density and synapses per neuron, *Brain Res.* **329:**195–203.

Uylings, H. B., Ruiz-Marcos, A., and Van Pelt, J., 1986a, The metric analysis of three-dimensional dendritic tree patterns: a methodological review, *J. Neurosci. Methods* **18:**127–151.

Uylings, H. B. M., Van Eden, C. G., and Hofman, M. A., 1986b, Morphometry of size/volume variables and comparison of their bivariate relations in the nervous system under different conditions, *J. Neurosci. Methods* **18:**19–37.

Uylings, H. B., and Van Pelt, J., 2002, Measures for quantifying dendritic arborizations, *Network* **13:**397–414.

Uylings, H. B., Van Pelt, J., and Verwer, R. W. H., 1989, Topological analysis of individual neurons, In: Capowski, J. J. (ed.), *Computer Techniques in Neuroanatomy*, New York: Plenum Press, pp. 215–239.

Valverde, F., 1970, The Golgi method. A tool for comparative structural analyses, In: Nauta, W. J. H., and Ebbeson, S. O. E. (eds.), *Contemporary Research Methods in Neuroanatomy*, Berlin: Springer, pp. 12–31.

Van Pelt, J., Van Ooyen, A., and Uylings, H. B. M., 2001, The need for integrating neuronal morphology databases and computational environments in exploring neuronal structure and function, *Anat. Embryol.* **204:**255–265.

Verwer, R. W. H., Van Pelt, J., and Uylings, H. B. M., 1992, An introduction to topological analysis of neurones, In: Stewart, M. G. (ed.), *Quantitative Methods in Neuroanatomy*, Chichester, England: John Wiley & Sons, pp. 295–323.

Vinkenoog, M., van den Oever, M. C., Uylings, H. B., and Wouterlood, F. G., 2005, Random or selective neuroanatomical connectivity. Study of the distribution of fibers over two populations of identified interneurons in cerebral cortex, *Brain Res. Brain Res. Protoc.* **14:**67–76.

Weibel, E. R., 1979, *Stereological Methods. Vol. I. Practical Methods for Biological Morphometry,* New York: Academic Press.

West, M. J., 1993, New stereological methods for counting neurons, *Neurobiol. Aging* **14:**275–285.

West, M. J., and Slomianka, L., 1998, Total number of neurons in the layers of the human entorhinal cortex, *Hippocampus* **8:**69–82.

West, M. J., Slomianka, L., and Gundersen, H. J. G., 1991, Unbiased stereological estimation of the total number of neurons in the subdivisions of the rat hippocampus using the optical fractionator, *Anat. Rec.* **231:**482–497.

Williams, R. W., and Rakic, P., 1988, Three-dimensional counting: an accurate and direct method to estimate numbers of cells in sectioned material, *J. Comp. Neurol.* **278:**344–352.

Zaborszky, L., Csordas, A., Buhl, D. L., Duque, A., Somogyi, J., and Nadasdy, Z., 2002, Computational anatomical analysis of the basal forebrain corticopetal system, In: Ascoli, A. (ed.), *Computational Neuroanatomy: Principles and Methods,* Totowa, NJ: Humana Press, pp. 171–197.

Zaborszky, L., Leranth, C., Makara, G. B., and Palkovits, M., 1975, Quantitative studies on the supraoptic nucleus in the rat: II. Afferent fiber connections, *Exp. Brain Res.* **22:**525–540.

17

Three-Dimensional Computerized Reconstruction from Serial Sections: Cell Populations, Regions, and Whole Brain

JAN G. BJAALIE and TRYGVE B. LEERGAARD

JAN G. BJAALIE AND TRYGVE B. LEERGAARD • Neural Systems and Graphics
Computing Laboratory, Centre for Molecular Biology and Neuroscience, and Department
of Anatomy, University of Oslo, N-0317 Oslo, Norway

Abstract: Three-dimensional (3D) reconstruction of data collected from serial sections brings together information that has been separated by sectioning. From 3D reconstructions, shapes of structures and complicated distributions of tissue elements can be analyzed to an extent not possible by inspection of individual sections. This chapter reviews preparatory steps that are needed to produce high-quality series of sections, procedures that are typically used to acquire data from the sections, alignment of the section data, use of global and local coordinate systems for comparison of data across experiments, and useful tools (visualization and others) for extracting information from 3D reconstructions. The examples shown are based on axonal tracing data from the rat cerebro–cerebellar system, made available via The Rodent Brain Workbench (www.rbwb.org).

Keywords: brain atlas, coordinate system, histology, section alignment, three dimensional, visualization

I. INTRODUCTION

The brain consists of numerous areas and regions that are intermingled and interconnected in a complex fashion. Serial sectioning followed by the use of various microscopic techniques is frequently used to study shapes, dimensions, and spatial distributions of objects at multiple levels of granularity. At the light microscopic level, the objects studied range from parts of cells, such as dendritic or axonal elements, via cell populations to regions, areas, and whole brains. A problem with the sectioning, however, is that information that is inherently three dimensional (3D) is separated. Furthermore, the orientation of the section plane influences the visualization, analysis, and interpretations of the data collected from the sections. Since many neurobiological problems rely on accurate mapping of information in space, 3D reconstruction of selected features from serial sections will often represent an advantage for data presentation, as well as for the implementation of analysis that will reveal essential features in the material. It is therefore expected that the use of 3D reconstruction will lead to a more complete understanding or spatial organization, influencing the transition from data to knowledge.

Computerized 3D reconstruction of data from serial sections has been employed for more than three decades (for references, see Bjaalie, 1991). In this method, data acquired from a stack of consecutive sections are brought together in registration. By use of surface- or volume-based rendering, features present in the 3D reconstruction can be visualized in a much more complete fashion than made possible by inspection of single sections only. An additional advantage posed by the 3D format is the opportunity, relying on well-designed software, to perform manipulations and analyses that could not have been performed at the single section level. Important examples include re-slicing at any chosen angle, which may reveal spatial relationships that are difficult to detect in the original plane of sectioning, and density measurements, used to identify important aspects of spatial distribution patterns. Many categories of quantitative analyses may be performed directly on single sections. But with 3D reconstruction from series of sections, such analyses tend to be more complete and accurate. Finally, the use of computerized 3D reconstruction offers opportunities for efficient comparisons of data from different experimental animals of the same species. The 3D digital data, if assembled carefully and according to standardized procedures, may be fitted to common coordinate systems and reused in novel combinations.

In this chapter, we will deal with 3D reconstruction of geometric representations, essentially *point and line coded data*. In our context, point data represent distributions of defined populations of cell bodies or axonal terminal fields. Line data may represent boundaries of territories identified by a specific label, the boundaries of areas or nuclei, or the pial or ventricular surfaces of the brain. Volume rendering, based on data present in a full 3D array of volume elements (voxels), is not dealt with (for considerations of such methods, see, e.g., Pommert *et al.*, 2002; Senft, 2002). We will describe some of the basic requirements for 3D reconstruction at the level of the biological material, before briefly discussing data acquisition procedures, and alignment of the acquired point and line coded data into a 3D reconstruction. Further, we will outline principle steps of visualization and analysis, using examples based on software that is generally available for the neuroscience community. Our experiences related to the use of local and global coordinate systems will be integrated in the descriptions. The examples given and illustrations shown are taken from system level investigations of neuronal populations and brain regions, more specifically from our investigations of the rat cerebro–cerebellar system.

II. EXAMPLE DATA

To illustrate and exemplify the different stages involved in 3D reconstruction of data collected from serial sections, with subsequent visualization and analysis, we have mostly used published data that are available from the database application *Functional Anatomy of the Cerebro–Cerebellar System in Rat* (FACCS), which is a part of the Rodent Brain Workbench (www.rbwb.org; see also Bjaalie *et al.*, 2005). The cerebro–cerebellar system is monosynaptically interrupted in the pontine nuclei, a major brain stem nuclear complex. The system originates in large parts of the cerebral cortex, reaches most parts of the cerebellum, and is one of the largest projection systems in the brain (Brodal, 1982; Brodal and Bjaalie, 1992; Ruigrok, 2004). The data available in FACCS are axonal tracing data. Axonal tracers were injected in the cerebral cortex and/or cerebellum and the distribution of the ensuing labeling in the pontine nuclei (anterogradely labeled cerebro–pontine axonal terminal fields, and retrogradely labeled ponto–cerebellar neuronal cell bodies) was mapped in 3D. The focus of the present examples and illustrations is on the cerebro–pontine projection, with data primarily from Leergaard *et al.* (2000a, b, 2004). The overall aims of these studies were to discover principles of topographical organization and overall structural design of the cerebro–ponto–cerebellar system, including the transformations of the maps from one part of the system to the next. At a more specific level, the aims were to elucidate organizations reminiscent of developmental principles, and organizations that may influence the nature of the information transfer in the system (for review, see Leergaard, 2003).

III. PREPARATORY STEPS: TISSUE BLOCKING AND HISTOLOGY

Brain material that is sectioned and submitted to 3D reconstruction may be produced under a variety of conditions and by use of diverse experimental procedures. The procedures include, e.g., perturbations imposed on the experimental animals, surgical manipulation with introduction of dyes or tracers, tissue fixation and extraction, and cutting of sections and application of histological procedures. In the context of 3D reconstruction, it is particularly important to produce a material of serial sections that is as complete as possible and without distortions in the individual sections. The quality of the sections relies on multiple factors, most of which are specific to the sort of tissue and type of investigation, and should be optimized within given limits. Considerations pertaining to the use of chemical tissue fixation, embedding, choice of microtome and section thickness, section sampling, and subsequent histological processing are well described elsewhere (see, e.g., Maunsbach and Afzelius, 1999; Romeis, 1968; Woods and Ellis, 1994). In the experimental material illustrated in Figs. 17.1–17.9 (data from Brevik *et al.*, 2001; Leergaard *et al.*, 2000a, b, 2004; Lillehaug *et al.*, 2002), brains were fixed by paraformaldehyde perfusion, cryoprotected with buffered sucrose, and sectioned at 50 μm on a freezing microtome. To minimize distortions, microtomy was performed carefully at stable temperatures. For some of the material (from the cerebellum, see Brevik *et al.*, 2001; Lillehaug *et al.*, 2002), tissue blocks were embedded in gelatin prior to sectioning. Series of sections that were incomplete or contained damaged sections were not used for reconstruction.

In this section, we will deal with the most fundamental and invariant preparatory steps, relevant in the context of 3D reconstruction. These steps include the dividing of the tissue into blocks and the subsequent handling of the tissue blocks, orientation of plane of sectioning, use of fiducials and block-face images, and estimates of shrinkage. General advice concerning preparation of histological material may be found in standard textbooks on histology and microscopic techniques (see, e.g., Romeis, 1968; Woods and Ellis, 1994).

A. Tissue Block Size

High-quality histological material is usually easier to produce from small tissue blocks. The smaller the tissue block, the smaller are the chances of having distortions in the sections. The investigator should therefore consider the possibility of cutting sections from several smaller tissue blocks rather than from one large block. If different regions of the brain are studied in separate tissue blocks, relationships between regions may nevertheless be studied. Thus, the application of local coordinate systems (outlined in section "Coordinate Systems," and in the Appendix) will not only facilitate comparisons across animal analysis, but also allow transfer of the data from the local coordinate systems onto common global coordinates.

B. Tissue Block Imaging

Prior to the sectioning, it is often useful to capture images of the tissue from standardized angles (Fig. 17.1A, B), before and after tissue blocking or tissue embedding. Such images may serve to extract positional information for a series of sections, as well as to measure tissue shrinkage due to the histological processing. Images of the tissue blocks may also serve to document positional information about, e.g., axonal tracer injection sites (Fig. 17.1D) that may be transferred to standard diagrams (Fig. 17.1E) and compared with microscope images (Fig. 17.4B).

C. Orientation of Section Plane

For most types of investigation, exact determination of the orientation of the section plane in relation to the whole brain and the tissue block facilitates subsequent registration and assembly of sections into a 3D reconstruction. The plane of sectioning is usually determined with a cut applied directly to the tissue, defining the block face to be placed on the microtome tissue platform, or by using a mold and an embedding medium (Paxinos and Watson, 1998, 2005; Swanson, 1999). Several strategies and tools may be used to ensure a defined and reproducible section plane. Section planes are usually chosen in relation to recognizable anatomical structures (e.g., midline of the brain, ventral surface of the brain, or others), or to reproduce

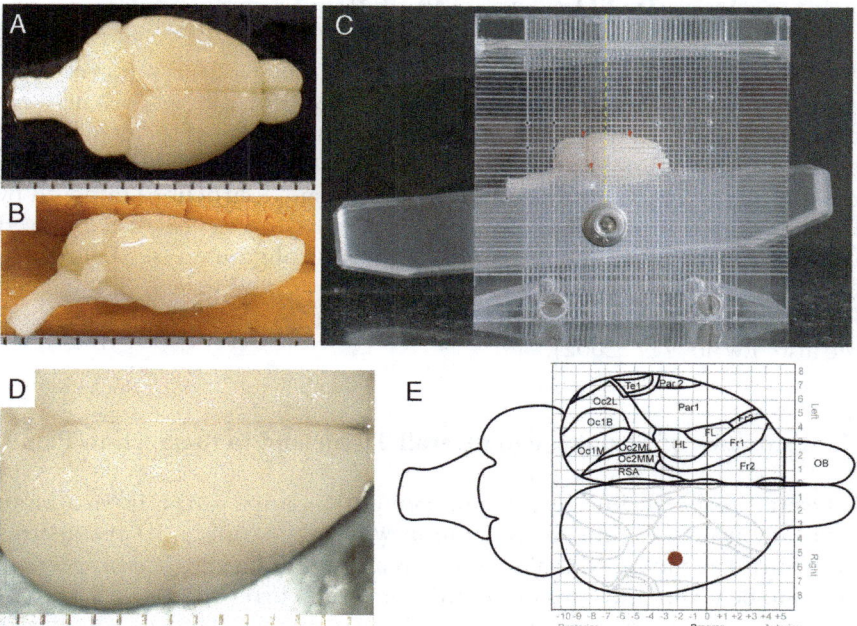

Figure 17.1. Preparation of paraformaldehyde fixed rodent brains for microtomy. (A) and (B) show isolated mouse brains photographed from standard dorsal (A) and lateral (B, C) angles. Various measurements from the images are used for verification of the obtained section angle, for assigning spatial coordinates to individual sections, and for adjustment of shape and proportion of subsequent 3D reconstructions. (C) shows a custom-made device for flexible definition of reproducible coronal/transverse section planes in rodent brains (developed by T. B. Leergaard and K. A. Rekdahl, University of Oslo, Norway). Isolated or embedded rodent brains are placed on a pivotable mount and oriented to align selected anatomical landmarks to a grid (red arrows), before a razor blade is inserted in a vertical slit (dashed yellow line) to apply a coronal cut to the brain in a guillotine fashion. The depicted mouse brain (C) is oriented to reproduce the coronal section plane used in the stereotaxic atlas of Paxinos and Franklin (2001). (D) shows an isolated rat brain in a view from dorsal. A focal injection of a fluorescent neuronal tracer is visible in the parietal cerebral cortex. (E) shows the position and extent of the tracer injection shown in (D), mapped onto a standard diagram of the rat cerebral cortex. Line spacing in the grids in (C) and (E) is 1 mm. (C) and (D) are from the database application FACCS (www.rbwb.org; see also Bjaalie *et al.*, 2005).

standardized angles (e.g., as used in published brain atlases). For example, use of commercially available brain blockers (such as blockers for rodent brains provided by David Kopf Instruments, Tujunga, CA) facilitates the reproduction of section planes used in stereotaxic rat and mouse brain atlases (Paxinos and Franklin, 2001; Paxinos and Watson, 1998, 2005). A similar customized device is shown in Fig. 17.1C. The angle of sectioning is typically chosen to reproduce those of previous investigations, or of standard atlases.

D. Fiducials and Block-Face Imaging

Fiducial markers are introduced artificially in the tissue as references for the alignment of sections. Classical fiducials are drilled holes, oriented orthogonally to the section plane, or notches in the surface of the tissue or the embedding medium. The use of fiducials in the tissue itself, if placed close to regions of interest, is not recommended, as tissue destructions imposed by the fiducial markers may be substantial. Other approaches can be employed to compensate for lack of fiducials during section alignment (see below). To facilitate section alignment, some investigators capture images of the tissue block face (of every section) during sectioning (Toga and Banerjee, 1993; see also, Ewald *et al.*, 2002).

E. Estimates of Shrinkage in Overall Tissue and in the Section Plane

Overall tissue shrinkage (or enlargement) relative to the in vivo brain is usually not taken into consideration when preparing a 3D reconstruction. Shrinkage in individual sections, as a result of histological processing, needs to be taken into consideration to ensure a correctly proportioned 3D reconstruction. Correct spatial proportions are obtained by subsequent size adjustment of the data derived from the sections.

IV. DATA ACQUISITION: USER-GUIDED AND AUTOMATIC PROCEDURES

To acquire simplified geometric representations (point and line coded data) of selected objects present in the sections, methods ranging from completely manual to nearly fully automatic may be employed. Data need to be recorded at sufficient resolution (using appropriate magnification) and accuracy across regions of interest ranging from small divisions of a brain nucleus (a few hundred micrometers) to entire sections (several centimeters). A basic approach is to use camera lucida drawings that are subsequently digitized (reviewed in Bjaalie, 1991). More sophisticated methods, briefly described below, include a semiautomatic method based on image-combining computerized microscopy (ICCM) and a more automated method combining digital camera technology with the use of image analysis software.

A. Image-Combining Computerized Microscopy

This method is currently available, e.g., in the Neurolucida system (Micro-BrightField Inc., Williston, VT), or in customized versions (Leergaard and Bjaalie, 1995). The principle of ICCM was first introduced by Glaser and van der Loos (Glaser *et al.*, 1979, 1983; Glaser and Glaser, 1990). The method is based on the mixing of a computer graphical drawing area with the image of the specimen. Movement of the microscope stage (controlled by stepping motors via the computer) is accompanied by a translation of the graphical image. The user thus obtains direct feedback during the data entry procedure. ICCM systems are usually optimized for high-resolution recording of numerous x, y (or x, y, z) coordinates across large brain regions. The precision of the data recording depends on the quality of the microscope optics, the stepping motors, and the resolution of the graphic feedback image (Glaser *et al.*, 1983). Carefully calibrated customized systems based on high-precision Märzhäuser microscope stages (Märzhäuser Wetzlar Gmbh, Wetzlar, Germany) have been reported to be able to reproduce spatial locations with a precision of ± 2 μm (Leergaard and Bjaalie, 1995). Higher precision levels have been reported for commercially available systems.

B. Image Analysis

More automated data acquisition methods have been employed, motivated by the need for procedures that are less dependent on human pattern recognition abilities and more suitable for high-throughput analysis. Thus, image analysis techniques have been applied to detect cytoarchitectonic boundaries (Amunts and Zilles, 2001; Grefkes *et al.*, 2001; Rademacher *et al.*, 2001; Schleicher and Zilles, 1990; Schmitt *et al.*, 2003; Wree *et al.*, 1982), to delineate the boundaries of labeled axonal plexuses (Lillehaug *et al.*, 2002; Schwarz and Möck, 2001; Schwarz and Their, 1995), and to identify tracer-labeled neurons and terminal fields of tracer-labeled axons (Lillehaug *et al.*, 2002). In principle, these methods apply filtering techniques to digital grayscale images. Elements of interest are identified with use of threshold values creating binarized images (Fig. 17.2; see also Lillehaug *et al.*, 2002). Various procedures, included in most image analysis software packages, are in turn used to distinguish between objects of interest and artifacts or noise. The end result of such image analysis based methods may, e.g., be lists of point coordinates representing the centers of identified objects, such as labeled cell bodies (Fig. 17.2; see also Lillehaug *et al.*, 2002).

C. Data Acquisition Exemplified

In our investigations of the cerebro–pontine system illustrated here, we used image-combining microscopy as outlined above to digitize contour lines for a number of structures (the ventral surface of the pons, the outlines of the pontine gray, the contours of the corticobulbar and corticospinal fiber tracts, the midline of the brain, and the outline of the fourth ventricle; see Figs. 17.3A and 17.4E, F) as a reference for the alignment

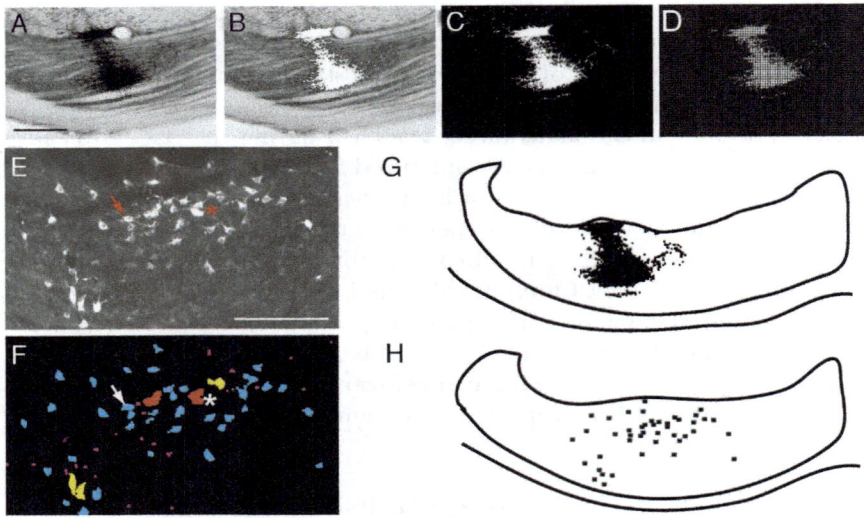

Figure 17.2. Automatic methods for data acquisition from serial sections through rat pontine nuclei (modified from Lillehaug *et al.*, 2002, with permission). *Automatic digitization of labeled fiber distributions:* (A) Labeled cerebro–pontine axonal plexuses are visible as dense and loose plexuses in an image of a transverse section through the pontine nuclei. The labeled fibers, appearing as black or shades of gray, are identified from the digitized gray-scale image by choosing a threshold value from minimum 0 (black) to a certain maximum level (dark gray), and represented as white (B). A binary image (C) is created from a chosen threshold value. By super-imposing a grid onto the binary image (D), it is possible to obtain *x, y* coordinates from squares containing more than a given number of white pixels, which in turn may be represented as black dots (G). *Automatic detection of labeled cell distributions:* Digital gray-scale image of transverse section through the pontine nuclei (E), show-ing pontocerebellar neurons labeled following injection of the fluorescent tracer rhodamine conjugated dextran amine into the cerebellar crus IIa. The digital image is first converted to a binary black and white image by applying a threshold value, and binary operators are applied to remove artifacts, fill closed gaps, and separate particles. (F) Unique colors are assigned to different area ranges, and spatial coor-dinates (*x, y*) are automatically assigned to the center of gravity of each particle, and may be represented by black dots in a graphic plot (H). Comparison of the original "raw image" with the analyzed image shows that most labeled neurons are recognized correctly. Some cells, however, were not counted as separate objects due to the merging of overlapping cells. Weakly labeled cells, cellular fragments, and small artifacts (pink) that gave rise to particle sizes below the defined size (area) range for labeled cells were excluded from the plot in (H). Scale bars: 200 μm.

of the sections. The density and distribution of anterogradely labeled ax-onal plexuses within the pontine nuclei were digitized semiquantitatively as points (Figs. 17.4C, E and 17.7; Leergaard *et al.*, 1995, 2000a, b, 2004). In areas with low densities of labeling, point coordinates were placed at regu-lar intervals along the length of single axons. In areas with dense labeling, a rough correspondence was sought between the density of labeling and the number of digitized points. When different investigators recorded data

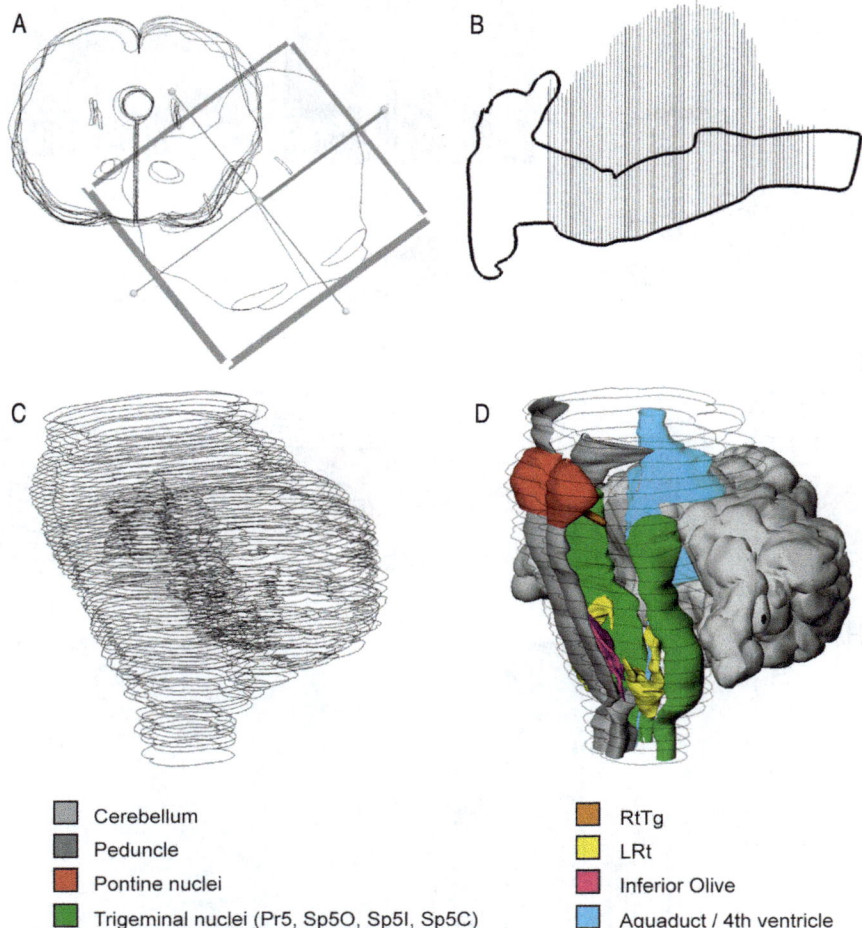

Cerebellum

Peduncle

Pontine nuclei

Trigeminal nuclei (Pr5, Sp5O, Sp5I, Sp5C)

RtTg

LRt

Inferior Olive

Aquaduct / 4th ventricle

Figure 17.3. Assembly and visualization of a 3D reconstruction of the rat brain stem and cerebellum (modified from Brevik *et al.*, 2001). (A) Series of digitized transverse brain sections are aligned interactively according to multiple anatomical landmarks. (B) The external boundary of the brain stem, digitized from a sagittal section through a different rat brain (bold contour line), is used as a template for fine adjustment of the dorsoventral alignment of the transverse sections. (C) Oblique lateral view of the complete aligned 3D reconstruction. Real-time rotation with inspection of the 3D reconstruction from multiple angles of view was used to facilitate alignment. (D) The outer boundaries of the cerebellum, the descending peduncle (corticobulbar and corticospinal tracts), the fourth ventricle, and precerebellar nuclei (pontine nuclei, trigeminal nuclei, reticulotegmental nucleus of the pons/RtTg, lateral reticular nucleus/LRt, and the inferior olive) are represented as solid surfaces.

from the same sections, some variation in the density of digitized points was observed, but spatial distribution patterns remained nearly identical. This approach allowed subsequent analyses of spatial distributions, density gradients, and overlap patterns (Figs. 17.5–17.9). A similar logic as above

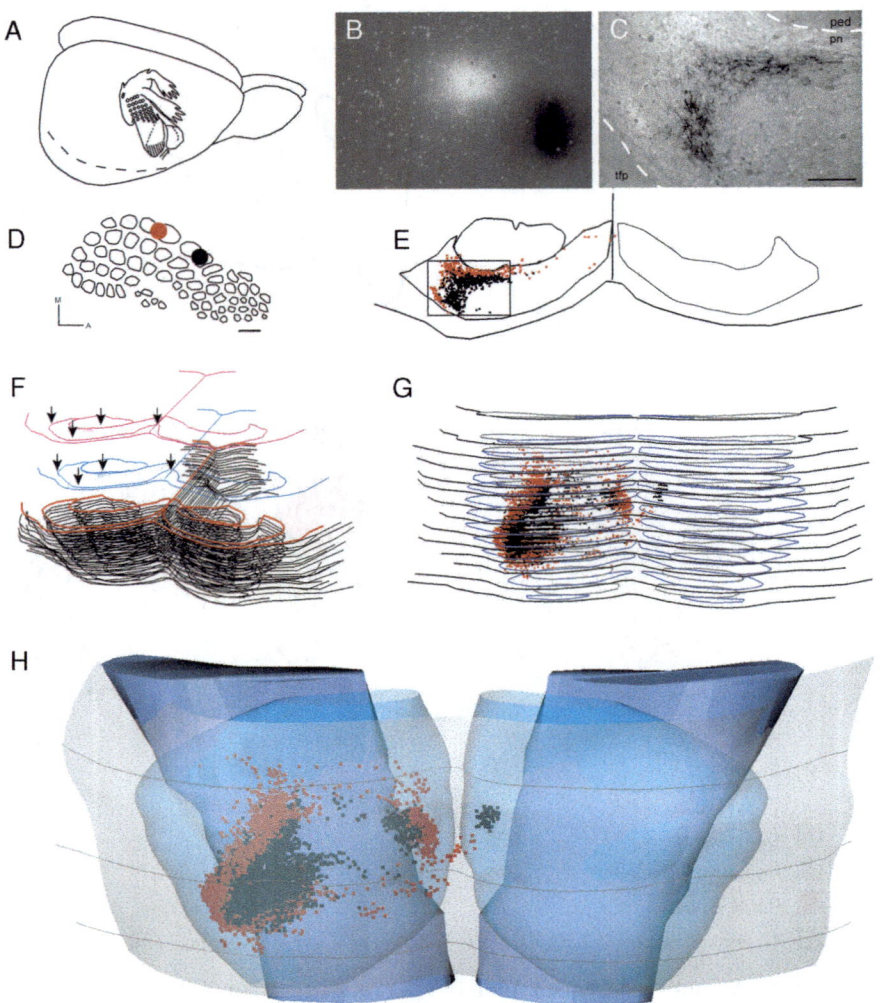

Figure 17.4. Methodological procedures from cerebral cortical injection of axonal tracer to assembly of 3D reconstruction of the rat pontine nuclei (modified from Bjaalie *et al.*, 2005). (A) Drawing of the right cerebral hemisphere of the rat with a cartoon representation of the somatotopic map of body surface representations found in the primary somatosensory cortex (SI, redrawn from Welker, 1971). The image in (B) shows two typical tracer injection sites (Fluoro-ruby, FR, with a bright appearance, and biotinylated dextran amine, BDA, seen as a black area) in a horizontal section through SI (data from Leergaard *et al.*, 2000a, case D46). The injected axonal tracers give rise to labeled axons in the pontine nuclei. The image in (C) is taken from a transverse section through the pontine nuclei and shows typical axonal clusters labeled with FR (bright) and BDA (black). The positions and sizes of the FR (red) and BDA (black) injection sites are indicated in a drawing of the SI whisker barrels (D), which represent the mystacial vibrissae arrangement on the contralateral mystacial pad (redrawn from Fabri and Burton, 1991). (E) Digitized transverse section [same section as in (C)] frame indicates position of the image shown in (C) in which labeled axonal plexuses are semiquantitatively represented

applies to investigations focusing on distribution of cell bodies, labeled with axonal tracing techniques or any other technique that would lead to labeling of a population of cells, or to studies of shapes and dimensions of brain areas and regions (for references based on the same analytical procedures as outlined here but with other data categories, see Brevik *et al.* 2001; Lillehaug *et al.*, 2002; Nikundiwe *et al.*, 1994; Vassbø *et al.*, 1999). For landmarks and boundaries, usually recorded as open or closed contour lines, the spacing of points and relationships of start and end points for each boundary through the stack of sections should be as consistent as possible to facilitate later surface modeling (see section "Primary Viewing of the Reconstruction").

V. ALIGNMENT OF DATA FROM SECTIONS: BUILDING THE RECONSTRUCTION

Since the data acquisition process is performed independently for each section, the spatial relationships between the geometric representations (points and lines) collected from the serial sections are disrupted. Data from each section level thus need to be registered to recover the original relative position prior to the sectioning. Depending on the software employed, manual, automatic, or combined methods may be used. Here we will only deal with the fundamental approaches independent of the particular implementation. Under most circumstances, the best end result is produced by the combined use of several approaches, as outlined below.

A. Fiducials and Block-Face Images

Given that the use of fiducial markers may be problematic, as discussed above (section "Orientation of Section Plane"), here we will only acknowledge that these markers represent a possible aid during alignment and emphasize that they must be used with care, in particular to avoid tissue damage. Block-face imaging has been reported to be of considerable aid

---

Figure 17.4. (*Cont.*) as dots (FR, red dots, BDA, black dots), together with lines representing selected anatomical boundaries. (F) Series of digitized sections are assembled to a 3D reconstruction by interactive alignment of multiple anatomical landmarks (arrows) while inspecting the growing 3D reconstruction (black sections) from multiple angles of view on the computer screen. (G) The complete 3D reconstruction with different anatomic structures displayed in different colors, in a view from ventral. (H) Visualization of the same reconstruction as in (G), with boundaries of the descending fiber tracts shown as solid surfaces and boundaries of the pontine nuclei and ventral brain stem surface as transparent surfaces. Scale bar in (C): 100 μm. Scale bar in (D): 1 mm. *Abbreviations*: A, anterior; M, medial; ped, peduncle; pn, pontine nuclei; tfp, transverse fibers of the pons.

during alignment of sections from large tissue blocks or whole brain (Toga and Banerjee, 1993; see also, Ewald *et al.*, 2002). With this method, the cut surface, or face, of the tissue block is digitally imaged prior to sectioning. An image is collected for each section level, and since the camera is fixed, the relative positions of all block-face images (and hence, sections) are thereby collected. To detect possible accidental movement of the camera during the sectioning procedure, calibration via other markers, such as a grid, is recommended.

B. Landmarks and Boundaries

The primary focus during data acquisition is usually to record data describing the region of interest, specific labeling patterns, or other features of relevance for the biological problem in question. In addition, it is essential to also record landmarks and boundaries for the purpose of alignment of data from the sections. It is particularly important to record distinguished landmarks close to the region of interest, but more remote landmarks may also be of value for the initial global orientation of the section drawings (see section "Example alignment"). Pial or ventricular surfaces, distinct boundaries of fiber tracts, the midline of the brain, and the most prominent architectonic boundaries represent examples of tissue elements that may provide useful guidance for the subsequent alignment of data from a series of sections.

C. Template Sections, Atlases, and Tomography Data

Template sections with an orientation perpendicular to the section plane used to generate the 3D reconstruction may be useful as a reference during the alignment procedure (Fig. 17.3B; Brevik *et al.*, 2001). The perpendicular section template serves as an aid to reconstruct the correct curvature and slope of the reconstructed brain region. Use of whole model templates, such as 3D digital atlases (Fig. 17.10; Leergaard *et al.*, 2003) or tomographic images of the intact specimen (or a similar specimen), also represents powerful strategies for the alignment of section data. A general discussion of the use of reference brains for registration is provided by Amunts and Zilles (in this volume).

D. Example Alignment

In the example reconstructions from the rat brain stem illustrated here, we used multiple landmarks and boundaries, as well as template sections. Thus, to build our brain stem reconstructions from series of transverse sections (Figs. 17.3A, B and 17.4E–G), we normally use the brain midline, the ventral surface of the brain stem (with particular emphasis on the groove of the basilary artery), and the floor of the fourth ventricle as global and initial landmarks, before conducting a more fine adjustment based on local landmarks, such as the external boundaries of the pontine nuclei and the descending corticobulbar and corticospinal pathways (peduncles). Irregularities in the surface, recognizable across sections, were also detected and used during the alignment, since these were often systematically repeated at the same locations through several adjacent sections. Furthermore, we used structures known to have a perpendicular orientation relative to the section plane: in the case of our brain stem reconstructions, the descending corticobulbar and corticospinal tracts (Figs. 17.3 and 17.4) were particularly useful. The overall straight course of these structures was faithfully reconstructed. Finally, to help perform an even finer alignment of the transverse series of sections, a series of sagittal sections through the brain stem of another brain (from an animal of the same strain, gender, and age as the reconstructed brain) were translated to the coordinate system of the reconstruction and used as a template to adjust the dorsoventral location of the sections.

VI. COORDINATE SYSTEMS

Implicitly, any 3D reconstruction consisting of x, y, z coordinate data occurs in a coordinate system. Coordinate systems determined by trivial factors, such as values recorded from a digitizing tablet or ranges of movement of a motorized microscope stage, are of limited value. Coordinate systems that refer to an established standard can, however, be of considerable use. Thus, transfer of all data collected in a given study to the same standardized coordinate system facilitates interindividual comparisons and extraction of organizational principles based on multiple experimental animals (for a general discussion, see Bjaalie, 2002).

3D reconstructions covering an entire brain require, needless to say, the use of a global coordinate system. Such coordinate systems are usually skull based and well defined in standard atlases for a given species (see, e.g., Paxinos and Franklin, 2001; Paxinos and Watson, 1998, 2005). But skull coordinates are not easily used, since standard sectioning procedures require that the brain is removed from the skull. Local coordinate systems, however, rely entirely on the use of landmarks and boundaries intrinsic to the brain and are therefore more easily implemented. It is recommended that 3D reconstructions covering only limited parts of the brain employ a standardized, local coordinate system, which in turn are mapped onto a global coordinate system defined in a standard atlas (see Appendix).

With the use of a local coordinate system, comparison of data collected across experimental animals is facilitated. If a small brain region is studied, and a defined rodent strain is chosen (to limit variations in shapes of the region studied), simple procedures can be employed to transfer data from multiple experimental animals to the same coordinate system. In each animal, the same landmarks and boundaries will have to be identified in order to establish the local coordinate system. Once this operation has been carried out, only adjustments for size differences (scaling) and orientation (rotation) are needed to overlay the data from different animals. In principle, the procedure is independent of section angle. In practice, it is considerably more convenient to perform the data transfer if the section angle is standardized (see section "Orientation of Section Plane"). This is primarily achieved by accurate determination of the section angle prior to sectioning (Fig. 17.1C) and secondarily by measurements from the sections. To this end, high-magnification images of the intact tissue block are useful.

For our investigations of the rat pontine nuclei, we have implemented a local coordinate system (Fig. 17.11; Brevik *et al.*, 2001, see also Leergaard *et al.*, 2000a, b, 2004), suitable for detailed analysis of experimental labeling patterns and presentation of neural tracing data in standardized diagrams (Figs. 17.5A, B, 17.6, and 17.9). This standardized coordinate system (detailed in section "Application of Local Coordinate System for the Rat Pontine Nuclei") consists of a bounding box oriented in parallel to the descending fiber tract at the level of the pontine nuclei, a plane corresponding to the brain stem midline, and boundaries corresponding to the extreme lateral, rostral, caudal, and ventral limits of the pontine nuclei. The dorsal boundary is defined indirectly, using the ventral surface of the peduncles as an intermediate landmark. It is applied individually to each 3D reconstruction, and subsequently size adjusted to fit the standardized coordinate system. In this way, point coordinate data from multiple animals are combined within the same spatial framework. Based on this coordinate system, we have made diagrams showing the outlines of the pontine nuclei in standard angles of view (Figs. 17.5, 17.6, and 17.9). These diagrams facilitate data presentation and comparison of results from different animals. By registering the local coordinate system to a global (skull-based) coordinate system, the local pontine positional values are readily translated into global (sterotaxic) values (see section "Translation Between Local and Global Coordinates").

VII. VISUALIZATION AND ANALYSES

In the present account, most of the descriptions are based on the program Micro3D, used in original research reports dealing with tract-tracing studies on insects (Berg *et al.*, 1998, 2005), chicken embryo (Diaz *et al.*, 2003), newborn rat (Leergaard *et al.*, 1995), adult rat (Leergaard *et al.*, 2000a, b, 2003, 2004; Zaborszky *et al.*, 2002), cat (Bajo *et al.*, 1999; Malmierca *et al.*, 1998), and monkey (Vassbø *et al.*, 1999). The Micro3D software has recently been made available for general use by the neuroscience community in a

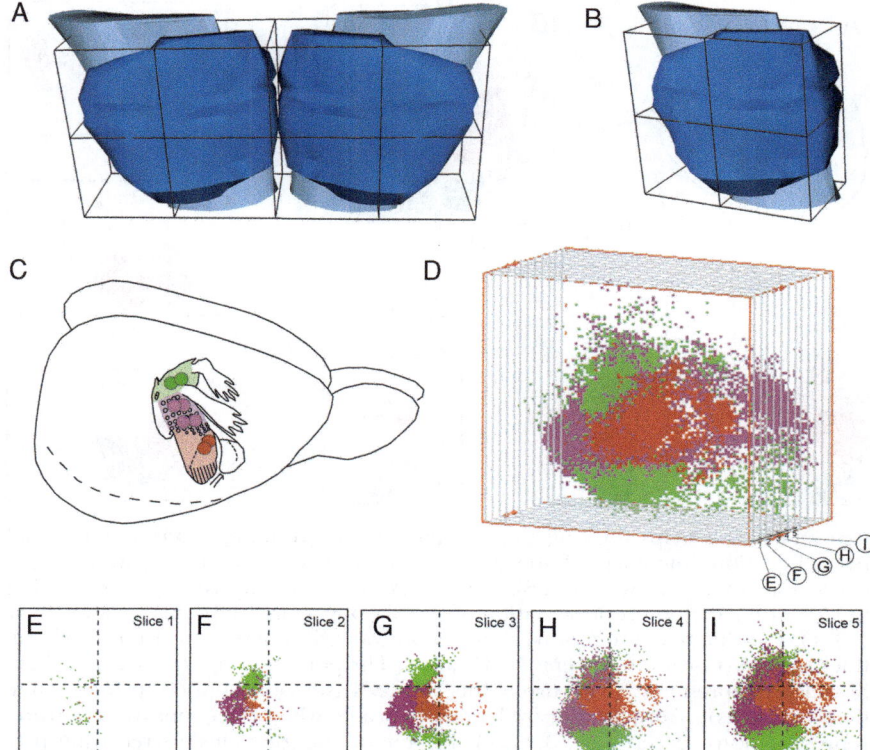

Figure 17.5. Standardized local coordinate system applied to 3D reconstruction of the rat pontine nuclei and used as basis for combined presentation and analysis of axonal tracing data obtained from different experimental animals [(C–I) are from Bjaalie *et al.*, 2005, with permission]. (A) shows a computer-generated 3D reconstruction of the pontine nuclei (dark blue solid surfaces) and cerebral peduncles (bright blue surfaces) in view from ventral. In (B), only the right pontine nuclei are shown in an oblique ventromedial view. The local pontine coordinate system is applied by size adjusting a cuboid bounding box to fit the defined nuclear boundaries (for further details, see section "Application of Local Coordinate System for the Rat Pontine Nuclei"). (C–I) show the spatial distribution of cerebro–pontine projections originating from eight different cortical sites in the primary somatosensory cortex. The eight injection sites were taken from six animals that had been submitted to tracing experiments in which the distribution of labeled cerebro–pontine axons was 3D reconstructed and transferred to the local pontine coordinate system (data from Leergaard *et al.*, 2000a, b). (C) The size and location of the eight tracer injection sites within the primary somatosensory cortex are shown in the drawing of the right cerebral hemisphere (modified from Welker, 1971). (D) Colored dots, representing labeled cerebro–pontine axons, are shown together in a simplified frame, representing the local coordinate system. (E–I) Spatial distribution patterns and topographic organization examined in consecutive 100-μm-thick slices oriented in parallel with the ventral surface of the pons, from ventral (E) to dorsal (I) at locations indicated in (D). The data sets were combined and analyzed using tools available in the database application FACCS (www.rbwb.org; Bjaalie *et al.*, 2005).

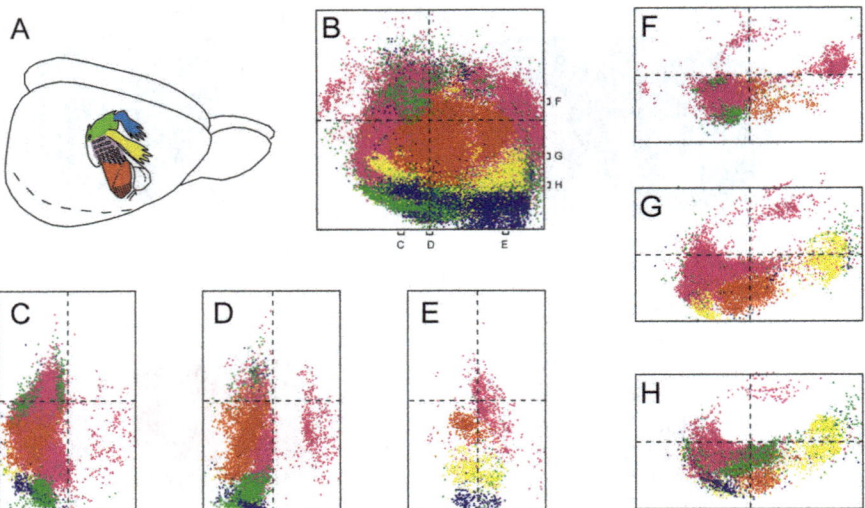

Figure 17.6. Dot maps showing the pontine labeling resulting from 54 individual tracer injections into major SI body representations in rat, color coded according to cortical site of origin (A) [from Bjaalie *et al.* (2005), with permission]. The drawing in (A), with a cartoon representation of the SI body map, is modified from Welker (1971). The dot maps are shown as projections viewed from ventral (B) or as 50-μm-thick sagittal (C–E) or transverse (F–H) slices. The pontine projections arising from SI whisker representations (purple dots, 40 cases) surround upper lip representations (red dots) externally and are largely colocated with projections from SI trunk representations (green dots, B, C, F). In general, pontine clusters receiving projections from different SI body parts have complementary shapes and close spatial relationships. In most slices through the models, patterns of interdigitating labeled clusters are visible. While the neighboring relationships of the 2D SI map are largely preserved in the pontine map, the 3D nature of this map allows introduction of new neighboring relationships.

version running under Linux (Bjaalie *et al.*, 2006; download available from www.nesys.uio.no and from the Rodent Brain Workbench, www.rbwb.org). Some illustrations (Figs. 17.5D–I and 17.6B–H) are produced with a suite of tools based on Java (Sun Microsystems Inc., Santa Clara, CA) made available via the database application FACCS (www.rbwb.org). The description below is generic only and does not deal with technical details, algorithms, or comparisons of software solutions. Several other comparable tools and approaches are available (see, e.g., Lohmann *et al.*, 1998; Maurin *et al.*, 1999; Van Essen *et al.*, 2001).

A. Primary Viewing of the Reconstruction

Rotation and zooming, performed in real time, with or without perspective viewing, represent primary manipulations that most users will perform immediately on a 3D reconstruction. The visualization of the geometry of points and lines in the 3D reconstruction relies on the chosen rendering parameters, such as point size, line thickness, and different colors for each object category (an object category in the present context is, e.g., different populations of labeled cells and different categories of boundaries of brain regions). Rendering of individual data points as spheres of variable diameter may also be useful.

A fundamental visualization approach that helps perceive the shapes and spatial relationships in the data is *surface modeling* based on contour line data, describing boundaries of brain regions or external (pial) or internal (ventricular) surfaces of the brain. Surface modeling from stacks of contour lines can be performed using various approaches, in principle, in two steps: first, a geometry of polygons is created and second, shading is applied to each polygon, resulting in the "resynthesis" of a surface based on the contour line data. The surface can typically be viewed as solid or transparent. The final surface may be either rough (with the surface tightly coupled to the original coordinate data, Fig. 17.5A, B), or smoothened (Figs. 17.3D and 17.4H). Methods range from the construction of a geometry of triangles directly based on the coordinates recorded along the contour lines (exemplified in Toga, 1990) to more complex parametric methods (Bjaalie *et al.*, 1997a; Geiger, 1993) and implicit methods (Shen *et al.*, 2004). Examples of point, line, and surface representations are given in Figs. 17.3–17.6, with the transition from stacks of contour lines to surfaces are exemplified in Fig. 17.4G, H.

Most data sets need to be edited and optimized to achieve an optimal surface modeling. Depending on the method employed, requirements may include the ordering of the coordinates describing each contour line, use of regular or irregular distances between points recorded along the contour lines, and use of few or many points along the contour lines. Requirements for the chosen surface modeling method should be taken into consideration before data acquisition is performed. The possibilities for editing the sequence and density of point coordinates in contour line data differ among software solutions. In general, independent of the method chosen, a good end result often follows from finding the right balance between the density of points used to represent the contour lines, and the spacing between the contour lines.

For combined visualization of points, lines, and surfaces, it is important to find a balance between object and background colors, and surface transparencies (Fig. 17.4H). Contour lines may be used to emphasize shape when surfaces are made smooth and transparent, and are especially useful when multiple layers of transparent surfaces are rendered (Fig. 17.4H). When data points representing labeled structures such as cells or axonal terminal fields, distributed through the thickness of a section, have been recorded without assigning individual z-values within the section (i.e., all data points from a given section are assigned the same z-value), randomized distribution of the data points within the thickness of the section may be used to avoid an artificial spacing of the data points when viewed "on edge."

In Fig. 17.4G, H, we exemplify the primary viewing of a reconstruction of the pontine nuclei from rat. The data shown represent the external brain surface at the level of the pons, nuclear boundaries, and landmarks, as well as patterns of axonal terminal fields resulting from dual tracer injections placed in specific, electrophysiologically, and architectonically defined divisions of the barrel cortex. The data show that the two injection sites generate topographically distinct labeling patterns (for further considerations, see Leergaard *et al.*, 2000a).

B. Slicing

Visualization and analysis based on a single plane of sectioning may lead to incomplete interpretations of spatial relationships. 3D reconstructions partly solve this problem by bringing together the information that was separated by the sectioning. But subdivision of the 3D reconstruction is often required to perform a detailed analysis and to extract essential information. Reslicing of the reconstruction at chosen thicknesses and orientations therefore represent a key analytical tool (Figs. 17.5 and 17.6). Such dynamic reslicing, independent of the original section plane, introduces a new dimension to the analysis of brain topography. In our example analysis from the rat pontine nuclei, the distribution and shape of the labeling patterns of cerebro–pontine terminal fields are analyzed to advantage with reslicing at variable angles (Fig. 17.6). When superimposing data from many experiments in the same local coordinate system for the pontine nuclei (see sections "Coordinate Systems" and "Application of Local Coordinate System for the Rat Pontine Nuclei"), reslicing turns out to be critical to reveal spatial relationships in the data. For somatosensory cerebro–pontine terminal fields, such an analysis has demonstrated the presence of concentrically organized shells or lamellae with spatial relationships in the cortical map being largely preserved in the pontine map (reviewed in Leergaard, 2003). Careful inspection and digital reslicing of the reconstructions from multiple angles of view have therefore been instrumental to fully understand the 3D distribution and shape of the labeling patterns (Leergaard *et al.*, 2000b, 2004; for another example, see Malmierca *et al.*, 1998).

C. Surface Modeling of Labeling Patterns

Spatial relationships among clouds of points representing patterns of labeling may also be shown to advantage with use of surface modeling. Such surface modeling, based on point data representing labeled structures such as cells and axonal terminal fields (and not contour line data representing distinct boundaries), may be achieved by manually digitizing a set of contour lines surrounding the dense regions of labeling (see, e.g., Leergaard *et al.*, 2000a, b; Malmierca *et al.*, 1998). This simple approach may work well for dense clusters of labeling. Automatic generation of a geometric representation surrounding the point clouds is obviously more attractive and has been shown for analysis of tract-tracing data in, e.g., Bjaalie *et al.* (1997a), as well as in a recent analysis of the rat pontine nuclei (Leergaard *et al.*, 2004). In the most recent example, a method based on scalar fields (generated by binning the point coordinates and estimating the relative density for each bin) was used. For visualization, the isosurfaces were extracted by marching cube-like algorithm (J.O. Nygaard, S. Gaure, C. Pettersen, H. Avlesen, and J.G. Bjaalie, in preparation; see also, Nadasdy and Zaborszky, 2001). In the example shown in Fig. 17.8C, D, isodensity surfaces were used to model point clouds representing axonal terminal fields originating in precisely defined parts of the primary motor and primary somatosensory cortices. Injection site details and labeling patterns as they occur in single sections are shown in Fig. 17.7. This analysis demonstrated to advantage the shapes of the terminal fields of labeling and aspects of the neighboring relationships (Leergaard *et al.*, 2004).

D. Analysis of Spatial Overlap

A useful approach for extracting quantitative information about the distribution of labeling deals with the visualization and analysis of spatial overlap between pairs of data. The problem raised is whether two populations of points are segregated, overlapping, or randomly distributed with respect to each other. If there is overlap, the overlap in turn needs to be quantitatively expressed.

The terms "segregation" and "overlap" are commonly used to describe spatial relationships, but they are not trivial to define, and different analytical models should be employed for different purposes (Bjaalie and Diggle, 1990; Bjaalie *et al.*, 1991; Diggle, 1986).

A typical problem confronted in investigations of system level connectivity in the brain is to characterize whether two populations of labeled neuronal elements are overlapping or segregated. To the extent that they are overlapping, interaction is more likely to occur also at a functional level. To estimate the amount of overlap in a material containing labeled axonal terminal fields, many investigators apply a grid to each section and then count the number of events (points of different categories representing labeling) in each bin in the grid before computing statistics from the data (Alloway *et al.*, 1999; He *et al.*, 1993; Leergaard *et al.*, 2000a, 2004). Figure 17.7E, F demonstrates such an analysis at the single section level for axonal tracing data. Figure 17.8E–H shows a similar analysis implemented at the level of complete 3D reconstructions. It shows the 3D distribution of bins expressing double labeling, defined as the presence of data points of both labeling categories at a defined threshold (Leergaard *et al.*, 2004). The results of such analyses are inherently related to the choice of bin size, and in case of axonal terminal fields, also to the data acquisition method

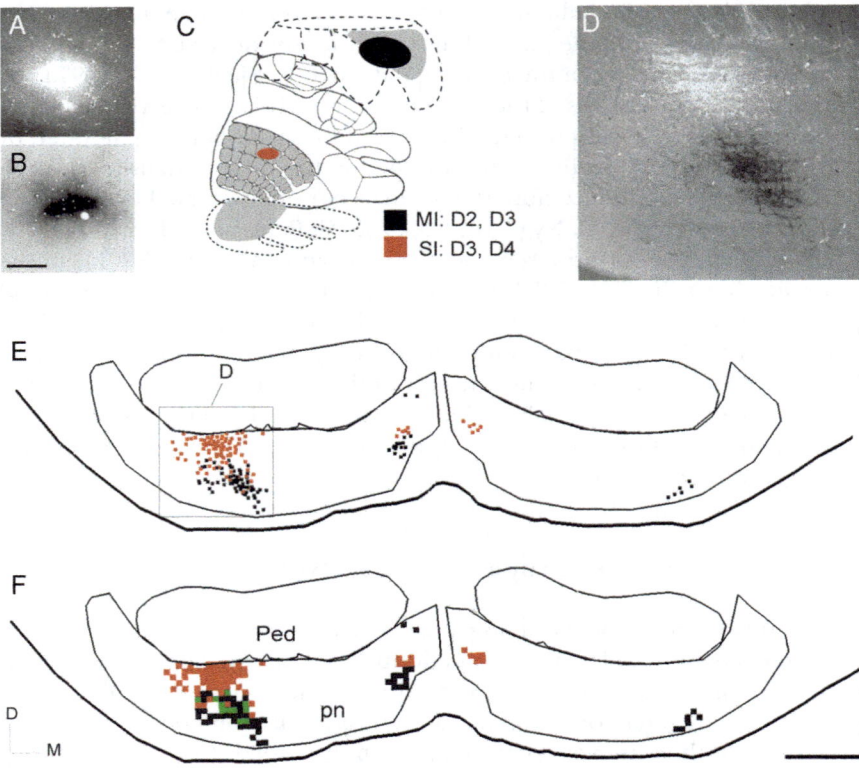

Figure 17.7. Digitization and 2D overlap analysis of two populations of cerebro–pontine axons labeled after dual injections of axonal tracers into homotopic whisker representations in the right primary motor (MI) and somatosensory (SI) cortices in rat (modified from Leergaard *et al.*, 2004). (A and B) Images of tangential sections through cortical layer V, showing a biotinylated dextran amine injection (BDA) in MI (A), and a Fluoro-ruby (FR) injection in SI (B). (C) The localization and extent of the injection sites are mapped onto a diagram of the rat sensorimotor cortex (for details, see Leergaard *et al.*, 2004). (D) Axonal plexuses, labeled with BDA (black) and FR (bright), are observed in a transverse section through the pontine nuclei. (E) Raw data plot of the section shown in (D). Anatomical landmarks and boundaries are represented by lines and the axonal labeling by black (BDA) and red (FR) dots. (F) Overlap analysis of the same section as shown in (E). The dot map is subdivided into 35-μm^2 bins, and the numbers of FR and BDA dots are counted in each bin. Bins containing at least one or more FR or BDA dots are colored red or black, respectively; those containing at least two of each type are colored green. The analysis shows that the axons arising from homotopic whisker representations in MI and SI are partly overlapping.

and the densities of points used to code a specific labeling intensity. The method exemplified here allows robust relative comparisons among cases, as well as 3D visualization of the shape and location of the estimated zones of overlap.

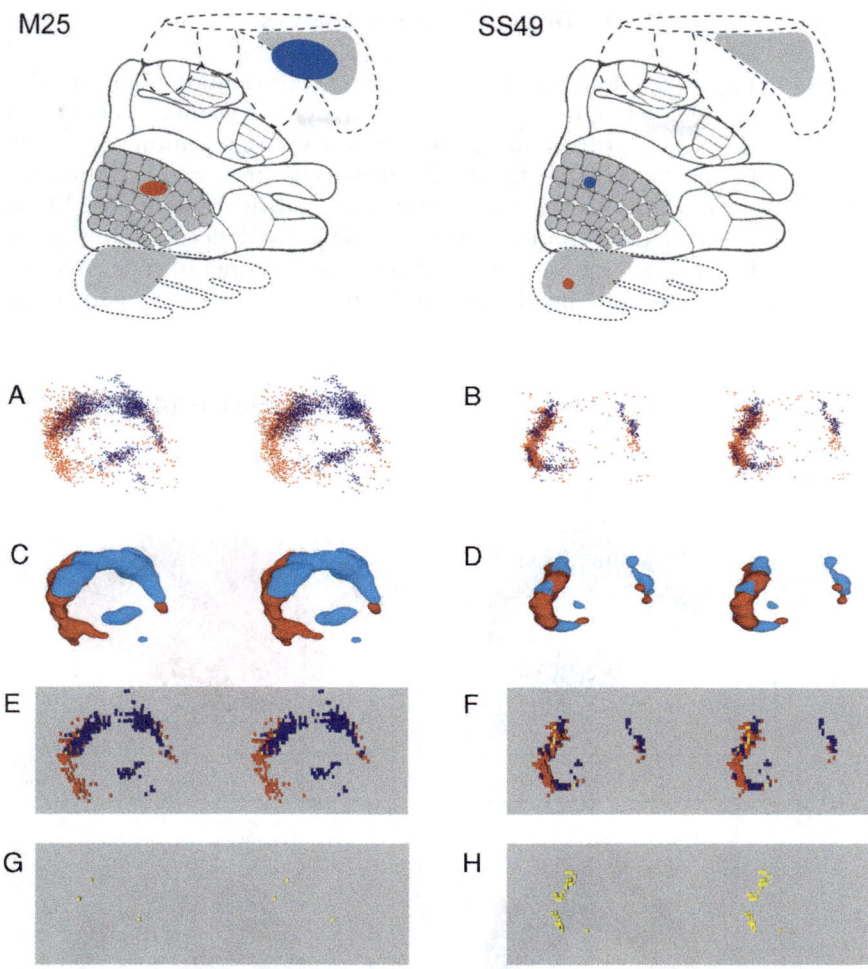

Figure 17.8. Computer-generated visualization (stereo-pairs showing the 3D distribution, shape, and relationship within the ipsilateral pontine nuclei) of cerebro-pontine projections labeled after axonal tracer injections into homotopic whisker representations in primary motor (MI), primary somatosensory (SI), and secondary somatosensory (SII) cortices. Left column shows dual injections into MI (blue) and SI (red). Right column shows dual injections into SI (blue) and SII (red) [modified from Leergaard *et al.* (2004), with permission]. The position and extent of the injections are indicated in the outline drawings of the cortical maps. To see 3D images (A–H), the viewer must cross the eye axis to let the stereo-pair of images merge. (A, B) Dot maps showing the distribution of cerebro–pontine axons labeled with biotinylated dextran amine (BDA, blue) and Fluoro-ruby (FR, red). (C, D) Isodensity surfaces surrounding the clusters of FR and BDA labeling, shown in view from ventral. (E, F) 3D distribution of the results of overlap analysis (for details, see Fig. 17.7), with red (FR) and blue (BDA), and yellow (FR and BDA) bins. The shape and distribution of labeled clusters originating from SI and SII are similar (B, D), while labeled clusters originating from MI have a more rostral and ventral distribution (A, C). The dual injections in SI and SII produce more overlap (F, H) than the dual injections in SI and MI (E, G).

E. Density Gradient Analysis

Another approach for extracting quantitative information about the distribution of the labeling is density gradient analysis. This method has been employed in our investigations of pontine nuclei organization (Fig. 17.9; Leergaard *et al.*, 2000b, 2004; for similar analysis of other brain regions, see Malmierca *et al.*, 1998; Vassbø *et al.*, 1999). The analysis shown in Fig. 17.9 is based on 2D "collapsed" projections of the 3D point data representing the labeling. The analysis may be repeated for different angles of view. A square grid is superimposed on the 2D map, and a gray or color level is assigned to

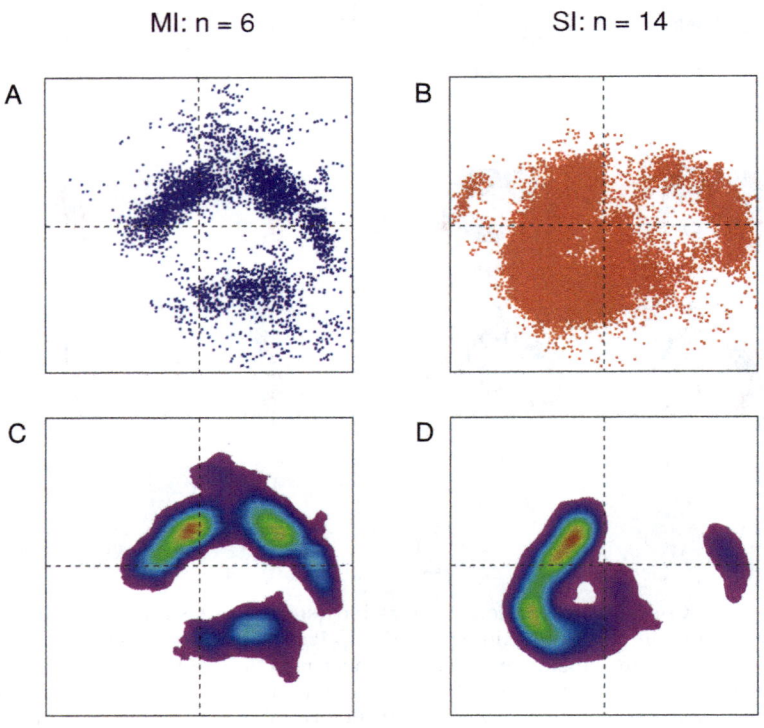

Figure 17.9. Dot maps (A, B) and density maps (C, D) showing the distribution of whisker-related cerebro–pontine projections from MI (A, C) and SI (B, D) [modified from Leergaard *et al.* (2004), with permission]. Data from 14 different tracing experiments (for details, see Leergaard *et al.*, 2004) have been accumulated in the same local pontine coordinate system and are shown in a view from ventral. Only the right half of the pontine coordinate system is shown. Dots represent the spatial distribution of labeled cerebro–pontine axons (A, B). The color gradient in (C) and (D) shows the highest densities in red and the lowest in violet. Densities < 5% of the maximum value are not shown. Overall, the distribution of high-density regions of labeling is different. Together, the SI and MI high-density regions form complementary parts of a ring-like volume. Scale bar: 500 μm.

each square, corresponding to the density of points within a user-defined radius centered on the square. Thereby, a grayscale (or color) coded density map is constructed. With the use of small grid size and short radius, it is possible to demonstrate changes in densities across short distances. The example shown in Fig. 17.9 uses data accumulated from several experiments and shows the overall differences in projection patterns of primary motor and primary somatosenory projections to the pontine nuclei. The density gradient analyses reveal that the high-density regions of the labeled projections are differently distributed, and that they together appear to form complementary parts of a ring-like volume (see also Leergaard *et al.*, 2004).

F. Stereo-Imaging

The classical (printed) journals have several limitations for visualization of 3D reconstructions. Perception of depth in 3D reconstructions, displayed in journal figures, may be achieved with the use of stereo-images. The stereoscopic effect is realized by the use of image pairs that have approximately 8° different vertical rotation. The viewer crosses the eye axis to merge the pair of images into a 3D image. A series of stereo images of point clouds, surface-modeled point clouds, and visualizations of spatial overlaps are shown in Fig. 17.8.

VIII. WHOLE BRAIN ATLASING

The development of 3D digital whole brain atlas systems represents an emerging opportunity for making more efficient use of data recorded from histological sections. Recent developments in the fields of informatics, imaging, and computerized microscopy make it technically feasible and scientifically attractive to create such atlas systems. The general motivation behind digital brain atlasing is to assign localization to data from the brain and to combine data of multiple categories and from different individuals in the same information structure. This will allow investigators to average comparable observations, describe variability, reuse data in new combinations, and overall to extract knowledge based on large data sets. Many data categories may be included in such atlas systems, including axonal tract-tracing data describing organization of projections and connections, as exemplified in the illustrations in the present chapter. At the level of the mouse brain, extensive efforts are currently undertaken to build whole brain atlas systems for gene expression data. Examples include the Allen Brain Atlas (Pennisi, 2003) and The Mouse Brain Atlas (MacKenzie-Graham *et al.*, 2003).

Local coordinate systems, recommended above to be used for 3D reconstructions of any region of the brain (smaller than the whole brain), can be translated to global coordinates employed in whole brain digital atlas systems. An example translation of a local coordinate system onto global coordinates is shown in section "Translation Between Local and Global Coordinates." Hence, once 3D reconstructions are prepared in a local coordinate system, the data contained in the reconstruction can in principle be translated to global coordinates and uploaded in databases that are accessed by whole brain atlas applications.

Here wee exemplify the potential use of a whole brain 3D digital atlas with an analysis of in vivo tracing of axonal trajectories and terminal fields using manganese-enhanced MRI (Leergaard *et al.*, 2003). Figure 17.10 illustrates a conceptual implementation of a 3D digital atlas template consisting of contour line data for the brain surface and selected internal structures, created by digitizing selected structures from a standard stereotaxic rat brain atlas (Swanson, 1999; Fig. 17.10A, B). This 3D template is used as a matrix for combining and segmenting MRI image volumes. The co-registration of the MR images with the atlas was achieved by matching multiple well-known anatomical structures, applying the same principles as described above for the alignment of data from serial sections. Since the identification of anatomical landmarks in MR images is variably difficult, depending on the employed scanning resolution and parameters, the use of a 3D atlas approach was particularly helpful for assessing the boundaries of several brain structures (Fig. 17.10I, J). In addition to aiding anatomical parcellation, the 3D atlasing approach was in this example fundamental for the demonstration of topographical organization in the corticothalamic system, based on in vivo tracing of pathways using manganese-enhanced MRI (Fig. 17.10C–J).

IX. SUMMARY OF ADVANTAGES AND LIMITATIONS

By assembling computerized 3D reconstructions from serial sections, complex spatial relationships may be visualized and analyzed in a more complete and versatile fashion than with the use of single sections only. A 3D reconstruction may be submitted to interactive inspection from any angle of view and sophisticated visualization and analysis including, as outlined above, digital re-slicing, surface modeling, analysis of spatial overlaps, density gradient analysis, and analysis of other complex spatial relationships between distributions recorded in the reconstruction. Altogether, the 3D reconstruction approach facilitates perception of complex architectonical features. Three-dimensional reconstructed data, as exemplified in the present chapter, are furthermore well suited for quantitative analysis.

A fundamental aspect of computerized 3D reconstruction is the opportunity to compare data from different experimental animals. This can in its simplest form be performed by (1) establishing local, standardized coordinate systems for the brain region of interest and (2) applying such coordinate systems to the 3D reconstructed data. This approach implies that data from different animals are size-adjusted and rotated to fit the chosen coordinate

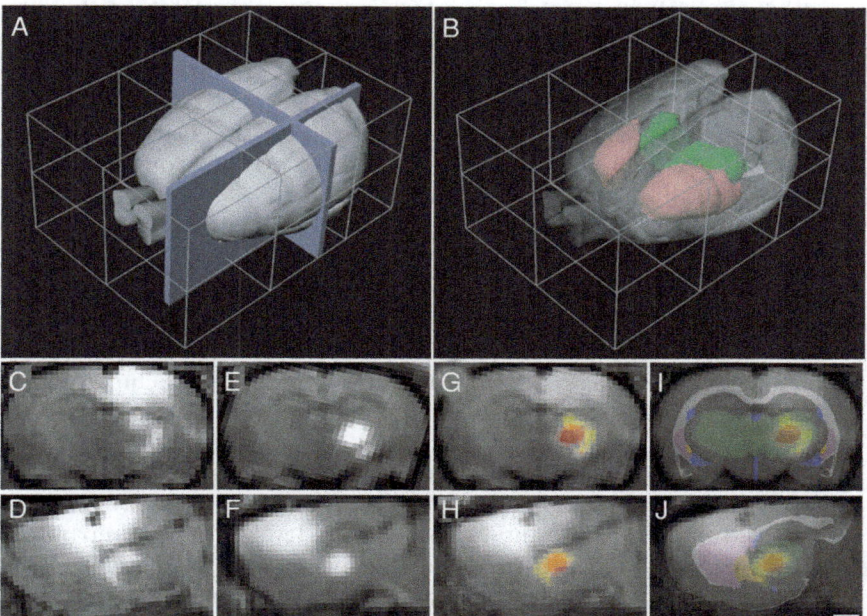

Figure 17.10. Digital rat brain atlasing applied to facilitate localization of manganese (Mn^{2+})-enhanced MRI signals, used to trace corticothalamic pathways [modified from Leergaard *et al.* (2003), with permission]. (A, B) 3D atlas reconstructed from serial section drawings available in the Swanson atlas of the rat brain (Swanson, 1999). In (A), the external surface of the brain is shown as a solid surface. Stereotaxic atlas coordinates are shown as a bounding box with selected reference lines. The position and orientation of the 200-μm-thick coronal and sagittal slices shown in (C–J) are indicated by solid blue boxes. In (B), the external surface is made transparent, and the boundaries of the caudate–putamen complex (pink) and the thalamus (green) are seen. (C–F) show coronal (C, E) and sagittal (D, F) T_1 weighted MRI slices, obtained with 390 $μm^3$ isotropic voxels, using a 3 T human clinical MRI scanner, about 10 h after focal injections of manganese chloride ($MnCl_2$) were made into medial (C, D) and lateral (E, F) parts of the somatosensory cortex in two adult rats, respectively. Different patterns of specific Mn^{2+} signal enhancement are detectable in cortico–cortical and cortico–subcortical pathways in the two animals. (G–J) By superimposing the MR volumes from the two cases onto the 3D digital atlas, it was possible to demonstrate topographically different labeling patterns. In G and H, yellow represents labeling in the thalamus following medical injection (also shown in C, D), and red labeling following lateral injection (also shown in E, F). Atlas regions are shown in (I–J) with different colors for different regions (bright gray, white matter; pink, caudate–putamen; orange, globus pallidus; green, thalamus).

system. More advanced warping, in particular needed for brain regions of highly variable shapes, requires the use of more advanced nonlinear transformations (see, e.g., Toga and Thompson, 1999). In our investigations of neural populations and sensory projection systems, use of 3D reconstructions, visualizations, and analysis has been critical for attaining a fuller understanding of topographic organization and complex spatial relationships among functionally defined neuronal elements (see, e.g., Bajo *et al.*, 1999; Bjaalie *et al.*, 1997b, 2005; Leergaard *et al.*, 1995, 2000a, b, 2004; Malmierca *et al.*, 1998; Nikundiwe *et al.*, 1994). Combined use of tomographic image data and geometric 3D reconstructions (Leergaard *et al.*, 2003) illustrates future possibilities for the construction of multidimensional 3D digital brain atlases, in which multiple modalities of brain data may be brought together and localized within a 3D atlas framework.

The technical and scientific limitations of 3D geometric reconstructions are found at several levels. First, the quality of the ensuing 3D models never exceeds the quality of the original tissue material and is further critically determined by the procedures used for data acquisition and subsequent alignment, as well as by the accuracy with which section angle, section position, and tissue shrinkage are monitored. The examples included in the present chapter are taken from small animal brains displaying limited individual variation. The level of complexity is considerably higher in studies of, e.g., primate or human cortical structures, and with the use of material with more artifacts and section distortions. Second, the 3D reconstruction method is overall time consuming, both at the level of tissue preparation and data acquisition. Robotic tissue processing and automated data acquisition/microscopy approaches (Herzig *et al.*, 2001; Visel *et al.*, 2004) may further help to rapidly acquire complete high-resolution images of entire sections at higher speeds. Automated procedures for acquiring the essential information from such images are being used (see, e.g., Lillehaug *et al.*, 2002). The 3D reconstruction process may also be further improved with the aid of tomographic templates. Technical improvement in resolution and quality of data obtained by tomographic imaging method will also presumably allow generation of data more suitable for detailed analysis. These contemporary developments offer new opportunities and will most likely influence the field substantially over the years to come.

APPENDIX

The descriptions in this section are modified from information posted in the database application FACCS (www.rbwb.org; see also Bjaalie *et al.*, 2005).

A. Application of Local Coordinate System for the Rat Pontine Nuclei

1. Purpose

The purpose is to facilitate interindividual comparison of data originating from different experimental animals.

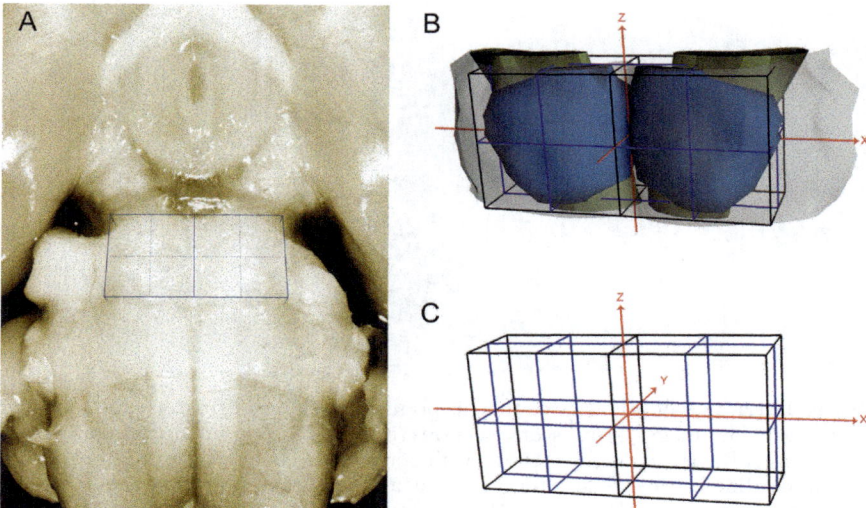

Figure 17.11. Local coordinate system for the rat pontine nuclei [from the database application FACCS (www.rbwb.org; Bjaalie *et al.*, 2005)]. (A) Image of the ventral surface of the rat brain with the standardized local coordinate system for the pontine nuclei. (B) The standardized local coordinate system shown in relation to the pontine nuclei (blue surfaces) and descending fiber tracts (dark gray surfaces). (C) The standardized local coordinate system shown in isolation.

2. Definition

The local coordinate system is a cuboid, referred to below as a bounding box, with anatomically defined orientation, center, and extent. In order to apply the coordinate system, using software tools such as those exemplified by Bjaalie *et al.* (2006), the following anatomical boundaries have to be identified from a stack of histological sections through the pontine nuclei and in the 3D reconstruction based on such sections (Fig. 17.12):

- ventral surface of corticobulbar and corticospinal fiber tracts (peduncle)
- rostralmost extent of pontine gray substance
- caudalmost extent of pontine gray substance
- ventralmost extent of pontine gray substance
- lateralmost extent of pontine gray substance
- midline of the brain.

The coordinate system is applied individually to the right and left pontine nuclei. A bilateral coordinate system is obtained by subsequently scaling the two sides so that all edges are aligned.

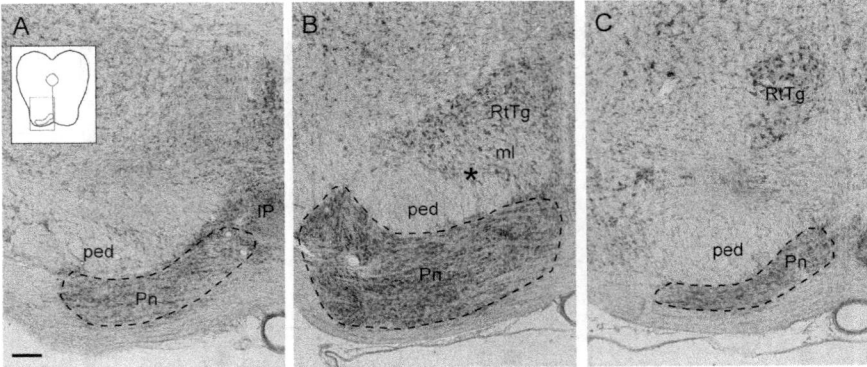

Figure 17.12. The boundaries of the rat pontine nuclei gray matter (dotted lines) determined in images of transverse sections through the pontine nuclei, stained for cresyl violet [from Brevik *et al.* (2001), with permission]. Small cell groups belonging to the pontine nuclei are also found inside, and along the dorsal aspect, of the descending fiber tracts. *Abbreviations:* IP, interpeduncular nuclei; ml, medial lemniscus; ped, peduncle (corticobulbar and corticospinal tracts); Pn, pontine nuclei; RtTg, reticulotegmental nucleus of the pons. Scale bar: 200 μm.

Orientation: The bounding box is oriented so that its surfaces are parallel or perpendicular to the midline of the brain and the long axis of the brain stem at the level of the pons (identified by the ventral surface of corticobulbar and corticospinal fiber tracts).

Extent toward rostral, caudal, ventral, and lateral: The extent of the bounding box in these directions is defined by the maximum extent of the pontine gray substance in the same directions.

Center point: The center point in each half of the bilateral coordinate system is placed at the ventral surface of the peduncle, halfway from rostral to caudal and halfway from the midline to the lateral end of the pontine nuclei.

Extent toward dorsal: The dorsal boundary of the pontine nuclei (pontine neurons located dorsal to the fiber tracts) is less distinct than the other boundaries. The dorsal extent of the bounding box, relative to the center point (see above), is therefore defined as the distance from the center point to the ventral boundary of the bounding box.

Origin: The origin of the local pontine coordinate system is placed at the intersection of the midline and a line connecting the left and right center points. All data coordinates made available through the FACCS database are related to this origin defined for a bilateral coordinate system. Note that the origin of the relative coordinates used in our presentation diagrams in several of the related publications for practical reasons is differently defined.

Standard size: Following application of the local pontine coordinate system to an individual animal, the coordinate system and related data may be size adjusted to a standard of choice in order to facilitate interindividual comparison. We have defined a standard based on the average size measured from the experimental animals included in the publication by Leergaard *et al.* (2000b). The bounding box is size adjusted to 2000 × 2000 × 1200 μm for each side (rostrocaudal × mediolateral × ventrodorsal distances), i.e., 2000 × 4000 × 1200 μm for both sides of the pontine nuclei.

3. Technical Implementation

a. Application to a 3D Reconstruction

A 3D reconstruction consists of a stack of digitized sections. Each digitized section contains point and line coded data representing labeling (points), boundaries, and landmarks (lines). The local coordinate system is applied to a 3D reconstruction as follows:

- *Orientation*: A cuboid bounding box is oriented along the long axis of the brain stem at the level of the pons.
- *Center point*: The center point is found at the ventral surface of the peduncle by measuring or slicing the reconstruction of the pontine nuclei halfway from rostral to caudal, and halfway from the midline to the lateral end.
- *Extent*: The boundaries of the box are adjusted to fit the (histologically defined) rostral, caudal, lateral, medial, and ventral limits of the pontine nuclei. The distance from the center point to the dorsal boundary is set to be equal to the distance from the center point to the ventral boundary.
- *Standard size*: Affine transformations (i.e., combinations of linear scaling, translation, and rotation) are used to bring individual coordinate systems (together with all related data) in register with a chosen standard coordinate system.
- *Application to 2D section images*: The local coordinate system may be applied to pontine nuclei section images if the position and angle of the sections are known and the above listed anatomical boundaries are identified (i.e., maximum rostrocaudal and mediolateral extent of pontine gray substance, and distance from center point to the maximum ventral extent of pontine gray substance). The coordinate system is applied using the above-described criteria.

b. Application to 2D Images

The local coordinate system may be applied to pontine nuclei section images if the position and angle of the sections are known and the above listed anatomical boundaries are identified (i.e., maximum rostrocaudal

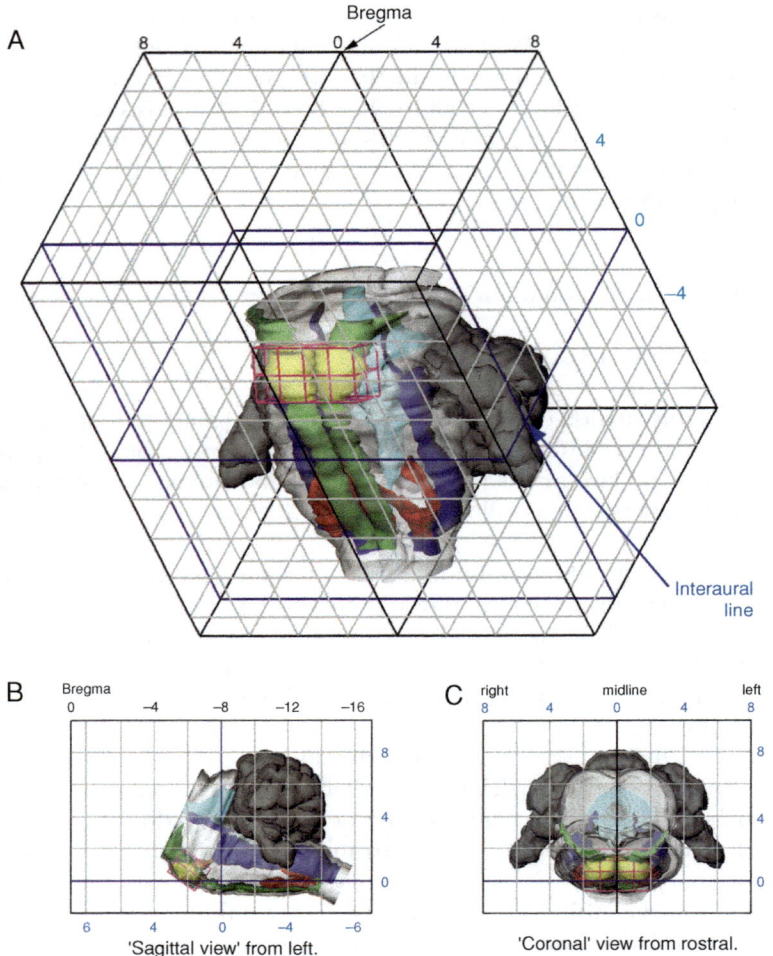

Figure 17.13. The standardized local coordinate system for the rat pontine nuclei (red bounding box) shown in relation to a 3D reconstruction of the brain stem (data from Brevik *et al.*, 2001), and the skull-based coordinate system of Paxinos and Watson (1998, 2005) in oblique lateral (A), sagittal (B), and coronal (C) views [modified from Bjaalie and Leergaard (2005)]. The external boundaries of the brain stem are shown as a transparent gray surface and other structures as solid surfaces (gray, cerebellum; yellow, pontine nuclei; green, descending corticobulbar and corticospinal tract; dark blue, trigeminal nuclei; red, lateral reticular nucleus; bright blue, aquaduct/fourth ventricle).

and mediolateral extent of pontine gray substance, and distance from center point to the maximum ventral extent of pontine gray substance). The coordinate is then applied using the criteria above for the application to a 3D reconstruction.

B. Translation Between Local and Global Coordinates

1. Purpose

The purpose is to translate data from the standardized local coordinate system for the pontine nuclei to a coordinate system for the whole brain (here exemplified by the skull-based stereotaxic coordinates of Paxinos and Watson, 1998).

2. Definitions

For the definition of standardized local coordinate system for the pontine nuclei, see above. For definition of skull-based stereotaxic coordinates, see Paxinos and Watson (1998, 2005; see also Swanson, 1999; Fig. 17.13).

3. Technical Description

Mediolateral levels are identical in the two coordinate systems and can thus be directly superimposed. The translation of rostrocaudal and dorsoventral coordinates from one coordinate system to the other is defined in Fig. 17.14, where the internal coordinate system for the pontine nuclei is applied to a 2D section image (see Appendix A.3.b *Application to 2-D images*; above) from the Paxinos and Watson (1998) stereotaxic atlas of the rat brain, allowing translation of rostrocaudal and dorsoventral coordinates between the two coordinate systems.

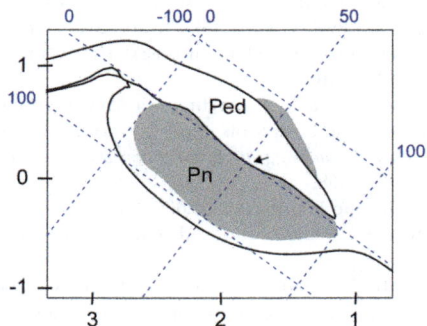

Figure 17.14. The local pontine nuclei coordinate system applied to the Paxinos and Watson (1998) stereotaxic atlas of the rat brain, at lateral level of 1.40 mm [modified from Brevik *et al.* (2001), with permission]. The dashed lines and blue numbers represent the local pontine coordinate system. The black numbers on the surrounding frame indicate the skull-based stereotaxic coordinates of Paxinos and Watson (1998). The *x*-axis shows rostrocaudal position in millimeters relative to the interaural line, while the *y*-axis indicates dorsoventral position relative to the interaural line.

ACKNOWLEDGMENTS. This work was supported by grants from The Research Council of Norway and EC grant QLG3-CT 2001-002256.

REFERENCES

Alloway, K. D., Crist, J., Mutic, J. J., and Roy, S. A., 1999, Corticostriatal projections from rat barrel cortex have an anisotropic organization that correlates with vibrissal whisking behavior, *J. Neurosci.* **19**:10908–10922.

Amunts, K., and Zilles, K., 2001, Advances in cytoarchitectonic mapping of the human cerebral cortex, *Neuroimaging Clin. N. Am.* **11**:151–169, vii.

Bajo, V. M., Merchán, M. A., Malmierca, M. S., Nodal, F. R., and Bjaalie, J. G., 1999, Topographic organization of the dorsal nucleus of the lateral lemniscus in the cat. *J. Comp. Neurol.* **407**:349–366.

Berg, B. G., Almaas, T. J., Bjaalie, J. G., and Mustaparta, H., 1998, The macroglomerular complex of the antennal lobe in the tobacco budworm moth *Heliothis virescens:* specified subdivision in four compartments according to information about biologically significant compounds, *J. Comp. Physiol. A* **183**:669–682.

Berg, B. G., Almaas, T. J., Bjaalie, J. G., and Mustaparta, H., 2005, Projections of male specific receptor neurons in the antennal lobe of the Oriental tobacco budworm moth, *Helicoverpa assulta.* A unique glomerular organisation among related species, *J. Comp. Neurol.* **486**:209–220.

Bjaalie, J. G., 1991, Three-dimensional computer reconstructions in neuroanatomy: basic principles and methods for quantitative analysis, In: Stewart, M. G. (ed.), *Quantitative Methods in Neuroanatomy,* Chichester: John Whiley & Sons Inc., pp. 249–293.

Bjaalie, J. G., 2002, Localization in the brain: new solutions emerging, *Nat. Neurosci. Rev.* **3**:322–325.

Bjaalie, J. G., Daehlen, M., and Stensby, T. V., 1997a, Surface modelling from biomedical data, In: Daehlen, M., and Tveito, A. (eds.), *Numerical Methods and Software Tools in Industrial Mathematics,* Boston: Birkhauser, pp. 9–26.

Bjaalie, J. G., and Diggle, P. J., 1990, Statistical analysis of corticopontine neuron distribution in visual areas 17, 18, and 19 of the cat, *J. Comp. Neurol.* **295**:15–32.

Bjaalie, J. G., Diggle, P. J., Nikundiwe, A., Karagülle, T., and Brodal, P., 1991, Spatial segregation between populations of ponto–cerebellar neurons: statistical analysis of multivariate spatial interactions, *Anat. Rec.* **231**:510–523.

Bjaalie, J. G., and Leergaard, T. B., 2005, Three-dimensional visualization and analysis of wiring patterns in the brain: experiments, tools, models, and databases, In: Koslow, S. H., and Subramaniam, S., (eds.), *Databasing the Brain: From Data to Knowledge,* Chichester: John Wiley & Sons Inc., pp. 349–368.

Bjaalie, J. G., Leergaard, T. B., and Pettersen, C., 2006, Micro3D: computer program for three-dimensional reconstruction, visualization, and analysis of neuronal populations and brain regions, *Int. J. Neurosci.,* in press.

Bjaalie, J. G., Leergaard, T. B., Lillehaug, S., Odeh, F., Moene, I., Kjode, J. O., and Darin, D., 2005, Database and tools for analysis of topographic organization and map transformations in major projection systems of the brain, *Neuroscience,* 136:618–696.

Bjaalie, J. G., Sudbø, J., and Brodal, P., 1997b, Corticopontine terminal fibres form small scale clusters and large scale lamellae in the cat, *Neuroreport* **8**:1651–1655.

Brevik, A., Leergaard, T. B., Svanevik, M., and Bjaalie, J. G., 2001, Three-dimensional computerised atlas of the rat brain stem precerebellar system: approaches for mapping, visualization, and comparison of spatial distribution data, *Anat. Embryol.* **204**:319–332.

Brodal, P., 1982, The cerebropontocerebellar pathway: salient features of its organization, In: Chan Palay, V., and Palay, S. (eds.), *The Cerebellum—New Vistas. Exp. Brain Res.,* Suppl. 6, Berlin and Heidelberg: Springer-Verlag, pp. 108–132.

Brodal, P., and Bjaalie, J. G., 1992, Organization of the pontine nuclei, *Neurosci. Res.* **13**:83–118.

Diaz, C., Glover, J. C., Puelles, L., and Bjaalie, J. G., 2003, The relationship between hodological and cytoarchitectonic organization in the vestibular complex of the 11-day chicken embryo, *J. Comp. Neurol.* **457**:87–105.

Diggle, P. J., 1986, Displaced amacrine cells in the retina of a rabbit: analysis of a bivariate spatial point pattern, *J. Neurosci. Methods* **18**:115–125.

Ewald, A. J., McBride, H., Reddington, M., Fraser, S. E., and Kerschmann, R., 2002, Surface imaging microscopy, an automated method for visualizing whole embryo samples in three dimensions at high resolution, *Dev. Dyn.* **225**:369–375.

Geiger, B., 1993, *Three-dimensional modelling of human organs and its application to diagnosis and surgical planning*, Institut National de Recherche en Informatique et en Automatique, Sophia-Antipolis, France, Report 2105.

Glaser, E. M., Gissler, M., and Van der Loos, H., 1979, An interactive camera lucida computer-microscope, *Soc. Neurosci. Abstr.* **5**:1697.

Glaser, J. R., and Glaser, E. M., 1990, Neuron imaging with Neurolucida—a PC-based system for image combining microscopy, *Comput. Med. Imaging Graph* **14**:307–317.

Glaser, E. M., Tagamets, M., McMullen, N. T., and Van der Loos, H., 1983, The image-combining computer microscope—an interactive instrument for morphometry of the nervous system, *J. Neurosci. Methods* **8**:17–32.

Grefkes, C., Geyer, S., Schormann, T., Roland, P., and Zilles, K., 2001, Human somatosensory area 2: observer-independent cytoarchitectonic mapping, interindividual variability, and population map, *Neuroimage* **14**:617–631.

He, S. Q., Dum, R. P., and Strick, P. L., 1993, Topographic organization of corticospinal projections from the frontal lobe: motor areas on the lateral surface of the hemisphere, *J. Neurosci.* **13**:952–980.

Herzig, U., Cadenas, C., Sieckmann, F., Sierralta, W., Thaller, C., Visel, A., and Eichele, G., 2001, Development of high-throughput tools to unravel the complexity of gene expression patterns in the mammalian brain, *Novartis Found. Symp.* **239**:129–146.

Leergaard, T. B., 2003, Clustered and laminar topographic patterns in rat cerebro–pontine pathways, *Anat. Embryol.* **206**:149–162.

Leergaard, T. B., Alloway, K. D., Mutic, J. J., and Bjaalie, J. G., 2000a, Three-dimensional topography of corticopontine projections from rat barrel cortex: correlations with corticostriatal organization, *J. Neurosci.* **20**:8474–8484.

Leergaard, T. B., Alloway, K. D., Pham, T. A., Bolstad, I., Hoffer, Z. S., Pettersen, C., and Bjaalie, J. G., 2004, Three-dimensional topography of corticopontine projections from rat sensorimotor cortex: comparisons with corticostriatal projections reveal diverse integrative organization, *J. Comp. Neurol.* **478**:306–322.

Leergaard, T. B., and Bjaalie, J. G., 1995, Semi-automatic data acquisition for quantitative neuroanatomy. MicroTrace—computer programme for recording of the spatial distribution of neuronal populations, *Neurosci. Res.* **22**:231–243.

Leergaard, T. B., Bjaalie, J. G., Devor, A., Wald, L. L., and Dale, A. M., 2003, In vivo tracing of major rat brain pathways using manganese-enhanced magnetic resonance imaging and three-dimensional digital atlasing, *NeuroImage* **20**:1591–1600.

Leergaard, T. B., Lakke, E. A., and Bjaalie, J. G., 1995, Topographical organization in the early postnatal corticopontine projection: a carbocyanine dye and 3-D computer reconstruction study in the rat, *J. Comp. Neurol.* **361**:77–94.

Leergaard, T. B., Lyngstad, K. A., Thompson, J. H., Taeymans, S., Vos, B. P., De Schutter, E., Bower, J. M., and Bjaalie, J. G., 2000b, Rat somatosensory cerebropontocerebellar pathways: spatial relationships of the somatotopic map of the primary somatosensory cortex are preserved in a three-dimensional clustered pontine map, *J. Comp. Neurol.* **422**:246–266.

Lillehaug, S., Oyan, D., Leergaard, T. B., and Bjaalie, J. G., 2002, Comparison of semi-automatic and automatic data acquisition methods for studying three-dimensional distributions of large neuronal populations and axonal plexuses, *Network* **13**:343–356.

Lohmann, K., Gundelfinger, E. D., Scheich, H., Grimm, R., Tischmeyer, W., Richter, K., and Hess, A., 1998, BrainView: a computer program for reconstruction and interactive visualization of 3D data sets, *J. Neurosci. Methods* **84**:143–154.

MacKenzie-Graham, A., Jones, E. S., Shattuck, D. W., Dinov, I. D., Bota, M., and Toga, A. W., 2003, The informatics of a C57BL/6J mouse brain atlas, *Neuroinformatics* **1**:397–410.

Malmierca, M. S., Leergaard, T. B., Bajo, V. M., Bjaalie, J. G., and Merchan, M. A., 1998, Anatomic evidence of a three-dimensional mosaic pattern of tonotopic organization in the ventral complex of the lateral lemniscus in Cat, *J. Neurosci.* **18**:10603–10618.

Maunsbach, A. B., and Afzelius, B. A., 1999, *Biomedical Electron Microscopy*, San Diego: Academic Press.

Maurin, Y., Banrezes, B., Menetrey, A., Mailly, P., and Deniau, J. M., 1999, Three-dimensional distribution of nigrostriatal neurons in the rat: relation to the topography of striatonigral projections, *Neuroscience* **91**:891–909.

Nadasdy, Z., and Zaborszky, L., 2001, Visualization of density relations in large-scale neural networks, *Anat. Embryol.* **204**:303–317.

Nikundiwe, A. M., Bjaalie, J. G., and Brodal, P., 1994, Lamellar organization of pontocerebellar neuronal populations. A multi-tracer and 3-D computer reconstruction study in the cat, *Eur. J. Neurosci.* **6**:173–186.

Paxinos, G., and Franklin, K. B. J., 2001, *The Mouse Brain in Stereotaxic Coordinates*, San Diego: Academic.

Paxinos, G., and Watson, C., 1998, *The Rat Brain in Stereotaxic Coordinates*, San Diego: Academic Press.

Paxinos, G., and Watson, C., 2005, *The Rat Brain in Stereotaxic Coordinates*, Burlington, MA: Academic Press.

Pennisi, E., 2003, Neuroscience. Mapping the brain's genes: a Microsoft dividend, *Science* **301**:1642.

Pommert, A., Tiede, U., and Höhne, K. H., 2002, Volume visualization, In: Toga, A. W., and Mazziotta, J. C. (eds.), *Brain Mapping. The Methods*, San Diego: Academic Press, pp. 707–723.

Rademacher, J., Burgel, U., Geyer, S., Schormann, T., Schleicher, A., Freund, H. J., and Zilles, K., 2001, Variability and asymmetry in the human precentral motor system. A cytoarchitectonic and myeloarchitectonic brain mapping study. *Brain* **124**:2232–2258.

Romeis, B., 1968, *Mikroskopische Technik*, München: R. Oldenbourg Verlag.

Ruigrok, T. J. H., 2004, Precerebellar nuclei and red nucleus, In: Paxinos, G. (ed.), *The Rat Nervous System*, San Diego, CA: Elsevier Academic Press, pp. 167–204.

Schleicher, A., and Zilles, K., 1990, A quantitative approach to cytoarchitectonics: analysis of structural inhomogeneities in nervous tissue using an image analyzer, *J. Microsc.* **157**(Pt 3):367–381.

Schmitt, O., Homke, L., and Dumbgen, L., 2003, Detection of cortical transition regions utilizing statistical analyses of excess masses, *Neuroimage* **19**:42–63.

Schwarz, C., and Möck, M., 2001, Spatial arrangement of cerebro–pontine terminals, *J. Comp. Neurol.* **435**:418–432.

Schwarz, C., and Their, P., 1995, Modular organization of the pontine nuclei: dendritic fields of identified pontine projection neurons in the rat respect the borders of cortical afferent fields, *J. Neurosci.* **15**:3475–3489.

Senft, S. L., 2002, Axonal navigation through voxel substrates: a strategy for reconstructing brain circuitry, In: Ascoli, G. (ed.), *Computational Neuroanatomy: Principles and Methods*, Totowa, NJ: Humana Press, pp. 245–270.

Shen, C., O'Brien, J. F., and Shewchuk, J. R., 2004, Interpolating and approximating implicit surfaces from polygon soup, In: *Proceedings of the ACM SIGGRAPH*, Los Angeles, CA: ACM Press.

Swanson, L. W., 1999, *Brain Maps: Structure of the Rat Brain*, Amsterdam: Elsevier.

Toga, A. W., 1990, Three-dimensional reconstruction, In: Toga, A. W. (ed.), *Three-Dimensional Imaging*, New York: Raven Press, pp. 251–272.

Toga, A. W., and Banerjee, P. K., 1993, Registration revisited, *J. Neurosci. Methods.* **48**:1–13.

Toga, A. W., and Thompson, P., 1999, An introduction to brain warping, In: Toga, A. W. (ed.), *Brain Warping*, San Diego: Academic Press, pp. 1–26.

Van Essen, D. C., Drury, H. A., Dickson, J., Harwell, J., Hanlon, D., and Anderson, C. H., 2001, An integrated software suite for surface-based analyses of cerebral cortex, *J. Am. Med. Inform. Assoc.* **8**:443–459.

Vassbø, K., Nicotra, G., Wiberg, M., and Bjaalie, J. G., 1999, Monkey somatosensory cerebrocerebellar pathways: uneven densities of corticopontine neurons in different body representations of areas 3b, 1, and 2, *J. Comp. Neurol.* **406**:109–128.

Visel, A, Thaller, C., and Eichele, G., 2004, GenePaint.org: an atlas of gene expression patterns in the mouse embryo, *Nucleic Acids Res.* **32**(Database issue): D552–D556.

Welker, C., 1971, Microelectrode delineation of fine grain somatotopic organization of (SmI) cerebral neocortex in albino rat, *Brain Res.* **26**:259–275.

Woods, A. E., and Ellis, R. C., 1994, *Laboratory Histopahtology: A Complete Reference*, Orlando, FL: Churchill Livingstone.

Wree, A., Schleicher, A., and Zilles, K., 1982, Estimation of volume fractions in nervous tissue with an image analyzer, *J. Neurosci. Methods.***6**:29–43.

Zaborszky, L., Csordas, A., Buhl, D., Duque, A., Somogyi, J., and Nadasdy, Z., 2002, Computational anatomical analysis of the basal forebrain corticopetal system, In: Ascoli, G. (ed.), *Computational Neuroanatomy: Principles and Methods*, Totowa, NJ: Humana Press, pp. 171–197.

18

Atlases of the Human Brain: Tools for Functional Neuroimaging

KATRIN AMUNTS and KARL ZILLES

KATRIN AMUNTS • Institute of Medicine, Research Center Jülich, D-52441 Jülich, Germany; Department of Psychiatry and Psychotherapy, RWTH Aachen University, Pauwelsstr. 30, D-52074 Aachen, Germany   KARL ZILLES • Institute of Medicine, Research Center Jülich, D-52441 Jülich, Germany; C. and O. Vogt Institute for Brain Research, Heinrich Heine University Düsseldorf, Universitätsstr. 1, D-40225 Düsseldorf, Germany; Brain Imaging Center West, Research Center Jülich, D-52441 Jülich, Germany

Abstract: Human brain atlases are frequently used tools for the analysis of data from functional imaging and neurophysiology studies. This chapter briefly reviews historical, two- and three-dimensional printed and electronic atlas systems. It emphasizes several key aspects of such atlases: spatial relationships of macro- and microstructures, their intersubject variability, definition of reference brains and spatial reference systems, linear and nonlinear registration procedures of data sets of individual brains to reference brains, and multimodal comparisons of structural and functional data in stereotaxic space. The Appendix outlines the method of generation of probabilistic cytoarchitectonic maps, and provides addresses of some of the electronic human brain atlases and software.

Keywords: cerebral cortex, cytoarchitecture, human brain mapping, neuroanatomy, probability maps

I. INTRODUCTION

Modern functional neuroimaging and neurophysiological techniques such as functional magnetic resonance tomography (fMRI), positron emission tomography (PET), receptor PET (rPET), single photon emission computed tomography (SPECT), and magnetoencephalography (MEG) provide detailed topographical information on human brain activity. Brain activity is hereby estimated via the blood-oxygen-level-dependent (BOLD) signal, blood flow and metabolism, distribution of receptor-binding sites, and neuronal signaling, respectively. The spatial resolution of these methods is in the millimeter range. In most cases, high-resolution anatomical MR images of the same brain are coregistered to enable topographical interpretation of imaging data. The spatial resolution of anatomical MRI enables a localization of the detected brain activities at a macroscopical level, e.g., gyri, sulci, and subcortical nuclei (e.g., basal ganglia, thalamus).

The size, precise location, shape, and extent of macroanatomical landmarks, however, differ significantly between the subjects as well as between the hemispheres (Duvernoy, 1991; Ono *et al.*, 1990; Paus *et al.*, 1996; Toga and Thompson, 2003). Brain macroscopy is not a stable feature over the whole life span, but changes during development and aging as well as in neurological and psychiatric disorders (Sowell *et al.*, 2003; Thompson *et al.*, 1998; Thompson and Toga, 2002). In order to compare anatomical and functional imaging data between different subjects (e.g., in group studies), it is necessary to transform the data into a common reference space. This space can be represented by an individual brain or an "average" brain. A common reference space also opens the possibility to integrate data concerning the microscopical structure of postmortem brains such as cytoarchitecture. Postmortem brains can be analyzed with maximal spatial resolution in the micrometer range. In contrast to the macroanatomical pattern of the brain with its highly variable sulci and gyri, microstructural parcellations, e.g., cytoarchitectonic areas, seem to be more closely related to brain function (Amunts *et al.*, 2004; Binkofski *et al.*, 2000; Bodegard *et al.*, 2000a; Geyer *et al.*, 1996, 2001; Horwitz *et al.*, 2003; Larsson *et al.*, 1999, 2002; Mazziotta *et al.*, 2001a, b; Naito *et al.*, 1999, 2000; Roland and Zilles, 1994, 1998; Uylings *et al.*, 2000; Young *et al.*, 2003; Zilles *et al.*, 1995).

Cytoarchitectonic parcellations represent one aspect of cortical (and subcortical) organization. These parcellations, however, can result in quite different maps, with differences in the number of areas, nomenclature, and hierarchies. The famous Brodmann map represents only one (well-known) of many (less well-known) maps. There is no general agreement about a "gold standard" of a cytoarchitectonic map. An extreme example of a cytoarchitectonic map is that of Bailey and von Bonin, who asked whether there is any objective basis for a detailed cytoarchitectonic map at all. They came to the final conclusion that ". . . vast areas are as closely similar in structure as to make any attempt at subdivisions unprofitable, if not impossible." As a consequence, their cytoarchitectonic map is based only on a parcellation into a few main types of cortical regions (Bailey and von Bonin, 1951). The other extreme can be seen in a map of Gerhardt who distinguished between "true" borders separating two different areas, gradual borders and likely borders within one area (Gerhardt, 1940). This approach resulted in a vast number of areas of the parietal cortex, which has not been reproduced until now (Zilles and Palomero-Gallagher, 2001). It seems to be clear, at least, that the problem of a reproducible and observer-independent definition of areal borders plays a key role (see also section "Method of the Generation of Cytoarchitectonic Probabilistic Maps").

Would it be, theoretically, sufficient to know the true cytoarchitectonic parcellation of the brain in order to disclose the relationship between brain structure and function? Is a cytoarchitectonic area the same as a cortical area? For some cortical regions, in particular primary sensory and motor area, this assumption seems to hold true. For many other areas, including higher associative ones, a correlation between a certain function and cytoarchitecture has not yet demonstrated. One aspect of this failure may be related to the nonadequate characterization of a brain function at the level of a cortical area. "Language" is clearly not a function of one cortical area, but includes different aspects, e.g., phonological and semantic processing, prosody, syntax, etc. involving different brain regions.

What, however, is a cortical area? This question provides an important argument for multimodal architectonic mapping, since not all subparcellations of the cerebral cortex constitute a cortical area. For example, the subdivision of areas V1 and V2 into blob and interblob regions (Livingstone and Hubel, 1987; Roe and Tso, 1995; Wong-Riley *et al.*, 1993) reflects differences in color and orientation selectivity (V1) and receptive field properties (V2), but these subdivisions do not constitute cortical areas. Additional examples are the somatotopy of the motor and somatosensory cortex, the tonotopical organization of the auditory cortex, each of which represents a functional segregation without representing an architectonic entity. The isolated analysis of only one aspect of cortical organization, without consideration of other mapping techniques, would lead to an over-parcellation of the cerebral cortex. Instead, we propose a multimodal approach which avoids this problem by providing an overview of the different hierarchical levels (e.g., cytoarchitectonic or receptor architectonic families of cortical areas) of the cortical organization.

The close relationship between brain function, connectivity, and architecture has been demonstrated in combined histological and electrophysiological experiments in monkeys (Luppino *et al.*, 1991; Matelli *et al.*, 1991; Tanji and Kurata, 1989). It is, therefore, possible to relate functional activations to microstructurally defined parcellations such as cytoarchitectonic areas of the cerebral cortex. Moreover, it has been shown that maps based on different histological and histochemical techniques frequently show a perfect spatial coincidence of areal borders; thus, corroborating the position of an areal border by multimodal imaging including receptor autoradiography (Zilles *et al.*, 2002a, b). Moreover, since a single receptor may not reveal all borders demonstrated by other markers, this finding can be used to define a family of neurochemically related areas by studying the regional pattern of one transmitter receptor and comparing its distribution with the maps revealed by other receptors or cytoarchitecture (see section "Microstructural Localization of Neural Functions"). We think that such a multimodal concept of cortical mapping improves and supplements classical cytoarchitectonic analysis.

An important perspective for a functionally relevant architectonical parcellation of the cortex arises from the integration of architectonic maps with recent PET, fMRI, and MEG studies in a common spatial reference system (see section "Microstructural Localization of Neural Functions") and databases (see sections "The Human Brain Atlas of the International Consortium of Human Brain Mapping as an Example of a Multimodal Brain Imaging Database," "Individual Reference Brain," and "Surface-Based Atlases and Flat Maps"). Thus, the analysis and evaluation of structural–functional relationship is a major goal of microstructural brain atlases. We here focus on such approaches based on microscopical data.

II. CLASSICAL BRAIN ATLASES

A. Brodmann's Map

One of the most widely used brain atlases is the cytoarchitectonic map of Brodmann (1909). Brodmann published descriptions of cortical areas based on cytoarchitectonic studies of the human brain. The book has been translated into English, and thus, became accessible to a broader readership (Brodmann, 1994). Brodmann subdivided the cortical surface into more than 50 areas. The areas were formally numbered in the sequence of their appearance in horizontal sections. This atlas did not consider subcortical nuclei and fiber tracts. He did not study the functions of the cortical areas, but provided some functional interpretations on the basis of the knowledge available at the beginning of the 20th century. Although Brodmann showed histological sections with labeled borders between cortical areas in his book and original articles, his final map is a schematic drawing of the left lateral and right medial surfaces of a "typical" brain. He was aware that this map was a simplification of his observations and noticed that "... a schematic drawing can reflect only the major spatial relationships, and therefore, precise topographical associations (i.e., between sulci and areal borders; remark of the authors) cannot be considered in general or only in a distorted manner; this is true in particular for all those cortical regions which have borders in the neighborhood of sulci and those regions which are located in the depth of such a cortical region" (Brodmann, 1908).

   Therefore, it is not possible to conclude from Brodmann's schematic drawing whether an actual areal border is located on the walls of a sulcus, or coincides with the bottom of the sulcus (Fig. 18.1). Even if one would accept this uncertainty, the identification of areal borders by sulcal landmarks can be very problematic, since many sulci are highly variable, and a small sulcus shown in Brodmann's map may not be present in another individual brain.

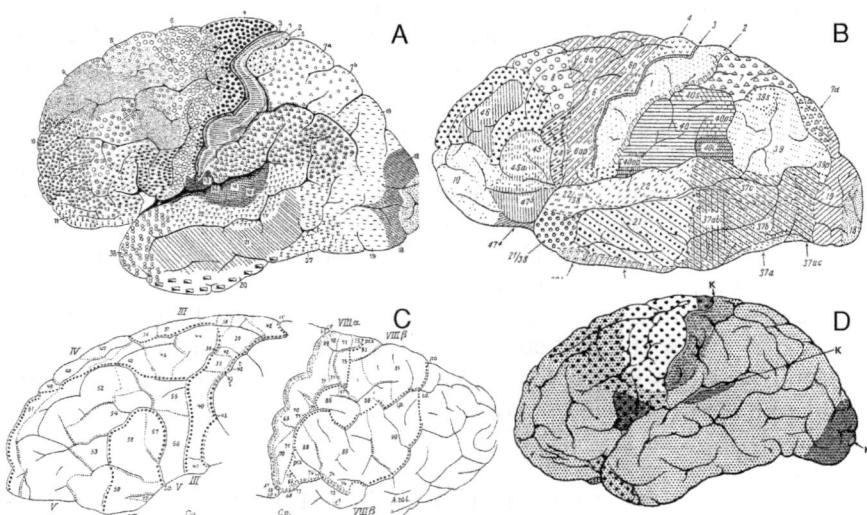

Figure 18.1. Historical cortical maps of (A) Brodmann (1909), (B) the Russian school (Sarkisov *et al.*, 1949), (C) Cecile and Oskar Vogt (1919), and (D) Bailey and von Bonin (1951). Note the different shapes and sulcal patterns, as well as the different number of areas.

B. Other Historical Maps

Although various complete cytoarchitectonic and myeloarchitectonic maps of the human cerebral cortex have also been published by other authors (e.g., Bailey and von Bonin, 1951; Campbell, 1905; Cecile and Oskar Vogt, 1919; Elliot Smith, 1907; Kleist, 1934); the Russian school (Sarkisov *et al.*, 1949); and von Economo and Koskinas, 1925, these maps did not achieve a comparably wide distribution and level of acceptance as did Brodmann's map. The comparison between these maps reveals differences with respect to (i) the underlying method of parcellation (e.g., cytoarchitecture, myeloarchitecture, unstained sections), (ii) the pattern of sulci and gyri of the particular brain used for the atlas, (iii) the location of areal borders with respect to sulci and gyri, and (iv) the number of the areas. An extreme position was held by Bailey and von Bonin (1951), who identified only four main cytoarchitectonic types in the isocortex. The Vogts, in contrast, found ~150 different myeloarchitectonic fields. They and their fellows defined dozens of transitional forms (Gerhardt, 1940; Riegele, 1931). In addition to differences between individual brains, one important reason of this discordances is presumably related to differing and highly observer-dependent criteria used for the delineation of the areas and the definition of their borders. This problem has only been mastered during the last years when statistical tools of cytoarchitectonic analysis have been introduced for an observer-independent definition of cortical borders (Annese *et al.*, 2004; Schleicher *et al.*, 1998, 1999, 2000; Schmitt *et al.*, 2003).

Starting with the development of techniques for functional imaging of the living human brain, another disadvantage of these classical printed, two-dimensional atlases became evident: They were not compatible with the three-dimensional imaging data. Moreover, the drawings of brains in the different historical maps showed differences in orientation, shape and size, and often they did not disclose all aspects of the brain surface in an optimal way (e.g., the insula, or the auditory cortex on the dorsal surface of the superior temporal gyrus). These problems, and the need of neurosurgeons to identify in space brain lesions, epileptic foci or targets for stereotaxic surgery, stimulated the development of a stereotaxic atlas systems, e.g., the atlas of Talairach and Tournoux (1988). Although originally published as a book, this atlas provides stereotaxic coordinates of cortical areas.

III. THREE-DIMENSIONAL ANATOMICAL BRAIN ATLASES

A. The Stereotaxic Atlas of Talairach and Tournoux

The atlas of Talairach and Tournoux (1988) provides two choices of defining the spatial position: (i) x-, y-, and z-coordinates in millimeter representing the distances from a line connecting the anterior and the posterior commissures (AC–PC line), the vertical line through the anterior commissure, and the midline between the hemispheres, and (ii) the concept of proportionality realized as a proportional grid system, where the dimensions of the grid system vary with the dimensions of the major axes of individual brains. Thus, each point of this atlas brain is defined by three coordinates (x, y, and z). The atlas brain is a single human postmortem brain, which has been sectioned sagittaly. From this series of sagittal sections, two further series of sections were reconstructed in the frontal and horizontal planes by point-to-point projections. The borders of fiber tracts, subcortical nuclei, and cortical areas were traced by comparison of macroanatomical features of the atlas brain with previous descriptions, e.g., by Brodmann (1909) for the cortical areas, and by Walker (1938), for the thalamus as well as by direct observations in the actual postmortem atlas brain (e.g., olfactory tract). It is notable, that the delineations of the cortical areas are not based on histological analysis of the atlas brain, but inferred as the authors emphasize by "the transfer of the cartography of Brodmann usually pictured in two-dimensional projections," [which] "sometimes possesses uncertainties" (Talairach and Tournoux, 1988).

Although originally created for neurosurgery, radiology, and neurology, the atlas rapidly gained an increasing importance for functional imaging studies such as PET and fMRI, since it is assumed that its application enables the comparison of brains and groups of brains within and between studies (Fox *et al.*, 1985; Friston, 1997). The atlas of Talairach and Tournoux is available in a computerized version, the "Talairach Daemon" (Lancaster *et al.*, 2001). It is also implemented as an anatomical reference system in various software packages, e.g., the widely used SPM software www.fil.ion.ucl.ac.uk/spm and the International Consortium of Human Brain Mapping (ICBM) viewer http://www.bic.mni.mcgill.ca/cgi/icbm_view/ (Fig. 18.2).

As compared to the classical architectonic maps, the atlas of Talairach and Tournoux

1. offers three-dimensional information on the topography of cortical areas, subcortical nuclei, and fiber tracts,
2. introduces a stereotaxic orientation using the AC–PC line as a reliable and easy-to-find anatomical landmark,
3. proposes a proportional grid system in order to compensate intersubject differences in brain shape and size, and
4. covers the whole cortical surface, several subcortical nuclei, and fiber tracts.

At the same time, however, several important problems remained unsolved:

1. The atlas is not based on microstructural analyses of the atlas brain, but on a vaguely defined method of transfer of microscopical (architectonic) information about cortical areas from the classical, schematic drawings of Brodmann to the atlas brain based on macroanatomical

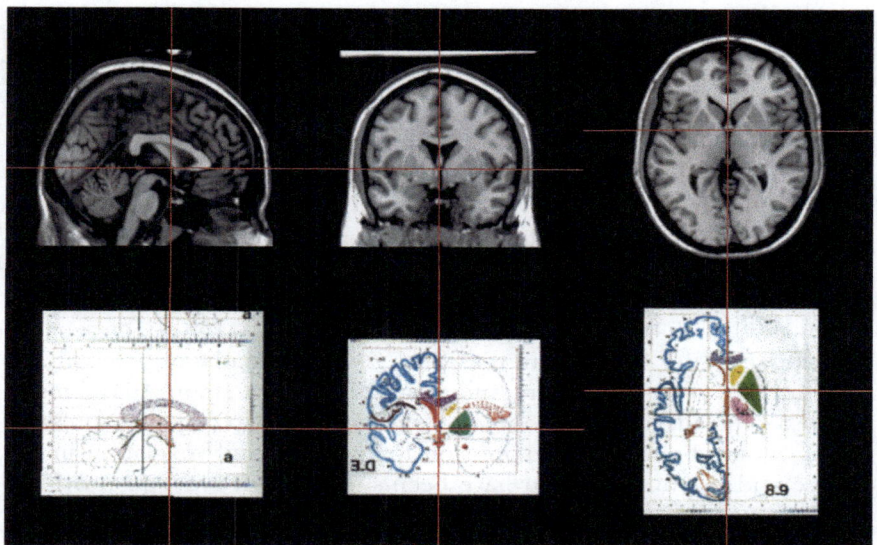

Figure 18.2. ICBM viewer (http://www.bic.mni.mcgill.ca/cgi/icbm_view/). Upper row: average of 27 T1-weighted scans of an individual brain ("MNI single-subject template"); lower row: from Talairach and Tournoux (1988).

similarities between brains. Such an approach, however, assumes an invariable correspondence between macroanatomical and microscopical features, which is true for the borders of only a few cortical areas (see below), but which cannot be accepted for the majority of other regions (e.g., the so-called higher associative areas of the parietal lobe).

2. The delineations of cortical areas are based on Brodmann's map. During the last years, however, it has been shown (see below) that Brodmann's map provides an oversimplified parcellation in some brain regions, or is insufficient in other regions. This is particularly important, e.g., in the extrastriate, visual region, and the intraparietal sulcus, where functional imaging studies and observations in nonhuman primates have already shown a lack of correspondence with the structural parcellations performed by Brodmann.

3. Borders between cortical areas are not indicated.

4. Intersubject variability in brain size, shape, and sulcal pattern as well as in microstructure is not considered (see below).

5. Interhemispheric asymmetry is not taken into account. Only the corticospinal tract "is represented on the right and the left in a different manner" in correspondence to the variations of the central sulcus in the two hemispheres (Talairach and Tournoux, 1988). The remaining structural asymmetries are not considered.

6. The atlas is based on a single postmortem brain, which shows macroanatomical deformation in comparison to in vivo brains. It appears, for example, more flat than most of the in vivo brains.

B. Anatomical Maps and Atlases Based on Other Modalities

Although the above-mentioned atlases use cytoarchitectonic information in nearly all cases, brain mapping is not limited to this modality. Maps can been created on the basis of the myeloarchitecture (Elliot Smith, 1907; Mai *et al.*, 2004; Vogt and Vogt, 1919), regional and laminar distribution of receptor binding sites of different neurotransmitters such as the glutamatergic, GABAergic, cholinergic, adrenergic, dopaminergic receptors (Zilles *et al.*, 1995, 1988b, 2002a, b, 2003; Zilles and Schleicher, 1993, 1995), immunohistochemical markers (Bidmon *et al.*, 1997; Geyer *et al.*, 2000a; Hayes and Lewis, 1992; Majocha *et al.*, 1985; Zilles *et al.*, 1991a), or gene expression, e.g., the Allen Brain Atlas (http://www.brainatlas.org/default.asp and http://www.brain-map.org/index.jsp). The Allen Brain Atlas aims to create a detailed, cellular-resolution, genome-wide map of gene expression in the mouse brain. Gene expression tomography and voxelation in the mouse and the human brain have been suggested in order to provide a high-throughput approach to map regional gene expression patterns in the brain (Singh and Smith, 2003).

The atlas of the human brain of Mai *et al.* (2004) combines a macroscopic part with a microscopic (myeloarchitectonic) part. The latter comprises 69 serial sections of one hemisphere. The macroscopic part is based on three brains, which were scanned using MRI before histological processing. In contrast to many other maps, this atlas shows a detailed parcellation of subcortical nuclei.

C. Fiber Tract Mapping

The development of diffusion tensor imaging and fiber tractography enable the mapping of fiber tracts in the living human brain (Barker, 2001; Basser and Jones, 2002; Coenen *et al.*, 2001; Conturo *et al.*, 1999; Ito *et al.*, 2002; Jones *et al.*, 2002; Krings *et al.*, 2001; Le Bihan *et al.*, 2001; Mori and van Zijl, 2002; Parker *et al.*, 2002; Poupon *et al.*, 2000; Powell *et al.*, 2004; Turner *et al.*, 1991). Diffusivity is known to depend upon the orientation of the principal axes of fiber tracts (for a review see, e.g., Basser and Jones, 2002). Fiber tract trajectories can be generated from a fluid velocity field. Several fiber tracts of the brain stem and spinal cord (Assaf *et al.*, 2000; Clark *et al.*, 1999; Wheeler-Kingshott *et al.*, 2002) have been traced. This method has also been successfully applied in neurological and psychiatric disorders, e.g., in multiple sclerosis (Assaf *et al.*, 2002) and schizophrenia (Ardekani *et al.*, 2003; Foong *et al.*, 2002; Iwasawa *et al.*, 1997). Artifacts may arise, however, when discrete, coarsely sampled, noisy, voxel-averaged direction field data are used. Further problems may occur if fiber tracking is applied to incoherently organized fiber tract (Basser and Jones, 2002) or if fibers merge, branch, or cross each other (Le Bihan *et al.*, 2001). It should be kept in mind that all presently available variations of this technique do not reveal anatomical connectivity in a strict sense, since this goal would require the demonstration of synaptic connectivity at an electron-microscopical level in combination with axonal tracing.

Although in vivo tracing studies have frequently been performed in experimental animals, this approach is not possible in the living human brain. Tracing studies in adult postmortem, pathologically altered brains have been reported using various staining techniques for degenerating nerve fibers (Clarke, 1994; Clarke and Miklossy, 1990; Di Virgilio and Clarke, 1997; Miklossy and van der Loos, 1991; Mufson *et al.*, 1990; Wiesendanger *et al.*, 2004) or polarized light microscopy (Axer and Keyserlingk, 2000). Mapping of fiber tracts has also been performed in immature brains, where the early myelinating fiber tracts can be easily distinguished from the unstained, later myelinating fiber tracts (Flechsig, 1920; Yakovlev and Lecours, 1967). All these studies have contributed important data on the topography of fiber tracts and their maturation during ontogeny. They did not, however, provide stereotaxic information concerning the course and extent of fiber tracts. Moreover, they did not consider intersubject variability and inter-hemispheric asymmetry, which are important aspects of brain atlases (see below).

Using a modified myelin stain, fiber tracts have been mapped in histological sections of 10 adult human brains, and probabilistic, three-dimensional maps of the optic radiation (Bürgel *et al.*, 1999), corticospinal tract (Rademacher *et al.*, 2001a), and the auditory system (Rademacher *et al.*, 2002) have been published Bürgel *et al.*, 2005 (Fig. 18.3). Compared to tracing studies in pathologically altered brains or in fetal brains, these

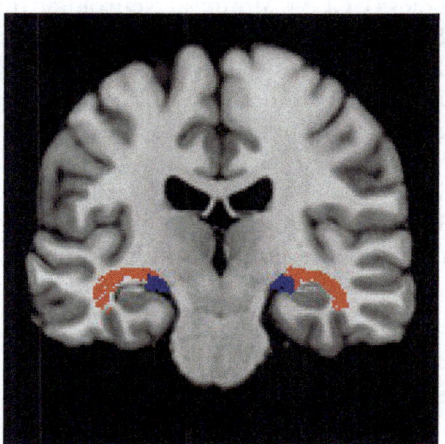

Figure 18.3. Thirty percent probability map of the optic radiation (red) and lateral geniculate body (blue). The fiber tract and the geniculate body have been traced in histological serial sections (thickness 20 μm) stained using the modified myelin stain of Heidenhain–Wölcke (Bürgel *et al.*, 1997, 1999) of 10 human postmortem brains and registered in the stereotaxic MNI reference space (Bürgel *et al.*, 2005).

maps display the normal and adult conditions. They took advantage of the high microscopical resolution, which enabled to trace fibers even if fibers abruptly change their direction, cross, or touch other fiber tracts ("kissing fibers"). The probabilistic fiber maps can also not reveal synaptic connectivity. These maps can be used, however, for evaluation of data from MR tractography as an independent measure.

An interesting extension of tractography has recently been developed. Thalamic subdivisions were classified according to the cortical region with which they show the highest connection probability (Behrens et al., 2003; Johansen-Berg et al., 2005). A further application led to the identification of the connectivity of SMA and pre-SMA (Johansen-Berg et al., 2004). This approach enabled the generation of population-based maps. The proposed parcellations are comparable to those found by invasive studies in the nonhuman primate brain. Parcellations of cortical areas and subcortical nuclei based on MR tractography is complementary to cyto-, receptor-, and myeloarchitectonic mapping, and add new information about fiber tracts and connectivity.

D. The Human Brain Atlas of the International Consortium of Human Brain Mapping as an Example of a Multimodal Brain Imaging Database

The primary goal of the International Consortium of Human Brain Mapping (ICBM) project (http://www.loni.ucla.edu/ICBM/) is the continuing development of a probabilistic reference system for the human brain (Mazziotta et al., 1995, 2001a). It comprises four core sites: University of California, Los Angeles (UCLA), Montreal Neurologic Institute (MNI), University of Texas at San Antonio, and Institute of Medicine, Research Center Jülich/Heinrich Heine University Düsseldorf. In addition, data acquisition sites in Asia and Europe contribute to this international consortium. Various modalities ranging from 3D tomographic images of in vivo and postmortem brains to histological preparations for cyto- or myeloarchitectonic observations, quantitative in vitro receptor autoradiography, and functional imaging using PET and fMRI have been included both from normal controls, different developmental stages, and pathologically altered brains (Toga, 2003). The atlas is based on the MNI template brain as standard reference space. It includes various data sets from ~7000 individuals.

IV. INTERSUBJECT VARIABILITY IN BRAIN ANATOMY

Human brains show an enormous intersubject variability of their anatomy. Variability comprises macroanatomical features (size, shape, sulcal pattern), the microstructure (e.g., cyto-, receptor-, and myeloarchitecture), and the relationship between microstructurally defined cortical areas and macroscopical landmarks.

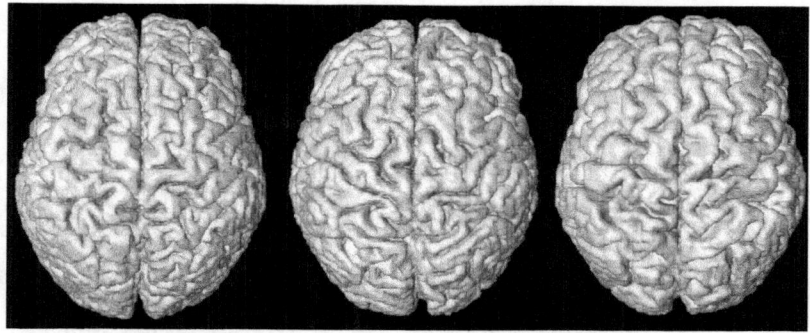

Figure 18.4. Three-dimensional reconstruction of the surface of three right-handed males (age between 30 and 40 years). Note the differences in the sulcal pattern between the brains.

A. Variability in Brain Macroscopy

Differences in brain shape between different ethnic populations or in absolute brain size within each gender as well as between female and male brains have been frequently described (Steinmetz *et al.*, 1995; Zilles *et al.*, 1997, 2001). Differences in shape between both hemispheres have been shown, e.g., for the occipital lobe (Falk *et al.*, 1991; LeMay and Kido, 1978; Zilles *et al.*, 1996) and the central region (Amunts *et al.*, 2000a).

Three-dimensional reconstructions of human brains using MRI at a resolution of ∼1 mm × 1 mm × 1 mm voxel size demonstrate the variability in sulcal patterns (Fig. 18.4). The degree of gyrification also differs considerably between subjects (Zilles *et al.*, 1988a). Although the large primary sulci, which develop early during ontogeny, are relatively constant in their presence and shape, the smaller secondary and tertiary sulci are highly variable. Sulcal variability (Thompson *et al.*, 1996) includes

1. the presence of a sulcus, e.g., the diagonal sulcus is present in ∼50% of the hemispheres (Amunts *et al.*, 1999);
2. the number of segments of a sulcus, e.g., the precentral sulcus can be split into two to four segments (Ono *et al.*, 1990);
3. the number and course of side branches and connections, e.g., the central sulcus can show up to four side branches and connections to either the postcentral or precentral sulci (Ono *et al.*, 1990);
4. the spatial location of sulci (Juch *et al.*, 2005);
5. shape of the endings, the width of sulci, and the depths (Van Essen, 2005).

The macroanatomical variability can be a confounding factor in many neuroimaging studies, e.g., group studies using fMRI or PET, if the aim is an analysis of corresponding brain regions. Various image registration tools (see below) can be applied in order to measure or eliminate

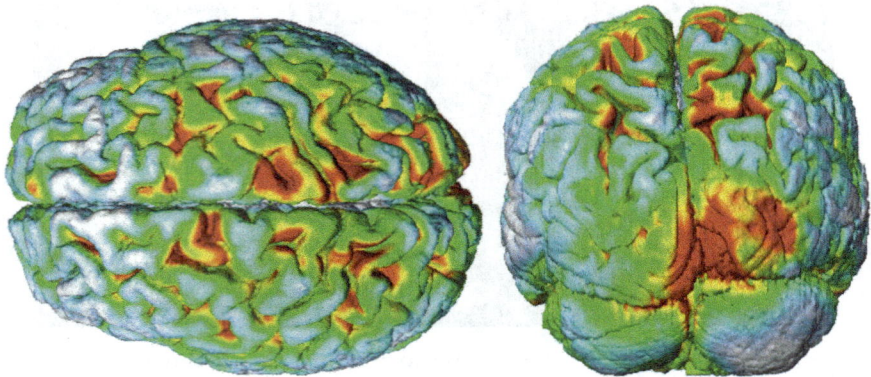

Figure 18.5. Intersubject differences between the brains of 24 healthy normal volunteers and an individual reference brain (all right-handed males). The 24 T1-weighted MR data sets have been warped to the individual target brain. As a first step, a linear affine transformation (scaling, rotation, translation) has been performed, which also normalizes the spatial orientation of all individual brains. As a next step, a nonlinear elastic deformation ("warping") has been applied. For each brain, deformation fields were calculated. The vectors of each of these fields indicate the direction and length of the elastic deformation (not the absolute distance between corresponding voxels of the original and the target brains) in every single voxel (Pieperhoff *et al.*, 2005). The 24 individual deformation fields were then averaged, and the averaged vector lengths were color coded. Red corresponds to a vector length of 5 mm, orange of 4 mm, yellow of 3 mm, green of 2 mm, and blue of 1 mm. Large differences in sulcal pattern and regional brain shape between the target brain and the sample correspond to large vector lengths, that is, the target brain differs maximally from the sample in those regions which are shown in red, and less in regions shown in green and blue. The figure demonstrates larger differences in shape in the occipital pole and lobe between the target brain and the actual sample as compared to the frontal lobe (Pieperhoff and Amunts, unpublished observations).

macroanatomical differences between individual brains. Image registration is, therefore, a necessary prerequisite for brain atlases which take the interindividual variability into account. Registration tools include (i) linear, affine transformation which removes and normalizes differences in shape and size, but retains left–right asymmetries as well as the individual sulcal pattern, and (ii) nonlinear transformations. After an ideal, nonlinear transformation of an MR data set of an individual brain to the reference brain of an atlas, brains are identical in size, shape, sulcal pattern, and asymmetry.

The information about the macrostructural, intersubject variability in a sample or the difference between an individual brain and the reference brain can be stored for further analysis and applications as "deformation field" (Fig. 18.5).

B. Microstructural Variability

The giant pyramidal cells (Betz-cells) in Brodmann's area 4 (primary motor cortex) may vary between 60 and 120 μm in height and 30 and 60 μm in width (von Economo and Koskinas, 1925). This is one of the numerous examples of intersubject variability in microstructure at the cellular level.

Here, we will focus on the microstructural variability at the level of cortical areas (e.g., size) and their laminar pattern (cytoarchitecture). Whereas some brains show a clearly visible laminar pattern with considerable differences in cell packing density between the layers of a cytoarchitectonically defined area, a much more blurred laminar pattern with minor differences in cell packing density between the layers can be found in the same cortical area of other brains. These cytoarchitectonic differences can only be analyzed using quantitative microscopical techniques, e.g., by registration of the volume fraction of cell bodies (gray level index, GLI; Schleicher and Zilles, 1990) vertical to the cortical surface between this surface and the cortex/white matter border. The results can be visualized as a profile curve which describes quantitatively the specific cytoarchitecture of an actual area in a single brain. The shapes of the profiles of the same area but from different brains can be parameterized as feature vectors, and the intersubject variability of these vectors can be analyzed using multivariate statistics. How large is the difference in cytoarchitecture between two different areas in one brain as compared to differences in cytoarchitecture of the same area between different brains? For example, it has been found that cytoarchitectonic differences in area 44 (one part of Broca's speech region) between 10 postmortem brains are greater than the mean difference between area 44 and 45 (other part of Broca's speech region) in the same sample of brains (Amunts et al., 1999).

Another aspect of intersubject variability is represented by the considerable variability in absolute and relative (with respect to the whole brain volume) size of a cytoarchitectonic area (Filimonoff, 1932; Kononova, 1935; Stensaas et al., 1974). This type of variability has a major impact on the construction of any brain atlas designed as a tool for the analysis of functional imaging data or stereotaxic neurosurgery, since the degree of variability differs between brain regions: whereas areas 44 and 45 of Broca's region vary in their absolute volumes up to a factor of 10 (Amunts et al., 1999), the hippocampus and the amygdala vary "only" by a factor of 2 (Amunts et al., 2005). Thus, the analysis of the intersubject variability is a major task for future brain atlases.

C. Cortical Areas and Macroanatomical Landmarks

The spatial positions of a cortical area and macroanatomical landmarks vary independently between brains (Zilles et al., 1997). A relatively strong association between macroanatomical landmarks and cortical areas has been found for primary cortical areas (Rademacher et al., 1993). The primary motor cortex (Brodmann's area 4) is always located in the anterior bank of the central sulcus, and the somatosensory cortex is consistently found in the posterior bank of this sulcus. The calcarine sulcus indicates the location of the primary visual cortex (areas V1 or Brodmann's area 17), and the primary auditory cortex is found on Heschl's gyrus (Fig. 18.1). However, the extent of area 4 in anterior direction (to premotor area 6) cannot be defined by any macroanatomical landmark. The same is true for the dorsal and ventral borders of area 17 (to the extrastriate cortical areas). For the majority of cortical areas, a sufficiently precise spatial relationship between the macroanatomical landmarks and the border of cortical areas cannot be established.

These findings do not imply a complete lack of spatial correlation between the location of a cortical area and gyri or sulci. In some cortical regions, gyri can be good indicators of the approximate location of a cortical area. For example, major parts of area 44 are consistently found on the free surface of the opercular part of the inferior frontal gyrus, and area 45 occupies usually the triangular part of this gyrus. However, when moving from the free surface of the triangular part into the depth of the horizontal branch of the Sylvian fissure and further to the surface of the orbital part of the inferior frontal gyrus, area 45 may reach the bottom of the sulcus and the free surface in some hemispheres, but not in others (Amunts et al., 1999, 2004). Consequently, the variability in the extent of one and the same area is low at the free surface of the opercular part, but high in the sulci. Additionally, these sulci may vary between brains (see also section "Variability in Brain Macroscopy").

It can be concluded, therefore, that any brain atlas based exclusively on macroanatomical landmarks for the identification of the position and extent of cortical areas may cause significant misinterpretations, e.g., of the cytoarchitectonic identity of activated cortical regions found in functional neuroimaging studies. There are two possible solutions of this problem:

1. Cyto- or myeloarchitectonic parcellations of the same living brain which has been studied by functional neuroimaging. This is only feasible if the parcellation can be performed by ultrahigh-resolution anatomical MR imaging. First attempts have been made already for selected regions of the cortex (Clark et al., 1992; Eickhoff et al., 2005d; Fatterpekar et al., 2002; Kruggel et al., 2003; Walters et al., 2003).
2. A probabilistic approach, which defines for each voxel of a reference space (or brain) the probability with which an actual cortical area, subcortical nucleus, or fiber tract is present in this voxel. As a consequence, it will not be possible to define unambiguously a certain voxel at a given stereotaxic location as the representation of one single anatomical structure (Fig. 18.6). Any voxel may have probabilities different from zero to belong to several cortical areas. This may not be a confounding factor in regions covered by a large single cortical area. In other regions, e.g., at the meeting point of several areas such as in the central operculum (primary and secondary motor cortex, primary and secondary somatosensory cortex, insula), a single voxel may represent several cortical areas with relatively similar probabilities.

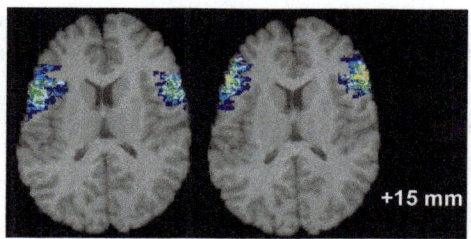

Figure 18.6. Cytoarchitectonic probabilistic maps of area 44 (left) and 45 (right). Orientation according to the AC–PC line. Location at $z = 15$ (i.e., 15 mm above the intercommissural line). Left hemisphere is left in the image. The probability is color coded. Yellow corresponds to an overlap in 7 out of 10 brains, orange to an overlap of 8 brains.

D. Image Registration Techniques to Eliminate Macroanatomical Variability

Image registration techniques can be used to transform one MR data set into another. Widely used synonyms of this procedure are "warping," "image transformation," "deformation," "spatial normalization," and "alignment." The data sets may come from one and the same subject (which has been scanned at several different occasions, e.g., during aging), or from different subjects. We here focus on registration tools which are used to transform MR data sets of different human brains to a common reference brain.

In this latter case, registration minimizes the macroanatomical differences between brains, but does not eliminate microstructural differences. A reference brain can be either an individual brain (Roland and Zilles, 1994), a digital brain phantom (Collins *et al.*, 1998), or an average brain (Collins *et al.*, 1994; Evans *et al.*, 1992, 1993).

Registration tools can be classified by the degrees of freedom, i.e., the number of free parameters, which define the registration and must be calculated for each individual brain. Here, we provide a few examples of frequently applied registration tools.

1. Affine Transformation

The rigid-body transformations are a subset of affine transformations. They have six degrees of freedom: three angles of rotation and three components of translation. Rigid-body transformation may change the location of a brain data set in space. This is necessary, for example, in order to align different brains to the AC–PC line. This type of transformation does not change the shape and size of the brains.

Other affine transformations have 12 degrees of freedom. In addition to the changes which can be performed by the rigid-body transformation, these transformations enable linear changes in brain size and shape due to scaling and sheering. They are frequently used as the initial step for a subsequent nonlinear transformation (see section "Nonlinear, Nonrigid Transformations").

2. Transformations of Medium Complexity

These transformations have several hundred up to several thousand degrees of freedom. Examples of this type of registration tools have been implemented in SPM (http://www.fil.ion.ucl.ac.uk/spm) and AIR (http://air.bmap.ucla.edu). The registration tool of SPM calculates a spatial transformation as linear combination of a given set of basic functions (Ashburner and Friston, 1999). Hereby, the coefficients of these linear combinations have to be determined. The tool implemented in AIR defines the transformation of a voxel using higher polynomials (Woods *et al.*, 1998).

These tools are frequently used for the spatial normalization of functional imaging data. They do not lead to a precise match of the anatomical MR data of different brains. Such a precise match is often not required due to the relatively low spatial resolution of functional imaging data. Moreover, SPM offers an average reference brain as a target brain for this transformation. Such average brain shows much less anatomical detail than a single subject brain.

3. Nonlinear, Nonrigid Transformations

Registration tools of this group are based on equations developed by scientists working on the physics of elastic bodies or fluids. These methods model the brain as an elastic body or a viscose fluid (Bajcsy and Kovacic, 1989; Christensen *et al.*, 1994; Gee *et al.*, 1993). A distance measure has to be defined in order to estimate the quality of the match between the transformed data set and the reference brain. The sum of the differences in gray values between the reference brain and the transformed brain is an example of such a measure. Frequently, nonlinear registrations are stepwise procedures using sequentially different levels of spatial resolution (Henn and Witsch, 2004b).

The registration tool based on a model which is known from linear elasticity physics requires the repeated solution of the Navier–Lamé equation which belongs to the second-order partial differential equations. The mathematical procedure is quite demanding, if we have to handle a three-dimensional data set with 256 voxels in each direction (as it is usually the case in an anatomical MR data set of a human brain), because the resulting system of equations has 24 million unknowns. The system can efficiently be solved using multiscale and multigrid approaches (Henn and Witsch, 2004a).

In general, registration tools based on elastic and fluid models are time-consuming (e.g., in the range of 3.5 h per brain on a recent Linux system). Furthermore, their application usually requires several steps of prealign-

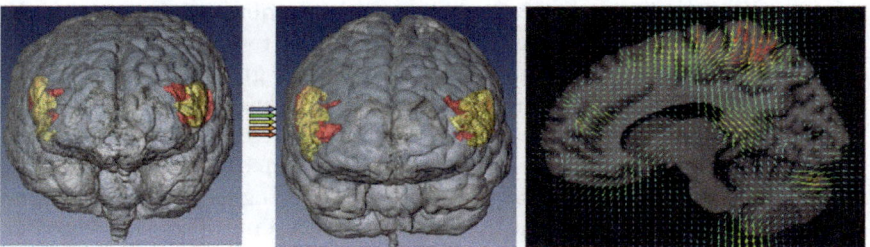

Figure 18.7. Three-dimensional reconstruction of a MR data set of a postmortem brain (left) with areas 44 (red) and 45 (yellow) before (left) and after (middle) nonlinear, elastic registration to a reference brain. Right: sagittal section of the brain after registration, in which the length and direction of the vectors of the transformation are visualized by color-coded arrows. Red tones correspond to large deformations, blue tones to small deformation (Amunts *et al.*, 2004).

ment, e.g., an affine, linear registration, and a gray-level normalization, which is further a quite critical step. Transformations of this kind, however, can result in a precise match at the single-voxel level. They have successfully been applied in order to warp both in vivo structural MR data sets and 3D reconstructed histological data sets (Fig. 18.7).

An inherent problem of nonlinear transformations is to distinguish a sufficient from an insufficient match. Theoretically, criteria such as the gray-value differences (see above) may be applicable. The question remains, however, whether homologue structures in different brains, e.g., sulci, have been matched. This seems to be no major problem for the large and consistently occurring primary sulci. It is a problem, however, when small side branches, highly variable small sulci (e.g., diagonal sulcus), or corresponding sulci with a different number of segments (e.g., the precentral sulcus of two brains with two and four segments, respectively) have to be warped to a reference brain with differing sulcal morphology. Moreover, the homology of smaller sulci is not always clear. The comparison of textbooks and atlases shows that different authors have classified the sulci of the brain in a different way e.g., compare the atlas of Duvernoy (1991) with that of Ono *et al.* (1990) with respect to the intraparietal and the lower postcentral sulci, or the occipital sulci). Observer-independent tools for the definition of sulci may help to define the sulci of an individual brain, but they cannot resolve principal morphological differences and the problem of sulcal homology. Anatomist/Brain VISA (http://anatomist/info/) is a software which supports segmentation, 3D reconstruction, and identification of sulci (Mangin *et al.*, 1995; Rivière *et al.*, 2002).

V. STANDARD REFERENCE BRAINS

The strategy for selecting the spatial reference system (e.g., standard reference brain) of an atlas depends on the actual answers to several major questions. Is it necessary to select an individual reference brain with the most "prototypical" anatomy? Or, is it better to generate a mean brain by averaging over a large population of brains? How to consider well-known differences in brain shape and size between sexes, groups of handedness, normal and pathologically altered brains? Do these differences require a standard brain for each of these groups separately? Or, is it irrelevant which type of standard reference brain has been chosen since all individual brains, irrespective of their actual macroanatomy, will be registered to this brain? Here, we can only discuss some aspects of the different strategies. Presently, no final answer to all the above questions is available.

A. Individual Reference Brain

How can we select an individual brain with macroanatomical features which are most similar to those of all other brains of a large sample? The size of such a brain can be precisely calculated by averaging the sizes (e.g., volumes) of all brains of the sample. The location and sizes of the subcortical nuclei (e.g., thalamus, striatum, red nucleus) and ventricles may also be used for identifying the most prototypical brain. The problem becomes extremely difficult, however, if we want to identify the most prototypical cortical ribbon in a single brain. This is due to the extremely variable gyrification pattern. Moreover, the degree of gyrification varies considerably between brains (Armstrong *et al.*, 1991). Thus, some cortical regions are located at the free surface, other regions are hidden in the depths of the sulci. The latter portion, however, includes two-third of the whole cortical surface (Zilles *et al.*, 1988a). Consequently, it is not possible to find homologue positions on the folded cortical surface in different brains on the basis of macroscopical information alone. It is also difficult if not impossible to identify a "mean" brain, if the intraparietal sulcus in one brain is represented by one large furrow whereas two separate segments represent this sulcus in another brain.

Roland and Zilles (1994) used 21 MRI data sets obtained with a 3D FLASH sequence at a 1.5 T Siemens Magnetom from healthy 20 to 30 years old, right-handed, male volunteers for the selection of a reference brain. The MR data sets were interactively peeled (removing meninges, skull, and vessels from brain tissue), and then scaled in coronal, sagittal, and horizontal directions until the deviations in brain volume among the 21 brains were minimal. The standard brain was selected as the brain which deviates the least in size and hemispheric shape from the 20 other brains. All possible combinations of two brains were run until a minimum deviant brain was found. This brain serves as the reference brain of the European Human Brain Database [ECHBD (Roland and Zilles, 1996) and the Neurogenerator (http://www.neurogenerator.org/about.htm; Roland *et al.*, 2001)].

An advantage of this approach is that formal criteria based on a similarity measure have been applied in order to define this reference brain. The brain, however, is typical in some regions, but less typical in others. For example, the occipital lobes are extraordinarily asymmetric. The left occipital pole protrudes into the space frequently covered by the right hemisphere (Amunts *et al.*, 2000b). Consequently, medially located activations in the left visual cortex may appear as activations right of the midsagittal plane when using this brain as a reference for the topographical interpretation of functional imaging data.

Another individual reference brain is the T1-weighted "single-subject MNI brain" of the Montreal Neurological Institute (Collins *et al.*, 1994; Evans *et al.*, 1993; Holmes *et al.*, 1998), which is widely used as a reference brain in functional imaging studies, e.g., in SPM (www.fil.ion.ucl.ac.uk/spm) and the ICBM viewer (http://www.bic.mni.mcgill.ca/cgi/icbm_view/). This brain has been selected by chance and is not supposed to be a minimum deviant brain of a larger sample of brains. Since the data set of this single brain represents an average of 27 scans, the contrast and the signal-to-noise ratio of the data set are excellent (Figs. 18.2 and 18.8). Moreover, it has the advantage that it is coregistered to the "average brain" of the MNI (see section "Average Brain"). Functional imaging studies have been registered to both the "single-subject MNI brain" and the "average brain" during the last years (Brett *et al.*, 2002).

B. Average Brain

The Montreal Neurological Institute created a series of brain templates which are averages over larger samples of normal brains. The ICBM has adopted these templates as spatial reference systems (see section "The Human Brain Atlas of the International Consortium of Human Brain Mapping as an Example of a Multimodal Brain Imaging Database"). One of the first templates was the MNI305 (Evans *et al.*, 1993). The current standard is known as the MNI152, which is based on 152 scans (Fig. 18.8). This template is available together with several commonly used functional imaging analysis

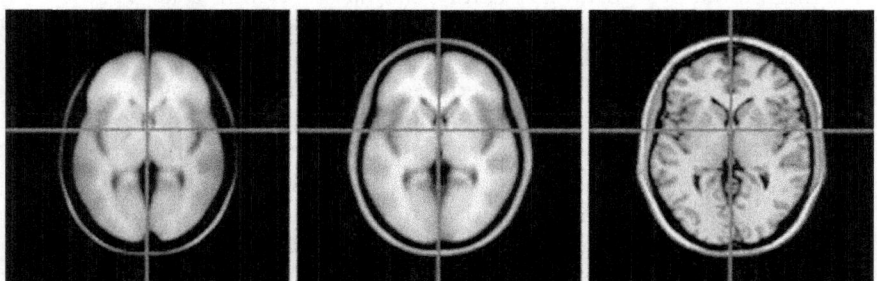

Figure 18.8. Average of 305 T1-weighted volumes (MNI305, left), average of 152 T1-weighted volumes (MNI152, center), and average of 27 T1-weighted scans of the single MNI brain (right) at $z = 0$ (MNI space). The single-subject brain volume is coregistered with the average volumes (http://www.bic.mni.mcgill.ca/cgi/icbm_view/).

packages, including SPM and FSL (http://www.fmrib.ox.ac.uk/fsl/). These two templates have been supplemented by templates from children aged between 4 and 18 years, 12 and 18 years, and 4 and 11 years, which are all based on a large number of scans. Although the MNI templates are supposed to be spatially oriented according to the Talairach atlas, they do not have precisely the same orientation, size, and shape. At the level of the interhemispheric fissure, the anterior commissure does not correspond to the origin $x = 0$, $y = 0$, and $z = 0$ in the Talairach atlas, but is shifted by 4 mm in rostrocaudal and dorsoventral directions. For more detailed information see Brett *et al.* (2002).

C. Alternative Approaches Based on Volume Data

It is also feasible to use any structural MR data set of an individual brain, which belongs to the sample of an fMRI group study, as reference brain (in analogy to the individual reference brain, see section "Individual Reference Brain"). This approach has the advantage of a correspondence of the data quality (resolution, contrast, scanner properties) between the "reference" brain and the individual brains. The results, however, cannot directly be compared with other studies using different reference brains.

Another possibility is the construction of a "synthetic" brain. An example of this approach is the digital brain phantom as proposed by Collins *et al.* (1998). The advantage of phantoms is the precise knowledge of its physical features. They can be used in order to test and to evaluate image analysis algorithms (e.g., tissue segmentation, correction of motion artifacts, partial volume effects, and many other methodical problems). A major problem of this approach is, however, the inevitable simplification of the real brain anatomy. The digital phantom of Collins *et al.* is based on the high-resolution single subject MNI template (see above). After preprocessing, the data set was classified with respect to different tissue types (gray and white matter, cerebrospinal fluid, fat, and background). Using this data set, different imaging modalities (MR, PET) can be simulated (Collins *et al.*, 1998).

Reference brains can be generated or selected for a special group of subjects. For example, if the brain shape of right-handed males has to be compared with that of right-handed females, it may be problematic to register both groups, e.g., to a right-handed male reference brain. The same is true when brains of patients have to be compared with brains of healthy subjects. The standard deviations will be different between both groups, and the results may be confounded. If the selected reference brain is a postmortem brain, it may provide detailed, microstructural information, but the selected brain may be distorted due to histological processing and artifacts. In vivo brains, in contrast, are free of such artifacts, but do not contain microstructural information.

The selection of an adequate reference brain depends on the neuroscientific question. Human brain anatomy and function vary across age and gender, and may depend on the individual health status and many other factors. In conclusion, there is no gold standard for the selection of a reference brain.

D. Surface-Based Atlases and Flat Maps

A strategy completely different from that of 3D reference brains based on volume data is the mapping of structural and functional data to a geometrically well-defined, two-dimensional surface (ellipsoidal or planar). Flat maps (planar maps) enable a simultaneous visualization of regions on the cortical surface as well as those buried in the depths of sulci while respecting surface topology (Van Essen *et al.*, 1998). Flat maps are part of widely distributed analysis tools of functional imaging data, e.g., developed by David Van Essen's laboratory (http://brainmap.wustl.edu/vanessen.html), Rainer Goebel's laboratory (http://www.brainvoyager.de/), Lohmann, von Cramon, 2000; Dale *et al.*, 1999, Fischl *et al.* 1999 (http://surfer.nmr.mgh.harvard.edu).

When flat maps are constructed, artificial cuts of the cortical surface must be introduced in order to enable the "flattening" of the complete cortical surface. The surface of a hemisphere is frequently cut along the bottom of sulci such as the calcarine and the cingulate sulci. The visual cortex is a good example of the advantages of flat maps. fMRI studies exploring the retinotopy of visual areas have shown that the occipital lobe can be subdivided into the striate and several extrastriate areas (Dougherty *et al.*, 2003; Sereno *et al.*, 1995; Shipp *et al.*, 1995; Tootell *et al.*, 1997). The flat maps demonstrate at the first glance that the early extrastriate areas (e.g., V2, V3v, VP, V4v, V5), which are hidden to some degree in the depths of sulci, surround the primary visual cortex in a belt-like arrangement. The registration of sulci, gyri, or cytoarchitectonically defined areas to ellipsoidal surfaces rather than planar (flat) maps may circumvent at least some of the problems caused by the artificial cuts necessary for the construction of flat maps.

Both the volume and the surface-based atlases can be seen as complementary approaches. The data can be transferred from one into the other, and vice versa. The population-average, landmark-, and surface-based (PALS) atlas of human cerebral cortex includes both volume and surface representations of the cortex (Van Essen, 2005). The environment includes a human brain database (http://sumsdb.wustl.edu:8081/sums/), tools for surface reconstruction (SurFit), visualization and analysis of surface maps (via Caret software), and other tools (Van Essen *et al.*, 2001). It enables the analysis of brain data sets both in a population-based reference system (Van Essen, 2005) and in an individual brain, the MNI single-subject brain (Van Essen, 2002, 2004; see also sections "Individual Reference Brain" and "Average Brain."). First attempts have been done to integrate our cytoarchitectonic probabilistic maps in this atlas (Fig. 18.9).

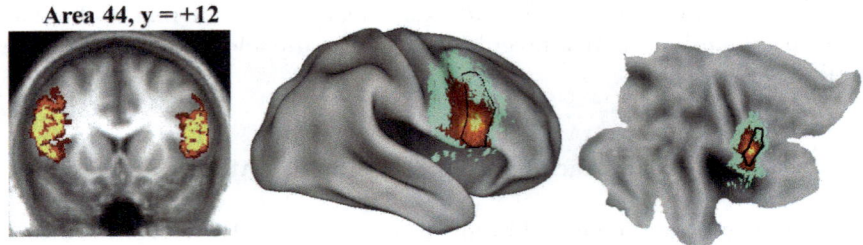

Figure 18.9. Probabilistic cytoarchitectonic map of area 44 (Amunts *et al.*, 1999, 2004) in coronal slice (overlaid on average sMRI for PALS-B12) plus lateral view and flat map of the PALS-B12 surface (Van Essen, 2005).

A major problem of the surface-based mapping is the construction of accurate surface reconstructions from 3D structural MRI data. For example, if the MR images have an isotropic resolution of 1 mm, and the cortical thickness varies between 2 and 4 mm, many voxels at the brain surface and the cortex/white matter border represent both (partial volume effect) parts of the cortex and parts of the pial surface plus cerebrospinal fluid and white matter, respectively. Voxels, which cover a sulcus and the adjoining gray matter on either side of the sulcus, "average" signals from the cerebrospinal fluid, the pial surface, and the cortical ribbon. Different strategies have been developed to handle such problems (http://white.stanford.edu/html/teo/mri/mri.html; http://v1/wustl.edu/software.html).

VI. MICROSTRUCTURAL LOCALIZATION OF NEURAL FUNCTIONS

A major goal of neuroimaging studies is to localize neural mechanisms, and hereby to understand structural–functional relationships. The studies, e.g., of Penfield (Penfield and Rasmussen, 1950) or of Ojemann (Ojemann and Mateer, 1979), aimed at this goal by using neurophysiological mapping techniques during neurosurgery. Such approaches are major steps forward to reveal functional mechanisms, but the anatomical localization of neuronal activities is necessarily restricted to the level of macroscopical landmarks. The scientifically most interesting microstructural–functional relationships can only be studied, however, by the registration of functional imaging data, macroscopical landmarks, and most importantly cyto-, myelo-, and/or chemoarchitectonic data in an identical spatial reference system, i.e., a multimodal brain atlas (Mazziotta *et al.*, 2001a).

Therefore, multimodal atlas systems have been designed which combine functional findings such as fMRI, PET, MEG data with anatomical findings such as in vivo structural MRI data and postmortem cyto-, myelo-, and receptor architectonic data from different brains. Meanwhile, scientists at the

Karolinska Institute in Stockholm (Per Roland and coworkers) and at the Research Center Jülich and the C. and O. Vogt Brain Research Institute, University of Düsseldorf (Karl Zilles, Katrin Amunts, Axel Schleicher and coworkers) have performed the combined microstructural–functional comparisons of this type in the human brain (Amunts *et al.*, 2004; Binkofski *et al.*, 2000, 2002; Bodegard *et al.*, 2000a,b; Ehrsson *et al.*, 2000; Eickhoff *et al.*, 2005a; Geyer *et al.*, 2001; Herath *et al.*, 2001; Horwitz *et al.*, 2003; Indefrey *et al.*, 2001; Kötter *et al.*, 2001; Larsson *et al.*, 1999, 2002; Naito *et al.*, 1999, 2000; Roland *et al.*, 2001; Roland and Zilles, 1994, 1996, 1998; Stöckel *et al.*, 2004; Wohlschläger *et al.*, 2005; Young *et al.*, 2003, 2004; Zilles *et al.*, 1995).

Presently, ~40 cytoarchitectonically defined cortical areas, 9 fiber tracts, and 16 nuclei/subnuclei have been mapped (Amunts *et al.*, 1999, 2000b, 2003, 2005; Bürgel *et al.*, 1999; Choi *et al.*, 2005; Eickhoff *et al.*, 2005b; Geyer, 2004; Geyer *et al.*, 1996, 1999, 2000b; Grefkes *et al.*, 2001; Morosan *et al.*, 2001; Rademacher *et al.*, 2001a, b, 2002; Zilles *et al.*, 2002a, 2003, 2004; Zilles and Palomero-Gallagher, 2001).

The combination of microstructural maps with functional imaging data has not only improved the precision of the localization of the functional activations, but led to new discoveries, e.g., subdivisions of the human primary motor cortex, Brodmann's area 4. Area 4 has been subdivided into an anterior and a posterior part, both subserving different functions (Geyer *et al.*, 1996). Functional maps show a much more detailed segregation of the regional organization of the human cerebral cortex than was expected from classical architectonic maps. This stimulated new architectonic studies. For example, human area VIP (Ventral IntraParietal area) in the intraparietal sulcus has been detected in a fMRI study, employing different moving stimuli (Bremmer *et al.*, 2001). This area cannot be found in Brodmann's classical map, but stimulated by the functional imaging study, the cytoarchitectonic correlate of VIP has now been delineated in histological sections of postmortem brains (Choi *et al.*, 2005). On the other hand, cytoarchitectonic data can generate working hypotheses for functional imaging studies. For example, the cytoarchitectonic mapping of the secondary somatosensory cortex has revealed a subdivision into four distinct cytoarchitectonic areas, for which no correlate can be found in Brodmann's map (Amunts *et al.*, 2003; Eickhoff *et al.*, 2005b). Functional imaging has elucidated the functional correlates of these architectonical subdivisions in more detail (Eickhoff *et al.*, 2005a).

Recently, a SPM toolbox (Fig. 18.10) has been developed (http://www.fz-juelich.de/ime/index.php), which is MATLAB based, integrated into the SPM2 software package (www.fil.ion.ucl.ac.uk/spm), and enables an integration of probabilistic cytoarchitectonic maps and results of functional imaging studies (Eickhoff *et al.*, 2005c). The toolbox includes the functionality for the construction of summary maps comprising probability of several cortical areas. This is enabled by an algorithmic definition of the most probable assignment of each voxel to one of these areas. Its main feature is to provide several measures defining the degree of correspondence

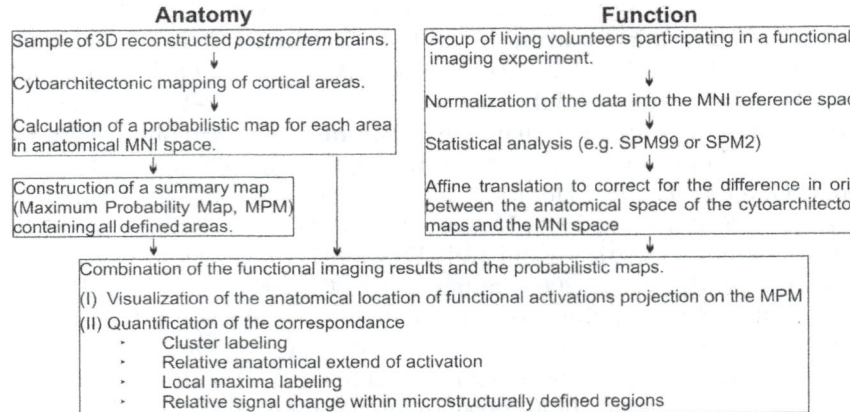

Anatomy	Function
Sample of 3D reconstructed *postmortem* brains.	Group of living volunteers participating in a functional imaging experiment.
↓	↓
Cytoarchitectonic mapping of cortical areas.	Normalization of the data into the MNI reference space.
↓	↓
Calculation of a probabilistic map for each area in anatomical MNI space.	Statistical analysis (e.g. SPM99 or SPM2)
↓	↓
Construction of a summary map (Maximum Probability Map, MPM) containing all defined areas.	Affine translation to correct for the difference in origin between the anatomical space of the cytoarchitectonic maps and the MNI space

Combination of the functional imaging results and the probabilistic maps.

(I) Visualization of the anatomical location of functional activations projection on the MPM

(II) Quantification of the correspondance
- Cluster labeling
- Relative anatomical extend of activation
- Local maxima labeling
- Relative signal change within microstructurally defined regions

Figure 18.10. Combination of microanatomical and functional data using the novel SPM toolbox (Eickhoff *et al.*, 2005c).

between architectonic areas and functional foci. The software together with the presently existing probability maps is available as open source software to the neuroimaging community. This new toolbox provides an easy-to-use tool for the integrated analysis of functional and microanatomical data in a common reference space.

The toolbox contains a summary map, which is computed for all hitherto published probabilistic maps. It defines the most likely cytoarchitectonic area for each voxel ("maximum probability map," MPM), and hereby enables the definition of nonoverlapping volumes of interest for each area (Fig. 18.11).

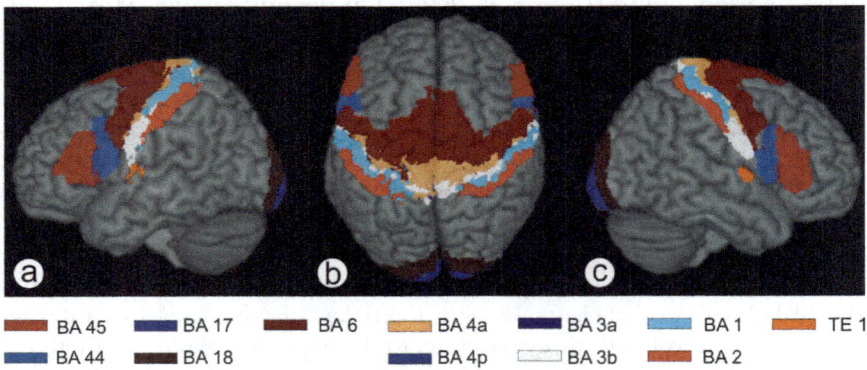

■ BA 45	■ BA 17	■ BA 6	■ BA 4a	■ BA 3a	■ BA 1	■ TE 1
■ BA 44	■ BA 18		■ BA 4p	☐ BA 3b	■ BA 2	

Figure 18.11. Surface rendering of the T1-weighted MNI reference brain and the MPMs of all presently published cortical areas, which were defined by observer-independent, quantitative cytoarchitectonic mapping. Only the surface extent of the different areas is shown. (A) Lateral view of the left hemisphere, (B) dorsal view, and (C) lateral view of the right hemisphere.

Ten years of experience with microstructural brain mapping suggests that the classical architectonic maps of the human brain must be modified. New areas have to be defined, and other cytoarchitectonic areas, e.g., Brodmann's area 19 of the extrastriate cortex, have to be replaced by more detailed parcellation schemes. Not only the number and location of architectonic areas have to be considerably changed, but also the exclusively descriptive organizational principles have to be enriched by more functional aspects. Hypotheses have been suggested which propose hierarchies of cortical areas, based on their structural or functional connectivity (Kötter *et al.*, 2001; Passingham *et al.*, 2002; Stephan *et al.*, 2000; Van Essen, 1985).

A novel promising, more functionally related chemoarchitectonic approach to brain mapping has recently been suggested. This approach is based on the receptor architecture of the cerebral cortex, i.e., the analysis of the regional and laminar distribution patterns of numerous receptor binding sites of various neurotransmitter systems. The receptor binding sites can be visualized by quantitative in vitro receptor autoradiography (Zilles, 1991; Zilles *et al.*, 1991b, 1995, 2002a; Zilles and Palomero-Gallagher, 2001). It has been demonstrated that the balance between the total binding values (averaged over all cortical layers of a cortical area) of various different receptors is a characteristic feature of a cortical area. This balance is like a "receptor fingerprint" of a cortical unit (Zilles *et al.*, 2002a). Moreover, it has been shown that areas of similar function show similar receptor fingerprints and differ from those with other properties (Zilles *et al.*, 2002a). Therefore, mathematical measures of similarity may help to define neurochemical and functional hierarchies and relationships of cortical areas.

VII. SUMMARY OF ADVANTAGES/LIMITATIONS OF CYTOARCHITECTONIC PROBABILISTIC MAPS

A. Advantages

1. Microstructural reference for data from imaging and neurophysiological studies of the living human brain for structural–functional analyses.
2. Link between different modalities of data (receptor architecture, myeloarchitecture, fMRI, MEG, PET, etc.) as well as between postmortem and in vivo data.
3. Maps can be coregistered to any individual MR data set of a brain, e.g., brains of patients.
4. Quantification of interhemispheric differences and intersubject variability of cortical areas ("probabilistic" maps).
5. Open system—new cortical areas can be considered (cortical areas, nuclei, and fiber tracts).
6. High-spatial resolution.

B. Limitations

1. The maps provide a probabilistic, but not an unambiguous anatomical classification of an individual in vivo MR data set (due to intersubject variability).
2. The maps do not cover the whole brain ("work in progress").
3. They represent a single aspect of microstructural organization, i.e., cytoarchitecture.
4. No maps available for the developing human brain and for the pathologically altered brain.

APPENDIX

A. Method of the Generation of Cytoarchitectonic Probabilistic Maps

1. The procedure is described in detail in Amunts *et al.* (2000b) and Zilles *et al.* (2002b).
2. Tissue: whole human postmortem brains fixed in buffered formalin (4%) or Bodian mixture (formalin, glacial acetic acid, and ethanol).
3. MR scan of the fixed postmortem brain, e.g., T1-weighted 3D FLASH sequence, flip angle 40°, repetition time of TR = 40 ms, echo time of TE = 5 ms for each image; spatial resolution was 1 mm × 1 mm × 1.17 mm, 8-bit gray-level resolution ("MR data set").
4. Embedding in paraffin.
5. Serial histological sectioning at a microtone for large sections in coronal, sagittal, or horizontal plane (6000–8000 sections, thickness 20 µm).
6. Each 60th blockface of the paraffin block (distance between images: 1.2 mm) is digitized via a CCD camera ("blockface images").
7. Each 60th section is stained for cell bodies (Merker, 1983). Neighboring sections are stained for myelin (Bürgel *et al.*, 1997; Gallyas, 1979) or Heidenhain–Wölcke (Burck, 1973).
8. Digitalization of the stained sections; ~130 per brain in coronal sections ("histological images").
9. Three-dimensional reconstruction of the histological images using the blockface images and the MR data set (Schormann *et al.*, 1995).
10. Definition of borders of cortical areas using an algorithmic approach based on the
    - measurement of the GLI as an indicator of the cell packing density in cortical regions (Schleicher and Zilles, 1990; Wree *et al.*, 1982);
    - generation of a sequence of GLI profiles, covering the cortical region of interest and reflecting the laminar distribution of the GLI from the surface of the cortex to the white matter border;

- extraction of features defining the shape of the GLI profiles (Schleicher *et al.*, 1999);
- calculation of multivariate distance measures for the quantification of differences between neighboring groups of profiles;
- definition of borders of cortical areas at those positions, at which the distance measure reaches a local maximum.

11. Definition of borders of subcortical nuclei and fiber tracts in histological sections.
12. Tracing (interactively) of borders of cortical areas, subcortical nuclei, and fiber tracts into the 3D reconstructed histological volume.
13. Registration of the complete volume data to standard reference space [e.g., ECHBD (Roland and Zilles, 1994) and MNI template (Evans *et al.*, 1993; Mohlberg *et al.*, 2003)] using different steps of preprocessing, linear, affine, and nonlinear, elastic transformations (Amunts *et al.*, 2004; Schormann and Zilles, 1998).
14. Registration of the delineated areas, nuclei, and fibers (=anatomical structures) to the standard reference space.
15. Superimposition of anatomical structures of 10 postmortem brains in the common reference space.
16. Probability maps (population maps, probabilistic maps) are generated which describe, for each voxel of the reference space, the relative frequency with which an anatomical structure was present in the sample of 10 brains. The frequency is color coded.

B. Useful Links and Atlas Tools (in alphabetical order)

1. AIR http://air.bmap.ucla.edu
2. Anatomist/Brain VISA http://anatomist/info/
3. Allan Atlas for gene expression (mouse brain) http://www.brainatlas.org/default.asp and http://www.brain-map.org/index.jsp
4. Brain Voyager http://www.brainvoyager.de/
5. Caret and surface-based atlas http://sumsdb.wustl.edu:8081/sums/
6. Cytoarchitectonic probabilistic maps http://www.fz-juelich.de/ime/index.php
7. Extraction of brain surfaces http://white.stanford.edu/html/teo/mri/mri.html; http://v1/wustl.edu/software.html
8. Free surfer http://surfer.nmr.mgh.harvard.edu
9. FSL http://www.fmrib.ox.ac.uk/fsl/
10. ICBM http://www.loni.ucla.edu/ICBM/
11. Neurogenerator http://www.neurogenerator.org/about.htm
12. SPM www.fil.ion.ucl.ac.uk/spm/
13. Talairach Daemon http://ric.uthscsa.edu/projects/talairachdaemon.html
14. Web site of the MNI, ICBM viewer http://www.bic.mni.mcgill.ca/cgi/icbm_view/

REFERENCES

Amunts, K., Eickhoff, S., and Zilles, K., 2003, Multimodal mapping of human cerebral cortex—individual variability, In: Ng, V., Barker, G. J., and Hendler, T. (eds.), *Psychiatric Neuroimaging*, Amsterdam, Berlin, Oxford, Tokyo, Washington, DC: IOS press, pp. 16–20.

Amunts, K., Jäncke, L., Mohlberg, H., Steinmetz, H., and Zilles, K., 2000a, Interhemispheric asymmetry of the human motor cortex related to handedness and gender, *Neuropsychologia* **38**:304–312.

Amunts, K., Kedo, O., Kindler, M., Pieperhott, P., Mohlberg, H., Shah, N. J., Habell, U., and Zilles, K., 2005, Cytoarchitectonic mapping of the amygdala, hippocampal region and entorhinal cortex: intersubject variability and probability maps. *Anat. Embryol.* (Berl.) **200**:343–352.

Amunts, K., Malikovic, A., Mohlberg, H., Schormann, T., and Zilles, K., 2000b, Brodmann's areas 17 and 18 brought into stereotaxic space—where and how variable? *Neuroimage* **11**:66–84.

Amunts, K., Schleicher, A., Bürgel, U., Mohlberg, H., Uylings, H. B. M., and Zilles, K., 1999, Broca's region revisited: cytoarchitecture and intersubject variability, *J. Comp. Neurol.* **412**:319–341.

Amunts, K., Weiss, P. H., Mohlberg, H., Pieperhoff, P., Gurd, J., Shah, J. N., Marshall, C. J., Fink, G. R., and Zilles, K., 2004, Analysis of the neural mechanisms underlying verbal fluency in cytoarchitectonically defined stereotaxic space—the role of Brodmann's areas 44 and 45, *Neuroimage* **22**:42–56.

Annese, J., Pitiot, A., Dinov, I. D., and Toga, A. W., 2004, A myelo-architectonic method for the structural classification of cortical areas, *Neuroimage* **21**:15–26.

Ardekani, B. A., Nierenberg, J., Hoptman, N. J., Javitt, D. C., and Lim, K. O., 2003, MRI study of white matter diffusion anisotropy in schizophrenia, *Neuroreport* **14**:2025–2029.

Armstrong, E., Curtis, M., Buxhoeveden, D. P., Gregoe, C., Zilles, K., Casanova, M. F., and McCarthy, W. F., 1991, Cortical gyrification in the rhesus monkey: a test of the mechanical folding hypothesis, *Cereb. Cortex* **1**:426–432.

Ashburner, J., and Friston, K. J., 1999, Nonlinear spatial normalization using basic functions, *Hum. Brain Mapp.* **7**:254–266.

Assaf, Y., Ben-Bashat, D., Chapman, J., Peled, S., Biton, I. E., Kafri, M., Segev, Y., Hendler, T., Korczyn, A. D., Graif, M., and Cohen, Y., 2002, High b-value q-space analyzed diffusion-weighted MRI: application to multiple sclerosis, *Magn. Reson. Med.* **47**: 115–126.

Assaf, Y., Mayk, A., and Cohen, Y., 2000, Displacement imaging of spinal cord using q-space diffusion-weighted MRI, *Magn. Reson. Med.* **44**:713–722.

Axer, H., and Keyserlingk, D. G., 2000, Mapping of fibre orientation in human internal capsule by means of polarized light and confocal scanning laser microscopy, *J. Neurosci. Methods* **94**:165–175.

Bailey, P., and von Bonin, G., 1951, *The Isocortex of Man*, Urbana: University of Illinois Press.

Bajcsy, R., and Kovacic, S., 1989, Multiresolution elastic matching, *Comput. Vis. Graph. Image Process* **46**:1–21.

Barker, G. J., 2001, Diffusion-weighted imaging of the spinal cord and optic nerve, *J. Neurol. Sci.* **186**:S45–S49.

Basser, P. J., and Jones, D. K., 2002, Diffusion-tensor MRI: theory, experimental design and data analysis—a technical review, *NMR Biomed.* **15**:456–467.

Behrens, T. E., Johansen-Berg, H., Woolrich, M. W., Smith, S. M., Wheeler-Kingshot, C. A., Boulby, P. A., Barker, G. J., Sillery, E. L., Sheehan, K., Ciccarelli, O., Thompson, A. J., Brady, J. M., and Matthews, P. M., 2003, Non-invasive mapping of connections between human thalamus and cortex using diffusion imaging, *Nat. Neurosci.* **6**:750–757.

Bidmon, H. J., Wu, J. Y., Godecke, A., Schleicher, A., Mayer, B., and Zilles, K., 1997, Nitric oxide synthase-expressing neurons are area-specifically distributed within the cerebral cortex of the rat, *Neuroscience* **81**:321–330.

Binkofski, F., Amunts, K., Stephan, K. M., Posse, S., Schormann, T., Freund, H. -J., Zilles, K., and Seitz, R. J., 2000, Broca's region subserves imagery of motion: a combined cytoarchitectonic and fMRI study, *Hum. Brain Mapp.* **11**:273–285.

Binkofski, F., Fink, G. R., Geyer, S., Buccino, G., Gruber, O., Shah, J. N., Taylor, J. G., Seitz, R. J., Zilles, K., and Freund, H. J., 2002, Neural activity in human primary motor cortex areas 4a and 4p of human motor cortex is modulated differentially by attention to action, *J. Neurophysiol.* **88**:519.

Bodegard, A., Geyer, S., Naito, E., Zilles, K., and Roland, P. E., 2000a, Somatosensory areas in men activated by moving stimuli: cytoarchitectonic mapping and PET, *Neuroreport* **11**:187–191.

Bodegard, A., Ledberg, A., Geyer, S., Naito, E., Zilles, K., and Roland, P. E., 2000b, Object shape differences reflected by somatosensory cortical activation in human, *J. Neurosci.* **20 RC 51**:1–5.

Bremmer, F., Schlack, A., Shah, N. J., Zafiris, O., Zilles, K., and Fink, G. R., 2001, Polymodal motion processing in posterior parietal and premotor cortex: a human fMRI study strongly implies equivalencies between humans and monkeys, *Neuron* **29**:287–296.

Brett, M., Johnsrude, I. S., and Owen, A. M., 2002, The problem of functional localization in the human brain, *Nat. Rev. Neurosci.* **3**:243–249.

Brodmann, K., 1908, Beiträge zur histologischen Lokalisation der Großhirnrinde. VI: die Cortexgliederung des Menschen, *J. Psychol. Neurol.* **X**:231–246.

Brodmann, K., 1909, *Vergleichende Lokalisationslehre der Großhirnrinde in ihren Prinzipien dargestellt auf Grund des Zellenbaues*, Leipzig: Barth JA.

Brodmann, K., 1994, *Brodmann's Localization in the Cerebral Cortex*, London: Smith Gordon.

Burck, H. -C., 1973, *Histologische Technik*, Stuttgart: Thieme-Verlag.

Bürgel, U., Amunts, K., Hoemke, L., Mohlberg, H., Gilsbach, J. M., Zilles, K., 2005, White matter fiber tracts of the humans brain: Three-dimensional mapping of microscopic resolution, topography and intersubject variability, *Neuroimage* Epub ahead of print.

Bürgel, U., Mecklenburg, I., Blohm, U., and Zilles, K., 1997, Histological visualization of long fiber tracts in the white matter of adult human brains, *J. Brain Res.* **3**:393–404.

Bürgel, U., Schormann, T., Schleicher, A., and Zilles, K., 1999, Mapping of histologically identified long fiber tracts in human cerebral hemispheres to the MRI volume of a reference brain: position and spatial variability of the optic radiation, *Neuroimage* **10**:489–499.

Campbell, A. W., 1905, *Histological Studies on the Localisation of Cerebral Function*, Cambridge: Cambridge University Press.

Choi, H.-J., Zilles, K., Mohlberg, H., Schleicher, A., Fink, G. R., and Amunts, K., 2005, Cytoarchitectonic mapping of the anterior ventral bank of the human intraparietal sulcus, *J. Comp. Neurol.*, in press.

Christensen, G. E., Rabbitt, R. D., and Miller, M. I., 1994, 3-D brain mapping using deformable neuroanatomy, *Phys. Med. Biol.* **39**:609–618.

Clark, C. A., Barker, G. J., and Tofts, P. S., 1999, Magnetic resonance diffusion imaging of the human cervical spinal cord in vivo, *Magn. Reson. Med.* **41**:1269–1273.

Clark, V. P., Courchesne, E., and Grafe, M., 1992, In vivo myeloarchitectonic analysis of human striate and extrastriate cortex using magnetic resonance imaging, *Cereb. Cortex* **2**:424.

Clarke, S., 1994, Modular organization of human extrastriate visual cortex: evidence from cytochrome oxidase pattern in normal and macular degeneration cases, *Eur. J. Neurosci.* **6**:725–736.

Clarke, S., and Miklossy, J., 1990, Occipital cortex in man: organization of callosal connections, related cyto- and myeloarchitecture, and putative boundaries of functional visual areas, *J. Comp. Neurol.* **298**:188–214.

Coenen, V. A., Krings, T., Mayfrank, L., Polin, R. S., Reinges, M. H. T., Thron, A., and Gilsbach, J. M., 2001, Three-dimensional visualization of the pyramidal tract in a neuronavigation system during brain tumor surgery: first experiences and technical note, *Neurosurgery* **49**:86–93.

Collins, D. L., Neelin, P., Peters, T. M., and Evans, A. C., 1994, Automatic 3D intersubject registration of MR volumetric data in standardized Talairach space, *J. Comput. Assis. Tomogr.* **18:**192–205.

Collins, D. L., Zijdenbos, A., Kollokian, V., Sled, J. G., Kabani, N. J., Holmes, C. J., and Evans, A. C., 1998, Design and construction of a realistic digital brain phantom, *IEEE Trans. Med. Imaging* **17:**463–468.

Conturo, T. E., Lori, N. F., Cull, T. S., Akbudak, E., Snyder, A. Z., Shimony, J. S., McKinstry, R. C., Burton, H., and Raichle, M. E., 1999, Tracking neuronal fiber pathways in the living human brain, *Proc. Natl. Acad. Sci.* **96:**10422–10427.

Dale, A. H., Fischl, B., and Sereno, H. I., 1999, Cortical surface based analysis. I: segmentation and surface reconstruction, *Neuroimage* **9:**179–194.

Di Virgilio, G., and Clarke, S., 1997, Direct interhemispheric visual input to human speech areas, *Hum. Brain Mapp.* **5:**347–354.

Dougherty, R. F., Koch, V. M., Brewer, A. A., Fischer, B., Modersitzky, J., and Wandell, B. A., 2003, Visual field representations and locations of visual areas V1/2/3 in human visual cortex, *J. Vis.* **3:**586–598.

Duvernoy, H., 1991, *The Human Brain. Surface, Three-Dimensional Sectional Anatomy and MRI,* Wien, New York: Springer.

Ehrsson, H. H., Naito, E., Geyer, S., Amunts, K., Zilles, K., Forssberg, H., and Roland, P. E., 2000, Simultaneous movements of upper and lower limbs are coordinated by motor representations that are shared by both limbs: a PET study, *Eur. J. Neurosci.* **12:**3385–3398.

Eickhoff, S., Amunts, K., Mohlberg, H., and Zilles, K., 2005a, The human parietal operculum: II. Stereotaxic maps and correlation with functional imaging results, *Cereb. Cortex*, Epub ahead of print.

Eickhoff, S., Schleicher, A., Zilles, K., and Amunts, K., 2005b, The human parietal operculum: I. Cytoarchitectonic mapping of subdivisions, *Cereb. Cortex*, Epub ahead of print.

Eickhoff, S., Stephan, K. E., Mohlberg, H., Grefkes, C., Fink, G. R., Amunts, K., and Zilles, K., 2005c, A new SPM toolbox for combining probabilistic cytoarchitectonic maps and functional imaging data, *Neuroimage* **25:**1325–1335.

Eickhoff, S., Walters, N., Schleicher, A., Egan, G. F., Zilles, K., Watson, J. D. G., and Amunts, K., 2005d, High-resolution MRI reflects myelo- and cytoarchitecture of human cerebral cortex, *Hum. Brain Mapp.* **24:**206–215.

Elliot Smith, G., 1907, A new topographical survey of the human cerebral cortex, being an account of the distribution of the anatomically distinct cortical areas and their relationship to the cerebral sulci, *J. Anat.* **41:**237–254.

Evans, A. C., Collins, D. L., Mills, S. R., Brown, E. D., Kelly, R. L., and Peters, T. M., 1993, 3D statistical neuroanatomical models from 305 MRI volumes, In: *Proceedings of the IEEE-NSS-MI Symposium*, pp. 1813–1817.

Evans, A. C., Collins, D. L., and Milner, B., 1992, An MRI-based stereotactic atlas from 250 young normal subjects, *Soc. Neurosci. Abstr.* **18:**408.

Falk, D., Hildebolt, C., Cheverud, J., Kohn, L. A. P., Figiel, G., and Vannier, M., 1991, Human cortical asymmetries determined with 3D MR technology, *J. Neurosci. Methods* **39:**185–191.

Fatterpekar, G. M., Naidich, T. P., Delamn, B. N., Aguinaldo, J. G., Gultekin, S. H., Sherwood, C. C., Hof, P. R., Drayer, B. P., and Fayad, Z. A., 2002, Cytoarchitecture of the human cerebral cortex: MR microscopy of excised specimens at 9.4 Tesla, *Am. J. Neuroradiol.* **23:** 1313–1321.

Filimonoff, I. N., 1932, Über die Variabilität der Großhirnrindenstruktur. Mitteilung II—Regio occipitalis beim erwachsenen Menschen, *J. Psychol. Neurol.* **44:**2–96.

Fischl, B., Sereno, M. I., and Dale, A. M., 1999, Cortical surface based analysis. II: inflation, flatterring, and a surface based coordinate system, *Neuroimage* **9:**195–207.

Flechsig, P., 1920, *Anatomie des menschlichen Gehirns und Rückenmarks auf myelogenetischer Grundlage,* Leipzig: Thieme.

Foong, J., Symms, M. R., Barker, G. J., Maier, M., Miller, D. G. H., and Ron, M. A., 2002, Investigating regional white matter in schizophrenia using diffusion tensor imaging, *Neuroreport* **13:**333–336.

Fox, P. T., Perlmutter, J. S., and Raichle, M., 1985, A stereotactic method of localization for positron emission tomography, *J. Comput. Assis. Tomogr.* **9**:141–153.

Friston, K. J., 1997, Testing for anatomically specified regional effects, *Hum. Brain Mapp.* **5**:133–136.

Gallyas, F., 1979, Silver staining of myelin by means of physical development, *Neurol. Res.* **1**:203–209.

Gee, J. C., Reivich, M., and Bajcsy, R., 1993, Elastically deforming an atlas to match anatomical brain images, *J. Comput. Assis. Tomogr.* **17**:225–236.

Gerhardt, E., 1940, Die Cytoarchitektonik des Isocortex parietalis beim Menschen, *J. Psychol. Neurol.* **49**:367–419.

Geyer, S., 2004, *The Microstructural Border Between the Motor and the Cognitive Domain in the Human Cerebral Cortex*, Berlin, Heidelberg: Springer.

Geyer, S., Ledberg, A., Schleicher, A., Kinomura, S., Schormann, T., Bürgel, U., Klingberg, T., Larsson, J., Zilles, K., and Roland, P. E., 1996, Two different areas within the primary motor cortex of man, *Nature* **382**:805–807.

Geyer, S., Matelli, M., Luppino, G., and Zilles, K., 2000a, Functional neuroanatomy of the primate isocortical motor system, *Anat. Embryol.* **202**:443–474.

Geyer, S., Schleicher, A., Schormann, T., Mohlberg, H., Bodegard, A., Roland, P. E., and Zilles, K., 2001, Integration of microstructural and functional aspects of human somatosensory areas 3a, 3b, and 1 on the basis of a computerized brain atlas, *Anat. Embryol.* **204**:351–366.

Geyer, S., Schleicher, A., and Zilles, K., 1999, Areas 3a, 3b, and 1 of human primary somatosensory cortex: I. Microstructural organisation and interindividual variability, *Neuroimage* **10**:63–83.

Geyer, S., Schormann, T., Mohlberg, H., and Zilles, K., 2000b, Areas 3a, 3b, and 1 of human primary somatosensory cortex: II. Spatial normalization to standard anatomical space, *Neuroimage* **11**:684–696.

Grefkes, C., Geyer, S., Schormann, T., Roland, P., and Zilles, K., 2001, Human somatosensory area 2: observer-independent cytoarchitectonic mapping, interindividual variability, and population map, *Neuroimage* **14**:617–632.

Hayes, T. L., and Lewis, D. A., 1992, Nonphosphorylated neurofilament protein and calbindin immunoreactivity in layer III pyramidal neurons of human neocortex, *Cereb. Cortex* **2**:56–67.

Henn, S., and Witsch, K., 2004a, A multigrid approach for minimizing a nonlinear functional for digital image matching, *Computing* **64**:339–348.

Henn, S., and Witsch, K., 2004b, Iterative multigrid regularization techniques for image matching, *SIAM J. Sci. Comput.* **23**:1077–1093.

Herath, P., Klingberg, T., Young, Y., Amunts, K., and Roland, P., 2001, Neural correlates of dual task interference can be dissociated from those of divided attention: an fMRI study, *Cereb. Cortex* **11**:796–805.

Holmes, C. J., Hoge, R., Collins, L., Woods, R., Toga, A. W., and Evans, A. C., 1998, Enhancement of MR images using registration for signal averaging, *J. Comput. Assis. Tomogr.* **22**:324–333.

Horwitz, B., Amunts, K., Bhattacharyya, R., Patkin, D., Jeffries, J., Zilles, K., and Braun, A. R., 2003, Activation of Broca's area during the production of spoken and signed language: a combined cytoarchitectonic mapping and PET analysis, *Neuropsychologia* **41**:1868–1876.

Indefrey, P., Brown, C. M., Hellwig, F., Amunts, K., Herzog, H., Seitz, R. J., and Hagoort, P., 2001, A neural correlate of syntactic encoding during speech production, *Proc. Natl. Acad. Sci.* **98**:5933–5936.

Ito, R., Mori, S., and Melhem, E. R., 2002, Diffusion tensor brain imaging and tractography, *Neuroimaging Clin. N. Am.* **12**:1–19.

Iwasawa, T., Matoba, H., Ogi, A., Kurihara, H., Saito, K., Yoshida, T., Matsubara, S., and Nozaku, A., 1997, Diffusion-weighted imaging of the human optic nerve: a new approach to evaluate optic neuritis in multiple sclerosis, *Magn. Reson. Med.* **38**:484–491.

Johansen-Berg, H., Behrens, T. E. J., Robson, M. D., Drobnjak, I., Rushworth, M. F. S., Brady, J. M., Smith, S. M., Higham, D. J., and Matthews, P. M., 2004, Changes in connectivity profiles define functionally distinct regions in human medial frontal cortex, *Proc. Natl. Acad. Sci.* **101:**13335–13340.

Johansen-Berg, H., Behrens, T. E. J., Sillery, E., Ciccarelli, O., Thompson, A. J., Smith, S. M., and Matthews, P. M., 2005, Functional–anatomical validation and individual variation of diffusion tractography-based segmentation of the human thalamus, *Cereb. Cortex* **15:**31–39.

Jones, D. K., Williams, S. C., Gasston, D., Horsefiled, M. A., Simmons, A., and Howard, R., 2002, Isotropic resolution diffusion tensor imaging with whole brain acquisition in a clinically acceptable time, *Hum. Brain Mapp.* **15:**216–230.

Juch, H., Zimine, I., Seghier, M. L., Lazeyras, F., and Fasel, J. H. D., 2005, Anatomical variability of the lateral frontal lobe surface: implication for intersubject variability in language neuroimaging, *Neuroimage* **24:**504–514.

Kleist, K., 1934, *Gehirnpathologie*, Leipzig: Barth.

Kononova, E. P., 1935, Structural variability of the cortex cerebri. Inferior frontal gyrus in adults (Russian), In: Sarkisov, S. A., and Filimonoff, I. N. (eds.), *Annals of the Brain Research Institute. Vol. I.*, Moscow-Leningrad: State Press for Biological and Medical Literature, pp. 49–118.

Kötter, R., Stephan, K. E., Palomero-Gallagher, N., Geyer, S., Schleicher, A., and Zilles, K., 2001, Multimodal characterisation of cortical areas by multivariate analyses of receptor binding and connectivity data, *Anat. Embryol.* **204:**333–350.

Krings, T., Coenen, V. A., Axer, H., Möller-Hartmann, W., Mayfrank, L., Weidemann, J., Kränzlein, H., Gilsbach, J. M., and Thron, A., 2001, In vivo 3D visualization of pyramidal tracts using anisotropic diffusion weighted magnetic resonance imaging, *Neurosci. Lett.* **307:**192–196.

Kruggel, F., Brückner, M. K., Arendt, T., Wiggins, C. J., and von Cramon, D. Y., 2003, Analyzing the neocortical fine-structure, *Med. Image Anal.* **7:**251–264.

Lancaster, J. L., Woldorff, M. G., Parsons, L. M., Liotti, M., Freitas, C. S., Rainey, L., Kochunov, P. V., Nickerson, D., Mikiten, S. A., and Fox, P. T., 2001, Automated Talairach atlas labels for functional brain mapping, *Hum. Brain Mapp.* **10:**120–131.

Larsson, J., Amunts, K., Gulyas, B., Malikovic, A., Zilles, K., and Roland, P. E., 1999, Neuronal correlates of real and illusory contour perception: functional anatomy with PET, *Eur. J. Neurosci.* **11:**4024–4036.

Larsson, J., Amunts, K., Gulyas, B., Malikovic, A., Zilles, K., and Roland, P. E., 2002, Perceptual segregation of overlapping shapes activates posterior extrastriate visual cortex in man, *Exp. Brain Res.* **143:**1–10.

Le Bihan, D., Mangin, J.-F., Poupon, C., Clark, C. A., Pappata, S., Molko, N., and Chabriat, H., 2001, Diffusion tensor imaging: concepts and applications, *J. Magn. Reson. Imaging* **13:**534–546.

LeMay, M., and Kido, D. K., 1978, Asymmetries of the cerebral hemispheres on computed tomograms, *J. Comput. Assis. Tomogr.* **2:**471–476.

Livingstone, M. S., and Hubel, D. H., 1987, Connections between layer 4B of area 17 and the thick cytochrome oxidase stripes of area 18 in the squirrel monkey, *J. Neurosci.* **7:**3371–3377.

Lohmann, G., and von Cramon, D. Y., 2000, Automated Labelling of the human cortical surface using sulcal bassins. *Med. Image Anal.* **4:**179–188.

Luppino, G., Matelli, M., Camarda, R. M., Gallese, V., and Rizzolatti, G., 1991, Multiple representations of body movements in mesial area 6 and the adjacent cingulate cortex: an intracortical microstimulation study in the macaque monkey, *J. Comp. Neurol.* **311:** 463–482.

Mai, J. K., Assheuer, J., and Paxinos, G., 2004, *Atlas of the Human Brain*, San Diego, London: Elsevier.

Majocha, R. E., Marotta, C. A., and Benes, F. M., 1985, Immunostaining of neurofilament protein in human postmortem cortex: a sensitive and specific approach to the pattern analysis of human cortical cytoarchitecture, *Can. J. Biochem. Cell Biol.* **63:**577–584.

Mangin, J.-F., Frouin, V., Bloch, I., Régis, J., and Lopez-Krahe, J., 1995, From 3D magnetic resonance images to structural representations of the cortex topography using topology preserving deformations, *J. Math. Imaging Vis.* **5:**297–318.

Matelli, M., Luppino, G., and Rizzolatti, G., 1991, Architecture of superior and mesial area 6 and the adjacent cingulate cortex in the macaque monkey, *J. Comp. Neurol.* **311:** 445–462.

Mazziotta, J. C., Toga, A. W., Evans, A. C., Fox, P. T., and Lancaster, J. L., 1995, A probabilistic atlas of the human brain: theory and rationale for its development, *Neuroimage* **2**:89–101.

Mazziotta, J., Toga, A., Evans, A., Fox, M., Lancaster, J., Zilles, K., Woods, R., Paus, T., Simpson, G., Pike, B., Holmes, C., Collins, L., Thompson, P., Macdonald, D., Iacoboni, M., Schormann, T., Amunts, K., Palomero-Gallagher, N., Geyer, S., Parsons, L., Narr, K., Kabani, N., LeGoualher, G., Boomsma, D., Cannon, T., Kawashima, R., and Mazoyer, B., 2001a, A probabilistic atlas and reference system for the human brain: international consortium for brain mapping (ICBM), *Phil. Trans. R. Soc. Lond. B* **356**:1293–1322.

Mazziotta, J., Toga, A., Evans, A., Fox, P., Lancaster, J., Zilles, K., Woods, R., Paus, T., Simpson, G., Pike, B., Holmes, C., Collins, L., Thompson, P., Macdonald, D., Iacoboni, M., Schormann, T., Amunts, K., Palomero-Gallagher, N., Geyer, S., Parsons, L., Narr, K., Kabani, N., Le Goualher, G., Feidler, J., Smith, K., Boomsma, D., Pol, H. H., Cannon, T., Kawashima, R., and Mazoyer, B., 2001b, A four-dimensional probabilistic atlas of the human brain, *J. Am. Med. Inform. Assoc.* **8**:401–430.

Merker, B., 1983, Silver staining of cell bodies by means of physical development, *J. Neurosci. Methods* **9**:235–241.

Miklossy, J., and van der Loos, H., 1991, The long-distance effects of brain lesions: visualization of myelinated pathways in the human brain using polarizing and fluorescence microscopy, *J. Neuropathol. Exp. Neurol.* **50**:1–15.

Mohlberg, H., Lerch, J., Amunts, K., Evans, A. C., and Zilles, K., 2003, Probabilistic cytoarchitectonic maps transformed into MNI space, In: *Proceedings of the Ninth International Conference of Neuroimage CD Rom on Functional Mapping of the Human Brain*, New York, 2003.

Mori, S., and van Zijl, P. C., 2002, Fibre tracking: principles and strategies—a technical review, *NMR Biomed.* **15**:468–480.

Morosan, P., Rademacher, J., Schleicher, A., Amunts, K., Schormann, T., and Zilles, K., 2001, Human primary auditory cortex: cytoarchitectonic subdivisions and mapping into a spatial reference system, *Neuroimage* **13**:684–701.

Mufson, E. J., Brady, D. R., and Kordower, J. H., 1990, Tracing neuronal connections in postmortem human hippocampal complex with the carbocyanine dye DiI, *Neurobiol. Aging* **11**:649–653.

Naito, E., Ehrsson, H. H., Geyer, S., Zilles, K., and Roland, P. E., 1999, Illusory arm movements activate cortical motor areas: a positron emission tomography study, *J. Neurosci.* **19**:6134–6144.

Naito, E., Kinomura, S., Geyer, S., Kawashima, R., Roland, P. E., and Zilles, K., 2000, Fast reaction to different sensory modalities activate common fields in the motor areas, but the anterior cingulate cortex is involved in the speed of reaction, *J. Neurophysiol.* **83**:1701–1709.

Ojemann, G., and Mateer, C., 1979, Human language cortex: localization of memory, syntax, and sequential motor-phoneme identification systems, *Science* **205**:1401–1403.

Ono, M., Kubik, S., and Abernathey, C. D., 1990, *Atlas of the Cerebral Sulci*, Stuttgart, New York: Thieme.

Parker, G. J. M., Stephan, K. E., Barker, G. J., Rowe, J. B., MacManus, D. G., Wheeler-Kingshot, C. A. M., Ciccarelli, O., Passingham, R. E., Spinks, R. L., Lemon, R. N., and Turner, R., 2002, Initial demonstration of in vivo tracing of axonal projections in the macaque brain and comparison with the human brain using diffusion tensor imaging and fast marching tractography, *Neuroimage* **15**:797–809.

Passingham, R. E., Stephan, K. E., and Kötter, R., 2002, The anatomical basis of functional localization in the cortex, *Nature* **3**:606–616.

Paus, T., Tomaiolo, F., Otaky, N., Macdonald, D., Petrides, M., Atlas, J., Morris, R., and Evans, A. C., 1996, Human cingulate and paracingulate sulci: pattern, variability, asymmetry, and probabilistic maps, *Cereb. Cortex* **6**:207–214.

Penfield, W., and Rasmussen, T., 1950, *The Cerebral Cortex of Man*, New York: Macmillan.

Pieperhoff, P., Mohlberg, H., and Amunts, K., 2005, Überlagerungen anatomischer und funktioneller Daten. In: Schneider, F., and Fink, G. R. (eds.), *Funktionelle Kernspintomographie in Psychiatrie und Neurologie*, Berlin: Springer.

Poupon, C., Clark, C. A., Frouin, V., Régis, J., Bloch, I., LeBihan, D., and Mangin, J.-F., 2000, Regularization of diffusion-based direction maps for the tracking of brain white matter fascicles, *Neuroimage* **12:**184–195.

Powell, H. W. R., Guye, M., Parker, G. J. M., Symms, M. R., Boulby, P., Koepp, M. J., Barker, G. J., and Duncan, J. S., 2004, Noninvasive in vivo demonstration of the connections of the human parahippocampal gyrus, *Neuroimage* **22:**740–747.

Rademacher, J., Bürgel, U., Geyer, S., Schormann, T., Schleicher, A., Freund, H.-J., and Zilles, K., 2001a, Variability and asymmetry in the human precentral motor system. A cytoarchitectonic and myeloarchitectonic brain mapping study, *Brain* **124:**2232–2258.

Rademacher, J., Bürgel, U., and Zilles, K., 2002, Stereotaxic localization, intersubject variability, and interhemispheric differences of the human auditory thalamocortical system, *Neuroimage* **17:**142–160.

Rademacher, J., Caviness, J., Steinmetz, H., and Galaburda, A. M., 1993, Topographical variation of the human primary cortices: implications for neuroimaging, brain mapping, and neurobiology, *Cereb. Cortex* **3:**313–329.

Rademacher, J., Morosan, P., Schormann, T., Schleicher, A., Werner, C., Freund, H.-J., and Zilles, K., 2001b, Probabilistic mapping and volume measurement of human primary auditory cortex, *Neuroimage* **13:**669–683.

Riegele, L., 1931, Die Cytoarchitektonik der Felder der Broca'schen Region, *J. Psychol. Neurol.* **42:**496–514.

Rivière, D., Mangin, J.-F., Papadopoulos-Orfanos, D., Martinez, J. M., Frouin, V., and Régis, J., 2002, Automatic recognition of cortical sulci of the human brain using a congregation of neural networks, *Med. Image Anal.* **6:**77–92.

Roe, A. W., and Tso, D. Y., 1995, Visual topography in primate v2: multiple representations across functional stripes, *J. Neurosci.* **15:**3689–3715.

Roland, P. E., Svensson, P., Lindeberg, T., Risch, T., Baumann, P., Dehmel, A., Fredriksson, J., Halldorson, H., Forsberg, L., Young, J., and Zilles, K., 2001, A database generator for human brain imaging, *Trends Neurosci.* **24:**562–564.

Roland, P. E., and Zilles, K., 1994, Brain atlases—a new research tool, *Trends Neurosci.* **17:**458–467.

Roland, P. E., and Zilles, K., 1996, The developing European computerized human brain database for all imaging modalities, *Neuroimage* **4:**S39–S47.

Roland, P. E., and Zilles, K., 1998, Structural divisions and functional fields in the human cerebral cortex, *Brain Res. Rev.* **26:**87–105.

Sarkisov, S. A., Filimonoff, I. N., and Preobrashenskaya, N. S., 1949, *Cytoarchitecture of the Human Cortex Cerebri (Russ.)*, Moscow: Medgiz.

Schleicher, A., Amunts, K., Geyer, S., Kowalski, T., Schormann, T., Palomero-Gallagher, N., and Zilles, K., 2000, A stereological approach to human cortical architecture: identification and delineation of cortical areas, *J. Chem. Neuroanat.* **20:**31–47.

Schleicher, A., Amunts, K., Geyer, S., Kowalski, T., and Zilles, K., 1998, An observer-independent cytoarchitectonic mapping of the human cortex using a stereological approach, *Acta Stereol.* **17:**75–82.

Schleicher, A., Amunts, K., Geyer, S., Morosan, P., and Zilles, K., 1999, Observer-independent method for microstructural parcellation of cerebral cortex: a quantitative approach to cytoarchitectonics, *Neuroimage* **9:**165–177.

Schleicher, A., and Zilles, K., 1990, A quantitative approach to cytoarchitectonics: analysis of structural inhomogeneities in nervous tissue using an image analyser, *J. Microsc.* **157:**367–381.

Schmitt, O., Hömke, L., and Dümbgen, L., 2003, Detection of cortical transition regions utilizing statistical analyses of excess masses, *Neuroimage* **19:**42–63.

Schormann, T., Dabringhaus, A., and Zilles, K., 1995, Statistics of deformations in histology and improved alignment with MRI, *IEEE Trans. Med. Imaging* **14:**25–35.

Schormann, T., and Zilles, K., 1998, Three-dimensional linear and nonlinear transformations: an integration of light microscopical and MRI data, *Hum. Brain Mapp.* **6:**339–347.

Sereno, M. I., Dale, A. M., Reppas, J. B., Kwong, K. K., Belliveau, J. W., Brady, T. I., Rosen, B. R., and Tootell, R. B. H., 1995, Borders of multiple visual areas in humans revealed by functional magnetic resonance imaging, *Science* **268**:889–893.

Shipp, S., Watson, J. D. G., Frackowiak, R. S. J., and Zeki, S., 1995, Retinotopic maps in human prestriate visual cortex: the demarcation of areas V2 and V3, *Neuroimage* **2**:125–132.

Singh, R. P., and Smith, D. J., 2003, Genome scale mapping of brain gene expression, *Biol. Psychiatry* **53**:1069–1074.

Sowell, E. R., Peterson, B. S., Thompson, P. M., Welcome, S. E., Henkenius, A. L., and Toga, A. W., 2003, Mapping cortical change across the human life span, *Nat. Neurosci.* **6**:309–315.

Steinmetz, H., Staiger, J. F., Schlaug, G., Huang, Y., and Jäncke, L., 1995, Corpus callosum and brain volume in women and men, *Neuroreport* **6**:1002–1004.

Stensaas, S. S., Eddington, D. K., and Dobelle, W. H., 1974, The topography and variability of the primary visual cortex in man, *J. Neurosurg.* **40**:755.

Stephan, K. E., Zilles, K., and Kötter, R., 2000, Coordinate-independent mapping of structural and functional data by objective relational transformation (ORT), *Phil. Trans. R. Soc. Lond. B* **355**:37–54.

Stöckel, M. C., Weder, B., Binkofski, F., Choi, H. -J., Amunts, K., Pieperhoff, P., Shah, N. J., and Seitz, R., 2004, Left and right parietal lobule in tactile object discrimination, *Eur. J. Neurosci.* **19**:1067–1072.

Talairach, J., and Tournoux, P., 1988, *Coplanar Stereotaxic Atlas of the Human Brain*, Stuttgart: Thieme.

Tanji, J., and Kurata, K., 1989, Changing concepts of motor areas of the cerebral cortex, *Brain Dev.* **11**:374–377.

Thompson, P. M., Moussai, J., Zohoori, S., Goldkorn, A., Khan, A. A., Mega, M. S., Small, G. W., Cummings, J. L., and Toga, A. W., 1998, Cortical variability and asymmetry in normal aging and Alzheimer's disease, *Cereb. Cortex* **8**:492–509.

Thompson, P. M., Schwartz, C., Lin, R. T., Khan, A. A., and Toga, A. W., 1996, Three-dimensional statistical analysis of sulcal variability in the human brain, *J. Neurosci.* **16**:4261–4274.

Thompson, P. M., and Toga, A. W., 2002, A framework for computational anatomy, *Comput. Vis. Sci.* **5**:13–34.

Toga, A., 2003, Brain atlases for the 21st century, In: Ng, V., Barker, G. J., and Hendler, T. (eds.), *Psychiatric Neuroimaging*, Ohmsha: IOS Press, pp. 9–15.

Toga, A. W., and Thompson, P. M., 2003, Mapping brain asymmetry, *Nat. Rev. Neurosci.* **4**: 37–48.

Tootell, R. B. H., Mendola, J. D., Hadjikhani, N. K., Ledden, P. J., Liu, A. K., Reppas, J. B., Sereno, M. I., and Dale, A. M., 1997, Functional analysis of V3A and related areas in human visual cortex, *J. Neurosci.* **17**:7060–7078.

Turner, R., Le Bihan, D., and Chesnick, A. S., 1991, Echo-planar imaging of diffusion and perfusion, *Magn. Reson. Med.* **19**:247–253.

Uylings, H. B. M., Sanz Arigita, E., de Vos, K., Smeets, W. J. A. J., Pool, C. W., Amunts, K., Rajkowska, G., and Zilles, K., 2000, The importance of a human 3D database and atlas for studies of prefrontal and thalamic functions, In: Uylings, H. B. M., van Eden, C. G., de Bruin, J. P. C., Feenstra, M. G. P., and Pennartz, C. M. A. (eds.), *Progress in Brain Research*, Vol. 126, Amsterdam: Elsevier Science BV, pp. 357–368.

von Economo, C., and Koskinas, G. N., 1925, *Die Cytoarchitektonik der Hirnrinde des erwachsenen Menschen*, Berlin: Springer.

Van Essen, D. C., 1985, Functional organization of primate visual cortex, In: Peters, A., and Jones, E. G. (eds.), *Cerebral Cortex*, London: Plenum Press, pp. 259–329.

Van Essen, D. C., 2002, Surface-based atlases of cerebellar cortex in human, macaque, and mouse. *Ann. Ny. Acad. Sci.* **978**:468–479.

Van Essen, D. C., 2004, Surface-based approaches to spatial localization and registration in primate cerebral cortex, *Neuroimage* **23**:s97–s107.

Van Essen, D. C., 2005, A population-average, landmark- and surface-based (PALS) atlas of human cerebral cortex, *Neuroimage* **28**:635–662.

Van Essen, D. C., Drury, H. A., Dickson, J., Harwell, J., Hanlon, D., and Anderson, C. H., 2001, An integrated software suite for surface-based analyses of cerebral cortex, *J. Am. Med. Inform. Assoc.* **8**:443–459.

Van Essen, D. C., Drury, H. A., Joshi, S., and Miller, M. I., 1998, Functional and structural mapping of human cerebral cortex: solutions are in the surfaces, *Proc. Natl. Acad. Sci.* **95**:788–795.

Vogt, C., and Vogt, O., 1919, Allgemeinere Ergebnisse unserer Hirnforschung, *J. Psychol. Neurol.* **25**:292–398.

Walker, A. E., 1938, *The Primate Thalamus*, Chicago: University of Chicago Press.

Walters, N., Egan, G. F., Kean, M., Jenkinson, M., Kril, J. J., and Watson, J. D. G., 2003, In vivo identification of human cortical areas using high resolution MRI: an approach to structure–function correlation, *Proc. Natl. Acad. Sci.* **100**:2981–2986.

Wheeler-Kingshott, C. A. M., Hickman, S. J., Parker, G. J. M., Ciccarelli, O., Symms, M. R., Miller, D. H., and Barker, G. J., 2002, Investigating cervical spinal cord structure using axial diffusion tensor imaging, *Neuroimage* **16**:93–102.

Wiesendanger, E., Clarke, S., Kraftsik, R., and Tardif, E., 2004, Topography of cortico-striatal connections in man: anatomical evidence for parallel organization, *Eur. J. Neurosci.* **20**:1915–1922.

Wohlschläger, A. M., Specht, K., Lie, C.-H., Mohlberg, H., Bente, K., Pietrzyk, U., Stöcker, T., Zilles, K., Amunts, K., and Fink, G. R., 2005, Linking retinotopic fMRI mapping and anatomical probability maps of human occipital areas V1 and V2, *Neuroimage* **15**:73–82.

Wong-Riley, M. T. T., Hevner, R. F., Cutlan, R., Earnest, M., Egan, R., Frost, J., and Nguyen, T., 1993, Cytochrome oxidase in the human visual cortex: distribution in the developing and the adults brain, *Vis. Neurosci.* **10**:41–58.

Woods, R. P., Grafton, S. T., Watson, J. D. G., Sicotte, N. L., and Mazziotta, J. C., 1998, Automated image registration: II. Intersubject validation of linear and nonlinear tools, *J. Comput. Assis. Tomogr.* **22**:153–165.

Wree, A., Schleicher, A., and Zilles, K., 1982, Estimation of volume fractions in nervous tissue with an image analyzer, *J. Neurosci. Methods* **6**:29–43.

Yakovlev, P. I., and Lecours, A. R., 1967, The myelogenetic cycles of regional maturation of the brain, In: Yakovlev, P. I. and Lecours, A. R. (eds.), *Regional Development of the Brain in Early Life*, Oxford: Blackwell Scientific Publishing, pp. 3–70.

Young, J. P., Geyer, S., Grefkes, C., Amunts, K., Morosan, P., Zilles, K., and Roland, P. E., 2003, Regional cerebral blood flow correlations of somatosensory Areas 3a, 3b, 1, and 2 in humans during rest: a PET and cytoarchitectural study, *Hum. Brain Mapp.* **19**: 183–196.

Young, J. P., Herath, P., Eickhoff, S., Choi, H.-J., Grefkes, C., Zilles, K., and Roland, P. E., 2004, Somatotopy and attentional modulation of the human parietal and opercular regions, *J. Neurosci.* **24**:5391–5399.

Zilles, K., 1991, Codistribution of receptors in the human cerebral cortex, In: Mendelsohn, F. A. O., and Paxinos, G. E. (eds.), *Receptors in the Human Nervous System*, San Diego: Academic Press, pp. 165–206.

Zilles, K., Armstrong, E., Schleicher, A., and Kretschmann, H. J., 1988a, The human pattern of gyrification in the cerebral cortex, *Anat. Embryol.* **179**:173–179.

Zilles, K., Dabringhaus, A., Geyer, S., Amunts, K., Qü, M., Schleicher, A., Gilissen, E., Schlaug, G., Seitz, R., and Steinmetz, H., 1996, Structural asymmetries in the human forebrain and the forebrain of non-human primates and rats, *Neurosci. Behav. Rev.* **20**:593–605.

Zilles, K., Eickhoff, S., and Palomero-Gallagher, N., 2003, The human parietal cortex: a novel approach to its architectonical mapping, *Adv. Neurol.* **93**:1–20.

Zilles, K., Hajos, F., Kalman, M., and Schleicher, A., 1991a, Mapping of glial fibrillary acidic protein-immunoreactivity in the rat forebrain and mesencephalon by computerized image analysis, *J. Comp. Neurol.* **308**:340–355.

Zilles, K., Kawashima, R., Dabringhaus, A., Fukoda, H., and Schormann, T., 2001, Hemispheric shape of European and Japanese brains: 3-D MRI analysis of intersubject variability, ethnical, and gender differences, *Neuroimage* **13**:262–271.

Zilles, K., and Palomero-Gallagher, N., 2001, Cyto-, myelo-, and receptor architectonics of the human parietal cortex, *Neuroimage* **14**:S8–S20.

Zilles, K., Palomero-Gallagher, N., Grefkes, C., Scheperjans, F., Boy, C., Amunts, K., and Schleicher, A., 2002a, Architectonics of the human cerebral cortex and transmitter receptor fingerprints: reconciling functional neuroanatomy and neurochemistry, *Eur. Neuropsychopharmacol.* **12**:587–599.

Zilles, K., Palomero-Gallagher, N., and Schleicher, A., 2004, Transmitter receptors and functional anatomy of the cerebral cortex, *J. Anat.* **205**:417–432.

Zilles, K., Schlaug, G., Matelli, M., Luppino, G., Schleicher, A., Qü, M., Dabringhaus, A., Seitz, R., and Roland, P. E., 1995, Mapping of human and macaque sensorimotor areas by integrating architectonic, transmitter receptor, MRI and PET data, *J. Anat.* **187**:515–537.

Zilles, K., and Schleicher, A., 1993, Cyto- and myeloarchitecture of human visual cortex and the periodical GABA-A receptor distribution, In: Gulyas, B., Ottoson, D., and Roland, P. E. (eds.), *Functional Organization of the Human Visual Cortex*, Oxford: Pergamon Press, pp. 111–120.

Zilles, K., Schleicher, A., 1995, Correlative imaging of transmitter receptor distributions in human cortex, In: Stumpf, W., and Solomon, H. (eds.), *Autoradiography and Correlative Imaging*, San Diego: Academic Press, pp. 277–307.

Zilles, K., Schleicher, A., Langemann, C., Amunts, K., Morosan, P., Palomero-Gallagher, N., Schormann, T., Mohlberg, H., Bürgel, U., Steinmetz, H., Schlaug, G., and Roland, P. E., 1997, A quantitative analysis of sulci in the human cerebral cortex: development, regional heterogeneity, gender difference, asymmetry, intersubject variability and cortical architecture, *Hum. Brain Mapp.* **5**:218–221.

Zilles, K., Schleicher, A., Palomero-Gallagher, N., and Amunts, K., 2002b, Quantitative analysis of cyto- and receptor architecture of the human brain, In: Mazziotta, J. C., and Toga, A. (eds.), *Brain Mapping: The Methods*, Amsterdam: Elsevier, pp. 573–602.

Zilles, K., Schleicher, A., Rath, M., and Bauer, A., 1988b, Quantitative receptor autoradiography in the human brain. Methodical aspects, *Histochemistry* **90**:129–137.

Zilles, K., Werner, L., Qü, M., Schleicher, A., and Gross, G., 1991b, Quantitative autoradiography of 11 different transmitter binding sites in the basal forebrain region of the rat—evidence of heterogeneity in distribution patterns, *Neuroscience* **42**:473–481.

19

Neuron and Network Modeling

GIORGIO A. ASCOLI and RUGGERO SCORCIONI

Abstract: Computer technology constitutes a formidable asset in the acquisition, manipulation, analysis, and modeling of neuroanatomical data. Single-cell arborizations can be digitally represented as a large number of connected cylinders. In this form, neuronal structure is amenable to three-dimensional (3D) rendering, extensive quantitative characterization, and computational modeling of biophysics,

GIORGIO A. ASCOLI • Krasnow Institute for Advanced Study and Psychology Department, George Mason University, 4400 University Drive, MS2A1 Fairfax, VA 22030-4444
RUGGERO SCORCIONI • Krasnow Institute for Advanced Study, George Mason University, 4400 University Drive, MS2A1 Fairfax, VA 22030-4444

electrophysiology, outgrowth, network connectivity, and dynamics. This chapter describes the state of the art in neuron and network modeling, with particular emphasis on the methods to acquire, analyze, and synthesize neuroanatomical data. Several commercial and freeware systems are available to reconstruct neuronal morphology in digital format, from a variety of preparations, either directly from the microscope or off-line from captured images. The resulting, increasing amount of digital data (and meta-data) can be archived and publicly distributed to maximize scientific impact. This database enables continuing efforts in modeling dendritic branching of neurons throughout the central nervous system, including cortex, cerebellum, and spinal cord. The experimental acquisition of complete axonal projections from single neurons poses additional challenges, which are only recently being overcome. The combination of dendritic and axonal reconstructions (or models), together with the surface and volumetric representation of the surrounding tissue, allows the computational derivation of synaptic connectivity. Taken together, such models constitute a powerful substrate for the implementation of large-scale, anatomically realistic neural networks. These advances can be instrumental for the cross-scale elucidation of the relationship between structure, activity, and function in the brain.

Keywords: algorithm, axon, computer, connectivity, dendrite, reconstruction, simulation

I. INTRODUCTION

The mammalian brain is often referred to as the "most complex object in the universe." Indeed, the sheer number of cells and their connections must be compounded with their exquisite organization, from the intricacy of dendritic and axonal branching, to the specificity of the interactions among neuronal classes. Facing such mighty complexity, neuroscientists have traditionally reverted to two levels of analysis. At the system level, descriptions are typically qualitative, with interactions among functional components simply tagged as "present" or "absent" (or perhaps "strong" and "weak"). In contrast, quantitative characterization accompanies the reductionist approach to investigate ever more "elementary" components, from individual cells to spines, synaptic densities, single receptors, their subunits, and individual amino acids. Can the rigorous biophysical knowledge of cellular and subcellular processes be synthesized at the network level?

Computer technology constitutes a formidable asset in the acquisition, manipulation, analysis, and modeling of neuroanatomical data. Models have played a fundamental role in most fields of science, including several sub-disciplines within neurobiology. Neuroanatomy has somehow lagged behind, as its cross-scale complexity prevented the intuitive development of abstract theories. This deadlock can be now solved by adopting hardware and software tools to render neuronal and network structures quantitatively accessible to our understanding. This chapter describes the state of the art in neuron and network modeling, with particular emphasis on the methods to acquire, analyze, and synthesize neuroanatomical data in digital format.

II. DIGITAL MORPHOMETRY OF SINGLE NEURONS

A. Computer Acquisition

Typical neuroanatomical experiments result in chemically processed tissue mounted on a microscope slide. The corresponding observable microscope image can only be further manipulated in a limited way. A key step toward the flexible, quantitative, and extensive analysis and modeling of these data consists of their computer acquisition or *digitization*. The resulting digital files represent data in numerical (machine-readable) format.

Among the essential elements of digital neuroanatomy (as of much of neuroscience) are individual neurons. Ramon y Cajal pioneered the use of *camera lucida*, or drawing tube, a system of mirrors mounted between the microscope oculars and the stage, which allows the precise hand-tracing of the specimen. With this method, still in use in many neuroscience laboratories, neuronal structure is captured on paper as a pencil drawing of its two-dimensional (2D) projection.

Dendritic and axonal trees can be described in digital form as a series of interconnected cylinders, each characterized by the three spatial coordinates of the end point, the diameter, and the identity of the cylinder they are attached to in the path to the soma. Several systems, alternative to camera lucida, have been developed to acquire neuronal morphology directly in digital form. The most widely adopted commercial system is MicroBright-Field's Neurolucida (www.microbrightfield.com). The Neurolucida setup includes a computer–microscope interface, a motorized stage, and a complete software suite (Glaser and Glaser, 1990). Similarly to the camera lucida system, the user sees the computer's monitor overlaid on the microscopic image. However, instead of drawing with a pencil on paper, the user virtually draws the structure of interest with mouse clicks, and the digital file is created in the computer memory in real time. Fine regulation of the focus is logged as depth, yielding precise spatial information in all three dimensions. The adjustable size of the mouse cursor determines the diameter of the structure being traced. In addition, extended structures can be followed continuously, thanks to the joystick-controlled horizontal movement of the stage, virtually paralleled by corresponding moves on the drawing screen. Neurolucida also allows the reconstruction of arborizations across multiple serial sections (Fig. 19.1A).

An alternative to the online reconstruction of neurons at the microscope is constituted by the semiautomated acquisition of a *stack* of (possibly tiled) digital images, serially ordered by their depth (or focal plane). Digital reconstruction of neuronal morphology can then be carried out off-line. This system allows lengthy reconstructions to be completed with minimal operation of the (usually expensive) microscope, including from perishable preparations such as fluorescence stains for confocal microscopy. In addition, image stacks can be postprocessed, e.g., by contrast optimization, filtering, and deconvolution.

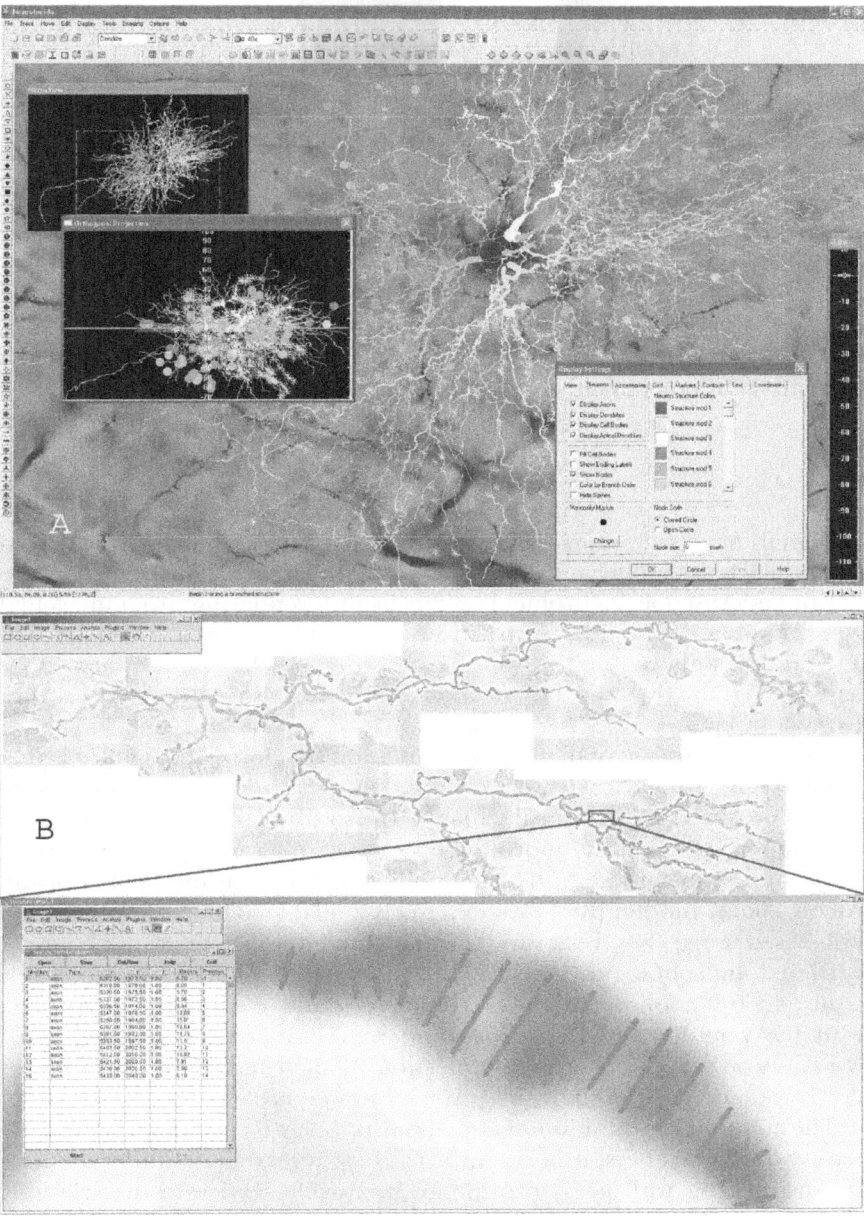

Figure 19.1. Computer-assisted digital reconstructions. (A) Screenshot from Micro-BrightField's Neurolucida® software for 3D neuron reconstruction. This image shows neuron reconstruction superimposed on a live image from the microscope. Additional windows provide accessory tools that assist with the neuron tracing are also shown. Neuron reconstruction by Robin Price. (Courtesy of MicroBrightField, Inc.) (B) Neuron_morpho ImageJ plug-in screenshot. The top window contains a cerebellar climbing fiber image from a Z-stack captured by optical microscopy. A portion of this image is enlarged in the bottom window, showing the semimanual tracing operation. Each reconstructed tracing point (red lines) is represented as one row in the inset window. Column entries for each point correspond to a progressive numerical identity, type, x, y, and z coordinates, radius, and identity of the parent segment.

An additional powerful software for the off-line digital reconstruction of neuronal morphology from image stacks is Neuron_morpho (Fig. 19.1B), a plug-in of the NIH-distributed imaging program ImageJ. Both ImageJ (http://rsb.info.nih.gov/ij) and Neuron_morpho (www.maths.soton.ac.uk/staff/D'Alessandro/morpho) are freely available and run on all JAVA-compatible platforms (including Windows, Linux, and MacOS). Another, less widely used, system for digital reconstruction uses polynomial interpolation to join 2D reconstructions from serial images (Wolf *et al.*, 1995). Several ongoing projects are also attempting to automate the digital reconstruction process by pattern recognition (e.g., He *et al.*, 2003; Rodriguez *et al.*, 2003), an extremely difficult but ultimately commanding step of progress in this field.

B. Digital Files

Digital neuronal morphologies can be displayed and inspected in "pseudo-3D," including angle views different from that originally imaged under the microscope. Neurolucida offers its own rendering program, called NeuroExplorer (Fig. 19.2). An extremely popular neuronal visualization and editing software tool is Cvapp (Cannon *et al.*, 1998), a freeware, JAVA-based program that can be run both locally or through a web browser (www.compneuro.org/CDROM/nmorph).

A major advantage of digital representation of neuronal structure is that virtually any geometrical feature captured by the cylinder-based description can be measured and statistically analyzed quickly, reliably, and precisely (see also Appendix). Over 50 morphometric functions can be extracted from single or multiple neuromorphological files with L-Measure (Scorcioni and Ascoli, 2001), another JAVA program freely available both for download and web-based usage (Fig. 19.2; www.krasnow.gmu.edu/L-Neuron).

Digital morphologies can be also used to implement anatomically realistic simulations of neuronal biophysics and electrophysiology, e.g., with the popular NEURON environment (Hines and Carnevale, 2001; www.neuron.yale.edu). A large collection of such models is available for download and use (Migliore *et al.*, 2003; http://senselab.med.yale.edu).

The computer acquisition of digital morphology is considerably labor intensive ($\sim$1 week-person per neuron). Thus, researchers willing to share reconstructions with peers provide an invaluable service to the scientific community (Gardner *et al.*, 2003). Several electronic collections of neuronal morphology are available for a variety of cell classes (reviewed in Ascoli, 2002a; Turner *et al.*, 2002). Although almost each archive of neuronal reconstructions comes with its own unique file format (Ascoli *et al.*, 2001a), these can be easily interconverted using tools such as Cvapp and L-Measure. Nevertheless, particular care must be taken in considering the lab-idiosyncratic morphological characteristics, which can derive from specific experimental conditions and protocols, hardware and software setups, and individual operators' bias (Scorcioni *et al.*, 2004).

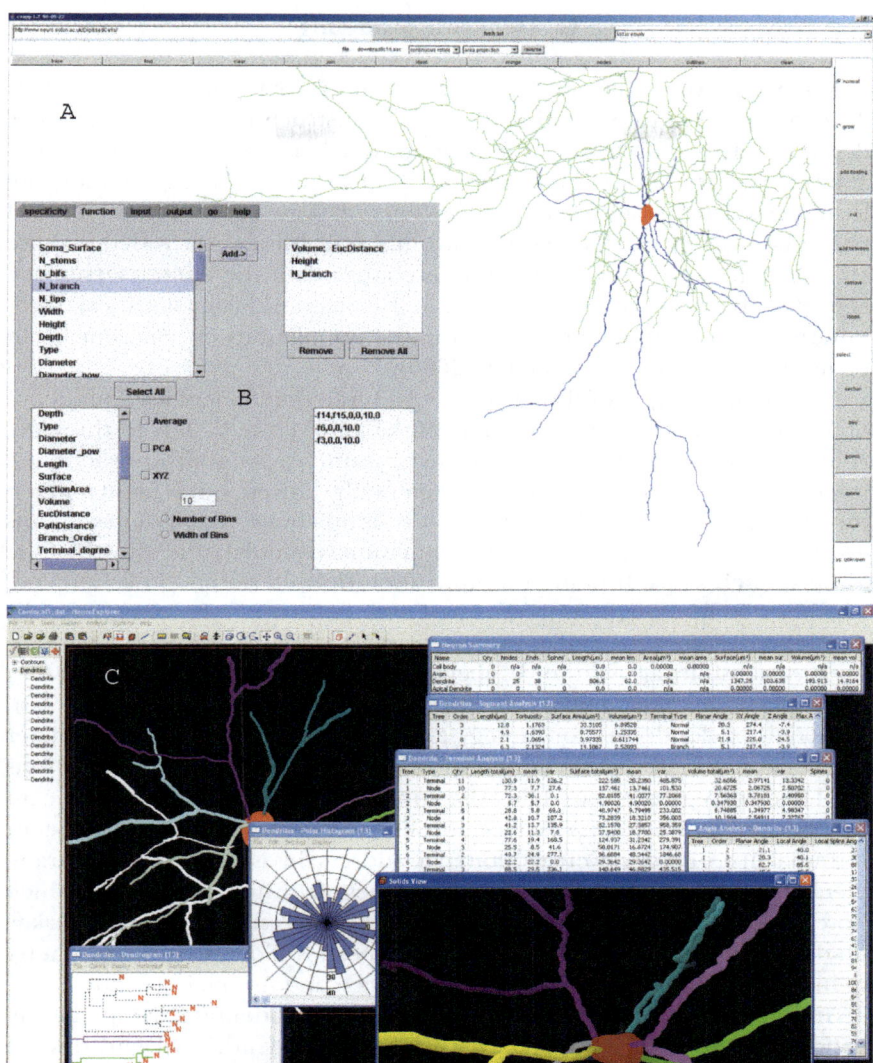

Figure 19.2. Electronic tools for rendering and analyzing neuronal morphology. (A) Cvapp display of a neuron from Markram's neocortical database (http://microcircuit.epfl.ch; LBC cell C300301B1 from layer 4). (B) Screenshot (inset) of the L-Measure web-based graphical user interface (see also Appendix). (C) Screenshot from MicroBrightField's NeuroExplorer (TM) software showing results of quantitative morphological analyses and an interactive 3D graphical representation of a reconstructed neuron. (Courtesy of MicroBrightField, Inc.)

C. Dendritic Modeling

A single neuron can be represented in digital form by tens of thousands of 3D coordinates of its branching neurites. This structure can be modeled by designing algorithms to generate synthetic neurons in virtual reality (Ascoli, 1999). The natural variability of neuronal anatomy within a given morphological class can be captured by *stochastic* simulations, in which nonidentical virtual cells are generated in different runs of the model (provided that the seed of random number generation is changed). If the parameters of the algorithm have a straightforward geometric meaning, their statistical distributions can be extracted directly from the populations of real neurons to be modeled (Ascoli and Krichmar, 2000).

A seminal example of this approach is Burke's diameter-based model of dendrogram geometry (reviewed in Burke and Marks, 2002). In this algorithm, dendrites sequentially elongate by a unitary length step, each time sampling diameter dependent, experimentally derived, bifurcation and termination probabilities. If a bifurcation is sampled, two daughter segments are attached at the next steps. If a termination is sampled, the growth of the given branch stops. If neither a bifurcation nor a termination is sampled, another dendritic segment is attached and the process repeats. Originally developed to describe spinal motoneurons, variations of this model have been successfully applied to cerebellar Purkinje cells (Ascoli *et al.*, 2001b) and hippocampal pyramidal cells (Donohue *et al.*, 2002) as well (Fig. 19.3).

Models of dendritic morphology exclusively based on branch diameter are generally under constrained. In other words, the simulated neurons tend to display greater variability than observed in the real cells. In a recent advancement, *hidden variables* (specifically, path distance and the number of terminal tips, or *degree*) were exploited to address this issue. All model parameters were made dependent on the local values of the hidden variables, which were updated at every step of the algorithm. The resulting hidden Markov model successfully captured all relevant properties of dendrogram geometry in hippocampal pyramidal cells (Samsonovich and Ascoli, in press).

An additional element in neuromorphological modeling is the spatial embedding of dendrograms, i.e., the 3D orientation of dendrites. Dendrites can be described as "pointing" in a given absolute direction, or in an orientation relative to the origin of their internal coordinates, i.e., the soma (Ascoli, 1999). A Bayesian method was recently introduced to measure the relative contribution of these various components of tropism from experimental data (Samsonovich and Ascoli, 2003). In all principal cell classes of the rat hippocampus, it was found that the major (and only statistically significant) component of systematic growth was away from the soma. Thus, the heavily polarized shape of hippocampal pyramidal and granule cells may be solely produced by the local orientation of the stems from the soma, which may be genetically determined. As a result, a simple two-parameter model (only specifying the amount of "push" away from the soma, and that of random deflection) can surprisingly capture the emergent shape of these cell classes

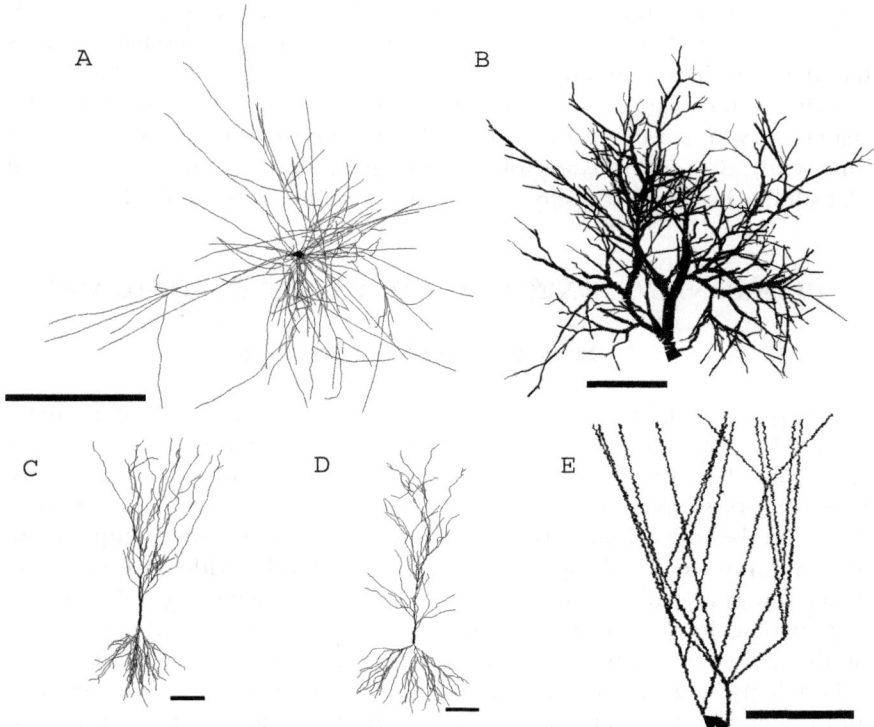

Figure 19.3. Simulated dendritic morphologies. (A) Diameter-based model of a spinal motoneuron. Scale bar: 1000 μm. (B) Diameter-based model of a cerebellar Purkinje cell. Scale bar: 50 μm. (C) Hidden Markov model of a hippocampal CA3 pyramidal neuron. Scale bar: 100 μm. (D) Hidden Markov model of a hippocampal CA1 pyramidal neuron. Scale bar: 100 μm. (E) Globally constrained model of a dentate granule cell. Scale bar: 100 μm.

(Fig. 19.3). Thus, the hidden Markov model of dendrograms and the algorithm of dendritic orientation together constitute a remarkably complete description of neuronal morphology, which was recently also applied to dentate granule cells (Samsonovich and Ascoli, in press).

An alternative model of granule cell morphology was developed based on global constraints such as the position of terminations along the principal component of the dendritic field (Winslow *et al.*, 1999). While the resulting shape coarsely reproduces the structure of real neurons (Fig. 19.3), algorithms of this type cannot be taken (even metaphorically) as *mechanistic* models of development, because real growing branches have access only to locally expressed and stored signals, and not to global information regarding the whole tree or the distal surrounding environment.

Nevertheless, algorithms based on the overall distribution of branching probability against the number of bifurcations, even if a global termination is externally imposed to the whole tree, can still be taken as *descriptive* models of development, if they capture the temporal dynamics of neuronal growth (Van Ooyen and Van Pelt, 2002). From this point of view, even local models based on dendritic diameter must be considered "hidden," since the parameter distributions are measured from adult shapes, and kept constant during virtual growth. An extensive review of computational models of neuronal outgrowth, with a discussion of the strengths, weaknesses, and biological plausibility, has been recently published (Donohue and Ascoli, 2004).

III. AXONAL CONNECTIVITY IN THE ELECTRONIC AGE

A. Semiautomated Vectorization

Computer-assisted digital reconstructions of dendritic trees, albeit time-consuming, have now become standard routine in modern cellular neuroanatomy laboratories. The axonal arborizations of projection neurons, however, are simply too huge to be digitized in the same fashion. Only a small number of such reconstructions have been successfully completed in what amounts to a truly heroic effort of dedicated individuals. Camera lucida tracings are still widely adopted to obtain a permanent graphic record of axonal projections from single neurons. How can these data be converted in digital form for improved analysis and modeling?

Pencil drawings can be computer-acquired with a high-resolution scanner, and a segment representation of the tracings can be obtained with freely available software (e.g., www.wintopo.com). Alternatively, camera lucida projections can be directly acquired in electronic form by using computer-interfaced tablets (e.g., Gras and Killman, 1983). In both cases, the result is a set of digitized but disjoined segments (for a technical comparison, see Ewart et al., 1989).

Using a tablet-acquired data set from Tamamaki et al. (1988) as a test bed, we have recently developed an algorithm to fully reconnect axonal arborizations in the same format as typically obtained with the techniques described in section "Computer Acquisition" (Fig. 19.4). This implementation (which can be also applied to scanned-in camera lucida paper-and-pencil drawings) simply follows a nearest-neighbor strategy, taking into account the average spread of axonal branches in the thickness of the serial sections (Scorcioni and Ascoli, in press).

Using this semiautomated vectorization procedure, we reconstructed eight complete axonal morphologies from individual neurons, including at least one for each of the principal cell classes of the hippocampal formation: entorhinal cortex layer II stellate cells, dentate gyrus granule cells, CA3, CA2, and CA1 pyramidal cells, and subicular neurons (Fig. 19.5). While the amount of effort required to hand-trace these large arborizations on paper or tablets is still quite considerable, the semiautomated image processing now makes it feasible to obtain larger collections of digital axonal data from each neuronal class.

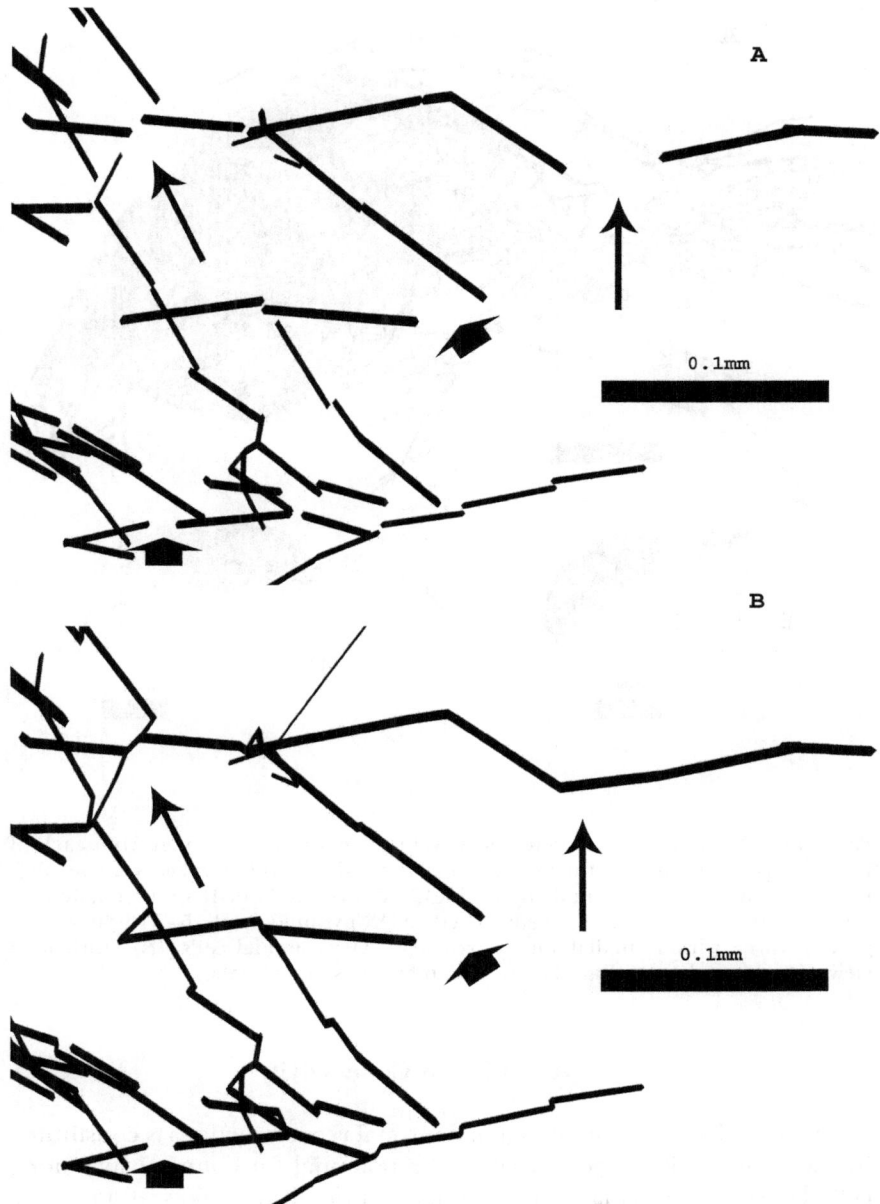

Figure 19.4. Reconstruction of axonal trees from manual tracings. (A) Raw vectorized data. Note various disconnected segments (arrows). (B) Algorithmically connected arborization. Long thin arrows indicate branches that have been joined by an additional segment. Short thick arrows indicate "true" gaps that should remain disconnected in the final reconstruction.

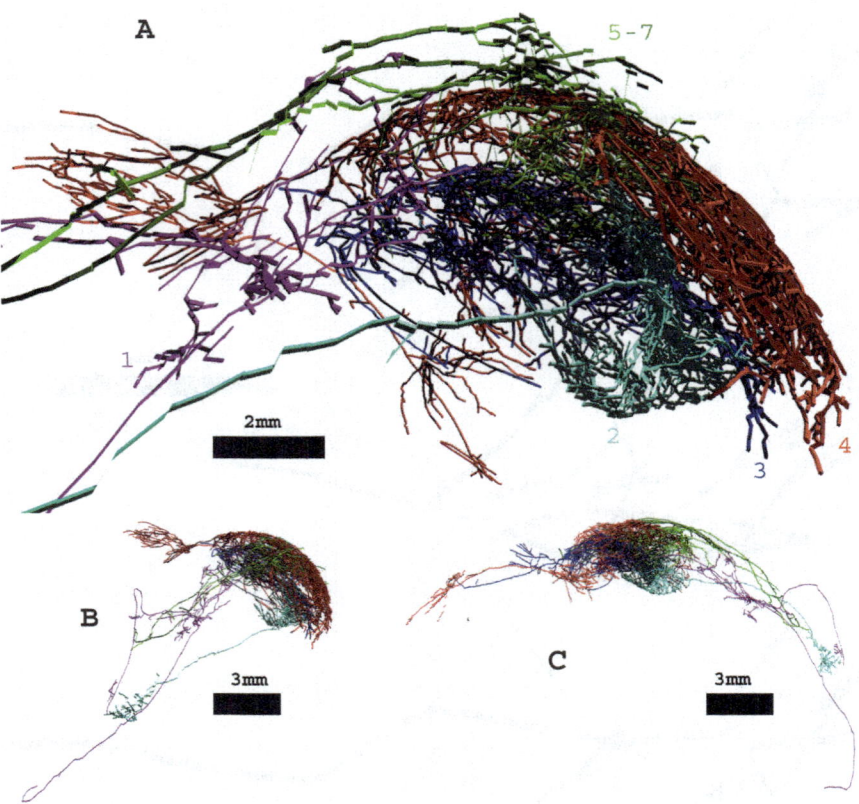

Figure 19.5. Montage of complete ipsilateral projections of one axon from each of the principal cell classes of the hippocampal formation. (A) Lateral view. 1 (purple): subicular pyramidal neuron; 2 (light blue): enthorhinal layer II spiny stellate cell; 3 (dark blue): CA3 pyramidal cell; 4 (red): CA2 pyramidal cell; 5–7 (light-to-dark green): three (distal, medial, and proximal) CA1 pyramidal cells. (B) Horizontal view. (C) Coronal view. (Raw data provided by Dr. N. Tamamaki.)

B. Derivation of Connectivity

A powerful application of digital neuronal reconstructions is constituted by the computational derivation of the potential for connectivity among cell classes. In particular, when axonal and dendritic arborizations share the same anatomical space, it is possible to identify a minimum interaction distance within which synapses could be established. A similar (and computationally equivalent) approach consists of defining a "sphere of influence" around neurites, and analyzing their overlaps. In either case, given an "input" and "output" arborization, it is possible to mathematically derive the number and spatial distribution of potential synapses (e.g., Kalisman *et al.*, 2003).

The relevant geometric parameters for this analysis (e.g., interaction distance) can be estimated from experimental data, such as the size of dendritic spines, the interbouton distance on axons, and the length of growth cones. It is important to stress that this approach yields an estimate of the *potential* for synaptic connectivity, rather than a direct number of synapses. This potential can be regarded as an upper limit of the number of synapses, or as the combinatorial pool of possible synapses, a subset of which is expressed at any given time. In light of the anatomical plasticity of synaptic connections (which are formed and eliminated continuously in at least some regions of the cortex), this measure can be physiologically relevant in the study of the cellular and network bases of learning and memory (Stepanyants *et al.*, 2004).

Potential connectivity is affected both by the intrinsic shape of afferent and efferent cells and by their spatial distribution and orientation. Thus, this type of analysis yields results that are cell-class specific, and can be used to compare different types of neurons within an anatomical region, inferring their possible functional roles (Stepanyants *et al.*, 2004). This approach has also been applied to elucidate the information processing in entire sensory pathways of model systems down to synaptic level (Jacobs and Pittendrigh, 2002). Alternatively, it is possible to estimate parameters of the spatial distribution of specific morphologies to ensure effective connectivity of the network (Costa and Manoel, 2003).

Information of system-level connectivity among brain regions is currently being collated in electronic databases, such as the Brain Architecture Management System (Bota and Arbib, 2004; http://brancusi.usc.edu/bkms). The advances in computational neuroanatomy described in this and previous sections of this chapter will soon make it possible to create web-based archives of neuronal connectivity at the cellular level (Ascoli and Atkeson, in press).

C. Models of Axonal Navigation

Projecting neurons navigate long distances toward their target before expressing full arborizations in the neuropil (see, e.g., Fig. 19.5). Therefore, computational models of axonal anatomy must include algorithmic descriptions of pathfinding in addition to intrinsic structural determinants. Much is known about the molecular correlates of axonal navigation (e.g., Donohue and Ascoli, 2004). Nonetheless, the theoretical understanding of these complex phenomena is still incomplete, and relatively little information has been so far integrated in computational models.

Senft and Ascoli (1999) proposed a phenomenological model in which axons navigated toward groups of neurons (or glia), turning and possibly bifurcating depending on their local orientation relative to their target, until they arrived within a given distance. At this point, axons started establishing synapses, again turning and bifurcating as necessary to optimally interact with their local postsynaptic counterparts. After making synaptic contact, the axonal branch in this model was temporarily inhibited from further synapsing. This "refractory period" was a critical parameter of the algorithm, which could discriminate among diverse axonal morphologies such as perforant pathways, sprouting neurites, and climbing fibers (long-, medium-, and short-lasting inhibition, respectively).

Earlier computational models concentrated on a mechanistic description of the biophysical processes underlying axonal movement, such as filopodial dynamics (Buettner, 1995). Several studies have focused on the mathematical description of chemical and cellular gradients as the main guiding cue for axons (e.g., Goodhill, 1998). Increasing attention in computational neuroanatomy is also being paid to the effect of competition (both for external targets and for internal metabolic resources) among axons (Van Ooyen and Van Pelt, 2002). A recent model integrated gradient navigation, axon–axon interaction, and the further influence of patterned activity (Yates *et al.*, 2004). Other relevant efforts include the attempt to describe neuritic navigation and connectivity in 2D with a cell automata formalism (Segev and Ben-Jacob, 2000), and the introduction of cell fate mechanisms in the computational description of axonal pathfinding (Eglen and Willshaw, 2002).

Notably, the relative scarcity of complete axonal reconstructions in digital format prevents a rigorous statistical comparison of the simulated axonal morphologies with the corresponding experimental data. In this sense, a wealth of useful data may become available when the resolution of Diffusion Tensor Imaging reaches the scale of individual axonal bundles (Mori, 2002).

IV. BOTTOM-UP NETWORK MODELING

A. System-Level Boundaries and Virtual Stereology

Axonal projections (and in some cases, dendritic trees as well) are typically affected by system-level geometric constrains, such as the shape of the afferent and efferent nuclei and regions. Thus, in order to fully characterize neuritic shape and neuropil connectivity, it is important to include in the model a digital representation of the relevant tissue and layer boundaries. These data can be acquired from neuroanatomical preparation in ways similar to those described in "Computer Acquisition." Suitable raw data include high-resolution ex vivo microscopic magnetic resonance imaging, or µ-MRI (see, e.g., Lester *et al.*, 2002), cytostructural boundaries traced from intracellular filling experiments (Fig. 19.6), or classic histochemical preparation such as Nissl or myelin stains.

From serially traced system-level boundaries, it is possible to compute continuous surfaces (rendered, e.g., as tiled triangles) and the corresponding volumes (list of internal voxels). Both representations carry important information. Surfaces often determine the orientation of axonal and

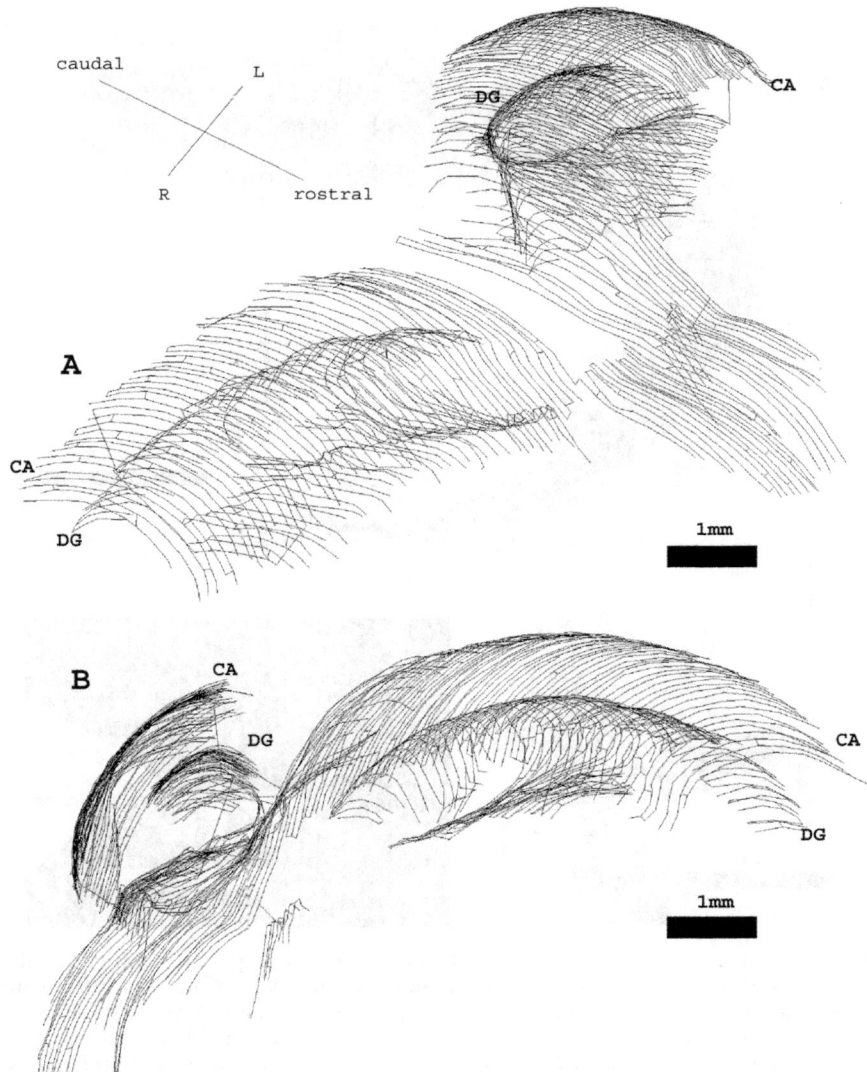

Figure 19.6. Cytostructural boundaries of the rat hippocampus traced from serial sections. (A) Dorsolateral view. Dentate gyrus and Ammon's horn are clearly visible in both the left and right hippocampi. (B) Mediocaudal view. One of the hippocampi is approximately displayed along its transversal axis, and the other one along its longitudinal axis. (Raw data provided by Dr. N. Tamamaki.)

dendritic arborization, while cell bodies, synapses, and branch coordinates can be virtually positioned in the appropriate volumes. Following this strategy, a large number of reconstructed (or simulated) neurons can be assembled in 3D (Scorcioni *et al.*, 2002) to "recreate" regional anatomy from cellular-level information (Fig. 19.7).

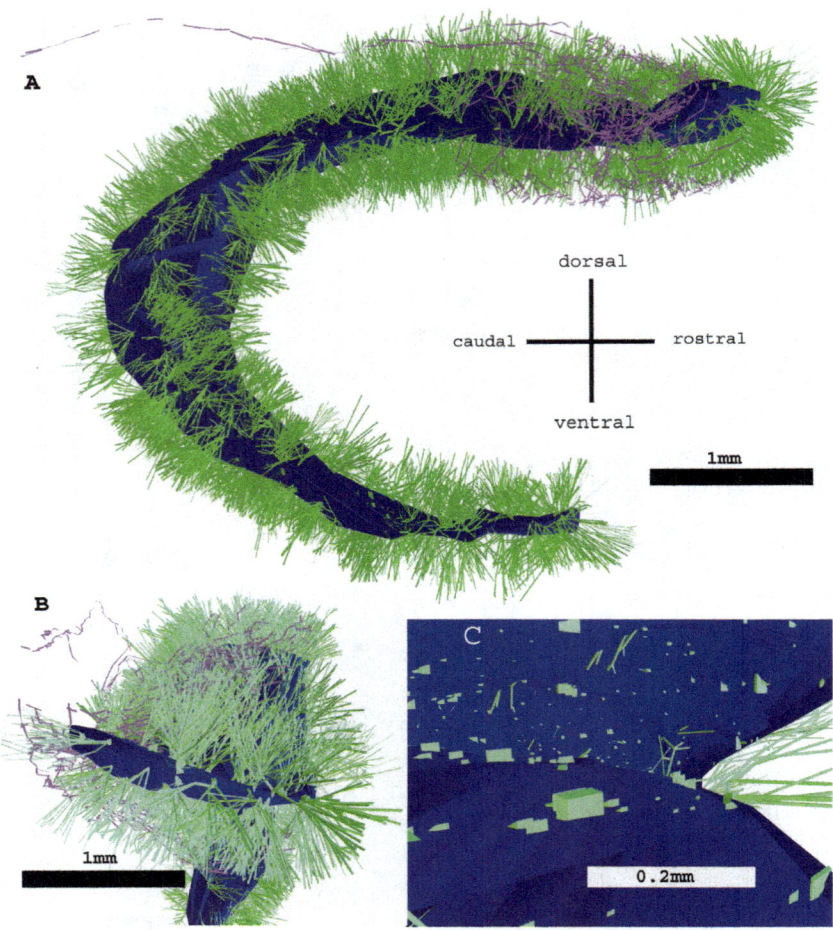

Figure 19.7. Large-scale model of the rat dentate gyrus showing one thousand granule cells (green) over a surface reconstruction of the cellular layer (blue), and the axon from a single spiny stellate cell (purple) projecting from layer II of the entorhinal cortex. (A) Dorsomedial view. (B) Detail on one of the dentate blade endings. (C) Detail from within the hilus. The volume is mostly empty as the granule cell axons (mossy fibers) are not included in the visualization.

Large-scale neuroanatomical models such as those displayed in Fig. 19.7 can be used to compute synaptic connectivity (see section "Derivation of Connectivity"), as well as to impose global constraints to models of neuronal morphology (see "Dendritic Modeling" and "Models of Axonal Navigation"). "Virtual slices" at arbitrary planes of orientation can be explored to foster intuition and guide electrophysiological and anatomical experiments. In addition, basic stereological properties, such as spatial occupancy and density of various subcellular components (e.g., dendrites and axons), can be derived for each layer and position in the virtual tissue. This approach has been applied to evaluate optimal motoneuron packing in the spinal cord (Burke and Marks, 2002), to quantitatively analyze the basal forebrain corticopetal system (Zaborszky *et al.*, 2002), and the somatosensory cerebro–cerebellar and ascending auditory pathways (Leergaard and Bjaalie, 2002). An important application of this line of study will consist of the inclusion of glia and blood vessels in considering the relationship between structure and function in neural systems.

B. Anatomically Realistic Neural Networks

Biophysical models of single-cell electrophysiology now routinely include a faithful description of neuronal morphology and account for the resulting functional compartmentalization (e.g., Lazarewicz *et al.*, 2002a; Migliore *et al.*, 2003). In contrast, most artificial neural networks have grown increasingly abstract, and retain almost none of the anatomical characteristics of the brain regions they are supposed to represent. Recent efforts, however, have concentrated on the development of anatomically realistic neural network models.

Small networks can be assembled "by hand" out of individual cell models, within the framework of existing modeling environments, such as NEURON (e.g., www.physiol.ucl.ac.uk/research/silver_a/neuroConstruct). Even with massively parallel supercomputers, however, simulation of activity dynamics at the level of subcellular electrophysiological mechanisms can be carried out only for a limited number of neurons. Nevertheless, simple, computationally efficient formalisms exist to capture essential neuronal dynamics (Izhikevich, in press), which can be in principle applied to real-scale network models. The problem remains to automatically and efficiently assemble a realistic anatomical network construct.

The cellular-level anatomy of the CA1 area of a hippocampal slice was simulated with the powerful ArborVitae software (Senft and Ascoli, 1999). Cell bodies for a variety of morphological classes were distributed in layers, subsequently warped to reproduce the natural folds of the rodent archicortex. Dendrites were then virtually grown in 3D using approaches as described in "Dendritic Modeling." Finally, axons were made to navigate toward and connect with targets specified according to the known wiring diagram of area CA1. The whole simulation could be run, displayed, and saved in a limited amount of time.

Recently, ArborVitae was augmented with the ability to "read in" digital representations of system-level experimental data of layer surfaces and regional volumes. Thus, cells can be distributed according to their precise spatial location, while axons' navigation can be simulated through a "real" voxel substrate (Senft, 2002). Figure 19.8 shows an example of this application to the simulation of a corticothalamic projection within an imaged human brain.

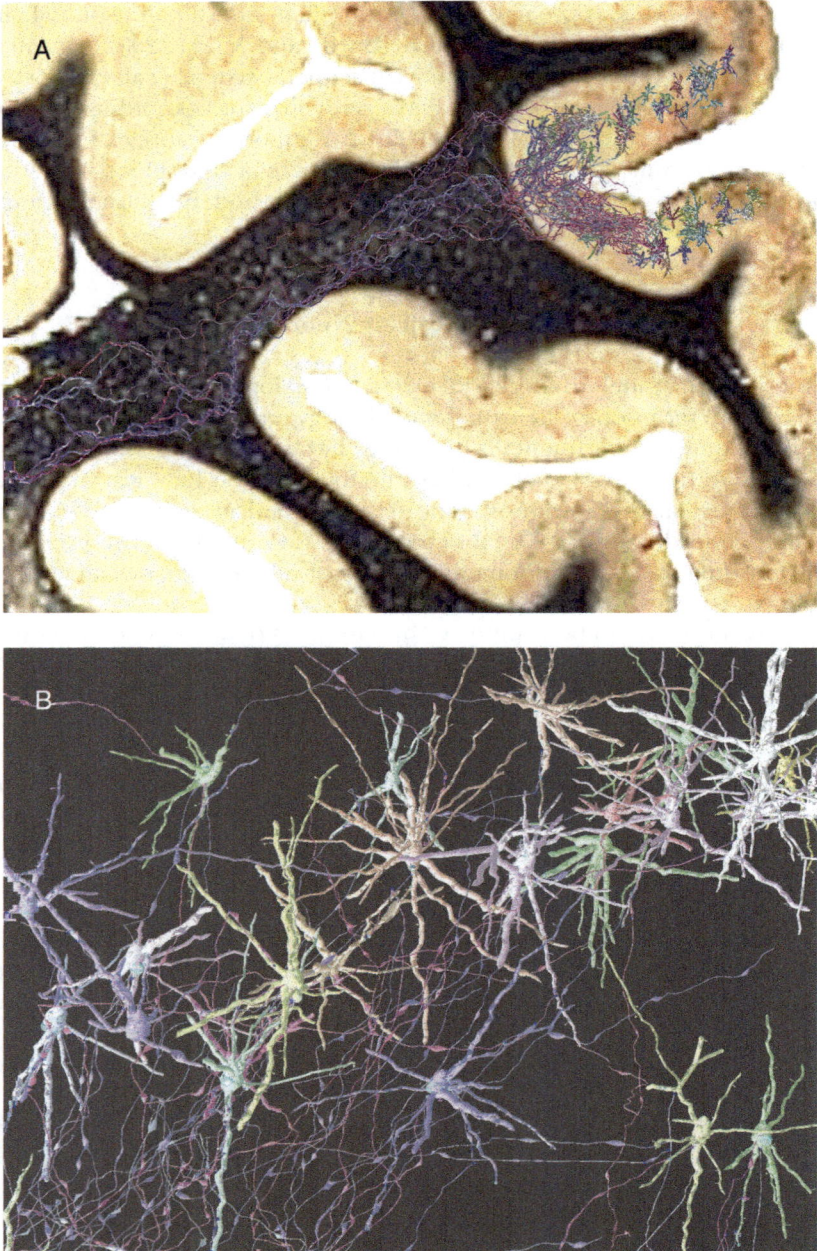

Figure 19.8. An ArborVitae model of human corticothalamic projections. (A) Reverse-contrast display of axons departing from a cortical sulcus and navigating through the white matter toward the thalamus. (B) Zoom-in on the gray matter, with several cortical neurons visible. Unique colors were assigned to each individual cell and its respective processes. (Courtesy of Dr. S. L. Senft.)

The results of ArborVitae simulations can be used to model (offline) the activity dynamics of neuronal populations. The results can then be reloaded for interactive display and analysis. In fact, once the specific connectivity among all cellular classes is obtained from anatomical data (see section "Derivation of Connectivity"), neural network dynamics can be run without the need to explicitly simulate and display the details of network structure (e.g., Ascoli and Atkeson, in press). Three-dimensional arrangements of neurons are however essential to reproduce system-level properties captured by imaging techniques, such as EEG and fMRI, as well as functional interactions with other complex components (e.g., glia).

V. PHYSIOLOGICAL RELEVANCE

A. Influence of Morphology on Neuronal Electrophysiology

Paradoxical as it may sound to the readers of this book, many theoretical neuroscientists question the need to consider anatomy in computational models of the nervous system. In fact, many early computational models of brain function, from the (sub)cellular to the system level, essentially approximated neuroanatomy away. Recent mounting evidence (especially from modeling studies), nevertheless, indicates that both neuronal morphology and network connectivity play a critical role in shaping activity (and thus, presumably, function).

At the cellular level, there is widespread consensus that the integrative properties of dendrites in most neuronal classes are sculpted by their active membrane properties (e.g., Migliore *et al.*, 2003). Voltage-dependent channels, however, are not always uniformly distributed throughout the dendritic trees, creating complex interactions that can only be investigated with numerical simulations. For example, the peculiar bursting activity of CA3 pyramidal cells can be obtained with two distinct channel distributions, but the corresponding subcellular mechanisms are drastically different (Lazarewicz *et al.*, 2002b).

If the distribution of channels is maintained uniform throughout the dendritic trees, nearly all of the known neocortical spiking patterns can simply derive from the different morphologies of the various neuronal classes (Mainen and Sejnowski, 1996). Specifically, the ability of action potentials to forward- and back propagate in the neuronal dendrites of various morphological classes is dramatically variable due to the geometrical difference alone (Vetter *et al.*, 2001). The topological structure of the trees also influences firing patterns (Van Ooyen *et al.*, 2002).

Even the natural variability among individual neurons *within* a morphological class can heavily affect spiking dynamics. When simulated with the same plausible distribution of membrane properties, a set of morphologically accurate CA3 pyramidal cells were shown to fire both regularly and irregularly, with a wide frequency range between 1 and 100 Hz (Krichmar *et al.*, 2002). Such a morphological control of electrophysiological behavior was robust with respect to the distribution of active channel and the simulation protocols (reviewed in Krichmar and Nasuto, 2002). Similarly, in a combined experimental and computational study, Schaefer *et al.* (2003) showed that temporal integration in neocortical pyramidal cells is affected by the proximal branching pattern in apical trees.

As these biophysical mechanisms relating structure and activity at the single neuron level are uncovered, it is important to critically consider the corresponding subtlety in the representation of digital anatomy in electrophysiological simulations (Lazarewicz *et al.*, 2002a). Depending on both experimental and simulation protocols, morphologies of the same class, reconstructed in different laboratories, can yield more disparate firing properties than can morphologies of different classes, reconstructed in the same laboratory (Scorcioni *et al.*, 2004).

B. Network Dynamics

Since morphology affects the intrinsic excitability of individual neurons, it can be expected to influence network dynamics as well. There are, however, multiple additional avenues of interaction between neuroanatomy and network activity. For example, the spatial location of the neurons and the axonal path length can determine the pattern of onset response latencies (e.g., Kotter *et al.*, 2002).

The typically laminated arrangement of fiber tracts in the cortex also determines the synaptic position in the dendritic layers. This in turn correlates with the electrotonic distance of the input signal from the soma. Even an oversimplified model of passive integration can illustrate that specifying this elementary level of anatomical information deeply changes nearly all dynamical aspects of a recurrent network (Ascoli, 2003). Similar conclusions can be drawn in less orderly and more abstract Hopfield-type networks (Costa *et al.*, 2003).

When multiple cell classes are considered within a subregion, the specific pattern of their connectivity also powerfully modulate the input–output function of neural network (Ascoli and Atkeson, in press). Thus, the function of a given subclass, and the robustness of its contribution to network activity, can be inferred individually for each neuronal class of the subregion. In this perspective, biological neural networks can be viewed as assemblies of functional motifs. The internal anatomy of each motif determines its specific function. Likewise, the anatomy of the motif assembly determines the overall network function.

It should be noted that neuronal structure correlates with and in fact determines all of the above characteristics (time delays, electrotonic distances, synaptic connectivity), as well as the intrinsic firing properties of each morphology class (and individual cell). Thus, the effects of several of these characteristics are likely to be strongly correlated in biological networks. It is parsimonious to hypothesize that various levels of biophysical organization, such as distribution of dendritic channels, branching pattern, layer position of synapses, and class-specific interconnectivity, coevolved to robustly express the desired network functions. This coevolution may also guarantee a certain degree of homeostatic balance in the resulting activity dynamics.

Given the intricacy of these interactions, computational modeling is extremely useful to separate (at least in silico) and quantitatively examine the contributions of each anatomical property to network dynamics. Anatomically realistic neural network models carry the potential to similarly investigate several other mechanisms of interaction between structure and activity. These include, but are not limited to, ephaptic interactions, intrinsic (or extrinsic) electric field modulation of neuronal firing, chemical inhomogeneity in the extracellular medium, glia buffering, and control. Finally, these effects should be expected to be compounded with the natural interindividual variability of neural connectivity, and the related functional and structural effects of (and on) synaptic plasticity.

C. Design Principles

A complementary approach to understanding the physiological relevance of neuron and network anatomy consists of the analysis of the possible principles underlying their structural design. For example, dendritic trees can be observed to increase their space occupancy (i.e., elongate and sprout additional branching) in response to deafferentiation, due, e.g., to lesioning of the presynaptic cell population (Shetty and Turner, 1999). This experimental observation could be interpreted by postulating that the principle behind, or goal of (some of the characteristics of), the shape of dendrites is the homeostatic formation of a given number of synapses. This hypothesis can be further tested by altering the number of synapses with a different extrinsic manipulation.

A popular assumption is that biological shape has evolved to *optimize* the expression of its intended function while *minimizing* the metabolic or structural cost, such as the total wiring length of input and output cables (Cherniak *et al.*, 2002). In particular, both the axonal branching pattern (Mitchison, 1992) and the volumetric ratio between axon and dendrites (Chklovskii *et al.*, 2002) appear to be close to optimal in the mammalian neocortex. The clustered organization of cortical connections reflect a key topological characteristic of small-world networks (Hilgetag and Kaiser, in press), resulting in highly efficient yield of *functional* connectivity despite a limited *physical* connectivity.

The same design principles can be investigated at larger scales. For example, the spatial placement of macroscopic components of the nervous system (functional subregions in the cortex, or ganglia in invertebrates) can be also compatible with optimal wiring and connectivity (e.g., Young and

Scannell, 1996). Similarly, the intricate spatial pattern of ocular columns may correspond to the optimization of the trade-off between coverage and continuity (Carreira-Perpinan and Goodhill, 2002).

VI. CONCLUSIONS AND FUTURE PERSPECTIVES

A central tenet of neuroinformatics is the digital representation and archiving of (in principle) all relevant information about the nervous system (Ascoli *et al.*, 2003). This goal may constitute a powerful basis for the creation of a large-scale, low-level model of brain structure, activity, and function. Neuroanatomy is leading the pack of success stories in neuroinformatics (Ascoli, 2002b). What is the rationale for envisioning a structural model of the brain, down to the detail of dendritic morphology (Samsonovich and Ascoli, 2002)?

Virtual experiments can be carried out quickly, reliably, safely, and inexpensively. In silico investigations can go beyond the boundary of wet lab technical and physical limits (e.g., simultaneously recording from millions of neurons). They allow the exploration of a large number of promising questions, and optimal experimental conditions (only the best of which to be implemented in a "real" experiment). Virtual experiments can also examine the theoretical effect of each model parameter separately by precisely reproducing all other initial conditions. Finally, they limit the use of ethically charged invasive procedures. A detailed, large-scale model of the mammalian brain will also foster scientific education both at the basic and advanced levels.

A simple estimation of the computational power necessary to handle the structure and activity of billions of neurons and trillions of synapses may lead to the conclusion that a truly realistic model of the brain at the cellular level is implausible in any foreseeable future. However, such a model could be dynamically computed piecewise with a multiscale strategy. When virtually recording the overall activity of the cortex, no detail may be necessary about dendritic spines in the cerebellum. Certainly, powerful computational and statistical techniques will be required to exploit such a large-scale model, yet such challenges in silico look less insurmountable than those faced when envisioning the corresponding experiments in real brains.

Devil's advocates will maintain that neuroscience is still too far from the reach of such a grand goal. While this argument is difficult to disagree with, it is relieving to look at the temporal growth of the GeneBank database. From the start of the Human Genome Project (1986), it took 6 years to establish acceptable guidelines, and almost 10 years to complete the yeast genome. Yet only 3 years after that, an entire human chromosome was mapped. It was only one additional year (2000) before all 23 human chromosomes were completed. The clearly exponential graph now looks dramatically flat until the incept of the quite recent boom. The creation of a realistic, large-scale human brain model may follow a similar pattern.

Although the computational power of hardware and software is increasing at an exponential rate, so is the amount of experimental data collected in biomedical sciences in general, and neuroscience in particular. In principle, much of these data need to be incorporated in the realistic model. Neuroinformatics started when experimental neuroscience was already a mature field. Is modeling catching up, or is the gap between relevant published data and corresponding computational simulations ever widening?

The answer to this question depends on what is meant by "relevant," i.e., what level of modeling computational neuroscientists are designing. This, in turn, is defined by the type of scientific explanation being sought, e.g., behavior in terms of genes, network rhythms in terms of neuronal spiking properties, synaptic strength in terms of calcium buffering. In this perspective, computational models become a constructive definition of our quantitative understanding of the structure, activity, and function of the nervous system. The "ultimate race" between experiments and models, then, is along the fine line dividing data and knowledge.

APPENDIX

A. Simple Extraction of Morphometric Parameters with L-Measure

Common measurements extracted from neuronal morphology include total neuronal length and minimum, average, and maximum dendritic diameter. As a first basic example, a step-by-step guide is given, showing how to extract these basic parameters with L-Measure. Only requirements are an Internet connection and a JAVA-enabled browser.

1. Connect to the online version of L-Measure at www.krasnow.gmu .edu/L-Neuron (click "L-Measure" in the left column, then click "On-line Version"). A security window will appear to ask for access permission, click "Yes."
2. From the panel "Function," select "Length" in the top left box, and then click "Add." A new measurement "Length" will appear in the list of functions to be measured on the right.
3. From the same panel, add function "Diameter."
4. From the panel "Input," open and add the neurons you wish to analyze and measure. You can freely download electronic neuronal sample from www.krasnow.gmu.edu/L-Neuron (click "Morphology Database").
5. From the panel "Go" click the "Go" button. L-Measure will list all extracted measurements in the bottom panel. For each selected function, L-Measure displays six values:

   a. Total sum: In case of "Length" this is the total sum of all segment lengths, which represents the first desired measurement.

    b. Compartments included: It lists how many compartments were included in this measurement.

    c. Compartments excluded.

    d. Minimum value: It reports the minimum value for the specified function. In the diameter row, this represents the second desired measurement.

    e. Average value: It reports the average value across all segments (the third desired measurement).

    f. Maximum value: In the diameter row, this represents the fourth desired measurement.

    g. Standard deviation.

B. A More Complex Example

This section illustrates a more complex example in which a Sholl diagram of length vs. Euclidean distance is generated from basal dendrites with a sphere radial increment of 50 μm. For this section, a pyramidal cell in SWC file format is required, which can be freely downloaded from www.krasnow.gmu.edu/L-Neuron (click "Morphology Database").

1. Connect to the online version of L-Measure at www.krasnow.gmu .edu/L-Neuron (click "L-Measure" in the left column).

2. From the "Specificity" panel, select the function "Type" and insert the value "3" followed by the "=" radio button. Then click the "Add" button (type = 3 in SWC files identifies basal dendrites).

3. From the top left box in the "Function" panel, select "Length."

4. From the bottom left box, select "Euclidean Distance."

5. Select "Width of Bins" from the bottom radio button, and insert a value of 50.

6. Click the "Add" button. A new function named "Length vs. Euclidean Distance" will be added to the right top panel.

7-A. In the "Input" panel, add the SWC neuronal reconstruction file.

8. In the "Go" panel, click the "Go" button. The bottom box will show the resulting measurement table.

To obtain a graphical representation of the Sholl diagram, additional steps are required together with a spreadsheet-like software, such as Microsoft Excel. To obtain an Excel compatible output, insert the following extra step into the above sequence:

7-B. In the "Output" panel, click the "Save As" button, choose a directory, and write "example.xls."

9. Double-click the produced ".xls" file. Microsoft Excel will automatically recognize and open the selected file.

10. Within Excel, select the first two rows.

11. Click the "Chart Wizard" button.
12. Select "XY scatterplot."
13. Press "finish."

ACKNOWLEDGMENTS. The authors are grateful to Drs. Stephen L. Senft and Nobuaki Tamamaki, and to MicroBrightField, Inc., for supplying material used in some of this chapter's illustrations. Support was provided by R01 grants NS39600 (jointly funded by NINDS, NIMH, and NSF under the Human Brain Project) and AG025633 as part of the NSF/NIH Collaborative Research in Computational Neuroscience Program.

REFERENCES

Ascoli, G. A., 1999, Progress and perspectives in computational neuroanatomy, *Anat. Rec.* **257**(6):195–207.

Ascoli, G. A., 2002a, Neuroanatomical algorithms for dendritic modelling, *Network* **13**(3):247–260.

Ascoli, G. A., 2002b, Computing the brain and the computing brain, In: Ascoli, G. A. (ed.), *Computational Neuroanatomy: Principles and Methods*, Totowa, NJ: Humana Press, pp. 3–26.

Ascoli, G. A., 2003, Passive dendritic integration heavily affects spiking dynamics of recurrent networks, *Neural Netw.* **16**:657–663.

Ascoli, G. A., and Atkeson, J. C., 2005, Incorporating anatomically realistic cellular-level connectivity in neural network models of the rat hippocampus, *Biosystems.* **79**:173–181.

Ascoli, G. A., De Schutter, E., and Kennedy, D. N., 2003, An information science infrastructure for neuroscience, *Neuroinformatics* **1**(1):1–2.

Ascoli, G. A., and Krichmar, J. L., 2000, L-Neuron: a modeling tool for the efficient generation and parsimonious description of dendritic morphology, *Neurocomputing* **32–33**:1003–1011.

Ascoli, G. A., Krichmar, J. L., Nasuto, S. J., and Senft, S. L., 2001a, Generation, description, and storage of dendritic morphology data, *Philos. Trans. R. Soc. Lond. B Biol. Sci.* **356**(1412):1131–1145.

Ascoli, G. A., Krichmar, J. L., Scorcioni, R., Nasuto, S. J., and Senft, S. L., 2001b, Computer generation and quantitative morphometric analysis of virtual neurons, *Anat. Embryol.* **204**(4):283–301.

Bota, M., and Arbib, M. A., 2004, Integrating databases and expert systems for the analysis of brain structures: connections, similarities, and homologies, *Neuroinformatics* **2**(1):19–58.

Buettner, H. M., 1995, Computer simulation of nerve growth cone filopodial dynamics for visualization and analysis, *Cell Motil. Cytoskeleton* **32**(3):187–204.

Burke, R. E., and Marks, W. B., 2002, Some approaches to quantitative dendritic morphology, In: Ascoli, G. A. (ed.), *Computational Neuroanatomy: Principles and Methods*, Totowa, NJ: Humana Press, pp. 27–48.

Cannon, R. C., Turner, D. A., Pyapali, G. K., and Wheal, H. V., 1998, An online archive of reconstructed hippocampal neurons, *J. Neurosci. Methods* **84**(1–2):49–54.

Carreira-Perpinan, M. A., and Goodhill, G. J., 2002, Development of columnar structures in visual cortex, In: Ascoli, G. A. (ed.), *Computational Neuroanatomy: Principles and Methods*, Totowa, NJ: Humana Press, pp. 337–358.

Cherniak, C., Mokhtarzada, Z., and Nodelman, U., 2002, Optimal-wiring models of neuroanatomy, In: Ascoli, G. A. (ed.), *Computational Neuroanatomy: Principles and Methods*, Totowa, NJ: Humana Press, pp. 71–82.

Chklovskii, D. B., Schikorski, T., and Stevens, C. F., 2002, Wiring optimization in cortical circuits, *Neuron* **34**(3):341–347.

Costa Lda, F., and Manoel, E. T., 2003, A percolation approach to neural morphometry and connectivity, *Neuroinformatics* **1**(1):65–80.

Costa Lda, F., Barbosa, M. S., Coupez, V., and Stauffer, D., 2003, Morphological Hopfield networks, *Brain Mind* **4**:91–105.

Donohue, D. E., and Ascoli, G. A., 2005, Models of neuronal outgrowth, In: Koslow, S. H., and Subramaniam, S. (eds.), *Databasing the Brain: From Data to Knowledge*, Wiley, New York, NY, pp. 303–323.

Donohue, D. E., Scorcioni, R., and Ascoli, G. A., 2002, Generation and description of neuronal morphology using L-Neuron: a case study, In: Ascoli, G. A. (ed.), *Computational Neuroanatomy: Principles and Methods*, Totowa, NJ: Humana Press, pp. 49–70.

Eglen, S. J., and Willshaw, D. J., 2002, Influence of cell fate mechanisms upon retinal mosaic formation: a modelling study, *Development* **129**(23):5399–5408.

Ewart, D. P., Kuzon, W. M., Jr., Fish, J. S., and McKee, N. H., 1989, Nerve fibre morphometry: a comparison of techniques, *J. Neurosci. Methods* **29**(2):143–150.

Gardner, D., Toga, A. W., Ascoli, G. A., Beatty, J. T., Brinkley, J. F., Dale, A. M., Fox, P. T., Gardner, E. P., George, J. S., Goddard, N., Harris, K. M., Herskovits, E. H., Hines, M. L., Jacobs, G. A., Jacobs, R. E., Jones, E. G., Kennedy, D. N., Kimberg, D. Y., Mazziotta, J. C., Miller, P. L., Mori, S., Mountain, D. C., Reiss, A. L., Rosen, G. D., Rottenberg, D. A., Shepherd, G. M., Smalheiser, N. R., Smith, K. P., Strachan, T., Van Essen, D. C., Williams, R. W., and Wong, S. T., 2003, Towards effective and rewarding data sharing, *Neuroinformatics* **1**(3):289–295.

Glaser, J. R., and Glaser, E. M., 1990, Neuron imaging with Neurolucida—a PC-based system for image combining microscopy, *Comput. Med. Imaging Graph.* **14**(5):307–317.

Goodhill, G. J., 1998, Mathematical guidance for axons, *Trends Neurosci.* **21**(6):226–231.

Gras, H., and Killmann, F., 1983, NEUREC—a program package for 3D-reconstruction from serial sections using a microcomputer, *Comput. Programs Biomed.* **17**(1–2):145–155.

He, W., Hamilton, T. A., Cohen, A. R., Holmes, T. J., Pace, C., Szarowski, D. H., Turner, J. N., and Roysam, B., 2003, Automated three-dimensional tracing of neurons in confocal and brightfield images, *Microsc. Microanal.* **9**(4):296–310.

Hilgetag, C. C., and Kaiser, M., 2004, Clustered organisation of cortical connectivity, *Neuroinformatics* **2**:353–360.

Hines, M. L., and Carnevale, N. T., 2001, NEURON: a tool for neuroscientists, *Neuroscientist* **7**(2):123–135.

Izhikevich, E. M., 2004, Which model to use for cortical spiking neurons? *IEEE Trans. Neural Netw.* **15**:1063–1070.

Jacobs, G. A., and Pittendrigh, C. S., 2002, Predicting emergent properties of neuronal ensembles using a database of individual neurons, In: Ascoli, G. A. (ed.), *Computational Neuroanatomy: Principles and Methods*, Totowa, NJ: Humana Press, pp. 151–170.

Kalisman, N., Silberberg, G., and Markram, H., 2003, Deriving physical connectivity from neuronal morphology, *Biol. Cybern.* **88**(3):210–218.

Kotter, R., Nielsen, P., Dyhrfjeld-Johnsen, J., Sommer, F. T., and Northoff, G., 2002, Multi-level neuron and network modeling in computational neuroanatomy, In: Ascoli, G. A. (ed.), *Computational Neuroanatomy: Principles and Methods*, Totowa, NJ: Humana Press, 359–382.

Krichmar, J. L., and Nasuto, S. J., 2002, The relationship between neuronal shape and neuronal activity, In: Ascoli, G. A. (ed.), *Computational Neuroanatomy: Principles and Methods*, Totowa, NJ: Humana Press, pp. 105–126.

Krichmar, J. L., Nasuto, S. J., Scorcioni, R., Washington, S. D., and Ascoli. G. A., 2002, Effects of dendritic morphology on CA3 pyramidal cell electrophysiology: a simulation study, *Brain Res.* **941**(1–2):11–28.

Lazarewicz, M. T., Boer-Iwema, S., and Ascoli, G. A., 2002a, Practical aspects in anatomically accurate simulations of neuronal electrophysiology, In: Ascoli, G. A. (ed.), *Computational Neuroanatomy: Principles and Methods*, Totowa, NJ: Humana Press, pp, 127–148.

Lazarewicz, M. T., Migliore, M., and Ascoli, G. A., 2002b, A new bursting model of CA3 pyramidal cell physiology suggests multiple locations for spike initiation, *Biosystems* **67**:129–137.

Leergaard, T. B., and Bjaalie, J. G., 2002, Architecture of sensory map transformations: axonal tracing in combination with 3-d reconstruction, geometric modeling, and quantitative analyses, In: Ascoli, G. A. (ed.), *Computational Neuroanatomy: Principles and Methods*, Totowa, NJ: Humana Press, pp. 199–218.

Lester, D. S., Hanig, J. P., and Pine, P. S., 2002, Quantitative neurotoxicity, In: Ascoli, G. A. (ed.), *Computational Neuroanatomy: Principles and Methods*, Totowa, NJ: Humana Press, pp. 383–400.

Mainen, Z. F., and Sejnowski, T. J., 1996, Influence of dendritic structure on firing pattern in model neocortical neurons, *Nature* **382**(6589):363–366.

Migliore, M., Morse, T. M., Davison, A. P., Marenco, L., Shepherd, G. M., and Hines, M. L., 2003, ModelDB: making models publicly accessible to support computational neuroscience, *Neuroinformatics* **1**(1):135–139.

Mitchison, G., 1992, Axonal trees and cortical architecture, *Trends Neurosci.* **15**(4):122–126.

Mori, S., 2002, Principle and applications of diffusion tensor imaging: a new MRI technique for neuroanatomical studies, In: Ascoli, G. A. (ed.), *Computational Neuroanatomy: Principles and Methods*, Totowa, NJ: Humana Press, pp. 271–292.

Rodriguez, A., Ehlenberger, D., Kelliher, K., Einstein, M., Henderson, S. C., Morrison, J. H., Hof, P. R., and Wearne, S. L., 2003, Automated reconstruction of three-dimensional neuronal morphology from laser scanning microscopy images, *Methods* **30**(1):94–105.

Samsonovich, A. V., and Ascoli, G. A., 2002, Towards virtual brains, In: Ascoli, G. A. (ed.), *Computational Neuroanatomy: Principles and Methods*, Totowa, NJ: Humana Press, pp. 425–436.

Samsonovich, A. V., and Ascoli, G. A., 2003, Statistical morphological analysis of hippocampal principal neurons indicates cell-specific repulsion of dendrites from their own cell, *J. Neurosci. Res.* **71**(2):173–187.

Samsonovich, A. V., and Ascoli, G. A., 2005, Statistical determinants of dendritic morphology in hippocampal pyramidal neurons: a hidden Markov model, *Hippocampus* **15**: 166–183.

Samsonovich, A. V., and Ascoli, G. A., 2005, Algorithmic description of hippocampal granule cell dendritic morphology, *Neurocomputing* **65–66**:253–260.

Schaefer, A. T., Larkum, M. E., Sakmann, B., and Roth, A., 2003, Coincidence detection in pyramidal neurons is tuned by their dendritic branching pattern, *J. Neurophysiol.* **89**(6):3143–3154.

Scorcioni, R., and Ascoli, G. A., 2001, Algorithmic extraction of morphological statistics from electronic archives of neuroanatomy, *Lect. Notes Comp. Sci.* **2084**:30–37.

Scorcioni, R., and Ascoli, G. A., 2005, Algorithmic reconstruction of complete axonal arborizations in rat hippocampal neurons, *Neurocomputing* **65–66**:15–22.

Scorcioni, R., Boutiller, J. M., and Ascoli, G. A., 2002, A real scale model of the dentate gyrus based on single-cell reconstructions and 3D rendering of a brain atlas, *Neurocomputing* **44–46**:629–634.

Scorcioni, R., Lazarewicz, M. T., and Ascoli, G. A., 2004, Quantitative morphometry of hippocampal pyramidal cells: differences between anatomical classes and reconstructing laboratories, *J. Comp. Neurol.* **473**(2):177–193.

Segev, R., and Ben-Jacob, E., 2000, Generic modeling of chemotactic based self-wiring of neural networks, *Neural Netw.* **13**(2):185–199.

Senft, S. L., 2002, Axonal navigation through voxel substrates: a strategy for reconstructing brain circuitry, In: Ascoli, G. A. (ed.), *Computational Neuroanatomy: Principles and Methods*, Totowa, NJ: Humana Press, pp. 245–270.

Senft, S. L., and Ascoli, G. A., 1999, Reconstruction of brain networks by algorithmic amplification of morphometry data, *Lect. Notes Comp. Sci.* **1606**:25–33.

Shetty, A. K., and Turner, D. A., 1999, Aging impairs axonal sprouting response of dentate granule cells following target loss and partial deafferentation, *J. Comp. Neurol.* **414**(2):238–254.

Stepanyants, A., Tamas, G., and Chklovskii, D. B., 2004, Class-specific features of neuronal wiring, *Neuron.* **43**(2):251–259.

Tamamaki, N., Abe, K., and Nojyo, Y., 1988, Three-dimensional analysis of the whole axonal arbors originating from single CA2 pyramidal neurons in the rat hippocampus with the aid of a computer graphic technique, *Brain Res.* **452**(1–2):255–272.

Turner, D. A., Cannon, R. C., and Ascoli, G. A., 2002, Web-based neuronal archives: neuronal morphometric and electrotonic analysis, In: Kotter, R. (ed.), *Neuroscience Databases—A Practical Guide*, Amsterdam: Elsevier, pp. 81–98.

Van Ooyen, A., Duijnhouwer, J., Remme, M. W. H., and Van Pelt, J., 2002, The effect of dendritic topology on firing patterns in model neurons, *Network* **13**:311–325.

Van Ooyen, A., and Van Pelt, J., 2002, Competition in neuronal morphogenesis and the development of nerve connections, In: Ascoli, G. A. (ed.), *Computational Neuroanatomy: Principles and Methods*, Totowa, NJ: Humana Press, pp. 219–244.

Vetter, P., Roth, A., and Hausser, M., 2001, Propagation of action potentials in dendrites depends on dendritic morphology, *J. Neurophysiol.* **85**(2):926–937.

Winslow, J. L., Jou, S. F., Wang, S., and Wojtowicz, J. M., 1999, Signals in stochastically generated neurons, *J. Comput. Neurosci.* **6**(1):5–26.

Wolf, E., Birinyi, A., and Pomahazi, S., 1995, A fast three-dimensional neuronal tree reconstruction system that uses cubic polynomials to estimate dendritic curvature, *J. Neurosci. Methods* **63**:137–145.

Yates, P. A., Holub, A. D., McLaughlin, T., Sejnowski, T. J., and O'Leary, D. D., 2004, Computational modeling of retinotopic map development to define contributions of EphA–ephrinA gradients, axon–axon interactions, and patterned activity, *J. Neurobiol.* **59**(1):95–113.

Young, M. P., and Scannell, J. W., 1996, Component-placement optimization in the brain, *Trends Neurosci.* **19**(10):413–415.

Zaborszky, L., Csordas, A., Buhl, D., Duque, A., Somogyi, J., and Nadasdy, Z., 2002, Computational anatomical analysis of the basal forebrain corticopetal system, In: Ascoli, G. A. (ed.), *Computational Neuroanatomy: Principles and Methods*, Totowa, NJ: Humana Press, pp. 171–198.

20

Functional Connectivity of the Brain: Reconstruction from Static and Dynamic Data

ZOLTAN NADASDY, GYORGY BUZSAKI, and LASZLO ZABORSZKY

ZOLTAN NADASDY • California Institute of Technology, Pasadena, CA 91125
GYORGY BUZSAKI AND LASZLO ZABORSZKY • Rutgers, The State University of
New Jersey, Newark, NJ 07102

Abstract: The central nervous system is a single complex network connecting each neuron through a number of synaptic connections. However, only a small fraction of the total connections functionally link neurons together. If the smallest multineuronal architecture within which functional links are established constitutes circuitry, then what are the basic operating principles of these circuitries from which we can understand both the composition and the dynamics of the larger networks? We argue that a finite class of circuitries, the "basic circuitries," can be identified as repeating structural motifs tightly associated with specific dynamics. Functional circuitries, however, cannot be derived from the static architecture simply because they do not obey structural borders. Fortunately, since the constituent neurons do act in synergy, we can infer from the dynamics the minimal structural conditions that constitute a circuitry. In this chapter, instead of giving a precise definition of the "basic circuitry," we outline a set of methods that may elucidate such a definition. We argue that since the concept of circuitry incorporates both dynamic and static features, understanding can be achieved through combining the structural and dynamic aspects of the available data. We review methods of extracting functional information from static data first. Next, we review methods of extracting structural information from dynamic data. Ideally, these two approaches should converge and define circuitry based on the fragile concept of functional connectivity.

Keywords: cell types, circuitry, databasing, functional connectivity, large-scale recording, population statistics

I. INTRODUCTION

The ultimate objective of neuronal tract-tracing is to reveal the functional architecture of the nervous system. Progress toward this objective must rely on a precise definition of the architecture in order to successfully explain and predict the activity flow within its circuitries. This inferential process, however, is rather limited since the reconstruction or prediction of putative dynamics based on an abstract network architecture requires simulations that are extremely sensitive to small variations of a large number of parameters. Therefore, we propose that the coapplication of anatomical and electrophysiological methods is essential for describing a functional architecture of the nervous system. Although the inferential process of the electrophysiology and the neuroanatomy are quite opposite in nature, they support and complement each other. To find a link between them, we introduce the "structural and dynamic compactness" criterion, that is, to determine the smallest multiple-neuronal cluster, which generates the shortest invariant activity pattern. This practical definition of "circuitry" is sufficient to introduce the problem; however, further qualification and classification must go beyond the structural and temporal "compactness criteria," one of the key challenges of research for the next decade. This chapter will review constituent elements and main organization principles of cortical circuitries in the "The Building Blocks: Neurons, Circuitries, and Assemblies" section, specifically circuitries of the isocortex and the hippocampus will be discussed (both addressed as cortex). Next we discuss a number of innovative applications of neuroanatomy and electrophysiology related to circuitries of the brain. We separately discuss methods related to morphological data in the "Static Data" section and physiological data in the "Dynamic Data" section. However, emphasis is put on the convergence and dialog between the two approaches in the "Concluding Remarks" Furthermore, we restrict our review to the rodent brain; however, the principles we outline can be generalized to the mammalian brain.

II. THE BUILDING BLOCKS: NEURONS, CIRCUITRIES, AND ASSEMBLIES

Despite the complexity and size differences between the invertebrate and the vertebrate nervous systems, both consist of large-scale repetition of compact architectural modules, which we denote as "basic circuitries." In order to define the basic circuitries within and across brains of different species, we first describe the constituent elements, the basic neuron types, and their specific connectivity pattern. The basic circuitry, which is uniform within a given brain structure, varies across different structures depending upon the computational needs. These basic circuitries, once we understand their dynamics, will enable us to infer the architecture from their activity pattern.

First, we need to make a distinction between structural and functional connectivity[1] and dynamics. The relationship between structural and functional connectivity is best understood if we decompose the large network of the nervous system into the smallest multineuronal information-processing subunit, or "motif." Motifs are conceived as small directed graphs of M-nodes within a large network. It has been shown through simulations that the number of structural motifs derived from anatomical connections of nervous systems in various species is smaller than that from random graphs (Sporns and Kötter, 2004). In direct contrast, when considering effective connections, real nervous systems show more functional motifs than do random graphs. This may suggest that the nervous system tends to maximize the number of functional motifs but minimize the number of structural motifs. However, when considering the diversity of activity patterns generated by networks of different functional motifs, it turns out that the number of dynamics is smaller than the number of functional motifs. Apparently, there is redundancy by which different functional motifs generate the same activity pattern (Prinz *et al.*, 2004). Therefore, dynamics may be more closely associated with architectures than to the functional connectivity. Defining circuitries by the dynamics they implement may allow us to reduce the necessary number of basic circuitries and simplify their classification. The price for this reduction is that the morphological composition of these circuitries may be rather complex.

---

[1] We also make a distinction between functional connectivity pertaining to connections between neurons as opposed to interareal connections inferred from functional imaging.

The prototypes of cortical circuitries are those that have been described in the hippocampus (Somogyi *et al.*, 1998). From architectural and developmental points of view, isocortical circuitries derive from the prototypical circuitries with a certain complexity added. Specifically, while the circuitries are relatively homogeneously distributed and coaligned in the hippocampus, the neocortical organization is complex and is conceived as an expansion of the hippocampal cytoarchitecture. This expansion involves three types of topological transformations. The first is a layer multiplication that adds a new set of interlaminar circuitries to the isocortex, nonexistent in the hippocampus. According to one view, the laminar structure of isocortex can be conceived as the unfolding of the hippocampus and superimposing of three subregions, the dentate, CA3, and CA1, as different layers but at the same time preserving the connections (Watts and Thomson, 2005). The second type of expansion of cortical development is a superposition of the same circuitry within the same cortical layer and often within the same volume, which makes reconstruction of the synaptic circuitry particularly difficult (Somogyi *et al.*, 1998). The third type of expansion is a radial specialization that forms columns and selective reciprocal tangential connections with functionally similar circuitries. This extension is responsible for creating the patchy functional architecture of the neocortex.

A. Classes of Neurons

Neuronal circuitries, at the lowest level, consist of two mutually exclusive classes of neurons, excitatory glutamatergic neurons (mainly pyramidal cells), and GABAergic inhibitory interneurons.[2] On the basis of an assessment from the CA1 area of the hippocampus, the pyramidal-to-interneuron ratio is 33:1 (Aika *et al.*, 1994). Principal neurons are the main excitatory projection neurons as they establish long-range connections and transfer information between different structures. In the cortex, pyramidal cells are reciprocally connected to the thalamus and to each other via axon collaterals. In spite of their diverse axonal projection patterns, pyramidal cells show a characteristic bipolar dendritic arborization which consists of apical and basal dendritic tufts. In contrast, most cortical interneurons incorporate a more diverse morphology. Interneurons establish differential reciprocal connections with other interneurons. The taxonomy of interneurons is still unresolved since the categories constructed based on histochemical staining, morphological features, termination sites, and firing patterns do not mesh. According to two extreme viewpoints on interneuronal diversity, cell types may either represent a finite set of discrete classes or blends of continuous feature distributions (Gupta *et al.*, 2000). Several inhibitory interneuron types have been classified solely based on morphological features, such as

---

[2] Among the few exceptions, the spiny stellate neurons in layer 4 are excitatory and considered interneurons.

basket, small basket, nest basket, axo-axonic, spiny stellate, aspiny stellate, Martinotti, double-bouquet, and a number of smaller classes (Gupta *et al.*, 2000). The inhibitory interneuron diversity seems to scale with the evolution of the mammalian brain. While brain structures with long evolution history, such as cerebellum, basal ganglia, and thalamus, consist of a few cell types, structures that specialized later, such as hippocampus and neocortex, show larger diversity. Moreover, interneurons show layer specificity in terminals (Freund and Buzsaki, 1996; Somogyi and Klausberger, 2005) and circuitry specificity. One speculation about interneuronal diversity is that it is the result of circuitry specialization. According to this view, GABAergic interneurons are added to the glutamatergic neurons to serve specific functional roles (Földy *et al.*, 2005). This is consistent with their different origin from pyramidal cells during the early development of the nervous system and a migration path orthogonal to that of the pyramidal cells (Rakic, 1995). The most accepted definition of interneuronal species takes several features into account, such as the postsynaptic target, the layer specificity, and the expression of species-characteristic markers (Freund and Buzsáki, 1996; Maccaferri and Lacaille, 2003; Somogyi and Klausberger, 2005).

B. Basic Circuitry

The next level of organization is the "basic circuitry."[3] All extrinsic and intrinsic glutamatergic pathways terminate on both pyramidal cells and GABAergic interneurons. Therefore, the basic circuitry is composed of pyramidal → pyramidal, pyramidal → interneuron, interneuron → pyramidal, and interneuron → interneuron functional units. The pyramidal–pyramidal excitatory connection is feed-forward if it connects two pyramidal cells between two areas or between subregions of the cortex (Fig. 20.1A). It is recurrent if axons project back to the same pyramidal cell population (Fig. 20.1B).[4]

Connections between neurons, in general, can be synaptic or electrically coupled (gap junction). Both pyramidal and interneurons can mutually connect through gap junctions that are instrumental for gamma and higher frequency band network synchronization. Interneurons seem to electrically couple only with the same subtypes (Gibson *et al.*, 1999; Tamás *et al.*, 2000). Conversely, connections between the same type of interneurons are often mutual, and the outcome of a steady-state input is an oscillation with zero-phase lag synchrony (empirical result and modeling: Destexhe and

---

[3] We would like to make a clear distinction between the concepts of basic circuitry and a "canonical microcircuit" of the neocortex (Douglas *et al.*, 1989). While the former denotes the basic information processing circuit involving only a few pyramidal and interneurons, the latter describes the basic architecture of an isocortical volume incorporating all six layers and all prototypical connections between the known neuron types.

[4] Very few autaptic axons, i.e., axons projecting back to the very same neuron, have been observed (Tamas *et al.*, 1997).

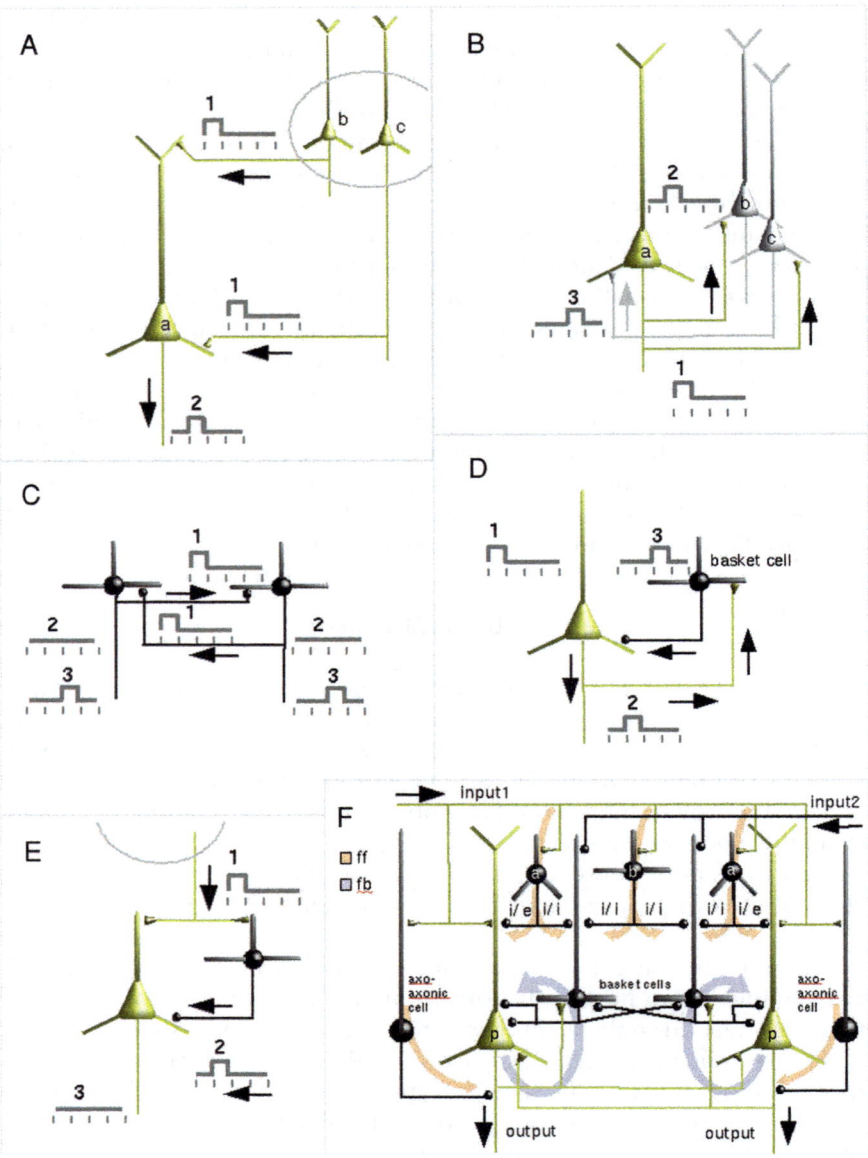

Figure 20.1. Most common examples of basic circuitry types. (A) Feed-forward excitatory connection involves excitatory projection from a different group of neurons
(*b*, *c*) terminating on the target neuron *a*. The excitatory input from (*b*, *c*) must
precede the action potentials in *a*. The relative timing of events is indicated by pulses
and numbers. The flow of action potentials is indicated by arrows. (Examples: Schaffer collateral system in the hippocampus, layer 2–3 pyramidal → layer 5 pyramidal
synapses in the isocortex.) (B) The recurrent or feedback excitation involves excitatory collaterals from *a* to other neurons of the same group (*b*, *c*). The excitatory input
from *a* must precede the action potential in *b* and *c*, which in turn may cause a second

Babloyantz, 1993; modeling: Wang and Rinzel, 1993; Vreeswijk, 1996). The other main class of basic circuitry is the connection between projection neurons (pyramidal) and interneurons. Within an interneuron–principal cell couple, the interneuron controls the probability of the principal cell generating an action potential (AP) for a given excitatory input. The interneuron–principal cell connection can be feed-forward (Fig. 20.1E) or recurrent (feedback) (Fig. 20.1D). It is feed-forward if the interneuron, activated by an excitatory input, has an inhibitory effect on the target pyramidal neuron

←———————————————————————————————————————

Figure 20.1. (*Cont.*) wave of excitatory postsynaptic potential on neuron *a*. (Examples: Hippocampal CA3 recurrent collateral system, recurrent connections between layer 2–3 pyramidal neurons in the isocortex.) (C) Inhibitory–inhibitory connection between two GABAergic neurons. The sequence of action is important in order to understand how mutual inhibition causes oscillation with zero-phase-lag synchrony. When both neurons inhibit each other (phase 1), they both will be released from inhibition at the same time (Phase 2). Then they both elicit action potentials (Phase 3), which in turn cause them to be inactive for the next period, and the cycle starts over. Assuming sufficient depolarizing driving force, the phases of mutual inhibition and disinhibition generate a self-sustaining oscillation within the interneuronal network. (Examples: basket cells in hilus, layer 2–3 interneurons in the isocortex.) (D) Recurrent or feedback inhibition. In this case, the excitatory collateral projects to interneurons, which in turn project back to the same neuron. The recurrent inhibition is evoked by an excitatory input on the glutamatergic neuron (Phase 1), which generates an action potential. The action acts on the interneuron (Phase 2) through a collateral axonal projection, which in turn evokes an inhibition and when backprojected to the glutamatergic neuron (Phase 3) causes suppression of action potentials in the next phase. This type of control is effective to decrease the probability of action potentials of pyramidal neurons. (Examples: interneuron—pyramidal connections in the CA3 area of the hippocampus and pyramidal cell—basket cell feedback inhibition in the CA area of the hippocampus.) (E) Feed-forward inhibition. Distant excitatory inputs often projects to interneurons which terminate on local pyramidal cells. Functionally, the distant excitatory input (Phase 1) precedes the interneuron's response (Phase 2), which causes a suppression in the target pyramidal cells (Phase 3). A classic example of this circuitry is the lateral inhibition, common contrast-enhancement mechanism in sensory structures. Small black spheres are GABAergic terminals. Cone-shaped terminals are glutamatergic. Principal cells are shown in yellow, and interneurons are shown black. (F) A highly reduced model of isocortical circuitry can be conceived as the combination of above described feed-forward (orange) and feedback (viola) circuitries within a shared volume. This circuit repeats in each layer of the isocortex, often in juxtaposition and in superimposition with the same circuit within the same volume. Excitatory and inhibitory inputs are arriving on the top from left as input 1 and right as input 2, respectively. Only a single layer is featured. Connections between pyramidal neurons (p) and four different GABAergic interneuron types are illustrated. The laminar segregation of their terminals relative to the pyramidal cell is emphasized. Specifically, while the feed-forward inhibitory interneurons (*a* and *b*; such as double-bouquet, neurogliaform, and bitufted cells) preferentially target apical dendrites, the basket cells terminate perisomatically on the pyramidal neurons. Axo-axonic cell terminals, in contrast, occupy the axon initial segment. In addition, interneurons establish extensive and laminar interconnections amongst each other. The reciprocal excitatory–excitatory connections are also extensive. (After Somogyi *et al.*, 1998.)

prior to the common excitatory input activating the pyramidal neuron directly. The connection is recurrent if the interneuron, activated through recurrent axon collaterals, feeds back to the same pyramidal cell with a delay. Note that this morphological classification includes a dynamic criterion, the relative timing of the inhibitory and excitatory inputs. Accordingly, since pyramidal cells and interneurons are mutually interconnected within a region, whether inhibition is feed-forward or feedback can only be determined based on the relative timing of inhibition and excitation. Methods to resolve timing relationships will be discussed in Section 4C–D.

Different interneuron classes localize in specific hippocampal and cortical layers, and they establish connections with specific layers. Their axon termination is coaligned with excitatory input. In addition to their layer-specific distribution, different interneurons terminate selectively on specific parts of the pyramidal neuron. A highly simplified model of isocortical circuitry is shown in Fig. 20.1F. Only one excitatory and one inhibitory input is illustrated. Three basic types of interneurons are featured: axo-axonic, feed-forward inhibitory to excitatory interneuron (e.g., bistratified cell), and feed-forward inhibitory–inhibitory interneurons (e.g., basket cells). The feed-forward interneurons preferentially terminate on the apical dendrites and middle range dendrites. Basket cells are part of the feedback, feed-forward, and reciprocal inhibitions. Their terminals preferentially target the pyramidal cell soma. Axo-axonic cell terminals occupy the axon initial segment of pyramidal cells. The combination of convergent feed-forward and feedback connections effectively imposes a complex temporal pattern of inhibition on pyramidal cells through spatially segregated gating of the excitatory input from basal or apical dendrites (Fig. 20.1C). It is assumed that a concerted action of different types of interneurons is able to impose a complex temporal pattern of hyperpolarizations on the pyramidal cell (Somogyi and Klausberger, 2005).

C. Vertical Organization of Circuitries

Both the laminar arborization of the dendritic tree as well as the laminar organization of axons are cell-type specific. Furthermore, the efficacy of synapses on the postsynaptic cell is highly dependent on the spatial localization of the terminals relative to the postsynaptic cell's morphology. The closer the terminal is to the axon initial segment, the more effective is the inhibitory conductance. Therefore, inhibitory interneurons, such as the axo-axonic cells in the isocortex and the basket cells in the hippocampus, can effectively suppress the pyramidal cell response regardless of the excitatory postsynaptic potential, while interneurons terminating on the apical dendrite can suppress the integration of excitatory postsynaptic potential selectively for a specific dendritic cluster or branch.

Two examples of how basic circuitry types combine within the same volume to form a functional unit are shown in Fig. 20.2A. The first is a complex

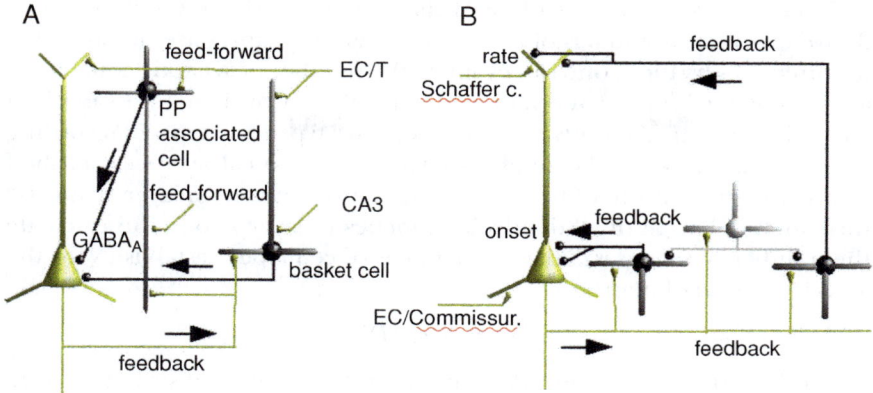

Figure 20.2. Hippocampal circuitries conceived as the composition of basic circuitries. (A) Combination of inhibitory feed-forward and feedback circuitries. The feed-forward circuitry is established through excitatory input connections from entorhinal cortex or thalamus terminating on perforant path (PP) associated cells that inhibit pyramidal cells near the soma. This is convergent with the recurrent inhibition through basket cells that are coaligned with the Schaffer collateral/comissural input and also terminate perisomatically. The design of the two convergent but spatially segregated inhibitory sources suggest coordinated inhibition on the pyramidal cells. (B) Another example of the hippocampal CA1 circuitries where the in vitro physiological response has been clarified. Spatially segregated recurrent (feedback) inhibition circuitries originate and act on the same pyramidal neuron. One is acting on the apical dendrites, the other is perisomatic. During stimulation through alveus, the inhibition rapidly shifts from the soma to the apical dendrites. While the somatic inhibition acts in a time-dependent fashion, the distant dendritic inhibition is prolonged and frequency dependent (Pouille and Scanziani, 2004).

feed-forward/feedback circuitry of hippocampal CA1 pyramidal cell, basket cell, and perforant-path-associated cell. Basket cell terminals are coaligned with the Shaffer collateral or comissural input, while the perforant-path-associated interneuron is coaligned with the entorhinal/thalamic input. The combination of the two types of inhibition, feed-forward and feedback, exerts an effective control over the integration in the pyramidal cell as they both act on the soma (Somogyi *et al.*, 1998). The other example, representing a combined feedback inhibition, is also from the CA1 area of the hippocampus (Fig. 20.2B). By using simultaneous somatic and dendritic recordings in vitro, a rapid shift from somatic feedback to dendritic feedback inhibition can be observed (Pouille and Scanziani, 2004). The somatic feedback most likely acts via basket cells. These observations suggest that two spatially disjoint circuitries process input onset time and prolonged rate in the same pyramidal cells separately. This simple circuitry, by utilizing the somatodendritic dynamics of inhibition, may enable pyramidal cells to allocate separate channels for processing time-encoded and firing-rate-encoded information.

From the layer specificity of dendritic and axonal distributions, one can derive quantitative models of circuitries. Even if synapses are not available for direct observation, one can estimate the number of synaptic contacts by applying Peters' rule (Peters and Feldman, 1976). Consider, for example, a cortical volume traced for two types of neurons, type i and type j. According to Peters' rule, given S_j^u the number of synapses in cortical layer u established by presynaptic neurons of type j, the number of neurons in layer u, and D^u the summed length of all dendritic branches in layer u, one can calculate the number of synapses S that all neurons of cell type j establish with the apical dendrite of neuron i:

$$S_{ij}^u = S_j^u d_i^u / D^u.$$

This formula is based on the assumption that synapses distribute evenly. Applying this formula, Binzegger et al. (2004) were able to calculate the number of synapses between inhibitory and excitatory neurons in the primary visual cortex of a cat. The calculation led to a few surprising revisions of the traditional circuitry diagram of area 17 (Gilbert, 1983; Gilbert and Wiesel, 1983; Szentagothai, 1978). For example, the most important circuitry in area 17 was believed to consist of a high-bandwidth sensory feed-forward pathway of X/Y afferents, originating from the dorsal lateral geniculate nuclei (LGN) and terminating in layer 4 spiny stellate cells. What changed this view was taking into consideration that layer 4 cells establish massive excitatory connections with layer 2/3, 5, 6 pyramidal cells and feedback to layer 4 through a recurrent loop. Based on a quantitative assessment of the synapses, it became evident that the feed-forward pathway comprises only 21% of all excitatory connections (Binzegger et al., 2004). In contrast, at least 34% of connections are intrinsic, i.e., establishing long-range horizontal connections within the given layer and interconnecting different columns (Fig. 20.3). Similar relationships were found among inhibitory–inhibitory and inhibitory–excitatory neurons.

The selectivity of excitatory and inhibitory connections appears to be circuitry and layer specific within the same cortical area. For example, feed-forward projections from pyramidal neurons preferentially target other pyramidal neurons; however, feedback connections mainly target interneurons (Watts and Thomson, 2005). More specifically, feed-forward projections from layer 4 to 3 and from layer 3 to 5 target pyramidal cells and to lesser degree interneurons. "Feedback" projections from layer 5 to 3 and from layer 3 to 4, on the other hand, mainly target interneurons.

Using one of the most powerful techniques, optical release of caged glutamate in combination with intracellular or multiple extracellular recordings, Callaway and colleagues demonstrated an intracolumnar fine-scale organization of layer 2/3 cortical neurons (Yoshimura et al., 2005). Specifically, adjacent pairs of layer 2/3 pyramidal neurons that are connected to each other share common input from layer 4. Conversely, those that are not connected share negligible input. In contrast with this fine-scale specificity, inhibitory or layer 5 excitatory input are all shared across layer 2/3 cells,

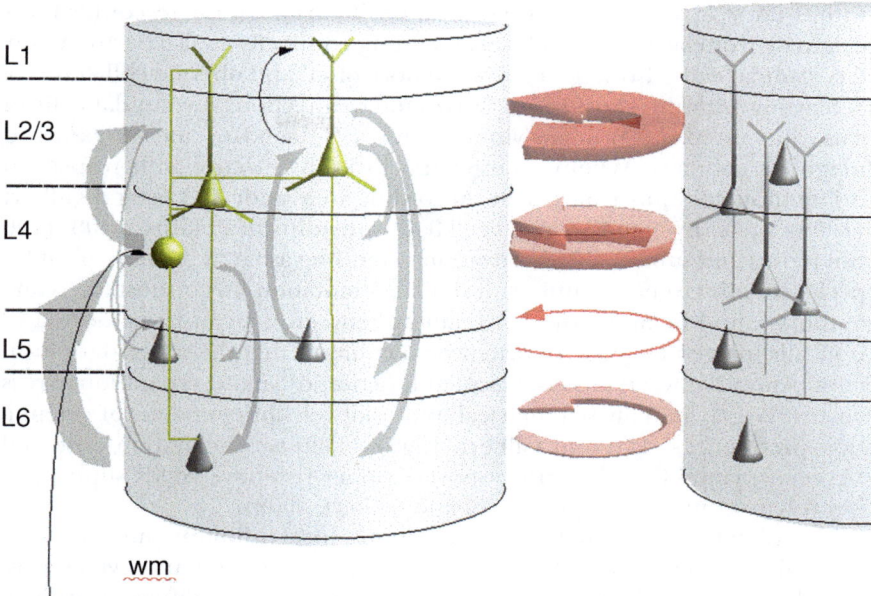

Figure 20.3. The cortical circuitry of excitatory connections based on synapse density in cat visual cortex area 17 (Binzegger *et al.*, 2004). The cylindrical volumes represent cortical columns. Arrows are the local excitatory connections. The arrow thickness is proportional to the number of synapses (total number of synapses between excitatory neurons = 13.6 × 1010). Gray arrows represent connections between layers. Pink arrows are within layer connections. Some of these connections project from other columns. Note that the within layer excitatory connections outnumber the feed-forward sensory connections (cortical layers numbered; wm, white matter).

regardless of whether they are connected or not. Whether this example of a layer-specific fine-scale organization of neurons represents a functional subnetwork independent of columnar compartmentalization remains to be investigated.

D. Horizontal Organization of Circuitries

We must make a clear distinction between the vertical specialization of circuitry and the horizontal (short and long range) associational connections. These associational connections interconnect different vertical circuitries, regardless of whether these circuitries cluster according to columns or not. The range of these horizontal connections vary from adjacent columns (50–200 μm) to different cortical areas or different hemispheres (few centimeters). Excitatory associational long-range connections localize broadly between layer 3 and layer 6. Horizontal excitatory axon collaterals and interneurons from layer 3 arborize in "patches" at distant excitatory targets within layer 2/3 and 5. Furthermore, excitatory long-range connections originate from layer 5 where local ascending connections also derive. Layer 6 pyramidal cells also send long-range horizontal and oblique collaterals.

Starting with the short-range horizontal connections, pyramidal cells in cats' and monkeys' primary visual cortex (V1) send excitatory axons to neighbor columns. Whether target specificity is selective with respect to orientation columns is not clear. According to a study by Das and Gilbert (1995), spikes from the neuron's cell body spreading in a radius of 400–1100 μm radius and subthreshold activation extending as far as 3.2 mm in cat V1 preferentially target columns with similar orientation. In contrast, the same authors found by measuring correlation between neuron pairs located in optically imaged maps of V1 orientation columns that the strength of local connections between cells is a graded function of lateral separation across cortex, largely radially symmetrical and relatively independent of orientation preferences (Das and Gilbert, 1999). Collinear facilitation observed between pyramidal cells with nonoverlapping receptive fields supports a key role in contour integration (Li and Gilbert, 2002).

Complementary to excitatory connections, local inhibitory interneurons may substantially contribute to the functional segregation and dynamic assembly of orientation columns. Selective suppression of different orientation domains of adjacent columns may enhance the contour integration. The first important difference between lateral inhibitory and excitatory connections is that inhibition has a shorter (250–500 μm) radius. Morphological reconstruction of large GABAergic basket cell axonal arborization and superposition on local orientation maps obtained by optical imaging have revealed selective targeting. Specifically, visual area 17 layer 4 clutch cells (a subtype of basket cells) arborize isotropically near their cell body within 50 μm, restricted to the nearest adjacent columns (Budd and Kisvarday, 2001). However, axons beyond this core show highly domain-specific topography (Kisvarday et al., 2002).

The long-distance horizontal connections, which extend beyond specific cortical subregions, form a massive cortico-cortical network. Probably the best studied such interareal network is the visual pathway where the most complete functional connectivity map is available. This map revealed a network of distributed hierarchical processing (Felleman and Van Essen, 1991) based on the systematic mapping of long-range cortical associational connections that have been published during the last few decades. It illustrates the enormous effort to extract connectivity information from published data. Mapping associative connections between other cortical areas, such as the prefrontal cortex (Kötter et al., 2001; Rempel-Clower and Barbas, 2000), is in progress. This type of mapping involves a combination of retrograde tracing with electrophysiology because the range of connections is at the super millimeter level and cannot be tracked from the same section. High-resolution functional and optical imaging applying voltage-sensitive fluorescent dyes with confocal or two-photon microscopy may substantially facilitate the functional verification of the connections, especially the horizontal associational and callosal connections (Chen-Bee *et al.*, 2000; Grinvald and Hildesheim, 2004; Petersen, in this volume).

E. Columnar Organization of the Neocortex

The organization module that integrates both vertical and horizontal connections is the cortical column. The concept of columnar architectonics of the cerebral cortex arose originally from the early physiological observations of Mountcastle (1957) of the vertical columnar organization of the somatosensory cortex. This was soon followed by an analogous architectural principle in the visual cortex found by Hubel and Wiesel (1959). The columnar architectonic principle of the cortex has received crucial support from studying the callosal and associational connections in primates by Goldman and Nauta (1977). The arborization spaces of callosal columns are one order of magnitude larger (300 μm–3 mm) as compared to the orientation columns of Hubel and Wiesel (1972). Even after transections of large parts of the corpus callosum, the distribution of degenerated fibers shows a discontinuous pattern: in coronal sections hourglass-shaped territories containing massive degeneration are alternating with areas containing little or no degenerated terminals (Zaborszky and Wolff, 1982). Cortico-cortical associational connections show inhomogeneous distribution pattern similarly to callosal columns (Zaborszky, 2002). The systematic studies by Burkhalter, Killackey, Malach, and more recently by Sakman and their colleagues (Coogan and Burkhalter, 1993; Johnson and Burkhalter, 1997; Koralek *et al.*, 1990; Lubke *et al.*, 2000; Malach, 1994; Paperna and Malach, 1991) in the rodent cortex and that in the prefrontal cortex in primates by Patricia Goldman-Rakic, Helen Barbas, and David Lewis (Barbas and Rempel-Clover, 1997; Goldman-Rakic, 1984; Pucak *et al.*, 1996) amply confirmed the columnar nature of associational connections that can be utilized to predict the hierarchical organization of cortico-cortical connections as shown in the often cited diagram of Van Essen (Felleman and Van Essen, 1991). The size of associational columns in primates compared with the size of the associational columns in rats show a remarkable congruence.

The idea of columnar organization of the neocortex[5] is part of a more general hypothesis of modular organization of the nervous system, a widely documented principle of design for both vertebrate and invertebrate brains (Szentagothai, 1983). Some of the main characteristics of the modular principle are summarized in a review by Liese (1990) and in a superb book on the anatomy and functions of cerebral cortex by Mountcastle (1998). The following features can be enlisted: (1) Modules are local networks of cells in any region of the CNS, containing one or more electrically compact

---

[5] A more detailed discussion of the columnar–modular organization of the cortex is beyond the scope of this chapter and the reader is referred to a recent review by Rockland (1998).

circuits active in a particular behavioral function. (2) Modules are dynamic entities—modules, repeated iteratively within each larger structure, function independently, or they may act together when combined in groups whose composition may vary from time to time. (3) Modules may differ in cell type and number in internal and external connectivity and in the mode of neuronal processing between different large entities; but within any single entity, like the neocortex, they have a basic similarity of internal design and operation, ranging in diameter from about 150 to 1000 μm. (4) The neighborhood relations between connected subsets of modules in different entities result in nested systems that serve distributed functions. A cortical area defined in classical cytoarchitectural terms may belong to more than one and sometimes to several such systems. (5) Modules may develop through ontogenesis and phylogenesis by duplication of homeobox genes (Allman, 1998).

Modules can often be anatomically differentiable from surrounding tissue, for example, associational or callosal columns in the cortex using tracing methods, the striosome–matrix compartments in the striatum using AChE histochemical reaction (Graybiel and Ragsdale, 1978), or the application of immunocytochemical and autoradiographic methods for the presence of various transmitters and receptors (Gerfen, 1985). AChE staining also delineates patches in the superior colliculus that represent special sites where information from various sensory modalities can be integrated (Chevalier and Mana, 2000). In other brain regions, such as the auditory nuclei (Malmierca *et al.*, 1998), the pontine gray (Leergaard *et al.*, 2000), or the basal forebrain (Zaborszky *et al.*, 2005b), computational anatomical methods helped to reveal a clustered, putatively modular organization, defined by patterns of connectivity. For the historical record, we reprint here in Fig. 20.4 Szentagothai's imaginative models about cortical modules in which he envisioned to place the elementary circuitries of the neocortex in repetitive spaces of callosal columns.

F. Synaptic Reconstruction

Precise assessment of the number of synapses has been done entirely by using electron microscopy (EM) in 2D sections (see Avendano, this volume). Full 3D reconstruction of neurons based on EM has only recently became available (Fiala and Harris, 2001; see also the chapter by Duque and Zaborszky, this volume). As a shortcut, using the correlation between dendritic spines and synapses, it is possible to estimate the number of synapses based on two-photon microscopic reconstruction of neurons (Yuste and Denk, 1995) and investigate the specificity of synaptic connections relative to random dendritic contacts (Kalisman *et al.*, 2005). These measurements provide the most reliable quantitative assessment that could guide further calculations on the bandwidth at different components of the circuitry, thus allowing reconstruction of the functional connectivity.

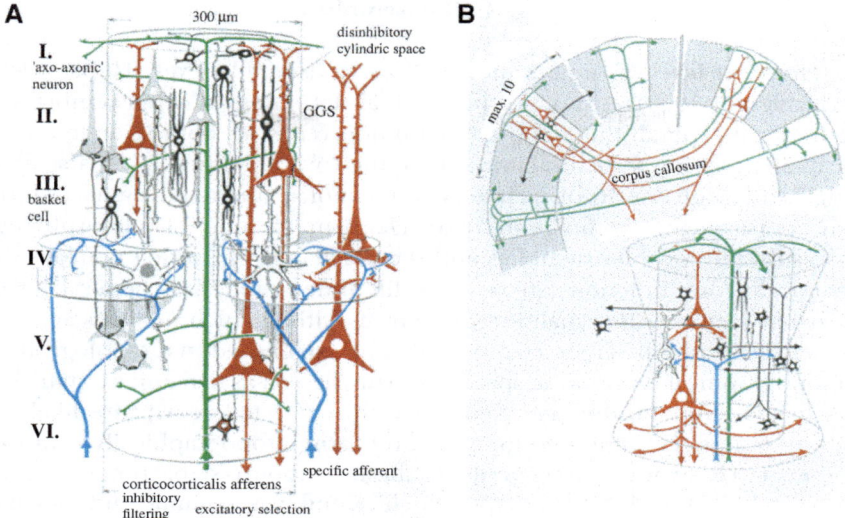

Figure 20.4. The callosal and associational columns. (A) Arrangement of neurons and local circuits in one cortical columnar module. Pyramidal cells: red; specific afferents: blue; corticocortical afferents: green; inhibitory neurons: solid black; TKN: spiny stellate excitatory neuron; GGS: inhibitory neuron (double-bouquet cell of Cajal) connected to other inhibitory neurons. The effect of the Martinotti cell in L VI spread upto L I. (B) Interconnections of associational and callosal columns. The output of each column originates from pyramidal cells, their terminal axonal arborizations are labeled green. Ipsilateral connections maximally up to 10 columns. The lower right scheme shows some of the dynamic features: in L I and VI the excitation expands the diameter of the column, in L IV, the inhibition shrinks the cylinder (Reprinted with permission from Szentagothai and Rethelyi, 2002.)

Combining voltage-sensitive Ca^{2+} imaging with two-photon microscopy it was possible to monitor activity of 100–1000 neurons from a 150×150 or 300×300 μm cortical volume of area 18 in cat and rat in vivo at a < 1-μm spatial resolution, allowing unambiguous cell body identification (Ohki *et al.*, 2005). By improving the stability and signal-to-noise ratio of the voltage-sensitive dyes, the temporal resolution of this technique can be reduced to sub-second rates to achieve a dynamic imaging of whole populations of neurons simultaneously.

For submicroscopic structures, such as gap junction, EM verification remains to be necessary (Fukuda and Kosaka, 2000). However, the combination of precise morphology-based modeling with physiological level simulations allows predictions about network dynamics to be made. For example, multicompartmental modeling of the cellular morphology of interneurons and pyramidal cells, including the number of synapses and various input currents, led Traub *et al.* (1999) to conjecture that gap junctions are necessary for high-frequency synchronization, commonly observed in the hippocampus and neocortex.

G. Cell Assemblies

The next level of organization is the cell assembly level (Hebb, 1949; Wickelgren, 1999). Although the original definition of cell assembly was purely conceptual, the empirical definition relies on both structural and dynamic criteria. Structural criteria are the "overlapping set," "sparse coding," and "high density of excitatory connections," and the dynamic criteria are "persistence," "completion," and "Hebbian learning." The morphological substrate of cell assemblies and the mechanism by which neurons dynamically form functional groups are still undetermined. In general, a cell assembly represents a coalition of neurons within which neurons act in a synergistic fashion. These coalitions can be established in a topological or a topographical basis in the spatial domain, as well as dynamic or static basis in the temporal domain. Neurons with correlated activity are likely to group together during development and form a topographically compact network. This type of cell assembly is static and supports stable functions over a period of time. The other type of cell assembly is dynamic and may not segregate into topographical cell clusters. This is typical in structures supporting flexible associative function between cells such as the hippocampus (Harris *et al.*, 2003) or interface structures with a large input/output divergence (higher sensory cortical areas). While topographical cell clustering is not a defining feature of cell assemblies, temporal synergy is. On the other hand, since temporal synergy often derives from the common input to the constituent neurons, topographical and temporal compactness are usually codetermined. Temporal compactness can be detected as coherence, while spatial compactness is a morphological feature. Since cells with overlapping dendritic arborization are likely to share input, they must respond to the common input with a synergistic fashion (see section "Statistical Modeling"). However, temporal compactness can be derived from the high interconnectedness of a group of neurons that generate the same spike pattern regardless of the input. Therefore, cell assemblies created by the common input must be distinguished from cell assemblies formed by the synergy of neurons. To consider all these possibilities of neuronal ensemble formation, one needs to combine anatomy with physiology. Fortunately, the combination of morphological and dynamical features results in a finite set and, as we argue, the lexicon of connection patterns and dynamics fall into several categories. For example, neurons with overlapping dendritic trees sharing a common input can either form a feed-forward or a recurrent network. If the output of these cells reliably reflects the temporal structure of the input, then the network is feed-forward, otherwise it is likely to be recurrent.

The ability to record simultaneous activity from a population of identifiable neurons brings up important questions concerning the relationship of anatomical connections between neurons ("structural connectivity"), the observed correlations between the activity of different neurons ("functional connectivity"), and the causal sequence ("effective connectivity"). It has been considered for many years that several neuronal configurations are compatible with the same observed firing correlations leading to the concept of a simplest "effective" mechanism that can account for the data (e.g., Aertsen *et al.*, 1989). Furthermore, the same observed activity pattern recorded from a group of neurons may be underlain by different network connectivity, suggesting that the robustness and self-organization of activity patterns are more important than the precise architecture (Prinz *et al.*, 2004). Much further work will need to be done before we can unequivocally specify the relationships between structural and functional connectivity, the number of their distinct configurations, and the potential benefit of redundancy at any of these levels.

III. STATIC DATA

A. Connectivity Matrix

The static architecture of neuronal information processing relies on the map of connectivity, i.e., the connectivity matrix. We will refer to this as "static connectivity" in contrast with the "functional connectivity," which represents active connections, equivalent with the map of information flow, defined by measuring spike-train to spike-train correlations (see section "Dynamic Data"). Theoretically, one can construct an immense matrix including all neurons as i indices and list the same neurons as j indices to represent all the monosynaptically connected neuron pairs with the number of synapses. Within this matrix, we should find numerous isolated hot spots representing major hubs and symmetric blocks indicating long-range connections as well as local reciprocal connections. More detailed analysis of the matrix would resolve the primary associations between basic neuron types (e.g., association of Purkinje and mossy cells in the cerebellum, or basket cells and granule cells in the dentate area of the hippocampus), those that we have described as circuitries. The next level of associations would indicate static assemblies, those that represent functionally related groups, such as ocular dominance columns of the visual cortex, barrels in the somatosensory cortex of the rat, or glomeruli of the olfactory bulb. The next level of associations would reveal brain regions, such as the thalamus, hippocampus, colliculus-superior, etc., including functionally distinct cortical areas. To construct such a matrix is beyond the current technological capabilities and may not be feasible at all. However, what is technically feasible is the regional mapping of connections. We refer to this as the connectivity map, a small part of the theoretical connectivity matrix limited to a specific structure. This is within a reach of the current technology, since a database like this could be developed incrementally (see section "Databases"). Detailed regional mapping and tracing of connections of various brain areas over a century have revealed the critical organizational features of these structures allowing to make generalizations and construct accurate models (see chapter by Ascoli and Scorcioni, in this volume).

B. The Importance of 3D Reconstruction

In order to create connectivity maps, first one has to identify the elements of connections, such as dendrites, axons, spines, and synapses. There are two basic approaches of extracting these structures from histochemically prepared slices. Both approaches are based on human operators to recognize these elements; however, data registration has been substantially accelerated by computer technology. One is based on image analysis and the other is based on vectorial tracing. Both methods start with application of specific markers (Amunts and Zilles, Ascoli and Scorcioni, and Bjaalie and Leergard, this volume). The goal of image analysis is to recognize elements of connections from images. To date, there are a number of image-enhancement methods, such as edge detection, contrast enhancement, and texture analysis, that can aid or make the recognition of different structures unsupervised (He *et al.*, 2003; Rodriguez *et al.*, 2003).

The other approach derives from the technique of camera lucida, which is based on manual tracing of outlines under the microscope at different levels of details, including cellular, population, and structure level. Originally, this method was introduced to trace the contours of cells and connections with maximal precision. Today, under computerized microscope control, the tracing of individual sections is still done by an operator; however, traces are registered with their X, Y, and focal depth coordinates to a database by computer. The computer encodes each contour segment by a vector in the 3D Cartesian coordinate system in addition to the coordinates indexed by the actual section under the microscope. By combining these sections, we can reconstruct the virtual 3D structure of an entire traced neuronal system. The reconstruction may involve interpolation between adjacent sections unless the sectioning was gapless. The result is a 3D vectorial representation of the cells that may include, besides cell bodies, the corresponding dendritic processes and eventually axons. One major advantage of 3D representation in a Cartesian coordinate system is that it enables one to view the data from different angles and virtually zoom in to any level of detail. Furthermore, using a standard stereotactic coordinate system, the 3D representation allows incremental development of a database by adding new elements. The most widely adopted commercial system for vectorial data acquisition is MicroBrightField's Neurolucida® (MicroBrightField Inc., VT; see Ascoli and Scorcioni, this volume).

The two types of data representation, image and vector, are fundamentally different and combining them is a major challenge in developing neuroanatomical databases for the future. Although the 3D vectorial representation is better suited for tracking neural processes across sections than the image format, it does not automatically identify the connections. Synaptic connections can only be verified with EM or physiology. Therefore, connections revealed by light microscopy are only putative and inferentially based on a set of critical attributes. These attributes are synaptic boutons, dendritic spines, or the proximity of axons and dendrites. A less reliable but still useful attribute is the overlapping dendritic arborization that indicates shared input source since en passant axons are likely to establish contact with all neurons with overlapping dendritic arbors.

Ideally, for a population database, one would like to map neurons along with their cell body, dendrites, and axons in relation to other neurons and structure outlines or other available morphological markers, such as cortical layers. Axons, however, are difficult to trace due to their small diameter and that they may traverse across multiple sections and depth planes. However, an intermediary solution for reconstruction of the axonal tree is to scan camera lucida paper-and-pencil drawings and apply an algorithm that follows a nearest-neighbor strategy (Ascoli and Scorcioni, this volume). Because the connectivity is not readily available from tracings, reconstruction of the *connectivity map* of large populations of neurons requires using a few inferential heuristics. Such heuristics are the following: (a) neurons that group together are likely to be functionally related; (b) neurons with overlapping dendritic arborization are likely sharing input; (c) spatial association of chemically identified cell types reflects functional synergy. Guided by these heuristics, methods have been developed to extract and visualize hidden association of neurons (Stepanyants *et al.*, 2004). The first group of methods is based on cell body distribution; the second is based on dendritic morphology.

C. Statistical Modeling

When a population of cell bodies have been selected with an unbiased sampling, traced, and registered in 3D, a valid question is whether the distribution of neurons suggests a pattern that further implicates functional relatedness. Obviously, the anatomical relationship is only suggestive and a functional relationship must be tested by physiological methods. The null hypothesis (H_0) is that neurons distribute evenly within the volume of interest. In contrast, if neurons cluster, we need to reject H_0, which again does not necessarily imply any functional relatedness among neurons of the same cluster. One algorithm of testing for homogeneity is the following:

1. For digitization, design a set of grids with linearly increasing grid sizes. This grid serves to partition the total volume of interest into voxels.
2. Apply the largest grid size that is smaller than the smallest possible cluster size.
3. Count the number of cell bodies within each voxel. Cell counts represent local densities.
4. Impose a density threshold. Voxels with local density larger than threshold can be visualized by surface rendering.
5. Employ the next larger grid size and recursively apply steps 3–5 until the grid is larger than the largest possible cluster size.

If H_0 is correct, then the average cell density within the filled voxels should increase linearly (that is, the total number of cells within the filled voxels should increase exponentially) with the grid size, while the number of filled voxels should also increase linearly. Deviation from the linear density change, for example, a stepwise increase in average cell density, indicates inhomogeneity. The precise cell density and grid size at which the largest stepwise increase of density occurs is the one that corresponds to the critical density and cluster size, respectively. This algorithm was applied to reveal hidden clustering of cell bodies within the basal forebrain cholinergic system (Nadasdy et al., submitted; Nadasdy and Zaborszky, 2001; Zaborszky et al., 2002, 2005b). Having the critical density and size determined, one can apply a grid size that fits to the cluster size and compute the local cell density within the voxels constructed by this grid (Fig. 20.5B). Next, select only those voxels where the density is equal to or larger than the critical density. These voxels delineate putative neuronal clusters (Fig. 20.5E). To visualize these clusters, apply a surface rendering on the selected voxels. To eliminate sharp edges between adjacent polygons, it is recommended to interpolate and smoothen the polygons before surface rendering is applied (Fig. 20.5F). For construction of volumetric database and isodensity surface rendering, see the Appendix..

Visualization of such clusters can place these clusters into the context of the macro-architecture (Fig. 20.5D–F). We can take the density analysis one step further and apply it to the association of different cell types. For example, if the brain sections were histochemically stained for choline acetyltransferase, parvalbumin, calretinin, and calbindin, then each of these markers will label a subpopulation of neurons specifically associated with a specific cell type. These populations can be considered either as four independent spatial distributions (H_0) or a coordinated distribution of four markers (H_A). One could ask the following question: Knowing the spatial distribution of each marker population, what is the probability of finding a parvalbumin and a cholinergic neuron within the same voxel by chance? The combined by-chance probability is the product of the local probabilities of occurrences of the two markers. If the observed coregistration probability is larger than the expected by-chance coregistration of markers, then we must reject H_0. Rejection of H_0 does not necessarily imply functional relatedness. Nonetheless, assuming that neurons within the same density cluster share common input, H_A is suggestive of a functional association of different cell types (Zaborszky et al., 2005b).

Dendritic morphology, when combined with cellular density data, further supports the inference from the spatial distribution to the connectivity map. Utilizing the overlapping dendritic volume as an indicator for shared input provides an additional attribute of functional cell clusters. To explore this option, we used a large database of basal forebrain cholinergic neurons ($n = 15,700$ neurons) and extracted an unbiased random sample of 750 neurons for dendritic morphology tracing. Next, we determined the 3D orientation of the main axis of the dendritic tree and, for each individual cell, we replaced the dendrite with a vector pointing from the cell body. The vector's orientation was identical to the orientation of the dendritic

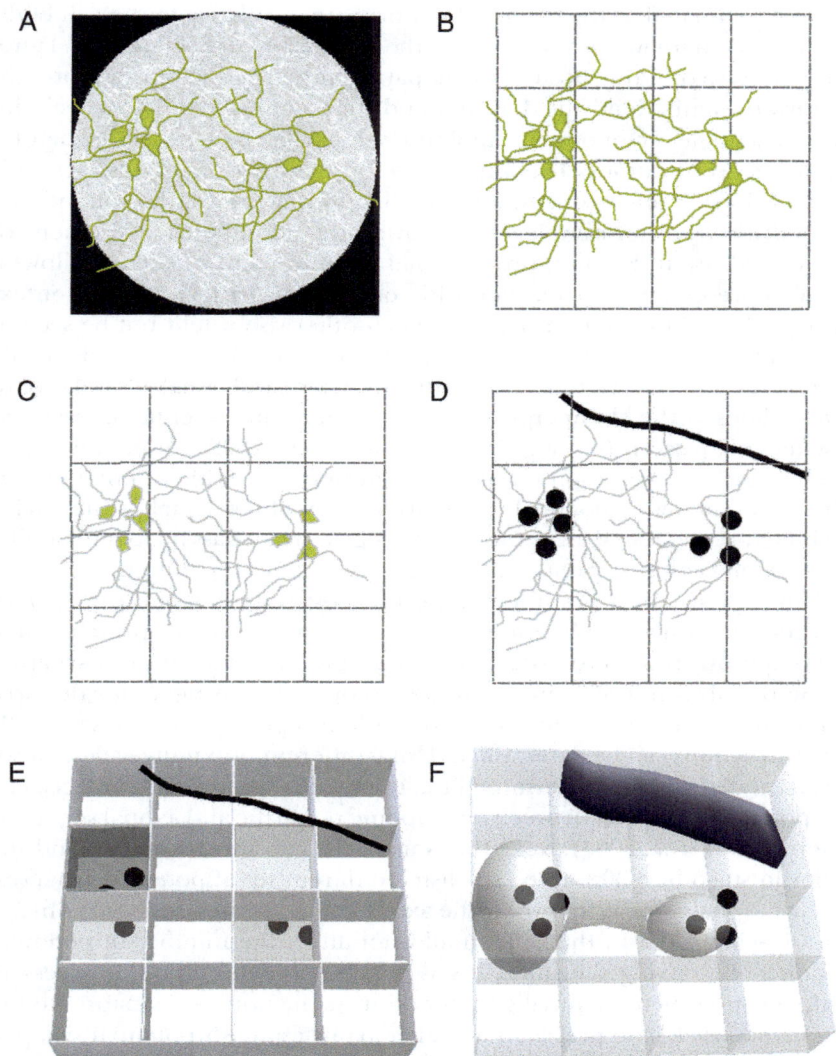

Figure 20.5. Visualization of neuron clusters by using isodensity surface rendering. (A) Microscopic image of immunohistochemically labelled neurons. (B) Camera lucida reconstruction of neuron outlines with processes. (C) A grid is defined that incorporates the size of clusters to be determined. Cell bodies are highlighted. (D) Cell bodies enter with their voxel coordinates. (E) The number of cell bodies within each 3D grid is determined and the local density is assigned to each voxel. (F) Voxels at which the density exceeds a certain threshold create a volume. This volume, encompassing the spatial structure of neuron clusters, can be visualized by surface rendering and superimposed on clusters obtained from other cell markers.

volume, estimated as the center of dendritic mass relative to the cell body, and vector length was proportional to the dimension of the dendrites. Here, one can use a different abstraction of parameters. The vector can represent dendritic length, dendritic density, dendritic volume, etc. Alternatively, instead of a single vector one can apply two vectors that better describe bipolar dendritic arborizations. The importance of dendritic vector abstraction is twofold. First, it allows one to estimate the overlapping of the volume that a dendritic tree can sample, thus providing a quantitative assessment of the shared input. Second, proportional enhancement of vectors allows a visual representation of the dendritic orientation pattern in the context of the cellular distribution pattern, which otherwise would not be seen at a true microscopic scale. When neurons represented with their dendritic vectors were projected to a 3D coordinate system and visualized relative to the outlines of the cholinergic column, the dendritic orientation revealed a systematic pattern (see Fig. 5 in Zaborszky *et al.*, 2002). Along the septal cholinergic volume, a spiral staircase pattern of dendritic orientation was observed that was orthogonal to the orientation of the septal column. This architecture suggested an optimal sampling of the en passant axons parallel to the septal column (Somogyi *et al.*, manuscript in preparation.

Although synapses with their respective postsynaptic target can reliably be identified only by EM, the presence of synaptic marker proteins such as synaptophysin or ProSAP2/Shank3 can give an estimation of synapses at the population level using light microscopy. Also, using a tour de force confocal regimen (see Wouterlood, this volume) putative synapses from 3D reconstructions can be determined. Using different proximity scales, Stepanyants and colleagues determined the number of "graphical" contacts, i.e., the potential synapses, between overlapping dendritic and axonal segments (Stepanyants *et al.*, 2004). Next, they shifted the coordinates of the dendritic arbor in silico by ~30 μm to establish a different set of potential synapses. Their rationale was as follows. If the axons and dendrites establish contacts on a by-chance basis, the shift should not affect the number of potential synapses. By varying the shifts, it was possible to determine the necessary shift beyond which the misalignment destroys the potential synapses to the by-chance level. The precision at which axons establish potential synaptic contacts was found to be consistent within the shaft dimension (0.4 μm) and dendritic spine dimension (2.0 μm). Considering this metric, it was also evident that the number of actual synapses represent only 10–30% of the potential synapses, indicating significant room for plasticity to convert potential synapses to actual (Stepanyants *et al.*, 2002, 2004). Exploration of the regional and laminar differentiation of potential synaptic contacts will hopefully elucidate further details of cortical circuitries. The use of computational and statistical methods could play a major role in exploiting the richness of data available from the 3D morphology and population scale reconstruction. Therefore, systematic construction and incremental combination of morphological data collected by different laboratories using a common data registration and database system is vital for these quantitative methods to gain momentum.

D. Databases

To be able to derive synaptic connectivity from static neuronal population data, we first need a representative sample of neurons which have been identified and mapped in 3D. More than 100 years of systematic mapping of a variety of structures and the morphological characterization of different neurons have amounted to a vast data source, which have been partially analyzed and the results published. Unfortunately, most of the original data are no longer available. It is now imperative to integrate independently generated data into one consistent database. The first problem is the data structure of choice to incorporate the full scale of complexity of static neuronal data. The language of the database must be vectorial to represent connections and flexible enough to accommodate new aspects. Born from this motivation, MorphML, a unified vectorial data structure has been designed for flexible representation of various neuromorphological objects from the subcellular level to the macroscopic scale morphology (http://www.morphml.org). The purpose of this data format is multifold. First, it enables incremental data integration across different laboratories and platforms. Second, it supports visualization of anatomical data as a rendering tool, designed specifically for this data format. Third, it provides a "core" data format that allows conversions between different morphological databases. Fourth, it supports the import of morphology data into dynamic modeling environments and simulation software such as Neuron (Hines and Carnevale, 1997) and Genesis (Bower and Beeman, 1998). Fifth, it could be maintained by industrial standard database management software (e.g., XML, IXL, Oracle). Currently, data conversion modules allow conversion of the Neurolucida data format to the 3D annotation system.

Arguably, it is now feasible to integrate different levels of morphological data and construct a detailed 3D representation of large population of neurons. With multiple-level data integration, it will soon be possible to reveal parts of the connectivity matrix by incremental development of the connectivity map based on morphological tracing. Furthermore, automatic data acquisition, either image analysis based or vectorial tracing, is expected to speed up data generation. Within the next 10 years, we anticipate that the complete 3D vectorial database of the rat brain will be available to address specific questions about hidden organization principles. A preliminary version of a database that allows integration and analysis of different 3D data sets collected at different levels and different laboratories that uses MorphML format can be viewed at http://www.ratbrain.org/ (Zaborszky et al., 2005a). More details on this database and a list of other databases or electronic brain atlases can be found in the Appendix.

IV. DYNAMIC DATA

A. Dynamic System Approach

The inductive approach to understand network dynamics as a multiplication of single neuron operations fails for multiple reasons. First, inferences about the connectivity matrix, a prerequisite of functional connectivity, could not be made even with the complete neuromorphometric database. Recovering the connectivity matrix would require mapping all synapses onto each neuron, which is impractical. Secondly, given that all synaptic connections are mapped, calculating the range of dynamics exhibited by a connectivity matrix is intractable. Specifically, when trying to solve all the differential equations describing channel dynamics in multicompartmental models of a population of neurons, the effect of small uncertainties and the sensitivity for small perturbations scales with iterations so quickly that it could render the state of the neurons unpredictable.

Fortunately, it is not necessary to know all the connections and solve the differential equations for each compartment in order to predict the dynamics of a given network of neurons. There is a shortcut. Simulations on 20 million versions of a three-cell model of the pyloric network suggest that the number of possible different dynamics is much less than the possible synaptic strength configurations (Prinz et al., 2004). Given the chance of reduction, theoretically we can classify circuitry architectures based on the invariant dynamics they generate. These dynamics at the mesoscopic[6] level may be robust enough to tolerate small differences in the subnetwork level organization (Freeman, 2000). Using this classification we can infer from the dynamics the architecture and vice versa. Recently, these dynamics became empirically permeable by using large-scale high-resolution recording of a population of neurons.

B. Large-Scale Recording of Neuronal Populations

1. Limitations and Constraints

To attain an empirical ground on large-scale network activity and capture the mesoscopic dynamics, we need to record as many neurons simultaneously as possible. Coincidentally, the number of neurons required to decode the location of a rat in an open space or the motion of the hand/arm on monkeys (both with better than 10-cm precision) is at least a hundred (Kemere et al., 2004; Musallam et al., 2004; Zhang et al., 1998). To record this quantity of neurons at a single spike temporal precision is now feasible by extracellular recording. Imaging techniques are approaching this scale and

---

[6] An intermediate level between microscopic and macroscopic (Freeman, 2000).

will soon complement the extracellular recording methods. Albeit the spatiotemporal resolution of the blood oxygenation level-dependent signal and the indirect link between neural activity and hemodynamic changes make fMRI less suitable to capture the functional architecture on the cellular scale, recent advances in voltage-sensitive dye (Petterson, this volume) and two-photon calcium imaging (Goldberg *et al.*, this volume) are promising techniques to study the spatiotemporal dynamics of hundreds of neurons in the living brain. We focus on the extracellular recording technique in this chapter; however, the statistical approach for recovering functional connectivity from the spike pattern of multiple neurons is indifferent to the data acquisition technique.

Concerning the traditional extracellular recording technique, it is important to consider the limitation of the electrode. To date, the most commonly used electrodes are sharp electrodes, microwire electrodes, tetrodes, multiple wire electrodes, and electrode arrays. Positioning electrodes individually imposes a serious constraint on the number of electrodes. Practically, remote control operated individually movable electrode microdrives can control up to 16 electrodes. This may be sufficient to record activity of up to 30–40 neurons. To attain a simultaneous recording of > 100 neurons requires chronically implanted electrodes. Several configurations have been devised. One is an array of parallel microwires, mounted together allowing about 400 μm spacing between the 72 μm-diameter wires (MicroProbe Inc., MD). Another option is using silicon-substrate, micromachined probes that come in various shapes and configurations and cause much less tissue displacement relative to microwires. Among other configurations, multiple shank silicon probes allow simultaneous multisite data acquisition from the same cell as well as from the different cell groups (Vetter *et al.*, 2004). The third technology is the Utah electrode (the "Utah array"), which is a penetrating array of electrodes mounted on a 4 mm × 4 mm base containing 100 silicon spikes that are up to 1.5 mm long and designed to be implanted in primate and human cortical structures (Nordhausen *et al.*, 1996).

Another constraint in extracellular recording is sampling rate. To resolve spike shapes, the sampling rate must exceed 20 kHz per electrode. Sampling rates higher than 25 kHz do not improve spike discrimination significantly (Nadasdy *et al.*, 1998). A simultaneous recording of 100 channels at this sampling rate would require 2 MHz multiplexing with the digitized data streamed to a hard disk. If the signal is digitized at 16 bits per channel (12 bits is sufficient, 16 bits is recommended), then the necessary data transfer rate is 32 MB/s, which can be easily transferred in standard network cards and streamed to a hard disk. However, we still have to be cautious with disk space since 2 h of continuous recording with this bandwidth takes a total (nonredundant) disk space of about 28 GB per session.

Another important constraint is the type and geometry of the electrode. Multiplying electrodes can be done by bundling multiple microwires together. The standard technique is to bundle four electrodes to make a tetrode. This technique multiplies the recording sites around a single neuron, improving spatial resolution and consequently spike discrimination (Buzsaki, 2004; Henze *et al.*, 2000). There is no substantial gain in neuron yield of bundling more than four microwires. Another way to increase the number of recording sites is by multiplying electrode arrays that are mounted separately. Several such electrode configurations have been deployed through the last 15 years. There are two basic strategies, concerning the type of scientific questions addressed. One strategy is to chronically implant as many electrodes as possible, where it is certain that a number of electrodes will not detect any signal from active neurons, simply because they cannot be individually positioned near the neurons. The alternative strategy is to make the individual electrodes independently movable. This technique is better suited for acute preparations where the research objective requires inserting electrodes in different locations every day for systematic mapping. Chronically implanted tetrodes are between these two strategies since the whole tetrode array can manually be lowered every day. Electrode configurations for chronic implants range from 16 (MicroProbe Inc., MD) to 100 (Utah probe, University of Utah, Salt Lake City, UT) recording sites, on a single connector. Many laboratories use custom-made electrode arrays. Electrode implants can be skull mounted or floating. To minimize movement artifacts, it is critical to keep the electrode and signal stable. The stability of floating electrodes is higher than that of skull mounted because the electrode moves with the brain.

Extracellular recording strategies have to consider several trade offs. In order to increase the neuronal yield of recordings, the obvious strategy is to multiply electrodes. However, the cost of multiplying electrodes is multifold. First, the tissue displacement increases with the number of electrodes. According to simulations, a vertically inserted 50-μm-diameter microwire electrode collides with 80% of the dendrites of the nearest recorded neuron (Claverol and Nadasdy, 2004). Second, the increased number of channels requires upgrading the data acquisition hardware to keep the sampling rate at least 20 kHz per channel and stream larger data blocks to a hard disk. The third factor is that the isolation of neurons from an increased number of channels substantially increases the volume and time of data processing. For example, a thorough manual spike sorting of a 2-h recording from eight channels may take 2 weeks in our practice. Considering that for publication it is necessary to process at least 30 such recordings and to replicate it at least on one other preparation, it would take more than 2 years of graduate student life with no weekends off to publish. Consequently, 100 channels would take at least 28 years to analyze which may eventually delay graduation.

2. Signal-Based Source Identification

A typical microwire electrode or tetrode can detect signals from neurons as far as 150 μm away (Fig. 20.6). Such a volume contains about 1100 neurons based on hippocampal cell density estimates (Henze *et al.*, 2000). The probability of detecting neurons decreases with the distance from the electrode. In the outer part of this volume no individual spikes are discernible primarily

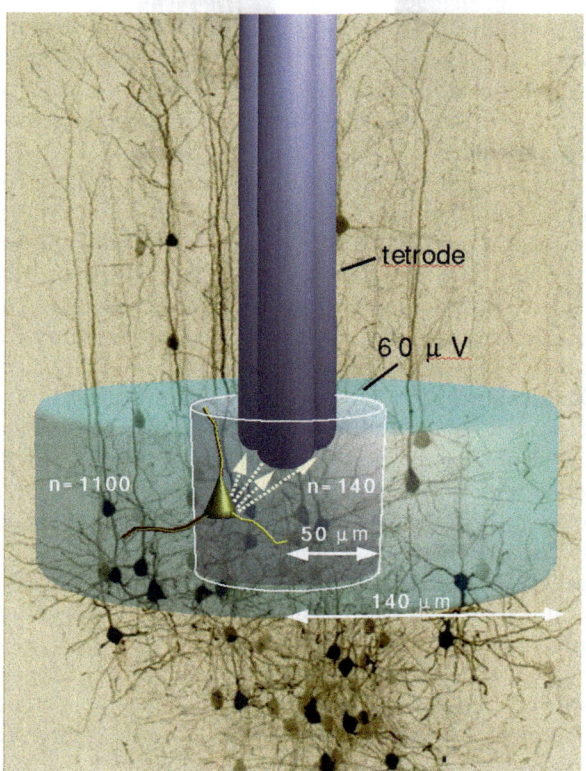

Figure 20.6. The simplified scheme of single-unit isolation from extracellular recording. The montage is an artistic rendering of a ~300-µm-wide cortical volume with the tip of a ~50-µm-diameter tetrode inserted in the vicinity of pyramidal cells. The broad blue cylinder represents the largest distance from the electrode within which spikes can be detected. This volume contains ~1100 neurons. The 50-µm-radius inner cylinder is the volume within which spikes are discernible. A pyramidal cell (highlighted) at the vicinity of the electrode projects differential signals to the four electrode tips as a function of distance. Since the voltage attenuates proportionally with the distance, the location of the source can be recovered from the differential amplitudes reaching the electrodes. Although this volume contains ~140 neurons, one can usually record up to 8 neurons. In the volume around the 50 µm radius, the signal amplitude drops below 60 mV. From this point we can consider the signal "multiunit," where individual spikes are still detectable but cannot be identified with discrete sources. From this range to the 140-µm range, the signal asymptotically converges to noise as spikes from an increasing number of neurons start to overlap. Numbers are based on hippocampal estimates (Henze *et al.*, 2000). (Background image is enhanced GFP-expressing lentivirus injected into the parenchyma of rat layer 2/3 somatosensory cortex from Brecht *et al.*, 2004.)

because spikes at average firing rate overlap in time and space, making the signal similar to $1/f$ type noise. Within the inner volume of 100 µm, spikes become discernible; however, due to the additive background noise, the amplitude variation is too large to classify neurons. It is only within a 50-µm radius volume around the electrode that spikes can reliably be associated with neurons. Intriguingly, based on a hippocampal cell density estimate, ~140 neurons should be detectable from this volume, we typically discriminate only a small fraction, on average less than 10 neurons. Whether this discrepancy is due to the silence of neurons or other attenuating factors, remains to be clarified.

In order to associate spikes with neurons, most spike-sorting methods use waveform-discrimination-based algorithms. The underlying assumption behind this group of methods is that the spike waveform differences correspond to different neurons. The biological justification for this assumption is the point source model of action potential generation. According to this model, action potentials going down the axon of the same neuron are almost identical if you measure it near the axon hillock or intracellularly from the soma. Consequently, the main source of variation of mixed waveforms is the distinct spatial location of the neurons relative to the electrode. However, the point source is affected by a number of other factors that need to be considered. For example, the intrinsic dynamics of spike generation (the interaction of back-propagating action potentials and dendritic spikes), the cell type (interneuron versus pyramidal cell), and ongoing local field oscillations substantially contribute to the spike waveforms (Buzsaki *et al.*, 1996) either by moving the point source or by adding a nonlocalized variance to the waveform.

An alternative approach in spike sorting is based on the spatial localization of the source. This principle was first utilized by the stereotrodes and further exploited by the tetrodes. Projecting simultaneous traces of spike waveforms (all data points or only peak-to-peak values), measured from two adjacent electrodes as *X* and *Y* coordinates against each other, results in clusters of points. These clusters reflect the differential amplitude ratios caused by the unequal distance of the neurons relative to the two electrodes. Using multiple recording sites within the critical 50–60 μm volume and knowing the spacing between the sites, one can triangulate the sources (Nadasdy *et al.*, 1998). Sources determined from the amplitude ratios of collinearly arranged recording sites discriminate between neurons more reliably than spike waveforms (Harris *et al.*, 2000). Silicon multiprobes further expand the "scope" of electrodes. For example, multiple shanks and collinearly arranged recording site geometry with 20 μm spacing provide precise localization of neurons as well as reliable segregation of different groups of neurons from each shank (Csicsvari *et al.*, 2003). The distance and geometry of recording sites can be optimized for a given structure of interest.

3. Spike Sorting

While electrodes and amplifiers allow recording from more than 100 channels simultaneously, isolation of single units from all of these channels is one of the main bottlenecks of data processing. Spike sorting is a critical step not only for isolating the sources of spike generation but also for classifying neurons. More research is needed to elucidate the relationship between action potential generation and current propagation in the extracellular space in order to maximize information available from the waveforms. For example, it has been reported that spikes generated by interneurons have a shorter polarity reversal (spike half width) than do pyramidal cells (Bartho *et al.*, 2004). The correlative inference from the average spike shape to the type of neuron can serve only as a putative classification considering that the waveform can be affected by many other factors. Combining evidences including firing rate statistics and cross-correlation analysis is necessary to classify neurons based on dynamic properties.

Spike-sorting methods consist of two main steps: feature extraction and clustering. The goal of feature extraction is to determine a set of discriminative features which, when projected to an *n*-dimensional space, best separate individual spikes. When spikes are classified based on waveforms, either the whole spike shape or several descriptive features are used (Abeles and Goldstein, 1977; Fee *et al.*, 1996, respectively, and review by Lewicki, 1998). In either case, the database usually contains redundant dimensions that are suboptimal for clustering. The most common method to reduce dimensionality of spike features is based on the principal component analysis (PCA). Using PCA, the main axes of "ideal" projection are determined by the largest two eigenvalues (Abeles and Goldstein, 1977). Although the largest eigenvalue projections maximize the global variance of the whole spike sample, they may not be optimal to separate the clusters. A more recent approach is based on clustering wavelet coefficients computed from the individual waveforms (Quian-Quiroga *et al.*, 2004).

It has been argued that the main causes of unit identification errors regardless of the method being used are (1) the additive noise derived from overlapping background activity, (2) the intrinsic amplitude modulation of the spikes due to subthreshold membrane oscillations, (3) somatodendritic back propagation of action potentials (Buzsaki, 2004), and (4) the human operator's ability to supervise the clustering (Harris *et al.*, 2000). The human operator's suboptimal performance and the time constraint are the main motivations to develop quasi or fully automatic and unsupervised spike-sorting tools. As the number of electrodes and channels increases, isolation of single units (spikes generated by a distinct neuron) "online" during the experiment simply cannot be performed reliably by a human operator. Such unsupervised spike-sorting methods have already been deployed (KlustaKwik, http://klustakwik.sourceforge.net/; WaveClus, Quian-Quiroga *et al.*, 2004). For a while, off-line spike sorting will remain necessary simply because the online classification is CPU intensive and requires more time to classify a channel than to record from it.

C. Reconstruction of Functional Connectivity

Spike sorting identifies spikes with a distinct neuron, thus allowing the reconstruction of multiple spike trains. The next major step is to recover the hidden functional architecture of these neurons, based on the temporal correlation between the spikes in different spike trains. Before this step is taken, we need to distinguish between sources of temporal correlations. Correlations in neuronal activity can be caused by the spatiotemporal structure of stimuli or can be derived from the functional connectivity of neurons.[7] The former one is referred as signal correlation and the latter one is referred as noise correlation. Usually both sources contribute to the actual activity pattern. To elucidate the functional connectivity, we must consider the intrinsic variance in neuronal activity, i.e., the variance that is independent of the input signal characteristics. Therefore, the input to the studied circuitry should either be kept constant or assumed to be evenly distributed. If each neuron is an ideal encoder, which encodes the stimulus independent of other neurons, then all correlations can be explained by the stimulus and there is no circuitry to uncover. In contrast, the typical scenario is that the input can only be partially recovered from the activity of neurons, suggesting that a nontrivial circuitry is involved where neurons are not independent. The answer for the general question "whether neurons encode the input independently from one another or not" likely depends on the studied structure. The complexity of dependence greatly affects the level of statistics efficient to recover the circuitry from the activity of neurons.

If neurons are highly interdependent, then recovering the circuitry is a complex and computationally NP-complete problem* where the computation time is an exponential function of the number of cells (spike trains). Therefore, the first practical step is to reduce the complexity to pairwise cross-correlations of spike processes and build the functional connectivity from the pairwise relationships. Several cross-correlational techniques have been developed, primarily between spike trains of two neurons. Nonetheless, higher order correlations are also practical to compute.

The first-order relationship between two spike trains is captured by the cross-correlation and cross-coherence of the two spike trains. For the calculation of the cross-correlation function see the Appendix. The graphical representation of cross-correlation is the cross-correlogram. The qualitative evaluation of cross-correlogram can tell (i) the independence of the two spike generation processes, (ii) the time precedence between the two processes, and (iii) the polarity of the effect (inhibitory/excitatory). What a cross-correlogram does not determine is whether the dependency between the two neurons is direct or indirect. The interpretation of cross-correlograms is illustrated in Fig. 20.7. Cross-correlograms are constructed by treating one spike train as the reference and the other train as the

---

[7] The method of reconstructing the stimulus from the activity is called the reverse correlation technique.

*NP-complete problems are decision problems verifiable in non-deterministic polynomial time.

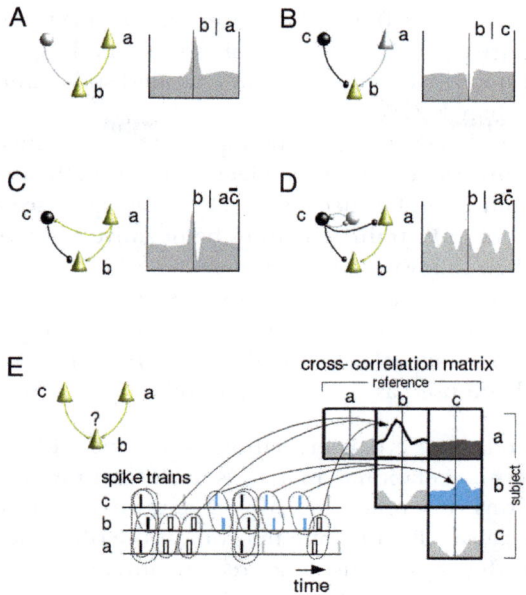

Figure 20.7. Inference from cross-correlations to circuitries. (A) A central peak with a few millisecond delay indicates a monosynaptic excitatory drive from neuron *a* to *b*. (B) A central through is suggestive of a monosynaptic inhibitory drive from *c* to *b*. (C) A central bipolar wave with positive peak followed by the through indicate a feed-forward excitatory drive (*a–b*), followed by recurrent inhibition (*a–c–b*). The opposite polarity order would suggest feed-forward inhibition combined with a non-monosynaptic recurrent excitation (not shown). (D) Periodicity in the cross-correlogram is consistent with an excitatory drive from *a* to *b*, modulated by an oscillatory input from a locally phase-coupled interneuron circuitry *c*. Alternatively, the same cross-correlogram can reflect two inhibitory interneurons firing in synchrony (not shown). (E) A fictitious example for a three-neuronal excitatory circuitry illustrates the limitation of cross-correlation analysis. Neurons *a* and *c* both terminate on *b*. It is evident from the cross-correlogram that *a* drives *b* with certain delay (represented by outlined area in the cross-correlational matrix and corresponding empty ticks in the spike train inset). The *c–b* cross-correlogram also indicates an excitatory drive between neurons *c* and *b* (blue area and blue ticks in the spike train inset). However, the flat *a–c* cross-correlogram suggests that spikes of neurons *a* and *c* coincide only by chance (gray area in the cross-correlogram). The dilemma, whether *a–b* and *c–b* drives are conjunctive or disjunctive, i.e., *a* and *c* drive must coincide to excite *b* or any of them is sufficient to drive *b*, cannot be resolved based solely on the pairwise cross-correlograms. One needs to look at the triple cross-correlogram and evaluate whether the frequency of $a - b - c$ triplets (indicated by black ticks in the spike train inset) is higher than chance or not. (This panel is modified from Nadasdy, 2000).

subject. The computational algorithm of the cross-correlogram goes as follows:

1. Define a time window W of interest. This is usually longer than the refractory period (3 ms) and smaller than the longest causal interactions

within a network (200 ms). Moreover, define a Δt precision of spike time measurement. Reasonably, this should be larger than the spike width and smaller than the expected smallest spike time precision, but not larger than $W/20$.

2. Then iteratively, take the next/first spike at the reference train and look for spike coincidences on the subject train within the time window centered around the reference spike. The time lag τ (see the Appendix) of the subject spike train can either be negative when preceding or be positive when following the reference.

3. Next, the bin of the cross-correlation histogram at τ is incremented.

4. After all subject spikes within W have been registered, move the time window to the next reference spike and repeat steps 2–3 until the last spike has been used as the reference spike.

The interpretation of cross-correlations is summarized in Fig. 20.7. The objective of cross-correlational analysis is to determine the interaction between two neurons (reference and subject neurons). Since correlation can occur due to a causal effect by one neuron on the other neuron, as well as due to a causal effect by an outside source on either of the two neurons, the reduced model of interactions must include at least three neurons. When considering the dynamics of a three neuronal (a, b, c) subnetwork based on the information available from two neurons, the possible cross-correlations are consistent with one of the following schemes: (i) excitatory drive from a to b (Fig. 20.7A); (ii) inhibitory drive from c to b (Fig. 20.7B); (iii) combined excitatory and inhibitory drives (Fig. 20.7C); and (iv) common oscillatory drive to a and b is also evident from cross-correlograms (Fig. 20.7D).

D. Effective Connectivity

Although action potentials are uniform in shape and generation site, the inputs that elicit them derive from many neurons firing at different times. In principle, a strong input connection has a larger contribution to a spike series than a weak input. In order to recover the functional connectivity between neurons from an extracellular recording, it is important to determine the most effective source of input for a given neuron. The tight temporal correlation between the input and the output spikes is the key to resolving which spike was caused by which neuron in the network, even when higher order (hidden) dependencies are involved.

Higher order dependencies, involving three or more neurons, are beyond the scope of the cross-correlation method. Consider the example illustrated by Fig. 20.7E. Let us assume that we know the connectivity between the three neurons (a, b, c), where neuron a and c both have an excitatory drive on neuron b. The excitatory dependencies between a–b and c–b are evident from the combined cross-correlograms (on the right). Nonetheless, we would like to know whether a single input from neuron a or c is sufficient to excite b or the two inputs must coincide to excite neuron b.[8] From pairwise cross-correlations this information is not available. To infer the "effective connectivity" from multiple spike trains involving more than two neurons requires higher order statistics.

Motivated by mapping the "effective connectivity," Gerstein and colleagues introduced the joint peristimulus time histogram (JPSTH) method (Aertsen *et al.*, 1989). The JPSTH was a generalization of cross-correlation method to spike triplets of different neurons that coincide within the time window W. The appropriate statistics were developed much later (Abeles and Gat, 2001; Baker *et al.*, 2001; Frostig *et al.*, 1990; Palm *et al.*, 1988). As an illustration of such statistical analysis on >2D dependency between spike trains, consider Fig. 20.7E again. The joint occurrence of an *abc* triplet is simplified by reducing it to the coincidences of two intervals, τ_{ab} and τ_{ac} interspike intervals. In a random spike process, we assume that the *acb* triplets are merely by-chance co-occurrences of the *ab* and *ac* intervals. Therefore, their joint probability is the product of the component probabilities. Otherwise, the triplets are not random coincidences:

$$H_0: P_{acb} = P\tau_{ab} \cdot P\tau_{ac}$$
$$H_A: P_{acb} \neq P\tau_{ab} \cdot P\tau_{ac}$$

If it turns out that $P_{acb} > (P\tau_{ab} \cdot P\tau_{ac})$, that is, the observed frequency of the triplet is higher than the product of the individual probabilities, then the *acb* triplets must be coordinated above chance coincidence of *ab* and *ac* intervals. The $P\tau_{ab}$ and $P\tau_{ac}$ probabilities are estimated from the pairwise cross-correlation functions; however, the two interspike-interval pools are not mutually exclusive because they both contain a fraction of the other interval pool. For example, the *ab* cross-correlations include the *abc* events and likewise the *ac* cross-correlations, consequently these events will be counted twice. As a result, the product of cross-correlation functions overestimates the expected probability of triplets. Therefore, the product must be corrected and renormalized to fit to the individual firing rates and cross-correlograms. This renormalization must conform not only with the $P\tau_{ab}P\tau_{ac}$ cross-correlations but also with the $P\tau_{bc}$. Because the true probability density functions are not directly available from the data, various computational methods have been developed for estimating it. The correct estimation of random coincidence is critical in order to determine the confidence interval for significance testing. Among those methods, we point the reader's attention to the method of calculating a "surprise function" (Palm *et al.*, 1988), spike jittering or "histogram blurring" technique (Abeles and Gat, 2001) and time resolved cross-correlation method (Baker *et al.*, 2001).

---

[8] Although there may be thousands of excitatory inputs terminating on a pyramidal cell, the above question, whether the two cells (a, c) belong to populations that have a conjunctive or disjunctive effect on neuron b, remains relevant.

An alternative approach to detecting hidden dependency time structures from spike trains is based on searching for joint events using pattern-searching algorithms. The rationale behind this approach is the following. When several neurons are recorded simultaneously, these neurons are part of a larger hidden network. We further assume that neurons are finite state machines and spikes are generated from neuron to neuron as a *causal sequence* of state transitions. If a sequence occurs more often than chance, we have a statistical basis to believe that the spike sequence follows a deterministic process, which is constrained by the network architecture (unless it was imposed by the temporal structure of the input to the structure which we do not consider here). If the circuitry favors one spike sequence over other sequences, then this sequence must recur more often than others. Consequently, the most invariant sequence must be related to the activity flow, which in turn, is dependent upon the circuitry. Thus, the objective of pattern searching spike trains is to determine repeating sequences and reconstruct their underlying functional connectivity. Various algorithms of searching the high dimensionality space of multiple spike trains for repeating motifs of spikes were introduced during the last 10 years (Abeles and Gerstein 1988; Nadasdy *et al.*, 1999; Lee and Wilson, 2002; Ikegaya *et al.*, 2004). The common feature of these algorithms is to define a template motif $S_n = \{t_i^{j1}, t_{i+1}^{j2}, t_{i+2}^{j3}, t_l^{nm}\}$ from the data, which is a vector of spike latency t of the ith spike generated by the jth neuron where $j = [1 \ldots n]$. Let us name this as an n sequence. Then the computer program performs iterative searches by comparing the template n sequence with each instance of n sequences on a data segment where the length of n sequence is equal to the dimension of searching space n. After every partial match between the template and the data, the dimension is incremented until n. When the search is complete for a given template, the template is replaced by a new n sequence from the data and the searching session starts over. After an exhaustive search of the whole database for all possible S_n sequences, the detected recurring n sequences can be rank ordered and those that occur more than chance are considered to be representing the predominant flow of activity within the network of the given set of neurons. As we have mentioned, the searching process is CPU intensive and computation time increases exponentially with the number of spike trains. A spike train database of 100 neurons would require a supercomputer to run the search.

Since dependency and causality are interchangeable terms, reconstruction of effective connectivity from dynamic data can be approached as recovering the hidden causal sequence of activity flow within the network. The formal definition of causality on time series was first introduced by Granger (1969) in the context of linear regression models of stochastic processes and since then it is referred as "Granger causality." Accordingly, if the variance of the prediction error from a time series $Y(t)$ is reduced by including the past measurements from time series $X(t)$, then the time series $X(t)$ "caused" the time series $Y(t)$. Independent of the Granger causality, numerous other attempts were made to quantify causality from neuronal data. To take a full advantage of multichannel data, such as simultaneously recorded spike trains, Kaminsky and Blinowska (1991) suggested a multivariate spectral measure to determine directional influence, defined as the "directed transfer function," which is fully compatible with Granger causality (Kaminsky *et al.*, 2001). Using the formalism of directed transfer functions, one can reconstruct the causal-dependency network within a population of neurons. It is important to note that Granger causality does not imply direct causal effect between the two spike trains (i.e., neurons). The effect can be mediated by other (hidden) neurons or distributed over the whole population of neurons, which would make the causal sequence untraceable. However, the question of direct causality can also be addressed within the framework of directed transfer functions (Kaminski *et al.*, 2001). Granger causality spectra have been successfully applied on local field potentials, clarifying interactions between different recording sites at specific frequency bands, (Brovelly *et al.*, 2003) and can be applied to point process spectra such as spike trains (Kaminski *et al.*, 2001).

A problem common in all these methods is the definition of confidence. To obtain confidence intervals for significance estimations, we need to know the probability density function of the given variable, such as triplet occurrence, sequence repetition, or causality. Since there is no a priori knowledge about the generative process of these variables, we must rely on Monte Carlo methods, such as the simulation of the spike generation process and the construction of surrogate spike trains (Nadasdy *et al.*, 1999). The limitation of the Monte Carlo approach is that we can never be sure that randomization of a variable leads to a biologically plausible null hypothesis or not. Consensus on these methods has yet to be achieved.

In summary, to recover the functional connectivity from a local cell assembly based on the dynamics, the following methods should be applied. Use electrodes with multiple recording sites at 20–50 μm apart allowing multiple spike measurements from the same neuron. Combining several such electrodes with at least 100–200 μm spacing between them allows the recording of nonoverlapping groups of cells. Configurable recording geometry is preferred to suit the electrode to the size and cytoarchitecture of the target tissue. For an optimal design, the potential tissue displacement and interference with normal neuronal functions should be considered. Using an unsupervised spike-sorting software, the classification of 3–7 neurons from a single electrode (shank or tetrode) is feasible, which scales up with the number of electrodes used. Cross-correlations can be applied to reveal first-order dependencies. Higher order statistics, essential for reconstructing the network dynamics, can be computed by joint probability and sequence-searching algorithms.

An example of the complete process of the reconstruction of effective connectivity is shown in Fig. 20.8. Cortical recording from the somatosensory cortex of an awake rat was performed by using a silicon microprobe (Buzsaki, 2004). For illustration purposes, the image of the electrodes was superimposed on the histological section using electrode traces as landmarks.

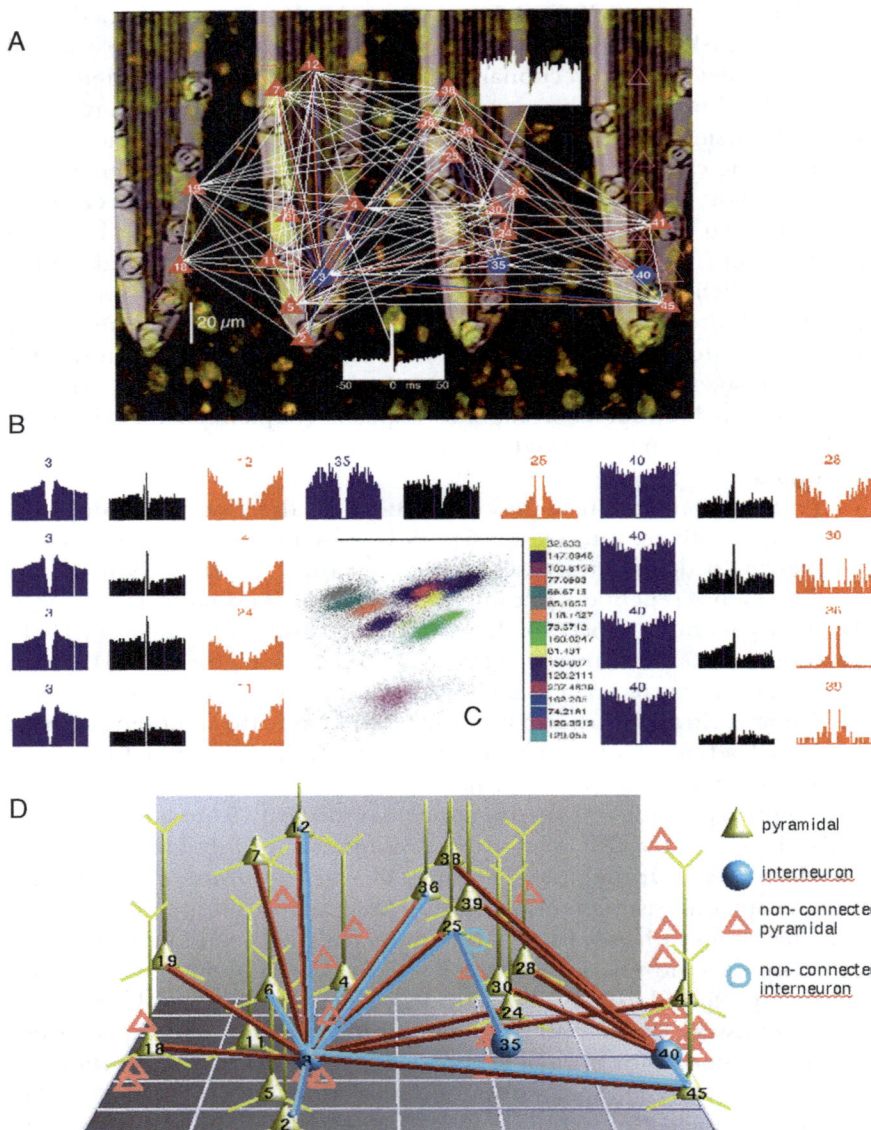

Figure 20.8. Reconstruction of effective connectivity between neurons based on the temporal coherence of spiking activity recorded extracellularly from the somatosensory cortex of a rat. (A) The theoretically possible network of neurons is illustrated by white lines (only a subset) connecting putative pyramidal (red triangles) and interneurons (blue circles). Neurons were isolated using spike sorting and localized by calculating the "center of mass" of the spike amplitudes. The putative locations enabled identification of neurons with their histological traces (background) and recovery of the electrode traces. The image of the multiple shank silicon electrode was aligned and superimposed on the traces. Spacing between adjacent recording sites of the electrode was 20 μm with 200 μm intershank distance (electrode shanks

The neurons isolated from each electrode were localized by calculating the center of mass of their spike amplitude. Possible functional connections were considered by pairing all these neurons and computing pairwise cross-correlograms of their spike times. Neurons with the largest modulation were selected as active nodes of the functional network, using the cross-correlation function. Interneurons and pyramidal cells were identified based on their firing rate and the polarity of the cross-correlation modulation. Reciprocal connections were also evident from the cross-correlograms. Note that not all of the recorded neurons are engaged in the network. The fact that certain neurons did not participate in the functional network supports the validity of detected connections. Using such heuristics, one can construct a network that is most consistent with the dynamics sampled by the cross-correlations. Although this network is rather incomplete and connections are only putative, it is the most reliable method to date for parsing the architecture of circuitries.

V. CONCLUDING REMARKS

After a century of extensive piece-by-piece brain mapping, finally all the main pathways of the central nervous system have been described and validated across different species. Although for most structures the cytoarchitecture has been defined and major cell types have been identified, the key principles that relate the microscopic architecture to activity patterns are still unknown. We argue that the conceptual bridge between these two levels is the dynamics of microcircuitries. The microcircuitry is a mesoscopic entity between neurons and macroscopic networks that generates a dynamic mosaic of functional cell assemblies by assigning neurons to different subnetworks, and these subnetworks are dynamically allocated to various functions. We reviewed classification of these microcircuitries as precursors for "basic circuitries." Although the catalog of basic circuitries is at the doorstep, to be able to make inferences from the architecture to the activity pattern and vice versa requires a coordinated anatomical and physiological approach that embraces both static and dynamic aspects of neuronal data acquisition.

Figure 20.8. (*Cont.*) were brought closer for illustration purpose). (B) Auto- and cross-correlation histograms featuring the coherency between the three interneurons (numbers 3, 35, and 40). The autocorrelograms for interneurons are purple, interneuron–pyramidal cell cross-correlograms are black, and pyramidal cell auto-correlograms are red. (C) The main PCA clusters show clear isolation of single units. Numbers at the right are the Mahalanobis distances for the given cluster. (D) Effective connections verified by the cross-correlational analysis among the theoretically possible network. Neurons whose connectedness could not be verified are marked by empty symbols. Note that most pyramidal cells receive inhibitory connections from the interneurons. A few of these connections are reciprocal. In contrast, no pyramidal–pyramidal connection was observed.

Basic circuitries are recognized as instrumental for filtering and generating specific activity patterns and oscillations. Their cellular composition must be specific for the given structures and consistent across different species. Plausibly, the evolving and developing brain multiplies these circuitries as basic information-processing units to construct large networks. The unique features of basic circuitries are the constituent neuron types, connections, and the specific activity patterns they implement. We argued that these activity patterns cannot be derived from the physiology of isolated neurons. Instead, the focus needs to shift from single-cell recording to the circuitry of ensembles of neurons, an approach that involves a combination of neuroanatomical and physiological methods.

Among these methods, this chapter discussed algorithms designed to extract information from a large population of neurons that is relevant for functional considerations, such as neuronal clusters. We also reviewed methods of collecting physiological data that capture the dynamic aspects of circuitry function, such as spike pattern generation. We highlighted on methods of the multiple electrode recordings and the analysis of action potential patterns generated by a population of neurons. We illustrated the inferential processes from the two opposite directions: one that proceeds from the structure to the dynamics and the other that proceeds from the dynamics to the structure. The two types of inferences complement each other and they should converge to a solution where the circuitry architecture is consistent with the activity pattern. The circuitry and activity pattern as a functional unit may be considered as a fundamental building block of the nervous system.

Concerning strategies of neuronal tract-tracing methods, two dominating tendencies, the expanding computational approach in neuroanatomical data analysis and combining existing methods, are broadly covered by different chapters of this volume. Specifically, using a sophisticated combination of tracing methods, the synaptology of any circuitry can be accurately determined (Sesack et al., this volume). Using extracellular, juxtacellular, and intracellular recording methods (Duque and Zaborszky and Sik, this volume), the anatomical features can be correlated with an electroencephalogram, multiunit activity, local field potential, and intrinsic membrane characteristics. Functional networks or entire brain regions are available for reconstruction (Bjaalie and Leergard, this volume) and statistical methods on testing structural hypotheses have already been developed (Bjaalie and Leergard, this volume). Furthermore, neural functions can be correlated with microstructural variation (Amunts and Zilles, this volume).

Last but not least, we emphasized the importance of 3D representation and the use of volumetric data structure that is a prerequisite for exploiting the information available from neuronal tracings. The third dimension is critical to complete the paradigm shift we are witnessing in three major fronts of research: (i) functional neuroimaging, (ii) computational data analysis and visualization, and (iii) data sharing and integration into incremental databases.

APPENDIX

A. Parametric Modeling of Neuroanatomical Data

1. Definition of Volumetric Database

A volumetric database consists of a 3D distribution of one or more variables. For the sake of simplicity, let us focus only on one variable. This variable must have spatial gradients that are continuous in space. Usually this variable is the density of a given feature. Therefore, the first step of constructing a volumetric database is to define this structural feature (such as the cell body, the dendritic segment, or the expression of certain marker), which has spatial density gradients. Before extracting this variable the space needs to be discretized. Discretization is generally achieved by defining a grid structure that is optimal for sampling the data. Reasonably, the grid size must be large enough to contain multiple instances of the given feature, but small enough to represent the fine-scale spatial variation of the variable. We refer to the 3D units of the grid as voxels. The voxel size determines the spatial resolution of the density of the variable. Voxels can also be considered as 3D bins within which the instances of the variable are counted. The voxels usually, but not necessarily, have equal edge lengths and each voxel is addressed by its Cartesian coordinates (x, y, and z).

2. Construction of Volumetric Data of Neuronal Density

The given camera lucida database (we recommend Neurolucida, however, it can be another 3D vectorial data type) must be exported as an ASCII file and parsed for cell bodies and structure outline coordinates. As a result, the data should contain only cell bodies $\{b_t\}$ of a given cell type t, represented individually by their single Cartesian point coordinates $\{p_{xyz}\}$ and section identifiers s. If the z coordinates were not recorded, the section identifier should be considered as the z coordinate.

$$b_t = p_{xyz} \qquad (20.1)$$

Conversely, since a given position could be occupied by only one cell body, the Cartesian coordinates together with the cell type completely define a cell b_t as x_b, y_b, z_b, and t. Structure outlines can be compiled to separate files. The medial, lateral, dorsal, and ventral extremes of the cell population should be taken as edges of a 3D framework to incorporate the region of interest. The total volume V occupied by neurons (neuronal space) is expressed as a vector $\vec{r}$ with minimal r density function:

$$V = \{\vec{r} : \rho_{\vec{r}} > 0\} \qquad (20.2)$$

For quantitization of local density differences, V was subdivided into "voxels" and defined as follows:

$$v = d_x, d_y, d_z, \qquad (20.3)$$

where d is the edge of the voxel. If the within section depth of the coordinates were not registered, the section index s can be used as d_z. Thus, voxel and cell definitions can be simplified as $vs = d_x, d_y$ and $b_i = p_{xys}$, respectively. The i, j, and k indices of a voxel containing a $b_i = p_{xys}$ cell body is calculated as

$$ i = d_x \left(\frac{x_p}{V_{\text{length}}} \right), \quad j = d_y \left(\frac{y_p}{V_{\text{width}}} \right), \quad k = s. \tag{20.4} $$

Cells are then counted in each voxel for each cell type separately, providing a local density function ρ:

$$ \rho_{(v_{ijk})} = \text{count}\,(v_{ijk}) = nb_i. \tag{20.5} $$

For visualization, we define a volume Ω based on a function of minimum density σ:

$$ \Omega = \left\{ v_{ijk} : \rho_{(v_{ijk})} \geq \sigma \right\}. \tag{20.6} $$

The anatomical distribution of high-density locations is often visualized by a manifold rendered around the volume and denoted as $\delta\Omega$. The manifold can be defined by surface rendering and carried out using commercially available visualization tools such as the Matlab software package (Math-Works, Natick, MA). After surface rendering, one must pay attention to convert the voxel coordinates back to stereotaxic coordinates or coordinates used by the data acquisition system to be able to superimpose the voxels on structure outlines.

Note that the critical step was the conversion of the 3D point-coordinate database, where the entries were the cells, to density data (Eq. 20.5), which became a volumetric database. In contrast to the original parametric database where cell bodies are represented by their x, y, and z coordinates, the entries of the volumetric database are cell density or cell count. The key advantage of the volumetric database is that it allows one to employ parametric statistical methods. Moreover, it enables one to slice data at any angle and visualize it from any point of view.

3. Differential Density 3D Scatter Plot

Valuable information can be extracted from the volumetric database that is not available from the 3D tracings of cells. To extract structural features and visualize them, we use the example of cell density. The density of neurons is indirectly related to functional clusters of the population because neurons close to one another are likely sharing input. We can apply a density threshold and highlight voxels with higher than threshold density. Taking the visualization one step further, we randomly choose a neuron from the selected voxels and visualize that neuron superimposed on the background of local density. Furthermore, we can select not only one but n number of neurons from a given voxel, where n is proportional to the density.

4. Isodensity Surface Rendering

The spatial organization of a large population of neurons can be very complex. Instead of visualizing each neuron, the global pattern of density differences may elucidate important structural principles that are not apparent from a large population of neurons. Surface rendering around high-density cell groups can capture this global pattern. We developed an algorithm that renders a surface around voxels where cell density is larger than a certain threshold. The procedure of discretization of the 3D space into voxels and calculating the per voxel cell density is described under section "Construction of Volumetric Data of Neuronal Densities." By applying different density thresholds, the sharp transition of neuronal density captured by two thresholds can be indicative of the size of a neuronal aggregate. The stepwise increase in density may be related to deviation of randomness that requires further statistical tests.

5. Isorelational Surface Rendering

In order to simplify the complex spatial relationship of large neuron populations and extract the associative relationship of different cell types, we can calculate the density ratio of the two cell types in each voxel. Highlighting the locations at which the density ratio exceeds a certain level reveals the spatial configuration of the cell-to-cell associations between different cell types. To illustrate this, first we constructed volumetric databases of density for each cell type and discretized the space by dividing it into voxels. For this, the different cell types must be carefully mapped to a common 3D coordinate system. If the different cell types were traced from different sections, it must be considered that inference of the joint density from separated sections may be affected by section distortion. If density changes between adjacent sections are negligible, the within-section z coordinates of different cell types can be collapsed into the same section s to obtain an estimate for the real joint density. The next step is to calculate local density ratios between the cell types for each voxel. When density ratios had been assigned to each voxel of the volumetric database, we imposed a predefined density ratio criterion and selected voxels that met this criterion for visualization. The volume of these voxels represented a specific numerical association between cell types. Voxels where the predefined density ratio of cell types had been established were surface rendered. Thus, cell bodies with density ratios larger than a specific value were covered by the surface. Conversely, voxels with density ratios smaller than the critical one were located outside of the surface. The unique feature of the "isorelational surface rendering" method is the visual representation of an abstract relationship that may be more important for understanding the functional architecture of neurons than the exact locations of cell bodies. For visualization purposes, a range of critical density values must be applied for testing the integrity of clouds and to make sure that there are no hollow spaces covered. Examples for analyses described under sections "Differential Density 3D Scatter Plot," "Isodensity Surface Rendering," and "Isorelational Surface Rendering" can be found in Nadasdy and Zaborszky (2001) and Zaborszky *et al.* (2005b).

B. Databases

1. List of Electronic Databases of Rodent Brains

Brain Maps	Swanson	Computer Graphic Files, Elsevier, 1992–1998
3D Rat Brain	Timsari *et al.* (1999)	http://www-hbp.usc.edu/Projects/3dAtlas.htm
The Rat Brain	Paxinos and Watson	Academic Press, 1998–2005 CD-ROM
Brain Browser	Bloom and Young	Academic Press, 1993
Rat Brain	Toga *et al.* (1995)	www.loni.ucla.edu/Research/Atlases/Rat/Atlas.html
Rat Brain	Nissanov and Bertrand (1998)	www.neuroterrain.org
IMGEM Mouse Brain	Sugaya *et al.*	http://sugaya.ucf.edu
Mouse Brain	Allen Project	www.brain-map.org
Mouse Brain Library	Reed *et al.* (1999)	www.mbl.org/mbl_main/atlas.html
Comp Mouse Brains	Hof *et al.*	http://www.neuroscion.com/laboratory/mousebrainmaps.html
Brain for Macintosh	Nissanov, Tretiak	http://ece.drexel.edu/ICVC
Mouse Atlas Project	Baldock *et al.*	http://genex.hgu.mrc.ac.uk
Rat-Brain Atlas	Pich, Danckaert	http://www.gwer.ch/qv/ratatlas/ratatlas.htm
Mouse MRI Images	Jacobs *et al.*	http://www.gg.caltech.edu/hbp/atlas.html
High-Resolution Mouse Atlas	Sidman *et al.*	http://www.hms.harvard.edu/research/brain/
Virtual RatBrain	Zaborszky *et al.*	http://www.ratbrain.org/
CCDB	Martone *et al.*	http://ccdb.ucsd.edu

2. Virtual Rat Brain Project (Zaborszky et al., 2005a; http://www.ratbrain.org/)

The *software architecture* is based on the NetBeans platform, which provides a modular component framework for Java applications. Layered on top of this are library modules for connecting to the database, dealing with MorphML, and viewing 3D data. The top layer is a collection of applications for warping and submitting data to the database, browsing database contents, and performing analysis on multiple data sets. Figure 20.9 shows the system architecture that consists of three major entities: (a) *the contributor* who does the data acquisition and registers the data into a standardized

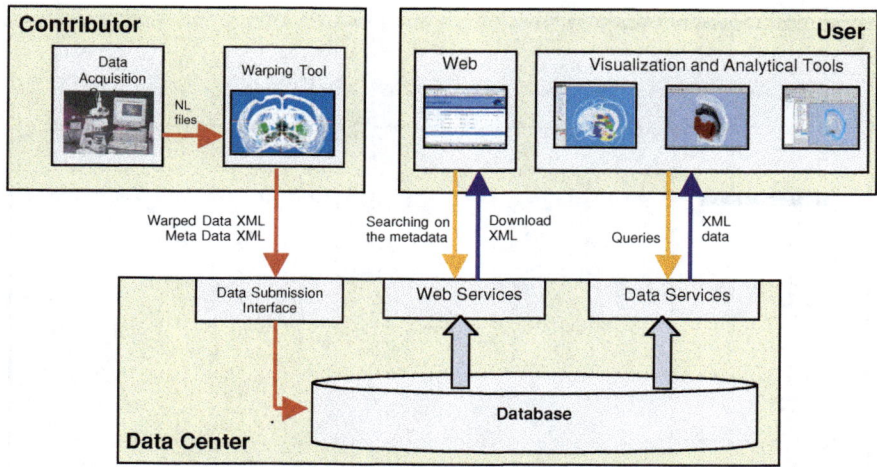

Figure 20.9. System architecture of the Virtual Rat Brain Project. For more details, see at website www.ratbrain.org.

coordinate space (reference brain) by using a special software (warping tool) and uploads it to the database; (b) *the data center* that stores and provides data services to the user; and (c) *the user* who access the data by using the analytical and visualization tools or web browser.

The *warping tool* was created to register data sets into a common coordinate space. The GUI was designed for easy spatial navigation and spatial point selection (Fig. 20.10). The loaded data sets are visualized in a 3D window. The program generates and visualizes three transparent normal section planes that give the basis of the spatial navigation. The user can move these planes along the x, y, and z axes and define points where the planes intersect. The GUI contains three additional 2D windows designed to show "virtual slices" corresponding to each plane. The registration process begins by loading the reference brain (from the database) and the data set to be warped (from the local disk) into the same 3D space. Then the user can define reference points in his/her data set and corresponding points in the reference brain. When all the reference point pairs are defined, the program creates an affine transformation and applies it to the experimental data set on user demand. The result of the action is a new separate data file that contains the experimental data transformed into the standard coordinate space of the reference brain. At the end of the warping procedure, the user is allowed to add descriptive data to the new data set. The tool is able to save the generated data on the local disk as well as upload it to the database. The tool is able to upload original, transformed, and descriptive data. The program currently supports MicroBrightField's Neurolucida files and MDPlot's Accustage files.

The *overlapping analysis tool* (Fig. 20.11) is suited to compare density and overlap between two or more cell populations. The program divides the

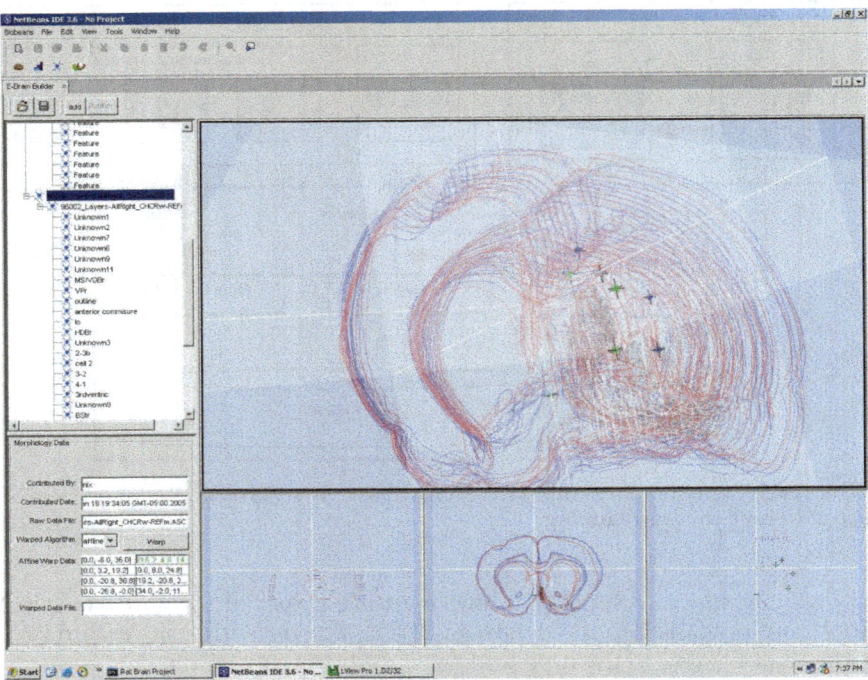

Figure 20.10. The warping tool is used for registering data onto the reference brain and submitting it to the database. Corresponding points for a newly acquired data set and the reference contours can be selected and used to construct a mapping function.

whole brain volume occupied by the data sets into boxes (bins, voxels) of a given size and counts the cell types within each box. If the cell number of each population within a box are equal to or above a certain threshold, the program shows this box in a different color indicating the spatial segregation or overlap. The user selects which cell populations to analyze. The program also makes density measurements comparing the cell numbers from different populations in each box. The program can open multiple data files for cross-brain analyses. A similar approach has been used by Alloway *et al.* (1999) and Leergaard *et al.* (2000). There are several outputs of the program including visualization of bin distributions and summarized cell and bin numbers in table and Microsoft® Excel format (Fig. 20.12; http://www.ratbrain.org/modules/Tools/).

C. Cross-Correlation Function

Consider spike processes as oscillations. Then, the temporal interaction between two cells can be captured as coherent oscillation of the two spike processes, which is quantified by cross-correlation and cross-coherence

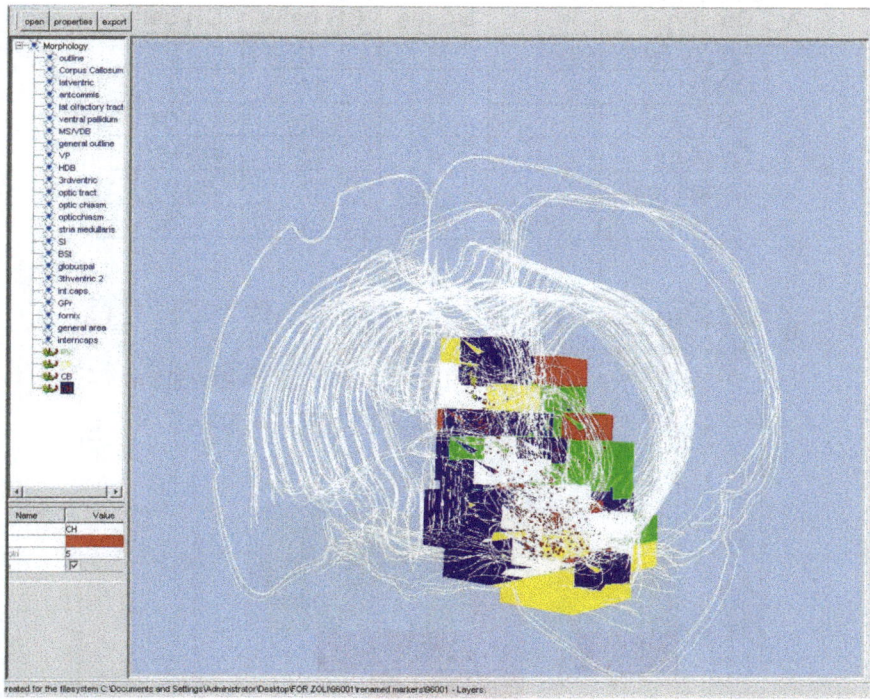

Figure 20.11. The overlap analysis tool allows comparison of overlap of population data sets. The tree display on the left shows named features in the reference brain and reference markers used for warping. The color of the bins is determined by the cell type that is represented by the highest number in the particular bin. Red: cholinergic; blue: calbindin; yellow: calretinin; green: parvalbumin; white: overlapping bins.

functions. The simple cross-correlation function is the product of two spike processes r_1 and r_2 with different time lags τ applied within a time window of stimulus or behavioral event S at time t:

$$CC_{12S}(\tau) = \langle r_{1S}(t + \tau) r_{2S}(t) \rangle_t. \qquad (20.7)$$

Since part of the coherent oscillations derive from stimulus induced nonstationarities of the firing rate, in practice we normalize the cross-correlation function by the shift predictor to get a shift-predictor-corrected cross-correlation function:

$$CC_{12S}(\tau) = \langle r_{1S}(t + \tau) r_{2S}(t) - m_{1S}(t + \tau) m_{2S}(t) \rangle_t, \qquad (20.8)$$

where m_{1S} and m_{2S} are the mean responses of the two neurons for stimulus S at time t. The stimulus-condition-independent component of cross-correlation can be obtained by averaging CC over S:

$$CC_{12}(\tau) = \langle CC_{12S}(t) \rangle_s. \qquad (20.9)$$

Bin Type	Boxes	CH cells	CR cells
Empty (with cells)	3	1	2
CH	8	64	1
CR	16	3	104
Overlap	6	26	43
Undefined	0	0	0
Sum	33	94	150

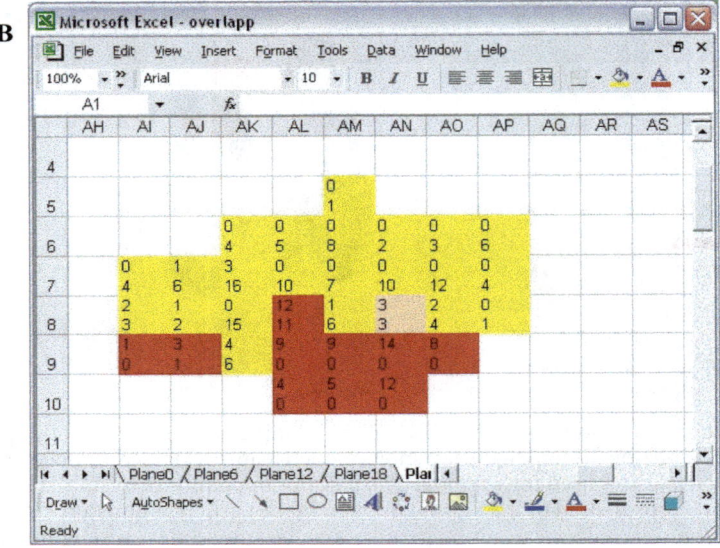

Figure 20.12. The outputs of the overlapping analysis tools in Excel format. (A) The table shows the bin numbers and the occupying cell numbers from a section. (B) The colored table shows the spatial distribution of the bins from the same section. The color of the bins is determined by the cell type that is represented by the highest number in the particular bin. This example is from the basal forebrain (horizontal limb of the diagonal band nucleus), and shows the locations in which the majority of cells are cholinergic (CH, red) or calretinin-containing (CR, yellow). In each bin the upper number represents the cholinergic cells and the lower number the calretinin-containing neurons. Beige: both cell types are represented by an equal number.

For the interpretation of cross-correlation function, see the text (see section "Reconstruction of Functional Connectivity").

ACKNOWLEDGMENTS. We thank Grant Mulliken and Rajan Bhattacharyya for valuable comments and suggestions and thank Peter Bartho for providing figure components for Fig. 20.8. The development of the "Virtual Rat Brain Project" was supported by NIH/NINDS grant NS023945 to Laszlo Zaborszky and funding from UK research councils (MRC and BBSRC) under the eScience program to Fred W. Howell. We are thankful to Peter Varsanyi and Ms. Nicola McDonnell for the database development.

REFERENCES

Abeles, M., and Gat, I., 2001, Detecting precise firing sequences in experimental data, *J. Neurosci. Methods* **107**(1–2):141–154; Erratum in *J. Neurosci. Methods* **112**(2):203.

Abeles, M., and Gerstein, G. L., 1988, Detecting spatiotemporal firing patterns among simultaneously recorded single neurons, *J. Neurophysiol.* **60**(3):909–924.

Abeles, M., and Goldstein, M. 1977, Multispike train analysis, *Proc. IEEE* **65**:762–773.

Aertsen, A. M., Gerstein, G. L., Habib, M. K., and Palm, G., 1989, Dynamics of neuronal firing correlation: modulation of "effective connectivity," *J. Neurophysiol.* **61**(5):900–917.

Aika, Y., Ren, J. Q., Kosaka, K., and Kosaka, T., 1994, Quantitative analysis of GABA-like-immunoreactive and parvalbumin-containing neurons in the CA1 regions of the rat hippocampus using a stereological method, the dissector, *Exp. Brain Res.* **99**:267–276.

Allman, J. M. 1998, *Evolving Brains*, New York: Scientific American Library, W. H. Freeman and Co.

Alloway, K. D., Crist, J., Mutic, J. J., and Roy, S. A., 1999, Corticostriatal projections from rat barrel cortex have an anisotropic organization that correlates with vibrissal whisking behavior, *J. Neurosci.* **19**(24):10908–10922.

Baker, S. N., Spinks, R., Jackson, A., and Lemon, R. N., 2001, Synchronization in monkey motor cortex during a precision grip task: I. Task-dependent modulation in single-unit synchrony, *J. Neurophysiol.* **85**(2):869–885.

Barbas, H., and Rempel-Clower, N., 1997, Cortical structure predicts the pattern of corticocortical connections, *Cereb. Cortex* **7**(7):635–646.

Bartho, P., Hirase, H., Monconduit, L., Zugaro, M., Harris, K. D., and Buzsaki, G., 2004, Characterization of neocortical principal cells and interneurons by network interactions and extracellular features, *J. Neurophysiol.* **92**(1):600–608.

Binzegger, T., Douglas, R. J., and Martin, K. A. C., 2004, A quantitative map of the circuit of cat primary visual cortex, *J. Neurosci.* **24**(39):8441–8453.

Bower, J. M., and Beeman, D., 1998, *The Book of GENESIS: Exploring Realistic Neural Models with the General Neural Simulation System*, 2nd ed., New York: Springer-Verlag.

Brecht, M., Fee, M. S., Garaschuk, O., Helmchen, F., Margrie, T. W., Svoboda, K., and Osten, P., 2004, Novel approaches to monitor and manipulate single neurons in vivo, *J. Neurosci.* **24**:9223–9227.

Brovelli, A., Ding, M., Ledberg, A., Chen, Y., Nakamura, R., and Bressler, S. L., 2004, Beta oscillations in a large-scale sensorimotor cortical network: directional influences revealed by Granger causality, *Proc. Natl. Acad. Sci. U. S. A.* **101**(26):9849–9854.

Budd, J. M., and Kisvarday, Z. F., 2001, Local lateral connectivity of inhibitory clutch cells in layer 4 of cat visual cortex (area 17), *Exp. Brain Res.* **140**(2):245–250.

Buzsaki, G., 2004, Large-scale recording of neuronal ensembles, *Nat. Neurosci.* **7**(5):446–451.

Buzsaki, G., Penttonen, M., Nadasdy, Z., and Bragin, A., 1996, Pattern and inhibition-dependent invasion of pyramidal cell dendrites by fast spikes in the hippocampus in vivo, *Proc. Natl. Acad. Sci. U. S. A.* **93**(18):9921–9925.

Chen-Bee, C. H., Polley, D. B., Brett-Green, B., Prakash, N., Kwon, M. C., and Frostig, R. D., 2000, Visualizing and quantifying evoked cortical activity assessed with intrinsic signal imaging, *J. Neurosci. Methods* **97**(2):157–173.

Chevalier, G., and Mana, S., 2000, Honeycomb-like structure of the intermediate layers of the rat superior colliculus, with additional observations in several other mammals: AChE patterning, *J. Comp. Neurol.* **419**(2):137–153.

Claverol, E., and Nadasdy, Z., 2004, Intersection of microwire electrodes with proximal CA1 stratum-pyramidal neurons at insertion for multiunit recordings predicted by a 3-D computer model, *IEEE Trans. Biomed. Eng.* **51**(12):2211–2216.

Coogan, T. A., and Burkhalter, A., 1993, Hierarchical organization of areas in rat visual cortex, *J. Neurosci.* **13**(9):3749–3772.

Csicsvari, J., Henze, D. A., Jamieson, B., Harris, K. D., Sirota, A., Bartho, P., Wise, K. D., and Buzsaki, G., 2003, Massively parallel recording of unit and local field potentials with silicon-based electrodes, *J. Neurophysiol.* **90**(2):1314–1323.

Das, A., and Gilbert, C. D., 1995, Long-range horizontal connections and their role in cortical reorganization revealed by optical recording of cat primary visual cortex, *Nature* **375**(6534):780–784.

Das, A., and Gilbert, C. D., 1999, Topography of contextual modulations mediated by short-range interactions in primary visual cortex, *Nature* **399**(6737):655–661.

Destexhe, A., and Babloyantz, A., 1993, A model of the inward current Ih and its possible role in thalamocortical oscillations, *Neuroreport* **4**(2):223–226.

Douglas, R., Martin, K., and Witteridge, D., 1989, A canonical microcircuit for neocortex. *Neural. Comput.* **1**:480–488.

Fee, M. S., Mitra, P. P., and Kleinfeld, D., 1996, Automatic sorting of multiple unit neuronal signals in the presence of anisotropic and non-Gaussian variability, *J. Neurosci. Methods* **69**(2):175–188; Erratum in *J. Neurosci. Methods* **71**(2):233.

Felleman, D. J., and Van Essen, D. C., 1991, Distributed hierarchical processing in the primate cerebral cortex. *Cereb. Cortex* **1**:1–47.

Fiala, J. C., and Harris, K. M., 2001, Extending unbiased stereology of brain ultrastructure to three-dimensional volumes, *J. Am. Med. Inform. Assoc.* **8**(1):1–16.

Földy, C., Dyhrfjeld-Johnsen, J., and Soltesz, I., 2005, Structure of cortical microcircuit theory, *J. Physiol.* **562.1**:47–54.

Freeman, W. J., 2000, Mesoscopic neurodynamics: from neuron to brain, *J. Physiol. (Paris)* **94**(5–6):303–322.

Freund, T. F., and Buzsaki, G., 1996, Interneurons of the hippocampus, *Hippocampus* **6**(4):347–470.

Frostig, R. D., Frostig, Z., and Harper, R. M., 1990, Recurring discharge patterns in multiple spike trains: I. Detection, *Biol. Cybern.* **62**(6):487–493.

Fukuda, T., and Kosaka, T., 2000, Gap junctions linking the dendritic network of GABAergic interneurons in the hippocampus, *J. Neurosci.* **20**(4):1519–1528.

Gerfen, C. R., 1985, The neostriatal mosaic: I. Compartmental organization of projections from the striatum to the substantia nigra in the rat, *J. Comp. Neurol.* **236**(4):454–476.

Gibson, J. R., Beierlein, M., and Connors, B. W., 1999, Two networks of electrically coupled inhibitory neurons in neocortex, *Nature* **402**(6757):75–79.

Gilbert, C. D., 1983, Microcircuitry of the visual cortex, *Annu. Rev. Neurosci.* **6**:217–247.

Gilbert, C. D., and Wiesel, T. N., 1983, Functional organization of the visual cortex. *Prog. Brain Res.* **58**:209–218.

Goldman, P. S., and Nauta, W. J., 1977, Columnar distribution of cortico-cortical fibers in the frontal association, limbic, and motor cortex of the developing rhesus monkey, *Brain Res.* **122**(3):393–413.

Goldman-Rakic, P. S., 1984, Modular organization of prefrontal cortex, *Trends Neurosci.* **7**:419–429.

Granger, C. W. J., 1969, Investigating causal relations by econometric methods and cross-spectral methods, *Econometrica*, **34**:424–438.

Graybiel, A. M., and Ragsdale, C. W., Jr., 1978, Histochemically distinct compartments in the striatum of human, monkeys, and cat demonstrated by acetylthiocholinesterase staining, *Proc. Natl. Acad. Sci. U. S. A.* **75**(11):5723–5726.

Grinvald, A., and Hildesheim, R., 2004, VSDI: a new era in functional imaging of cortical dynamics, *Nat. Rev. Neurosci.* **5**(11):874–885.

Gupta, A., Wang, Y., and Markram, H., 2000, Organizing principles for a diversity of GABAergic interneurons and synapses in the neocortex, *Science* **287**(5451):273–278.

Harris, K. D., Csicsvari, J., Hirase, H., Dragoi, G., and Buzsaki, G., 2003, Organization of cell assemblies in the hippocampus, *Nature* **424**(6948):552–556.

Harris, K. D., Henze, D. A., Csicsvari, J., Hirase, H., and Buzsaki, G., 2000, Accuracy of tetrode spike separation as determined by simultaneous intracellular and extracellular measurements, *J. Neurophysiol.* **84**(1):401–414.

He, W., Hamilton, T. A., Cohen, A. R., Holmes, T. J., Pace, C., Szarowski, D. H., Turner, J. N., and Roysam, B., 2003, Automated three-dimensional tracing of neurons in confocal and brightfield images, *Microsc. Microanal.* **9**(4):296–310.

Hebb, D. O., 1949, *Organization of Behavior*, New York: Wiley.

Henze, D. A., Borhegyi, Z., Csicsvari, J., Mamiya, A., Harris, K. D., and Buzsaki, G., 2000, Intracellular features predicted by extracellular recordings in the hippocampus in vivo, *J. Neurophysiol.* **84**(1):390–400.

Hines, M. L., and Carnevale, N. T., 1997, The NEURON simulation environment, *Neural Comput.* **9**:1179–1209.

Hubel, D. H., and Wiesel, T. N., 1959, Receptive fields of single neurones in the cat's striate cortex, *J. Physiol.* **148**:574–591.

Hubel, D. H., and Wiesel, T. N., 1972, Laminar and columnar distribution of geniculo-cortical fibers in the macaque monkey, *J. Comp. Neurol.* **146**(4):421–450.

Ikegaya, Y., Aaron, G., Cossart, R., Aronov, D., Lampl, I., Ferster, D., and Yuste, R., 2004, Synfire chains and cortical songs: temporal modules of cortical activity, *Science* **304**(5670): 559–564.

Isseroff, A., Schwartz, M. L., Dekker, J. J., and Goldman-Rakic, P. S., 1984, Columnar organization of callosal and associational projections from rat frontal cortex, *Brain Res.* **293**(2):213–223.

Johnson, R. R., and Burkhalter, A., 1997, A polysynaptic feedback circuit in rat visual cortex, *J. Neurosci.* **17**(18):7129–7140.

Kalisman, N., Silberberg, G., and Markram, H., 2005, The neocortical microcircuit as a tabula rasa, *Proc. Natl. Acad. Sci. U. S. A.* **102**(3):880–885.

Kaminski, M., Ding, M., Truccolo, W. A., and Bressler, S. L., 2001, Evaluating causal relations in neural systems: granger causality, directed transfer function and statistical assessment of significance, *Biol. Cybern.* **85**(2):145–157.

Kaminski, M. J., and Blinowska, K. J., 1991, A new method of the description of the information flow in the brain structures, *Biol. Cybern.* **65**:203–210.

Kemere, C., Shenoy, K. V., and Meng, T. H., 2004, Model-based neural decoding of reaching movements: a maximum likelihood approach, *IEEE Trans. Biomed. Eng.* **51**(6):925–932.

Kisvarday, Z. F., Ferecsko, A. S., Kovacs, K., Buzas, P., Budd, J. M., and Eysel, U. T., 2002, One axon-multiple functions: specificity of lateral inhibitory connections by large basket cells, *J. Neurocytol.* **31**(3–5):255–264.

Koralek, K. A., Olavarria, J., and Killackey, H. P., 1990, Areal and laminar organization of corticocortical projections in the rat somatosensory cortex, *J. Comp. Neurol.* **299**(2):133–150.

Kötter, R., Hilgetag, C. C., and Stephan, K. E., 2001, Connectional characteristics of areas in Walker's map of prefrontal cortex, *Neurocomputing* **38–40**:741–746.

Lee, A. K., and Wilson, M. A., 2002, Memory of sequential experience in the hippocampus during slow wave sleep, *Neuron* **36**(6):1183–1194.

Leergaard, T. B., Alloway, K. D., Mutic, J. J., and Bjaalie, J. G., 2000, Three-dimensional topography of corticopontine projections from rat barrel cortex: correlations with corticostriatal organization, *J. Neurosci.* **20**(22):8474–8484.

Leise, E. M. 1990, Modular construction of nervous system: a basic principle of design for invertebrates and vertebrates, *Brain Res. Rev.* **15**:1–23.

Lewicki, M. S., 1998, A review of methods for spike sorting: the detection and classification of neural action potentials, *Network* **9**(4):R53–R78 (Review).

Li, W., and Gilbert, C. D., 2002, Global contour saliency and local collinear interactions, *J. Neurophysiol.* **88**(5):2846–2856.

Lubke, J., Egger, V., Sakmann, B., and Feldmeyer, D., 2000, Columnar organization of dendrites and axons of single and synaptically coupled excitatory spiny neurons in layer 4 of the rat barrel cortex, *J. Neurosci.* **20**(14):5300–5311.

Maccaferri, G., and Lacaille, J. C., 2003, Interneuron diversity series: hippocampal interneuron classifications—making things as simple as possible, not simpler, *Trends Neurosci.* **26**:564–571.

Malach, R., 1994, Cortical columns as devices for maximizing neuronal diversity, *Trends Neurosci.* **17**(3):101–104 (Review).

Malmierca, M. S., Leergaard, T. B., Bajo, V. M., Bjaalie, J. G., and Merchan, M. A., 1998, Anatomic evidence of a three-dimensional mosaic pattern of tonotopic organization in the ventral complex of the lateral lemniscus in cat, *J. Neurosci.* **18**(24):10603–10618.

Mountcastle, V. B., 1957, Modality and topographic properties of single neurons of cat's somatic sensory cortex, *J. Neurophysiol.* **20**(4):408–434.

Mountcastle, V. B., 1998, *Perceptual Neuroscience: The Cerebral Cortex*, Harvard University Press, Cambridge, MA.

Musallam, S., Corneil, B. D., Greger, B., Scherberger, H., and Andersen, R. A., 2004, Cognitive control signals for neural prosthetics, *Science*, **305**(5681):258–262.

Nadasdy, Z., 2000, Time sequences and their consequences, *J. Physiol. (Paris)* **94**:505–524.

Nadasdy, Z., Csicsvari, J., Penttonen, M., Hetke, J., Wise, K., and Buzsáki, G., 1998, Extracellular recording and analysis of neuronal activity: from single cells to ensembles, In: Eichenbaum, H. B., and Davis, J. L. (eds.), *Neuronal Ensembles: Strategies for Recording and Decoding*, New York: Wiley.

Nadasdy, Z., Hirase, H., Czurko, A., Csicsvari, J., and Buzsaki, G., 1999, Replay and time compression of recurring spike sequences in the hippocampus, *J. Neurosci.* **19**(21):9497–9507.

Nadasdy, Z., and Zaborszky, L., 2001, Visualization of density relations in large-scale neural networks, *Anat. Embryol. (Berl.)* **204**(4):303–317.

Nordhausen, C. T., Maynard, E. M., and Normann, R. A., 1996, Single unit recording capabilities of a 100 microelectrode array, *Brain Res.* **726**(1–2):129–140.

Ohki, K., Chung, S., Ch'ng Y. H., Kara, P., and Reid, R. C., 2005, Functional imaging with cellular resolution reveals precise micro-architecture in visual cortex, *Nature* **433**(7026):597–603.

Palm, G., Aertsen, A. M., and Gerstein, G. L., 1988, On the significance of correlations among neuronal spike trains, *Biol. Cybern.* **59**(1):1–11.

Paperna, T., and Malach, R., 1991, Patterns of sensory intermodality relationships in the cerebral cortex of the rat, *J. Comp. Neurol.* **308**(3):432–456.

Peters, A., and Feldman, M. L., 1976, The projection of the lateral geniculate nucleus to area 17 of the rat cerebral cortex: I. General description, *J. Neurocytol.* **5**:63–84.

Peters, A., and Payne, B. R., 1993, Numerical relationships between geniculocortical afferents and pyramidal cell modules in cat primary visual cortex, *Cereb. Cortex* **3**:69–78.

Pouille, F., and Scanziani, M., 2004, Routing of spike series by dynamic circuits in the hippocampus, *Nature* **429**(6993):717–723.

Prinz, A. A., Bucher, D., and Marder, E., 2004, Similar network activity from disparate circuit parameters, *Nat. Neurosci.* **7**(12):1287–1288.

Pucak, M. L., Levitt, J. B., Lund, J. S., and Lewis, D. A., 1996, Patterns of intrinsic and associational circuitry in monkey prefrontal cortex, *J. Comp. Neurol.* **376**(4):614–630.

Quian-Quiroga, R., Nadasdy, Z., and Ben-Shaul, Y., 2004, Unsupervised spike detection and sorting with wavelets and superparamagnetic clustering, *Neural Comput.* **16**(8):1661–1687.

Rakic, P., 1995, Radial versus tangential migration of neuronal clones in the developing cerebral cortex, *Proc. Natl. Acad. Sci. U. S. A.* **92**(25):11323–11327.

Rempel-Clower, N. L., and Barbas, H., 2000, The laminar pattern of connections between prefrontal and anterior temporal cortices in the Rhesus monkey is related to cortical structure and function, *Cereb. Cortex* **10**(9):851–865.

Rockland, K. S., 1998, Complex microstructures of sensory cortical connections, *Curr. Opin. Neurobiol.* **8**(4):545–551 (Review).

Rodriguez, A., Ehlenberger, D., Kelliher, K., Einstein, M., Henderson, S. C., Morrison, J. H., Hof, P. R., and Wearne, S. L., 2003, Automated reconstruction of three-dimensional neuronal morphology from laser scanning microscopy images, *Methods* **30**(1):94–105.

Somogyi, P., and Klausberger, T., 2005, Defined types of cortical interneurone structure space and spike timing in the hippocampus, *J. Physiol. (London)* **562**:9–26.

Somogyi, P., Tamas, G., Lujan, R., and Buhl, E. H., 1998, Salient features of synaptic organisation in the cerebral cortex. *Brain Res. Brain Res. Rev.* **26**(2–3):113–135.

Sporns, O., and Kötter, R., 2004, Motifs in brain networks, *PLoS Biol.* **2**(11):e369.

Stepanyants, A., Hof, P. R., and Chklovskii, D. B., 2002, Geometry and structural plasticity of synaptic connectivity, *Neuron* **34**(2):275–288.

Stepanyants, A., Tamas, G., and Chklovskii, D. B., 2004, Class-specific features of neuronal wiring, *Neuron* **43**(2):251–259.

Szentagothai, J., 1978, The neuron network of the cerebral cortex: a functional interpretation. *Proc. R. Soc. Lond. B.* **201**:219–248.

Szentagothai, J., 1983, The modular architectonic principle of neural centers, *Rev. Physiol. Biochem. Pharmacol.* **98**:11–61.

Tamas, G., Buhl, E. H., Lorincz, A., and Somogyi, P., 2000, Proximally targeted GABAergic synapses and gap junctions synchronize cortical interneurons, *Nat. Neurosci.* **3**(4):366–371.

Tamas, G., Buhl, E. H., and Somogyi, P., 1997, Massive autaptic self-innervation of GABAergic neurons in cat visual cortex, *J. Neurosci.* **17**(16):6352–6364.

Traub, R. D., Jefferys, J. G. R., and Whittington, M. A., 1999, *Fast Oscillations in Cortical Circuits*, Bradford Books, MIT Press, Cambridge, MA.

Vetter, R. J., Williams, J. C., Hetke, J. F., Nunamaker, E. A., and Kipke, D. R., 2004, Chronic neural recording using silicon-substrate microelectrode arrays implanted in cerebral cortex, *IEEE Trans. Biomed. Eng.* **51**(6):896–904.

Vreeswijk, C. V., 1996, Partial synchronization in populations of pulse-coupled oscillators, *Phys. Rev. E Stat. Phys. Plasmas Fluids Relat. Interdiscip. Topics* **54**(5):5522–5537.

Wang, X. J., and Rinzel, J., 1993, Spindle rhythmicity in the reticularis thalami nucleus: synchronization among mutually inhibitory neurons, *Neuroscience* **53**(4):899–904.

Watts, J., and Thomson, A. M., 2005, Excitatory and inhibitory connections show selectivity in the neocortex, *J. Physiol. (London)* **562**:89–97.

Wickelgren, W. A., 1999, Webs, cell assemblies, and chunking in neural nets: introduction, *Can. J. Exp. Psychol.* **53**(1):118–131.

Yoshimura, Y., Dantzker, J. L., and Callaway, E. M., 2005, Excitatory cortical neurons form fine-scale functional networks, *Nature* **433**(7028):868–873.

Yuste, R., and Denk, W., 1995, Dendritic spines as basic functional units of neuronal integration, *Nature* **375**(6533):682–684.

Zaborszky, L., 2002, The modular organization of brain systems. Basal forebrain: the last frontier, *Prog. Brain Res.* **136**:359–372.

Zaborszky, L., Buhl, D. L., Pobalashingham, S., Bjaalie, J. G., and Nadasdy, Z., 2005b, Three-dimensional chemoarchitecture of the basal forebrain: spatially specific association of cholinergic and calcium binding protein-containing neurons, *Neuroscience*, **136**(3):697–713.

Zaborszky, L., Csordas, A., Buhl, D. L., Duque, A., Somogyi, J., and Nadasdy, Z., 2002, Computational anatomical analysis of the basal forebrain corticopetal system, In: Ascoli, G. A. (ed.), *Computational Neuroanatomy: Principles and Methods*, New Jersey: Humana Press.

Zaborszky, L., Varsanyi, P., Mc Donnell, N. C., and Howell, F. W., 2005a, 3D cellular database of the rat brain: integrated anatomical data, *Soc. Neurosci. Abstr.* **31**. Program No. 570.3.*2005 Abstract Viewer/Itinerary Planner*. Washington, DC: Society for Neuroscience, 2005.Online.

Zaborszky, L., and Wolff, J. R., 1982, Distribution patterns and individual variations of callosal connections in the albino rat, *Anat. Embryol. (Berlin)* **165**(2):213–232.

Zhang, K., Ginzburg, I., McNaughton, B. L., and Sejnowski, T. J., 1998, Interpreting neuronal population activity by reconstruction: unified framework with application to hippocampal place cells, *J. Neurophysiol.* **79**(2):1017–1044.

Index

Printed by Books on Demand, Germany